Medicinal Plants - Recent Advances in Research and Development

Hsin-Sheng Tsay • Lie-Fen Shyur
Dinesh Chandra Agrawal
Yang-Chang Wu • Sheng-Yang Wang
Editors

Medicinal Plants - Recent Advances in Research and Development

Springer

Editors
Hsin-Sheng Tsay
Department of Applied Chemistry
Chaoyang University of Technology
Taichung, Taiwan

Dinesh Chandra Agrawal
Department of Applied Chemistry
Chaoyang University of Technology
Taichung, Taiwan

Sheng-Yang Wang
College of Agriculture and Natural Resources, Department of Forestry
National Chung Hsing University
Taichung, Taiwan

Lie-Fen Shyur
Agricultural Biotechnology Research Center
Taipei, Taiwan

Yang-Chang Wu
School of Pharmacy and Chinese Herbal Medicine Research Center
China Medical University
Taichung, Taiwan

ISBN 978-981-10-1084-2 ISBN 978-981-10-1085-9 (eBook)
DOI 10.1007/978-981-10-1085-9

Library of Congress Control Number: 2016956887

Printed on acid-free paper

This Springer imprint is published by Springer Nature
The registered company is Springer Nature Singapore Pte Ltd.
The registered company address is: 152 Beach Road, #22-06/08 Gateway East, Singapore 189721, Singapore

This book is dedicated to commemorate the 70th birthday of Professor Dr. Hsin-Sheng Tsay, who has made an immense contribution to the research on medicinal plants in Taiwan and has trained and mentored thousands of young minds in the techniques of plant tissue culture and beyond.

Preface

The editors thought of bringing out this book given the importance of medicinal plants in human health and vigorous ongoing efforts worldwide in the research and development of medicinal plants. The book contains a plethora of information about resources and conservation; biosynthesis and metabolic engineering; biotechnological tools including bioreactor technology; phytochemical research; herbal medicines and plant-derived agents in cancer prevention and therapy, in metabolic syndrome management, and in the modulation of immune-related disorders; and toxicology of medicinal plants. The book consists of 20 chapters, mostly review articles written by experts in their respective fields.

Chapter 1 consists of ten case studies of promising new drug discovery resulting from the Chinese herbal medicine (CHM)-derived products. In each case study, the active principles of CHM have been elucidated. Chapter 2 describes the ethnopharmacological role of *Scutellaria lateriflora* and its medicinal applications in detail. Also, it contains a summary of the mechanism of action attributed to the neuroprotection and other pharmacological actions associated with *S. lateriflora*. Chapter 3 provides an overview of drug adulteration and evaluation of herbal products with a special reference to the approaches in adulterant detection and regulatory perspectives to control such malpractices. Also, adulteration in slimming phytotherapeutic formulations and PDE-5 inhibitors in herbal formulations have been discussed. Chapter 4 contains an overview of the application and challenges and success of various candidate markers used in the DNA barcoding of plants. The development of multilocus and tiered approaches along with the new frontier areas for application of DNA barcoding of medicinal plants and its products has been analyzed in detail. Chapter 5 summarizes the current development in "omics" approaches in the investigation of selected pharmacologically bioactive compounds from three major classes of secondary metabolites, terpenoids, alkaloids, and phenolics, which are being used as drugs in the prevention or therapy of human diseases. Chapter 6 describes the research findings on in vitro plant regeneration, genetic transformation, and molecular characterization of some of the genes involved in the bacoside biosynthesis in the memory booster plant *Bacopa monniera* (Brahmi). Chapter 7 pertains to the metabolic engineering of isoprenoid pathway using squalene synthase as a tool to enhance secondary metabolite contents in *Withania somnifera* (Ashwagandha) or commonly known as Indian ginseng. Chapter 8 reviews the reports on in vitro methodologies, the use of different elicitors, gene functions, genetic modifications, and expression profiling for a better understanding of and to enhance the constituents in *Salvia miltiorrhiza*. Chapter 9 reviews the work carried out on in vitro propagation, somatic embryogenesis, and controlling hyperhydricity (vitrification) in cultures of selected medicinal plant species in Taiwan. Chapter 10 describes the commercial production of paclitaxel, 10-deacetylbaccatin III (10-DAB), and camptothecin (CPT) by the cultivation of plants by farming and production of raw materials and target compounds by cell or hairy root cultures in bioreactors. Chapter 11 provides comprehensive information on the biological effects and pharmacological importance of lucidone, an active constituent of fruits and leaves of *Lindera erythrocarpa*. Chapter 12 briefly describes various aspects of pharmacokinetics, which need to be addressed to generate reliable data on safety and efficacy of herbal drugs. Chapter 13 reviews the role of green tea in cancer prevention and therapy. Chapter 14 presents a review on the traditional Chinese medicine (TCM) oncology theory and its approach toward cancer, therapeutic effects, and various anticancer compounds obtained from TCM herbal plants with the hope of providing a better understanding of the role of drugs in the treatment of cancer. Chapter 15 outlines the results of various scientific studies on *Angelica* species that were reported to have anticancer and antitumor activities. Chapter 16 illustrates the clinical utilization of Chinese herbal medicines acting on rheumatoid arthritis (RA), elucidates their mechanism of action, analyses their limitations and problems, and discusses their development and application prospects. Chapter 17 describes the status, health sector responses, role of traditional medical practitioners, medicinal plant use, and its challenges in the management of noncommunicable diseases in Uganda. Chapter 18 explores possible immune molecular targets of disease-modifying antirheumatic herbal agents and discusses their role in the management of arthritic conditions. Chapter 19 describes the causes and pathophysiology of diabetes, its perspective in Ayurveda, and Ayurvedic plants having antidiabetic potential. Finally, Chapter 20 presents a brief overview of the mechanism of hepatotoxicity and highlights the nexus between herbal medicines and liver injury. Also, it discusses the potential ways that can assure quality check of herbal medicines through the imposition of regulatory laws, databases, and resorting to toxicogenomics.

The editors hope that this compendium of review articles will be very useful as a reference book for advanced students, researchers, academics, business houses, and all people concerned with medicinal plants.

Taichung, Taiwan — Hsin-Sheng Tsay
Taipei, Taiwan — Lie-Fen Shyur
Taichung, Taiwan — Dinesh Chandra Agrawal
May 2016 — Yang-Chang Wu
Sheng-Yang Wang

Acknowledgements

The editors wish to thank all the contributors to this book who took the time to write their valuable manuscripts. Without their contributions, this book would not have been possible.

The editors wish to place on record special thanks to one of the editors of this book Professor Agrawal for initiating the book, handling the correspondence, and managing it from the start to finish. Without his untiring efforts, this book would not have seen the light of the day.

The editors sincerely thank Springer for publishing this book and all the Springer staff members who were ever ready to answer our queries about the book. We would especially like to place on record our appreciation and thanks to Ms. Aakanksha Tyagi, who from the day one handled the correspondence very efficiently and provided all needed help to make this book a reality.

Peer reviewing is an important component of technical writing. The editors would like to express their sincere thanks to all the peer reviewers who took time out from their busy schedules and provided their critical comments on several manuscripts. It helped us to maintain a certain standard of the chapters. We would especially like to thank Professor Dhanasekaran of the Department of Drug Discovery and Development, Harrison School of Pharmacy, Auburn University, Auburn, USA; Dr. Neha Patel, Division of Plant Sciences, Research School of Biology, College of Medicine, Biology and Environment, the Australian National University, Canberra ACT 0200, Australia; Dr. Sushim Gupta, Bacterial Epidemiology and Antimicrobial Resistance Research Unit, USDA, ARS, Athens, GA, USA; Dr. Rishi Kishore Vishwakarma, Le CBS-Centre de Biochimie Structurale-CNRS, UMR 5048-UM-INSERM U 105429 rue de Navacelles 34090, Montpellier, France; Dr. Manish Nivsarkar, Director, Department of Pharmacology and Toxicology, BV Patel Pharmaceutical Education and Research Development (PERD) Centre, Ahmedabad, India; and Dr. Pathirage Kamal Perera, Institute of Indigenous Medicine, University of Colombo, Rajagiriya, Sri Lanka, for their help in peer reviewing some of the manuscripts.

One of the editors, Professor Agrawal, wishes to thank his spouse Mrs. Manju Agrawal for being a great support throughout the book and never complained about his working at home at late hours and weekends. Also, he wishes to thank his daughters Ms. Somya and Ms. Neha for their help in proofreading some of the manuscripts.

Contents

About the Editors and Contributors

About the Editors

Hsin-Sheng Tsay, Ph.D. Professor/Dr. Hsin-Sheng Tsay is a renowned researcher and teacher. He completed his Ph.D. in agronomy from the National Taiwan University about 40 years ago. He worked with Professor Toshio Murashige, University of California, Riverside, for his Ph.D. on anther culture of tobacco. Later, he served at the Taiwan Agricultural Research Institute (TARI) for about 25 years and became head of the Agronomy Department. In TARI, his research pertained to anther culture of rice. There, he also worked on asparagus, papaya, sweet potato, and bamboo. For the last 25 years, Prof. Tsay has been working with medicinal plants indigenous to Taiwan and China. For the last 14 years, he has been working at the Chaoyang University of Technology. There he served as dean of the College of Science and Engineering and director and chair professor of the Graduate Institute of Biotechnology. Professor Tsay has transferred about 20 technologies pertaining to the functional foods for commercialization. He has published 265 research papers and has guided 23 Ph.D. and more than 100 master's students.

Professor Tsay has made a significant contribution to the biotechnology of medicinal plants. During his career, he has organized about 80 plant tissue culture training workshops for international and national researchers. Of these, about 45 tissue culture workshops were conducted for local researchers, professors, vocational school teachers, students, and farmers. These were sponsored by the National Science Council and the Council of Agriculture, Taiwan. About 35 workshops were supported by the International Cooperation and Development Fund (ICDF), and the National Science Council, Taiwan. Participants (more than 500) in these workshops came from about 40 countries. Professor Tsay was invited by 13 countries to conduct plant tissue culture training workshops. He has a galaxy of students across the globe.

During his career, Prof. Tsay has won several national and international awards, including the "National Science Council Outstanding Research Award" for three times. He is on the editorial boards of several international journals and serves as a reviewer for several SCI journals of plant biotechnology.

Lie-Fen Shyur, Ph.D. Professor/Dr. Lie-Fen Shyur is currently serving as a research fellow and deputy director in the Agricultural Biotechnology Research Center, Academia Sinica, Taiwan, and also holding adjunct or joint professorships at five academic institutions in Taiwan. She has participated as editorial board member and invited referee for many international scientific journals and governmental and academic committees.

Dr. Shyur's lab research foci include (1) research and development of phytomedicines and their derived phyto-agents for prevention or therapy of inflammatory diseases, including cancers, septic shock, and hepatitis, (2) elucidating biosynthesis pathway of pharmacologically bioactive compounds in medicinal plants, and (3) industrial enzyme biotechnology. Her research achievements include publication of lab results in high-caliber and reputed journals, such as *Pharmacology and Therapeutics*, *Cancer Research*, *Molecular and Cancer Therapeutics*, *Molecular Oncology*, *Current Opinion in Chemical Biology*, *Molecular Medicine*, *Journal of Biological Chemistry*, *Environmental Science &Technology*, *Journal of Medicinal Chemistry*, and others. In addition, Dr. Shyur has obtained more than 25 international patents and two national awards, namely, the 2014 Silver Award of the National Invention and Creation and the 10th National Innovation Award (2013). A few of technology licensing and cooperative research and development agreement (CRADA) and projects were/are proceeded with local biotech or pharmaceutical companies. The ultimate goal of Dr. Shyur's lab research is to develop agricultural or plant-derived agents for human health-care or bio-industrial applications.

Dinesh Chandra Agrawal, Ph.D. Professor/Dr. Agrawal graduated in 1976 from the Aligarh Muslim University (Central Govt. University) and after that obtained his Ph.D. degree in 1982 from the Garhwal University (renamed as Hemwati Nandan Bahuguna Garhwal University and became Central Govt. University since 2009). His Ph.D. research work on the physiological aspects of *Pinus caribaea* was carried out at the Forest Research Institute, Dehradun (Central Govt. Institute). After his Ph.D. degree, Professor Agrawal has more than 34 years of research experience in plant biotechnology of diverse species including medicinal plants. After serving for more than 31 years, in 2013, he retired as a chief scientist and professor of biological sciences from the CSIR-National Chemical Laboratory, Pune, the top ranking institute in chemical sciences under the umbrella of Council of Scientific and Industrial Research (CSIR), Ministry of Science and Technology, Govt. of India. Currently, he is working as a professor in the Department of Applied Chemistry, Chaoyang University of Technology (CYUT), Taiwan.

While in CSIR-NCL, Professor Agrawal worked as a coordinator and project leader of several research projects funded by the Govt. of India. He has more than 155 research articles to his credit on different aspects of plant biotechnology including medicinal plants. Also, he has written two books on "laboratory record writing." More than 35 M.Tech./M.Sc. and 7 Ph.D. students have completed their thesis work under his guidance.

Professor Agrawal has been bestowed several prestigious awards and fellowships such as the Alexander von Humboldt Fellowship (Germany), DBT Overseas Associateship (USA), British Council Scholar (UK), European Research Fellow (UK), and INSA Visiting Scientist. During these fellowships, he had opportunities to work in the USA, Germany, and the UK. Also, he had research collaboration with UMR Vigne et Vins, INRA, Centre de Recherché Colmar, France.

Professor Agrawal has served as a member of the editorial board of *Medicinal and Aromatic Plant Abstracts*, NISCAIR, Govt. of India, and has reviewed a large number of research papers for several SCI journals on plant biotechnology. For more than 10 years, he has been a member of the executive committee of the Humboldt Academy, Pune Chapter, and held the position of treasurer.

Yang-Chang Wu, Ph.D. Professor/Dr. Yang-Chang Wu was born in Chiayi, Taiwan, in 1951. He obtained his Ph.D. in pharmacognosy from the College of Pharmacy, Kaohsiung Medical University (KMU), Taiwan, in 1986. After that, he joined the group of Prof. Yoshimasa Hirata at Meijo University, Japan, as a postdoctoral researcher from 1986 to 1987. Later, he joined the laboratory of Prof. Kuo-Hsiung Lee for further postdoctoral research at the University of North Carolina (UNC), Chapel Hill, USA. There he worked on the various synthetic approaches toward natural products and medicinal chemistry.

In 1990, he became professor at the College of Pharmacy at KMU and director of the Graduate Institute of Natural Products (GINP) in 1992. Later, he served as the dean of the Office of Research and Development at KMU, from 2006 to 2009. Attributed to his significant contribution to research on natural products, he was selected as the chair professor and vice-president of the Graduate Institute of Integrated Medicine and College of Chinese Medicine at China Medical University (CMU), Taiwan, from 2010 to 2012. Since 2012, he was appointed as the chair professor and vice-president as well as dean of the School of Pharmacy, CMU, Taiwan.

In 2007, he was awarded by the Wang Ming-Ning Foundation for outstanding merit and high scholastic achievement to medical and pharmaceutical research. In 2009, he received the National Science Council Outstanding Research Award in Taiwan. He also received the Outstanding Medical and Pharmaceutical Technology

Award in 2010 by the TienTe Lee Biomedical Foundation, Taiwan. Professor Wu is known for his expertise in the area of translational research on Chinese herbal medicine, functional food, and new drug development.

Professor Wu has served as an editorial board member of 6 journals and as a referee for about 30 journals. Also, he is an outstanding member of the American Society of Pharmacognosy (ASP) and ten more other associations. So far, Prof. Wu has published more than 510 research articles in SCI journals along with the authorship in several book chapters. He has been granted more than 30 patents and is in cooperation with more than 20 industry-academic organizations. Professor Wu has transferred six patent/technologies (including one new drug R&D tech transfer) to industry.

Sheng-Yang Wang, Ph.D. Dr. Sheng-Yang Wang is a distinguished professor at the Department of Forestry, National Chung Hsing University, Taiwan. Dr. Wang obtained his Ph.D. degree from the National Taiwan University. He is one of the well-known phytochemists in Taiwan and has expertise in the qualitative and quantitative determination of natural products by chromatography and spectroscopy. He has published more than 120 scientific articles so far. Dr. Wang has obtained several important scientific awards in his academic career, including "Excellent Research Award" for Young Faculty of National Chung Hsing University (2004); Dr. Ta-You Wu Memorial Award established by the National Science Council (2004), Taiwan; Academia Award of the Chinese Forestry Association and Top 10 Outstanding Agricultural Specialists of Taiwan in 2011; Academic Award, Forest Products Association of ROC; Academician Shang-Fa Academy Award for Distinguished Young Scholars in 2012; Development Program of Industrialization for Agricultural Biotechnology (DPIAB) Award for Outstanding Industry-University Collaboration Projects in 2013; Silver Award of the National Invention and Creation 2014; and Gold Award in "Seoul International Invention Fair 2014." The research areas of interest in Dr. Wang's laboratory are (1) development of new methodology in the isolation and structural elucidation of natural products, (2) phytomedicine investment of indigenous plants of Taiwan, and (3) functional genomic study of woody plants.

Contributors

Dinesh Chandra Agrawal Department of Applied Chemistry, Chaoyang University of Technology, Taichung, Taiwan

Godwin Anywar Department of Plant Sciences, Microbiology and Biotechnology, College of Natural Sciences, Makerere University, Kampala, Uganda

Uriel Bachrach Department of Molecular Biology, Hebrew University-Hadassah Medical School, Jerusalem, Israel

Barbara Kemppainen College of Veterinary Medicine, Auburn University, Auburn, AL, USA

Dipita Bhakta-Guha Cellular Dyshomeostasis Laboratory (CDHL), SASTRA University, Thanjavur, Tamil Nadu, India

Mili Bhargava Biotechnology Division, CSIR-Central Institute of Medicinal and Aromatic Plants, Lucknow, India

Hung-Chi Chang Department of Applied Chemistry, Chaoyang University of Technology, Taichung, Taiwan

Department of Golden-Ager Industry Management, Chaoyang University of Technology, Taichung, Taiwan

Shu-Hwa Chang Taiwan Forestry Research Institute, Taipei, Taiwan, Republic of China

Chia-Chen Chen Department of Chinese Pharmaceutical Sciences and Chinese Medicine Resources, China Medical University, Taichung, Taiwan

Fang Chen Department of Botany, Key Laboratory of Bio-Resources and Eco-Environment Ministry of Education, College of Life Sciences, Sichuan University, Chengdu, People's Republic of China

Fen-Hui Chen Taiwan Forestry Research Institute, Taipei, Taiwan, Republic of China

Emily Chin-Fun Chen Department of Applied Chemistry, Chaoyang University of Technology, Taichung, Taiwan

Hsiao-Hang Chung Department of Horticulture, National Ilan University, Yilan, Taiwan

Ranjeet Prasad Dash Department of Drug Discovery and Development, Harrison School of Pharmacy, Auburn University, Auburn, AL, USA

Department of Pharmacology and Toxicology, B. V. Patel Pharmaceutical Education and Research Development (PERD) Centre, Ahmedabad, Gujarat, India

Muralikrishnan Dhanasekaran Department of Drug Discovery and Development, Harrison School of Pharmacy, Auburn University, Auburn, AL, USA

Thomas Efferth Institute of Pharmacy and Biochemistry, Johannes Gutenberg University, Mainz, Germany

Qiuyan Guo Institute of Chinese Materia Medica, China Academy of Chinese Medical Sciences, Beijing, China

Sushim Kumar Gupta Bacterial Epidemiology and Antimicrobial Resistance Research Unit, USDA, ARS, Athens, GA, USA

Department of Applied Chemistry, Chaoyang University of Technology, Taichung, Taiwan

Cheng-Kuen Ho Taiwan Forestry Research Institute, Taipei, Taiwan, Republic of China

Sneha Joshi College of Veterinary Medicine, Auburn University, Auburn, AL, USA

Esezah Kakudidi Department of Plant Sciences, Microbiology and Biotechnology, College of Natural Sciences, Makerere University, Kampala, Uganda

Esther Katuura Department of Plant Sciences, Microbiology and Biotechnology, College of Natural Sciences, Makerere University, Kampala, Uganda

Juliet Kiguli Department of Community Health & Behavioral Sciences, College of Health Sciences, Makerere University, Kampala, Uganda

Claude Kirimuhuzya Western Campus, Department of Pharmacology, Kampala International University, Bushenyi, Uganda

Bashir M. Khan Plant Tissue Culture Division, CSIR-National Chemical Laboratory, Pune, Maharashtra, India

Uma Kumari Plant Tissue Culture Division, CSIR- National Chemical Laboratory, Pune, Maharashtra, India

Chao-Lin Kuo Department of Chinese Pharmaceutical Sciences and Chinese Medicine Resources, China Medical University, Taichung, Taiwan

Kuo-Hsiung Lee Natural Products Research Laboratories, UNC Eshelman School of Pharmacy, University of North Carolina, Chapel Hill, NC, USA

Chinese Medicine Research and Development Center, China Medical University and Hospital, Taichung, Taiwan

Na Lin Institute of Chinese Materia Medica, China Academy of Chinese Medical Sciences, Beijing, China

A. Lingesh Department of Pharmaceutical Analysis, National Institute of Pharmaceutical Education and Research, Hyderabad (NIPER-H), Hyderabad, Telangana, India

Madhukar Lohani Department of Drug Discovery and Development, Harrison School of Pharmacy, Auburn University, Auburn, AL, USA

Aiping Lu School of Chinese Medicine, Hong Kong Baptist University, Kowloon, Hong Kong

Manasi Malgaonkar National Research Institute of Basic Ayurvedic Sciences, Pune, India

Mohammed Majrashi Department of Drug Discovery and Development, Harrison School of Pharmacy, Auburn University, Auburn, AL, USA

Xia Mao Institute of Chinese Materia Medica, China Academy of Chinese Medical Sciences, Beijing, China

Amira Mira Department of Pharmacognosy, Faculty of Pharmacy, Mansoura University, Mansoura, Egypt

Susan L. Morris-Natschke Natural Products Research Laboratories, UNC Eshelman School of Pharmacy, University of North Carolina, Chapel Hill, NC, USA

Vanisree Mulabagal Department of Civil Engineering, Auburn University, Auburn, AL, USA

Soma Narasimha Murthy National Research Institute of Basic Ayurvedic Sciences, Pune, India

Katie Musgrove Natural Products Research Laboratories, UNC Eshelman School of Pharmacy, University of North Carolina, Chapel Hill, NC, USA

Manish Nivsarkar Department of Pharmacology and Toxicology, B. V. Patel Pharmaceutical Education and Research Development (PERD) Centre, Ahmedabad, Gujarat, India

Karuppaiya Palaniyandi Department of Botany, Key Laboratory of Bio-Resources and Eco-Environment Ministry of Education, College of Life Sciences, Sichuan University, Chengdu, People's Republic of China

Mansi Patel Department of Drug Discovery and Development, Harrison School of Pharmacy, Auburn University, Auburn, AL, USA

Neha Patel Division of Plant Sciences, Research School of Biology, College of Medicine, Biology and Environment, The Australian National University, Canberra, ACT, Australia

Parth Patel Molecular Microbial Biochemistry Laboratory, Department of Biochemistry, The Maharaja Sayajirao University of Baroda, Vadodara, Gujarat, India

David Paul Department of Pharmaceutical Analysis, National Institute of Pharmaceutical Education and Research Hyderabad (NIPER-H), Hyderabad, Telangana, India

Sharad Pawar National Research Institute of Basic Ayurvedic Sciences, Pune, India

Pathirage Kamal Perera Institute of Indigenous Medicine, University of Colombo, Rajagiriya, Sri Lanka

Sanchita Biotechnology Division, CSIR-Central Institute of Medicinal and Aromatic Plants, Lucknow, India

N. Satheeshkumar Department of Pharmaceutical Analysis, National Institute of Pharmaceutical Education and Research Hyderabad (NIPER-H), Hyderabad, Telangana, India

K.J. Senthil Kumar Department of Forestry, National Chung Hsing University, Taichung, Taiwan

Ashok Sharma Biotechnology Division, CSIR-Central Institute of Medicinal and Aromatic Plants, Lucknow, India

Poonam Sharma Division of Biochemical Sciences, CSIR-National Chemical Laboratory, Pune, India

Kuniyoshi Shimizu Division of Systematic Forest and Forest Products Sciences, Department of Agro-Environmental Sciences, Faculty of Agriculture, Graduate School of Kyushu University, Kyushu, Japan

Amey Shirolkar National Research Institute of Basic Ayurvedic Sciences, Pune, India

Lie-Fen Shyur Agricultural Biotechnology Research Center, Academia Sinica, Taipei, Taiwan

Swati Srivastava Biotechnology Division, CSIR-Central Institute of Medicinal and Aromatic Plants, Lucknow, India

Yi-Chang Sung Agricultural Biotechnology Research Center, Academia Sinica, Taipei, Taiwan

Hsin-Sheng Tsay Department of Applied Chemistry, Chaoyang University of Technology, Taichung, Taiwan

Department of Agronomy, National Chung Hsing University, Taichung, Taiwan

Rishi K. Vishwakarma Plant Tissue Culture Division, CSIR-National Chemical Laboratory, Pune, India

Le CBS-Centre de Biochimie Structurale-CNRS, UMR 5048-UM-INSERM U 1054, Montpellier, France

Shenghua Wang Department of Botany, Key Laboratory of Bio-Resources and Eco-Environment Ministry of Education, College of Life Sciences, Sichuan University, Chengdu, People's Republic of China

Sheng-Yang Wang Department of Forestry, National Chung Hsing University, Taichung, Taiwan

Agricultural Biotechnology Research Center, Academia Sinica, Taipei, Taiwan

Zohara Yaniv Department of Genetic Resources, Institute of Plant Sciences, Agricultural Research Organization, The Volcani Center, Bet Dagan, Israel

Yanqiong Zhang Institute of Chinese Materia Medica, China Academy of Chinese Medical Sciences, Beijing, China

Yu Zhao Natural Products Research Laboratories, UNC Eshelman School of Pharmacy, University of North Carolina, Chapel Hill, NC, USA

Chapter 1
Chinese Herbal Medicine-Derived Products for Prevention or Treatment of Diseases Affecting Quality of Life

Kuo-Hsiung Lee, Susan L. Morris-Natschke, Yu Zhao, and Katie Musgrove

Abstract Chinese herbal medicine (CHM) has been used for several thousands of years to treat human illness, which makes CHM the best source to provide valuable and unique information for modern drug discovery and development. Development of CHM products as adjunct therapies to augment the efficacy and offset the toxicity of Western medicine is an excellent approach for rapid advancement into US FDA-approved new drugs. Developing CHM products as high-quality dietary supplements must particularly emphasize standardization through qualitative and quantitative quality controls on single herbs and multiple herbs of the prescription formulas by using the most advanced scientific technology, especially toxicity profile testing. A combination of advanced medicinal chemistry and natural products chemistry, coupled with cutting-edge life science technology, will play a very important role for converting CHM products, especially the pure single active principles, through modification and synthesis into clinical trial candidates very efficiently and effectively. This chapter presents ten case studies of promising new drug discovery resulting from CHM-derived products: compounds from *Curcuma longa* (Jiang Huang), *Antrodia camphorata* (Chang-ku), *Apium graveolens* (Han Qin), *Momordica charantia* (Ku Gua), *Monascus purpureus* (Hong Chi), *Astragalus membranaceus* (Huang Chi), *Scutellaria* decoction (Huang Chin Tang), *Eucommia ulmoides* (Tu Chung), *Ligusticum wallichii* (Chuan Chiung), and *Lycium barbarum* (Kou Chi Tzu).

K.-H. Lee (✉)
Natural Products Research Laboratories, UNC Eshelman School of Pharmacy, University of North Carolina, Chapel Hill, NC 27599-7568, USA

Chinese Medicine Research and Development Center, China Medical University and Hospital, Taichung, Taiwan
e-mail: khlee@unc.edu

S.L. Morris-Natschke • Y. Zhao • K. Musgrove
Natural Products Research Laboratories, UNC Eshelman School of Pharmacy, University of North Carolina, Chapel Hill, NC 27599-7568, USA

H.-S. Tsay et al. (eds.), *Medicinal Plants - Recent Advances in Research and Development*, DOI 10.1007/978-981-10-1085-9_1

Keywords Chinese herbal medicine (CHM) • Drug discovery and development • Disease prevention and treatment • Medicinal chemistry • Natural products chemistry • Traditional Chinese medicine (TCM)

Abbreviations

Ach	Acetylcholine
AIDS	Acquired immunodeficiency syndrome
ANS	1-anilino-8-naphthalene-sulfonate
ATP	Adenosine triphosphate
BDFI	Bioactivity-directed fractionation and isolation
CD	Circular dichroism
CHM	Chinese herbal medicine
COX	Cyclooxygenase
CRF	Cancer-related fatigue
DNA	Deoxyribonucleic acid
EDHF	Endothelium-derived hyperpolarizing factor
EGFR-TKI	Epidermal growth factor receptor tyrosine kinase inhibitor
FDA	Food and Drug Administration
FRET	Förster-type fluorescence resonance energy transfer
FTIR	Fourier transform infrared spectroscopy
GAP	Good agriculture practice
GCP	Good clinical practice
GI	Gastrointestinal
GLP	Good laboratory practice
GMP	Good manufacturing practice
GRAS	Generally recognized as safe
HBV	Hepatitis B virus
HCV	Hepatitis C virus
HIV	Human immunodeficiency virus
HIV-1	Human immunodeficiency virus type 1
HPLC	High-pressure liquid chromatography
HPV	Human papilloma virus
IC_{50}	Half maximal inhibitory concentration
IL	Interleukin
IND	Investigational New Drug
iNOS	Inducible nitric oxide synthase
IOIP	Idiopathic orbital inflammatory pseudotumors
IV	Intravenous
LBP	*Lycium barbarum* polysaccharide
MALDI-TOF MS	Matrix-assisted laser desorption/ionization time-of-flight mass spectrometry
*mc*IRBP	*Momordica charantia* insulin receptor (IR)-binding protein

MCP	Monocyte chemoattractant protein
MOEA	Ministry of Economic Affairs
NBP	3-*n*-butylphthalide
NOS	Nitric oxide synthase
RCT	Randomized controlled trial
RMR	Red mold rice
SFDA	State Food and Drug Administration
TCM	Traditional Chinese medicine
TNF	Tumor necrosis factor
US	United States
WHO	World Health Organization

1.1 Introduction

Generations of Chinese people have used Chinese herbal medicine (CHM), the most important medicine of traditional Chinese medicine (TCM), for thousands of years for disease prevention and treatment. CHM and their active principles and derivatives provide a broad and profound base for the discovery of effective and safe dietary supplements, in addition to new medicine for the prevention or treatment of diseases affecting quality of life. The symbiotic relationship between the ancient practice of CHM, with its thousands of years of experience, and the modern technology and scientific advancements of today has proven to be the key to effective new drug discovery and development.

The knowledge base of CHM is both vast and diverse. Chinese medicines are all derived from natural products, with roughly 90 % coming from plants. Approximately 5000 plant species with therapeutic value have been identified, many of them used as *Min Chien Yao*, or folk drugs. About 500 of them are commonly prescribed by doctors of Chinese medicine based on a series of systemic and self-contained theories (Chung Yao, *Chinese Materia Medica*). This systematic approach is necessary because most CHM-derived products involve multicomponent processed herbal formulations. Although CHM does not provide the same scale and scope of modern-day clinical trials, they no doubt offer a vast knowledge base from which to gain valuable insights relevant to current chronic health issues. The combination of CHM and modern scientific practices has the potential to lead to both new clinical trial candidates and adjunct therapies for current Western medicine treatments (Lee et al. 2013).

In 2013, the authors detailed several case studies of modern drug discovery from CHM, including ephedrine/pseudoephedrine from *Ephedra sinica* (Ma Huang), indirubin from *Indigofera tinctoria* (Qing Dai – Dang Gui Long Hui Wan), and artemisinin from *Artemisia annua* (Qing Hao) (Lee et al. 2013). It should be noted that Tu Youyou from the China Academy of Chinese Medical Sciences, Beijing, was recognized in 2011 with the Lasker-DeBakey Clinical Medical Research Award and in 2015 together with Drs. Satoshi Ōmura and William C. Campbell, with the Nobel Prize in Physiology or Medicine, based on her outstanding contribution for the discovery of artemisinin in malaria therapy. Tu Youyou is credited with the inspiration and diligence to probe the ancient CHM literature for an applicable solution to malaria, which ultimately led to the identification of artemisinin/qinghaosu as the effective active principle. Dr. Yi Zhao of Guangxi Traditional Chinese Medical University, China, also made a significant contribution with regard to the elucidation of the pharmacological effect, as well as the mechanism of action of artemisinin/qinghaosu as evidenced in his academic papers entitled *Studies on Artemisia annua L./Qinghaosu* (Zhao et al. 1986, 1987; Zhao 2011). Overall, the discovery of artemisinin serves as the best example of producing a world-class new drug from ancient CHM via modern medicinal chemistry studies.

1.1.1 Current Areas of Interest for CHM-Derived Drugs

Chronic diseases such as diabetes, arthritis, heart disease, high blood pressure, etc. can be debilitating to patients and can drastically alter their quality of life. CHM has the potential to offer solutions to many of these modern chronic diseases. Some of the current areas of interest for scientists focusing on CHM-derived drug discovery are as follows: antioxidant and antiaging activity, blood pressure-lowering effects, hypolipidemic action, blood sugar-lowering effects, antiallergic functions, and antiarthritis properties. While we highlight quite a few CHM-derived drugs in our case studies, there are other notable discoveries concerning the treatment and prevention of these devastating chronic diseases.

In the coming years, CHM will lead to many new clinical trial drug candidates. The ancient practices of CHM combined with modern scientific evidence have clearly demonstrated that CHM can be a promising treatment modality for chronic diseases to greatly improve quality of life for patients. This combination of ancient and modern-day principles may also lead to a higher probability of success and a more efficient scientific process. This chapter presents ten case studies of promising new drug discovery resulting from CHM-derived products: compounds from *Curcuma longa* (Jiang Huang), *Antrodia camphorata* (Chang-ku), *Apium graveolens* (Han Qin), *Momordica charantia* (Ku Gua), *Monascus purpureus* (Hong Chi), *Astragalus membranaceus* (Huang Chi), *Scutellaria* decoction (Huang Chin Tang), *Eucommia ulmoides* (Tu Chung), *Ligusticum wallichii* (Chuan Chiung), and *Lycium barbarum* (Kou Chi Tzu).

1.2 Bringing TCM to Mainstream

1.2.1 Obstacles for Bringing TCM to Mainstream

Lack of standardization is a major obstacle to the development of CHM as world-class dietary supplements and new medicines. Unless herbal products can be guaranteed to be efficacious and safe with validated quality and consistency, they cannot be patented or tested in clinical trials. Even their commercialization and marketing as dietary supplements are weakened without such assurances. Four issues to be addressed in bringing CHM products into the mainstream pharmaceutical market are as follows: *high quality*, *high consistency*, *high safety*, and *high efficacy*. Good methods of quality control should be applied at all stages, including plant growth, production, processing, and storage. In addition, the products should be validated as being from the correct plant species, plant strain, and plant part as well as uniform from manufacturer to manufacturer and from batch to batch. The products must be free of toxicity and contaminants (e.g., heavy metals) with all of their constituents characterized. Possible interactions with other drugs must be determined, and controlled clinical trials should be performed to prove efficacy.

1.2.2 Methods for Bringing TCM to Mainstream

Two main areas for development of CHM products are as adjunct therapies to Western medicine and for prevention or treatment of diseases that are difficult to be treated satisfactorily with Western medicine. Strategies for development of CHM-based world-class new drugs should emphasize using *medicinal chemistry approaches* to study the active principals of CHM and to modify the lead compound into suitable clinical trial candidates. Furthermore, *mechanism of action studies* on active principles, active extracts, and effective formulas of CHM can provide needed scientific proof and future directions for new drug research. Also, it would be extremely helpful to establish *international collaborative platforms* for development of CHM-based new drugs.

Two examples of new botanical drugs already approved by the US FDA are Veregen (MediGene, Germany) in October 2006 and Fulyzaq (Salix Pharmaceuticals, USA) in December 2012. These products were indicated for use against human papilloma virus (HPV) and noninfectious diarrhea in adult patients with HIV/AIDS on antiretroviral therapy, respectively. Their major constituents are kunecatechins (component mixture from a green tea extract) and crofelemer (an extract from *Croton lechleri*), respectively.

A national program for developing world-class new botanical drugs was initiated by the Ministry of Economic Affairs (MOEA), Taiwan, during 2000–2005 (Lee 2015). The seven-person committee (Committee for Promotion of Chinese Herbal Medicine Industry and Technology) received FDA approval for four Investigational

New Drug (IND) applications for partially purified Chinese herbal medicine products for clinical trials, established a platform technology to produce high-quality herbal medicine products with cutting-edge methodology for quality control, and founded excellent preclinical and IND-related infrastructures, including good agriculture, laboratory, manufacturing, and clinical practice (GAP, GLP, GMP, and GCP). Both the software and hardware of these infrastructures were set up to international standards (Lee 2015).

Medicinal chemistry is the art of combining *chemistry* and *biology* for optimal drug discovery, and these two research areas are complementary, just like *yin* and *yang* of TCM. The discovery of new bioactive compounds depends on valid biological assays, and new chemistry can make the discovery of new biological targets possible.

CHM-derived world-class new drugs and high-quality dietary prescriptions can come from three sources: herbal formulas, extract fractions, and single herbs. As mentioned above, the quality control of herbal formulas is critical, while any modification and improvement should be followed by reevaluation of efficacy and toxicity. In all cases, herbal formulas should continue to be used according to the conformation dictated by TCM diagnosis and principles. Both herbal formulas and single herbs from those formulas can be subjected to pharmacological testing as well as bioactivity-directed fractionation and isolation (BDFI) to discover active fraction mixtures and active single natural product lead compounds, respectively. After the structures of the active lead compounds are elucidated, an optimized lead can be pursued through an iterative cycle that includes the design of modified analogs, synthesis of these analogs, bioactivity screening, and analysis of results. Various preclinical studies to discern mechanism of action and other pharmacological properties (solubility, pharmacokinetics, etc.) are important to make sure that a lead is a viable clinical trial candidate. The goals of the lead development process are to improve pharmacological profiles by increasing activity, decreasing toxicity, or circumventing metabolic, pharmacokinetic, solubility, or drug-resistance problems. The following case studies will highlight the rationale, diversity, and strength of these processes as well as introduce several examples of CHM-derived drugs used now or in the future to treat or prevent diseases affecting quality of life.

1.3 Case Studies Highlighting CHM-Derived Drugs Used to Treat/Prevent Diseases Affecting Quality of Life

1.3.1 Curcuma longa *(Turmeric)*

1.3.1.1 Introduction

Curcuma longa (Chinese: Jiang Huang 薑黃) is a rhizomatous herbaceous perennial member of the Zingiberaceae family. More commonly known as turmeric, it is native to tropical Southeast Asia and now cultivated in the tropical and subtropical regions of the world, especially in India and China. Turmeric has a long historical use as a traditional medicine. In Ayurveda medicine, turmeric is primarily used as a treatment for inflammatory conditions. In TCM, it is used to treat biliary disorders, anorexia, cough, diabetes, wounds, hepatic disorders, and rheumatism. It has also been used as a sinusitis stimulant, aspirant, carminative, emmenagogue, astringent, detergent, and diuretic (Gupta et al. 2015). Besides its medicinal use, turmeric has long been part of the daily diet in Asian countries and has not been shown to cause any toxicity. Turmeric powder and many other extracts from the rhizomes were found to possess versatile bioactivity, including wound-healing, anti-inflammatory, hypolipidemic, cytotoxicity, antiprotozoan, antibacterial, antifungal, and antifertility effects (Chattopadhyay et al. 2004).

1.3.1.2 Chemical Constituents

To date, over 200 compounds have been identified from turmeric. These compounds belong to various structural types including diarylheptanoids (curcuminoids), diarylpentanoids, phenylpropene, phenolic compounds, monoterpenes, sesquiterpenes, diterpenes, triterpenoids, alkaloid, sterols, and fatty acids (Anonymous 1999b).

Curcuminoids with an aryl-C7-aryl skeleton, include curcumin (curcumin I, **1**, Table 1.1), demethoxycurcumin (curcumin II, **2**, Table 1.1), bisdemethoxycurcumin (curcumin III, **3**, Table 1.1), *p,p′*-dihydroxydicinnamoyl methane, *p*-hydroxycinnamoyl-feruloylmethane, dihydrocurcumin, etc. Sesquiterpenes include *ar*-turmerone (**4**, Table 1.1), a-turmerone (**5**, Table 1.1), β-turmerone (**6**, Table 1.1), curlone, etc. Dried turmeric rhizomes usually contain 1.5–5 % essential oils. Compounds **4**–**6** are the major sesquiterpenes of the essential oils, and these compounds may account for at least 40 % of essential oils of turmeric rhizomes.

The pharmaceutical products of turmeric are dried whole rhizomes, ground turmeric, turmeric oils, turmeric oleoresin, and curcumin. The quality control of turmeric has been thoroughly reviewed (Li et al. 2011). Curcuminoids are the main active compounds. They are primarily accumulated in turmeric rhizomes (3–15 %) with curcumin (**1**, Table 1.1) as the principal constituent. The contents of curcuminoids in turmeric rhizomes often vary with varieties, locations, sources, and cultivation conditions. According to the Indian Pharmacopoeia in 1996, dried turmeric rhizomes should contain not less than 1.5 % of curcumin (**1**) (w/w) (India 1996). The Pharmacopoeia of the People's Republic of China (2005) requires no less than 1.0 % of curcumin (**1**) content (w/w) in dried turmeric rhizomes (Anonymous 2005). The Thai Herbal Pharmacopoeia recommended that dried turmeric should contain no less than 6 % of turmeric oil (v/w) and 5 % of total curcuminoids (w/w) (Thaikert and Paisooksantivatana 2009). The WHO (World Health Organization) suggests that not less than 4.0 % of volatile oil, and not less than 3.0 % of curcuminoids should be present in turmeric (Anonymous 1999d). The quality control of rhizomes, powders, and extract products was reported through defining and verifying the presence of curcumin (**1**), demethoxycurcumin (**2**), and bisdemethoxycurcumin (**3**) and determining their concentrations by HPLC chromatogram (Li et al.

Table 1.1 Structures of selected bioactive compounds from plant species in the case studies

Case study	Natural species	Examples of active compounds or modified derivatives
1	*Curcuma longa*	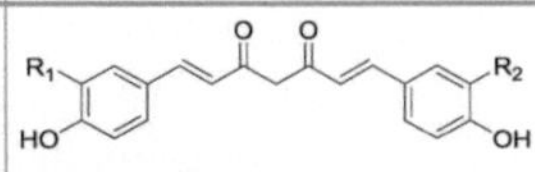 **1** (Curcumin) $R_1 = OCH_3$, $R_2 = OCH_3$ **2** (Demethoxycurcumin) $R_1 = OCH_3$, $R_2 = H$ **3** (Bisdemethoxycurcum) $R_1 = H$, $R_2 = H$ **7** (JC-9 = ASC-J9) $R_1 = OCH_3$, $R_2 = CH_3$ 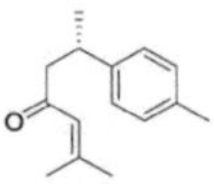 **4** (*ar*-Turmerone) 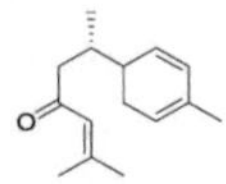 **5** (α-Turmerone) 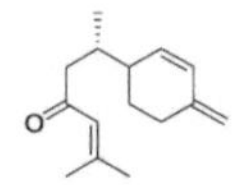 **6** (*β*-Turmerone)
2	*Antrodia camphorata*	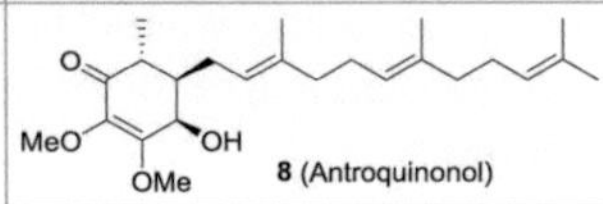 **8** (Antroquinonol)
3	*Apium graveolens*	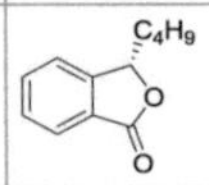 **9** (*S*-(-)-3-*n*-Butylphthalide, *S*-(-)-NBP)
4	*Momordica charantia*	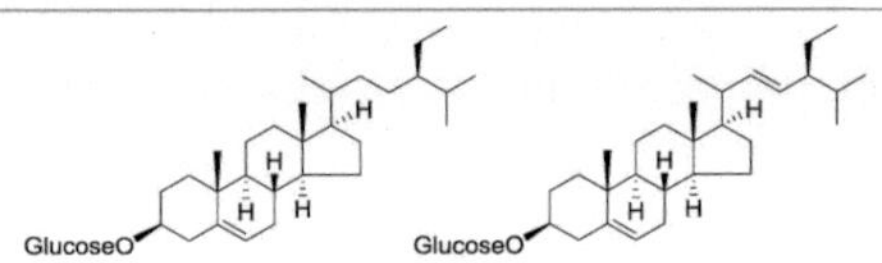 **10** (Charantin: mixture of sitosteryl glucoside (left) & stigmasteryl glucoside (right))
5	*Monascus purpureus*	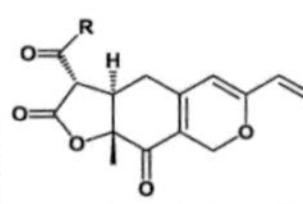 **11** (Monascin) $R = C_5H_{11}$ **12** (Ankaflavin) $R = C_7H_{15}$ 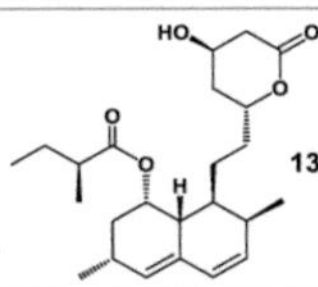 **13** (Monacolin K = Lipitor)
6	*Astragalus membranaceus*	Immunostimulating polysaccharides
7	*Scutellaria baicalensis* (one of four herbs in Huang Chin Tang	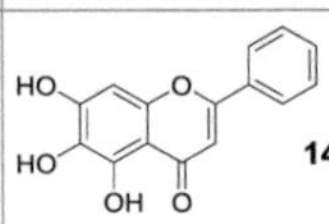 **14** (Baicalein)
8	*Eucommia ulmoides*	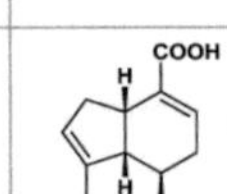 **15** (Geniposidic acid) 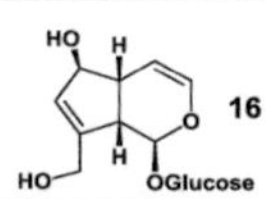 **16** (Aucubin)
9	*Ligusticum wallichii*	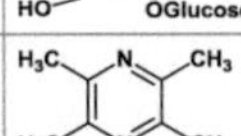 **17** (Tetramethylpyrazine = ligustrazine = chuanxiongzine)
10	*Lycium barbarum*	*Lycium barbarum* polysaccharides (LBPs)

2011). Turmeric oils and oleoresins with various promising activities have been marketed globally. To control the quality of these products, Ar-turmerone (**5**), turmerone (**6**), and *β*-turmerone (**7**) are usually employed as chemical markers, for example, a minimum of 40 % of these compounds in turmeric oils and oleoresins are required (Li et al. 2011).

1.3.1.3 Bioactivity

The primary active compound of turmeric is curcumin (**1**, Table 1.1), which is also responsible for turmeric's vibrant yellow color. Curcumin was isolated in 1815, obtained in crystalline form in 1870, and ultimately identified as 1,7-bis-(4-hydroxy-3-methoxyphenyl)-1,6-heptadiene-3,5-dione-(1E,6E) (Miłobdzka and Lampe 1910). Chemically, curcumin is a diarylheptanoid, in which two ferulic acid fragments are connected by a methylene bridge. The β-diketone moiety of curcumin can undergo keto–enol tautomerism (Fig. 1.1) and exists entirely in the enol form in both solution and solid phases (Pedersen et al. 1985).

Extensive studies have indicated that curcumin possesses versatile bioactivity, including anticarcinogenic, immunomodulatory, antioxidant, anti-inflammatory, anti-angiogenesis, anticancer, chemopreventive, anti-Alzheimer's disease, anti-thrombotic, antimalarial, anti-rheumatoid arthritis, anti-HIV, wound healing, anti-hepatotoxic, anti-psoriasis, hypoglycemic, and antihyperlipidemic effects. Thus, curcumin has therapeutic potential against a wide range of diseases, such as

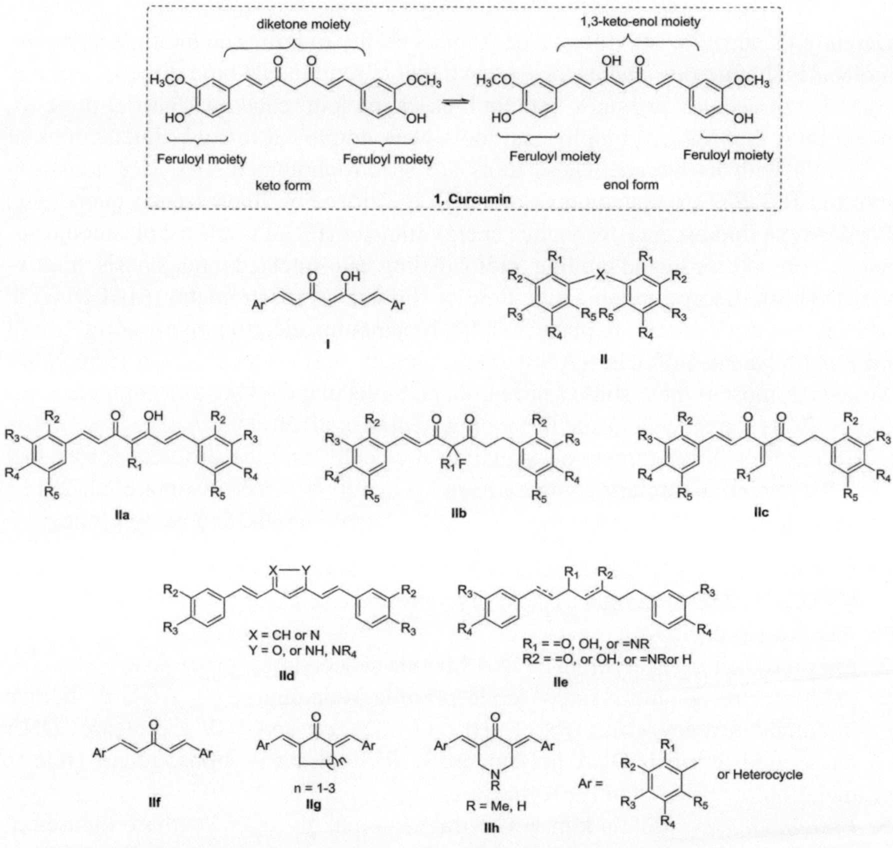

Fig. 1.1 Tautomerism of curcumin and design of different analogs

inflammatory, lung, neurological, liver, metabolic, autoimmune, and cardiovascular diseases, as well as cancers. Nowadays, around 65 clinical trials are ongoing using curcumin for various diseases, especially for anti-inflammatory, chemoprevention, or cancer therapy (Bairwa et al. 2014; Goel et al. 2008; Hsu and Cheng 2007).

Curcumin has been testified to be nontoxic even at high dosages. To date, no toxicity has been found in any animal and human studies of curcumin. It has been classified as "generally recognized as safe" (GRAS) by the National Cancer Institute (Anand et al. 2007; Sharma et al. 2001). However, several pharmacokinetic and pharmacodynamic studies on curcumin have indicated that it is rapidly metabolized, conjugated in the liver, and excreted in the feces, thus having limited systemic bioavailability (Cheng et al. 2000; Ireson et al. 2001). Research studies have shown that the administration doses as high as 8 g of curcumin per day to human subjects resulted in only an average peak serum concentration of ~1.77 μM of curcumin (Anand et al. 2007; Sharma et al. 2001).

1.3.1.3.1 Curcumin Is a Multi-target Natural Compound

Curcumin's versatile activities come from its ability to influence multiple signaling molecules. Numerous studies have shown that curcumin can bind directly to many signaling molecules, enzymes, protein kinases, protein reductases, carrier proteins, metal ions, etc. Various biophysical tools were employed to study interactions of curcumin with its targets. These tools are spectrophotometry, Fourier transform infrared (FTIR), circular dichroism (CD) spectroscopy, fluorescence quenching, Förster-type fluorescence resonance energy transfer (FRET), surface plasmon resonance, competitive ligand binding, radiolabeling, site-directed mutagenesis, matrix-assisted laser desorption/ionization time-of-flight mass spectrometry (MALDI-TOF MS), immunoprecipitation, phage display biopanning, electron microscopy, 1-anili no-8-naphthalene-sulfonate (ANS) displacement, and co-localization techniques. Moreover, most of these studies have utilized molecular docking as a computational tool to study the mode and site of binding (Gupta et al. 2011).

Curcumin's direct targets on signaling molecules and the different forces that bind the curcumin–protein complex have been well reviewed (Gupta et al. 2011). According to this review, the targets of curcumin fall into the following groups:

1. *Inflammatory molecules*: tumor necrosis factor (TNF)-α, cyclooxygenase (COX)-1, COX-2, human a1-acid glycoprotein (AGP), and myeloid differentiation protein 2 (MD-2);
2. *Enzymes*: histone acetyltransferases, histone deacetylases, glyoxalase I, xanthine oxidase, proteasomes, sarco (endo)plasmic reticulum Ca^{2+} ATPase, human immunodeficiency virus type 1 (HIV-1) integrase and HIV-1 protease, DNA methyltransferase 1, DNA polymerase l, ribonuclease A, lipoxygenase, matrix metalloproteinases, and lysozyme;
3. *Protein kinases*: protein kinase C, viral sarcoma, glycogen synthase kinase-3β, ErbB2 (HER2/neu), and phosphorylase kinase;

4. *Protein reductases*: thioredoxin reductase and aldose reductase;
5. *Carrier proteins*: casein, albumin, fibrinogen, β-lactoglobulin, and immunoglobulin;
6. *Other proteins*: cell survival proteins such as Bcl-2, cytoskeletal proteins such as FtsZ, and homotetrameric proteins such as transthyretin; and
7. *Tubulin*.

1.3.1.3.2 Chemical Modification Study on Curcumin

With an aim to improve the bioactivity as well as bioavailability while retain a similar safety profile to curcumin, numerous curcumin analogs have been obtained through total synthesis or semi-synthesis (Bairwa et al. 2014; Lin et al. 2006a, b). Based on the structural features of curcumin, the modifications can be grouped into two series. In series I (Fig. 1.1), modifications were focused on the phenyl moiety. Various groups were substituted on the phenyl ring, or the phenyl rings were replaced by various heterocycles. In series II (Fig. 1.1), various linkers were used to connect the two phenyl ring moieties. According to the different type of the linkers, the modifications could be further grouped as: substitutions on the 1,6-heptadiene moiety (**IIa**, **IIb**, **IIc**, Fig. 1.1), forming a heterocycle on the 1,6-heptadiene moiety (**IId**, Fig. 1.1), altering the conjugation of the 1,6-heptadiene moiety (**IIe**, Fig. 1.1), and changing the β-diketone to a monocarbonyl moiety (**IIf**, **IIg**, **IIh**, Fig. 1.1). The analogs produced were assayed for their antioxidant, anti-inflammatory, cytotoxic, anti-malarial, anti-HIV, antitrypanosomal, or antileishmanial effects. The results showed that the modifications can affect the bioactivity, and some new derivatives have exhibited improved activity as well as drug profile. However, no common structure–activity relationship (SAR) correlations could be concluded from these modifications.

1.3.1.3.3 Preclinical and Clinical Research on Curcumin and Its Derivatives in the Areas of Inflammation, Chemoprevention, and Cancer

Extensive research conducted on cell cultures, animal models, and clinical trials indicated that curcumin may have potential as a therapeutic agent in inflammatory bowel disease, pancreatitis, arthritis, and chronic anterior uveitis. Research shows that curcumin can interact with numerous molecular targets involved in inflammation. Curcumin modulates the inflammatory response by downregulating the activity of cyclooxygenase-2 (COX-2), lipoxygenase, and inducible nitric oxide synthase (iNOS) enzymes, inhibiting the production of the inflammatory cytokines tumor necrosis factor-alpha (TNF–α), interleukin (IL)-1, -2, -6, -8, and -12; monocyte chemoattractant protein (MCP); and migration inhibitory protein and downregulating mitogen-activated and Janus kinases.

Several clinical trials of curcumin were conducted for various inflammatory conditions, including postoperative inflammation, arthritis, uveitis, inflammatory pseudotumors, dyspepsia, irritable bowel syndrome, inflammatory bowel disease, pancreatitis, and *Helicobacter pylori* infection. Most studies have been promising and warranted further research on curcumin's therapeutic value for treatment of inflammatory conditions (Jurenka 2009). For example, studies proved curcumin to be superior to phenylbutazone and placebo in reducing spermatic cord edema, which resulted from surgery for inguinal hernia or hydrocele (Satoskar et al. 1986). In a preliminary double-blind, randomized, controlled trial (RCT), curcumin given at 1,200 mg daily was effective in improving joint swelling, morning stiffness, and walking time (Deodhar et al. 1980). In a clinical trial involving 32 patients with anterior uveitis, after administrating 375 mg curcumin for 12 weeks, visual acuity and aqueous flare improvements and decrease in keratic precipitates were observed (Lal et al. 1999). Curcumin was also used for idiopathic orbital inflammatory pseudotumors (IOIP). In a small study of eight patients with IOIP, 375 mg curcumin three times daily was given for 6–22 months, until complete regression of symptomology was achieved. Patients were followed for 2 years and assessed at 3-month intervals. Only five patients completed the study, but four completely recovered on curcumin therapy (Lal et al. 2000). Moreover, in several clinical trials and a pilot study, curcumin demonstrated anti-inflammatory activity and therapeutic benefit in various gastrointestinal conditions, including dyspepsia, *Helicobacter pylori* infection, peptic ulcer, irritable bowel syndrome, Crohn's disease, and ulcerative colitis. Besides the examples mentioned above, numerous clinical trials are ongoing to explore the effect of curcumin on various inflammatory conditions. Jurenka has published an excellent review on this topic (Jurenka 2009).

It is well acknowledged that proinflammatory states are related to tumor promotion. Accordingly, the chemopreventive activity of curcumin has also been widely explored. Preclinical cancer research has shown that curcumin can inhibit carcinogenesis in various cancer types, such as colorectal, pancreatic, gastric, prostate, hepatic, breast, and oral cancers, as well as leukemia, and at various stages of carcinogenesis (Aggarwal et al. 2003). Several animal studies demonstrated that curcumin can inhibit all three stages of carcinogenesis: initiation, promotion, and progression. In the initiation and promotion stages, curcumin modulates transcription factors controlling phases I and II detoxification of carcinogens; downregulates proinflammatory cytokines, free-radical-activated transcription factors, and arachidonic acid metabolism via cyclooxygenase and lipoxygenase pathways; and scavenges free radicals. In the promotion and progression stages of carcinogenesis, curcumin decreases frequency and size of tumors and induces apoptosis via suppression of NF-kB and AP-1 in several cancer types (Chan 1995; Hong et al. 2004; Jurenka 2009; Kawamori et al. 1999; Singh and Aggarwal 1995). Currently, there are at least three ongoing clinical trials exploring the preventive benefits of curcumin as therapy to treat patients with adenomatous polyps at risk for colorectal cancer (Jurenka 2009).

In cancer therapy clinical trials, curcumin alone, as well as combinations with other agents such as bioperine and ashwagandha, was used to treat rectal cancer, pancreatic cancer, multiple myeloma cancer, osteosarcoma, or oral mucositis (Jurenka 2009). For example, one clinical trial indicated that curcumin can stabilize disease progression in patients with advanced pancreatic cancer. Twenty-one patients received 8 g of curcumin daily until the disease progressed. One patient achieved disease stabilization for 18 months (Dhillon et al. 2008).

JC-9 (**7**, also known as ASC-J9, Table 1.1), a synthetic derivative of curcumin, exhibited potent activity against PC-3 (IC_{50} 1.1 μM) and LNCaP (IC_{50} 1.3 μM) cell lines (Lin et al. 2006a, b). Furthermore, JC-9 was found to be active in vivo against hepatocellular carcinoma and bladder cancer (Ma et al. 2008; Miyamoto et al. 2007). The mechanism of action study indicated that JC-9 enhances androgen receptor degradation (Shi et al. 2009). Another study indicated that JC-9 and its derivatives can overcome EGFR-TKI lung adenocarcinoma drug resistance and reduce EGFR-TKI-induced GI adverse effects (Wada et al. 2015). Therefore, JC-9 is a potential clinical trial candidate for treating prostate cancer, liver cancer, bladder cancer, and other cancers. JC-9 was licensed by AndroScience Corp. (San Diego, CA) and succeeded in phase 2 clinical trials against acne in 2014. Moreover, antiprostate clinical trials with JC-9 are being planned (Itokawa et al. 2008; Lee et al. 2008).

1.3.1.4 Conclusion

Turmeric has been used in traditional medicine since ancient times with a wide spectrum of therapeutic areas. Enormous amounts of data from preclinical and clinical research have confirmed turmeric's benefits as described in traditional medicine. However, clinical applications of turmeric as a botanical drug can be made only after extensive research on its bioactivity, mechanism of action, pharmacotherapeutics, and toxicity, as well as developing standard QC. Curcumin is the major biologically active constituent of turmeric. In spite of numerous reports showing its putative mechanism(s), multiple molecular targets, and wide range of therapeutic applications, curcumin has not yet been approved for treatment of any human disease, even though it has been reported to be safe for humans at gram dosages. The main obstacle to utilizing curcumin therapeutically is its poor systemic bioavailability. Also, because curcumin can bind to multiple signaling molecules, it is rather difficult to determine which target is causing the desired effect in a certain disease. Therefore, further study should be focused on designing additional different curcumin analogs that will be more effective and better absorbed, as well as on developing a clearer understanding of the actual functional meaning of direct interactions of curcumin with various specific targets. Such discoveries will help to elevate this fascinating natural product as a therapeutic agent for the treatment of human diseases.

1.3.2 Antrodia camphorata

1.3.2.1 Introduction

Antrodia camphorata (syn. *Antrodia cinnamomea*, *Taiwanofungus camphoratus*) is a unique basidiomycete fungus of the Fomitopsidaceae family. It is a parasite that affects the inner cavity of *Cinnamomum kanehirae* Hayata (Lauraceae) (bull camphor tree), which grows only in Taiwan with a distribution in broad-leaved forests at an altitude of 200–2,000 m (Chang and Chou 1995). *A. camphorata*, also called "Niu-chang-chih" (牛樟芝), "Chang-chih," "Niu-chang-ku," or "Chang-ku" in Taiwan, is used to lessen the discomforts caused by alcohol drinking or exhaustion and is believed to be beneficial to preserve human vitality and promote longevity. Moreover, it has been used in TCM to treat a number of illnesses including diarrhea, abdominal pain, hypertension, itchy skin, viral infection, stomatitis, diabetes mellitus, nephritis, proteinuria, liver cirrhosis, hepatoma, influenza, car sickness, heat-related fever, and motion sickness (Geethangili and Tzeng 2011; Wu and Ryvarden 1997). Therefore, in Taiwan, *A. camphorata* is also called as "ruby in mushroom."

Production of *A. camphorata* can occur by gathering mature fruiting bodies of wild *A. camphorata*, wood or solid-state cultivation, and liquid-state fermentation. Mature fruiting bodies of the wild *A. camphorata* are now close to extinction due to their very slow growth rate (1 or more years) and overharvesting (Huang et al. 2012).

1.3.2.2 Chemical Constituents

Currently, around 200 compounds have been identified from *A. camphorata*. These compounds have been summarized in four reviews (Ao et al. 2009; Geethangili and Tzeng 2011; Huang et al. 2012; Lu et al. 2013). Generally, the compounds isolated from *A. camphorata* include triterpenoids, diterpenoids, monoterpenes, steroids, lignans, benzoquinones, benzenoids, maleic/succinic acid derivatives, fatty acids and their esters, polysaccharides, etc. Terpenoids are the main chemical constituents of fruiting bodies of *A. camphorata*. Among the over 40 isolated terpenoids, most of them are lanostane (**III**, Fig. 1.2) and ergostane (**IV**, Fig. 1.2) triterpenes.

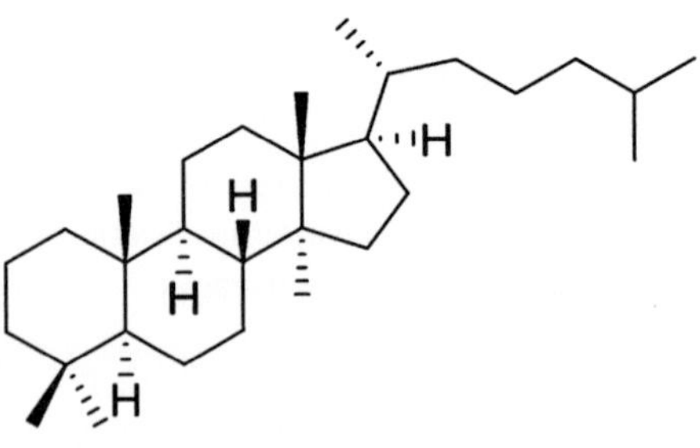

III, Lanostane

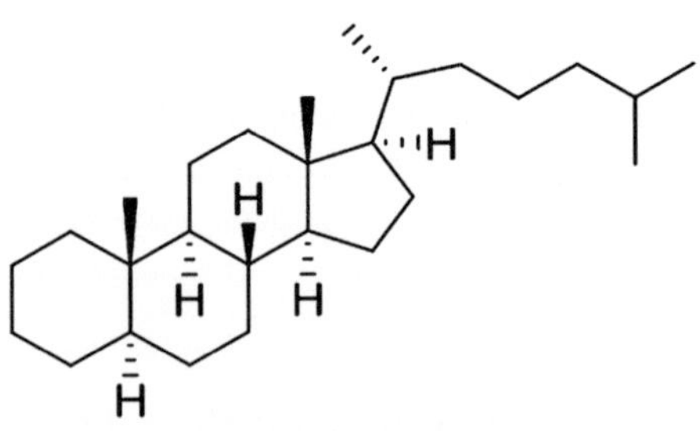

IV, Ergostane

Fig. 1.2 Most common triterpene skeletons found in *Antrodia camphorata*

1.3.2.3 Pharmacological Effects

Various crude extracts of *A. camphorata*, such as ethanol/methanol, chloroform, ethyl acetate, and aqueous extracts, have been tested for their pharmacological effects in vivo and in vitro in different modes. These research efforts indicated that *A. camphorata* possesses versatile bioactivities, including anticancer activity, anti-inflammatory and immunomodulatory effects, inhibition of hepatitis B virus (HBV) replication, antioxidant properties, antimicrobial activity, hepatoprotective effects, prevention of liver fibrosis, neuroprotective activity, antihypertensive effects, vaso-relaxation properties, antihyperlipidemic activity, and cardiovascular effects (Ao et al. 2009; Geethangili and Tzeng 2011; Huang et al. 2012; Lu et al. 2013).

Beside the crude extract, various pure natural products from *A. camphorata* have also been widely studied and found to exhibit cytotoxic, neuroprotective, anti-inflammatory, anti-insecticidal, anti-HBV, and anti-HCV effects. For example, several triterpenoids, such as camphoratins B–F and zhankuic acid A, showed moderate to potent cytotoxicity, with EC_{50} values ranging from 0.3 to 3 μM against KB and KB-VIN human cancer cell lines. Moreover, camphoratins F and J, as well as zhankuic acid A and their related compounds, were also found to exhibit anti-inflammatory NO production inhibition activity with IC_{50} values of less than 5 μM and were more potent than the nonspecific NOS inhibitor nitro-L-arginine methyl ester (Shi et al. 2011; Wu et al. 2010).

Antroquinonol (**8**, Table 1.1), antroquinonol B, antroquinonol C, and related compounds are ubiquinone derivatives isolated from mycelia, fruiting bodies, or both of *A. camphorata*. Antroquinonol was reported to exhibit cytotoxic activities against cancer cell lines MCF-7, MDA-MB-231, Hep 3B, Hep G2, DU145, and LNCaP with IC_{50} values ranging from 0.13 to 6.09 μM (Lee et al. 2007). Golden Biotechnology Corp. of Taiwan patented the application of antroquinonol and its related compounds for inhibiting the growth of breast, lung, hepatic, and prostate cancers (Liu et al. 2008). Antroquinonol (Hocena®) has completed a phase 1 study in the USA (https://clinicaltrials.gov/ct2/show/NCT01134016), and a phase 2 study is planned to treat non-small cell lung cancer (https://clinicaltrials.gov/ct2/show/NCT02047344). It was granted Orphan Drug Designation by the US FDA for pancreatic cancer (January 21, 2015), acute myeloid leukemia (April 30, 2015), and hepatocellular cancer (July 23, 2015).

1.3.2.4 Conclusion

The pharmacological effects of an extract or isolates from *A. camphorata* have been widely studied for decades. However, most studies have been performed in vitro and in vivo with animals. Next, more clinical studies are expected to confirm the therapeutic benefit of *A. camphorata*. Although over 200 compounds have been isolated, little research regarding modification and synthesis of active compounds has been conducted. Therefore, continued medicinal chemistry study on *A. camphorata* may help to provide more promising leads for drug development. Also, it is urgent to establish standard quality control and methods for developing the crude extract of *A. camphorata* as a botanical drug that could be accepted worldwide.

1.3.3 Apium graveolens

1.3.3.1 Introduction

Apium graveolens Linn. (celery, Han Qin, 旱芹, Umbelliferae), belonging to the family Apiaceae, is a common annual herb widely cultivated in temperate zones. Its leaf stalks are consumed as a popular vegetable. In various forms, such as fresh herb, stalk, leaves, seeds, seed oil, and oleoresin, celery is used to flavor foods. Celery stalks have shown a broad bioactivity spectrum, including antihyperlipidemic, antihypertensive, and memory enhancement effects (Anonymous 2009). Its seeds are used for treatment of bronchitis, asthma, and diseases of the liver and spleen (Satyavati and Raina 1976).

1.3.3.2 Chemical Constituents

The chemical constituents of the whole herb are represented by psoralen, bergapten, isopimpinellin, luteolin, apiin, chrysoeriol, apigenin, quercetin, and volatile oils including d-limonene, myrcene, isobutyric acid, valeric acid, 3-isobutylidenephthalide, 3-isovalidenephthalide, and cis-3-hexen-1-yl-pyruvate. Celery seeds contain various bioactive compounds, including phthalides, coumarins, flavonoids, sesquiterpenoids, and aromatic glucosides. The seed oil is a valuable product in both the flavor and fragrance industries. 3-*n*-Butylphthalide, 3-*n*-butyl-4,5-dihydrophthalide (sedanenolide), and sedanolide are reported to be the major flavor components of the seed oil. The volatile oils of the leaves were reported to contain octene-4,5-dione, 2-isopropyloxyethane, sabinyl acetate, 1,4-butanediol, seselin, rutaretin, celereoin, celeroside, apiumoside, isoquercitrin, vellein, apiumetin, nodakenin, myristic acid, 8-hydroxy-5-methoxypsoralen, umbelliferone, and nodakenetin (Anonymous 1999a).

1.3.3.3 Bioactivity

Celery has been used as a medicinal plant for its aphrodisiac, anthelmintic, antispasmodic, carminative, diuretic, laxative, sedative, stimulant, and tonic effects. The medicinal parts of the plant are the roots, leaves, and seeds. Products of celery are also used for blood purification, for regulating bowel movements for evacuation, for glandular stimulation, and as a cure for gallstones and kidney stones. Celery seeds are implicated in arthritic pain relief and for treating rheumatic conditions and gout. Celery is also effective at lowering blood pressure due to the 3-*n*-butyl phthalide constituent, which has been demonstrated to relax the smooth muscles that line blood vessels (Anonymous 1999a).

Various extracts or fractions of celery were reported to possess antioxidant activity, anti-inflammatory effect, hepatoprotective activity, antimicrobial activity, and inhibition of blood platelet aggregation. Coumarins isolated from celery were reported to help prevent free radicals from damaging cells, thus decreasing the mutations that increase the potential for cells to become cancerous (Sowbhagya 2014).

S-(-)-3-*n*-butylphthalide [*S*-(-)-3-butyl-1(3*H*)-isobenzofuranone, *S*-(-)-NBP, **9**, Table 1.1] was isolated from celery seeds. Pharmacological studies indicated that s-(-)-NBP, as well as synthesized (±)-NBP, has anti-ischemic effects. When stroke-prone spontaneously hypertensive rats were pretreated with (±)-NBP, the onset of stroke was delayed, life span was prolonged, and the neurological deficit score was decreased. (±)-NBP was also reported to ameliorate brain edema and blood–brain barrier damage in middle cerebral artery-occluded rats. (±)-NBP may have neuroprotective effects, since it improved mitochondria dysfunction. S-(-)-NBP and (±)-NBP also exhibited an inhibitory effect on thrombus formation in rats (Zhu et al. 2004). (±)-NBP has been developed as an anti-cerebral ischemic drug and has completed or is ongoing in phase 2–4 clinical trials in China (https://clinicaltrials.gov/ct2/show/NCT02149875, NCT01405248, NCT00724724, and NCT02594995). The results from the completed clinical trials confirmed its effect for treatment of mild and moderate acute ischemic stroke. In phase 2 trials, the total rate of efficacy was 70.3 % among the 91 effective cases. Phase 3 trials indicated that, in 282 effective cases, the total rate of efficacy was 63.9 %. In phase 4 trials, among the two groups with 305 and 1,147 effective cases, the efficacy ratios in the two groups were 78.4 % and 78.2 %, respectively. (±)-NBP was approved by the State Food and Drug Administration (SFDA) of China in 2005 for the treatment of ischemic stroke.

1.3.3.4 Conclusion

Celery and the natural products isolated from celery have exhibited great health benefits. The approval of (±)-NBP by the China SFDA for treating ischemic stroke is an excellent accomplishment. Nevertheless, further study on (±)-NBP-related compounds is still needed for meeting the approval from the US FDA, so that the world-class anti-stroke drugs can be obtained.

1.3.4 Momordica charantia

1.3.4.1 Introduction

In TCM, plants with a bitter flavor and cold property have long been used in the treatment of diabetes and high blood sugar (Chen et al. 2015). *Momordica charantia* (Chinese: ku gua 苦瓜), known as bitter melon or bitter cucumber, is a common edible vegetable, although all parts of the plant have also been used medicinally. The plant is a tropical and subtropical vine belonging to the family Cucurbitaceae, which produces a warty or ridged oblong fruit. Although it is bitter, the rind of the green fruit is often eaten cooked in stir fries in Chinese cooking or curries in Southeastern dishes. When the fruit is fully ripe, the rind is extremely bitter, but the red pith is sweet. In Japanese Kampo medicine, bitter melon is used for the depletion of yin and, thus, to treat yin stages of chronic diseases such as diabetes (Rister 1999). Indeed, the antidiabetic potential of this plant species has gained significant scientific attention (Broadhurst et al. 2000; Chaturvedi 2012; Grover and Yadav 2004; Horax et al. 2010; Hsu et al. 2011; Wang et al. 2012), particularly regarding the hyperglycemic effects of various chemical components.

1.3.4.2 Chemical Constituents

Components of this plant include flavonoids, isoflavones, anthroquinones, glucosinolates, steroidal saponins, glycol alkaloids, cucurbitane triterpenoids (e.g., charantin (**10**, Table 1.1)), and polypeptides (e.g., polypeptide-p) (Hamid et al. 2015; Hung et al. 2012; Kaur et al. 2016; Medagama and Bandara 2014; Snee et al. 2011; Tan et al. 2008). As many of these compounds have a bitter taste (Drewnowski and Gomez-Carneros 2000), studies have been conducted to determine how to best mask the bitter taste, if wider acceptance and consumption of *M. charantia* is to be promoted for medicinal value (Snee et al. 2011). The phytochemistry of the plant, particularly regarding antidiabetic constituents, has been well reviewed (Grover and Yadav 2004; Joseph and Jini 2013; Raman and Lau 1996).

1.3.4.3 Bioactivity

Bitter melon has been used as a folk medicine for numerous conditions, including stomach and gastrointestinal disorders (e.g., to treat colic in infants and peptic ulcers in adults as well for carminative, laxative, or purgative effects) (Grover and Yadav 2004). However, in traditional uses and modern medical studies (Joseph and Jini 2013; Raman and Lau 1996), the plant has been mostly appreciated for its ability to relieve diabetes by improving glucose metabolism (Hasan and Khatoon 2012; Leung et al. 2009), although it has also been studied for anticancer, antiviral (including HIV, herpes, polio), analgesic, anti-inflammatory, and hypotensive effects (Grover and Yadav 2004; Snee et al. 2011).

Among the compounds listed above, cucurbitane-type triterpenoids have been linked to AMP-activated protein kinase activity as a possible hypoglycemia mechanism (Tan et al. 2008), while polypeptide-p mimics the action of human insulin and might be used as a plant-based insulin replacement (Paul and Raychaudhuri 2010).

An exciting recent development is the discovery of a potential novel antidiabetic protein. *Momordica charantia* insulin receptor (IR)-binding protein (*mc*IRBP) is a novel insulin receptor (IR)-binding polypeptide with a distinct binding site from that of insulin (Lin et al. 2014). However, *mc*IRBP also enhances glucose uptake in cells and clearance in normal and diabetic mice, likely via stimulation of similar biopathways as insulin and regulation of genes affecting glucose and lipid metabolism (Lin et al. 2014).

1.3.4.4 Conclusion

Momordica charantia offers an alternative treatment to both injectable insulins, the only treatment for type 1 diabetes and current oral type 2 antidiabetic drugs, including sulfonylureas/insulinotropics, biguanides, α-glucosidase inhibitors, thiazolidinediones, and gliptins. Further studies and strictly controlled clinical trials are needed to evaluate thoroughly the potential of *mc*IRBP to treat diabetes, a life-altering metabolic disease, which can lead to blindness, renal disease, lower extremity amputations, and even death.

1.3.5 Monascus purpureus

1.3.5.1 Introduction

Red mold rice (RMR) is produced from ordinary rice by fermentation with the yeast species *Monascus purpureus* (Chinese: 紅麴). In East Asia, RMR is used to flavor, color, and preserve food and medicinally to improve blood circulation and digestion, as well as to treat diabetes (Heber et al. 1999; Li 1596).

1.3.5.2 Chemical Constituents

The fermentation process produces various secondary polyketide metabolites (Li 1596; Ma et al. 2000), including monascin (**11**, Table 1.1), ankaflavin (**12**, Table 1.1), and several monacolins.

1.3.5.3 Bioactivity

The hypolipidemic and hypocholesterolemic effects of these compounds have been noted for many years (Endo 1979; Lee et al. 2010). Indeed monacolin K (**13**, Table 1.1) is the same compound now widely marketed as the cholesterol-lowering drug lovastatin. RMR has also been studied for its anti-obesity effects, which have been linked to its lipolytic activity and subsequent prevention of body fat accumulation (hypertrophy and hyperplasia of adipose tissue), as well as mild appetite suppression (Chen et al. 2008). Finally, RMR fermented with *M. purpureus* strain NTU 568 [found to produce a higher content of the active components (Shi and Pan

2011)] has been developed as a commercial functional food to prevent Alzheimer's disease (AD) (Lee and Pan 2011a, b).

1.3.5.4 Conclusion

These fermented products hold promise for preventive medicine, particularly decreasing heart disease risk. Continued studies are aimed at identification of the neuroprotective metabolites in RMR, further evaluation of their mechanism of action, and structural modifications to improve pharmacological profiles as potential drugs for neurodegenerative disorders, including AD and Parkinson's disease (Lin et al. 2015).

1.3.6 Astragalus membranaceus

1.3.6.1 Introduction

The roots of *Astragalus membranaceus* (Chinese: Huang Chi 黃耆) (known in TCM for sweet flavor, slightly warm property) are used in various TCM formulas to counteract symptoms associated with a deficiency of chi, e.g., low energy, lack of strength, anorexia, slow healing, etc. (Batch 2010). As the authors have previously provided a detailed description of this herb (Lee et al. 2013), this section will focus only on the studies of PG2, an IV injection used to alleviate cancer-related fatigue and improve the quality of life for cancer patients.

1.3.6.2 Chemical Constituents

Polysaccharide immunostimulatory principles were developed as PG2 by Pharmagenesis in the USA and PhytoHealth Corporation of Taiwan based on the initial advice of Dr. K.H. Lee (Lee et al. 1993).

1.3.6.3 Bioactivity

In a randomized, double-blind placebo-controlled phase 2/3 clinical study (https://clinicaltrials.gov/ct2/show/NCT00523107), patients with advanced cancer who experienced moderate to severe cancer-related fatigue (CRF) had a higher response rate when they received PG2 compared with patients who received a placebo (normal saline). This study concluded that PG2 could be effective and safe for managing CRF in advanced cancer patients, without major or irreversible toxicities (Chen et al. 2012). Subsequently, PG2 was approved for clinical use in treating CRF by the Taiwan Department of Health in April 2011, particularly in cancer patients who developed severe fatigue after receiving chemotherapy. In another clinical trial, a combination treatment with PG2 and platinum-based chemotherapy in non-small cell lung cancer patients led to improved quality of life indices, including management of fatigue, nausea, vomiting, pain, and appetite loss (Kuo et al. 2015). Recruitment is currently ongoing for a phase 4 clinical trial (https://clinicaltrials.gov/ct2/show/NCT01720550) to evaluate the use of different doses of PG2 treatment for fatigue improvement in advanced cancer patients who are under standard palliative care in a hospice setting. Currently, all clinical studies have been performed with an IV injectable form of PG2.

1.3.6.4 Conclusion

In addition to the clinical studies, gene expression profiling showed that PG2 product batches were of high quality and consistency as well as functionally equivalent regarding their effects on the immune and hematopoietic systems (Kuo et al. 2015). A bioinformatics-based approach provided a quantitative measurement for the quality and consistency of herbal medicines and revealed new roles (e.g., immune modulation) for PG2 in cancer treatment. In the same study, PG2 and doxorubicin acted synergistically on induced cell death in HL-60 cells, raising the possibility of a novel antitumor function for PG2. Also, clinical trials (http://www.phytohealth.com.tw/upimages/newarrival1121228011612.pdf) are being performed with PG2 in Taiwan for potential use in treating stroke, both hemorrhagic and ischemic.

1.3.7 Huang Chin Tang

1.3.7.1 Introduction

Huang Chin Tang (or *Scutellaria* decoction) has been used as a Chinese herbal medicine for over 1,800 years to treat various GI issues, including diarrhea, nausea, vomiting, and abdominal cramps, as well as fever and headaches.

1.3.7.2 Chemical Constituents

Huang Chin Tang contains the four herbs shown below in a 3:2:2:2 ratio.

- *Scutellaria baicalensis* (Chinese: Huang Qin 黃芩) (common: skullcap): part used – root, main purpose – to regulate digestive and intestinal functions
- *Paeonia lactiflora* (Chinese: Bai Shao 芍藥) (common: white peony): part used – root, main purpose – combines with *Glycyrrhiza* to relax abdominal cramps
- *Glycyrrhiza uralensis* (Chinese: Zhi Gan Cao 炙甘草) (common: licorice): part used – root, main purpose – combines with *Paeonia* to relax abdominal pain and

cramps as well as to harmonize all ingredients and diminish side effects due to the more harsh components, such as *Scutellaria*

- *Ziziphus jujuba* (Chinese: Da Zao 大棗) (common: jujube, Chinese date): part used – fruit, main purpose – to stimulate digestive functions

Thus, this Chinese medicine formulation is a complex mixture containing numerous compounds, unlike Western drugs, which usually contain only a single compound. Liquid chromatography and tandem mass spectrometry have been used to identify various chemicals, including flavonoids, triterpene saponins, and monoterpene glycosides, as well as assign them to the four individual herbs from PHY906, a carefully prepared, consistent, quality-controlled formulation of Huang Chin Tang for cancer therapy (Ye et al. 2007). Flavonoids, such as baicalein (**14**, Table 1.1), are major active compounds from *Scutellaria*.

1.3.7.3 Bioactivity

While cancer chemotherapeutic drugs can lengthen the life of a cancer patient, their toxicities and side effects can also severely diminish the quality of that life. PHY906 showed promising results in phase 1 clinical trials in 2010 (Saif et al. 2010) and provided relief from chemotherapy-induced gastrointestinal toxicity, especially severe diarrhea. While PHY906 did not affect the initial intestinal DNA damage caused by irinotecan, it did encourage regeneration of the intestinal progenitor or stem cells and several signaling components, ultimately restoring the intestinal epithelium. Other simultaneous action modes were also evident (Lam et al. 2010). Notably, in addition to reducing irinotecan-induced toxicities, PHY906 could potentiate the drug's anticancer activity and exert favorable actions in multiple cancers treated with numerous chemotherapy drugs acting by different mechanisms (Liu and Cheng 2012). In a 2014 phase 2 clinical trial, a combination of PHY906 and capecitabine, a prodrug of 5-FU, was found to be a viable option for patients with advanced pancreatic cancer treated previously with gemcitabine (Saif et al. 2014). In the small cohort, quality-of-life indices of hand-to-foot syndrome and diarrhea were improved, and median overall survival was extended, especially in two patients with partial responses (69 and 83 weeks) compared with the previously reported 2 months for other second-line salvage therapies (Kang and Saif 2008). In a recently reported study, PHY906 potentiated the anti-hepatoma activity of the drug sorafenib by multiple mechanisms, particularly attraction of macrophages with a higher M1/M2 (tumor rejection) expression (Lam et al. 2015).

1.3.7.4 Conclusion

Based on the studies above, this CHM formula shows great promise as an adjuvant cancer therapy.

1.3.8 Eucommia ulmoides

1.3.8.1 Introduction

Eucommia ulmoides (Chinese: Tu Chung 杜仲) is a unique tree called the hardy or hard rubber tree and originating in central China. The most medically useful plant parts are the stem bark and leaves, used to make Tu Chung tea. This tea is considered to be useful in blood pressure regulation and weight control, as well as to rejuvenate skin elasticity and strengthen the lower back and knees. *Eucommia* is, thus, known as a “longevity herb” with its continued use thought to slow the aging process by improving body metabolism and strengthening the muscular skeletal system. *Eucommia* is well known in the traditional medicines of China, Japan, and Korea and is mentioned in the Chinese classic *Shennong Bencaojing* Yang (1998). It has analgesic (especially for pain in the lower back and knees), tonic (particularly for right kidney yang), and hypotensive actions. Regarding the former action, *Eucommia* is likely best known as part of the TCM arthralgia remedy *Duhuo Jisheng Tang* and is used to treat rheumatoid arthritis, rheumatic back pain, and sciatica (Dharmananda 1999, 2000). In most traditional formulations, *Eucommia* is a relatively minor ingredient in a large combination of herbs, e.g., it is 7 % of *Duhuo Jisheng Tang*; however, together with *Dipsacus asperoides*, it is a major component of *Duzhong Wan*, a TCM for treating lower back pain during pregnancy (Huang et al. 2014).

1.3.8.2 Chemical Constituents

Numerous chemical constituents and pharmacological activities of *Eucommia ulmoides* have been described (Deyama et al. 2001; Lee et al. 2012). The main chemical components of interest about the traditional use of Tu Chung are iridoid glycosides, including geniposidic acid and aucubin, and lignans, such as pinoresinol glucoside and diglucoside.

1.3.8.3 Bioactivity

Various lignans have been linked to antihypertensive and vasodilating effects (Deyama et al. 2001; Luo et al. 2010; Sih et al. 1976). In mechanistic studies (Jin et al. 2008), the vasodilation induced by an aqueous extract of *Eucommia ulmoides* was found to depend on endothelium-derived hyperpolarizing factor (EDHF) with the activation of K^+ channels. In addition, muscarinic Ach receptor agonists could also play a role in lowering blood pressure via vasodilation. Geniposidic acid (**15**, Table 1.1) also has antihypertensive effects (Deyama et al. 2001), while aucubin (**16**, Table 1.1) and other iridoids are likely responsible for the anti-inflammatory effects of *Eucommia*. Aucubin inhibits the arachidonic acid pathway,

TNF-α-induced responses, and NF-kB activation (Benito et al. 2000; Park 2013), which may partly explain the herbal use in the treatment of arthritis (Wang et al. 2015b), as well as indicate a potential for improving obesity-induced atherosclerosis (Park 2013). Aucubin and geniposidic acid promoted collagen synthesis (Li et al. 1998), and geniposidic acid increased the skin moisture content of UV-damaged skin in hairless mice (Jimbo et al. 2015). Due to the important pharmacological effects of geniposidic acid and aucubin, a high-performance liquid chromatography – tandem mass spectrometry method has been developed to determine these two compounds in rat plasma, particularly for the pharmacokinetic study after oral administration of *Eucommia* herbal tea (Zhang et al. 2014). Due to its medicinal values, geniposidic acid is used as a food additive in Japan (Zhang et al. 2014).

Neuroprotective properties have also been ascribed to *Eucommia ulmoides* aqueous bark extracts, implying potential application in the prevention or treatment of Alzheimer's and other neurodegenerative diseases (Kwon et al. 2012, 2014). *Eucommia* leaves have also been made into a health beverage (Yue et al. 1999).

1.3.8.4 Conclusion

Eucommia ulmoides is a TCM recommended for vitality enhancement and longevity. In addition to fairly common plant flavonoids, it also contains less common lignans and iridoid compounds. More scientific studies are needed to verify the plant's suggested role as a "longevity herb" in preventing bone loss, inducing fat loss, and reducing elevated blood pressure and triglycerides. However, its potential as a nutritional supplement is fairly high, because, unlike many other plants, it is active at relatively low oral dosages.

1.3.9 Ligusticum wallichii *(or* Cnidium officinale*)*

1.3.9.1 Introduction

In TCM, the rhizomes from *Ligusticum wallichii* or *L. chuanxiong* (Chinese: Chuan Chung or Chuan Xiong 川芎), plant species in the carrot family (Umbelliferae), are well known for both medicinal (to treat/prevent disease) and edible (to provide nutrition) uses. In Japanese Kampo medicine, the roots of *Cnidium officinale* are used for the same purposes. Commonly, the plant is known as a lovage root and is particularly used as an important herb in blood tonics, for treating pain from headache, rheumatic arthralgia, or traumatic injury and to relieve gynecological problems. Si Wu Tang (four-substance decoction) is one of the most famous gynecological TCM formulas containing *Ligusticum*, as well as *Angelica sinensis* (Dang Gui), *Rehmannia* (Sheng Di Huang), and *Paeonia alba* (Bai Shao) (Zhou and Qu 2009). This formula is used for treating menstrual disorders, such as amenorrhea and dysmenorrhea.

1.3.9.2 Chemical Constituents

Ligusticum roots contain other important bioactive compounds, including alkaloids, particularly tetramethylpyrazine; phthalides, such as ligustilide and senkyunone; organic acids, such as ferulic acid; anthraquinones, such as chrysophanic acid; polysaccharides; ceramides; and cerebrosides (Li et al. 2012b).

1.3.9.3 Bioactivity

Pharmacological studies have shown that *Ligusticum* TCM formulas and particularly its antihypertensive constituent tetramethylpyrazine (**17**, Table 1.1), also known as ligustrazine or chuanxiongzine, can increase coronary circulation and decrease oxygen consumption, increase myocardial contractility and decrease heart rate, and cause vasodilation and lower blood pressure (Lee 2015). A review (Ran et al. 2011) compiled the current scientific research on the chemistry and pharmacology of *L. chuanxiong*, pointing out its valuable therapeutic properties as well as medical potential. The pharmacological targets and pharmacokinetics of *L. chuanxiong*, as well as other TCM formulas used for the treatment of cardiovascular and cerebrovascular disease, have been studied with the aims of clarifying the traditional uses and modernizing these herbal medicines (Li et al. 2012a; Zeng et al. 2013).

1.3.9.4 Conclusion

As *Ligusticum* is widely used in TCM for treating migraine, the herb's effects on the blood–brain barrier have been studied (Wang et al. 2015a). The alkaloid tetramethylpyrazine has also shown in vivo nootropic and neuroprotective effects in rat models (Kong et al. 2015; Lu et al. 2014; Wu et al. 2013), as well as in vitro actions against neuro-inflammation (Kim et al. 2014), giving it a possible use in treatment of Alzheimer's or other neurodegenerative diseases, which could greatly improve quality of life for many elderly adults.

1.3.10 Lycium barbarum

1.3.10.1 Introduction

Lycium barbarum (Chinese: Kou Chi Tzu 枸杞子) is a species of boxthorn in the family Solanaceae. It produces a bright red-orange oblong fruit, known as a wolfberry or goji berry. In China, these sweet-tasting berries are celebrated in an annual festival each August in Ningxia province, China, the main location of their cultivation. The berries can be eaten either raw or dried; made into juice, wine, or tea; or processed into powders, tablets, and tinctures (Cheng et al. 2015). In TCM, the berries (*Fructus lycii*) are regarded to benefit the liver and kidney, replenish vital essence (chi), and improve eyesight. The root bark (*Cortex lycii*) is also used to alleviate fever and to treat night sweats, pneumonia, and cough (Potterat 2010; Yao et al. 2011).

1.3.10.2 Chemical Constituents

L. barbarum fruits are rich in polysaccharides/proteoglycans (5–8 % of the dried fruit), termed LBPs, from six carbohydrates: galactose, glucose, rhamnose, arabinose, mannose, and xylose. Other major compounds are the coumarin scopoletin and glycosylated vitamin C (2-*O*-β-D-glucopyranosyl-L-ascorbic acid) (Tang et al. 2012). Water-soluble LBPs can be separated by an extraction process to remove more lipid-soluble constituents, including carotenoids such as zeaxanthin (Cheng et al. 2015). The LBPs are considered to be the most important, bioactive components (Shan et al. 2011). Other components in the fruit are carbohydrates, carotenoids, flavonoids, betaine, cerebroside, β-sitosterol, amino acids, trace elements, vitamins, etc. (Amagase and Farnsworth 2011). Detailed phytochemical reviews are available (Potterat 2010; Yao et al. 2011).

1.3.10.3 Bioactivity

In TCM, plants from the genus *Lycium*, including *L. barbarum* and the related species *L. chinense*, have been used for tonic, aphrodisiac, hepatoprotective, antiseptic, hypotensive, and hypoglycemic properties (Anonymous 1999c; Potterat 2010; Yao et al. 2011). The traditional “yin”-nourishing effect can be associated with immunoregulating actions. The LBPs exert both indirect effects on immune modulation and direct effects on cytoprotection, antiaging, and neuromodulation (Chang and So 2008). Such scientific evidence provides some support for the folk medicine use of wolfberries to increase longevity (Potterat 2010; Yao et al. 2011). Also, a recent report found that wolfberries can enhance the protective effect of influenza vaccine in older mice (Du et al. 2014). Similar support has also been found for the eye health benefits, e.g., LBPs have shown neuroprotective effects on retinal ganglion cells (Mi et al. 2013). LBPs have also been linked to immune-enhancing effects by increasing expression of interleukin-2 and tumor necrosis factor-alpha (Gan et al. 2003), as well as anticancer effects, such as apoptotic induction, stimulating interest in *Lycium* as a cancer therapy agent/adjuvant (Tang et al. 2012). The scientific studies related to the above bioactivities, as well as metabolism stimulation, cardiovascular benefits, diabetes, and antioxidant effects, have been well described in several reviews (Amagase and Farnsworth 2011; Cheng et al. 2015; Jin et al. 2013).

1.3.10.4 Conclusion

L. barbarum undoubtedly has nutritional value for public health, and its chemical components have shown various biological effects in cellular and animal studies, as well as some human clinical studies. Indeed, the Natural Standard Research Collaboration recently reported an evidence-based review to compile folkloric precedent with consolidated safety and efficacy data from the scientific literature (Ulbricht et al. 2015). However, further studies should be performed to truly clarify detailed mechanisms, possible synergistic effects, and ensured safety profiles, such as contraindication with other food and medications. The general public needs to be told about not only the potential health benefits but also the difference between health claims and structure–function claims regarding the potential of *Lycium*, as well as other products, as a functional food (Lasekan 2014).

1.4 Summary

Chinese herbal medicine (CHM) has been used for many thousands of years to both treat and prevent human disease. This invaluable knowledge makes CHM the best resource for modern scientists in their efforts to develop new drug research and future clinical trial candidates. Chronic disease seriously affects the quality of life of patients, and many ancient CHM-derived remedies could hold the answer to alleviating many of those diseases' most debilitating symptoms, as well as preventing those diseases at the outset and treating the disease once contracted. Development of CHM products as adjunct therapies to augment the efficacy and offset the toxicity of Western medicine is an excellent approach for rapid advancement into US FDA-approved new drugs. Developing CHM products as high-quality dietary supplements must particularly emphasize standardization through qualitative and quantitative quality controls. A combination of advanced medicinal chemistry and natural products chemistry, coupled with cutting-edge life science technology, will play a very important role for converting CHM products, especially the pure single active principles, through modification and synthesis into clinical trial candidates very efficiently and effectively. This mixture of current Western medicine and the ancient therapies of CHM could indeed hold promising answers for the treatment of chronic diseases to greatly improve quality of life of patients, making CHM of the upmost importance in future drug discovery and development. As evidenced by the ten case studies outlined above, the active principles of CHM have been elucidated. Accordingly, there is a major need to develop these compounds into world-class medicines with great efficacy and low toxicity, making them superior to the current Western medicine therapies currently available.

Acknowledgments The authors would like to acknowledge support from three NIH grants: CA017625 and CA177584 from the National Cancer Institute and AI033066 from the Institute of Allergy and Infectious Diseases awarded to K.-H. Lee. Support in part was also received from the

Taiwan Ministry of Health and Welfare Clinical Trial Center of Excellence, Taiwan (MOHW103-TDU-B-212-113002).

References

Aggarwal BB, Kumar A, Bhari AC (2003) Anticancer potential of curcumin: preclinical and clinical studies. Anticancer Res 23:363–398

Amagase H, Farnsworth NR (2011) A review of botanical characteristics, phytochemistry, clinical relevance in efficacy and safety of *Lycium barbarum* fruit (Goji). Food Res Int 44:1702–1717. doi:10.1016/j.foodres.2011.03.027

Anand P, Kunnumakkara AB, Newman RA, Aggarwal BB (2007) Bioavailability of curcumin: problems and promises. Mol Pharm 4:807–818. doi:10.1021/mp700113r

Anonymous (1999a) State Administration of Traditional Chinese Medicine. Chinese Materia Medica, vol 5. Shanghai Scientific and Technical Publishers, Shanghai, pp 906–908

Anonymous (1999b) State Administration of Traditional Chinese Medicine. Chinese Materia Medica, vol 8. Shanghai Scientific and Technical Publishers, Shanghai, pp 631–498

Anonymous (1999c) State Administration of Traditional Chinese Medicine of the People's Republic of China, 'Chinese Herbs', vol 19. Shanghai Science and Technology Press, Shanghai, p 267

Anonymous (1999d) WHO monographs on selected medicinal plants. World Health Organization, Geneva

Anonymous (2005) The Pharmacopoeia of People's Republic of China. China, E.C.O.T.P.O.P.S.R.O. Chemical Industry Press, Beijing

Anonymous (2009) Dictionary of Chinese Traditional Medicine, vol 1. Scientific Technology Press, Shanghai, pp 1560–1561

Ao ZH, Xu ZH, Lu ZM, Xu HY, Zhang XM, Dou WF (2009) Niuchangchih (*Antrodia camphorata*) and its potential in treating liver diseases. J Ethnopharmacol 121:194–212. doi:10.1016/j.jep.2008.10.039

Bairwa K, Grover J, Kania M, Jachak SM (2014) Recent developments in chemistry and biology of curcumin analogues. RSC Adv 4:13946–13978. doi:10.1039/c4ra00227j

Batch PA (2010) Prescription for nutritional healing: a practical A-to-Z reference to drug-free remedies using vitamins, minerals, herbs and food supplements, 4th edn. Avery Penguin Putnam, New York

Benito PB, Díaz Lanza AM, Silván Sen AM, De Santos Galindez J, Fernandez Matellano L, Sanz Gómez A, Abad Martínez MJ (2000) Effects of some iridoids from plant origin on arachidonic acid metabolism in cellular systems. Planta Med 66:324–328

Broadhurst CL, Polansky MM, Anderson RA (2000) Insulin-like biological activity of culinary and medicinal plant aqueous extracts in vitro. J Agric Food Chem 48:849–852. doi:10.1021/jf9904517

Chan MMY (1995) Inhibition of tumor necrosis factor by curcumin, a phytochemical. Biochem Pharmacol 49:1551–1556. doi:10.1016/0006-2952(95)00171-U

Chang TT, Chou WN (1995) *Antrodia cinnamomea* sp. nov. on *Cinnamomum kanehirai* in Taiwan. Mycol Res 99:756–758

Chang RCC, So KF (2008) Use of anti-aging herbal medicine, *Lycium barbarum*, against aging-associated diseases. What do we know so far? Cell Mol Neurobiol 28:643–652. doi:10.1007/s10571-007-9181-x

Chattopadhyay I, Biswas K, Bandyopadhyay U, Banerjee RK (2004) Turmeric and curcumin: biological actions and medicinal applications. Curr Sci (Bangalore) 87:44–53

Chaturvedi P (2012) Antidiabetic potentials of *Momordica charantia*: multiple mechanisms behind the effects. J Med Food 15:101–107. doi:10.1089/jmf.2010.0258

Chen WP, Ho BY, Lee CL, Lee CH, Pan TM (2008) Red mold rice prevents the development of obesity, dyslipidemia and hyperinsulinemia induced by high-fat diet. Int J Obes 32:1694–1704. doi:10.1038/ijo.2008.156

Chen HW, Lin IH, Chen YJ, Chang KH, Wu MN, Su WH, Huang GC, Lai YL (2012) A novel infusible botanically-derived drug, PG2, for cancer-related fatigue: a phase II double-blind, randomized placebo-controlled study. Clin Invest Med 35:E1–E11

Chen H, Guo J, Pang B, Zhao L, Tong X (2015) Application of herbal medicine with bitter flavor and cold property on treating diabetes mellitus. Evid Based Complement Alternat Med 2015:529491. doi:10.1155/2015/529491

Cheng AL, Hsu CH, Lin JK, Hsu MM, Ho YF, Shen TS, Ko JY, Lin JT, Lin BR, Ming-Shiang W, Yu HS, Jee SH, Chen GS, Chen TM, Chen CA, Lai MK, PU YS, Pan MH, Wang YJ, Tsai CC, Hsieh CY (2000) Phase I clinical trial of curcumin, a chemopreventive agent, in patients with high-risk or pre-malignant lesions. Anticancer Res 21:2895–2900

Cheng J, Zhou ZW, Sheng HP, He LJ, Fan XW, He ZX, Sun T, Zhang X, Zhao RJ, Gu L, Cao C, Zhou SF (2015) An evidence-based update on the pharmacological activities and possible molecular targets of *Lycium barbarum* polysaccharides. Drug Des Dev Ther 9:33–78. doi:10.2147/DDDT.S72892

Deodhar S, Sethi R, Srimal R (1980) Preliminary study on antirheumatic activity of curcumin (diferuloyl methane). Indian J Med Res 71:632–634

Deyama T, Nishibe S, Nakazawa Y (2001) Constituents and pharmacological effects of *Eucommia* and Siberian ginseng. Acta Pharmacol Sin 22:1057–1070

Dharmananda S (1999) Chinese herbal therapy for sciatica and lumbago. 1999, START Group Manuscripts, Portland

Dharmananda S (2000) An analysis of Chinese herb prescriptions for rheumatoid arthritis. 2000, START Group Manuscripts, Portland

Dhillon N, Aggarwal BB, Newman RA, Wolff RA, Kunnumakkara AB, Abbruzzese JL, Ng CS, Badmaev V, Kurzrock R (2008) Phase II trial of curcumin in patients with advanced pancreatic cancer. Clin Cancer Res 14:4491–4499. doi:10.1158/1078-0432.CCR-08-0024

Drewnowski A, Gomez-Carneros C (2000) Bitter taste, phytonutrients, and the consumer: a review. Am J Clin Nutr 72:1424–1435

Du X, Wang J, Niu X, Smith D, Wu D, Meydani SN (2014) Dietary wolfberry supplementation enhances the protective effect of flu vaccine against influenza challenge in aged mice. J Nutr 144:224–209. doi:10.3945/jn.113.183566

Endo A (1979) Monacolin K, a new hypocholesterolemic agent produced by a *Monascus* species. J Antibiot (Tokyo) 32:852–854. doi:10.7164/antibiotics.32.852

Gan L, Zhang SH, Liu Q, Xu HB (2003) A polysaccharide-protein complex from *Lycium barbarum* upregulates cytokine expression in human peripheral blood mononuclear cells. Eur J Pharmacol 471:217–222. doi:10.1016/S0014-2999(03)01827-2

Geethangili M, Tzeng YM (2011) Review of pharmacological effects of *Antrodia camphorata* and its bioactive compounds. Evid Based Complement Alternat Med 2011:212641. doi:10.1093/ecam/nep108

Goel A, Kunnumakkara AB, Aggarwal BB (2008) Curcumin as "Curecumin": from kitchen to clinic. Biochem Pharmacol 75:787–809. doi:10.1016/j.bcp.2007.08.016

Grover JK, Yadav SP (2004) Pharmacological actions and potential uses of *Momordica charantia*: a review. J Ethnopharmacol 93:123–132. doi:10.1016/j.jep.2004.03.035

Gupta SC, Prasad S, Kim JH, Patchva S, Webb LJ, Priyadarsini IK, Aggarwal BB (2011) Multitargeting by curcumin as revealed by molecular interaction studies. Nat Prod Rep 28:1937–1955. doi:10.1039/c1np00051a

Gupta AK, Mishra R, Lal RK (2015) Genetic resources, diversity, characterization and utilization of agronomical traits in turmeric (*Curcuma longa* L.). Ind Crop Prod 77:708–712. doi:10.106/j.indcrop.2015.09.030

Hamid K, Alqahtani A, Kim MS, Cho JL, Cui PH, Li CG, Groundwater PW, Li GQ (2015) Tetracyclic triterpenoids in herbal medicines and their activities in diabetes and its complications. Curr Top Med Chem 15:2406–2430. doi:10.2174/1568026615666150619141940

Hasan I, Khatoon S (2012) Effect of *Momordica charantia* (bitter gourd) tablets in diabetes mellitus: type 1 and type 2. Prime Res Med (PROM) 2:72–74

Heber D, Yip I, Ashley JM, Elashoff DA, Elashoff RM, Go VL (1999) Cholesterol-lowering effects of a proprietary Chinese red-yeast rice dietary supplement. Am J Clin Nutr 69:231–236

Hong J, Bose M, Ju J, Ryu JH, Chen X, Sang S, Lee MJ, Yang CS (2004) Modulation of arachidonic acid metabolism by curcumin and related β-diketone derivatives: effects on cytosolic phospholipase A2, cyclooxygenases and 5-lipoxygenase. Carcinogenesis 259:1671–1679. doi:10.1093/carcin/bgh165

Horax R, Hettiarachchy N, Over K, Chen P, Gbur E (2010) Extraction, fractionation and characterization of bitter melon seed proteins. J Agric Food Chem 58:1892–1897. doi:10.1021/jf902903s

Hsu CH, Cheng AL (2007) Clinical studies with curcumin. In: Aggarwal BB, Surh YJ, Shishodia S (eds) The molecular targets and therapeutic uses of curcumin in health and disease. Springer, New York, pp 471–480

Hsu C, Hsieh CL, Kuo YH, Huang CJ (2011) Isolation and identification of Cucurbitane-type triterpenoids with partial agonist/antagonist potential for estrogen receptors from *Momordica charantia*. J Agric Food Chem 59:4553–4561. doi:10.1021/jf200418g

Huang CH, Chen YJ, Lin CC (2012) *Antrodia cinnamomea* (*A. camphorata*, Neu Chang Chin): an exceptional polypore mushroom with potential antitumor and immunomodulatory effects. Curr Top Nutraceuticals Res 10:61–74

Huang YX, Liu EW, Wang L, Huo Y, Wang Q, Olaleye O, Wang T, Gao XM (2014) LC/MS/MS determination of pharmacokinetic studies of six compounds in rat plasma following oral administration of the single and combined extracts of *Eucommia ulmoides* and *Dipsacus asperoides*. Chin J Nat Med 12:0469–0476. doi:10.1016/S1875-5364(14)60073-X

Hung HY, Qian K, Morris-Natschke SL, Hsu CS, Lee KH (2012) Recent discovery of plant-derived anti-diabetic natural products. Nat Prod Rep 29:580–606. doi:10.1039/c2np00074a

India. Ministry of Health and Family Welfare (1996) Indian Pharmacopoeia, Controller of Publications: Delhi, 1996

Ireson C, Orr S, Jones DJ, Verschoyle R, Lim CK, Luo JL, Howells L, Plummer S, Jukes R, Williams M, Steward WP, Gescher A (2001) Characterization of metabolites of the chemopreventive agent curcumin in human and rat hepatocytes and in the rat in vivo, and evaluation of their ability to inhibit phorbol ester-induced prostaglandin E2 production. Cancer Res 61:1058–1064

Itokawa H, Shi Q, Akiyama T, Morris-Natschke SL, Lee KH (2008) Recent advances in the investigation of curcuminoids. Chin Med 3:1–13. doi:10.1186/1749-8546-3-11

Jimbo N, Kawada C, Nomura Y (2015) Herb extracts and collagen hydrolysate improve skin damage resulting from ultraviolet-induced aging in hairless mice. Biosci Biotechnol Biochem 79:1624–1628. doi:10.1080/09168451.2015.1046362

Jin X, Otonashi-Satoh Y, Sun P, Kawamura N, Tsuboi T, Yamaguchi Y, Ueda T, Kawasaki H (2008) Endothelium-derived hyperpolarizing factor (EDHF) mediates endothelium-dependent vasodilator effects of aqueous extracts from *Eucommia ulmoides* Oliv. leaves in rat mesenteric resistance arteries. Acta Med Okayama 62:319–325

Jin M, Huang Q, Zhao K, Shang P (2013) Biological activities and potential health benefit effects of polysaccharides isolated from *Lycium barbarum* L. Int J Biol Macromol 54:16–23. doi:10.1016/j.ijbiomac.2012.11.023

Joseph B, Jini D (2013) Antidiabetic effects of *Momordica charantia* (bitter melon) and its medicinal potency. Asian Pac J Trop Dis 3:93–102

Jurenka JS (2009) Anti-inflammatory properties of curcumin, a major constituent of *Curcuma longa*: a review of preclinical and clinical research. Altern Med Rev 14:141–153

Kang SP, Saif MW (2008) Optimal second line treatment options for gemcitabine refractory advance pancreatic cancer patients. Can we establish standard of care with available data? J Pancreas (Online) OP 9:83–90

Kaur P, Garg V, Gulait M, Singh SK (2016) Oral delivery of antidiabetic polypeptide-k: journey so far and the road ahead. Curr Drug Deliv 13(2):236–244

Kawamori T, Lubet R, Steele VE, Kelloff GJ, Kaskey RB, Rao CV, Reddy BS (1999) Chemopreventive effect of curcumin, a naturally occurring anti-inflammatory agent, during the promotion/progression stages of colon cancer. Cancer Res 59:597–601

Kim M, Kim SO, Lee, Lee JH, Jung WS, Moon SK, Kim YS, Cho KH, Ko CN, Lee EH (2014) Tetramethylpyrazine, a natural alkaloid, attenuates pro-inflammatory mediators induced by amyloid β and interferon-γ in rat brain microglia. Eur J Pharmacol 741:504–511. doi:10.1016/j.ejphar.2014.06.037

Kong X, Zhong M, Su X, Qin Q, Su H, Wan H, Liu C, Wu J, Shang H, Zhang Y, Lin N (2015) Tetramethylpyrazine promotes migration of neural precursor cells via activating the phosphatidylinositol 3-kinase pathway. Mol Neurobiol. doi:10.1007/s12035-015-9551-1

Kuo YL, Chen CH, Chuang TH, Hua WK, Lin WJ, Hsu WH, Chang PM, Hsu SL, Huang TH, Kao CY, Huang CY (2015) Gene expression profiling and pathway network analysis predicts a novel antitumor function for a botanical-derived drug, PG2. Evid Based Complement Alternat Med 2015:917345. doi:10.1155/2015/917345

Kwon SH, Kim MJ, Ma SX, You IJ, Hwang JY, Oh JH, Kim SY, Kim HC, Lee SY, Jang CG (2012) *Eucommia ulmoides* Oliv. bark protects against hydrogen peroxide-induced neuronal cell death in SH-SY5Y cells. J Ethnopharmacol 142:337–345. doi:10.1016/j.jep.2012.04.010

Kwon SH, Ma SX, Hong SI, Kim SY, Lee SY, Jang CG (2014) *Eucommia ulmoides* Oliv. bark attenuates 6-hydroxydopamine-induced neuronal cell death through inhibition of oxidative stress in SH-SY5Y cells. J Ethnopharmacol 152:173–182. doi:10.1016/j.jep.2013.12.048

Lal B, Kapoor AK, Asthana OP, Agrawal PK, Prasad R, Kumar P, Srimal RC (1999) Efficacy of curcumin in the management of chronic anterior uveitis. Phytother Res 13:318–322. doi:10.1002/(SICI)1099-1573(199906)13:4<318::AID-PTR445>3.0.CO;2-7

Lal B, Kapoor A, Agrawal OP, Asthana OP, Srimal RC (2000) Role of curcumin in idiopathic inflammatory orbital pseudotumours. Phytother Res 14:443–447. doi:10.1002/1099-1573(200009)14:6<443::AID-PTR619>3.0.CO;2-V

Lam W, Bussom C, Guan F, Jiang Z, Zhang W, Gullen EA, Liu SH, Cheng YC (2010) The four-herb Chinese medicine PHY906 reduces chemotherapy-induced gastrointestinal toxicity. Sci Transl Med 2:45ra59. doi:10.1126/scitranslmed.3001270

Lam W, Jiang Z, Guan F, Huang X, Hu R, Wang J, Bussom S, Liu SH, Zhao H, Yen Y, Cheng YC (2015) PHY906(KD018), an adjuvant based on a 1800-year-old Chinese medicine, enhanced the anti-tumor activity of Sorafenib by changing the tumor microenvironment. Sci Rep 5:9384. doi:10.1038/srep09384

Lasekan O (2014) Exotic berries as a functional food. Curr Opin Clin Nutr Metab Care 17:589–595. doi:10.1097/MCO.0000000000000109

Lee KH (2015) Chinese herbal medicine-derived products for prevention or treatment of disease affecting quality of life. Plenary lecture presented at the 2nd international conference of traditional and complementary medicine on health. Taipei, Taiwan, 26 October 2015

Lee CL, Pan TM (2011a) Composition and method for prevention and treatment of Alzheimer's disease. Taiwan Patent I341204

Lee CL, Pan TM (2011b) Red mold fermented products and Alzheimer's disease: a review. Appl Microbiol Biotechnol 91:461–469. doi:10.1007/s00253-011-3413-1

Lee KH, Tao-Wiedman TW, Wu H (1993) Pharmagenesis Advisory Board Meeting

Lee TH, Lee CK, Tsou WL, Liu SY, Kuo MT, Wen WC (2007) A new cytotoxic agent from solid-state fermented mycelium of *Antrodia camphorata*. Planta Med 13:1412–1415. doi:10.1055/s-2007-990232

Lee KH, Lin L, Shih CCY, Su CY, Ishida J, Ohtsu H, Wang HK, Itokawa H, Chang CS (2008) US Patent 7,355,081

Lee CL, Kung YH, Wu CL, Shu YW, Pan TM (2010) Monascin and ankaflavin act as novel hypolipidemic and high-density lipoprotein cholesterol-raising agents in red mold dioscorea. J Agric Food Chem 58:9013–9019. doi:10.1021/jf3042402

Lee KH, Morris-Natschke SL, Qian K, Dong Y, Yang X, Zhou T, Belding E, Wu SF, Wada K, Akiyama T (2012) Recent progress of research on herbal products used in traditional Chinese medicine: the herbs belonging to the divine husbandman's herbal foundation canon (Shen Nong Ben Cao Jing). J Trad Compl Med 2:6–26

Lee KH, Morris-Natschke SL, Brattlie J, Xie J, Belding E (2013) Modern drug discovery from Chinese Materia Medica used in traditional Chinese medicine. In: Adams JD, Lien EJ (eds) Traditional Chinese medicine: scientific basis for its use. RSC Publishing, Cambridge, pp 81–134

Leung L, Birtwhistle R, Kotecha J, Hannah S, Cuthbertson S (2009) Anti-diabetic and hypoglycaemic effects of *Momordica charantia* (bitter melon): a mini review. Br J Nutr 102:1703–1708. doi:10.1017/S0007114509992054

Li SZ (1596) Pen Tsao Kang Mu (Compendium of Materia Medica Systematic Pharmacopoeia) China

Li Y, Kamo S, Metori K, Koike K, Che QM, Takhashi S (1998) The promoting effects of geniposidic acid and aucubin in *Eucommia ulmoides* Oliv leaves on collagen synthesis. Biol Pharm Bull 21:1306–1310

Li SY, Yuan W, Deng G, Wang P, Yang P, Aggarwal BB (2011) Chemical composition and product quality control of turmeric (*Curcuma longa* L.). Pharm Crop 2:28–54. doi:10.2174/2210290601102010028

Li B, Xu X, Wang X, Yu H, Li X, Tao W, Wang Y, Yang L (2012a) A systems biology approach to understanding the mechanism of action of Chinese herbs for treatment of cardiovascular disease. Int J Mol Sci 13:13501–13520. doi:10.3390/ijms131013501

Li W, Tang Y, Chen Y, Duan JA (2012b) Advances in the chemical analysis and biological activities of chuanxiong. Molecules 17:10614–10651. doi:10.3390/molecules170910614

Lin L, Shi Q, Nyarko AK, Bastow KF, Wu CC, Su CY, Shih CC, Lee KH (2006a) Antitumor agents. 250. Design and synthesis of new curcumin analogues as potential anti-prostate cancer agents. J Med Chem 49:3963–3972. doi:10.1021/jm051043z

Lin L, Shi Q, Su CY, Shih CC, Lee KH (2006b) Antitumor agents 247. New 4-ethoxycarbonylethyl curcumin analogs as potential antiandrogenic agents. Bioorg Med Chem 14:2527–2534. doi:10.1016/j.bmc.2005.11.034

Lin HY, Ho TY, Lin C, Li CC, Hsiang CY (2014) A novel insulin receptor-binding protein from *Momordica charantia* enhances glucose uptake and glucose clearance in vitro and in vivo through triggering insulin receptor signaling pathway. J Agric Food Chem 62:8952–8961. doi:10.1021/jf3042402

Lin CM, Lin YT, Lin RD, Huang WJ, Lee MH (2015) Neurocytoprotective effects of aliphatic hydroxamines from lovastatin, a secondary metabolite from *Monascus*-fermented red mold rice, in 6-hydroxydopamine (6-OHDA)-treated nerve growth factor (NGF)-differentiated PC12 cells. ACS Chem Neurosci 6:716–724. doi:10.1021/cn500275k

Liu SH, Cheng YC (2012) Old formula, new Rx: the journey of PHY906 as cancer adjuvant therapy. J Ethnopharmacol 140:614–623. doi:10.1016/j.jep.2012.01.047

Liu SY, Kuo MT, Wen WC (2008) US Patent No. 7,385,088

Lu MC, El-Shazly M, Wu TY, Du YC, Chang TT, Chen CF, Hsu YM, Lai KH, Chiu CP, Chang FR, Wu YC (2013) Recent research and development of *Antrodia cinnamomea*. Pharmacol Ther 139:124–156. doi:10.1016/j.pharmthera.2013.04.001

Lu C, Zhang J, Shi X, Miao S, Bi L, Zhang S, Yang Q, Zhou X, Zhang M, Xie Y, Miao Q, Wang S (2014) Neuroprotective effects of tetramethylpyrazine against dopaminergic neuron injury in a rat model of Parkinson's disease induced by MPTP. Int J Biol Sci 10:350–357. doi:10.7150/ijbs.8366

Luo LF, Wu WH, Zhou YJ, Yan J, Yang GP, Ouyang DS (2010) Antihypertensive effect of *Eucommia ulmoides* Oliv. extracts in spontaneously hypertensive rats. J Ethnopharmacol 129:238–243. doi:10.1016/j.jep.2010.03.019

Ma J, Li Y, Ye Q, Li J, Hua Y, Ju D, Zhang D, Cooper R, Chang M (2000) Constituents of red yeast rice, a traditional Chinese food and medicine. J Agric Food Chem 48:5220–5225. doi:10.1021/jf000338c

Ma WL, Hsu CL, Wu MH, Wu CT, Wu CC, Lai JJ, Jou YS, Chen CW, Yeh S, Chang C (2008) Androgen receptor is a new potential therapeutic target for the treatment of hepatocellular carcinoma. Gastroenterology 135:947–955. doi:10.1053/j.gastro.2008.05.046

Medagama AB, Bandara R (2014) The use of complementary and alternative medicines (CAMs) in the treatment of diabetes mellitus: is continued used safe and effective? Nutr J 13:102. doi:10.1186/1475-2891-13-102

Mi XS, Zhong JX, Chang RCC, So KF (2013) Research advances on the usage of traditional Chinese medicine for neuroprotection in glaucoma. J Integr Med 11:233–240. doi:10.3736/jintegrmed2013037

Miłobdzka J, Lampe V (1910) Zur kenntnis des curcumins. Ber Deutsch Chem Ges 43:2163–2170

Miyamoto H, Yang ZM, Chen YT, Ishiguro H, Uemura H, Kubota Y, Nagashima Y, Chang YJ, Hu YC, Tsai MY, Yeh S, Messing EM, Chang C (2007) Promotion of bladder cancer development and progression by androgen receptor signals. J Natl Cancer Inst 99:558–568. doi:10.1093/jnci/djk113

Park KS (2013) Aucubin, a naturally occurring iridoid glycoside inhibits TNF-α-induced inflammatory responses through suppression of NF-kB activation in 3T3-L1 adipocytes. Cytokine 62:407–412. doi:10.1016/j.cyto.2013.04.005

Paul A, Raychaudhuri SS (2010) Medicinal uses and molecular identification of two *Momordica charantia* varieties – a review. eJBio 6:43–51

Pedersen U, Rasmussen PB, Lawesson SO (1985) Synthesis of naturally occurring curcuminoids and related compounds. Liebigs Ann Chem 8:1557–1569. doi:10.1002/jlac.198519850805

Potterat O (2010) Goji (*Lycium barbarum* and *L. chinense*: phytochemistry, pharmacology and safety in the perspective of traditional uses and recent popularity. Planta Med 76:7–19. doi:10.1055/s-0029-1186218

Raman A, Lau C (1996) Anti-diabetic properties and phytochemistry of *Momordica charantia* L. (Cucurbitaceae). Phytomedicine 2:349–362. doi:10.1016/S0944-7113(96)80080-8

Ran X, Ma L, Peng C, Zhang H, Qin LP (2011) *Ligusticum chuanxiong* Hort: a review of chemistry and pharmacology. Pharm Biol 49:1180–1189. doi:10.3109/13880209.2011.576346

Rister R (ed) (1999) Japanese herbal medicine: the healing art of Kampo. Avery Publishing Group, New York, p 31

Saif MW, Lansigan F, Ruta S, Lamb L, Mezes M, Elligers K, Grant N, Jiang ZL, Liu SH, Chang YC (2010) Phase I study of the botanical formulation PHY906 with capecitabine in advanced pancreatic and other gastrointestinal malignancies. Phytomedicine 17:161–169. doi:10.1016/j.phymed.2009.12.016

Saif MW, Li J, Lamb L, Kaley K, Elligers K, Jiang Z, Bussom S, Liu SH, Cheng YC (2014) First-in-human phase II trial of the botanical formulation PHY906 with capecitabine as second-line therapy in patients with advance pancreatic cancer. Cancer Chemother Pharmacol 73:373–380. doi:10.1007/s00280-013-2359-7

Satoskar R, Shah S, Shenoy S (1986) Evaluation of anti-inflammatory property of curcumin (diferuloyl methane) in patients with postoperative inflammation. Int J Clin Pharmacol Ther Toxicol 24:651–654

Satyavati GV, Raina MK (1976) Indian Council of Medical Research. In Medicinal plants of India, vol 1. New Delhi, p 107

Shan X, Zhou J, Ma T, Chai Q (2011) *Lycium barbarum* polysaccharides reduce exercise-induced oxidative stress. Int J Mol Sci 12:1081–1088. doi:10.3390/ijms12021081

Sharma RA, McLelland HR, Hill KA, Ireson CR, Euden SA, Manson MM, Pirmohamed M, Marnett LJ, Gescher AJ, Steward WP (2001) Pharmacodynamic and pharmacokinetic study of oral *Curcuma* extract in patients with colorectal cancer. Clin Cancer Res 7:1894–1900

Shi YC, Pan TM (2011) Beneficial effects of *Monascus purpureus* NTU 568-fermented products: a review. Appl Microbiol Biotechnol 90:1207–1217. doi:10.1007/s00253-011-3202-x

Shi Q, Shih CY, Lee KH (2009) Novel anti-prostate cancer curcumin analogues that enhance androgen receptor degradation activity. Anti Cancer Agents Med Chem 9:904–912. doi:10/2174/187152009789124655

Shi LS, Chao CH, Shen DY, Chan HH, Chen CH, Liao YR, Wu SJ, Leu YL, Shen YC, Kuo YH, Lee EJ, Qian K, Wu TS, Lee KH (2011) Biologically active constituents from the fruiting body of *Taiwanofungus camphoratus*. Bioorg Med Chem 19:677–683. doi:10.1016/j.bmc.2010.10.032

Sih CJ, Ravikumar PR, Huang FC, Tuckner C, Whitlock H Jr (1976) Isolation and synthesis of pinoresinol diglucoside, a major antihypertensive principle of Tu Chung (*Eucommia ulmoides*, Oliver). J Am Chem Soc 98:5412–5413. doi:10.1021/ja00433a070

Singh S, Aggarwal BB (1995) Activation of transcription factor NF-kB is suppressed by curcumin (diferuloylmethane). J Biol Chem 270:24995–25000. doi:10.1074/jbc.270.42.24995

Snee LS, Nerurkar VR, Dooley DA, Efird JT, Shovic AC, Nerurkar PV (2011) Strategies to improve palatability and increase consumption intentions for *Momordica charantia* (bitter melon): a vegetable commonly used for diabetes management. Nutr J 10:78. doi:10.1186/1475-2891-10-78

Sowbhagya HB (2014) Chemistry, technology, and nutraceutical functions of celery (*Apium graveolens* L.): an overview. Crit Rev Food Sci Nutr 54:389–398. doi:10.1080/10408398.2011.586740

Tan MJ, Ye JM, Turner N, Hohnen-Behrens C, Ke CQ, Tang CP, Chen T, Weiss HC, Gesing ER, Rowland A, James DE, Ye Y (2008) Antidiabetic activities of triterpenoids isolated from bitter melon associated with activation of the AMPK pathway. Chem Biol 15:263-–273. doi:10.1016/j.chembiol.2008.01.013

Tang WM, Chan E, Kwok CY, Lee YK, Wu JH, Wan CW, Chan RY, Yu PH, Chan SW (2012) A review of the anticancer and immunomodulatory effects of *Lycium barbarum* fruit. Inflammopharmacology 20:307–314. doi:10.1007/s10787-011-0107-3

Thaikert R, Paisooksantivatana Y (2009) Variation of total curcuminoids content, antioxidant activity and genetic diversity in turmeric (*Curcuma longa* L.) collections. Kasetsart J (Nat Sci) 43:507–518

Ulbricht C, Bryan JK, Costa D, Culwell X, Giese N, Isaac R, Nummy K, Pham T, Rapp C, Rusie E, Weissner W, Sindsor RC, Woods J, Zhou S (2015) An evidence-based systematic review of goji (*Lycium* spp.) by the Natural Standard Research Collaboration. J Diet Suppl 12:184–240. doi:10.3109/19390211.2014.904128

Wada K, Lee JY, Hung HY, Shi Q, Lin L, Zhao Y, Goto M, Yang PC, Kuo SC, Chen HW, Lee KH (2015) Novel curcumin analogs to overcome EGFR–TKI lung adenocarcinoma drug resistance and reduce EGFR–TKI-induced GI adverse effects. Bioorg Med Chem 23:1507–1514. doi:10.1016/j.bmc.2015.02.003

Wang X, Sun W, Cao J, Qu H, Bi X, Zhao Y (2012) Structures of new triterpenoids and cytotoxicity activities of the isolated major compounds from the fruit of *Momordica charantia* L. J Agric Food Chem 60:3927–3933. doi:10.1021/jf204208y

Wang Q, Shen L, Ma S, Chen M, Lin X, Hong Y, Liang X, Feng Y (2015a) Effects of *Ligusticum chuanxiong* and *Gastrodia elata* on blood-brain barrier permeability in migraine rats. Pharmazie 70:421–426

Wang SN, Xie GP, Qin CH, Chen YR, Zhang KR, Li X, Wu Q, Dong WQ, Yang J, Yu B (2015b) Aucubin prevents interleukin-1 beta induced inflammation and cartilage matrix degradation via inhibition of NF-kB signaling pathway in rat articular chondrocytes. Int Immunopharmacol 24:408–415. doi:10.1016/j.intimp.2014.12.029

Wu SH, Ryvarden L (1997) *Antrodia camphorata* ("niu-chang-chih"), new combination of a medicinal fungus in Taiwan. Bot Bull Acad Sin Taipei 38:273–276

Wu SJ, Leu YL, Chen CH, Chaho CH, Shen DY, Chan HH, Lee EJ, Wu TS, Wang YH, Shen YC, Qian K, Bastow KF, Lee KH (2010) Camphoratins A–J, potent cytotoxic and anti-inflammatory triterpenoids from the fruiting body of *Taiwanofungus camphoratus*. J Nat Prod 73:1756–1762. doi:10.1021/np1002143

Wu W, Yu X, Luo XP, Yang SH, Zheng D (2013) Tetramethylpyrazine protects against scopolamine-induced memory impairments in rats by reversing the cAMP/PKA/CREB pathway. Behav Brain Res 253:212–216. doi:10.1016/j.bbr.2013.07.052

Yang SZ (translator) (1998) The Divine Farmer's Materia Medica. Blue Poppy Press, Boulder, CO

Yao X, Peng Y, Xu LJ, Li L, Wu QL, Xiao PG (2011) Phytochemical and biological studies of *Lycium* medicinal plants. Chem Biodivers 8:976–1010. doi:10.1002/cbdv.201000018

Ye M, Liu SH, Jiang Z, Lee Y, Tilton R, Cheng YC (2007) Liquid chromatography/mass spectrometry analysis of PHY906, a Chinese medicine formulation for cancer therapy. Rapid Commun Mass Spectrom 21:3593–3607. doi:10.1002/rcm.2832

Yue H, Hu X, Su K, Zhai W (1999) Development of healthy beverage from *Eucommia ulmoides* foliage. Chin Med Herbs 22:385–387

Zeng M, Pan L, Qi S, Cao Y, Zhu H, Guo L, Zhou J (2013) Systematic review of recent advances in pharmacokinetics of four classical Chinese medicines used for the treatment of cerebrovascular disease. Fitoterapia 88:50–75. doi:10.1016/j.fitote.2013.04.006

Zhang L, Ma YL, Liu Y, Zu YB (2014) Development and validation of high liquid performance chromatography-tandem mass spectrometry method for simultaneous determination of geniposidic acid and aucubin in rat plasma for pharmacokinetic study after oral administration of Du-zhong tea extract. J Chromatogr B Anal Technol Biomed Life Sci 963:62–69. doi:10.1016/j.jchromb.2014.04.033

Zhao Y (2011) Studies on Qinghao and the Qinghaosus. Traditional Chinese Medical College, China Press, Beijing

Zhao Y, Hanton WK, Lee KH (1986) Antimalarial agents. 2. Artesunate, an inhibitor of cytochrome oxidase activity in *Plasmodium berghei*. J Nat Prod 49:139–142. doi:10.1021/np50043a018

Zhao Y, Hall IH, Oswald CB, Yokoi T, Lee KH (1987) Antimalarial agents. III. Mechanism of action of artesunate against Plasmodium berghei infection. Chem Pharm Bull 35:2052–2061

Zhou J, Qu F (2009) Treating gynaecological disorders with traditional Chinese medicine: a review. Afr J Tradit Complement Altern Med 6:494–517

Zhu XZ, Li XY, Liu J (2004) Recent pharmacological studies on natural products in China. Eur J Pharmacol 500:221–230. doi:10.1016/j.ejphar.2004.07.027

Chapter 2
Ethnopharmacological Importance of Western Medicinal Herb, *Scutellaria lateriflora*

Madhukar Lohani, Mansi Patel, Mohammed Majrashi, Sneha Joshi, Barbara Kemppainen, Vanisree Mulabagal, and Muralikrishnan Dhanasekaran

Abstract *Scutellaria lateriflora* also known as American skullcap is a well-known medicinal herb which has been used in traditional medicine and Western materia medica for years. *S. lateriflora* has exhibited plethora of pharmacological effects. The plant has been shown to affect the central nervous system, peripheral nervous system, and other organs such as gastrointestinal tracts, liver, sexual organs, endocrine system, blood, kidney, and muscles. However, chief use of this Western medicinal herb is for the central nervous system pathologies. *S. lateriflora* has primarily reduced the symptoms associated with hyperarousal disorders. The hyperarousal disorders are seizures, insomnia, pain, spasms, and anxiety. Further, the plant parts are used in various botanical formulations and are currently on the US market. In this article, we describe ethnopharmacological role of *S. lateriflora* and its medicinal applications in detail. Additionally, we have summarized the mechanism of action attributed to the neuroprotection and other pharmacological actions associated with *S. lateriflora*.

Keywords Anxiolytics • Ethnopharmacology • Hyperarousal disorders • Immunomodulatory • Neuroprotection • *Scutellaria lateriflora* • Western herbal medicines

M. Lohani • M. Patel • M. Majrashi • M. Dhanasekaran (✉)
Department of Drug Discovery and Development, Harrison School of Pharmacy, Auburn University, Auburn, AL 36849, USA
e-mail: dhanamu@auburn.edu

S. Joshi • B. Kemppainen
College of Veterinary Medicine, Auburn University, Auburn, AL 36849, USA

V. Mulabagal
Department of Civil Engineering, Auburn University, Auburn, AL 36849, USA

H.-S. Tsay et al. (eds.), *Medicinal Plants - Recent Advances in Research and Development*, DOI 10.1007/978-981-10-1085-9_2

Abbreviations

5-HT	Serotonin
5-HT_7	Serotonin-7
ACTH	Adrenocorticotropic hormone
AMPARs	α-Amino-3-hydroxy-5-methyl-4-isoxazolepropionic acid receptors
AP-1	Activated protein 1
BDNF	Brain-derived neurotrophic factor
BDZ	Benzodiazepine
CNS	Central nervous system
COX	Cyclooxygenase
CREB	Cyclic AMP-responsive element-binding protein
CRH	Corticotropin-releasing hormone
DPPH	2,2-Diphenyl-1-picrylhydrazyl
GABA	Gamma-aminobutyric acid
GABAR	Gamma-aminobutyric acid receptor
IFN-γ	Interferon-γ
IL	Interleukin
iNOS	Nitric oxide synthase
NF-IL6	Nuclear factor IL-6
NF-kB	Nuclear factor k-light-chain enhancer of activated B cells
NMDAR	N-Methyl-D-aspartate
PrP	Prion protein
ROS	Reactive oxygen species
S. lateriflora	*Scutellaria lateriflora*
SOD	Superoxide dismutase
TGF	Transforming growth factor
TLR2	Toll-like receptor 2
TNF-α	Tumor necrosis factor-α
WHM	Western herbal medicine

2.1 Introduction

Medicinal plants (herbal medicines and botanicals) are accepted forms of healthcare products that are indicated for the prophylactic, therapeutic, and cosmetic purposes globally. Traditional medicines derived from various curative plants due to their cultural and social influence, prolonged use, healing efficacy, and minimal adverse reactions have lately gained incredible attention for their use as a novel alternative medicine for the treatment and protection of broad spectrum of diseases (Kumar et al. 2012). Traditional medicines are natural substances (plant components or chemical constituents) obtained from a botanical with minimal or no industrial handling/processing. Treatment with traditional medicine has shown to elicit substantial pharmacological responses to treat disease states/pathological conditions locally or globally. Traditional medicines are widely classified based on their source of geographical origin, "Eastern or Western" medicines. Eastern herbal medicine considers health and disease as the two sides of a coin, where health is considered as a balanced state versus disease as an unbalanced state. Western herbal medicines develop a remedy by hypothetical deduction and divide health from disease, and the main importance is on the individual's body. The Western method tends to change the environment, whereas the Eastern approach prefers to adapt to the environment. Western herbal medicine uses a complex science system approach for its treatment (Morin 2008). Complex science system gives an understanding of living systems as complete systems representing multifaceted incorporations of organized subsystems or parts linked to and surrounded in their environments (Koithan et al. 2012). As per the National Cancer Institute (NCI dictionary), the Western medicine is described as "A system in which medical doctors and other healthcare professionals (such as nurses, pharmacists, and therapists) treat symptoms and diseases using drugs, radiation, or surgery. Also called allopathic medicine, biomedicine, conventional medicine, mainstream medicine, and orthodox medicine." Furthermore, Macmillan dictionary describes it as "the type of medical treatment that is the most popular in North America and Western European countries, based on the use of drugs and surgery to treat symptoms (signs of illness). In these countries, other types of medical treatment are called alternative medicine or complementary medicine." Consequently, the Western medicine comprehends all categories of conventional health remedies comprising the operational/surgical procedures, radiation and treatment for cancer, herbal interventions, and physiotherapy. The conventional and established lines of cures are acupuncture, stretch, homeopathy, oriental practices, and alternative medicines. The alternative cure or medicines are body touch (massage, chiropractic, tai chi, yoga), mind motivation (meditation, hypnosis), senses stimulation (art therapy, rhythmic movement of the body for a music–dance), acupuncture, light therapy (depression), Ayurveda, homeopathy, naturopathy, diet and supplements (herbs, nutrition), Chinese or Oriental medicine, and Western herbal medicine (WHM).

WHM is particularly established on Anglo-American traditional herbal system (Casey et al. 2007). Even though the theory and practice of WHM was built on traditional descriptive models of health, presently, it has been validated by continuous practice, with modern scientific methods and experiments (Braun et al. 2013; Snow 2015). WHM uses naturally occurring root, bark, flower, or other parts to formulate the herbal extract, from single or several plant constituents for the treatment of the diseases or disorders (Tilburt and Kaptchuk 2008; Niemeyer et al. 2013). Currently, several countries such as Australia, Canada, New Zealand, the UK, the USA, and Western Europe practice WHM to significantly improve health care and the quality of human health (Wills and Stuart 2004; Nissen and Evans 2012; Niemeyer et al. 2013; Snow 2015). More specifically, psychiatric conditions such as mood, anxiety, and sleep disorders are known to be treated with WHM (Kessler et al. 2005; Sarris et al. 2011). The widely used WHM are agrimony, American ginseng, bilberry, black cohosh, burdock, cat's claw, chamomile, *Echinacea*, kava kava, Maca root, milk thistle, *Rhodiola rosea*, saw palmetto, Siberian ginseng, skullcap, St. John's wort, Suma root, and Valerian root. These WHM have unique and defined mechanisms of action leading to significant effects in the body (Table 2.1).

Table 2.1 List of widely used Western health medicines and their use

Medicinal herb	Common name	Uses
Agrimonia eupatoria	Agrimony	Alleviates indigestion
		Analgesic and inflammatory effects (reduces pain and inflammation)
		Reduces infection (sore throat and bladder infections)
		Remedy for arthritis and rheumatism
Panax quinquefolius	American ginseng	Cardiac stimulant (improves cardiovascular activity)
		Enhances memory and concentration
		Neuroprotection in Alzheimer's and Parkinson's disease
		Stimulates CNS, combats depression, stress, and fatigue
		Treatment of anorexia nervosa (stimulates appetite)
		Treats insomnia (induces sleep)
Vaccinium myrtillus	Bilberry	Effective against diabetic retinopathy
		Improves vision (night blindness, cataracts, glaucoma)
		Reduces mild inflammation of the throat and mouth
Actaea racemosa	Black cohosh	Antifatigue
		Modulates estrogen production (relieves the symptoms of PMS, menstrual cramps, and problems associated with menopause)
Arctium lappa	Burdock	Anticancer activity
		Antibacterial properties
		Antifungal effect
		Beneficial in dermal pathologies
		Boosts immune system
		Helpful in digestive problems
		Source of vitamins and minerals
		Used as blood purifier
Uncaria tomentosa	Cat's claw	Antibacterial properties
		Anticancer activity
		Anticonvulsant
		Antifungal effect
		Antihistamine (allergies)
		Anti-inflammatory effect (osteoarthritis)
		Antioxidant (neuroprotection)
		Immune stimulant
		Reduces acne
		Slows neurodegeneration (Alzheimer's disease progression)

(continued)

Table 2.1 (continued)

Medicinal herb	Common name	Uses
Matricaria chamomilla	Chamomile	Alleviates cold symptoms
		Anxiolytic effects (treats various anxiety disorders)
		Anti-inflammatory (reduces aches, pains, menstrual cramps)
		Alleviates various gastrointestinal tract-GIT disorders (nausea, irritable bowel syndrome, abdominal pain)
Echinacea	Purple coneflowers	Anti-inflammatory
		Antioxidant
		Boosts immune system (reduces the duration and symptoms of colds and flu and fights viral, fungal, and bacterial infections)
		Used as blood purifier
Piper methysticum	Kava kava	Alleviates various gastrointestinal tract-GIT disorders (reduces stomach upset)
		Anxiolytic effects
		Asthma
		Beneficial in arthritis and rheumatism
		Diuretic
Lepidium meyenii	Maca root	Increases energy, strength, and stamina
		Improves cognitive functioning
		Nootropic effect (advances mental clarity and memory, enhances learning and mental ability)
		Reduces anxiety, stress, and depression
		Restores hormonal balance (normalizes progesterone, estrogen, and testosterone, decreases mood swings and hot flashes)
Silybum marianum	Milk thistle	Anticancer activity
		Anti-mushroom poisoning
		Beneficial in spleen disorders
		Eases heartburn
		Hepatic protection (valuable in liver diseases)
		Normalizes blood sugar levels
		Treats gallbladder disorders
		Treats seasonal allergies
Rhodiola rosea	Roseroot	Antioxidant effect
		Energy enhancer (increases ATP production, increases the capacity for physical work, shortens recovery time after strenuous activity)
		Improves concentration, memory, and attention span
		Induces emotional calming
		Neuroprotective effect
		Reduces the symptoms of menopause
		Relieves fatigue

(continued)

Table 2.1 (continued)

Medicinal herb	Common name	Uses
Serenoa repens	Saw palmetto	Builds body tissues
		Cough suppressant
		Digestive aid
		Diuretic
		Shrinks enlarged prostate
		Sleep inducer
Eleutherococcus senticosus	Siberian ginseng	Anti-inflammatory effect
		Enhances mental concentration, memory, athletic performance, and learning ability
		Increases male and female fertility
		Reduces stress and fatigue
		Relieves the symptoms of fibromyalgia and chronic fatigue syndrome
		Relieves the symptoms of menopause and menstrual disorders
		Restores energy
		Treats Alzheimer's disease
		Treats insomnia
Scutellaria	Skullcap	Antiallergic (reduces allergies, inflammation)
		Anti-atherosclerosis effect
		Beneficial in Parkinson's disease
		Decreases anxiety (anxiolytic)
		Effective against upper respiratory and urinary tract infections
		Helpful in smoking cessation
		Lowers cholesterol
		Reduces fever
		Reduces symptoms of drug and alcohol withdrawal
		Useful in depression and exhaustion
Hypericum perforatum	St. John's wort	Beneficial in treatment of burns, bruises, and sores
		Promotes wound healing
		Reduces jaundice
		Suppresses cough
		Treats depression, anxiety, and nervous exhaustion
Hebanthe eriantha	Suma root	Anticancer activity
		Antidiabetic activity
		Balances the systems of the body
		Decreases fatigue
		Increases appetite, energy, and stamina
		Improves physical and mental endurance
		Reduces the symptoms of menopause
Valeriana officinalis	Valerian root	Reduces panic attacks
		Relieves headache
		Relieves muscle and joint pain
		Valuable in insomnia and anxiety treatment

Genus *Scutellaria* belonging to the Lamiaceae (mint) family comprises of over 360 species. This genus is widely distributed in Europe, the USA, and East Asia. Plants of this genus were used in local folk medicine for years. Even though 360 species are reported, *Scutellaria lateriflora* (American skullcap) and *Scutellaria baicalensis* (Chinese skullcap) are the two well-studied species for their pharmacological potential and are a pivotal source of traditional herbal products for medicinal and nutritional purposes (Zhang et al. 2009). *Scutellaria baicalensis* (Chinese skullcap) is a traditional native Chinese botanical that has been consumed as a prophylactic herb and used as a remedial herbal product for cardiovascular diseases, hemorrhagic disorders and has neuroprotective effects resulting in enhanced memory. *S. lateriflora* is deep green to olive green and is commonly found in the swampy woods and meadows of North America (Bergeron et al. 2005; Upton and Dayu 2012). *S. lateriflora* derives its common name, skullcap, from the helmet-shaped upper lid of the seedpods (Bergeron et al. 2005). For over 200 years, *S. lateriflora*, a native plant of North America and Canada, has been highly valued in traditional Western medicines (Joshee et al. 2002). This species has been traditionally used by Native Americans and known as one of nature's most treasured "nervines" (Millspaugh 1974). *S. lateriflora* is also referred as a women's herb by Native Americans, to stimulate menstruation in women and also for helping discharge of placenta after childbirth (Joshee et al. 2002). *S. lateriflora* has been consumed mainly to treat hyperarousal disorders. It is used for the treatment of several nervous disorders, anxiety, hysteria, phobias, panic attacks, sedation, tension, depression, sleep disorders, stress, cancer, and anticonvulsant properties for several centuries (Brock et al. 2013; Makino et al. 2008; Wakabayashi and Yasui 2000; Zhang et al. 2009). Various commercial products of *S. lateriflora* exist as various dosage forms and are currently available for healing and nontherapeutic purposes. Formulations made from aerial parts of *S. lateriflora* are sold as teas, tinctures, tablets, and capsule forms (Li et al. 2012a). In the Pharmacopoeia and National Formulary of the USA, the dried aerial parts of *S. lateriflora* are noted as sedative/nerve tonics. *S. lateriflora* has shown to exert antispasmodic (St. Vitus' dance, muscle spasms), analgesic (neuralgia, fibromyalgia), anticonvulsant (seizure disorder, mild Tourette's syndrome), antianxiety (various anxiety disorders), sedative, appetite stimulant (anorexia nervosa) activity and reduces withdrawal symptoms associated with barbiturates and tranquilizers (Wilczańska-Barska et al. 2012; Gao et al. 2001; Foster and Tyler 1999; Gafner et al. 2003; Awad et al. 2003).

Complex science system gives an understanding of living systems as a whole, explains the complete organization, and represents the multifaceted incorporations of organized subsystems and links it with their environments (Koithan et al. 2012). It has been established that the potent use of dietary antioxidants displays cognitive-enhancing effect, psychostimulant activity, and antidepressant properties (Bouayed 2010; Bouayed et al. 2009). Hence, dietary supplements of herbal medicine rich in antioxidants could be a potentially effective and novel approach in the treatment of a wide range of neurological disorders (Dhanasekaran et al. 2007). In this article, we review the bioactive compounds and the ethnopharmacological importance of medicinal herb, *S. lateriflora*.

2.2 Bioactive Compounds in *S. lateriflora*

Bioactive compounds differ based on their chemical structures and functions. Thus, they are mainly classified based on the above concepts. Commonly, flavonoids exist in all herbs and botanicals. It is eminent that each botanical synthesizes these substances as a defense mechanism; however, historical use and recent investigations have shown to have significant therapeutic and cosmetic values. There are numerous well-characterized bioactive substances and phytochemicals. Literature on the chemical investigation of *S. lateriflora* indicated the presence of flavonoids, polyphenols, diterpenoids, iridoid glycosides, alkaloids, and phytosterols (Bruno et al. 1998; Awad et al. 2003; Gafner et al. 2003; Bergeron et al. 2005; Marsh et al. 2014). However, the major bioactive compounds reported in *S. lateriflora* are the flavonoids, namely, chrysin, baicalein, baicalin, wogonin, and wogonoside (Bergeron et al. 2005). These flavonoids are considered as biomarkers in commercial products of *S. lateriflora* used by herbal practitioners. The chemical structures of major flavonoids are shown in Fig. 2.1. In addition to these bioactive compounds, the roots and the leaves of *S. lateriflora* are reported to contain dihydrobaicalin, ikonnikoside I, scutellarin, oroxylin A-7-O-glucuronide, and 2′-methoxy-chrysin-7-O-glucuronide (Gafner et al. 2003; Bergeron et al. 2005; Marsh et al. 2014). *Scutellaria* species are a rich source of phenolic compounds, which exhibit a broad spectrum of biological activities (Choi et al. 2005; Kim et al. 2009; Shang et al. 2010; Lohani et al. 2013). Phenols such as caffeic acid, cinnamic acid, p-coumaric acid, and ferulic acid were found in the ethanoic extracts of *S. lateriflora* (Upton and Dayu 2012).

Flavonoids and amino acids (23 % of dry weight of the extract) present in the *S. lateriflora* mainly account for the anxiolytic activities (Bergeron et al. 2005; Sarris et al. 2011; Awad et al. 2003; Brock et al. 2012). Baicalin and dihydrobaicalin (14 % and 4 %) are the chief flavonoids present in the ethanol extract of *S. lateriflora*. The other flavonoids present in *S. lateriflora* are norwogonin-7-O-glucuronide, wogonin-7-O-glucuronide, 5,7-dihydroxy-6,8-dimethoxy flavone-7-O-glucuronide, dihidrooroxylin A-7-O-glucuronide, galangin-7-O-glucuronide, and 5,6,7-trihydro xy-flavanone-7-O-glucuronide (Li et al. 2012a). Among the amino acids, gamma-aminobutyric acid (GABA) is the major component (0.55 %) followed by glutamine (0.34 % of dry weight in the leaves and stem). δ-Cadinene, calamenene, β-elemene, α-cubebene, α-humulene, and α-bergamotene sesquiterpenoids are the main components identified in *S. lateriflora* oil. Studies have found that the flavonoids found in *S. lateriflora*, especially baicalein, were able to cross the blood–brain barrier and eventually reached the CNS (Tsai et al. 2002), whereas baicalin may act as a prodrug (Eiden et al. 2012).

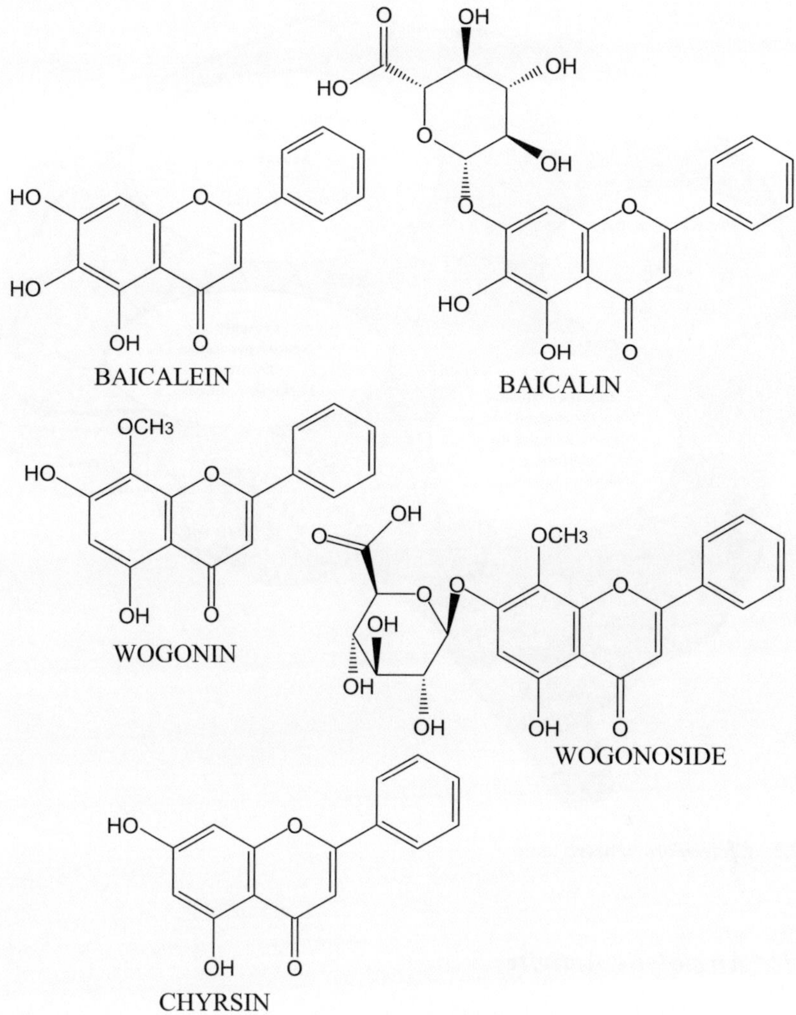

Fig. 2.1 Chemical structures of major flavonoids in *S. lateriflora*

2.3 Ethnopharmacological Properties of *S. lateriflora*

S. lateriflora is associated with a wide array of pharmacological actions. It is considered as a very important medicinal herb in traditional medicine and Western materia medica. Therapeutic activities exhibited by *S. lateriflora* have been summarized in this section in detail. Also, for better understanding, an overview of biological activities associated with *S. lateriflora* is shown in Figs. 2.2 and 2.3 and Table 2.1.

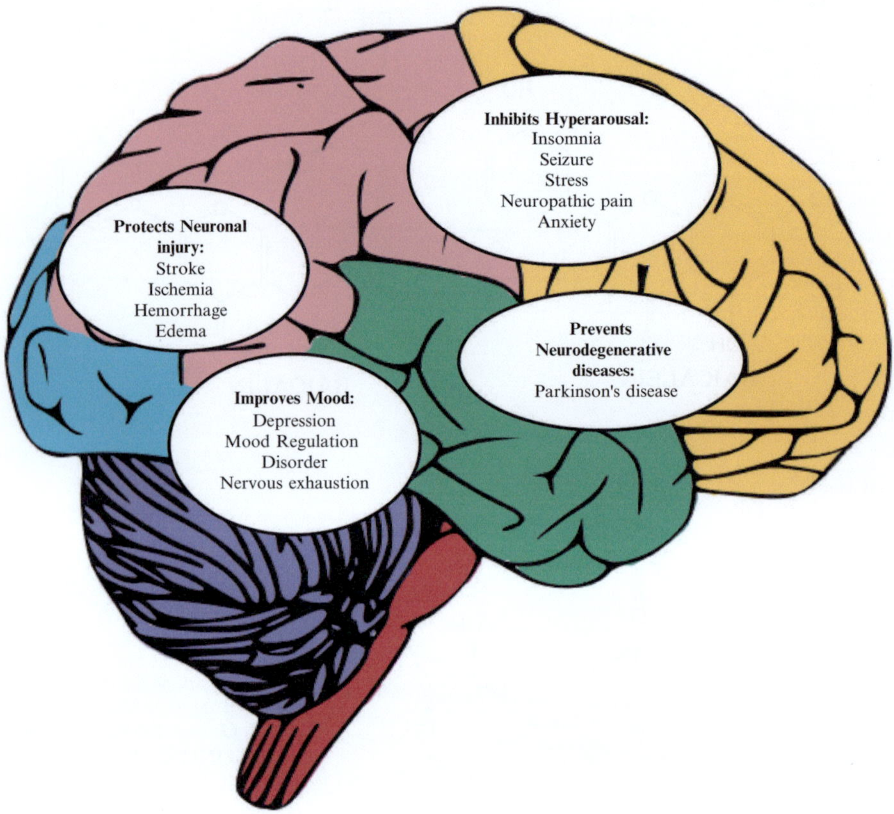

Fig. 2.2 *S. lateriflora*: a nerve tonic

2.3.1 Anxiolytic Activity

Globally, stress, depression, and anxiety are the most widespread psychiatric conditions. These mental health conditions are considered as harmful emotional experiences and are related to biochemical, cognitive, behavioral, and psychological changes (Saki et al. 2014). Hyperarousal is the major neuronal mechanistic alteration underlying most of the above pathological conditions. Hyperarousal occurs because of decreased inhibitory neurotransmission (GABA) and increased excitatory neurotransmission (glutamate). Anxiety is known to be one of the most prevalent neuropsychological disorders around the world. Concerning the global prevalence, there is a significant variation (10 % of the population in North America, Western Europe, and Australia/New Zealand compared to about 8 % in the Middle East and 6 % in Asia). Practical and functional parameters such as sex, age, values (cultural), conflict, and economic status explain the reason for the variability of occurrence. However, various statistical analyses point that one in 13 persons in the

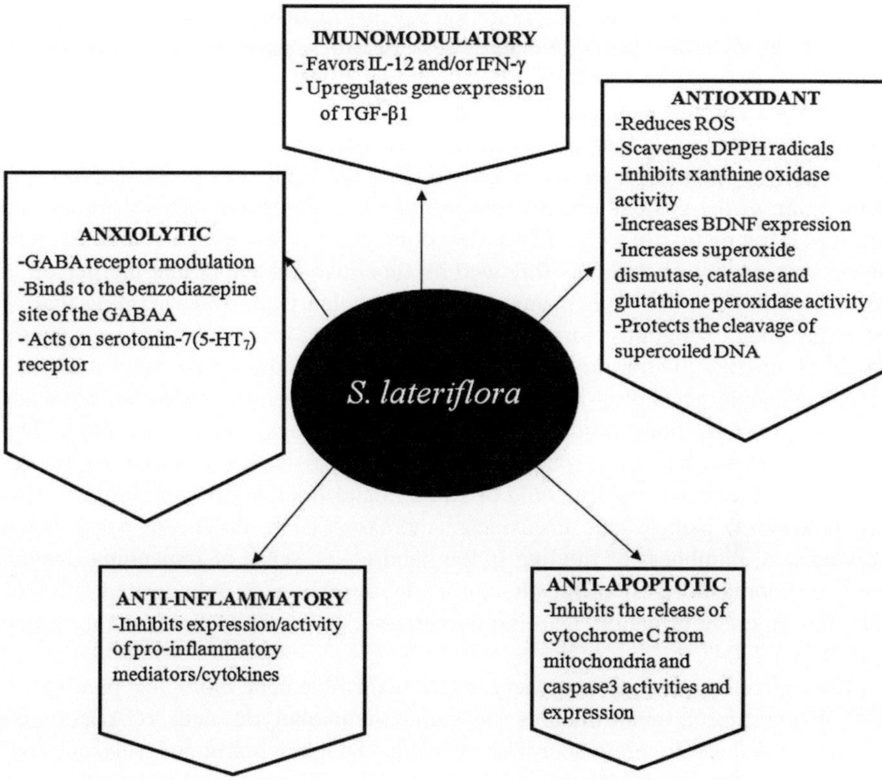

Fig. 2.3 Pharmacological activities exhibited by *S. lateriflora*

world suffers with an anxiety-like disorder. The lifetime occurrence of anxiety is about 18 % in the population of the USA. Out of these 18 %, 22.8 %, i.e., 4.1 % of the US population, are classed as severe anxiety cases (Kessler et al. 2005). Anxiety is prevalent more in female than in males (Serpell and Jagoe 1995). Anxiety is a normal reaction to stress and commonly caused by perception of real or perceived danger that threatens the security of an individual. Anxiety is a broad or common term for several disorders that cause nervousness, fear, apprehension, and worrying. These anxiety disorders affect the natural feelings and behavior and thus can manifest real physical symptoms in daily life. With regard to neuroanatomy and neurophysiology related to stress, the limbic–medial prefrontal circuit (amygdala, insula, hypothalamus, periaqueductal gray) appears to play an important role in the regulation of emotion. Processing of negative emotions (fear) is done by the cortical and subcortical areas. With respect to the neurochemical alteration in stress, GABA, serotonin (5-HT), and norepinephrine play a vital role in the etiology, progression, and pathology of the disease. Endocrine alterations are also very important in the manifestation of stress. Hence, these transmitters/mediators are the primary target to reduce the symptoms associated with stress-related disorders.

Anxiety disorders are the most common psychiatric disorders, including generalized anxiety disorder, post-traumatic stress disorder, panic disorder, obsessive–compulsive disorder, and phobias (social, simple, mixed, etc.). Generalized anxiety disorder is a chronic, nonspecific anxiety usually more prevalent in females. It is mainly caused due to a long period (more than 6 months) of worry/stress occurred in life events. Panic disorder arises abruptly and the symptoms peaks due to overstimulation of the sympathetic nervous system. Post-traumatic stress disorder is a result of a traumatic life event. Obsessive–compulsive disorder is a combination of distressing, repetitive thoughts followed by repetitive behaviors that the person is compelled to do to relive their anxiety. Social phobia is an overwhelming fear of everyday social situations. Simple phobia is a strong, irrational fear to a specific object or situation. Other types of anxiety are drug withdrawal (substance abuse)-associated anxiety and drug-induced anxiety (anticonvulsants, antibiotics, bronchodilators, steroids, nonsteroidal anti-inflammatory drugs, stimulants, etc.). The common symptoms of anxiety are insomnia, headache, backache, sweating, hyperventilation (breathing rapidly), cold or sweaty hands/feet, dry mouth, fatigue, feeling powerless, panic, fear, uneasiness, increased heart rate/heart palpitations, nervousness, numbness or tingling in the hands/ feet, sense of impending danger, panic or doom, shortness of breath, confusion, trembling, trouble concentrating or thinking, frequent urination, learning impairment, and memory dysfunction (Lavie and Milani 2004).

The main etiopathology of anxiety is genetic involvement, biological predisposition, hyperarousal, abnormalities in gamma-aminobutyric acid (GABAergic), adrenergic and serotonergic neurotransmission, change in hormone levels, and oxidative stress (Brawman-Mintzer and Lydiard 1997). Hyperarousal is a condition in which patients display an amplified psychological and physiological tension, such as reduction in pain tolerance, excessive startle and frighten responses, insomnia (sleeplessness), fatigue, and accentuation of personality traits. Decrease in the serotonergic (5HT) and GABAergic neurotransmission and increase in adrenergic and excitatory neurotransmission are normally seen in patients suffering from anxiety. This is typically related to an increased fear and sympathetic arousal due to over activation of the thalamus, hypothalamus, cortex, and amygdala. An increase in the sympathetic surge leads to panic attacks causing tachycardia (increased heart rate), shortness of breath, chest pain, and elevated stress. The hyperactivity of the brain and the loss of "feel good" factors are due to the diminishing activity of serotonin. In anxiety, the amygdala increases the production of norepinephrine in the hypothalamus; the pituitary gland increases corticotropin-releasing hormone (CRH) and adrenocorticotropic hormone (ACTH) production, respectively.

Glucose plays as a principal precursor for GABA production in vivo. Pyruvate and other amino acids also can act as precursors for GABA production. The first step in the GABA shunt is the transamination of α-ketoglutarate, formed from glucose metabolism in the Krebs cycle by GABA α-oxoglutarate transaminase into l-glutamic acid. Glutamic acid decarboxylase catalyzes the decarboxylation of glutamic acid to form GABA. GABA is present in high concentrations (millimolar) in many brain regions. These concentrations are about 1000 times higher than concentrations of the classical monoamine neurotransmitters in the same regions. GABA interneurons are abundant in the brain, with 50 % of the inhibitory synapses in the brain being GABA mediated. GABA is mainly present in the hypothalamus, hippocampus, cerebral cortex, and cerebellar cortex. GABAergic neurons (medium spiny) present in the basal ganglia send axons to the internal and external segment of the globus pallidus as well as the substantia nigra pars reticulata. Medium spiny neurons in the corpus striatum (nucleus caudatus, nucleus putamen, and nucleus accumbens) receive cortical, thalamic, and brain stem inputs. They play a key role in initiating and controlling movements of the body, limbs, and eyes. GABA is the major inhibitory neurotransmitter. GABA binds to $GABA_A$, $GABA_B$, and $GABA_C$ receptors. GABA counteracts the action of excitatory neurotransmitters such as acetylcholine and glutamate and thus promotes relaxation. GABAergic dysfunction in the brain causes the neurons to become overexcited, leading to excessive muscle contraction, hyperarousal, and a sense of restlessness. γ-Aminobutyric acid type A receptors that contain the δ subunit (δ$GABA_A$ receptors) are expressed in multiple types of neurons throughout the central nervous system (CNS), where they generate a tonic conductance that shapes neuronal excitability and synaptic plasticity. GABA regulates a variety of important behavioral functions, including memory, nociception, and anxiety, and may also modulate neurogenesis. Hence, $GABA_A$ receptors are considered to be novel therapeutic targets for the treatment of anxiety, memory dysfunction, pain, insomnia, and mood disorders. These receptors are highly responsive to benzodiazepines, barbiturates (sedative–hypnotic drugs), general anesthetics, and neuroactive steroids. A further remarkable feature of $GABA_A$ receptors is that their expression levels are highly dynamic and fluctuate substantially during development, and in response to physiological changes, including stress and the reproductive cycle. With respect to other etiologies, oxidative stress and inflammation can also contribute to the pathology. There is a correlation between oxidative stress and anxiety disorder in patients suffering from obsessive–compulsive disorder and panic disorder, and it is considered to be a characteristic feature in anxiety (Bouayed et al. 2009). Oxidative stress is possibly involved with pathogenesis and a risk factor for behavioral abnormalities and psychological disorders (Ng et al. 2008). Moreover, inflammation also plays roles in the pathogenesis of neuropsychological disorders, such as anxiety.

Managing anxiety is extremely important as depression also can be associated with the disease progression (Stein et al. 1990; Van Ameringen et al. 1991). At this time, there are pharmacological and non-pharmacological therapies available in the treatment of anxiety. Some of the therapies are adrenergic, GABAergic, and serotonergic drugs. The adrenergic drugs (sympatholytic drugs decreasing adrenergic neurotransmission) are clonidine (alpha-2 agonist), prazosin (alpha-1 adrenergic antagonist), and propranolol (beta-1/2 adrenergic antagonist). GABAergic drugs are benzodiazepines, barbiturates, and gabapentin. The other drugs used are tricyclic antidepressants (clomipramine and imipramine), selective norepinephrine reuptake inhibitors (venlafaxine), atypical antidepressants (nefazodone and mirtazapine), selective serotonin reuptake inhibitors, monoamine oxidase inhibitors, 5-HT_{1a} agonist (buspirone), and cyproheptadine (anti-serotonergic). The anticonvulsant pregabalin and a typical antipsychotic-quetiapine) can also reduce anxiety (Katzman 2009). There are innumerable adverse effects and substantial drug interactions with the current therapies. The most common anxiolytic drug therapy is the GABAergic drug, benzodiazepines. Common adverse effects due to the above mentioned therapies include drowsiness, dizziness, sedation, psychomotor impairment, memory disruption, vertigo, dysarthria, ataxia, impaired psychomotor function, anterograde amnesia, decrease in blood volume, gaining weight, aggression or paradoxical anxiety, and mortality. It is known that the prolonged use of benzodiazepine has a range of contraindications such as addiction and memory loss (Woods et al. 1987; Lader 1999; Sadock et al. 2000).

Anxiety, stress, and related disorders are complications frequently treated in the herbal medicinal clinics around the world (del Mundo et al. 2002). Conventionally, various herbal medicines have been used for managing anxiety and stress (Sarris et al. 2011). *S. lateriflora* (American skullcap) is a vital herb since ancient times in folklore medicinal practices for the treatment of anxiety- and stress-related disorders (King and Felter 1909). In WHM, *S. lateriflora* is mainly indicated for anxiety, sleeplessness, and several types of spasms. Additionally, recent clinical studies have also validated the therapeutic benefits of *S. lateriflora* in other anxiety-related disorders (Brock et al. 2013; Wolfson and Hoffmann 2003; Greenfield and Davis 2004). A clinical study in anxiety patients by Wolfson and Hoffmann has shown that *S. lateriflora* exhibited anxiolytic activity with minimal loss of cognition (Wolfson and Hoffmann 2003). In this clinical study, interestingly, a mild decline in cognition with no adverse reactions was shown by *S. lateriflora*. This indicated that this WHM could be a potential anxiolytic medicinal herb with minimal side effects (Wolfson and Hoffmann 2003). Researches have concluded that in animal models, flavonoids found in American skullcap (either in purified form or extracted from) and other species of *Scutellaria* exhibit their anxiolytic activities by modulating gamma-aminobutyric acid receptor function (Hanrahan et al. 2011). Few more studies inferred that bioactive flavonoids (baicalein, baicalin, and wogonin) possess affinities for the benzodiazepine-binding site of $GABA_A$ receptors and are responsible for the anxiolytic activity exhibited by *S. lateriflora* (Hui et al. 2000; Liao et al. 1998; Hanrahan et al. 2011; Bergeron et al. 2005). In addition, GABA and glutamine are also found in the aerial parts of *S. lateriflora*, which are also known to add to its anxiolytic activity (Bergeron et al. 2005; Sarris 2007; Zhang et al. 2009). In conclusion, commercial preparations of *S. lateriflora* have been proven to possess therapeutic benefits in anxiety-related disorders such as nervous exhaustion, muscular tension, hysteria, hypertension, tremors, convulsions, neuralgia, insomnia, and headache, with no evidence of toxicity in a double-blind, placebo-controlled clinical trial (Wolfson and Hoffmann 2003; Li et al. 2009).

With regard to the individual flavonoid present in *S. lateriflora*, they also have exhibited significant effect on the GABAergic neurotransmission. In vitro studies have established baicalein, a weak benzodiazepine receptor ligand, which has been known to possess anxiolytic and sedative effects arbitrated by its interaction with $GABA_A$ non-benzodiazepine sites. Additionally, baicalin has proven to have selective partial antagonism in the $GABA_A$ receptor (de Carvalho et al. 2011).

Wogonin exhibited anxiolytic effects by interacting with benzodiazepine receptors in the GABAergic system (Hui et al. 2002). In a few studies, it was found that the binding affinity for the benzodiazepine site was low for flavone glucuronides such as baicalin and wogonin-7-O-glucuronide (Wang et al. 2002). However, flavones with a 2′-hydroxyl group have higher affinity for such receptors. Particularly, oroxylin A displayed highest affinity to the $GABA_A$ receptor, followed by wogonin (Hui et al. 2002), whereas baicalein was not very active (Wang et al. 2002; Huen et al. 2003). Ligands binding to the benzodiazepine site of the $GABA_A$ receptor reduce the probability of action potentials caused by excitatory neurotransmitters such as glutamate, adrenaline and noradrenaline, associated with anxiety and stress (Rabow et al. 1995; Paladini et al. 1999). It has also been found that the extract of *S. lateriflora* subdued the activity of GABA transaminase. GABA transaminase which contributes to the anxiolytic activity was preferentially inhibited (de Carvalho et al. 2011). Additionally, it is postulated that since the amount of glutamine (known to have anxiolytic effect when extracted from *Valeriana*) found in *S. lateriflora* is relatively high, it may also add to the anxiolytic activity of American skullcap (Bergeron et al. 2005).

Additionally, other study on *S. lateriflora* extract has shown to have positive effects on the mood alteration (Brock et al. 2013). It has been proposed that *S. lateriflora* plays a role in the modulation of GABA and serotonin leading to improved mood and behavior (Wolfson and Hoffmann 2003; Gafner et al. 2003; Awad et al. 2003). A study by Gafner et al. (2003) found that major flavonoids isolated from *S. lateriflora* have shown binding affinity to $5\text{-}HT_7$ receptors, confirming the therapeutic potential of *S. lateriflora* for altering negative mood conditions. Phytochemicals such as scutellarein and ikonnikoside of *S. lateriflora* have shown to possess benzodiazepine and $5\text{-}HT_7$ receptor binding affinity in the brain (Liao et al. 1998; Hui et al. 2000; Gafner et al. 2003). In another study, it was found that baicalin, scutellarin, wogonin, lateriflorein, ikonnikoside I, and dihydrobaicalin from whole aerial parts of *S. lateriflora* extract bind to $5\text{-}HT_7$ receptor, proposing the probable of *S. lateriflora* for reduction of negative mood states (Gafner et al. 2003; Awad et al. 2003; Zhang et al. 2009). Serotonin $5\text{-}HT_7$ receptors are found in the thalamus, hypothalamus, and hippocampus. Upon activation by serotonin, $5\text{-}HT_7$ sets off a series of events and releases stimulatory G protein, G_s from the G protein-coupled receptor complex. G_s consequently activates adenylate cyclase that leads to an increase in the cAMP. $5\text{-}HT_7$ is associated with the antidepressant action, regulation of circadian rhythms, thermoregulation, learning and memory, mood, and smooth muscle relaxation. Novel, new, and atypical antipsychotic and antidepressants have high affinity for $5\text{-}HT_7$ receptors. It is thought that $5\text{-}HT_7$ receptor plays a role in the pathogenesis and prevention of psychiatric disorders and other pathological processes of the nervous system like sleep disorders, depression, anxiety, schizophrenia, and migraine. Various antidepressants such as tricyclics and selective serotonin reuptake inhibitors induce c-*fos* expression in a way consistent with $5\text{-}HT_7$ receptor activation within the suprachiasmatic nucleus. The effect on c-*fos* expression is inhibited after continuous treatment with antidepressants. Moreover, chronic treatment with an antidepressant drug leads to a downregulation of $5\text{-}HT_7$ receptor binding. It has been recently proposed that as an alternative therapy for depression, the combination of atypical antipsychotic with antidepressants could be used as a novel approach (Hedlund 2009). Thus, *S. lateriflora* can be a novel and potent WHM for other mental disorders.

Ionotropic glutamate receptors such as α-amino-3-hydroxy-5-methyl-4-isoxazole propionic acid receptors (AMPARs) and N-methyl-D-aspartate receptors (NMDARs) and gamma-aminobutyric acid receptor in neurons play important roles in synaptic plasticity (Bergeron et al. 2005; Awad et al. 2003; Xu et al. 2006; Brock et al. 2013). Activation of NMDAR and AMPAR leads to basal excitatory synaptic transmission and formation of synaptic plasticity like long-term potentiation (LTP) mechanisms (Malenka 1994; Lynch 2004; Citri and Malenka 2008). LTP is one of the major cellular mechanisms that is related to improved learning and memory (Bliss and Collingridge 1993; Cooke and Bliss 2006). Moreover, in contrast to anxiolytic BDZs that are known to block LTP, *S. lateriflora* extract treated hippocampal brain slice induced and maintained LTP, thereby proposing that *S. lateriflora* extract had no significant adverse effects on memory functions (del Cerro et al. 1992). The neurotonic effects of *S. lateriflora* extract are summarized in Fig. 2.2.

S. lateriflora has been traditionally used as an herbal remedy for the treatment of stress and anxiety and therefore been tested on human volunteers for its effects on mood. *S. lateriflora* displays effective antioxidant properties by exhibiting its anecdotal anxiolytic effects in various experimental animal and human models. A pilot survey among herbal medicine practitioners was conducted by Brock et al. (2012) to determine the efficacy of *S. lateriflora* using human volunteers suffering from various mental alterations or various other disorders and symptoms (depression, anxiety, and anxiety-like disorders such as insomnia, sleep-related disorders, frustration, menopausal mood swings, despondency, neuralgia, nerve weakness, emotional instability, shock, feelings of not coping, tinnitus, debility, hot flashes triggered by stress, low mood, grand mal, attention deficit hyperactivity disorder, obsessive–compulsive disorder, "liver heat rising"). It was found that *S. lateriflora* decreased the level of anxiety and depression; improved sleep quality; made them feel calmer, relaxed, and less irritable; elevated the mood; increased energy; helped them to focus; and decreased stress, tension, or nervousness. Such patients also had reduced stress- and anxiety-related physical symptoms like headache, digestive disturbance, premenstrual symptoms, inflammatory dermal conditions, twitches and spasms, and cardiovascular problems such as hypertension, tachycardia, and palpitations (Brock et al. 2012). Additionally, treatment with *S. lateriflora* did not display adverse reactions, dependability, or rebound excitability in human subjects suffering from mild anxiety (Brock et al. 2013). In a double-blind, placebo-controlled study carried on the healthy human volunteers, the aerial parts of *S. lateriflora* have also been reported to possess dose-dependent anxiolytic effects and have better level of relaxation, with no sign of toxicity (Upton and Dayu 2012; Brock et al. 2013; Wolfson and Hoffmann 2003).

2.3.2 Antioxidant Properties

Even though oxygen is an important element needed by different cells, its disruption in metabolism can lead to the production of oxygen-derived toxin and active reactive oxygen species (ROS) in cells and can lead to toxicity. Neurons are highly vulnerable to ROS. This scenario was noticed in several neurological disorders (Halliwell 2006; Ng et al. 2008; Bouayed et al. 2009). Bioactive compounds found in several medicinal plants scavenge or neutralize toxic free radicals and thereby suppress oxidative stress. Research studies have concluded that *S. lateriflora* exhibits significant radical-scavenging activities (Bochořáková et al. 2003; Wojcikowski et al. 2007; Lohani et al. 2013). Polyphenols, which are one of the major components of *S. lateriflora*, are also known to play a role in its antioxidant properties. The ethanolic extract of *S. lateriflora* containing flavonoids, polyphenols and glutathione suppressed the oxygen containing nucleophilic t-butyl hydroperoxide-induced lipid peroxidation and generation of reactive oxygen species (ROS) in the brain. Particularly, baicalin is known to have an effective lipoxygenase inhibitory activity. Lipoxygenase plays a vital role in the pathogenesis of diseases caused due to oxidative stress and inflammation (Chen et al. 2001). In rodents with type 2 diabetes mellitus, baicalin increased the activities of superoxide dismutase (SOD), catalase, and glutathione peroxidase (Waisundara et al. 2011; Guo et al. 2011). Baicalin also revealed its antioxidant activity by inhibiting the lipid peroxidation in the rat brain and kidneys (Ng et al. 2000). Moreover, baicalin also acts as a scavenger of 2,2-diphenyl-1-picrylhydrazyl (DPPH) radicals and inhibits xanthine oxidase activity in a concentration-dependent manner (Guo et al. 2011). Baicalein, the non-glycosidic form of baicalin, is also known to possess free radical-scavenging properties by reducing the ROS production (Park 2003). Baicalein reduces ROS by decreasing free radicals and/or indirectly inducing antioxidant enzymes – SOD and catalase activity (Shieh et al. 2000). In a rodent animal model of cerebral ischemia, it was established that baicalin significantly protected against the neuronal cell damage by decreasing the malondialdehyde levels and increasing SOD activities and glutathione level in the hippocampal cells. Baicalin also stimulated the brain-derived neurotrophic factor (BDNF) expression and suppressed the expression of caspase-3 at mRNA and protein levels (Cao et al. 2011). *S. lateriflora* extract had exhibited cytoprotection in various cells like H19-7 cells (hippocampal cells, as model to regulate memory and anxiety behavior) (Fournier and Duman 2013) and PC12 cells (dopaminergic cells) (Veglio et al. 2003). *S. lateriflora* extract had neuroprotection against hydrogen peroxide-induced toxicity by diminishing the ROS generation (Lohani et al. 2013).

In order to maintain the homeostasis of a cell, proper folding and functions of major macromolecules, such as DNA, are required. It has been concluded that increase in the oxidative stress results in the damage to DNA that can lead to cell death (Tharakan et al. 2005; Uttara et al. 2009). In a study conducted by Wojcikowski et al. (2007), *S. lateriflora* protected the cleavage of supercoiled DNA by hydroxyl radical mechanism. To maintain the cell homeostasis, the antioxidative enzymes like SOD and catalase play an important role in neutralizing ROS (Huang et al. 2012; Mian and XiQiang 2009). Increased activities of SOD and catalase by *S. lateriflora* extract are mainly due to the presence of flavonoids: baicalin and wogonin (Cheng et al. 2011; Waisundara et al. 2011; Wu et al. 2011; Wen et al. 2013). Some of the known markers of apoptosis are caspase-3, caspase-9, and cytochrome C proteins (Eves et al. 2002; Pugazhenthi et al. 2003). *S. lateriflora* extract also inhibited the expression of such proapoptotic markers in H19-7 cells, thereby implicating antiapoptotic property. Among flavonoids, chrysin inhibited the release of cytochrome C from mitochondria and caspase-3 activities, whereas baicalin suppressed caspase-3 expression (Liu et al. 2006; Izuta et al. 2008; Cao et al. 2010; Lagoa et al. 2011). The cyclic AMP-responsive element-binding protein (CREB) is a transcription factor which is posttranslationally activated. CREB plays a vital role as an antioxidant and has also been associated in neuronal growth and survival. BDNF in the brain is an important factor that acts as a probable antioxidant mediator (Chan et al. 2010) and helps in the regulation of neurogenesis and neuroplasticity (Lee and Son 2011). Decrease in the expression of BDNF and CREB is markers of the neurons that have undergone cell death due to oxidative stress (Yang et al. 2004; Somoza et al. 2010; Chan et al. 2010; Sarvestani et al. 2013). *S. lateriflora* extract due to the presence of its flavonoids wogonin and baicalein (Jeon et al. 2010) exhibits its neuroprotective role by enhancing the expression of BDNF but does affect the expression of CREB on the oxidative stress-induced cells.

In our laboratory, we have evaluated antioxidant property of *S. lateriflora* extract for the activity of its bioactive compounds on the mouse brain tissue homogenate by using various biochemical assays such as protein assay, Folin–Ciocalteu method, and 2,2-diphenyl-1-picrylhydrazyl (DPPH) assay. In our experiment, we found that the biochemical components such as proteins, glutathione, polyphenols, and flavonoids were higher in the ethanolic extracts of *S. lateriflora* than the aqueous extracts. This assay revealed that both the aqueous and ethanolic extracts of *S. lateriflora* reduced the DPHH radicals. Lipid peroxidation and generation of ROS production in the mouse brain tissue homogenate were induced by oxygen-containing nucleophilic t-butyl hydroperoxide. In our experiments, we also demonstrated that only the ethanolic extract of *S. lateriflora* but not aqueous extract reduced lipid peroxidation. This was mainly due to the antioxidant polyphenols and glutathione, present in *S. lateriflora*. Previous studies have established that the inhibition of robust lipoxygenase is caused by baicalin (Chen et al. 2001) which may add to the lipid peroxidation inhibitory properties of *S. lateriflora*. DNA damage is caused due to the oxidative stress which eventually leads to cell death (Tharakan et al. 2005; Uttara et al. 2009). Plasmid DNA was used to determine the effects of *S. lateriflora* against oxidative stress-induced DNA fragmentation. Our experiments demonstrated that the ethanolic extract of *S. lateriflora* protected the cleavage of supercoiled plasmid DNA triggered by hydroxyl radicals generated by hydrogen peroxide/UV exposure. The antioxidant property of *S. lateriflora* was further studied by Wojcikowski et al. (2007). This study was based on the oxygen radical absorbance capacity.

2.3.3 Anti-inflammatory Properties

Inflammation is known to play a key role in the initiation and progression of various disease states. Neurons are very susceptible to pro-inflammatory cytokines resulting in acute and chronic injury. Inflammation is associated with numerous neurological and neurodegenerative disorders. Neuroinflammation is also seen in various anxiety disorders. Cyclooxygenase and lipoxygenase are responsible for the production of inflammatory mediators which are also known to play a major role in the pathogenesis of a broad spectrum of diseases. Pro-inflammatory cytokines (PICs) that can cross the blood–brain barrier (BBB) from systemic circulation (Maes et al. 2012; Banks and Erickson 2010; Quan and Banks 2007) cause the loosening of the tightness of the junctions of capillary endothelial cells. Such PICs that enter the brain lead to the activation of microglia causing localized inflammation in the central nervous system (Qin et al. 2007). Microglia (macrophages) found in the central nervous system alters arousal by producing inflammatory mediators (Wager-Smith and Markou 2011). Inflammation may also modulate brain signals through vagus nerves and the neuronal axis carrying viscerosensory signals to specific brain areas involved in depression and anxiety. PICs affect synthesizing and degrading neurotransmitter enzyme activity, alter neurotransmitter content, and modulate neurotransmission (Gaykema et al. 2009; Gaykema and Goehler 2009). For instance, PICs alter the metabolism of tryptophan (directly) and serotonin (indirectly) by affecting indoleamine 2,3-dioxygenase, an enzyme responsible for the degradation. This results in the accelerated production of kynurenine and quinolinic acid which are known neurotoxins that cause the damage to the regular mitochondrial respiratory mechanisms and induce oxidative stress in neurons leading to anxiety and depression (Leonard and Maes 2012; Moylan et al. 2013). Increased levels of interleukin-6 (IL-6) and lower levels of serum cortisol are observed in the patients suffering from anxiety. A wide range of researches have also implicated that the psychological stress manifested by abnormal psychophysiological responses is due to the increase in the production of pro-inflammatory cytokines, tumor necrosis factor-alpha (TNF-α), interleukin-6 (IL-6), and interferon-gamma (IFN-γ), along with the decrease in the production of anti-inflammatory cytokines interleukin-10 (IL-10) and interleukin-4 (IL-4) (Maes et al. 2011). So, several reports have proven the roles of inflammation in the pathology of anxiety disorders.

Phytochemicals such as flavonoids are gaining copious beneficial attention due to its proven ability to prevent neuronal injury in both animal and human models by suppressing inflammation and oxidative stress. *S. lateriflora* has been traditionally used for the treatment of inflammatory disorders. Pharmacological studies clearly established that the anti-inflammatory activity of *S. lateriflora* is mainly attributed to the presence of flavonoids (wogonin, baicalein, and baicalin) which inhibits either the expression or activity of pro-inflammatory mediators (Li et al. 2009). It is thought that baicalein, present in *S. lateriflora*, specifically possesses the anti-inflammatory properties and is known to play an important role in neuroprotection. Few studies have confirmed that wogonin directly suppresses the prostaglandin synthesis (especially prostaglandin E_2) by inhibiting cyclooxygenase (COX-1 and COX-2) expression or activity (Wakabayashi and Yasui 2000; Chi et al. 2003; Gafner et al. 2003; Gurung et al. 2009; Lee and Kim 2010). Baicalin, baicalein, and wogonin can also induce 15-lipoxygenase (LOX) inhibitory activity (Butenko et al. 1993; Cui et al. 2010; Li et al. 2012c; Song et al. 2014). Controlled cortical impact injury induced neurotoxicity (Chen et al. 2008). Baicalein administration significantly improved the neuronal functional recovery and reduced neuronal degeneration within 24 h post-cortical injury. The neuroprotection was attributed to the significant decline in the levels of pro-inflammatory cytokines, tumor necrosis factor-α (TNF-α), interleukin-1β (IL-1β), and IL-6 in the brain (Chen et al. 2008). In another study conducted by Liu et al. (2010), baicalein (intravenous administration) protected against cerebral ischemia (permanent middle cerebral artery occlusion) in Sprague–Dawley rats (cerebral ischemia model). Baicalein downregulated the expression of pro-inflammatory enzymes, 12/15-LOX and phospholipase A (cPLA2), thereby inhibiting the production of pro-inflammatory cytokines in cerebral ischemia. In various in vitro studies, the anti-inflammatory properties of flavonoids (baicalin, baicalein, and wogonin) present in *S. lateriflora* inhibited 5-LOX and 12-LOX (Kimura et al. 1985; You et al. 1999). Various flavonoids have diverse inhibitory profile on the different pro-inflammatory enzyme activities. Baicalein, in particular, inhibited 5-LOX with a greater affinity than COX, whereas wogonin inhibited COX (Upton and Dayu 2012). Moreover, baicalein exerts its robust neuroprotective effects by blocking the activation of microglia in LPS-induced injury in dopaminergic neurons (Li et al. 2005). The excessive production of TNF-α, nitrous oxide, and superoxide radicals induced by the stimulation of LPS was inhibited by baicalein (Li et al. 2005, 2012b) and wogonin (Chiu et al. 2002; Lee et al. 2003a; Piao et al. 2004).

In the brain, major causes of apoptosis/programmed cell death are due to oxidative stress and inflammation. It is known that the oxygen–glucose deprivation inhibits the protein expression of toll-like receptor 2 (TLR-2) and results in neuronal cell death (via apoptosis). Studies have shown that baicalin, baicalein, wogonoside, and wogonin exhibit antiapoptotic effects (Li et al. 2012c). Nakamura et al. (2003) observed that in human retinal pigment epithelial cell lines (ARPE-19), dexamethasone, baicalein, and wogonin but not baicalin significantly inhibited the production of IL-6 and IL-8 mRNA induced by IL-1β. The activity of the transcription factor, nuclear factor k-light-chain enhancer of activated B cells (NF-kB), had been reduced by wogonin and baicalein, which may elucidate various anti-inflammatory and antitumor activities. Additionally, baicalein had a significantly higher inhibitory effect on the DNA-binding activity of transcription factor, nuclear factor IL-6 (NF-IL6), than NF-kB or activated protein 1 (Chen and Greene 2004). The binding activities of NF-kB were suppressed by wogonin, thereby indicating that such mechanism of wogonin may result in the inhibition of IL-6 and IL-8 mRNA (Nakamura et al. 2003). *S. lateriflora* increased gene expression in mouse macrophage in a dose-dependent manner (Pan and Halper 2003; Chuang et al. 2005), thereby resulting in the reduction of systemic and localized inflammatory responses. These results in turn support that various flavonoids found in *S. lateriflora* have an anti-inflammatory effect by the suppression of pro-inflammatory cytokines. Furthermore, baicalin, chrysin, and baicalein present in *S. lateriflora* support the production of anti-inflammatory cytokines like transforming growth factor-β1 (TGF-β1).

2.3.4 Immunomodulatory Activity

Immunomodulation is the modification and regulation of the immune response to a preferred level by immunopotentiation, immunosuppression, or induction of immunologic tolerance. The regulation of immune responses with various factors or immune-related agents such as monoclonal antibodies, cytokines, glucocorticoids, immunoglobulins, ultraviolet light, and plasmapheresis has shown to alter cellular or humoral immunity. In alternative medicine, the use of vitamins, minerals, natural foods, traditional medicines, or other nutrients promotes vigor and health or prevents degenerative or malignant diseases. The immune system plays a vital role in limiting the array of adverse effects caused as a result of infections or inflammation. Historically, flavonoids found in *S. lateriflora*, either in pure form or extracted from other plants of genus *Scutellaria*, are known to modulate immune function. *S. lateriflora*, for centuries, has been in the treatment of fever. Europeans were the first to identify this therapeutic potential in the year 1787 (Brock et al. 2012). The main cytokines produced by the innate immune cells are IL-12 and interferon-γ (IFN-γ) which play an important role in inducing cell-mediated immune response. Additionally, IFN-γ and IL-12 are also known to activate the cell-mediated immunity by stimulating type 1 and type 2 T-helper cells (Th1 and Th2) and CD4+ and CD8+ cells. A broad spectrum of immunomodulatory bioactive compounds have been found in the genus *Scutellaria* (Shang et al. 2010). Specifically, flavonoids from *S. lateriflora* extract are associated with immunomodulatory activities and offer protection against apoptosis-induced disruptions in a cell. Studies have shown that herbal extracts comprising baicalein and/or wogonin induce the production of IL-12 and/or IFN-γ production from leukocytes (Lim 2004). Moreover, lymphocytic proliferation is stimulated by wogonin (Ohtake et al. 2005). Flavonoids from *S. lateriflora* are also known to improve protection against infection by augmenting the concentration of IgA in the gut (Lim 2004).

Immune cells (T cells, B cells, and macrophages) secrete cytokine, transforming growth factor (TGF)-β1, which is responsible for inhibiting the secretion and activities of other cytokines like IFN-γ, TNF-α, and several interleukins. TGF-β1 plays a role in reducing the levels of the surface expression of cytokine receptors, including IL-2 receptor, thereby downregulating the immune cell activity. It can upsurge the expression of specific T cell cytokines and promote proliferation of such cells, especially from the underdeveloped immune cells. TGF-β1 plays a role in stimulating the B cell apoptosis. Moreover, it controls the expression of antibodies, MHC class II proteins and transferrin on both immature and mature B cells (Letterio and Roberts 1998; Lebman and Edmiston 1999). TGF-β1 is considered to be an effective immunosuppressive cytokine and has been present in the tumor and serum of most glioma patients. Baicalin, baicalein, and chrysin were found to play a role in the upregulation of TGF-β1 gene expression on macrophage (RAW 264.7) cells in a dose-dependent manner (Wang et al. 2006). Experimental study on rat malignant glioma cell line F98 demonstrated that flavonoid wogonin significantly inhibited TGF-β1-induced regulatory T cell activity in malignant gliomas. SMAD is an intracellular signal transducer and transcriptional modulator which is activated by TGF-β and activin type 1 receptor kinases. When TGF-β protein binds to a receptor on the cell surface, it activates a group of related SMAD proteins (including the SMAD3 protein also known as mothers against decapentaplegic homolog 3). These SMAD proteins bind together to form a protein complex that translocates to the nucleus. In the nucleus, the SMAD protein complex binds to specific areas of DNA which controls the activity of specific genes and regulates cell proliferation. By regulating gene activity and controlling cell proliferation, the SMAD3 protein helps as a transcription factor and tumor suppressor. The binding of tetracycline response element in the promoter region of many genes is regulated by TGF-β and the formation of SMAD3 or SMAD4 complex which initiates transcription. SMAD3/SMAD4/JUN/FOS complex formation occurs at the AP-1/SMAD site to regulate TGF-β-mediated transcription. Additionally, wogonin also inhibited IL-10 in regulatory T cell and suppressed Smad-3, GSK-3β, and ERK1/2 signaling pathways. Furthermore, wogonin enhanced phosphorylation of p38 MAPK, thereby indicating that it could impede T cells' response to TGF-β1 by modulating both SMAD and non-SMAD signaling pathways (Dandawate et al. 2012).

2.3.5 *Anticancer Property*

A study conducted by Shi et al. (2015) indicated that the flavonoid scutellarein found in *S. lateriflora* possessed anticancer potential. In this study, it was demonstrated that scutellarein could hinder the formation and the metastasis of tumor in HT1080 human fibrosarcoma cells. The rate of proliferation of such cancer cells was significantly inhibited through the initiation of programmed cell death. Additionally, in an in vivo study, it was found that scutellarein significantly decreased the volume and weight of the tumor cells. Overexpressed matrix metalloproteinases (an extracellular proteinase enzyme, MMPs) MMP-2, MMP-9, and MMP-14 assist the migration and invasion of tumor cells by altering cellular microenvironment (Coussens et al. 2002). *S. lateriflora* reduced the expression of MMP-2, MMP-9, and MMP-14 and initiated apoptosis of HT1080 cells by subduing the proliferation and thus reducing metastasis. In this experiment, it was also found that scutellarein effectively suppressed nuclear translocation and activation of nuclear factor NF-kB. NF-kB is known to be associated with various aspects of tumorigenesis, such as survival of cancer cell, its proliferation, and increase in the metastatic potential of tumor cells (Perkins 2012). Inhibition of the activity of NF-kB may further lead to the attenuation of the expression of MMP-2, MMP-9, and MMP-14 (which are known to be activated by NF-kB) further hindering the metastasis of tumor cells (Shi et al. 2015). Rat C6 glial cells are an experimental model used for the study of neoplastic gliomas. These cells provide an overview to illuminate the mechanisms of tumor invasion and angiogenesis. Due to the above properties, these cells have been used to study the etiology and to evaluate various therapies. Kim et al. (2001) found that wogonin inhibited the activation-induced C6 glial cells by inhibiting NF-kB-mediated inducible nitric oxide synthase (iNOS), thereby suppressing the production of NO. The various pharmacological activities of *S. lateriflora* are summarized in Fig. 2.3.

2.3.6 Other Pharmacological Effects

In addition to the above-discussed anxiolytic, antioxidant, anti-inflammatory, antitumor properties, the flavonoids from *S. lateriflora* also display anticonvulsant, anti-angiogenesis, antibacterial, antiviral, muscle relaxant, and other pharmacological properties (Dandawate et al. 2012; Parajuli et al. 2009; Shang et al. 2012). The herb has been used since ancient times for the prevention and treatment of hydrophobia caused due to rabies, nervousness, irritability, restlessness, and hysteria and for relieving indications of inflammation in patients with arthritis or fevers (Brock et al. 2013). In the modern times, *S. lateriflora* is used in the treatment of insomnia; digestive disorders such as colic, diarrhea, and heartburn; post-stroke paralysis; atherosclerosis; hyperlipidemia; allergies; skin conditions; and menstrual disorder. Extracts of *S. lateriflora* suppressed the replication of prion protein, thereby delaying the onset of prion disease in mice, and thus can be further studied in the therapeutic treatment of transmissible spongiform encephalopathies (Eiden et al. 2012). In the rodent models of acute seizures, *S. lateriflora* has proven to have an anticonvulsant activity (Zhang et al. 2009). In an in vitro study on mammalian renal tubular epithelial or fibroblast cells, *S. lateriflora* extract due to its antioxidant effects protected against renal injuries and urinary tract pathologies (Wojcikowski et al. 2007). *S. lateriflora* also had inhibitory properties against sucrose and maltase activities thereby indicating its probable use as an antidiabetic drug. Wogonin is known to possess a strong antiviral activity mainly against the *Hepatitis B virus*. Additionally, it has hepatoprotective effects (Gao et al. 1999; Marsh et al. 2014). Table 2.2 below illustrates various parts and different flavonoids and explains the actions of *S. lateriflora*.

Some flavonoids found in *S. lateriflora* are known to have spasmolytic effects on smooth muscle. It is postulated in an in vitro study that one of the mechanisms by which baicalein inhibits the contraction of muscles in bronchi or coronary arteries is via inhibition of cAMP phosphodiesterase (Nikaido et al. 1988). Kuhn and Winston (2001) in the herbal therapy and supplements stated that antispasmodic effects exhibited by *S. lateriflora* can be beneficial to reduce tremors in Parkinson's disease, restless leg syndrome, temporomandibular joint (TMJ) pain, bruxism, and physical symptoms of Tourette's syndrome. The ethanol extract of *S. lateriflora* is also known to have antimicrobial activity against *Cladosporium cucumerinum*,

Table 2.2 Ethnopharmacological properties of *S. lateriflora*

Plant part used	Medical uses	Mode of action	References
Whole plant	Beneficial in anxiety, nervous exhaustion, and sleep disorder (insomnia)	Binds to $GABA_A$ receptor, inhibits GABA reuptake, at high concentration, induces the release of GABA	Awad et al. (2003)
	Anxiolytic effect	Binds to $GABA_A$ receptor	Wolfson and Hoffmann (2003)
	Useful in anxiety (insomnia) and stress-related disorders	Activates the benzodiazepine-binding site of $GABA_A$ receptors	Upton and Dayu (2012)
	Beneficial in anxiety and stress-related disorders	Increase GABAergic neurotransmission	King and Felter (1909)
	Used in depression	Affects 5-HT7 receptors	Gafner et al. (2003)
	Used in the treatment of anxiety and sleeplessness	Blocks acute production of NO_2, inhibits expression of iNOS and COX-2 genes	Chen et al. (2001)
	To maintain the homeostasis of the cell	Protects the cleavage of supercoiled DNA induced by hydroxyl radicals	Wojcikowski et al. (2007)
Aerial parts	Reduces anxiety, insomnia, and stress	Binds to the benzodiazepine-binding site of GABA-α receptors	Brock et al. (2012)
	Positive effects on the mood	Binds to the benzodiazepine-binding site of GABA-α receptor, affects 5-HT neurotransmission	Brock et al. (2013)
	Anticonvulsant and anxiolytic activity	Binds to GABA-α receptor, inhibits GABA reuptake; GABA and glutamine are also found in the plant	Zhang et al. (2009)
	Neuroprotective effect	Scavenges (DPPH) radicals, reduces reactive oxygen species	Lohani et al. (2013)
	Antitumor activity	Induces apoptosis and cell cycle arrest at G1/G2 without affecting primary or nonmalignant cells	Parajuli et al. (2009)

(continued)

Table 2.2 (continued)

Plant part used	Medical uses	Mode of action	References
Pure compound (Baicalin)	Anxiolytic effect	GABA receptor modulation	Upton and Dayu (2012)
	Anxiolytic activity	Activates the benzodiazepine-binding site of GABA-α receptors	Xu et al. (2006)
	Neuroprotective in cerebral ischemia	Inhibits NMDA receptor function and PKCα activation	Liu et al. (2005)
	Neuroprotective activity	Attenuates GLU/NMDA-increased intracellular calcium	Lee et al. (2003a, b)
	Neuroprotective effect	Alleviates the increase in reactive oxygen species (ROS) production and alterations in cellular morphology	de Oliveira et al. (2015)
	Neuroprotective activity	Antagonizes the increase in intracellular free-calcium concentration [$Ca^{2+}{}_{i}$]	Gao et al. (2001)
	Alleviate neuropathic pain	Inhibits histone deacetylase 1 (HDAC1)	Cherng et al. (2014)
	Beneficial effect on adult neurogenesis, neuroprotection, improvement of cognitive function and emotional regulation	Increases the number of both class I and class II doublecortin-positive neurons	Jiang et al. (2013)
	Beneficial in stroke	Activates the cyclinD-mothers against decapentaplegic homolog 3 (SMAD3)-transforming growth factor (TGF-beta) network pathway	Li et al. (2013)
	Reduce brain edema after intracerebral hemorrhage	Inhibits NF-kB activation and reduces IL-1β and IL-6 production, as well as blood–brain barrier permeability	Zhou et al. (2014)
	Neuroprotection in cerebral ischemia, antioxidative effects	Inhibits xanthine and NADPH oxidases and increases superoxide dismutase	Cheng et al. (2013)
	Beneficial in stroke	Inhibits the inflammatory reaction in neuron damage, and toll-like receptor 2/4 (TLR) might be its targets	Li et al. (2012b)
	Parkinson's disease	Inhibits iron accumulation in different brain regions and has a protective effect on dopaminergic neurons	Xiong et al. (2012)

(continued)

Table 2.2 (continued)

Plant part used	Medical uses	Mode of action	References
Pure compound (wogonin)	Anxiolytic effect	Activates the benzodiazepine-binding site of GABA-α receptors	Hui et al. (2002) Tai et al. (2006)
	Antioxidant, neuroprotective effect	Inhibits the oxidative neuronal damage and lipid peroxidation and exhibited radical-scavenging activity	Cho and Lee (2004)
	Helpful in stroke	Inhibits induction of inflammatory mediators such as iNOS and TNF-α	Cho and Lee (2004)
	Reduces brain edema after traumatic brain injury	Attenuates the TLR4/ NF-kB-mediated inflammatory response	Chen et al. (2012)
	Neuroprotective effect	Inhibits activation-induced death of C6 glial cells by suppressing NO production, through inhibition of NF-kB-mediated iNOS induction	Kim et al. (2001)
	Neuroprotective effect	Inhibits the inflammatory activation of microglia	Lee et al. (2003a)
	Antitumor activity	Inhibits of TGF-β1 activity	Dandawate et al. (2012)

Escherichia coli, and Candida albicans (Bergeron et al. 1996). Wogonin inhibited *Vibrio cholerae* and *Staphylococcus aureus* (Upton and Dayu 2012).

Neuropathic pain as per the International Association for the Study of Pain (IASP) is caused due to a primary lesion or dysfunction of the nervous system. It is a chronic condition which leads to obstinate pain symptoms. It has been established that the manifestation of pain-associated genes in sensory neurons, immune cells, and glial cells is related to the cause and persistence of neuropathic pain. It is postulated that epigenetic modifications regulate the inflammatory process. Epigenetic modifications are the modifications caused in the gene expression without changing the DNA sequence and control the transcription and expression of pro- or antinociceptive genes. The main epigenetic alterations are DNA methylation and posttranslational histone modifications such as acetylation, methylation, and phosphorylation. Histone deacetylases have revealed that they play a role in the suppression of expression of cytokines such as IL-1b and TNF-α. Such PICs binding to their receptors can modify epigenetic processing. Further, the peripheral chronic inflammatory pains are also modulated by histone methylation, which results in the recruitment of NF-kB to pro-inflammatory genes. The acetylation of histones is catalyzed by histone acetyltransferase and removed by histone deacetylase. Inhibition of histone deacetylase prevents the removal of acetyl groups from histones, thereby improving the inflammatory symptoms caused by pain models (Cherng et al. 2014).

Symptoms of neuropathic pain include thermal and tactile hyperalgesia. Initially, neuropathic pain was thought to be linked with peripheral nerve dysfunction. However, neuropathy caused by diabetes mellitus, spinal stenosis, and injuries to the brain or spinal cord can also lead to chronic neuropathic pain, which relates the role of CNS. Treatment options for neuropathic pain include pharmacological and non-pharmacological methods. The pharmacological methods are opioids, antidepressants, GABAergic drugs (benzodiazepines), and anticonvulsants. The non-pharmacological ones are psychological management, physical methods, and sometimes surgery. Baicalein in its pure form is known to alleviate the symptoms caused by neuropathic pain by inhibiting the activity of histone deacetylase I enzyme. The activities of histone acetyltransferase and histone deacetylase balance the acetylation levels in the cells. Pure compound baicalin also has a beneficial effect on adult neurogenesis and improves cognitive functions by increasing the amount of doublecortin-positive neurons. Doublecortin is a microtubule-associated protein articulated by neuronal precursor cells and immature neurons in embryonic and adult cortical configurations. Doublecortin is immensely expressed in the focal cortical infarct surroundings and is a marker of the immature neurons in neurogenic niches. Strangely, the number of doublecortin-positive cells is clearly associated with the rescue of functional deficiency after stroke (Jiang et al. 2013). Baicalin also relives stroke symptoms by activating the SMAD3 and TGF-β pathways (Li et al. 2013). Baicalin plays a role in the reduction of brain edema caused after stroke by hindering NF-kB activation and reducing the production of IL-1β and IL-6 cytokines (Zhou et al. 2014).

2.4 Conclusions

Merging of new and novel modern pharmacological therapies with the orthodox traditional (Eastern and Western) medicines and other healing remedies (complementary and alternative medicine, integrative medicine) could be the future holistic approach for treating various disorders. *S. lateriflora* contains nutritionally rich phytochemicals, which exhibit substantial human and animal health remedial properties (antioxidant, antiapoptotic, and anti-inflammatory). These actions couple with their effects on neurochemical and endocrinal systems have shown to decrease hyperarousal-related disorders. Furthermore, it also exhibits anticancer action, decreases infection, improves gastrointestinal functions, and reduces inflammation. It can thus be postulated that *S. lateriflora* could be used as one of the potential experimental herbal drugs that could be further evaluated for its therapeutic potential in the treatment of various peripheral and neurological disorders.

References

Awad R, Arnason JT, Trudeau V, Bergeron C, Budzinski JW, Foster BC, Merali Z (2003) Phytochemical and biological analysis of skullcap (*Scutellaria lateriflora* L.): a medicinal plant with anxiolytic properties. Phytomedicine 10:640–649. doi:10.1078/0944-7113-00374

Banks WA, Erickson MA (2010) The blood-brain barrier and immune function and dysfunction. Neurobiol Dis 37:26–32. doi:10.1016/j.nbd.2009.07.031

Bergeron D, Audette M, Lambert RD (1996) Regulation of leukocyte interleukin 2 and interleukin 2 receptor gene expression by rabbit blastocoelic fluid. J Reprod Fertil 106:143–151

Bergeron C, Gafner S, Clausen E, Carrier DJ (2005) Comparison of the chemical composition of extracts from *Scutellaria lateriflora* using accelerated solvent extraction and supercritical fluid extraction versus standard hot water or 70% ethanol extraction. J Agric Food Chem 53:3076–3080. doi:10.1021/jf048408t

Bliss TV, Collingridge GL (1993) A synaptic model of memory: long-term potentiation in the hippocampus. Nature 361:31–39. doi:10.1038/361031a0

Bochořáková H, Paulová H, Slanina J, Musil P, Táborská E (2003) Main flavonoids in the root of *Scutellaria baicalensis* cultivated in Europe and their comparative antiradical properties. Phytother Res 17:640–644. doi:10.1002/ptr.1216

Bouayed J (2010) Polyphenols: a potential new strategy for the prevention and treatment of anxiety and depression. Curr Nutr Food Sci 6:13–18. doi:10.2174/157340110790909608

Bouayed J, Rammal H, Soulimani R (2009) Oxidative stress and anxiety: relationship and cellular pathways. Oxid Med Cell Longev 2:63–67

Braun LA, Spitzer O, Tiralongo E, Wilkinson JM, Bailey M, Poole SG, Dooley M (2013) Naturopaths and Western herbalists' attitudes to evidence, regulation, information sources and knowledge about popular complementary medicines. J Aust Tradit Soc 19:246–248

Brawman-Mintzer O, Lydiard RB (1997) Biological basis of generalized anxiety disorder. J Clin Psychiatry 58:16–25

Brock C, Whitehouse J, Tewfik I, Towell T (2012) The use of *Scutellaria lateriflora*: a pilot survey amongst herbal medicine practitioners. J Herb Med 2:34–41. doi:10.1016/j.hermed.2012.04.005

Brock C, Whitehouse J, Tewfik I, Towell T (2013) Identity issues surrounding American skullcap (*Scutellaria lateriflora*) and an optimised high performance liquid chromatography method to authenticate commercially available products. J Herb Med 3:57–64. doi:10.1016/j.hermed.2013.02.001

Bruno M, Cruciata M, Bondi ML, Piozzi F, de la Torre MC, Rodríguez B, Servettaz O (1998) Neo-clerodane diterpenoids from *Scutellaria lateriflora*. Phytochemistry 48:687–691. doi:10.1016/S0031-9422(97)01032-7

Butenko IG, Gladtchenko SV, Galushko SV (1993) Anti-inflammatory properties and inhibition of leukotriene C4 biosynthesis in vitro by flavonoid baicalein from *Scutellaria baicalensis* georgy roots. Agents Actions 39 Spec No:C49–C51

Cao Y, Li G, Wang Y, Fan Z, Yu D, Wang Z, Bi Y (2010) Neuroprotective effect of baicalin on compression spinal cord injury in rats. Brain Res 1357:115–123. doi:10.1016/j.brainres.2010.07.108

Cao Y, Mao X, Sun C, Zheng P, Gao J, Wang X, Min D, Sun H, Xie N, Cai J (2011) Baicalin attenuates global cerebral ischemia/reperfusion injury in gerbils via anti-oxidative and anti-apoptotic pathways. Brain Res Bull 85:396–402. doi:10.1016/j.brainresbull.2011.05.002

Casey MG, Adams J, Sibbritt D (2007) An examination of the prescription and dispensing of medicines by Western herbal therapists: a national survey in Australia. Complement Ther Med 15:13–20. doi:10.1016/j.ctim.2005.10.008

Chan SHH, Wu C-WJ, Chang AYW, Hsu K-S, Chan JYH (2010) Transcriptional upregulation of brain-derived neurotrophic factor in rostral ventrolateral medulla by angiotensin II: significance in superoxide homeostasis and neural regulation of arterial pressure. Circ Res 107:1127–1139. doi:10.1161/CIRCRESAHA.110.225573

Chen LF, Greene WC (2004) Shaping the nuclear action of NF-kappaB. Nat Rev Mol Cell Biol 5:392–401. doi:10.1038/nrm1368

Chen YC, Shen SC, Chen LG, Lee TJ, Yang LL (2001) Wogonin, baicalin, and baicalein inhibition of inducible nitric oxide synthase and cyclooxygenase-2 gene expressions induced by nitric oxide synthase inhibitors and lipopolysaccharide. Biochem Pharmacol 61:1417–1427

Chen SF, Hsu CW, Huang WH, Wang JY (2008) Post-injury baicalein improves histological and functional outcomes and reduces inflammatory cytokines after experimental traumatic brain injury. Br J Pharmacol 155:1279–1296. doi:10.1038/bjp.2008.345

Chen CC, Hung TH, Wang YH, Lin CW, Wang PY, Lee CY, Chen SF (2012) Wogonin improves histological and functional outcomes, and reduces activation of TLR4/NF-kb signaling after experimental traumatic brain injury. PLoS ONE. doi:10.1371/journal.pone.0030294

Cheng F, Song W, Guo S, Zhong X, Lu Y, Wang Q (2011) Protective effect of baicalin on H_2O_2 injured rats' primary cortical neurons. China J Tradit Chin Med Pharm 0–16

Cheng F, Lu Y, Zhong X, Song W, Wang X, Sun X, Qin J, Guo S, Wang Q (2013) Baicalin's therapeutic time window of neuroprotection during transient focal cerebral ischemia and its antioxidative effects *in vitro* and *in vivo*. Evid Based Complement Altern Med 2013:120261. doi:10.1155/2013/120261

Cherng CH, Lee KC, Chien CC, Chou KY, Cheng YC, Hsin ST, Lee SO, Shen CH, Tsai RY, Wong CS (2014) Baicalin ameliorates neuropathic pain by suppressing HDAC1 expression in the spinal cord of spinal nerve ligation rats. J Formos Med Assoc 113:513–520. doi:10.1016/j.jfma.2013.04.007

Chi YS, Lim H, Park H, Kim HP (2003) Effects of wogonin, a plant flavone from *Scutellaria* radix, on skin inflammation: *in vivo* regulation of inflammation-associated gene expression. Biochem Pharmacol 66:1271–1278. doi:10.1016/S0006-2952(03)00463-5

Chiu JH, Lay IS, Su MY, Chiu HL, Chiu AC, Lui WY, Wu CW (2002) Tumor necrosis factor-producing activity of wogonin in RAW 264.7 murine macrophage cell line. Planta Med 68:1036–1039. doi:10.1055/s-2002-35669

Cho J, Lee HK (2004) Wogonin inhibits excitotoxic and oxidative neuronal damage in primary cultured rat cortical cells. Eur J Pharmacol 485:105–110

Choi YA, Kang OH, Park HJ, Tae J, Kim DK, Kang CS, Choi SC, Yun KJ, Choi SJ, Nah YH, Kim YH, Bae KH, Lee YM (2005) Effect of processed *Scutellaria baicalensis* on dextran sulfate sodium-induced colitis in mice. Int J Mol Med 16:667–672

Chuang HN, Wang JY, Chiu JH, Tsai TH, Yeh SF, Fu SL, Lui WY, Wu CW (2005) Enhancing effects of *Scutellaria baicalensis* and some of its constituents on TGF-β1 gene expression in RAW 264.7 murine macrophage cell line. Planta Med 71:440–445. doi:10.1055/s-2005-864140

Citri A, Malenka RC (2008) Synaptic plasticity: multiple forms, functions, and mechanisms. Neuropsychopharmacology 33:18–41. doi:10.1038/sj.npp.1301559

Cooke SF, Bliss TVP (2006) Plasticity in the human central nervous system. Brain 129:1659–1673. doi:10.1093/brain/awl082

Coussens LM, Fingleton B, Matrisian LM (2002) Matrix metalloproteinase inhibitors and cancer: trials and tribulations. Science 295:2387–2392. doi:10.1126/science.1067100

Cui L, Zhang X, Yang R, Liu L, Wang L, Li M, Du W (2010) Baicalein is neuroprotective in rat MCAO model: role of 12/15-lipoxygenase, mitogen-activated protein kinase and cytosolic phospholipase A2. Pharmacol Biochem Behav 96:469–475. doi:10.1016/j.pbb.2010.07.007

Dandawate S, Williams L, Joshee N, Rimando AM, Mittal S, Thakur A, Lum LG, Parajuli P (2012) *Scutellaria* extract and wogonin inhibit tumor-mediated induction of Treg cells via inhibition of TGF-β1 activity. Cancer Immunol Immunother 61:701–711. doi:10.1007/s00262-011-1130-3

De Carvalho RSM, Duarte FS, de Lima TCM (2011) Involvement of GABAergic non-benzodiazepine sites in the anxiolytic-like and sedative effects of the flavonoid baicalein in mice. Behav Brain Res 221:75–82. doi:10.1016/j.bbr.2011.02.038

De Oliveira MR, Nabavi SF, Habtemariam S, Erdogan Orhan I, Daglia M, Nabavi SM (2015) The effects of baicalein and baicalin on mitochondrial function and dynamics: a review. Pharmacol Res 100:296–308. doi:10.1016/j.phrs.2015.08.021

Del Cerro S, Jung M, Lynch G (1992) Benzodiazepines block long-term potentiation in slices of hippocampus and piriform cortex. Neuroscience 49:1–6. doi:10.1016/0306-4522(92)90071-9

Del Mundo WFB, Shepherd WC, Marose TD (2002) Use of alternative medicine by patients in a rural family practice clinic. Fam Med 34:206–212

Dhanasekaran M, Tharakan B, Holcomb LA, Hitt AR, Young KA, Manyam BV (2007) Neuroprotective mechanisms of ayurvedic antidementia botanical Bacopa monniera. Phytother Res 21:965–969. doi:10.1002/ptr.2195

Eiden M, Leidel F, Strohmeier B, Fast C, Groschup MH (2012) A medicinal herb *Scutellaria lateriflora* inhibits PrP replication in vitro and delays the onset of prion disease in mice. Front Psychiatry. doi:10.3389/fpsyt.2012.00009

Eves EM, Boise LH, Thompson CB, Wagner AJ, Hay N, Rosner MR (2002) Apoptosis induced by differentiation or serum deprivation in an immortalized central nervous system neuronal cell line. J Neurochem 67:1908–1920. doi:10.1046/j.1471-4159.1996.67051908.x

Foster S, Tyler VE (1999) Tyler's honest herbal: a sensible guide to the use of herbs and related remedies, 4th edn. Haworth Herbal Press, New York

Fournier NM, Duman RS (2013) Illuminating hippocampal control of fear memory and anxiety. Neuron 77:803–806. doi:10.1016/j.neuron.2013.02.017

Gafner S, Bergeron C, Batcha LL, Reich J, Arnason JT, Burdette JE, Pezzuto JM, Angerhofer CK (2003) Inhibition of [3H]-LSD binding to 5-HT7 receptors by flavonoids from *Scutellaria lateriflora*. J Nat Prod 66:535–537. doi:10.1021/np0205102

Gao Z, Huang K, Yang X, Xu H (1999) Free radical scavenging and antioxidant activities of flavonoids extracted from the radix of *Scutellaria baicalensis* Georgi. Biochim Biophys Acta, Gen Subj 1472:643–650. doi:10.1016/S0304-4165(99)00152-X

Gao Z, Huang K, Xu H (2001) Protective effects of flavonoids in the roots of *Scutellaria baicalensis* Georgi against hydrogen peroxide-induced oxidative stress in HS-SY5Y cells. Pharmacol Res 43:173–178. doi:10.1006/phrs.2000.0761

Gaykema RPA, Goehler LE (2009) Lipopolysaccharide challenge-induced suppression of Fos in hypothalamic orexin neurons: their potential role in sickness behavior. Brain Behav Immun 23:926–930. doi:10.1016/j.bbi.2009.03.005

Gaykema RPA, Daniels TE, Shapiro NJ, Thacker GC, Park S-M, Goehler LE (2009) Immune challenge and satiety-related activation of both distinct and overlapping neuronal populations in the brainstem indicate parallel pathways for viscerosensory signaling. Brain Res 1294:61–79. doi:10.1016/j.brainres.2009.07.076

Greenfield J, Davis JM (2004) Medicinal herb production guide. Skullcap (*Scutellaria lateriflora* L). North Carolina Consortium on Natural Medicines and Public Health

Guo X, Chen B, Luo L, Zhang X, Dai X, Gong S (2011) Chemical compositions and antioxidant activities of water extracts of chinese propolis. J Agric Food Chem 59:12610–12616. doi:10.1021/jf202818p

Gurung SK, Lim H, Kim HP, Park HI (2009) Structural alterations of wogonin significantly reduce the inhibitory activity against COX-2 catalyzed PGE_2 production from LPS-induced RAW 264.7 cells. Biomol Ther 17:418–421. doi:10.4062/biomolther.2009.17.4.418

Halliwell B (2006) Oxidative stress and neurodegeneration: where are we now? J Neurochem 97:1634–1658. doi:10.1111/j.1471-4159.2006.03907.x

Hanrahan JR, Chebib M, Johnston GAR (2011) Flavonoid modulation of GABAA receptors. Br J Pharmacol 163:234–245. doi:10.1111/j.1476-5381.2011.01228.x

Hedlund PB (2009) The 5-HT7 receptor and disorders of the nervous system: an overview. Psychopharmacology (Berl) 206:345–354. doi:10.1007/s00213-009-1626-0

Huang TT, Zou Y, Corniola R (2012) Oxidative stress and adult neurogenesis—effects of radiation and superoxide dismutase deficiency. Semin Cell Dev Biol 23:738–744. doi:10.1016/j.semcdb.2012.04.003

Huen MSY, Leung JWC, Ng W, Lui WS, Chan MNS, Wong JT-F, Xue H (2003) 5,7-Dihydroxy-6-methoxyflavone, a benzodiazepine site ligand isolated from *Scutellaria baicalensis* Georgi, with selective antagonistic properties. Biochem Pharmacol 66:125–132

Hui KM, Wang XH, Xue H (2000) Interaction of flavones from the roots of *Scutellaria baicalensis* with the benzodiazepine site. Planta Med 66:91–93. doi:10.1055/s-0029-1243121

Hui KM, Huen MSY, Wang HY, Zheng H, Sigel E, Baur R, Ren H, Li ZW, Wong JT-F, Xue H (2002) Anxiolytic effect of wogonin, a benzodiazepine receptor ligand isolated from *Scutellaria baicalensis* Georgi. Biochem Pharmacol 64:1415–1424

Izuta H, Shimazawa M, Tazawa S, Araki Y, Mishima S, Hara H (2008) Protective effects of Chinese propolis and its component, chrysin, against neuronal cell death via inhibition of mitochondrial apoptosis pathway in SH-SY5Y cells. J Agric Food Chem 56:8944–8953. doi:10.1021/jf8014206

Jeon SJ, Bak HR, Seo JE, Kwon KJ, Kang YS, Kim HJ, Cheong JH, Ryu JH, Ko KH, Shin CY (2010) Synergistic increase of BDNF release from rat primary cortical neuron by combination of several medicinal plant-derived compounds. Biomol Ther 18:39–47. doi:10.4062/biomolther.2010.18.1.039

Jiang X, Xu J, Zou D, Yang L, Wang Y (2013) Baicalin influences the dendritic morphology of newborn neurons in the hippocampus of chronically stressed rats. Neural Regen Res 8:133–136. doi:10.3969/j.issn.1673

Joshee N, Patrick TS, Mentreddy RS, Yadav AK, Janick J, Whipkey A (2002) Skullcap: potential medicinal crop. In: Janick J, Whipkey A (eds) Trends in new crops and new uses. ASHS Press, Alexandria, pp 580–586

Katzman MA (2009) Current considerations in the treatment of generalized anxiety disorder. CNS Drugs 23:103–120. doi:10.2165/00023210-200923020-00002

Kessler RC, Chiu WT, Demler O, Merikangas KR, Walters EE (2005) Prevalence, severity, and comorbidity of 12-month DSM-IV disorders in the National Comorbidity Survey Replication. Arch Gen Psychiatry 62:617–627. doi:10.1001/archpsyc.62.6.617

Kim H, Kim YS, Kim SY, Suk K (2001) The plant flavonoid wogonin suppresses death of activated C6 rat glial cells by inhibiting nitric oxide production. Neurosci Lett 309:67–71

Kim EH, Shim B, Kang S, Jeong G, Lee J, Yu Y-B, Chun M (2009) Anti-inflammatory effects of *Scutellaria* baicalensis extract via suppression of immune modulators and MAP kinase signaling molecules. J Ethnopharmacol 126:320–331. doi:10.1016/j.jep.2009.08.027

Kimura Y, Okuda H, Arichi S (1985) Studies on Scutellariae Radix; XIII 1. Effects of various flavonoids on arachidonate metabolism in leukocytes. Planta Med 51:132–136. doi:10.1055/s-2007-969427

King J, Felter H (1909) King's American dispensatory. Ohio Valley Company, Ohio

Koithan M, Bell I, Niemeyer K, Pincus D (2012) A complex systems science perspective for whole systems of complementary and alternative medicine research. Forsch Komplementmed 19(1):7–14. doi:10.1159/000335181

Kuhn MA, Winston D (2001) Herbal therapy and supplements: a scientific and traditional approach. Lippincott Williams and Wilkins, Philadelphia

Kumar H, More SV, Han SD, Choi JY, Choi DK (2012) Promising therapeutics with natural bioactive compounds for improving learning and memory-a review of randomized trials. Molecules 17:10503–10539. doi:10.3390/molecules170910503

Lader MH (1999) Limitations on the use of benzodiazepines in anxiety and insomnia: are they justified? Eur Neuropsychopharmacol 9(Suppl 6):S399–S405

Lagoa R, Graziani I, Lopez-Sanchez C, Garcia-Martinez V, Gutierrez-Merino C (2011) Complex I and cytochrome c are molecular targets of flavonoids that inhibit hydrogen peroxide production by mitochondria. Biochim Biophys Acta Bioenerg 1807:1562–1572. doi:10.1016/j.bbabio.2011.09.022

Lavie CJ, Milani RV (2004) Prevalence of anxiety in coronary patients with improvement following cardiac rehabilitation and exercise training. Am J Cardiol 93:336–339. doi:10.1016/j.amjcard.2003.10.015

Lebman DA, Edmiston JS (1999) The role of TGF-beta in growth, differentiation, and maturation of B lymphocytes. Microbes Infect 1:1297–1304
Lee JH, Kim GH (2010) Evaluation of antioxidant and inhibitory activities for different subclasses flavonoids on enzymes for rheumatoid arthritis. J Food Sci 75:H212–H217. doi:10.1111/j.1750-3841.2010.01755.x
Lee EG, Son H (2011) Adult hippocampal neurogenesis and related neurotrophic factors. BMB Rep 44:421–421. doi:10.5483/BMBRep.2011.44.6.421
Lee H, Kim YO, Kim H, Kim SY, Noh HS, Kang SS, Cho GJ, Choi WS, Suk K (2003a) Flavonoid wogonin from medicinal herb is neuroprotective by inhibiting inflammatory activation of microglia. FASEB J 17:1943–1944. doi:10.1096/fj.03-0057fje
Lee HH, Yang LL, Wang CC, Hu SY, Chang SF, Lee YH (2003b) Differential effects of natural polyphenols on neuronal survival in primary cultured central neurons against glutamate- and glucose deprivation-induced neuronal death. Brain Res 986:103–113
Leonard B, Maes M (2012) Mechanistic explanations how cell-mediated immune activation, inflammation and oxidative and nitrosative stress pathways and their sequels and concomitants play a role in the pathophysiology of unipolar depression. Neurosci Biobehav Rev 36:764–785. doi:10.1016/j.neubiorev.2011.12.005
Letterio JJ, Roberts AB (1998) Regulation of immune responses by TGF-β. Annu Rev Immunol 16:137–161. doi:10.1146/annurev.immunol.16.1.137
Li FQ, Wang T, Pei Z, Liu B, Hong JS (2005) Inhibition of microglial activation by the herbal flavonoid baicalein attenuates inflammation-mediated degeneration of dopaminergic neurons. J Neural Transm 112:331–347. doi:10.1007/s00702-004-0213-0
Li S, Hong S, Shepardson NE, Walsh DM, Shankar GM, Selkoe D (2009) Soluble oligomers of amyloid Beta protein facilitate hippocampal long-term depression by disrupting neuronal glutamate uptake. Neuron 62:788–801. doi:10.1016/j.neuron.2009.05.012
Li HY, Hu J, Zhao S, Yuan ZY, Wan HJ, Lei F, Ding Y, Xing DM, Du LJ (2012a) Comparative study of the effect of baicalin and its natural analogs on neurons with oxygen and glucose deprivation involving innate immune reaction of TLR2/TNF α. J Biomed Biotechnol. doi:10.1155/2012/267890
Li HY, Yuan ZY, Wang YG, Wan HJ, Hu J, Chai YS, Lei F, Xing DM, Du LJ (2012b) Role of baicalin in regulating Toll-like receptor 2/4 after ischemic neuronal injury. Chin Med J (Engl) 125:1586–1593. doi: 10.3760/cma.j.issn.0366-6999.2012.09.012
Li XX, He GR, Mu X, Xu B, Tian S, Yu X, Meng FR, Xuan ZH, Du GH (2012c) Protective effects of baicalein against rotenone-induced neurotoxicity in PC12 cells and isolated rat brain mitochondria. Eur J Pharmacol 674(2–3):227–233. doi: 10.1016/j.ejphar.2011.09.181
Li H, Zhang Y, Yu Y, Li B, Chen Y, Wu H, Wang J, Li J, Xiong X, He Q, Tian J, Wang Z, Wang J (2013) Systemic revealing pharmacological signalling pathway networks in the hippocampus of ischaemia-reperfusion mice treated with baicalin. Evid Based Complement Alternat Med. doi:10.1155/2013/630723
Liao JF, Wang HH, Chen MC, Chen CC, Chen CF (1998) Benzodiazepine binding site-interactive flavones from *Scutellaria baicalensis* root. Planta Med 64:571–572. doi:10.1055/s-2006-957517
Lim BO (2004) Efficacy of wogonin in the production of immunoglobulins and cytokines by mesenteric lymph node lymphocytes in mouse colitis induced with dextran sulfate sodium. Biosci Biotechnol Biochem 68:2505–2511. doi:10.1271/bbb.68.2505
Liu LY, Wei EQ, Zhao YM, Chen FX, Wang ML, Zhang WP, Chen Z (2005) Protective effects of baicalin on oxygen/glucose deprivation- and NMDA-induced injuries in rat hippocampal slices. J Pharm Pharmacol 57:1019–1026. doi:10.1211/0022357056622
Liu P, Wang J, Li Q, Xu F, Wang Z, Xu H, Liu Z, Zhang X (2006) Effect of baicalin on HSP70 expression of hippocampal neurons in focal brain ischemia-reperfusion injury rats. Yao Xue Xue Bao 41:619–624
Liu C, Wu J, Xu K, Cai F, Gu J, Ma L, Chen J (2010) Neuroprotection by baicalein in ischemic brain injury involves PTEN/AKT pathway. J Neurochem 112:1500–1512. doi:10.1111/j.1471-4159.2009.06561.x

Lohani M, Ahuja M, Buabeid MA, Dean S, Dennis S, Suppiramaniam V, Kemppainen B, Dhanasekaran M (2013) Anti-oxidative and DNA protecting effects of flavonoids-rich *Scutellaria lateriflora*. Nat Prod Commun 8:1415–1418

Lynch M (2004) Analysis of the presynaptic signalling mechanisms underlying the inhibition of LTP in rat dentate gyrus by the tyrosine kinase inhibitor, genistein. Hippocampus 14:4–4. doi:10.1002/hipo.10147

Maes M, Kubera M, Obuchowiczwa E, Goehler L, Brzeszcz J (2011) Depression's multiple comorbidities explained by (neuro)inflammatory and oxidative and amp; nitrosative stress pathways. Neuro Endocrinol Lett 32:7–24

Maes M, Twisk FNM, Kubera M, Ringel K (2012) Evidence for inflammation and activation of cell-mediated immunity in Myalgic Encephalomyelitis/Chronic Fatigue Syndrome (ME/CFS): increased interleukin-1, tumor necrosis factor-α, PMN-elastase, lysozyme and neopterin. J Affect Disord 136:933–939. doi:10.1016/j.jad.2011.09.004

Makino T, Hishida A, Goda Y, Mizukami H (2008) Comparison of the major flavonoid content of *S. baicalensis*, *S. lateriflora*, and their commercial products. J Nat Med 62:294–299. doi:10.1007/s11418-008-0230-7

Malenka RC (1994) Synaptic plasticity in the hippocampus: LTP and LTD. Cell 78:535–538

Marsh Z, Yang T, Nopo-Olazabal L, Wu S, Ingle T, Joshee N, Medina-Bolivar F (2014) Effect of light, methyl jasmonate and cyclodextrin on production of phenolic compounds in hairy root cultures of *Scutellaria lateriflora*. Phytochemistry 107:50–60. doi:10.1016/j.phytochem.2014.08.020

Mian C, Xi QZ (2009) Research and application of superoxide dismutase in clinic. Food Drug 11:44–47

Millspaugh CF (1974) American medicinal plants Charles F Millspaugh. Dover, New York

Morin E (2008) On complexity. Hampton Press, Cresskill

Moylan S, Gustavson K, Karevold E, Øverland S, Jacka FN, Pasco JA, Berk M (2013) The impact of smoking in adolescence on early adult anxiety symptoms and the relationship between infant vulnerability factors for anxiety and early adult anxiety symptoms: the TOPP Study. PLoS ONE 8:e63252. doi:10.1371/journal.pone.0063252

Nakamura N, Hayasaka S, Zhang XY, Nagaki Y, Matsumoto M, Hayasaka Y, Terasawa K (2003) Effects of baicalin, baicalein, and wogonin on interleukin-6 and interleukin-8 expression, and nuclear factor-kb binding activities induced by interleukin-1β in human retinal pigment epithelial cell line. Exp Eye Res 77:195–202. doi:10.1016/S0014-4835(03)00116-7

Ng TB, Liu F, Wang ZT (2000) Antioxidative activity of natural products from plants. Life Sci 66:709–723

Ng F, Berk M, Dean O, Bush AI (2008) Oxidative stress in psychiatric disorders: evidence base and therapeutic implications. Int J Neuropsychopharmacol 11:851–876. doi:10.1017/S1461145707008401

Niemeyer K, Bell IR, Koithan M (2013) Traditional knowledge of Western herbal medicine and complex systems science. J Herb Med 3:112–119. doi:10.1016/j.hermed.2013.03.001

Nikaido T, Ohmoto T, Sankawa U, Tomimori T, Miyaichi Y, Imoto Y (1988) Inhibition of adenosine 3′,5′-cyclic monophosphate phosphodiesterase by flavonoids II. Chem Pharm Bull (Tokyo) 36:654–661

Nissen N, Evans S (2012) Exploring the practice and use of Western herbal medicine: perspectives from the social science literature. J Herb Med 2:6–15. doi:10.1016/j.hermed.2012.02.001

Ohtake N, Yamamoto M, Takeda S, Aburada M, Ishige A, Watanabe K, Inoue M (2005) The herbal medicine Sho-saiko-to selectively inhibits CD8+ T-cell proliferation. Eur J Pharmacol 507:301–310. doi:10.1016/j.ejphar.2004.11.037

Paladini AC, Marder M, Viola H, Wolfman C, Wasowski C, Medina JH (1999) Flavonoids and the central nervous system: from forgotten factors to potent anxiolytic compounds. J Pharm Pharmacol 51:519–526

Pan H, Halper J (2003) Cloning, expression, and characterization of chicken transforming growth factor β4. Biochem Biophys Res Commun 303:24–30. doi:10.1016/S0006-291X(03)00300-0

Parajuli P, Joshee N, Rimando AM, Mittal S, Yadav AK (2009) *In vitro* antitumor mechanisms of various *Scutellaria* extracts and constituent flavonoids. Planta Med 75:41–48. doi:10.1055/s-0028-1088364

Park S (2003) Antioxidative properties of baicalein, component from *Scutellaria baicalensis* Georgi and its application to cosmetics (I). J Korean Ind Eng Chem 1:657–665

Perkins ND (2012) The diverse and complex roles of NF-kB subunits in cancer. Nat Rev Cancer 12:121–132. doi:10.1038/nrc3204

Piao HZ, Jin SA, Chun HS, Lee J-C, Kim W-K (2004) Neuroprotective effect of wogonin: potential roles of inflammatory cytokines. Arch Pharm Res 27:930–936. doi:10.1007/BF02975846

Pugazhenthi S, Nesterova A, Jambal P, Audesirk G, Kern M, Cabell L, Eves E, Rosner MR, Boxer LM, Reusch JE-B (2003) Oxidative stress-mediated down-regulation of bcl-2 promoter in hippocampal neurons. J Neurochem 84:982–996. doi:10.1046/j.1471-4159.2003.01606.x

Qin L, Wu X, Block ML, Liu Y, Breese GR, Hong J-S, Knapp DJ, Crews FT (2007) Systemic LPS causes chronic neuroinflammation and progressive neurodegeneration. Glia 55:453–462. doi:10.1002/glia.20467

Quan N, Banks WA (2007) Brain-immune communication pathways. Brain Behav Immun 21:727–735. doi:10.1016/j.bbi.2007.05.005

Rabow LE, Russek SJ, Farb DH (1995) From ion currents to genomic analysis: recent advances in GABAA receptor research. Synapse 21:189–274. doi:10.1002/syn.890210302

Sadock B, Sadock V, Ruiz P (2000) Comprehensive textbook of psychiatry, 7th edn. Williams and Wilkins, Philadelphia

Saki K, Bahmani M, Rafieian-Kopaei M (2014) The effect of most important medicinal plants on two important psychiatric disorders (anxiety and depression)-a review. Asian Pac J Trop Med 7:S34–S42. doi:10.1016/S1995-7645(14)60201-7

Sarris J (2007) Herbal medicines in the treatment of psychiatric disorders: a systematic review. Phytother Res 21:703–716. doi:10.1002/ptr.2187

Sarris J, Panossian A, Schweitzer I, Stough C, Scholey A (2011) Herbal medicine for depression, anxiety and insomnia: a review of psychopharmacology and clinical evidence. Eur Neuropsychopharmacol 21:841–860. doi:10.1016/j.euroneuro.2011.04.002

Sarvestani NN, Khodagholi F, Ansari N, Farimani MM (2013) Involvement of p-CREB and phase II detoxifying enzyme system in neuroprotection mediated by the flavonoid calycopterin isolated from Dracocephalum kotschyi. Phytomedicine 20:939–946. doi:10.1016/j.phymed.2013.03.013

Serpell J, Jagoe J (1995) Early experience and the development of behaviour. The domestic dog: Its evolution, behaviour, and interactions with people, 1st edn. Cambridge University Press, Cambridge

Shang X, He X, He X, Li M, Zhang R, Fan P, Zhang Q, Jia Z (2010) The genus *Scutellaria* an ethnopharmacological and phytochemical review. J Ethnopharmacol 128:279–313. doi:10.1016/j.jep.2010.01.006

Shang Q, Liu Z, Chen K, Xu H, Liu J, Shang Q, Liu Z, Chen K, Xu H, Liu J (2012) A systematic review of Xuezhikang, an extract from red yeast rice, for coronary heart disease complicated by dyslipidemia. Evid Based Complement Altern Med 2012:1–18. doi:10.1155/2012/636547

Shi X, Chen G, Liu X, Qiu Y, Yang S, Zhang Y, Fang X, Zhang C, Liu X (2015) Scutellarein inhibits cancer cell metastasis in vitro and attenuates the development of fibrosarcoma *in vivo*. Int J Mol Med 35(1):31–38. doi:10.3892/ijmm.2014.1997

Shieh DE, Liu LT, Lin CC (2000) Antioxidant and free radical scavenging effects of baicalein, baicalin and wogonin. Anticancer Res 20:2861–2865

Snow J (2015) Context effects in western herbal medicine: fundamental to effectiveness. Explor J Sci Heal 12:55–62. doi:10.1016/j.explore.2015.10.004

Somoza R, Juri C, Baes M, Wyneken U, Rubio FJ (2010) Intranigral transplantation of epigenetically induced BDNF-secreting human mesenchymal stem cells: implications for cell-based therapies in Parkinson's disease. Biol Blood Marrow Transplant 16:1530–1540. doi:10.1016/j.bbmt.2010.06.006

Song L, Yang H, Wang HX, Tian C, Liu Y, Zeng XJ, Gao E, Kang YM, Du J, Li HH (2014) Inhibition of 12/15 lipoxygenase by baicalein reduces myocardial ischemia/reperfusion injury via modulation of multiple signaling pathways. Apoptosis 19:567–580. doi:10.1007/s10495-013-0946-z

Stein MB, Tancer ME, Gelernter CS, Vittone BJ, Uhde TW (1990) Major depression in patients with social phobia. Am J Psychiatry 147:637–639. doi:10.1176/ajp.147.5.637

Tai MC, Tsang SY, Chang LYF, Xue H (2006) Therapeutic potential of Wogonin: a naturally occurring flavonoid. CNS Drug Rev 11:141–150. doi:10.1111/j.1527-3458.2005.tb00266.x

Tharakan B, Dhanasekaran M, Manyam BV (2005) Antioxidant and DNA protecting properties of anti-fatigue herb Trichopus zeylanicus. Phytother Res 19:669–673. doi:10.1002/ptr.1725

Tilburt JC, Kaptchuk TJ (2008) Herbal medicine research and global health: an ethical analysis. Bull World Health Organ 86:594–599

Tsai TH, Liu SC, Tsai PL, Ho LK, Shum AYC, Chen CF (2002) The effects of the cyclosporin A, a P-glycoprotein inhibitor, on the pharmacokinetics of baicalein in the rat: a microdialysis study. Br J Pharmacol 137:1314–1320. doi:10.1038/sj.bjp.0704959

Upton R, Dayu RH (2012) Skullcap *Scutellaria lateriflora* L.: an American nervine. J Herb Med 2:76–96. doi:10.1016/j.hermed.2012.06.004

Uttara B, Singh AV, Zamboni P, Mahajan RT (2009) Oxidative stress and neurodegenerative diseases: a review of upstream and downstream antioxidant therapeutic options. Curr Neuropharmacol 7:65–74. doi:10.2174/157015909787602823

Van Ameringen M, Mancini C, Styan G, Donison D (1991) Relationship of social phobia with other psychiatric illness. J Affect Disord 21:93–99

Veglio F, Morello F, Morra Di Cella S, Del Colle S, Rabbia F, Mulatero P (2003) Recent advances in diagnosis and treatment of pheochromocytoma. Minerva Med 94:267–271

Wager-Smith K, Markou A (2011) Depression: a repair response to stress-induced neuronal microdamage that can grade into a chronic neuroinflammatory condition? Neurosci Biobehav Rev 35:742–764. doi:10.1016/j.neubiorev.2010.09.010

Waisundara VY, Siu SY, Hsu A, Huang D, Tan BKH (2011) Baicalin upregulates the genetic expression of antioxidant enzymes in Type-2 diabetic Goto-Kakizaki rats. Life Sci 88:1016–1025. doi:10.1016/j.lfs.2011.03.009

Wakabayashi I, Yasui K (2000) Wogonin inhibits inducible prostaglandin E(2) production in macrophages. Eur J Pharmacol 406:477–481

Wang Z, Song D, Berger TW (2002) Contribution of NMDA receptor channels to the expression of LTP in the hippocampal dentate gyrus. Hippocampus 12:680–688. doi:10.1002/hipo.10104

Wang JY, Chuang HN, Chiu JH, Fu SL, Tsai TH, Tsou AP, Hu CP, Chi CW, Yeh SF, Lui WY, Wu CW, Chou CK (2006) Effects of *Scutellaria baicalensis* Georgi on macrophage-hepatocyte interaction through cytokines related to growth control of murine hepatocytes. Exp Biol Med (Maywood) 231:444–455

Wen YF, Zhao JQ, Bhadauria M, Nirala SK (2013) Baicalin prevents cadmium induced hepatic cytotoxicity, oxidative stress and histomorphometric alterations. Exp Toxicol Pathol 65:189–196. doi:10.1016/j.etp.2011.08.005

Wilczańska-Barska A, Królicka A, Głód D, Majdan M, Kawiak A, Krauze-Baranowska M (2012) Enhanced accumulation of secondary metabolites in hairy root cultures of *Scutellaria lateriflora* following elicitation. Biotechnol Lett 34:1757–1763. doi:10.1007/s10529-012-0963-y

Wills RBH, Stuart DL (2004) Generation of high quality Australian skullcap products. A report for the rural industries research and development corporation, RIRDC. Australian Government, Barton

Wojcikowski K, Stevenson L, Leach D, Wohlmuth H, Gobe G (2007) Antioxidant capacity of 55 medicinal herbs traditionally used to treat the urinary system: a comparison using a sequential three-solvent extraction process. J Altern Complement Med 13:103–109. doi:10.1089/acm.2006.6122

Wolfson P, Hoffmann DL (2003) An investigation into the efficacy of *Scutellaria lateriflora* in healthy volunteers. Altern Ther Health Med 9:74–78

Woods JH, Katz JL, Winger G (1987) Abuse liability of benzodiazepines. Pharmacol Rev 39:251–413

Wu Y, Kim Y, Wu J, Yu X, Hao Y (2011) Effects of wogonin on inflammation-related factors in alveolar macrophages infected with influenza virus. Chinese J Pathophysiol 27:533–538

Xiong P, Chen X, Guo C, Zhang N, Ma B (2012) Baicalin and deferoxamine alleviate iron accumulation in different brain regions of Parkinson's disease rats. Neural Regen Res 7:133–136. doi:10.3969/j.issn.1673

Xu Z, Wang F, Tsang SY, Ho KH, Zheng H, Yuen CT, Chow CY, Xue H (2006) Anxiolytic-like effect of baicalin and its additivity with other anxiolytics. Planta Med 72:189–192. doi:10.1055/s-2005-873193

Yang EJ, Yoon J-H, Chung KC (2004) Bruton's tyrosine kinase phosphorylates cAMP-responsive element-binding protein at serine 133 during neuronal differentiation in immortalized hippocampal progenitor cells. J Biol Chem 279:1827–1837. doi:10.1074/jbc.M308722200

You KM, Jong HG, Kim HP (1999) Inhibition of cyclooxygenase/lipoxygenase from human platelets by polyhydroxylated/methoxylated flavonoids isolated from medicinal plants. Arch Pharm Res 22:18–24. doi:10.1007/BF02976430

Zhang Z, Lian X, Li S, Stringer JL (2009) Characterization of chemical ingredients and anticonvulsant activity of American skullcap (*Scutellaria lateriflora*). Phytomedicine 16:485–493. doi:10.1016/j.phymed.2008.07.011

Zhou QB, Jin YL, Jia Q, Zhang Y, Li LY, Liu P, Liu YT (2014) Baicalin attenuates brain edema in a rat model of intracerebral hemorrhage. Inflammation 37:107–115. doi:10.1007/s10753-013-9717-9

Chapter 3
Liquid Chromatography–Mass Spectrometry (LC–MS): Approaches to Adulterant Detection in Herbal Products

N. Satheeshkumar, David Paul, and A. Lingesh

Abstract An ever increasing global demand for herbal-based medicines in recent years has resulted in a serious public concern over its quality and safety. The consumption of adulterated herbal products can lead to unexpected and unpredictable pharmacological responses in the human body and various health-related risks. Adulteration of herbal products is a global concern and poses a major challenge for analytical laboratories to detect and characterize them. Liquid chromatography coupled with mass spectrometry (LC–MS) is widely used for screening of adulterants, mainly due to their high selectivity and sensitivity, which is crucial for analyzing complex natural product samples. The present article consists of an overview of drug adulteration and evaluation of herbal products with a special reference to the approaches in adulterant detection and regulatory perspectives to control such malpractices. Also, adulteration in slimming phytotherapeutic formulations, PDE-5 inhibitors in herbal formulations has been discussed.

Keywords Adulterants • Herbal slimming products • Liquid chromatography–mass spectrometry (LC–MS) • PDE-5 inhibitors

Abbreviations

APCI	Atmospheric pressure chemical ionization
APPI	Atmospheric pressure photoionization
AYUSH	Ayurveda, yoga, unani, siddha, homoeopathy
BMI	Body mass index
CAGR	Compound annual growth rate
CE	Capillary electrophoresis
CID	Collision-induced dissociation

N. Satheeshkumar (✉) • D. Paul • A. Lingesh
Department of Pharmaceutical Analysis, National Institute of Pharmaceutical Education and Research, Hyderabad (NIPER-H), Balanagar, Hyderabad 500037, Telangana, India
e-mail: satish@niperhyd.ac.in

H.-S. Tsay et al. (eds.), *Medicinal Plants - Recent Advances in Research and Development*, DOI 10.1007/978-981-10-1085-9_3

DDMS	N-di-desmethyl sibutramine
DMS	N-mono-desmethyl sibutramine
EFSA	European Food Safety Authority
EMA	European Medicines Agency
ESI	Electrospray ionization
EU	European Union
FT-ICR	Fourier transform ion cyclotron resonance
GC	Gas chromatography
GIA	Global industry analysts
HPLC	High-pressure or high-performance liquid chromatography
HPTLC	High-performance thin-layer chromatography
HRMS	High-resolution mass spectrometric
IMS	Ion mobility spectrometry
IT	Ion trap
IT–TOF	Ion trap–time of flight
LC–IR	Liquid chromatography–infrared spectroscopy
LC–MS	Liquid chromatography–mass spectrometry
LC–MS/MS	Liquid chromatography–tandem mass spectrometry
LC–NMR	Liquid chromatography–nuclear magnetic resonance spectroscopy
LTQ–orbitrap	Linear ion trap quadrupole–orbitrap
LTQ–FT	Linear ion trap–Fourier transform ion cyclotron resonance
MRM	Multiple reaction monitoring
OTC	Over-the-counter
PDE-5	Phosphodiesterase type 5
QAQC	Quality assurance and quality control
Q–FT-ICR	Quadrupole–Fourier transform ion cyclotron resonance
Q-IT	Quadrupole ion trap
QqQ	Triple quadrupole
Q–TOF	Quadrupole–time of flight
SATCM	State administration of traditional Chinese medicine
SIM	Selected ion monitoring
SRM	Selective reaction monitoring
TCM	Traditional Chinese medicine
TIM	Traditional Indian medicine
TOF	Time of flight
USFDA	United States food and drug administration
WHO	World Health Organization

3.1 Introduction

Since ancient times, plants have been a prime source of herbal medicines for the treatment of several human diseases. Medicinal plants are most abundant in tropical countries and are relied upon to support, promote, retain and regain human health worldwide. According to the World Health Organization (WHO), up to 80 % of people living in developing countries still depend principally on medicines from herbal sources for their healthcare concerns (WHO 2005). Systems of herbal medicines and formulations that are widely used in national healthcare systems around the world include Ayurvedic medicine, Chinese medicine, Unani medicine, Western herbal medicine, Japanese Kampo medicine and Tibetan Buddhist medicine. These traditional medicines are characterized in several ways under various jurisdictions around the world and are commonly sold either as a prescription or over-the-counter (OTC) medicines. It was estimated that about 25 % of all pharmaceuticals are directly or indirectly derived from higher plants (Bandaranayake 2006). In contrast to synthetic drugs and its analogues, herbal medicines are perceived to be safe and harmless due to their natural origin. Ascending health consciousness, growing health concern had boosted the attention towards preventive healthcare including the use of herbal products and supplements. According to Global Industry Analysts (GIA), the global herbal supplements and remedies market is forecasted to reach US $115 billion by the year 2020, accelerated by ageing population and increasing consumer awareness (GIA 2015). According to an estimate by WHO, the current requirement of herbal products is US $14 billion a year, and by the year 2050, it has a potential of US $5 trillion in trade (Aneesh et al. 2009).

Europe is the biggest market for the herbal health supplements, whereas Asia-Pacific is advancing at the fastest rate, largely owing to China and India with double-digit compound annual growth rate (CAGR) through 2017 (GIA 2015). India has 2.4 % of the world's area with 8 % of global biodiversity. In India, 80 % of the total population stays in the rural community and utilizes herbal medicines for various ailments. About 25,000 effective herbal-based formulations are utilized in traditional medicine in India, and more than two million practitioners are using the traditional therapeutic approaches for healthcare. It is estimated that more than 8000 plus manufacturing units are involved in the production of natural health products and traditional plant-based formulations in India, which requires tonnes of herbals annually as raw material. The government of India also has extended encouragement and support for the'traditional Indian medicine' (TIM) sector by establishing a separate department for the Indian systems of medicine and homoeopathy now known as AYUSH (ayurveda, yoga, unani, siddha, homoeopathy) in March 1995 to promote indigenous systems (Aneesh et al. 2009). In recent years, several reports have claimed the presence of synthetic drugs or modified analogues of approved or non-approved drugs in herbal remedies (Al-Safi et al. 2008; Bandaranayake 2006; Rocha et al. 2016; Starr 2015; Venhuis et al. 2009; WHO 1998). These adulterants are mixed with herbal products to increase the efficacy of the herbal products. The adulteration can give rise to the manifestation of various unpredictable responses in the human body, either due to the isolated effects or due to its drug-herbal interactions in the formulations (Liang et al. 2006). In many cases, the resulting side effects of adulterants could even be life threatening (George 2011).

The herbal formulations and supplements which are most regularly adulterated with pharmaceuticals are usually those advertised for treatment of various conditions such as obesity or erectile dysfunction, slimming agents and chronic illnesses (diabetes mellitus, hypertension, arthritis etc.) (Kanan et al. 2009; Li et al. 2010; Lu et al. 2009; Balayssac et al. 2009; Rocha et al. 2016). Another set of products which are frequently adulterated is nutritional supplements targeting bodybuilders and sports personalities (De Cock et al. 2001).

Adulteration is a global problem and poses a major challenge for analytical laboratories to identify, detect and to characterize adulterants. The practice of adulteration violates the laws of regulated countries because those herbal formulations are registered in disagreement with their real compositions (Deconinck et al. 2013; Poornima 2010). However, several studies have reported certain synthetic chemicals as adulterants in herbal products, representing a huge risk for public health (Cianchino et al. 2008; Holzgrabe and Malet-Martino 2011). This chapter covers an overview of drug adulteration in herbal products, evaluation of herbal products with special reference to the 'liquid chromatography–mass spectrometry' (LC–MS) approaches in adulterant detection and regulatory perspectives to control such malpractices. Also, adulterants in slimming phytotherapeutic formulations, PDE-5 inhibitors in herbal formulations have been discussed.

3.2 Drug Adulteration

An adulteration is an act of substituting the original crude drug partially or fully with other materials which are either free from or inferior in therapeutic and chemical properties or addition of low grade or spoiled drugs or entirely different drug similar to that of original drug substituted with an intention of making a profit (Kamboj 2012). Some manufacturers include synthetic pharmaceuticals in their products marketed as 'herbal medicine' or 'dietary supplement', to enhance the effect of their products. The adulteration of herbal drugs is a serious issue in the herbal industry, and it has caused a severe concern in the commercial use of phytoproducts and supplements. Adulteration takes place in two ways: (i) direct or intentional adulteration and (ii) indirect or unintentional adulteration.

3.2.1 Direct or Intentional Adulteration

Direct or intentional adulteration includes practices in which herbal drug is substituted partially or fully with other inferior products. Due to the morphological similarity with the authentic plants, much poor quality or inferior herbal materials are used as adulterants, which may or may not have any therapeutic values. For example, beeswax mixed with the coloured paraffin wax, papaya seed mixed with black pepper seed, citral to citrus oil products, liquorice powder mixed with olive powder and certain pharmaceuticals and its analogues are also used as adulterants (Ahmad et al. 2014; Kamboj 2012; Mukerjee 2002; Prakash et al. 2013).

3.2.2 Indirect or Unintentional Adulteration

This kind of adulteration may occur without any bad intention of the manufacturer or supplier. These products enter the market without any proper evaluation. Factors like growth conditions, geographical source, processing technology and storage conditions can interfere with the quality of the herbal product. For example, there may be a degradation of glycosides by enzymatic hydrolysis if digitalis leaves are dried above 65 °C. Use of such sub-standard material affects the quality of the herbal product (Ahmad et al. 2014; Kamboj 2012; Mukerjee 2002; Prakash et al. 2013).

3.3 Limitations in Screening of Herbal Products and Supplements

Standardization of an herbal product has to start from the cultivation stage itself. Analysis of raw materials for authentication, preliminary evaluation, chromatographic profiles, pesticides residue and heavy metal detection also need to be carried out. Herbal and its extract may contain several known and unknown components and some of which may be present in trace amounts, hence requirement of a sensitive tool for their detection and quantification poses a challenge to the analyst (Fan et al. 2006; Luis Cuadros-Rodríguez et al. 2016; Tistaert et al. 2011).

Screening and structure elucidation of novel adulterants and analogues which have not been previously reported in herbal products make the analytical task challenging. Additionally, the adulteration of herbal product with a single drug, a producer may use multiple moieties in trace quantity to achieve the desired effect and simultaneously make the process of detection more difficult. The complexity of matrixes of interest and the possible presence of very minute levels of adulterants, highly sensitive and selective analytical techniques thrust the need for their screening and estimation (Patel et al. 2014; Venhuis and De Kaste 2012).

3.4 Drug Evaluation

The confirmation of herbal drug identity and estimation of quality and purity is known as its evaluation. The drug from herbal sources generally can be evaluated or identified by the following five methods: organoleptic evaluation, morphological evaluation, macroscopic evaluation, microscopic evaluation and physical and chemical evaluations. Bioassays are used to evaluate the strength of drug in its preparation (Busse 2000; Kamboj 2012).

3.4.1 Analytical Evaluation

In general, quality control in pharmacopoeias is based on three important terminologies like identity, purity and content or assay. The content is the most critical one to assess, because, in most herbal formulations, the active constituents will be known or unknown and may be present in trace amount. Criteria like physical constants, moisture, ash content and solvent residues are also needed to be evaluated to prove identity and purity. The botanical identity of the herbals is of prime importance in establishing the quality control of herbal drugs and supplements. When the active constituents are known or unknown, a vast set of modern analytical approaches can be employed for the standardization of quality assurance and quality control (QAQC) (Busse 2000; Jordan et al. 2010; Kunle et al. 2012).

3.4.1.1 Chromatography and Chemical Fingerprints of Herbal Medicines

The determination of most of the phytoconstituents of herbal products is essential to guarantee the reliability of pharmacological and clinical research, to explore their bioactivities and possible side effects of active compounds. It enables to improve product quality and to enhance regulatory need. The chromatographic techniques, such as high-pressure or high-performance liquid chromatography (HPLC), gas chromatography (GC), capillary electrophoresis (CE) and high-performance thin-layer chromatography (HPTLC), can be applied for quality assessment of herbal products (Jordan et al. 2010; Kunle et al. 2012; Sahoo et al. 2010; Tistaert et al. 2011). According to the phytoequivalence concept developed in Germany, the chromatographic fingerprint-like chemical profile should be constructed and compared with the profile of a clinically proven reference product to ensure consistency of herbal products (Bauer 1998; Veit and Gaedcke 2004). Chromatography offers very powerful separation tools for complex chemical components in herbal fingerprints produced by the chromatographic instruments. It represents a relatively good illustration of various phytoconstituents of herbals and its products.

3.4.1.2 Liquid Chromatography–Mass Spectrometry

Adulteration of phytotherapeutic products and supplements has been discussed in several reports. Therefore, the development of instrumental analytical methodologies for detection of adulterants is of great relevance. Advances in the hyphenated instrumental techniques like liquid chromatography–tandem mass spectrometry (LC–MS/MS) have now made it possible to detect synthetic drugs and their structural analogues as adulterants selectively and with the highest sensitivity (Haneef et al. 2013). Over the past years, LC has been extensively applied in analysis and adulterant screening of herbals and its products. Since it is easy to learn and can be used to separate and analyze almost all the compounds in the herbal medicines by various detectors. LC chromatographic fingerprints can be utilized for the documentation of complete herbal extracts with adequate data that makes the on-line qualitative analysis possible. The qualitative analysis or structure elucidation of the chemicals in herbal drug and its products with simple HPLC is not possible. Therefore, hyphenated analytical techniques like LC–IR, LC–MS and LC–NMR have to be utilized for this kind of analysis (Haneef et al. 2013; Jalili et al. 2015; Nikam et al. 2012; Rasheed et al. 2013).

LC–MS is an analytical methodology that hyphenates the separation capabilities of LC with the mass analysis power of mass spectrometry. LC–MS is a powerful technique utilized for several applications requiring high sensitivity and selectivity with maximal accuracy. In the past years, several reports on LC–MS analysis of herbal medicines and products have been published. This trend reflects the advantages of LC–MS in solving complex problems in herbal products. There are mainly three primary ionization methodologies: electrospray ionization (ESI), atmospheric pressure chemical ionization (APCI) and atmospheric pressure photoionization (APPI). The ESI is primarily applied for the analysis of charged species, while APCI approach is commonly used for uncharged or difficult to charge species. Both are soft ionization techniques compatible with the most chromatographic separations. The LC–DAD–MS technique takes the advantage of chromatographic separation and detection using both DAD and MS. By utilizing hyphenated techniques, one could identify the chromatographic peaks and directly perform the on-line comparative study. With these advantages, LC–MS has now become the most employed approach for high throughput for the rapid screening and analysis of phytochemical constituents and adulterants in botanical products. In the analysis of the chemical components in herbal drugs, LC–MS technique is also used for the separation and identification of a variety of similar structural compounds and unknown adulterants, as an important qualitative tool. Screening and characterization of novel analogues as adulterants can also be carried out with the advanced LC–MS techniques that focus on marker fragment ions (Gross 2004; Lau et al. 2003; Luis Cuadros-Rodríguez et al. 2016; Rocha et al. 2016). Thus, LC–MS has become the most sensitive and reliable methodology for quality control and standardization of the herbal products.

Nowadays tandem mass spectrometry is used all over the world for the screening of adulterants in herbal medicines. Mass spectrometry consists of sample inlet

Table 3.1 Operational modes of triple quadrupole mass spectrometers

Modes	Q1	Q3	Applications
Full scan	Scanned	Scanned	1. Allows to see all ions contained in a sample
	No CID gas in Q2		
Product scan	Fixed	Scanned	1. Structural studies
			2. Identification of unknowns
	CID gas in Q2		3. Confirmation by spectral matching
Precursor scan	Scanned	Fixed	1. Structurally similar analyte detection, which produces a common fragment ion
	CID gas in Q2		
Neutral loss scan	Scanned	Scanned	1. Structurally similar analytes detection that eliminates or gain a common neutral molecule or selected common chemical group on collision
	CID gas in Q2		
Multiple reaction monitoring (MRM)	Fixed	Fixed	1. Highly specific detection and sensitivity detection of a single or multiple targeted analytes
	CID gas in Q2		

Q1 quadrupole 1, *Q2* quadrupole 2 (collision cell), *Q3* quadrupole 3, *CID* collision-induced dissociation

device, an ionization device, an ion path and detector. Every mass spectrometer also requires necessary auxiliary components with a device for signal processing with a combination of firmware and computer software. An MS/MS has two mass-selective devices, which is arranged in tandem that picks the precursor ion of the respective analyte, which is then passed into a collision cell, and very low flow of a collision gas is utilized for the disintegration of the precursor ions to several product ions. The second filter (selective device) behind the collision cell scans the product ions that are generated according to their respective *m/z* ratio. The term triple quadrupole mass spectrometer is used as synonymous to two quadrupole mass filters and collision cell, because it may represent a quadrupole (Dawson 2013; Holcapek et al. 2012). LC–MS/MS instrumentation allows the analyst to perform within the limits of sensitivity and mass resolution. Different kinds of experimental modes like full-scan mode used to identify the precursor ion, product ion scan mode to determine the fragments *m/z* ratios, precursor ion scan mode to confirm that only the molecule of interest gives rise to the *m/z* product ion and neutral loss scan mode for checking the related compounds family. Selective reaction monitoring (SRM) and multiple reaction monitoring (MRM) methodologies are most commonly used in sensitive and specific quantitation with LC–MS/MS. This reality has been fostered in large part by the tremendous revolution in computer processing technology, storage capabilities and retrieval of analytically processed data (Cappiello 2006; Dawson 2013). Operational modes of triple quadrupole mass spectrometers are presented in Table 3.1.

Newer hybrid MS instruments (LTQ–FT and LTQ–orbitrap) and processing tool packages (Thermo Fischer Scientific, USA) offer the option of multiple recording of MS signal intensities and identification. Tactical approaches for signal normalization, correcting variations in LC performances, are developed and employed with the help of fully automated computational software platforms for the label-free MS/MS analysis (Kee et al. 2015). On a conventional LC–MS/MS, the search for unknown adulterants requires multiple injections. One injection in full-scan LC/MS mode and second injection for targeted LC–MS/MS experiments. This step increases the time of the analyst to construct MS/MS method and to obtain MS data. The survey scan feature of the Waters Xevo TQ-MS instrument allows intelligent switching from LC–MS to LC–MS/MS data modes in a single run, thus bringing high throughput analysis. In ScanWave mode, duty cycle improvements result in a signal enhancement in scanning acquisition modes, which facilitates the detection of low-level adulterants. Consequently, a wide dynamic range of scan survey programmes, the rapid sampling rate for MS/MS generation, along with boosted delectability and fragmentation, will generate quality spectra for low-abundance species for adulterants present in herbal drugs (Patel et al. 2015). Advancement in analytical instrumentations had led to the introduction of rapid and effective separation techniques. MS systems capable of detecting, confirming and performing structural elucidation of multiple specific or non-specific analytes are available.

The wide range of data acquisition modes of MS data can be utilized for the detection and screening of known and unknown pharmaceutical adulterants in herbal remedies. Selected ion monitoring (SIM), an acquisition mode, is used with quadrupole mass analyzers to perform the analysis selectively. Significantly higher selectivity and sensitivity can be gained in selected reaction monitoring (SRM) and multiple reaction monitoring (MRM) modes through detection of characteristic product ions, formed by collision-induced dissociation (CID) of the precursor ion. The triple quadrupole instruments operated in MRM mode and the recording of one or more transitions for each target compound have been commonly employed LC–MS strategy and approach for targeted screening and quantification of adulterants in herbal products and supplements. The technology enhancement has revolutionized sensitivity, speed, reliability and performance delivered by these instruments. The QTRAP, LC–MS/MS systems of AB Sciex featuring the multicomponent IonDrive™ technology and advanced with a novel IonDrive High Energy Detector+, pushes the limits of LC–MS/MS quantitation superior to the past models in herbal adulteration screening. Performing MRM and scanning with high sensitivity LC–MS/MS system enable an analyst to identify, characterize and quantitate known and unknown adulterants more quickly and easily (Patel et al. 2015). The list of various LC–MS instruments based on mass separation methods has been given in Table 3.2.

Considering the European Union (EU) requirements for the performance of analytical methods for adulterants in herbal products and supplements and its result's interpretation, monitoring of at least two MRM transitions along with an evaluation of their relative intensities is recommended for screening and identification of an analyte (Jordan et al. 2010). The number of transitions monitored for each adulterant which can be simultaneously analyzed to obtain acceptable sensitivity is limited by the time, i.e. time spent for acquisition for MRM transition. The evolution of the hybrid technology of MS system facilitated with a support consisting of drug databases and data in the scientific manuscripts can become advanced and effective tools for screening of adulteration of herbal products and supplements. This advance has made it possible to combine various scanning modes to analyze and interpret both MRM data as well as product ion scan spectra in a single LC–MS run. Library-

Table 3.2 The list of various LC–MS instruments based on mass separation methods

Ion transmission instruments	Scanning	Magnetic sector
		Double focusing
		Quadrupole (Q)
	Non-scanning	Time of flight (TOF)
Trap instruments	Q-trap	
	Ion trap (IT)	
	Fourier transform ion cyclotron resonance (FT-ICR)	
MS/MS systems	Tandem	Triple quadrupole (QqQ), TOF–TOF
	Hybrid	Q–IT, Q–TOF, IT–TOF, Q–FT-ICR, LTQ–FT and LTQ–orbitrap

Q–IT quadrupole–ion trap, *Q–TOF* quadrupole–time of flight, *IT–TOF* ion trap–time of flight, *Q–FT-ICR* quadrupole–Fourier transform ion cyclotron resonance, *TOF–TOF* time of flight–time of flight, *LTQ–FT* linear ion trap–Fourier transform ion cyclotron resonance, *LTQ–orbitrap* linear ion trap–orbitrap

based screening and quantitation are possible with this hybrid technology, which can be utilized for synthetic pharmaceuticals and its analogues in herbal remedies (Chen et al. 2009). The UNIFI, an adulterant screening application, detects the existence of any synthetic compounds and then confirms the adulterant through an automatic reviewing process. It helps an analyst to draw a conclusion quickly. Figure 3.1 displays the flow of the analytical process involved in generalized screening approach. It integrates all the steps like data acquisition, peak picking, library searching, fragment ion confirmation and report generation in a systematic way. An additional provision of structure elucidation tools allows further scanning on unknown peaks that did not show matches from the adulterants library in the database (AB Sciex 2015).

In-house constructed product ion spectra library can be utilized for the concentration identity confirmation of undeclared drugs based on the areas of MRM traces (Chen et al. 2009). The collision energy setting in product ion scan experiments has to be optimized because it applies to all the analytes, and it may be troublesome to obtain data suitable for identification. To solve this issue, it is advisable to perform fragmentations at low, medium and high collision energy values and use averaged mass spectra rather than those obtained under single collision energy. High-resolution mass spectrometric (HRMS) instruments such as Fourier transform ion cyclotron resonance (FT-ICR), time of flight (TOF) or orbital ion trap (IT) provide high sensitivity in full-scan mode and accurate mass measurements. It can be utilized for the estimation of the elemental composition of detected ions and to enhance the selectivity of measurement by extracting the ion chromatograms of known or unknown compounds with narrow mass windows. Even though TOF instruments may provide accurate mass, but the disadvantage of TOF mass analyzers lies in the quantitative applications (Holcapek et al. 2012). The pros and cons of LC–MS instrument used for quantitation (Sargent 2013) are summarized in Table 3.3.

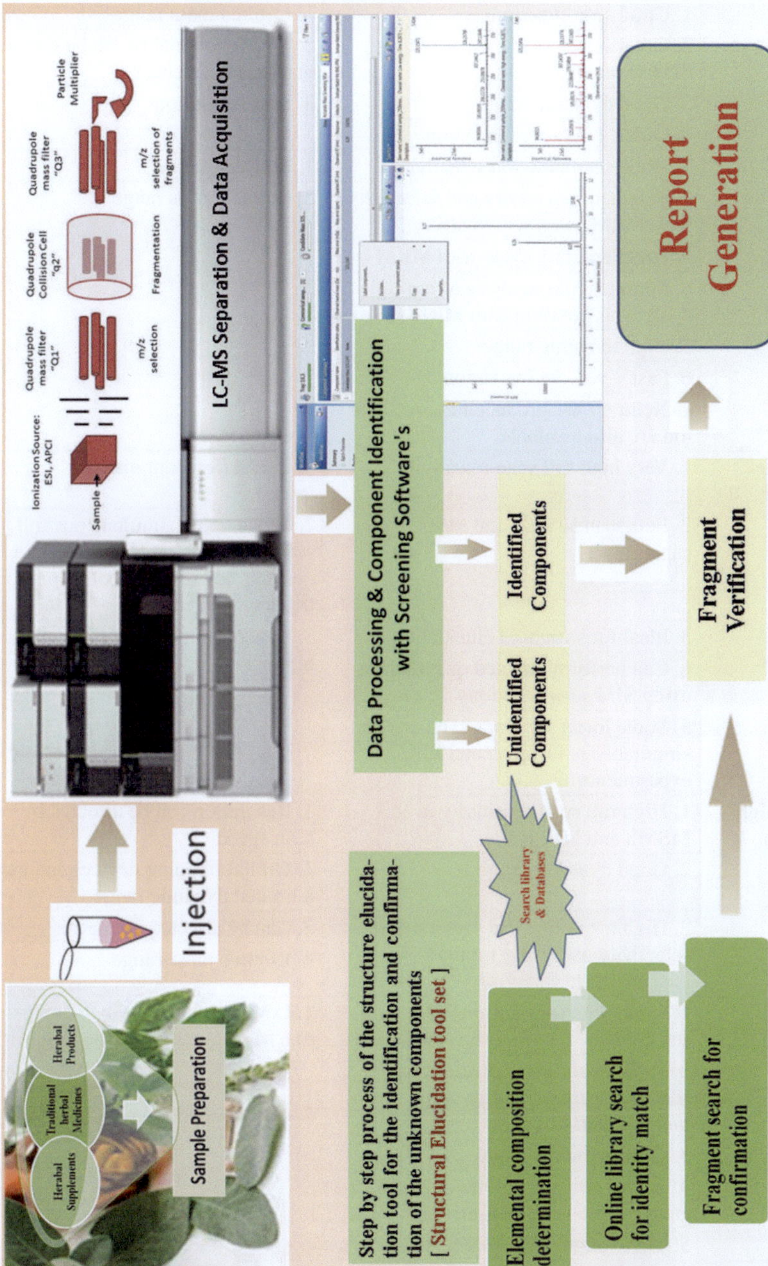

Fig. 3.1 A streamlined analytical process for adulterant screening in herbal products

Table 3.3 Summary of pros and cons of common types of LC–MS instruments

LC–MS type	Pros	Cons
Single quadrupole	1. Good scan function	1. Limited mass range
	2. Good selectivity/sensitivity using SIM scanning	2. SIM functionality can be prone to matrix interferences
	3. Good dynamic range	3. Low resolution
	4. Quick + ve/− ve ionization	
Triple quadrupole	1. Good scan function sensitivity	1. Low resolution
	2. Excellent sensitivity and selectivity with MRM, even with matrix	2. Limited mass range
	3. Excellent duty cycle with MRM	
	4. Simultaneous analysis of multiple analytes is possible with MRM	
	5. High dynamic range	
	6. Fast + ve/− ve ionization	
	7. Neutral loss, product and precursor ion are also available	
Ion trap (low resolution)	1. Very high full scan sensitivity	1. Can suffer from matrix interferences
	2. Full-scan MSMS and MS^n capability	2. When doing simultaneous full scan and MRM acquisitions duty cycle is slower when compared to a triple quadrupole
	3. Ideal for structural elucidation	3. Low resolution, but can run at higher resolution
	4. Can perform targeted quantitation using SIM scan functions	
	5. Some linear ion traps can perform simultaneous full scan and MRM experiments	
Ion trap (high resolution)	1. High full scan sensitivity in MS, MSMS and MS^n mode	1. Resolution can be affected by scan speed
	2. Good dynamic range	2. Orbital trapping devices can have a limited dynamic range
	3. High resolution can provide good selectivity using exact mass measurement	3. Can be affected by matrix
		4. Limited mass range
TOF (high resolution)	1. Good scan functionality and sensitivity	1. Lower sensitivity when compared to a triple quadrupole running MRM
	2. High resolution, provides high selectivity through exact mass measurement	
	3. Good dynamic range	
	4. Ability to get quantitation on multiple analytes in a single acquisition	
	5. Mass range: 20,000 *m/z*	

(continued)

Table 3.3 (continued)

LC–MS type	Pros	Cons
qTOF (high resolution)	1. Good full scan sensitivity	1. Lower sensitivity when compared to a triple quadrupole running MRM
	2. Good MS/MS scan functions	
	3. High resolution, providing high degree of selectivity via exact mass measurement	
	4. Good dynamic range	
	5. Ability to get quantitation on multiple analytes in a single run	
	6. Mass range more than 20,000 *m/z*	
	7. Resolution not affected by increased scan speed	

3.5 Adulteration of Herbal Slimming Products with Synthetic Drugs

Nowadays obesity is growing as a worldwide public health issue that affects millions of people (Correia 2008). The WHO estimates that more than 1.4 billion adults are overweight, and at least 500 million people are obese. Obesity is defined regarding body mass index (BMI). For normal people, it ranges between 25 and 30 kg m^{-2}. A person with BMI greater than or equal to 30 kg m^{-2} is considered as overweight (Keding et al. 2013; WHO 2016). Given the social and medical impact of being obese, as well as the issues in making diet control and physical exercises, it is not a wonder that patients often turn to over-the-counter (OTC) proprietary weight-loss products (Wang et al. 2008). Misguiding advertisements of some products have boosted the demand for phyto-formulations for obesity treatment. Furthermore, the beauty standards set by the influence of media has also made the population have enhanced demand on herbal slimming products. The consumption of herbal products and supplements claiming the potential to be used as slimming agents has enhanced markedly in the past decade. Herbal products and supplements that claim to have a weight-loss effect are marketed globally with the help of the internet facility (Ancuceanu et al. 2013; Ozdemir et al. 2013). These slimming products claimed to be natural, but the reports of adulterations with drugs for the treatment of obesity and constipation is increasing day by day (Al-Safi et al. 2008; Ancuceanu et al. 2013; Phattanawasin et al. 2012; Russo et al. 2016). Stimulants, laxatives and diuretics are also reported to use as adulterants in slimming formulations (Fraser and Wen 1998). Anorexics derived from amphetamines are also more frequently found in the phytoproducts used for slimming purposes. Fenproporex and fluoxetine were also identified as adulterants. Amfepramone, fenproporex and diazepam have been identified in commercialized slimming phytoproducts (Almeida et al. 2000; De Carvalho et al. 2011). Ephedrines, synephrine and caffeine used as stimulants were also reported as the most used drugs for slimming purposes (Viana et al. 2016). A pictorial representation of adulteration of herbal products with slimming agents

Fig. 3.2 A representation of adulteration of herbal products with synthetic slimming agents

is depicted in Fig. 3.2. USFDA-approved anti-obesity drugs sibutramine and orlistat have been more frequently found adulterants in slimming phytotherapeutics. In 2010, the European Medicines Agency (EMA) recommended the withdrawal of sibutramine by the European Union because of increased cardiovascular risks like tachycardia and arterial hypertension (Reeuwijk et al. 2014; Cohen and Ernst 2010; Muller et al. 2009). Psychiatric symptoms like psychosis and hypomania were also reported at the dose levels of sibutramine (Reeuwijk et al. 2014; Waszkiewicz et al. 2012). Also, adverse effects of N-mono-desmethyl sibutramine (DMS) and N-di-desmethyl sibutramine (DDMS) have been documented (Chen et al. 2009; Vidal and Quandte 2006). Another adulteration on slimming herbal product has again reported the presence of sibutramine as an undeclared product in a Chinese phytotherapeutic formulation (Jung et al. 2006).

Given these reports, quantitative monitoring of pharmaceutical agents present in weight-loss products is needed (Lee et al. 2013; Kim et al. 2014). An LC–MS strategy for the screening and quantification of sibutramine and its analogues in herbal supplements has been developed (Stahnke et al. 2012). LC–MS-based method for the identification of rimonabant polymorphs and sibutramine analogues in counterfeit ‘Acomplia’ and its imitation products was reported by Venhuis et al. (2011). A method for detection and estimation of sibutramine and its analogues, phenolphthalein, fenfluramine and orlistat in slimming herbal formulations with LC–MS instrumentation has been developed (Wang et al. 2008). An LC–MS/MS screening method was developed for studying the presence of synthetic adulterants and its quantitation in herbal remedies (Bogusz et al. 2006). This method also screened adulterants of different pharmacological effects including weight-reducing compounds (Bogusz et al. 2006). An LC–linear ion trap–(QTRAP)–MS method was used to screen multiple adulterants agents including slimming agents (Chen et al. 2009). A review on adulterants used in slimming phytotherapeutic products in addition to their analytical approaches was published by De Carvalho et al. (2011). According to another report, the herbal supplements marketed and exported from China contained adulterants like fenfluramine, phenolphthalein, sibutramine and its metabolites and orlistat (Lai et al. 2007). A sensitive, specific and accurate LC–ESI–MS method was developed for the simultaneous estimation of six adulterants in weight-loss products and supplements. They are, namely, fenfluramine, phenolphthalein, N-di-desmethyl sibutramine, N-mono-desmethyl sibutramine, sibutramine and orlistat (Wang et al. 2008). In this study, six matrix sources were considered, and samples from dietary fibre, tea, protein, cereal, vitamin and Chinese herb extracts were included. Jung et al. (2006) presented an analytical method of liquid chromatography coupled with triple quadrupole and TOF mass spectrometry for sibutramine. Song et al. (2006) determined orlistat in health food for controlling body weight by LC–MS method. By this methodology, sibutramine, its two metabolites and one analogue were successfully identified in an herbal product for weight loss. Khazan et al. (2014) identified and determined synthetic pharmaceuticals namely, sibutramine, phenolphthalein, phenytoin, bumetanide and rimonabant as adulterants in eight common herbal weight-loss products. Zeng et al. (2015) reported an LC–MS/MS method for simultaneous determination of 40 compounds including bisacodyl, sibutramine and its metabolites, etc. used for weight-loss effects.

3.6 Adulteration with PDE-5 Inhibitors and Its Analogues for Sexual Enhancement

The phosphodiesterase type 5 (PDE-5) inhibitor drugs are mainly used for the treatment of erectile dysfunction (ED) in males (Venhuis and De Kaste 2012). Synthetic PDE-5 inhibitors such as avanafil, sildenafil citrate, mirodenafil, tadalafil, udenafil and vardenafil and their analogues are the most commonly used drugs in herbal products as adulterants (Alp et al. 2013; Campbell et al. 2013; Park and Lee 2012). There are certain restrictions on the prevalent use of PDE-5 inhibitors, as these drugs require prescriptions by physicians (Goker et al. 2012). These restrictions have led interested persons to seek out herbal products and supplements claiming to enhance sexual performance. It is clearly evident by the recent dramatic increase in the consumption of health products for the aphrodisiac effects (Lee et al. 2013).

In fact, more than 46 PDE-5 inhibitors and analogues have recently been detected and isolated as adulterants in herbal supplements and products (Alp et al. 2013; Kee et al. 2012; Lee et al. 2013). Because these products are illegally promoted as natural and effective therapy for the enhancement of male sexual performance, consumers are seriously exposed to various health risks. It is known that PDE-5 inhibitors can drastically lower blood pressure when taken along with certain nitrate-containing drugs. It creates the possibility of a serious, life-threatening event for individuals who take these drugs. Increased cardiovascular risks like heart attack, stroke, chest pain, etc. are associated with the usage of these drugs in people with heart problems.

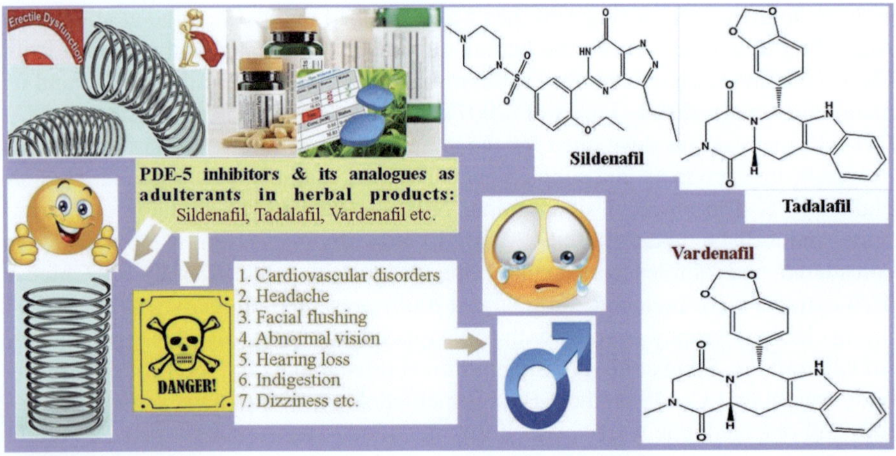

Fig. 3.3 Adulteration and its consequences of herbal products with PDE-5 inhibitors and analogues

Other major side effects associated with PDE-5 inhibitors and its analogues are headache, facial flushing, indigestion, dizziness, abnormal vision and hearing loss (Campbell et al. 2013; Rocha et al. 2016). Therefore, research into the monitoring and quantification of such pharmaceutical agents present in these products is needed to uphold human safety (Wang et al. 2012). Most of the published studies used HPLC in conjunction with diode array detector (PDA or DAD) and chemiluminescence detector for the screening of PDE-5 inhibitors in diverse samples (Di et al. 2011; Jalili et al. 2015). However, these methods have several drawbacks, as not only acquisition times are longer than the MS but also it is difficult to identify a variety of synthetic PDE-5 inhibitors and their analogues accurately. An LC–ESI–MS/MS method was reported for PDE-5 inhibitors and their analogues detection and was applied for qualitative confirmation and quantitative determination in herbal products. The use of LC–MS/MS detected adulteration of sildenafil, tadalafil and vardenafil in herbal products (Woo et al. 2013). Most of the synthetic PDE-5 inhibitor analogues synthesized by structural modification require a comparison with their respective MS/MS spectra for unequivocal identification (Lee et al. 2013; Venhuis et al. 2011). Hou et al. (2006) by using HR–ESI–MS instrument, found a new acetildenafil analogue from product intended as a dietary supplement. An LC–MS method was reported for the quantification of PDE-5 inhibitors adulterated in herbal food supplements (Radu et al. 2015). Kee et al. (2015) applied orbitrap mass spectrometry approach to differentiate isomeric sildenafil and thiosildenafil-like analogues used for the adulteration in herbal dietary supplements. A highly sensitive, selective and robust one-shot analysis LC–DAD–QTOF method for simultaneous screening and confirmation of target and non-target PDE-5 inhibitor analogues within a single chromatographic run in counterfeit herbal products was developed by Bortolini et al. (2015). They used a database of 82 PDE-5 inhibitor analogues and implemented an automatic non-target analysis of the above compounds. A representation of adulteration and its consequences in herbal products with PDE-5 inhibitors and analogues is shown in Fig. 3.3.

3.7 Regulatory Guidelines

WHO has developed guidelines for the quality assurance and quality control (QAQC) of herbal drugs which provide a description of the techniques and approaches needed for the cultivation and collection of medicinal plants. However, there is a gap between available knowledge and its implementation for various quality standards, because the majority of producers of herbal drugs are not aware of WHO's guidelines and continue their work without any quality control measures and SOPs, which affects the quality of phytoproducts (WHO 1998, 2005).

Herbal dietary supplements can be labeled as having certain healthful and nutritional properties but cannot make therapeutic claims. In 2015, the US Food and Drug Administration (USFDA) issued new guidelines for botanical drug products industry. The agency recommended a combination of tests and controls to ensure the identity, purity, quality, strength, potency and consistency of botanical products (USFDA 2015). Multiple tests for drug substance and drug product (e.g. spectroscopic and/or chromatographic fingerprints, chemical assay of characteristic markers and biological assay), raw materials and process controls (e.g. strict quality controls for the botanical raw materials and adequate in-process controls) and process validation (especially for the drug substance) are some of the tests and controls (Liu et al. 2015). The European Food Safety Authority (EFSA) works as an independent source for the scientific advice to the European commission to adopt legislation on the issues concerned with herbal supplements. Novel portable, fast screening spectrometric instruments such as ion mobility spectrometry (IMS) have been utilized to screen herbal products and supplements at a few FDA labs to prevent adulterated products from coming into the hands of consumers. FDA inspectors do it at ports of entry (Dunn et al. 2011; Song et al. 2014). In China, traditional Chinese medicine (TCM) includes Chinese *materia medica* (for Chinese herbs plus animal parts and minerals) and is regulated under the authority of the State Administration of Traditional Chinese Medicine (SATCM).

3.8 Conclusions

Herbal medicines and products play a significant role in the healthcare system of many developing countries and are rapidly gaining worldwide popularity among the consumers. From the reports, it is evident that the presence of undeclared synthetic drugs and its analogues in many herbal formulations is alarmingly increasing in the global market. The increasing reports of adulteration of botanical products and supplements represent a major concern to both consumers and regulatory agencies worldwide. However with recent advancements in hyphenated techniques and sophisticated analytical instruments and methodology, detection of adulteration can be made selective, accurate and sensitive. There is an urgent need for technical sets of quality standards for development and production of herbal formulations with necessary regulatory guidelines for commercialization. All countries must adopt multidisciplinary approaches to combat the threat of adulteration of herbal products to increase the confidence among consumers. The trend towards enhanced usage of LC–MS/MS based applications in synthetic adulterant screening and quantification in herbal products. The use of other analytical techniques along with most powerful LC–MS/MS instruments providing orthogonal information will likely be needed for unambiguous identification, elucidation and estimation of novel adulterants in herbal products. Therefore, the development and application of broad scope of all the screening analytical methodologies must be utilized to stop the menace of adulteration.

Acknowledgments The authors are thankful to the Science and Engineering Research Board (SERB), DST, New Delhi, India, for the financial support for this research work under 'Start-Up Research Grant' (Young Scientists) (NO. SB/YS/LS-82/2013). The authors would also like to thank Dr. Ahmed Kamal, Project Director, NIPER, Hyderabad, for his constant support and encouragement.

References

AB Sciex. http://sciex.com/products/mass-spectrometers/qtrap-systems. Accessed 20 Dec 2015

Ahmad I, Ahmad Khan MS, Cameotra SS (2014) Quality assessment of herbal drugs and medicinal plant products. Encycl Anal Chem:1–17. doi:10.1002/9780470027318.a9946

Al-Safi AS, Ayoub NM, Ayoub AM, Al-Momany E, Al-Doghim I, Al-Balas M, Alkofahi AS, Aboul-Enein FH, Aboul-Enein BH (2008) Public awareness of the abuse of herbs and drugs to decrease body weight: a novel national survey in Jordan. J Public Health 16:205–213. doi:10.1007/s10389-007-0166-5

Almeida AE, Ribeiro ML, Polese L (2000) Determination of amfepramone hydrochloride, fenproporex and diazepam in so called "natural" capsules used in the treatment of obesity. J Liq Chromatogr Relat Technol 23(7):1109–1118. doi:10.1081/JLC-100101512

Alp M, Coskun M, Goker H (2013) Isolation and identification of a new sildenafil analogue adulterated in energy drink: propoxyphenyl sildenafil. J Pharm Biomed Anal 72:155–158. doi:10.1016/j.jpba.2012.09.017

Ancuceanu R, Dinu M, Arama C (2013) Weight loss food supplements: adulteration and multiple quality issues in two products of Chinese origin. Farmacia 61(1):28–44

Aneesh TP, Hisham M, Sekhar MS, Manjusree M, Deepa TV (2009) International market scenario of traditional Indian herbal drugs in India is declining. Int J Green Pharm 3(3):184–190. doi:10.4103/0973-8258.56271

Balayssac S, Trefia S, Gilard V, Malet-Martino M, Martino R, Delsuc MA (2009) 2D and 3D DOSY 1 H NMR, a useful tool for analysis of complex mixtures: application to herbal drugs or dietary supplements for erectile dysfunction. J Pharm Biomed Anal 50:602–612. doi:10.1016/j.jpba.2008.10.034

Bandaranayake WM (2006) Quality control, screening, toxicity and regulation of herbal drugs. In: Iqbal A, Farrukh A (eds) Modern phytomedicine: turning medicinal plants into drugs. Wiley, New Jersey, pp 25–57

Bauer R (1998) Quality criteria and standardization of phytopharmaceuticals: can acceptable drug standards be achieved? Ther Innov Regul Sci 32(1):101–110. doi:10.1177/009286159803200114

Bogusz MJ, Hassan H, Al-Enazi E, Ibrahim Z, Al-Tufail M (2006) Application of LC-ESI-MS-MS for detection of synthetic adulterants in herbal remedies. J Pharm Biomed Anal 41:554–564. doi:10.1016/j.jpba.2005.12.015

Bortolini C, Pivato A, Bogialli S, Pastore P (2015) One-shot analysis of PDE-5 inhibitors and analogues in counterfeit herbal natural products using an LC-DAD-QTOF system. Anal Bioanal Chem 407(20):6207–6216. doi:10.1007/s00216-015-8801-4

Busse W (2000) The significance of quality for efficacy and safety of herbal medicinal products. Ther Innov Regul Sci 34(1):15–23. doi:10.1177/009286150003400102

Campbell N, Clark JP, Stecher VJ, Thomas JW, Callanan AC, Donnelly BF, Goldstein I, Kaminetsky JC (2013) Adulteration of purported herbal and natural sexual performance enhancement dietary supplements with synthetic phosphodiesterase type 5 inhibitors. J Sex Med 10(7):1842–1849. doi:10.1111/jsm.12172

Cappiello A (ed) (2006) Advances in LC-MS instrumentation, vol 72. Elsevier, Amsterdam

Chen Y, Zhao L, Lu F, Yu Y, Chai Y, Wu Y (2009) Determination of synthetic drugs used to adulterate botanical dietary supplements using QTRAP LC-MS/MS. Food Addit Contam Part A Chem Anal Control Expo Risk Assess 26(5):595–603. doi:10.1080/02652030802641880

Cianchino V, Acosta G, Ortega C, Martínez LD, Gomez MR (2008) Analysis of potential adulteration in herbal medicines and dietary supplements for the weight control by capillary electrophoresis. Food Chem 108(3):1075–1108. doi:10.1016/j.foodchem.2007.11.042

Cohen PA, Ernst E (2010) Safety of herbal supplements: a guide for cardiologists. Cardiovasc Ther 28(4):246–253. doi:10.1111/j.1755-5922.2010.00193.x

Correia D (2008) Voltammetric determination of benzodiazepines and diethylpropion as adulterants in phytotherapic formulations. Master thesis, Universidade Federal de Santa Maria, Brazil. Accessed 21 Dec 2015

Dawson PH (2013) Quadrupole mass spectrometry and its applications. Elsevier, Amsterdam

de Carvalho LM, Martini M, Moreira AP, de Lima AP, Correia D, Falcao T, Garcia SC, de Bairros AV, do Nascimento PC, Bohrer D (2011) Presence of synthetic pharmaceuticals as adulterants in slimming phytotherapeutic formulations and their analytical determination. Forensic Sci Int 204(1–3):6–12. doi:10.1016/j.forsciint.2010.04.045

de Cock KJS, Delbeke FT, Van EP, Desmet N, Roels K, De Backer P (2001) Detection and determination of anabolic steroids in nutritional supplements. J Pharm Biomed Anal 25:843–852. doi:10.1016/S0731-7085(01)00396-X

Deconinck E, Sacre PY, Courselle P, De Beer JO (2013) Chromatography in the detection and characterization of illegal pharmaceutical preparations. J Chromatogr Sci 51(8):791–806. doi:10.1093/chromsci/bmt006

Di Y, Zhao M, Nie Y, Wang F, Lv J (2011) A high-performance liquid chromatography: chemiluminescence method for potential determination of vardenafil in a dietary supplement. J Anal Methods Chem 2011:1–6. doi:10.1155/2011/982186

Dunn JD, Gryniewicz-Ruzicka CM, Kauffman JF, Westenberger BJ, Buhse LF (2011) Using a portable ion mobility spectrometer to screen dietary supplements for sibutramine. J Pharm Biomed Anal 54:469–474. doi:10.1016/j.jpba.2010.09.017

Fan XH, Cheng YY, Ye ZL, Lin RC, Qian ZZ (2006) Multiple chromatographic fingerprinting and its application to the quality control of herbal medicines. Anal Chim Acta 555(2):217–224. doi:10.1016/j.aca.2005.09.037

Fraser DB, Wen KC (1998) Chinese herbal medicines-manufacturing flaws and misuse. Forensic Sci Rev 10(2):67–80. doi: FSRv10n2-4

George P (2011) Concerns regarding the safety and toxicity of medicinal plants-an overview. J Appl Pharm Sci 1(6):40–44

GIA (2015) Global Industry Analysts (GIA), Inc., February 2015. Report on the global herbal supplements and remedies market trends drivers & projections. http://www.strategyr.com/marketresearch/herbalsupplementsandremediesmarkettrends.asp. Accessed 24 Dec 2015

Goker H, Coşkun M, Ayhan Kılcıgil G (2012) Chromatographic separation and identification of sildenafil and yohimbine analogues illegally added in herbal supplements. In: Sasikumar D (ed) Biochemistry, genetics and molecular biology. InTech Online, pp 978–953

Gross JRH (2004) Mass spectrometry: a textbook. Springer Science & Business Media, Germany

Haneef J, Shaharyar M, Husain A, Rashid M, Mishra R, Siddique NA, Pal M (2013) Analytical methods for the detection of undeclared synthetic drugs in traditional herbal medicines as adulterants. Drug Test Anal 5(8):607–613. doi:10.1002/dta.1482

Holcapek M, Jirasko R, Lísa M (2012) Recent developments in liquid chromatography -mass spectrometry and related techniques. J Chromatogr A 1259:3–15. doi:10.1016/j.jpba.2005.03.040

Holzgrabe U, Malet MM (2011) Analytical challenges in drug counterfeiting and falsification -the NMR approach. J Pharm Biomed Anal 55:679–687. doi:10.1016/j.jpba.2010.12.017

Hou P, Zou P, Low MY, Chan E, Koh HL (2006) Structural identification of a new acetildenafil analogue from pre-mixed bulk powder intended as a dietary supplement. Food Addit Contam Part A Chem Anal Control Expo Risk Assess 23(9):870–875. doi:10.1080/02652030600803856

Jalili R, Miraghaei S et al (2015) Detection of corticosteroid compounds and phosphodiesterase inhibitors (PDH-5) as counterfeit in herbal products available in Iranian market by HPLC method. J Rep Pharm Sci 4(1):75–81

Jordan SA, Cunningham DG, Marles RJ (2010) Assessment of herbal medicinal products: challenges, and opportunities to increase the knowledge base for safety assessment. Toxicol Appl Pharm 243(2):198–216. doi:10.1016/j.taap.2009

Jung J, Hermanns-Clausen M, Weinmann W (2006) Anorectic sibutramine detected in a Chinese herbal drug for weight loss. Forensic Sci Int 161(2–3):221–222. doi:10.1016/j.forsciint.2006.02.052

Kamboj A (2012) Analytical evaluation of herbal drugs. INTECH Open Access Publisher, Croatia

Kanan S, AbuYousef IA, Kanan S, Abu-Yousef IA, Gunasekar C, Abdo N, Narasimhan S (2009) Detection and quantification of synthetic drugs in herbal slimming formula. Eur J Sci Res 34:348–357

Keding GB, Msuya JM, Maass BL, Krawinkel MB (2013) Obesity as a public health problem among adult women in rural Tanzania. Glob Health Sci Pract 1(3):359–371. doi:10.9745/GHSP-D-13-00082

Kee CL, Ge X, Koh HL, Low MY (2012) Isolation and characterization of propoxyphenyl linked sildenafil and thiosildenafil analogues in health supplements. J Pharm Biomed Anal 70:265–272. doi:10.1016/j.jpba.2012.07.008

Kee CL, Ge X, Low MY (2015) Application of Orbitrap-mass spectrometry to differentiate isomeric sildenafil-and thiosildenafil-like analogues used for the adulteration of dietary supplements. Food Addit Contam Part A Chem Anal Control Expo Risk Assess 32(10):1737–1748. doi:10.1080/19440049.2015.1062923

Khazan M, Hedayati M, Kobarfard F, Askari S, Azizi F (2014) Identification and determination of synthetic pharmaceuticals as adulterants in eight common herbal weight loss supplements. Iran Red Crescent Med J 16(3):1–6. doi:10.5812/ircmj.15344

Kim HJ, Lee JH, Park HJ, Cho SH, Cho S, Kim WS (2014) Monitoring of 29 weight loss compounds in foods and dietary supplements by LC-MS/MS. Food Addit Contam Part A Chem Anal Control Expo Risk Assess 31(5):777–783. doi:10.1080/19440049.2014.888497

Kunle OF, Egharevba HO, Ahmadu PO (2012) Standardization of herbal medicines-a review. Int J Biodivers Conserv 4(3):101–112. doi:10.5897/IJBC11.163

Lai T, Hu X, Liu X (2007) Investigation of banned additives in healthy foods for weight control. Chin J Food Hyg 19:366–337. doi:10.1080/02652030801946553

Lau AJ, Holmes MJ, Woo SO, Koh HL (2003) Analysis of adulterants in a traditional herbal medicinal product using liquid chromatography- mass spectrometry- mass spectrometry. J Pharm Biomed Anal 31:401–406. doi:10.1016/S0731-7085(02)00637-4

Lee JH, Kim NS, Han KM, Kim SH, Cho S, Kim WS (2013) Monitoring by LC-MS/MS of 48 compounds of sildenafil, tadalafil, vardenafil and their analogues in illicit health food products in the Korean market advertised as enhancing male sexual performance. Food Addit Contam Part A Chem Anal Control Expo Risk Assess 30(11):1849–1857. doi:10.1080/19440049.2013.826878

Li N, Cui M, Lu X, Qin F, Jiang K, Li F (2010) A rapid and reliable UPLC-MS/MS method for the identification and quantification of fourteen synthetic anti-diabetic drugs in adulterated Chinese proprietary medicines and dietary supplements. Biomed Chromatogr 24:1255–1261. doi:10.1002/bmc.1438

Liang Q, Qu J, Guoan L, Yiming W (2006) Rapid and reliable determination of illegal adulterant in herbal medicines and dietary supplements by LC/MS/MS. J Pharm Biomed Anal 40:305–311. doi:10.1016/j.jpba.2005.07.035

Liu S-H, Chuang WC, Lam W, Jiang Z, Cheng YC (2015) Safety surveillance of traditional Chinese medicine: current and future. Drug Saf 38(2):117–128. doi:10.1007/s40264-014-0250-z

Lu Y, Chen J, Chen B, Yao S (2009) Simultaneous determination of 8 illegal synthetic drugs in antihypertensive dietary supplements by high-performance liquid chromatography-mass spectrometry. Se Pu 27(1):44–49

Luis CR, Cristina RS et al (2016) Chromatographic fingerprinting: an innovative approach for food identification and food authentication- a tutorial. Anal Chim Acta 909:9–23. doi:10.1016/j.aca.2015.12.042

Mukerjee PK (2002) Quality control of herbal drugs: an approach to the evaluation of botanicals. Business Horizons, New Delhi

Muller D, Weinmann W, Hermanns Clausen M (2009) Chinese slimming capsules containing sibutramine sold over the internet: a case series. Dtsch Arztebl Int 106(13):218–222. doi:10.3238/arztebl.2009.0218

Nikam PH, Kareparamban J, Jadhav A, Kadam V (2012) Future trends in standardization of herbal drugs. JAPS 2(6):38–44. doi:10.7324/JAPS.2012.2631

Ozdemir B, Sahin I, Kapucu H, Celbis O, Karakoc Y, Erdogan S, Onal Y (2013) How safe is the use of herbal weight-loss products sold over the internet? Hum Exp Toxicol 32:101–106. doi:10.1177/0960327112436407

Park S, Lee JG (2012) Determination of PDE-5 inhibitors and appetite suppressants in adulterated dietary supplements using LC/PDA and LC/MS. Food Addit Contam Part B Surveill 5(1):29–32. doi:10.1080/19393210.2012.656706

Patel DN, Li L, Kee CL, Ge X, Low MY, Koh HL (2014) Screening of synthetic PDE-5 inhibitors and their analogues as adulterants: analytical techniques and challenges. J Pharm Biomed Anal 87:176–190. doi:10.1016/j.jpba.2013.04.037

Patel DN, Lirui Q Qiao L, Yuk J, Yu K (2015) Waters application note on screening synthetic adulterants from herbal and dietary supplements (HDS) using the natural products application solution with UNIFI. http://www.waters.com/webassets/cms/library/docs/720005396en.pdf. Accessed 23 Dec 2015

Phattanawasin P, Sotanaphun U, Sukwattanasinit T, Akkarawaranthorn J, Kitchaiya S (2012) Quantitative determination of sibutramine in adulterated herbal slimming formulations by TLC-image analysis method. Forensic Sci Int 219(1–3):96–100. doi:10.1016/j.forsciint.2011.12.004

Poornima B (2010) Adulteration and substitution in herbal drugs a critical analysis. Int J Res Ayurveda Pharm 1(1):8–12

Prakash O, Jyoti, Kumar A, Kumar P, Manna NK (2013) Adulteration and substitution in Indian medicinal plants: an overview. J Med Plants Stud 1(4):127–132

Radu GL, Popescu AM, Niculae CG, Onisei T (2015) Identification by liquid chromatography-mass spectrometry of herbal food supplements adulterated with PDE-5 inhibitors. Rev Chim 66(1):1–5

Rasheed NMA, Nagaiah K, Waheed MA (2013) Recent analytical techniques in quality control of indigenous system of medicine. Ann Phytomed 2:44–58

Reeuwijk NM, Venhuis BJ, de Kaste D, Hoogenboom RL, Rietjens IM, Martena MJ (2014) Active pharmaceutical ingredients detected in herbal food supplements for weight loss sampled on the Dutch market. Food Addit Contam Part A Chem Anal Control Expo Risk Assess 31:1783–1793. doi:10.1080/19440049.2014.958574

Rocha T, Amaral JS, Oliveira MBPP (2016) Adulteration of dietary supplements by the illegal addition of synthetic drugs: a review. Compr Rev Food Sci F 15:43–62. doi:10.1111/1541-4337.12173

Russo G, Barbato F, Grumetto L (2016) A validated LC/UV method for the determination of four adulterating drugs in herbal slimming capsules. J Pharm Biomed Anal 117:436–445. doi:10.1016/j.jpba.2015.09.027

Sahoo N, Manchikanti P, Dey S (2010) Herbal drugs: standards and regulation. Fitoterapia 81:462–471. doi:10.1016/j.fitote.2010.02.001

Sargent M (ed) (2013) Guide to achieving reliable quantitative LC-MS measurements. RSC Analytical Methods Committee. ISBN 978-0- 948926-27-3

Song X, Xiaolan Z, Chen B, Yao SZ (2006) Determination of orlistat in health food for controlling body weight by high-performance liquid chromatography/electrospray spectrometry. J Hyg Res 35(3):355–366

Song F, Monroe D, El-Demerdash A, Palmer C (2014) Screening for multiple weight loss and related drugs in dietary supplement materials by flow injection tandem mass spectrometry and their confirmation by liquid chromatography tandem mass spectrometry. J Pharm Biomed Anal 88:136–143. doi:10.1016/j.jpba.2013.08.031

Stahnke H, Kittlaus S, Kempe G, Alder L (2012) Reduction of matrix effects in liquid chromatography-electrospray ionization – mass spectrometry by dilution of the sample extracts: how much dilution is needed? Anal Chem 84(3):1474–1482. doi:10.1021/ac202661j

Starr RR (2015) Too little, too late: ineffective regulation of dietary supplements in the United States. Am J Public Health 105(3):478–485. doi:10.2105/AJPH.2014.302348

Tistaert C, Dejaegher B, Heyden YV (2011) Chromatographic separation techniques and data handling methods for herbal fingerprints: a review. Anal Chim Acta 690(2):148–161. doi:10.1016/j.aca.2011.02.023

USFDA (2015) Botanical drug development guidance for industry draft August 2015. http://www.fda.gov/downloads/Drugs/GuidanceComplianceRegulatoryInformation/Guidances/UCM458484.pdf. Accessed 23 Dec 2015

Veit M, Gaedcke F (2004) Herbals: new specific and general monographs. Qual Mov Dyn Eur Pharmacopoeia 4:69

Venhuis BJ, De Kaste D (2012) Towards a decade of detecting new analogues of sildenafil, tadalafil, and vardenafil in food supplements: a history, analytical aspects and health risks. J Pharm Biomed Anal 69:196–208. doi:10.1016/j.jpba.2012.02.014

Venhuis BJ, Zwaagstra ME et al (2009) Trends in drug substances detected in illegal weight-loss medicines and dietary supplements. A 2002–2007 survey and health risk analysis. RIVM rapport. http://www.rivm.nl/en/Documentsandpublications/Scientific/Reports/2009/maart/Trendsindrugsubstancesdetectedinillegalweightlossmedicinesanddietarysupplements A2002–2007surveyandhealthriskanalysis. Accessed 26 Dec 2015

Venhuis BJ, Vredenbregt MVK, Kaun N, Maurin JK, Fijałek Z, de Kaste D (2011) The identification of rimonabant polymorphs, sibutramine, and analogues of both in counterfeit acomplia bought on the internet. J Pharm Biomed Anal 54:21–26. doi:10.1016/j.jpba.2010.07.043

Viana C, Zemolin GM, Müller LS, Dal Molin TR, Seiffert H, de Carvalho LM (2016) Liquid chromatographic determination of caffeine and adrenergic stimulants in food supplements sold in Brazilian e-commerce for weight loss and physical fitness. Food Addit Contam Part A Chem Anal Control Expo Risk Assess 33(1):1–9. doi:10.1080/19440049.2015.1112040

Vidal C, Quandte S (2006) Identification of a sibutramine-metabolite in patient urine after intake of a "pure herbal" Chinese slimming product. Ther Drug Monit 28(5):690–692. doi:10.1097/01.ftd.0000245392.33305.b0

Wang J, Chen B, Yao S (2008) Analysis of six synthetic adulterants in herbal weight-reducing dietary supplements by LC-electrospray ionization-MS. Food Addit Contam Part A Chem Anal Control Expo Risk Assess 25(7):822–830. doi:10.1080/02652030801946553

Wang R, Burnett AL, Heller WH, Omori K, Kotera J, Kikkawa K, Yee S, Day WW, DiDonato K, Peterson CA (2012) Selectivity of avanafil, a PDE5 inhibitor for the treatment of erectile dysfunction: implications for clinical safety and improved tolerability. J Sex Med 9(8):2122–2129. doi:10.1111/j.1743-6109.2012.02822.x

Waszkiewicz N, Zalewska-Szajda B, Zalewska B, Szajda SD, Simonienko K, Zalewska A, Szulc A, Ładny JR, Zwierz K et al (2012) Sibutramine-induced mania as the first manifestation of bipolar disorder. BMC Psychiatry 12:43. doi:10.1186/1471-244X-12-43

Woo H, Kim JW, Han KM, Lee JH, Hwang IS, Lee JH, Kim J, Kweon SJ, Cho S, Chae KR, Han SY, Kim J (2013) Simultaneous analysis of 17 diuretics in dietary supplements by HPLC and LC-MS/MS. Food Addit Contam Part A Chem Anal Control Expo Risk Assess 30(2):209–217. doi:10.1080/19440049.2012.738939

World Health Organization (1998) Quality control methods for medicinal plants materials. WHO, Geneva. http://apps.who.int/iris/bitstream/10665/41986/1/9241545100.pdf. Accessed 24 Dec 2015

World Health Organization (2005) Quality control methods for medicinal plants materials. WHO, Geneva. http://apps.who.int/medicinedocs/pdf/s7918e/s7918e.pdf. Accessed 24 Dec 2015

World Health Organization (2016) Obesity and overweight. http://www.who.int/mediacentre/factsheets/fs311/en. Accessed 7 Sep 2016

Zeng Y, Xu Y, Kee CL, Low MY, Ge X (2015) Analysis of 40 weight loss compounds adulterated in health supplements by liquid chromatography quadrupole linear ion trap mass spectrometry. Drug Test Anal. doi:10.1002/dta.1846

Chapter 4
DNA Barcoding of Medicinal Plants

Swati Srivastava, Sanchita, Mili Bhargava, and Ashok Sharma

Abstract DNA barcoding is a useful technique for diversity analysis whereby a standardized region of DNA is used for the identification of a species or a taxonomic group of organisms. These standard regions used for identification are called the DNA barcode. These are small sequences of the entire genome. In plants, DNA barcoding has application in phylogenetic analysis, authentication, inter- and intraspecific diversity, classification into wild and cultivated genotypes, the study of phylogeographical patterns, and in the detection of adulteration. The barcode loci, i.e., the DNA regions used for the identification are able to discriminate the closely related species and identify new cryptic species as well. Depending on the taxon and complexity of the species, different barcode loci are used for the purpose. In animals, the universal DNA barcode, i.e., mitochondrial cytochrome c oxidase I (COI) gene is used for species discrimination. However, this gene cannot be used for plants due to its limited divergence. Thus, its use is limited only to some algae. Efforts are going on to find suitable universal barcode loci for plants. Since the last decades, matK, rbcL, trnH-psbA, ITS, trnL-F, 5S-rRNA, and 18S-rRNA candidate regions are being used as DNA barcodes in plants. The article provides an overview of the use of these candidate regions through different approaches which have gained importance due to the challenges in DNA barcoding of plants. The development of multilocus and tiered approaches along with the new frontier areas for application of this technique has been analyzed in detail.

Keywords DNA barcoding • Forensics • Herbal drugs • Medicinal plants • Multilocus • Single locus • Tiered

---

S. Srivastava • Sanchita • M. Bhargava • A. Sharma (✉)
Biotechnology Division, CSIR-Central Institute of Medicinal and Aromatic Plants, Lucknow 226015, India
e-mail: ashoksharma@cimap.res.in

H.-S. Tsay et al. (eds.), *Medicinal Plants - Recent Advances in Research and Development*, DOI 10.1007/978-981-10-1085-9_4

Abbreviations

BOLD	Barcode of Life Data Systems
CBOL	Consortium for the Barcode of Life
iBOL	International Barcode of Life Project
WHO	World Health Organization

4.1 Introduction

Medicinal herbs have been used in Indian medicinal systems since ancient times. Currently, authentication of medicinal plants is a big issue. Trading of medicinal plants and its products worldwide is estimated at around US$60 billion, and an annual turnover of Ayurvedic medicine in the international market is about Rs. 3500 crores (US$813 million) (Biswas and Biswas 2014). According to the World Health Organization's (WHO) guidelines, authenticity, purity, and safety are important aspects of standardization and evaluation of traditional medicines. Due to commercialization and increased demand of Ayurvedic herbs, safety, quality, and assurance are big issues (Chan 2003). Taxonomists are busy in naming and annotating the huge number of organisms constituting the biodiversity. A large variety of species are measured annually. However, there still remains an enormous diverseness to be explored. The current scenario of extinction and conservation rates of biodiversity is also a serious concern (Costello et al. 2013).

In the last decade, DNA sequences are being extensively used in the biological process analysis such as phylogenetic analysis of organism identification. Among various approaches, the DNA barcoding was proposed to overcome the problems faced in the traditional taxonomy (Hebert et al. 2003a, b). This approach has succeeded in the identification of already existing as well as unknown species. In this technique, a standard region of DNA known as "DNA barcode" is used for the biodiversity analyses. Different regions of DNA are used as markers for DNA barcoding. Two main characteristics of a good marker are its universality and high resolution (Hollingsworth et al. 2011). The universality of any region refers to the applicability of the chosen DNA barcode to a large number of organisms. High-resolution ability implies that the markers must discriminate the closely related species. For efficient discriminatory power, a marker must show high interspecific and low intraspecific divergence. This distinction between inter- and intraspecific distances is known as the "DNA barcoding gap." The DNA barcoding is a widely used technique for quick and accurate identification of species (Bhargava and Sharma 2013). COI (cytochrome oxidase I) is the universal barcode marker in animals (Hebert et al. 2003a). However for plants, it has remained elusive (Li et al. 2015; Kress and Erickson 2008). COI has shown a good success rate in animals but in plants, due to limited divergence, it cannot be used. There has been much debate about the regions to be used as barcodes for plants (Hollingsworth et al. 2011).

Presently, DNA barcodes work on the standard gene of any locus as well as in whole chloroplast genome in plants. Although different kinds of genome-based strategies are developed for the identification, DNA barcoding is the most powerful tool. Approximately 300,000 plant species are available worldwide. The identification and classification of such a vast range of plants may be a difficult task for taxonomists. DNA barcoding helps in a rapid and accurate identification of plant species (Costion et al. 2011). DNA-based methods are more suitable as compared to proteins and RNAs because DNA is available in all the tissues of the organisms is more stable, and remains unaffected by external factors. The species discrimination in plants is difficult because of a higher level of gene tree paraphyly (Fazekas et al. 2008). matK, rbcL, trnH-psbA, ITS, trnL-F, 5S-rRNA, and 18S-rRNA are majorly used markers for plants with regard to their discrimination capacity (Table 4.1). Cowan and Fay (2012) have described the major challenges associated with DNA barcoding of plants. However, the studies on plant barcoding are increasing consistently due to its capability of identifying the unknown samples. A general concept of the formation of DNA barcode has been explained in Fig. 4.1.

4.2 History of DNA Barcoding and Success Stories

Two international initiatives are operating for the DNA barcoding analyses, viz., the International Barcode of Life project (iBOL) and Consortium for the Barcode of Life (CBOL). iBOL is the biggest biodiversity genomics initiative ever undertaken. Its mission is to maintain and update the barcode reference library, Barcode of Life

Table 4.1 Major candidate regions used for DNA barcoding of plants

Gene	Location/genome	Type	No. of GenBank accessions (plants)	No. of GenBank species (plants)
matK[a]	Plastid	Protein coding	~61,172	~33,089
rbcL[a]	Plastid	Protein coding	~86,806	~38,164
rpoB[a]	Plastid	Protein coding	~8328	~4399
rpoC1[a]	Plastid	Protein coding	~8650	~4787
trnH-psbA[a]	Plastid	IGS	~15,304	~4912
atpF-H[a]	Plastid	IGS	~370	~295
psbK-I[a]	Plastid	IGS	~3223	~1789
trnL-F	Plastid	Intron + IGS	~16,608	~7344
5S rRNA	Nuclear	Structural RNA	~9954	~2290
18S rRNA	Nuclear	Structural RNA	~40,291	~20,150
ITS	Nuclear	Transcribed spacers+5.8S rRNA	~179,520	~70,085

[a]CBOL proposed seven candidate barcodes

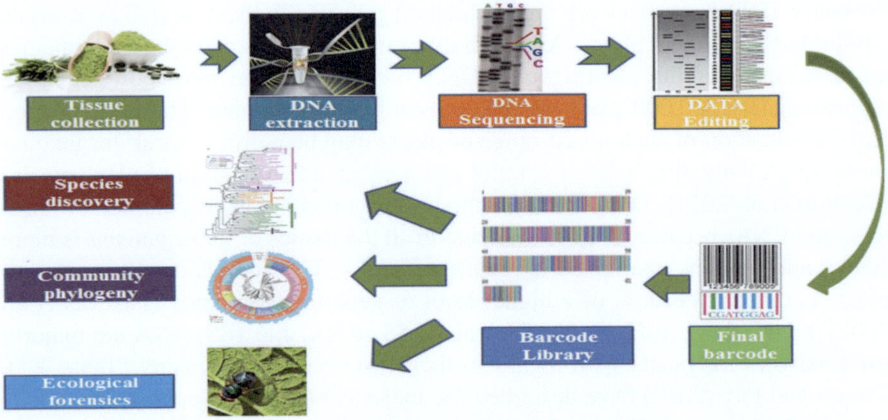

Fig. 4.1 Schematic representation of the formation of DNA barcodes

Data systems (BOLD), and further establishment as a robust resource for animal, plant, and fungal DNA barcodes (Ratnasingham and Hebert 2007). The work of the iBOL association is carried out by its constituent nodes, comprised of many counties grouped into separate operating teams. On the other hand, CBOL established in 2004 is functioning for DNA barcoding as a world methodology for the identification of plants and animals of earth's biodiversity. As far as the application of the technique is concerned, recently its usefulness has been explored in forensic botany to resolve the legal questions. Plant identifications at crime scenes are important in the criminal investigations. Every environment has a unique combination of pollens, suggesting the type of place where the crime took place. To overcome the problem related to ancient forensic botany, DNA barcoding could be a promising technique in several cases (Ward et al. 2005, 2009; Tsai et al. 2006, 2008; Ferri et al. 2009).

4.3 Status of DNA Barcoding of Medicinal Plants

Adulteration is a major problem in the herbal plant material market. Therefore, authentication and standardization is the prerequisite to minimize the unfair trade. According to the World Health Organization (WHO) total international seasoning, the drug market is calculated as US$62 billion and is anticipated to grow to the extent of US$5 trillion by the year 2050. The total available barcodes represent 363,584 sequences from 50,039 species. The criteria of DNA barcoding, i.e., minimum sequence length of 500 bp and more than three organisms per single species have been convincing by 13,761 species (Sarwat and Yamdagni 2015). However, most of these (98 %) are animal species. In January 2009, iBOL started with the target to collect barcodes for 5 million species in first 5 years. The scientists from

25 countries have contributed to this initiative (Hajibabaei et al. 2006). The DNA barcoding project has the goal of the reference library development that might provide data even for very low taxonomic level with short and specific DNA fragments. The major efforts are underway for barcoding of medicinal and aromatic plants worldwide (Cowan and Fay 2012; Elliott and Jonathan 2014). However, very little work has been reported for barcoding of Indian medicinal plants. India carries 7–8 % of world biodiversity with excessive resources of medicinal plants (45,500 approx.). Out of those, 8,000 plant species are of medicative worth, and 960 species are considered in a trade. Out of that, 178 species have a yearly consumption of more than 100 metric tons (Aneesh et al. 2009; Efferth and Greten 2012). The demand of the medicinal plants at industrial level is higher due to its global growth within the herbal industries. Thus, the Indian market is a center of herbs with the calculable trade of US$140 million annually. The botanical and natural ingredients export worldwide was more or less US$33 billion throughout 2010, and it was expected to reach US$93 billion by 2015 according to the December 2011 bulletin of Market News Service. The export of Indian medicinal plants and their products is estimated to be about $0.2 billion. In addition to the international trade, there is a considerable volume of international exchange of medicinal plants in India with a turnover of $1.6–$1.8 billion (Mishra et al. 2015). Total world seasoning herbal market is of the scale of $60 billion yearly with India's contribution of 2.5 %. Thus, in spite of having an extensive heritage of Ayurvedic literature and a good variety of medicinal plant species, India is still struggling with the potential market demand (Mishra et al. 2015). For increasing the India's share in the global herbal market, the improvement in quality control, standardization, scientific ways of production, and analysis of business products is necessary. The standardized mass produce of herbal products tested scientifically would not only maintain the efficacy of the herbals but also offers a competing edge to other medicines. China is presently leading the efforts on DNA barcoding of medicinal plants and has developed the database of DNA barcodes (Lou et al. 2010). Some reports are also available for DNA barcoding of Indian medicinal plants (Parveen et al. 2012; Ghosh et al. 2013). Due to importance and demand of herbal raw materials and products, the herbal industry suffers from substitution and adulteration of medicinal plants with its closely related species. Adulteration and mixing cause major changes in formulation and are also considered as illegal practices. The efficacy of any drugs/herbal product decreases when the herb is adulterated, and sometimes it could be lethal if it is substituted with toxic adulterants. The correct formulation is important for the medicinal herb to be effective. The main source of income of herbalist is the trading of medicinal plants. The economic constraints might offer an incentive for herbalists to substitute rare ingredients with cheaper and a lot of pronto offered species. Due to the illegal over-trading of medicinal plants, many plant species have become endangered in India. Therefore, to avoid these practices, some identification tags are required to detect plant materials. DNA barcoding is a useful tool for the discrimination of raw materials of medicinal plants.

4.4 Different Approaches of DNA Barcoding in Plants

Due to the complexity of DNA barcoding in plants, may it be amplification, sequencing, or a significant "barcoding gap," the technique has demanded ample attention toward the improvement of the methodology of the identification process. Thus, approaches like combining multiple barcodes at the totally different taxonomic group or multiple combinations of barcodes in a tiered fashion, such as a particular combination of one taxonomic group followed by a more robust combination at the next level, have recently gained importance. These approaches are mentioned below to introduce to the readers with the current methods in DNA barcoding in plants.

4.4.1 Single-Locus Approach

Due to the differences in the efficiency of barcoding markers in discriminating plants of different families, individual markers have been comparatively evaluated in a number of families (Gao et al. 2010a, b; Hollingsworth et al. 2009; Li et al. 2012; Muellner et al. 2011; Pettengill and Neel 2010). matK is the nearest plant analogue to COI, the animal DNA barcode. It typically provides high resolution, leading to good species identification as a result of its speedily evolving coding fragment among the plastid genome (Lahaye et al. 2008). However, the disadvantage of this barcode marker is due to unavailability of universal primer sets for all taxa. It creates a problem in PCR resulting in low PCR amplification particularly in non-angiosperms (Kress and Ericsson 2007; CBOL Plant Working Group 2009). As compared to matK, the barcode marker rbcL (ribulose-1,5-bisphosphate carboxylase/oxygenase large subunit) is easy to amplify and sequence. It is an important candidate for plant DNA barcoding even though its discriminatory power is not as good as matK. matK and rbcL have been suggested to be the core DNA barcodes for plants (Hollingsworth et al. 2009). Other than these, the plastid intergenic spacer trnH-psbA is also used as a supplementary DNA barcode. It has higher species discrimination success and variable intergenic spacers in plants (CBOL Plant Working Group 2009; Liu et al. 2012a, b). The main focus related to this locus includes the high frequency of mononucleotide repeats that cause simplex reads (Devey et al. 2009) and thus hamper the recovery of bidirectional sequences. Another common event encountered during this region is the microinversion. Microinversions in trnH-psbA have been studied in various angiosperms (Whitlock et al. 2010). Thus, the advanced design of trnH-psbA makes it tough to use as a barcode (Storchova and Olson 2007; Hao et al. 2010). However, the additional characters for species discrimination are provided by uncorrected microinversions (Jeanson et al. 2011). Although demonstrated a positive impact on species, discrimination by manually correcting inverted sequences has been demonstrated (Whitlock et al. 2010). A comprehensive analysis of the utility of trnH-psbA and its mixture has been studied (Pang et al. 2012). Many researchers have been concerned with the utilization of nuclear internal transcribed spacer (ITS) region in the form of the standard barcode and have recommended one of the core barcodes for seed plants (Li et al. 2015). Subsequent to this study, part of this region, ITS2 was also suggested to be a novel barcode for both plants and animals (China Plant BOL Group 2011; Yao et al. 2010; Chen et al. 2010). On the basis of its performance in phylogenetic studies (Baldwin 1992), resolving power of ITS was not underestimated; however, the following three major problems were encountered in its use.

4.4.1.1 Sequencing

One of the main limitations for nuclear ITS is its recovery, since the amplification and sequencing are difficult for this region (Kress et al. 2005). An alternative to it is the use of ITS2 that is less complicated to work with due to the small length of target fragment which makes amplification and sequencing easier.

4.4.1.2 Paralogous Gene Copies

Nuclear ITS fragment is available in multiple copies in the cells. Concerted evolution of multiple copies leads to divergence co-occurring in the individuals (Alvarez and Wendel 2003; Bailey et al. 2003). This can lead to rendering the sequences unreadable resulting in messy sequences. This paralogy phenomenon may lead to misidentification of samples because depending upon the variant sequence, the species will be identified subsequently. However, the identification of region, compared to other markers, is not compromise due to the availability of paralogous copies (Hollingsworth et al. 2011).

4.4.1.3 Fungal Contamination

The fungal ITS regions represent similarity with their plant correspondent. The primers considered for amplification as well as sequencing are very similar. Thus, the fungal DNAs are amplified by accident in several cases, particularly for the plants containing fungal endophytes. This can lead to misidentification of samples. Therefore, no matter the quantity of primer sets obtainable for this explicit barcode region, amplification and sequencing have been hard for numerous samples (Gonzalez et al. 2009).

4.4.2 Alternate Regions as DNA Barcodes

Besides the core plastid markers (matK, rbcL), the supplementary trnH-psbA, and ITS regions, there are other plastid protein-coding genes (rpoB, rpoC1), plastid intergenic spacers (atpF-H, psbK-I), and low-copy number genes being tested for identification in several families (Pillon et al. 2013).

4.4.3 Multilocus Approach

The multilocus approach is an adequate method for DNA barcoding of plants with good discrimination success (Kress and Erickson 2007; Fazekas et al. 2008; Newmaster et al. 2008; CBOL Plant Working Group 2009). The practice of using multiple barcodes has emerged in view of the unsatisfactory performance by individual loci. The high discrimination-related results could be obtained through combining the universality, discriminatory power, and amplification success of each locus. Multilocus combinations also promote high clade support values in monophyly-based identification of samples as in the case of Nyssaceae (Wang et al. 2012). Any plant barcode can be a combination of two or more locus. One of them may be a conservative coding region like rbcL and the other, a rapidly evolving noncoding region. The noncoding trnL intron and trnL-F intergenic spacer (IGS) are recommended for situations involving extremely degraded tissue (Taberlet et al. 2007). In bryophytes, the power of this region has been tested (Quandt et al. 2004; Stech et al. 2013). Thus, trnL-F and trnL regions were further used successfully for distinguishing the mysterious aquatic fern gametophyte (Li et al. 2009). Even a project on two-locus DNA barcode for plants (matK + rbcL) has been proposed by the CBOL Plant Working Group (2009). In some cases, the combination of three loci has failed to improve the discrimination better than two-locus barcodes in few cases (Wang et al. 2012). To avoid expenses of using a three loci combination for large data sets, the two-locus barcode was accepted as the standard barcode for land plants (CBOL 2009). In the case of two-locus, the preserved coding loci align well with the taxa of a community sample to determine deep phylogenetic branches. The hypervariable region of the DNA barcodes can align with ease in the subclades of closely related species (Kress et al. 2009). The complementation of rbcL gene and the noncoding trnH-psbA spacer region has been demonstrated (Fazekas et al. 2008). In contrast to CBOL, they suggested the use of more than two regions because of the decreased discrimination identified in barcoding analyses with three or more regions. This concept would also be beneficial when some of the loci recovered are of bad quality. Another efficient work using rbcL in combination with trnL-F for ferns has also been demonstrated and shown great potential for species discrimination (de Groot et al. 2011). The composition vector (CV) approach (Qi et al. 2004) has been described as an efficient method for analyzing rRNA data sets. The changed CV methodology incorporates an adjustable weighted algorithmic program for the vector distance as per the magnitude relation of sequence length found between a pair of taxa within the candidate genes. Recently, changed CV approach has been applied for studying huge multigene datasets for plant DNA barcoding (Li et al. 2012).

4.4.4 Tiered Approach

Combining barcode markers for discriminating species is robust and has high support values. A newer approach for combining the barcode, known as the tiered approach, is also evolving. Although vigorous efforts are going on to find suitable universal loci for plants, and there may be one in future, but relying on a single locus for plants will still be a bad choice. This is because of hybridization and introgression observed in some group of plants. Therefore, rather depending on maternally inherited genes, using a combination of both coding and noncoding regions in a stepwise manner will be the favorable and logical approach. It permits an unknown sample to be allotted at a taxon level, where a successful pair of primers can be targeted. Among a small group of taxa, the samples are aligned first followed by final assignment. In a specific taxonomic group, only a few studies have tested this methodology (Newmaster et al. 2006; Xiang et al. 2011). The first tier coding region, common in plants, has been used for differentiation at a definite taxonomic level, followed by a lot of variable second tier coding or noncoding region at the species level. Alignments at first tier (coding regions) would decrease the problem of aligning more divergent genera using noncoding regions at the second tier. So, under a common first tier sequence, the dataset will be properly organized to perform well at the next level of resolution. The method also preserves the efficiency of the previous multilocus approach since the complement regions for a group can still be used in this method. The rbcL has been considered as a primary tier barcode (Newmaster et al. 2006). Although it is the most identified plastid coding region in GenBank, covering a majority of groups and thus can work as a platform for comparison of different plastid genes. rbcL was analyzed to see how well it resolves congeneric species. This marker might be used for resolving congeneric species (85 % cases), so it should be used as the core first tier locus, followed by a choice of a secondary locus at the second tier. The method, therefore, provides flexibility in the choice of the next locus after a standard common region is used at a particular level. Similarly, this approach has been supported by Xiang et al. (2011) and also recommended that the use of matK at the generic level with further resolution at the second tier needs to be explored with a suitable second tier locus.

4.5 Bioinformatics Approaches

Bioinformatics play an important role in DNA barcoding analyses. The DNA barcoding processes depend on the availability of information in the form of data. If the data are available for query, we can use bioinformatics tools for the analysis of

Table 4.2 Different software and tools used for DNA barcoding

Categories of software and tools	Name of software and tools	References
Alignment-free and character based	BRONX (Barcode Recognition Obtained with Nucleotide eXposés)	Little (2011), Zhang et al. (2008), and Bertolazzi et al. (2009)
	BPSI (Back Propagation Species Identification)	
	BLOG (Barcoding with LOGic formulas)	
psbA-trnH based	PTIGS-Idit	Liu et al. (2011)
Distance based	TaxI	Steinke et al. (2005)
Character based	CAOS (Characteristic Attribute Organization System)	Sarkar et al. (2008)
Oligonucleotide frequency based	Oligonucleotide Frequency Barcode Generator (OFBG)	Tyagi et al. (2010)
Simultaneous sequence and structure alignment	4SALE	Seibel et al. (2006)
CBC detection	CBCAnalyzer	Wolf et al. (2005)
OTU clustering and annotation	jMOTU and taxonerator	Jones et al. (2011) and Kumar et al. (2011)
	CLOTU	
Identification of new barcode markers and associated PCR primers	ecoPrimers	Riaz et al. (2011)
Analysis of discrimination capacity of individual markers	TaxonGap	Slabbinck et al. (2008)
DNA barcoding based on Bayesian phylogenetics	Statistical Assignment Package (SAP)	Lou et al. (2010)
Analysis of resultant data	OTUbase	Beck et al. (2011), Brown et al. (2012) and Little (2010)
	SPIDER (SPecies IDentity and Evolution in R)	
	Barcode Quality Index (B)	

barcode data. After the collection of corresponding query sequences, sequence analysis and phylogenetic construction are performed. Sequence analysis basically involves the query and reference dataset sequence alignments. Some of the MSA programs ClustalW, T-Coffee, and MUSCLE, etc. are used for sequence analysis. In silico innovation approaches for DNA barcoding have been developed on the basis of compensatory base changes (CBCs) (Wolf et al. 2005), operational taxonomic units (OTUs) (Slabbinck et al. 2008), DNA metabarcoding (Riaz et al. 2011), locus-specific tools (Liu et al. 2011), tool for representing barcode symbology (Liu et al. 2012a, b), neural network techniques (Zhang et al. 2008), machine learning (Zhang et al. 2012a, b), data mining (Bertolazzi et al. 2009), composition vector (Kuksa and Pavlovic 2009), etc. The available software and tools analyzing the barcode data are given below (Table 4.2).

4.6 Limitations

4.6.1 The Absence of Universal Barcode and Selection of Appropriate Barcode Region

In DNA barcoding, the universality of the barcode is still a big problem. It is difficult to attain the universality of barcode due to the insufficient information of genetic variation in the less-studied taxonomic group. This problem is majorly found in plants as compared to animals. The differentiation and identification of species relying on interspecific variation among DNA sequences are due to the resolution capability of a barcode. Thus, there is a challenge in defining a good quality barcode consisting of a small and variable DNA sequence flanked by conserved regions (Hebert et al. 2003b; Moritz and Cicero 2004; Rubinoff et al. 2006a; Ficetola et al. 2010). The most important task of DNA barcoding is the identification of universal primers amplifying fragments with high resolution. However, it has been argued that a single short fragment will be sufficient to discriminate the organism at species level identification (Ficetola et al. 2010; Rubinoff et al. 2006a; Moritz and Cicero 2004). The single-locus DNA barcodes lack adequate variation in the closely connected taxonomic group, so for the identification of plants, no loci are available (Li et al. 2015).

4.6.2 Error Found in DNA Barcoding when Mitochondrial Sequences Are Used

DNA barcoding faced limitation due to the presence of the same copy of a gene of interest in the mitochondrial genome because of heteroplasmy in mtDNA, bacterial infection biasing, nuclear integration, and introgression in mtDNA. The duplication of a gene, i.e., if a portion of cytochrome oxidase I (COI) are duplicated in a given species, typical PCR may amplify these fragments. Thus, it will not be clear whether the paralogous copy had diverged from duplication of COI (Campbell and Barker 1999; Song et al. 2008). The heteroplasmy is the combination of more than one type of mitochondrial genome in a species. The overestimation of the quantity of distinctive species in barcoding results due to occurrence of co-amplification in divergent heteroplasmic copies of mtDNA (Rubinoff et al. 2006b; Song et al. 2008; Fišer Pečnikar and Buzan 2014; Magnacca and Brown 2010; Moulton et al. 2010; Valentini et al. 2009; Acs et al. 2010; Hurst and Jiggins 2005). The bacterial infection found in mtDNA due to the maternally inherited symbionts can cause linkage disequilibrium, and each individual becomes infected with such symbionts. These symbionts among closely connected species break the species barrier by conjugation followed by selective sweep leading to the identical mtDNA sequences in different species (Song et al. 2008; Whitworth et al. 2007). The nuclear integration of mtDNA creates error for barcoding. Nuclear mitochondrial pseudogenes (numts) are a nonfunctional duplication of mtDNA in the nucleus and occur in the major clades of eukaryotes. The presence of numts in the nuclear region creates a problem in DNA barcode data library construction and species identification. The potential existence of COI numts causes a major problem to DNA barcoding (Bensasson et al. 2001; Richly and Leister 2004; Song et al. 2008; Zhang and Hewitt 1996). The introgression in mtDNA also creates a problem for barcoding. Introgression is the process of transfer of a gene from one species into the gene pool of other species through recurrent backcrossing of an interspecific hybrid with one among its parents. It causes confusion in species boundaries between evolutionary lineages (phylogenies) that might commonly be divergent (Rubinoff 2006). In meta-analysis of phylogenetic studies, it was found that over 20 % of the studies lineages present problem due to mtDNA introgression (Ballard and Whitlock 2004; Fišer Pečnikar and Buzan 2014; Rubinoff 2006; Valentini et al. 2009; Vences et al. 2005; Acs et al. 2010; Hurst and Jiggins 2005; Machado and Hey 2003). There are limitations of using mtDNA in infer species boundaries with the retention of ancestral polymorphism, male-biased gene flow, and selection on any mtDNA nucleotide (the whole genome is one linkage group). The introgression along with hybridization and paralogy results in the transfer of mtDNA gene copies to the nucleus (Hebert et al. 2004; Ballard and Whitlock, 2004; Bensasson et al. 2001). These factors in mtDNA create a problem for both animal and plant DNA barcoding.

4.6.3 Lack of Comprehensive Reference Database

DNA barcoding is affected due to incomplete a priori identification of specimen in the reference database. The conflictions are created in data assessment; different laboratories work on the same taxa and explain different nomenclatures of the same species through morphological identification (Becker et al. 2011; Collins and Cruickshank 2013). If the reference database is not comprehensive, it will create misidentification of the taxa (Meyer and Paulay 2005; Valentini et al. 2009). DNA barcoding faces limitations when the selected individual represents to every taxon within the reference database. The unknown specimen taken from undescribed biodiversity causes problems in the identification (Fišer Pečnikar and Buzan 2014; Rubinoff 2006). The reference sequences from taxonomically verified specimen lead to the validity of DNA barcoding. In the absence of the reference data, DNA barcoding will face limitations and challenges (Ajmal et al. 2014). DNA barcoding will also face difficulty when the query sequence lacks its target in the reference database. Therefore, the barcoding-based identification of the query at the species level fails (Nielsen and Matz 2006; Virgilio et al. 2010). The reference sequences are verified from voucher specimen that is documented by experienced taxonomists. Due to lack of reference database, there will be no authentic library for recently identified query sequences. As a result, there will be a large quantity of legacy data in the form of sequences that are available in GenBank. These will not be used as a barcode. Thus, DNA barcoding does not improve the speed of cataloging the life on earth (Taylor and Harris 2012; Peterson et al. 2007).

4.6.4 Lack of Statistical Solution

DNA barcoding is a useful tool for the identification of unknown species. For this methodology, the threshold values providing a distinction between intraspecific and interspecific variation values are required. If the unknown sequence differs from the nearest reference sequence by a variation above the threshold, the organism containing the sequences will belong to a specific species, suggesting its classification needs additional investigation. The wide range of overlap between intra- and interspecific divergence values creates major problems. These overlaps seem comparatively restricted and far from the respective average values. Thus, only the mean values for intra- and interspecific comparisons of closely connected sibling species are required (Desalle 2006; Hebert et al. 2003a; Prendini 2005; Rubinoff 2006; Taylor and Harris 2012; Valentini et al. 2009; Vences et al. 2005; Casiraghi et al. 2010; Frézal and Leblois 2008). The use of a different threshold considering the tenfold rule (gap corresponds to a generic ten times the value of intraspecific divergence) has been proposed. This law has been extensively criticized (Meyer and Paulay 2005; Moritz and Cicero 2004; Matz and Nielsen 2005; Nielsen and Matz 2006; Valentini et al. 2009). To overcome this difficulty, currently it is predicted that the interspecific sequence divergence should increase to the threshold of 2 or 3 % dissimilarity. This threshold has been set on the basis of experimental proof observation of sequence variations among congeneric species (Hebert et al. 2003b). This approach might be simple to neglect the inconclusive or inaccurate results. Thus, there is a requirement of for statistical strategies when a sampled query sequence is the same as the specific database sequence to proof a species assignment of the query (Nielsen and Matz 2006). The strong assumptions based on the population genetics of the analyzed species revealed the statistical uncertainty in DNA barcoding (Nielsen and Matz 2006). The unrealistic assumption of excellent sequence identity at intraspecies level is abandoned. Thus, with not creating population genetic assumptions, the DNA barcoding is not possible (Acs et al. 2010; Hickerson et al. 2006). It has been observed that with robust population subdivision within species, the species assignment might fail due to the underlying demographics that have not been modeled capably. Another case is sequence sampled from a subpopulation with no gene flow with any of the population listed in the database. The DNA barcoding statistical methods which are used here do not categorize the query sequence as a member of the parental species, even though taxonomists would identify it as belonging to it. So, DNA barcoding might fail as a result of the unrecognition of taxonomical units corresponding to a population that is reproductively isolated and additionally if centered on a range of nucleotide changes (K) as a statistics within the hypothesis-based mostly approach. Nielsen and Matz (2006) have urged that a procedure that examines a number of nucleotide changes only between a query sequence and its best match within the information match in the database is not optimal. During this series for DNA barcoding, two ways, K-test and Bayesian check, are developed from a perspective of applied mathematics genetics. The performance of K-test faces drawback if some species were missing from the information and such behavior might lead to incorrect assignment of queries derived from these "unrecorded species." On the other hand, Bayesian theorem is used as an advantage upon the K- test in terms of accuracy and ability to face the negligence with more than one sequence per species in the database. However, this methodology is also with difficulties of significant phylogenetic assumptions and species level assumptions which are not always correct. Therefore, the convenience of a full Bayesian theorem might not eliminate the necessity for illation procedures with controlled frequentist (hypothesis-based) properties (Nielsen and Matz 2006). Still, DNA barcoding faces the problem to check the clear hypothesis meaning alternative of inappropriate or suboptimal analytical technique because of confusion on the objectives of the study.

4.6.5 Limitation of Distance-Based and Tree-Building Method Used in DNA Barcoding

In some reports, it is noted that DNA barcoding fails in the form of taxonomic approach because it does not recover correct species tree (Hebert and Gregory 2005; Will and Rubinoff 2004; Rubinoff et al. 2006a). Some criticism has arisen due to distance-based and character-based methods. Some reports have mentioned that the distance-based method should not be used for DNA barcoding, as it is a phenetic measure and is not appropriate for species identification (Casiraghi et al. 2010; Desalle 2006; Meyer and Paulay 2005). In the distance-based method, NJ tree acts as a standard part of the procedure for DNA barcoding (Casiraghi et al. 2010; Collins and Cruickshank 2013). But, there is a good documentation about the poor performance of NJ trees on the basis of trial and error and theoretical (Collins and Cruickshank 2013; Little 2011; Meier et al. 2006; Virgilio et al. 2010; Zhang et al. 2012b). The inappropriate use of NJ trees for identification can decrease the effectiveness of DNA barcoding. This will ensue either mtDNA paraphyly or misidentification of species independently. The NJ trees do not seem to be resolved by exploitation the other tree inference ways (Desalle 2006; Desalle 2007; DeSalle et al. 2005; Lowenstein et al. 2009; Rubinoff et al. 2006a; Taylor and Harris 2012; Austerlitz et al. 2009; Collins and Cruickshank 2013; Kerr et al. 2009; Little 2011; Little and Stevenson 2007; Lowenstein et al. 2009; Srivathsan and Meier 2012; Virgilio et al. 2010; Zhang et al. 2012b; Collins et al. 2012; Collins and Cruickshank 2013; Will et al. 2005). On the other hand, the character-based methodology are used to identify the nucleotide combinations (Collins and Cruickshank 2013). The character-based methodology have failed to break into the most stream of DNA barcoding (Savolainen et al. 2005). However, currently, DNA barcoding via tree-based approach did not stop at distance vs. character based approach. Avoiding any tree-building analysis due to its impression of inferring phylogenies and relationships with single-gene tree is well known as a problem of phylogenetics (DeSalle et al. 2005; Taylor and Harris 2012). Generally, the phylogenetic technique has been proposed as a data analysis in order to overcome the limitations of the threshold-based methodology in DNA barcoding. However, the application of these threshold-based approaches leads to some problem in a study on the relationship between DNA barcoding and molecular phylogeny. DNA barcoding is not a phylogenetic reconstruction. Still, these methods are being used along with the debate in phylogeny and identification in the area of DNA barcoding (Casiraghi et al. 2010; Moritz and Cicero 2004; Vogler and Monaghan 2007). The bootstrap resampling can further decrease the already low identification success rates associated with NJ trees (Brown et al. 2012; Collins and Cruickshank 2013; Fujita et al. 2012; Meyer and Paulay 2005; Monaghan et al. 2009; Puillandre et al. 2012; Virgilio et al. 2012; Zhang et al. 2012a). The use of bootstrap resampling in DNA barcoding studies creates confusion between species discovery and specimen identification. Bootstrapping in this situation also helps in addressing the problem with NJ trees, such as taxon-order bias and tied trees (Lowenstein et al. 2009; Meier et al. 2008). Use of bootstrap value as a cutoff for correct identification severely compromises the efficacy of a reference library and exacerbates the previously outlined weaknesses of using tree-based methods in general (Collins et al. 2012; Collins and Cruickshank 2013; Zhang et al. 2012b).

4.6.6 Limitation in Available Bioinformatics Tools and Algorithm

The biases occurred in methods used for the original cohort of DNA barcoding are being replicated by various studies and assisted by the analytical tools obtainable from the BOLD. A character-based tool, i.e., BLOG, has been made along with BOLD. But, presently it is available only on the Barcode of Life Data Portal (BDP) instead of various BOLD websites. The current popular methods could be a product of routine instead of wise selection. This means a systematic appraisal of taxa has not been capitalized by the barcoding movement. For DNA barcoding, easy-to-handle tools are required for species discrimination and identification. These tools use pairwise global alignment or alignment-free and automated selection of data partitions of an alignable group of samples (CBOL Plant Working Group 2009; Chu et al. 2009; Kress et al. 2009; Kuksa and Pavlovic 2009; Hollingsworth et al. 2011). Microinversions are common in noncoding regions leading to multiple groupings of samples.

4.6.7 *Absence of Effective Bioinformatics Pipeline for DNA Barcoding*

DNA barcoding is being used to recognize and identify the unknown species. Thus, despite its present limitations, the barcoding method provides a pipeline for the survey of biodiversity, a crucial task for prioritizing conservation efforts, given the present extinction crisis (Taylor and Harris 2012; Valentini et al. 2009). As mentioned earlier, there is a huge amount of sequence information stored in GenBank for which there are no voucher specimens, excluding this sequence from use as a barcode. During the DNA establishment of a barcoding reference library, there will be different unsampled taxa varying the depth of sample coverage for some markers. So, it is necessary to develop the bioinformatics framework having the access to select the sets of samples, directly comparable for a given set of markers. The integration of analytical approaches into a single easy-to-use workflow is required to provide comparable bioinformatics support for multi-marker barcoding in animals and plants (Hollingsworth et al. 2011; Bhargava and Sharma 2013).

4.7 Successful Uses of DNA Barcoding in Medicinal Plants

In many studies from 2003 to 2016, the results of DNA barcoding can provide accurate identification of many medicinal plant materials that are not morphologically distinguishable. DNA barcoding has found its applications in several areas like forensic science, biosecurity (Armstrong and Ball 2005), tracing of illegal trading of organisms (Galimberti et al. 2014), and pharmaceutical and herbal industries, among others (Gantait et al., 2014). When there is an insufficient morphological or anatomical data for the identification of a sample, a stretch of DNA sequence might be helpful in identifying a species. Samples with multiple fragments can now provide multiple species identification, giving a clear picture of habitat that offers a critical clue to the investigators (Ferri et al. 2015). DNA barcoding works with different identification fields and gives more accurate results of medicinal plant identification, i.e., DNA barcode identification with chemical analysis (Palhares et al. 2015) and next-generation sequencing (Shokralla et al. 2014). The plant materials are frequently encountered in criminal investigations but often overlooked as potential evidence. A forensic investigation that seeks to match evidence to a particular plant would require an updated database of samples. This requires the collection and genotyping of many samples from or near the crime scene. The law enforcement officers and attorneys are not very much familiar with the science of botany. So, the important plant-based evidence is often overlooked. Development of a robust DNA barcode database with highly authenticated sequence information will greatly contribute to the future of forensic botany (Ferri et al. 2009). Hallucinogenic compounds are pharmacological agents banned in most of the states or countries. They cause changes in perception, thought, emotion, and consciousness. Such kind of plants producing hallucinogens has been detected using DNA barcoding technique in some of the forensic studies (Murphy and Bola 2013; Ogata et al. 2013). Various DNA barcodes available in medicinal plants till date are listed below (Table 4.3).

Table 4.3 List of available DNA barcodes of medicinal plants

S.N.	Species	Family	DNA barcode region	Reference
1.	*Acanthopanacis cortex*	Araliaceae	ITS2	Han et al. (2016a)
2.	*Acanthopanacis cortex*	Araliaceae	ITS2	Zhao et al. (2015)
3.	*Aconitum*	Ranunculaceae	psbA-trnH	He et al. (2010)
4.	*Acori Tatarinowii Rhizoma*	Araceae	ITS2	Han et al. (2016a)
5.	*Andrographis paniculata*	Acanthaceae	rbcL	Osathanunkul et al. (2016)
6.	*Andrographis paniculata*	Acanthaceae	ITS2, rpoC1, trnH-psbA	Aziz et al. (2015)
7.	*Angelica* spp.	Apiaceae	matK, rbcL,ITS, ITS2, psbA-trnH	Yuan et al. (2015) and Techen et al. (2014)
8.	*Astragalus* spp.	Fabaceae	matK, rbcL, ITS	Xiao et al. (2011) and Techen et al. (2014)
9.	*Boerhavia* spp.	Nyctaginaceae	ITS, ITS2,	Selvaraj et al. (2012) and Techen et al. (2014)
10.	*Brugmansia*, *Datura*	Solanaceae	ITS2	Wu et al. (2015)
11.	*Bupleuri radix*	Apiaceae	ITS2	Han et al. (2016a)
12.	*Butea superb*	Fabaceae	matK	Wiriyakarun et al. (2013)
13.	*Cassia* species	Fabaceae	rbcL	Sheth and Thaker (2015)
14.	*Centella asiatica*	Apiaceae	ITS2, rpoC1, trnH-psbA	Aziz et al. (2015)
15.	*Citrus* spp.	Rutaceae	matk	Penjor et al. (2013)
16.	*Cleome*	Cleomaceae	matK, rbcL, ITS1	Tamboli et al. (2016)
17.	*Clinacanthus nutans*	Acanthaceae	ITS2, rpoC1, trnH-psbA	Aziz et al. (2015)
18.	*Cosmos caudatus*	Asteraceae	ITS2, rpoC1, trnH-psbA	Aziz et al. (2015)
19.	*Cymbidium*	Orchidaceae	ITS2	Sharma et al. (2012)
20.	*Cynanchum auriculatum*	Apocynaceae	trnL-F	Han et al. (2016b)
21.	*Cynanchum wilfordii*	Asclepiadaceae	trnL-F	Han et al. (2016b)
22.	*Dalbergiae Odoriferae* Lignum	Fabaceae	ITS2	Han et al. (2016a)
23.	*Dendrobium* spp.	Orchidaceae	psbA-trnH	Yao et al. (2009)

(continued)

Table 4.3 (continued)

S.N.	Species	Family	DNA barcode region	Reference
24.	*Dipsacus* spp.	Caprifoliaceae	ITS	Vickruck et al. (2011) and Techen et al. (2014)
25.	*Gentiana*	Gentianaceae	matK + ITS	Liu et al. (2016a)
26.	*Ginseng* genus	Araliaceae	matK, rbcL, ITS, psbA-trnH,	Dong et al. (2014),
			rpoB, rpoC1, ITS2	Wallace et al. (2012) and Liu et al. (2016b)
27.	*Ginseng radix*	Araliaceae	ITS2	Han et al. (2016b)
28.	*Hedyotis diffusa* Willd.	Rubiaceae	ITS	Sun et al. (2011)
29.	*Hypericum* spp.	Hypericaceae	ITS	Newmaster et al. (2013)
30.	*Illicium*	Schisandraceae	ITS + trnH-psbA	Zhang et al. (2015a)
31.	*Inulae flos*	Compositae	ITS2	Han et al. (2016b)
32.	*Justicia gendarussa*	Acanthaceae	ITS2, rpoC1, trnH-psbA	Aziz et al. (2015)
33.	*Lonicera* spp.	Caprifoliaceae	matK, rbcL, ITS, psbA-trnH, trnL-F	Techen et al. (2014)
34.	*Lonicerae japonicae* Flos	Caprifoliaceae	ITS2	Han et al. (2016b)
35.	*Meconopsis* spp.	Papaveraceae	ITS	Techen et al. (2014)
36.	*Mentha aquatica* L.	Lamiaceae	rbcL, matK, trnH-psbA, rpoB	De Mattia et al. (2011)
37.	*Mentha spicata* L.	Lamiaceae	rbcL, matK, trnH-psbA, rpoB	De Mattia et al. (2011)
38.	*Mucuna collettii*	Fabaceae	matK	Wiriyakarun et al. (2013)
39.	*Murraya koenigii*	Rutaceae	ITS2, rpoC1, trnH-psbA	Aziz et al. (2015)
40.	*Ochradenus* spp.	Resedaceae	ITS, rpoB, rpoC1	Techen et al. (2014)
41.	*Ocimum gratissimum*	Lamiaceae	rbcL, matK, trnH-psbA, rpoB	De Mattia et al. (2011)
42.	*Ocimum basilicum* L.	Lamiaceae	rbcL, matK, trnH-psbA, rpoB	De Mattia et al. (2011)

(continued)

Table 4.3 (continued)

S.N.	Species	Family	DNA barcode region	Reference
43.	*Ocimum tenuiflorum* L.	Lamiaceae	rbcL, matK, trnH-psbA, rpoB	De Mattia et al. (2011)
44.	*Origanum heracleoticum* L.	Lamiaceae	rbcL, matK, trnH-psbA, rpoB	De Mattia et al. (2011)
45.	*Origanum majorana* L.	Lamiaceae	rbcL, matK, trnH-psbA, rpoB	De Mattia et al. (2011)
46.	*Orthosiphon stamineus*	Lamiaceae	ITS2, rpoC1, trnH-psbA	Aziz et al. (2015)
47.	*Paris* spp.	Melanthiaceae	ITS2	Zhu et al. (2010)
48.	*Persicaria odorata*	Polygonaceae	ITS2, rpoC1, trnH-psbA	Aziz et al. (2015)
49.	*Phyllanthus niruri*	Phyllanthaceae	ITS2, rpoC1, trnH-psbA	Aziz et al. (2015)
50.	*Phyllanthus* spp.	Phyllanthaceae	psbA-trnH	Srirama et al. (2010)
51.	*Pinelliae Tuber*,	Araceae	matK, rbcL	Moon et al. (2016)
	Arisaematis Rhizoma			
52.	*Piper betel*	Piperaceae	ITS2, rpoC1, trnH-psbA	Aziz et al. (2015)
53.	*Piper sarmentosum*	Piperaceae	ITS2, rpoC1, trnH-psbA	Aziz et al. (2015)
54.	*Plectranthus asirensis*	Lamiaceae	rps16, rpoB	Al-Qurainy et al. (2014)
55.	*Polygonum multiflorum*	Polygonaceae	trnL-F	Han et al. (2016a)
56.	*Pueraria candollei*	Fabaceae	matK	Wiriyakarun et al. (2013)
57.	*Radix Astragali*	Fabaceae	ITS	Zheng et al. (2014)
58.	*Radix Rubi Parvifolii*	Gentianaceae	ITS2	Han et al. (2016b)
59.	*Rehmannia* spp.		ITS	Techen et al. (2014)
60.	*Rhodiola*	Crassulaceae	ITS	Zhang et al. (2015b)
61.	*Rhododendron*	Ericaceae	psbA-trnH	Chen et al. (2012)
62.	*Rhododendron* spp.	Ericaceae	matK, rbcL,ITS, ITS2, psbA-trnH	Yan et al. (2015), Chen et al. (2012) and Tsai et al. (2012)
63.	*Rhubarb*	Polygonaceae	matK	Xu et al. (2013)
64.	*Rosmarinus officinalis* L.	Lamiaceae	rbcL, matK, trnH-psbA, rpoB	De Mattia et al. (2011)
65.	*Rubus* spp.	Rosaceae	ITS, psbA-trnH, trnL-F	Newmaster et al. (2013)

(continued)

Table 4.3 (continued)

S.N.	Species	Family	DNA barcode region	Reference
66.	*Ruta* spp.	Rutaceae	ITS, rpoB, rpoC1	Al-Qurainy et al. (2011)
67.	*Sabia* spp.	Sabiaceae	matK, rbcL, psbA-trnH	Techen et al. (2014)
68.	*Salvia divinorum*	Lamiaceae	trnL	Murphy and Bola (2013)
69.	*Salvia officinalis* L.	Lamiaceae	rbcL, matK, trnH-psbA, rpoB	De Mattia et al. (2011)
70.	*Salvia rutilans*	Lamiaceae	rbcL, matK, trnH-psbA, rpoB	De Mattia et al. (2011)
71.	*Salvia sclarea*	Lamiaceae	rbcL, matK, trnH-psbA, rpoB	De Mattia et al. (2011)
72.	*Salvia uliginosa*	Lamiaceae	rbcL, matK, trnH-psbA, rpoB	De Mattia et al. (2011)
73.	*Sambucus chinensis*	Acanthaceae	ITS2, rpoC1, trnH-psbA	Aziz et al. (2015)
74.	*Isatis indigotica*	Cruciferae	ITS2	Chen et al. (2014)
75.	*Scutellaria baicalensis*	Lamiaceae	psbA-trnH	Guo et al. (2011)
76	*Scutellaria* spp.	Lamiaceae	matK, rbcL, psbA-trnH	Techen et al. (2014)
77.	*Senna* spp.	Fabaceae	psbA-trnH	Pansa Monkheang (2011)
78.	*Smilax* spp.	Smilacaceae	psbA-trnH	Techen et al. (2014)
79.	*Solanum* spp.	Solanaceae	matK, rbcL, ITS, psbA-trnH, trnL-F	Zhang et al. (2013) and Techen et al. (2014)
80.	*Thymus vulgaris* L.	Lamiaceae	rbcL, matK, trnH-psbA, rpoB	De Mattia et al. (2011)
81.	*Tulipa edulis*	Liliaceae	matK	Ma et al. (2014)
82.	*Uncaria*	Rubiaceae	ITS2	Zhang et al. (2015c)
83.	*Uyghur*	Apiaceae	ITS2	Fan et al. (2015)
84.	*Vitex* spp.	Lamiaceae	matK	Phoolcharoen and Sukrong (2013)

4.8 Conclusions

Over the past 12 years, DNA barcoding has been attracting a lot of interest all over the world. Researchers working in this field are busy in finding a more superior and desirable universal DNA barcode for an efficient conservation of the biodiversity. Since a major problem of barcoding lies in the case of plants, the research carried out so far in this area has been reviewed including the futuristic approaches. In the present chapter, various candidate markers used in plants and a number of barcoding reports have been summarized. Although the CBOL proposed seven candidate barcodes belonging to the plastid region, the proposed supplementary loci, i.e., nuclear-transcribed spacer regions ITS1 and ITS2, have a number of GenBank submissions of their respective sequences owing to its easy amplification due to high copy number. rbcL and matK (both plastid genes) come next followed by 18S rRNA (nuclear structural RNA), trnL-F (intron + IGS), and trnH-psbA (IGS), respectively. Since higher substitution rates are observed in plant nuclear genes than plastid genes, ITS is more in use and also acts as a supplementary marker. But once the choice of the locus is made, the approach of single-locus, multilocus, or tiered needs consideration. Based on the literature review, it can be inferred that multilocus and tiered approaches resulted in higher success rates than the single-locus approach if proper combinations of loci and selection of loci for each tier are done carefully.

Acknowledgment The financial assistance under BTISnet program of DBT, New Delhi, is gratefully acknowledged.

References

Acs Z, Challis RJ, Bihari P, Blaxter M, Hayward A, Melika G, Csóka G, Pénzes Z, Pujade-Villar J, Nieves-Aldrey JL, Schönrogge K, Stone GN (2010) Phylogeny and DNA barcoding of inquiline oak gall wasps (Hymenoptera: Cynipidae) of the Western Palaearctic. Mol Phylogenet Evol 55:210–225. doi:10.1016/j.ympev.2009.12.004

Ajmal AM, Gyulai G, Hidvégi N, Kerti B, Al Hemaid FMA, Pandey AK, Lee J (2014) The changing epitome of species identification – DNA barcoding. Saudi J Biol Sci 21:204–231. doi:10.1016/j.sjbs.2014.03.003

Al-Qurainy F, Khan S, Tarroum M, Al-Hemaid FM, Ali MA (2011) Molecular authentication of the medicinal herb Rutagraveolens (Rutaceae) and an adulterant using nuclear and chloroplast DNA markers. Genet Mol Res 10:2806–2816. doi:10.4238/2011.November.10.3

Al-Qurainy F, Khan S, Nadeem M, Tarroum M, Al-Ameri A (2014) Selection of DNA barcoding loci and phylogenetic study of a medicinal and endemic plant, plectranthus asirensis J.R.I. Wood from Saudi Arabia. Genet Mol Res 13:6184–6190. doi:10.4238/2014.August.7.31

Álvarez I, Wendel JF (2003) Ribosomal ITS sequences and plant phylogenetic inference. Mol Phylogenet Evol 29:417–434

Aneesh T, Hisham M, Sekhar M, Madhu M, Deepa T (2009) International market scenario of traditional Indian herbal drugs-India declining. Int J Green Pharm 3:184–190. doi:10.4103/0973-8258.56271

Armstrong KF, Ball SL (2005) DNA barcodes for biosecurity: invasive species identification. Philos Trans R Soc B Biol Sci 360:1813–1823. doi:10.1098/rstb.2005.1713

Austerlitz F, David O, Schaeffer B, Bleakley K, Olteanu M, Leblois R, Veuille M, Laredo C (2009) DNA barcode analysis: a comparison of phylogenetic and statistical classification methods. BMC Bioinformatics 10(Suppl 14):S10. doi:10.1186/1471-2105-10-S14-S10

Aziz NAA, Ahmad MI, Naim DM (2015) Molecular DNA identification of medicinal plants used by traditional healers in Malaysia. Genet Mol Res 14:15937–15947. doi:10.4238/2015. December.7.5

Bailey CD, Carr TG, Harris SA, Hughes CE (2003) Characterization of angiosperm nrDNA polymorphism, paralogy, and pseudogenes. Mol Phylogenet Evol 29:435–455

Baldwin BG (1992) Phylogenetic utility of the internal transcribed spacers of nuclear ribosomal DNA in plants: An example from the Compositae. Mol Phylogenet Evol 1:3–16

Ballard JWO, Whitlock MC (2004) The incomplete natural history of mitochondria. Mol Ecol 13:729–744

Beck D, Settles M, Foster JA (2011) OTUbase: an R infrastructure package for operational taxonomic unit data. Bioinformatics 27:1700–1701. doi:10.1093/bioinformatics/btr196

Becker S, Hanner R, Steinke D (2011) Five years of FISH-BOL: brief status report. Mitochondrial DNA 22(Suppl 1):3–9. doi:10.3109/19401736.2010.535528

Bensasson D, Zhang DX, Hartl DL, Hewitt GM (2001) Mitochondrial pseudogenes: evolution's misplaced witnesses. Trends Ecol Evol 16:314–321

Bertolazzi P, Felici G, Weitschek E (2009) Learning to classify species with barcodes. BMC Bioinforma 10(Suppl 14):S7. doi:10.1186/1471-2105-10-S14-S7

Bhargava M, Sharma A (2013) DNA barcoding in plants: evolution and applications of *in silico* approaches and resources. Mol Phylogenet Evol 67:631–641. doi:10.1016/j.ympev.2013.03.002

Biswas K, Biswas R (2014) DNA molecular markers based authentication of herbal drugs – a review. IJPRS 3:581–593

Brown SDJ, Collins RA, Boyer S, Lefort MC, Malumbres-Olarte J, Vink CJ, Cruickshank RH (2012) Spider: an R package for the analysis of species identity and evolution, with particular reference to DNA barcoding. Mol Ecol Resour 12:562–565. doi:10.1111/j.1755-0998.2011.03108.x

Campbell NJ, Barker SC (1999) The novel mitochondrial gene arrangement of the cattle tick, Boophilus microplus: fivefold tandem repetition of a coding region. Mol Biol Evol 16:732–740

Casiraghi M, Labra M, Ferri E, Galimberti A, De Mattia F (2010) DNA barcoding: a six-question tour to improve users' awareness about the method. Brief Bioinform 11:440–453. doi:10.1093/bib/bbq003

CBOL Plant Working Group (2009) A DNA barcode for land plants. Proc Natl Acad Sci U S A 106:12794–12797. doi:10.1073/pnas.0905845106

Chan K (2003) Some aspects of toxic contaminants in herbal medicines. Chemosphere 52:1361–1371. doi:10.1016/S0045-6535(03)00471-5

Chen S, Yao H, Han J, Liu C, Song J, Shi L, Zhu Y, Ma X, Gao T, Pang X, Luo K, Li Y, Li X, Jia X, Lin Y, Leon C (2010) Validation of the ITS2 region as a novel DNA barcode for identifying medicinal plant species. PLoS One 5, e8613. doi:10.1371/journal.pone.0008613

Chen K, Liu Y, Zhang L, Liu Z, Luo K, Chen S (2012) Species identification of Rhododendron (Ericaceae) using the chloroplast deoxyribonucleic acid PsbA-trnH genetic marker. Pharmacogn Mag 8:1–29. doi:10.4103/0973-1296.93311

Chen S, Zhu Z, Ma H, Yang J, Guo Q (2014) DNA barcodes for discriminating the medicinal plant Isatisindigotica Fort. (Cruciferae) and its adulterants. Biochem Syst Ecol 57:287–292. doi:10.1016/j.bse.2014.08.007

China Plant BOL Group (2011) Comparative analysis of a large dataset indicates that internal transcribed spacer (ITS) should be incorporated into the core barcode for seed plants. Proc Natl Acad Sci U S A 108:19641–19646. doi:10.1073/pnas.1104551108

Chu KH, Xu M, Li CP (2009) Rapid DNA barcoding analysis of large datasets using the composition vector method. BMC Bioinforma 10(Suppl 14):S8. doi:10.1186/1471-2105-10-S14-S8

Collins RA, Cruickshank RH (2013) The seven deadly sins of DNA barcoding. Mol Ecol Resour 13:969–975. doi:10.1111/1755-0998.12046

Collins RA, Armstrong KF, Meier R, Yi Y, Brown SDJ, Cruickshank RH, Keeling S, Johnston C (2012) Barcoding and border biosecurity: identifying cyprinid fishes in the aquarium trade. PLoS One 7, e28381. doi:10.1371/journal.pone.0028381

Costello MJ, May RM, Stork NE (2013) Can we name earth's species before they go extinct? Science 339:413. doi:10.1126/science.1237381

Costion C, Ford A, Cross H, Crayn D, Harrington M, Lowe A (2011) Plant DNA barcodes can accurately estimate species richness in poorly known floras. PLoS One 6, e26841. doi:10.1371/journal.pone.0026841

Cowan RS, Fay MF (2012) Challenges in the DNA barcoding of plant material. Methods Mol Biol Clifton NJ 862:23–33. doi:10.1007/978-1-61779-609-8_3

de Groot GA, During HJ, Maas JW, Schneider H, Vogel JC, Erkens RH (2011) Use of rbcL and trnL-F as a two-locus DNA barcode for identification of NW-European ferns: an ecological perspective. PLoS One 6(1), e16371. doi:10.1371/journal.pone.0016371

De Mattia F, Bruni I, Galimberti A, Cattaneo F, Casiraghi M, Labra M (2011) A comparative study of different DNA barcoding markers for the identification of some members of Lamiacaea. Food Res Int 44:693–702. doi:10.1016/j.foodres.2010.12.032

Desalle R (2006) Species discovery versus species identification in DNA barcoding efforts: response to Rubinoff. Conserv Biol J Soc Conserv Biol 20:1545–1547. doi:10.1111/j.1523-1739.2006.00543.x

Desalle R (2007) Phenetic and DNA taxonomy; a comment on Waugh. BioEssays News Rev Mol Cell Dev Biol 29:1289–1290. doi:10.1002/bies.20667

DeSalle R, Egan MG, Siddall M (2005) The unholy trinity: taxonomy, species delimitation and DNA barcoding. Philos Trans R Soc B 360:1905–1916. doi:10.1098/rstb.2005.1722

Devey DS, Chase MW, Clarkson JJ (2009) A stuttering start to plant DNA barcoding: microsatellites present a previously overlooked problem in non-coding plastid regions. Taxon 58:7–15

Dong W, Liu H, Xu C, Zuo Y, Chen Z, Zhou S (2014) A chloroplast genomic strategy for designing taxon specific DNA mini-barcodes: a case study on ginsengs. BMC Genet 15:138. doi:10.1186/s12863-014-0138-z

Efferth T, Greten HJ (2012) Medicinal and aromatic plant research in the 21st century. Med Aromat Plant 1e:110. doi:10.4172/2167-0412.1000e110

Elliott TL, Jonathan Davies T (2014) Challenges to barcoding an entire flora. Mol Ecol Resour 14(5):883–891. doi:10.1111/1755-0998.12277

Fan C, Li X, Zhu J, Song J, Yao H (2015) Endangered Uyghur medicinal plant *Ferula* identification through the second internal transcribed spacer. Evid Based Complement Alternat Med 2015:1–6. doi:10.1155/2015/479879

Fazekas AJ, Burgess KS, Kesanakurti PR, Graham SW, Newmaster SG, Husband BC, Percy DM, Hajibabaei M, Barrett SC (2008) Multiple multilocus DNA barcodes from the plastid genome discriminate plant species equally well. PLoS One 7, e2802. doi:10.1371/journal.pone.0002802

Ferri G, Alù M, Corradini B, Beduschi G (2009) Forensic botany: species identification of botanical trace evidence using a multigene barcoding approach. Int J Legal Med 123:395–401

Ferri G, Corradini B, Ferrari F, Santunione AL, Palazzoli F, Alu M (2015) Forensic botany II, DNA barcode for land plants: which markers after the international agreement? Forensic Sci Int Genet 15:131–136. doi:10.1016/j.fsigen.2014.10.005

Ficetola GF, Coissac E, Zundel S, Riaz T, Shehzad W, Bessière J, Taberlet P, Pompanon F (2010) An *in silico* approach for the evaluation of DNA barcodes. BMC Genomics 11:434. doi:10.1186/1471-2164-11-434

Fišer Pečnikar Ž, Buzan EV (2014) 20 years since the introduction of DNA barcoding: from theory to application. J Appl Genet 55:43–52. doi:10.1007/s13353-013-0180-y

Frézal L, Leblois R (2008) Four years of DNA barcoding: current advances and prospects. Infect Genet Evol J Mol Epidemiol Evol Genet Infect Dis 8:727–736. doi:10.1016/j.meegid.2008.05.005

Fujita MK, Leaché AD, Burbrink FT, McGuire JA, Moritz C (2012) Coalescent-based species delimitation in an integrative taxonomy. Trends Ecol Evol 27:480–488. doi:10.1016/j.tree.2012.04.012

Galimberti A, Labra M, Sandionigi A, Bruno A, Mezzasalma V, De Mattia F (2014) DNA barcoding for minor crops and food traceability. Adv Agric 2014:1–8. doi:10.1155/2014/831875

Gantait S, Debnath S, Nasim Ali M (2014) Genomic profile of the plants with pharmaceutical value. 3 Biotech 4:563–578. doi:10.1007/s13205-014-0218-9

Gao T, Yao H, Song J, Zhu Y, Liu C, Chen S (2010a) Evaluating the feasibility of using candidate DNA barcodes in discriminating species of the large Asteraceae family. BMC Evol Biol 10:324. doi:10.1186/1471-2148-10-324

Gao T, Yao H, Song J, Liu C, Zhu Y, Ma X, Pang X, Xu H, Chen S (2010b) Identification of medicinal plants in the family Fabaceae using a potential DNA barcode ITS2. J Ethnopharmacol 130:116–121. doi:10.1016/j.jep.2010.04.026

Ghosh S, Mahadani P, Sharma G (2013) Identification of ethnomedicinal plants (Rauvolfioideae: Apocynaceae) through DNA barcoding from northeast India. Pharmacogn Mag 9:255. doi:10.4103/0973-1296.113284

Gonzalez MA, Baraloto C, Engel J, Mori SA, Pétronelli P, Riéra B, Roger A, Thébaud C, Chave J (2009) Identification of amazonian trees with DNA barcodes. PLoS One 4, e7483. doi:10.1371/journal.pone.0007483

Guo X, Wang X, Su W, Zhang G, Zhou R (2011) DNA barcodes for discriminating the medicinal plant Scutellaria baicalensis (Lamiaceae) and its adulterants. Biol Pharm Bull 34:1198–1203

Hajibabaei M, Smith MA, Janzen DH, Rodriguez JJ, Whitfield JB, Hebert PDN (2006) A minimalist barcode can identify a specimen whose DNA is degraded: BARCODING. Mol Ecol 6:959–964. doi:10.1111/j.1471-8286.2006.01470.x

Han EH, Cho K, Goo Y, Kim M, Shin YW, Kim YH, Lee SW (2016a) Development of molecular markers, based on chloroplast and ribosomal DNA regions, to discriminate three popular medicinal plant species, Cynanchumwilfordii, Cynanchumauriculatum, and Polygonummultiflorum. Mol Biol Rep 43:323–332. doi:10.1007/s11033-016-3959-1

Han J, Pang X, Liao B, Yao H, Song J, Chen S (2016b) An authenticity survey of herbal medicines from markets in China using DNA barcoding. Sci Rep 6:18723. doi:10.1038/srep18723

Hao DC, Chen SL, Xiao PG (2010) Sequence characteristics and divergent evolution of the chloroplast psbA-trnH noncoding region in gymnosperms. J Appl Genet 51:259–273

He J, Wong KL, Shaw PC, Wang H, Li DZ (2010) Identification of the medicinal plants in Aconitum L. by DNA barcoding technique. Planta Med 76:1622–1628. doi:10.1055/s-0029-1240967

Hebert PDN, Gregory TR (2005) The promise of DNA barcoding for taxonomy. Syst Biol 54:852–859. doi:10.1080/10635150500354886

Hebert PDN, Cywinska A, Ball SL, deWaard JR (2003a) Biological identifications through DNA barcodes. Proc R Soc B Biol Sci 270:313–321. doi:10.1098/rspb.2002.2218

Hebert PDN, Ratnasingham S, deWaard JR (2003b) Barcoding animal life: cytochrome c oxidase subunit 1 divergences among closely related species. Proc Biol Sci 270(Suppl 1):96–99. doi:10.1098/rsbl.2003.0025

Hebert PDN, Penton EH, Burns JM, Janzen DH, Hallwachs W (2004) Ten species in one: DNA barcoding reveals cryptic species in the neotropical skipper butterfly *Astraptes fulgerator*. Proc Natl Acad Sci U S A 101:14812–14817. doi:10.1073/pnas.0406166101

Hickerson MJ, Meyer CP, Moritz C (2006) DNA barcoding will often fail to discover new animal species over broad parameter space. Syst Biol 55:729–739. doi:10.1080/10635150600969898

Hollingsworth ML, Clark AA, Forrest LL, Richardson J, Pennington RT, Long DG, Cowan R, Chase MW, Gaudeul M, Hollingsworth PM (2009) Selecting barcoding loci for plants: evaluation of seven candidate loci with species-level sampling in three divergent groups of land plants. Mol Ecol Resour 9:439–457

Hollingsworth PM, Graham SW, Little DP (2011) Choosing and using a plant DNA barcode. PLoS One 6:e19254

Hurst GDD, Jiggins FM (2005) Problems with mitochondrial DNA as a marker in population, phylogeographic and phylogenetic studies: the effects of inherited symbionts. Proc Biol Sci 272:1525–1534. doi:10.1098/rspb.2005.3056

Jeanson ML, Labat JN, Little DP (2011) DNA barcoding: a new tool for palm taxonomists? Annals Bot 108:1445–1451

Jones M, Ghoorah A, Blaxter M (2011) JMOTU and taxonerator: turning DNA barcode sequences into annotated operational taxonomic units. PLoS One 6, e19259. doi:10.1371/journal.pone.0019259

Kerr KC, Birks SM, Kalyakin MV, Red'kin YA, Koblik EA, Hebert P (2009) Filling the gap – COI barcode resolution in eastern Palearctic birds. Front Zool 6:29. doi:10.1186/1742-9994-6-29

Kress WJ, Erickson DL (2007) A two-locus global DNA barcode for land plants: the coding rbcL gene complements the non-coding trnH-psbA spacer region. PLoS One 2:e508

Kress WJ, Erickson DL (2008) DNA barcodes: genes, genomics, and bioinformatics. Proc Natl Acad Sci U S A 105:2761–2762

Kress JW, Wurdack JK, Zimmer AE, Weigt LA, Janzen DH (2005) Use of DNA barcodes to identify flowering plants. Proc Natl Acad Sci U S A 102:8369–8374. doi:10.1073/pnas.0503123102

Kress WJ, Erickson DL, Jones FA (2009) Plant DNA barcodes and a community phylogeny of a tropical forest dynamics plot in Panama. Proc Natl Acad Sci U S A 106:18621–18626

Kuksa P, Pavlovic V (2009) Efficient alignment-free DNA barcode analytics. BMC Bioinforma 10(Suppl 14):S9. doi:10.1186/1471-2105-10-S14-S9

Kumar S, Carlsen T, Mevik BH, Enger P, Blaalid R, Shalchian-Tabrizi K, Kauserud H (2011) CLOTU: an online pipeline for processing and clustering of 454 amplicon reads into OTUs followed by taxonomic annotation. BMC Bioinforma 12:182. doi:10.1186/1471-2105-12-182

Lahaye R, van der Bank M, Bogarin D, Warner J, Pupulin F, Gigot G, Maurin O, Duthoit S, Barraclough TG, Savolainen V (2008) DNA barcoding the floras of biodiversity hotspots. Proc Natl Acad Sci U S A 105:2923–2928. doi:10.1073/pnas.0709936105

Li FW, Tan BC, Buchbender V, Moran RC, Rouhan G, Wang CN, Quandt D (2009) Identifying a mysterious aquatic fern gametophyte. Plant Syst Evol 281:77–86. doi:10.1007/s00606-009-0188-2

Li CP, Yu ZG, Han GS, Chu KH (2012) Analyzing multi-locus plant barcoding datasets with a composition vector method based on adjustable weighted distance. PLoS One 7:e42154

Li X, Yang Y, Henry RJ, Rossetto M, Wang Y, Chen S (2015) Plant DNA barcoding: from gene to genome. Biol Rev Camb Philos Soc 90:157–166. doi:10.1111/brv.12104

Little DP (2010) A unified index of sequence quality and contig overlap for DNA barcoding. Bioinformatics 26:2780–2781. doi:10.1093/bioinformatics/btq507

Little DP (2011) DNA barcode sequence identification incorporating taxonomic hierarchy and within taxon variability. PLoS One 6:e20552. doi:10.1371/journal.pone.0020552

Little DP, Stevenson DW (2007) A comparison of algorithms for the identification of specimens using DNA barcodes: examples from gymnosperms. Cladistics 23:1–21. doi:10.1111/j.1096-0031.2006.00126.x

Liu C, Liang D, Gao T, Pang X, Song J, Yao H, Han J, Liu Z, Guan X, Jiang K, Li H, Chen S (2011) PTIGS-IdIt, a system for species identification by DNA sequences of the psbA-trnH intergenic spacer region. BMC Bioinforma 12(Suppl 13):S4. doi:10.1186/1471-2105-12-S13-S4

Liu Y, Zhang L, Liu Z, Luo K, Chen S, Chen K (2012a) Species identification of Rhododendron (Ericaceae) using the chloroplast deoxyribonucleic acid PsbA-trnH genetic marker. Pharmacogn Mag 8:29–36. doi:10.4103/0973-1296.93311

Liu C, Shi L, Xu X, Li H, Xing H, Liang D, Jiang K, Pang X, Song J, Chen S (2012b) DNA barcode goes two-dimensions: DNA QR code web server. PLoS One 7:e35146. doi:10.1371/journal.pone.0035146

Liu J, Yan HF, Ge XJ (2016a) The use of DNA barcoding on recently diverged species in the genus Gentiana (Gentianaceae) in China. PLoS One 11:e0153008. doi:10.1371/journal.pone.0153008

Liu Y, Wang X, Wang L, Chen X, Pang X, Han J (2016b) A nucleotide signature for the identification of American ginseng and its products. Front Plant Sci 7:1–319. doi:10.3389/fpls.2016.00319
Lou SK, Wong KL, Li M, But P, Tsui S, Shaw PC (2010) An integrated web medicinal materials DNA database: MMDBD (Medicinal Materials DNA Barcode Database). BMC Genomics 11:402. doi:10.1186/1471-2164-11-402
Lowenstein JH, Amato G, Kolokotronis SO (2009) The real maccoyii: identifying tuna sushi with DNA barcodes – contrasting characteristic attributes and genetic distances. PLoS One 4:e7866. doi:10.1371/journal.pone.0007866
Ma H, Zhu Z, Zhang X, Miao Y, Guo Q (2014) Species identification of the medicinal plant Tulipa edulis (Liliaceae) by DNA barcode marker. Biochem Syst Ecol 55:362–368. doi:10.1016/j.bse.2014.03.038
Machado CA, Hey J (2003) The causes of phylogenetic conflict in a classic Drosophila species group. Proc Biol Sci 270:1193–1202. doi:10.1098/rspb.2003.2333
Magnacca KN, Brown MJF (2010) Mitochondrial heteroplasmy and DNA barcoding in Hawaiian Hylaeus (Nesoprosopis) bees (Hymenoptera: Colletidae). BMC Evol Biol 10:174. doi:10.1186/1471-2148-10-174
Matz MV, Nielsen R (2005) A likelihood ratio test for species membership based on DNA sequence data. Philos Trans R Soc B 360:1969–1974. doi:10.1098/rstb.2005.1728
Meier R, Shiyang K, Vaidya G, Ng PKL (2006) DNA barcoding and taxonomy in Diptera: a tale of high intraspecific variability and low identification success. Syst Biol 55:715–728. doi:10.1080/10635150600969864
Meier R, Zhang G, Ali F (2008) The use of mean instead of smallest interspecific distances exaggerates the size of the "barcoding gap" and leads to misidentification. Syst Biol 57:809–813. doi:10.1080/10635150802406343
Meyer CP, Paulay G (2005) DNA barcoding: error rates based on comprehensive sampling. PLoS Biol 3, e422. doi:10.1371/journal.pbio.0030422
Mishra P, Kumar A, Nagireddy A, Mani DN, Shukla AK, Tiwari R, Sundaresan V (2015) DNA barcoding: an efficient tool to overcome authentication challenges in the herbal market. Plant Biotechnol 14(1):8–21. doi:10.1111/pbi.12419
Monaghan MT, Wild R, Elliot M, Fujisawa T, Balke M, Inward DJG, Lees DC, Ranaivosolo R, Eggleton P, Barraclough TG, Vogler AP (2009) Accelerated species inventory on Madagascar using coalescent-based models of species delineation. Syst Biol 58:298–311. doi:10.1093/sysbio/syp027
Moon BC, Kim WJ, Ji Y, Lee YM, Kang YM, Choi G (2016) Molecular identification of the traditional herbal medicines, Arisaematis Rhizoma and Pinelliae tuber and common adulterants via universal DNA barcode sequences. Genet Mol Res 15. doi:10.4238/gmr.15017064
Moritz C, Cicero C (2004) DNA barcoding: promise and pitfalls. PLoS Biol 2:e354. doi:10.1371/journal.pbio.0020354
Moulton MJ, Song H, Whiting MF (2010) Assessing the effects of primer specificity on eliminating numt coamplification in DNA barcoding: a case study from orthoptera (Arthropoda: Insecta). Mol Ecol Resour 10:615–627. doi:10.1111/j.1755-0998.2009.02823.x
Muellner AN, Schaefer H, Lahaye R (2011) Evaluation of candidate DNA barcoding loci for economically important timber species of the mahogany family (Meliaceae). Mol Ecol Resour 11:450–460
Murphy TM, Bola G (2013) DNA identification of Salvia divinorum samples. Forensic Sci Int Genet 7:189–193
Newmaster SG, Fazekas AJ, Ragupathy S (2006) DNA barcoding in land plants: evaluation of rbcL in a multigene tiered approach. Can J Bot 84:335–341
Newmaster SG, Fazekas AJ, Steeves RAD, Janovec J (2008) Testing candidate plant barcode regions in the Myristicaceae. Mol Ecol Res 8:480–490
Newmaster SG, Grguric M, Shanmughanandhan D, Ramalingam S, Ragupathy S (2013) DNA barcoding detects contamination and substitution in north american herbal products. BMC Med 11:1–222. doi:10.1186/1741-7015-11-222

Nielsen R, Matz M (2006) Statistical approaches for DNA barcoding. Syst Biol 55:162–169. doi:10.1080/10635150500431239

Ogata J, Uchiyama N, Kikura-Hanajiri R, Goda Y (2013) DNA sequence analyses of blended herbal products including synthetic cannabinoids as designer drugs. Forensic Sci Int 227:33–41. doi:10.1016/j.forsciint.2012.09.006

Osathanunkul M, Suwannapoom C, Khamyong N, Pintakum D, Lamphun SN, Triwitayakorn K, Osathanunkul K, Madesis P (2016) Hybrid analysis (barcode-high resolution melting) for authentication of Thai herbal products, Andrographispaniculata (Burm.f.) Wall.ex Nees. Pharmacogn Mag 12:71–75. doi:10.4103/0973-1296.176112

Palhares RM, Gonçalves Drummond M, Dos Santos AFBB, Pereira Cosenza G, das GraçasLinsBrandão M, Oliveira G (2015) Medicinal plants recommended by the world health organization: DNA barcode identification associated with chemical analyses guarantees their quality. PLoS One 10:e0127866. doi:10.1371/journal.pone.0127866

Pang X, Luo H, Sun C (2012) Assessing the potential of candidate DNA barcodes for identifying non-flowering seed plants. Plant Biol (Stuttg) 14:839–844

Pansa M (2011) Species diversity, usages, molecular markers and barcode of medicinal Senna species (Fabaceae, Caesalpinioideae) in Thailand. J Med Plants Res 5:6173. doi:10.5897/JMPR11.1075

Parveen I, Singh HK, Raghuvanshi S, Pradhan UC, Babbar SB (2012) DNA barcoding of endangered Indian *Paphiopedilum* species: DNA barcoding of Indian *Paphiopedilum* species. Mol Ecol Resour 12:82–90. doi:10.1111/j.1755-0998.2011.03071.x

Penjor T, Yamamoto M, Uehara M, Ide M, Matsumoto N, Matsumoto R, Nagano Y (2013) Phylogenetic relationships of Citrus and its relatives based on matK gene sequences. PLoS One 8, e62574. doi:10.1371/journal.pone.0062574

Peterson AT, Moyle RG, Nyári AS, Robbins MB, Brumfield RT, Remsen JV (2007) The need for proper vouchering in phylogenetic studies of birds. Mol Phylogenet Evol 45:1042–1044. doi:10.1016/j.ympev.2007.08.019

Pettengill JB, Neel MC (2010) An evaluation of candidate plant DNA barcodes and assignment methods in diagnosing 29 species in the genus *Agalinis* (Orobanchaceae). Am J Bot 97:1391–1406

Phoolcharoen W, Sukrong S (2013) Molecular analysis of *Vitex* species using candidate DNA barcoding and PCR-RFLP of the matK gene for authentication of *Vitex glabrata*. Nat Prod Commun 8:125–128

Pillon Y, Johansen J, Sakishima T, Chamala S, Barbazuk WB, Roalson EH, Price DK, Stacy EA (2013) Potential use of low-copy nuclear genes in DNA barcoding: a comparison with plastid genes in two Hawaiian plant radiations. BMC Evol Biol 13:35. doi:10.1186/1471-2148-13-35

Prendini L (2005) Comment on "Identifying spiders through DNA barcodes". Can J Zool 83:498–504. doi:10.1139/z05-025

Puillandre N, Lambert A, Brouillet S, Achaz G (2012) ABGD, automatic barcode gap discovery for primary species delimitation. Mol Ecol 21:1864–1877. doi:10.1111/j.1365-294X.2011.05239.x

Qi J, Wang B, Hao BL (2004) Whole proteome prokaryote phylogeny without sequence alignment: a k-string composition approach. J Mol Evol 58:1–11

Quandt D, Stech M (2004) Molecular evolution of the trnTUGU-trnFGAA region in Bryophytes. Plant Biol (Stuttg) 6:545–554

Ratnasingham S, Hebert PDN (2007) BOLD: the barcode of life data system (www.barcodinglife.org). Mol Ecol Notes 7:355–364

Riaz T, Shehzad W, Viari A, Pompanon F, Taberlet P, Coissac E (2011) ecoPrimers: inference of new DNA barcode markers from whole genome sequence analysis. Nucleic Acids Res 39:e145. doi:10.1093/nar/gkr732

Richly E, Leister D (2004) NUMTs in sequenced eukaryotic genomes. Mol Biol Evol 21:1081–1084. doi:10.1093/molbev/msh110

Rubinoff D (2006) Utility of mitochondrial DNA barcodes in species conservation. Conserv Biol J Soc Conserv Biol 20:1026–1033

Rubinoff D, Cameron S, Will K (2006a) Are plant DNA barcodes a search for the Holy Grail? Trends Ecol Evol 21:1–2. doi:10.1016/j.tree.2005.10.019

Rubinoff D, Cameron S, Will K (2006b) A genomic perspective on the shortcomings of mitochondrial DNA for "barcoding" identification. J Hered 97:581–594. doi:10.1093/jhered/esl036

Sarkar IN, Planet PJ, Desalle R (2008) caos software for use in character-based DNA barcoding. Mol Ecol Resour 8:1256–1259. doi:10.1111/j.1755-0998.2008.02235.x

Sarwat M, Yamdagni MM (2015) DNA barcoding, microarrays and next generation sequencing: recent tools for genetic diversity estimation and authentication of medicinal plants. Crit Rev Biotechnol 1–13. doi:10.3109/07388551.2014.947563

Savolainen V, Cowan RS, Vogler AP, Roderick GK, Lane R (2005) Towards writing the encyclopedia of life: an introduction to DNA barcoding. Philos Trans R Soc B 360:1805–1811. doi:10.1098/rstb.2005.1730

Seibel PN, Müller T, Dandekar T, Schultz J, Wolf M (2006) 4SALE–a tool for synchronous RNA sequence and secondary structure alignment and editing. BMC Bioinforma 7:498. doi:10.1186/1471-2105-7-498

Selvaraj D, Shanmughanandhan D, Sarma RK, Joseph JC, Srinivasan RV, Ramalingam S (2012) DNA barcode ITS effectively distinguishes the medicinal plant Boerhavia diffusa from its adulterants. Genom Proteom Bioinforma 10:364–367. doi:10.1016/j.gpb.2012.03.002

Sharma SK, Dkhar J, Kumaria S, Tandon P, Rao SR (2012) Assessment of phylogenetic interrelationships in the genus Cymbidium (Orchidaceae) based on internal transcribed spacer region of rDNA. Gene 495:10–15. doi:10.1016/j.gene.2011.12.052

Sheth BP, Thaker VS (2015) Identification of a herbal powder by deoxyribonucleic acid barcoding and structural analyses. Pharmacogn Mag 11:570–574. doi:10.4103/0973-1296.172963

Shokralla S, Gibson JF, Nikbakht H, Janzen DH, Hallwachs W, Hajibabaei M (2014) Next-generation DNA barcoding: using next-generation sequencing to enhance and accelerate DNA barcode capture from single specimens. Mol Ecol Resour 14:892–901. doi:10.1111/1755-0998.12236

Slabbinck B, Dawyndt P, Martens M, De Vos P, De Baets B (2008) TaxonGap: a visualization tool for intra- and inter-species variation among individual biomarkers. Bioinformatics 24:866–867. doi:10.1093/bioinformatics/btn031

Song H, Buhay JE, Whiting MF, Crandall KA (2008) Many species in one: DNA barcoding overestimates the number of species when nuclear mitochondrial pseudogenes are coamplified. Proc Natl Acad Sci U S A 105:13486–13491. doi:10.1073/pnas.0803076105

Srirama R, Senthilkumar U, Sreejayan N, Ravikanth G, Gurumurthy BR, Shivanna MB, Sanjappa M, Ganeshaiah KN, Uma Shaanker R (2010) Assessing species admixtures in raw drug trade of Phyllanthus, a hepato-protective plant using molecular tools. J Ethnopharmacol 130:208–215. doi:10.1016/j.jep.2010.04.042

Srivathsan A, Meier R (2012) On the inappropriate use of Kimura-2-parameter (K2P) divergences in the DNA-barcoding literature. Cladistics 28:190–194. doi:10.1111/j.1096-0031.2011.00370.x

Stech M, Veldman S, Larrain J, Muñoz J, Quandt D, Hassel K, Kruijer H (2013) Molecular species delimitation in the *Racomitrium canescens* complex (Grimmiaceae) and implications for DNA barcoding of species complexes in mosses. PLoS One 8(1), e53134. doi:10.1371/journal.pone.0053134

Steinke D, Vences M, Salzburger W, Meyer A (2005) TaxI: a software tool for DNA barcoding using distance methods. Philos Trans R Soc B 360:1975–1980. doi:10.1098/rstb.2005.1729

Storchová H, Olson MS (2007) The architecture of the chloroplast psbA-trnH non-coding region in angiosperms. Plant Syst Evol 268:235–256

Sun YL, Wang D, Yeon MH, Kim DH, Kim HG, Hong SK (2011) Molecular identification of medicinal herbs, Oldenlandia diffusa and Oldenlandia corymbosa based on nrDNA ITS region sequence. J Plant Biotechnol 38(4):301–307

Taberlet P, Coissac E, Pompanon F, Gielly L, Miquel C, Valentini A, Vermat T, Corthier G, Brochmann C, Willerslev E (2007) Power and limitations of the chloroplast trnL (UAA) intron for plant DNA barcoding. Nucleic Acids Res 35, e14. doi:10.1093/nar/gkl938

Tamboli AS, Patil SM, Gholave AR, Kadam SK, Kotibhaskar SV, Yadav SR, Govindwar SP (2016) Phylogenetic analysis, genetic diversity and relationships between the recently segregated species of *Corynandra* and *Cleoserrata* from the genus *Cleome* using DNA barcoding and molecular markers. C R Biol 339:123–132. doi:10.1016/j.crvi.2016.02.005

Taylor HR, Harris WE (2012) An emergent science on the brink of irrelevance: a review of the past 8 years of DNA barcoding. Mol Ecol Resour 12:377–388. doi:10.1111/j.1755-0998.2012.03119.x

Techen N, Parveen I, Pan Z, Khan IA (2014) DNA barcoding of medicinal plant material for identification. Curr Opin Biotechnol 25:103–110. doi:10.1016/j.copbio.2013.09.010

Tsai LC, Yu YC, Hsieh HM, Wang JC, Linacre A, Lee JC (2006) Species identification using sequences of the trnL intron and the trnL-trnF IGS of chloroplast genome among popular plants in Taiwan. Forensic Sci Int 164:193–200

Tsai LC, Yu YC, Hsieh HM (2008) Bidens identification using the non coding regions of chloroplast genome and nuclear ribosomal DNA. Forensic Sci Int Genet 2:35–40

Tsai CC, Chen CH, Chou CH (2012) DNA barcodes reveal high levels of morphological plasticity among Rhododendron species (Ericaceae) in Taiwan. Biochem Syst Ecol 40:169–177. doi:10.1016/j.bse.2011.10.013

Tyagi A, Bag SK, Shukla V, Roy S, Tuli R (2010) Oligonucleotide frequencies of barcoding loci can discriminate species across kingdoms. PLoS One 5:e12330. doi:10.1371/journal.pone.0012330

Valentini A, Pompanon F, Taberlet P (2009) DNA barcoding for ecologists. Trends Ecol Evol 24:110–117. doi:10.1016/j.tree.2008.09.011

Vences M, Thomas M, Bonett RM, Vieites DR (2005) Deciphering amphibian diversity through DNA barcoding: chances and challenges. Philos Trans R Soc B 360:1859–1868. doi:10.1098/rstb.2005.1717

Vickruck JL, Rehan SM, Sheffield CS, Richards MH (2011) Nesting biology and DNA barcode analysis of *Ceratina dupla* and *C. mikmaqi* and comparisons with *C. calcarata* (Hymenoptera: Apidae: Xylocopinae). Can Entomol 143:254–262. doi:10.4039/n11-006

Virgilio M, Backeljau T, Nevado B, De Meyer M (2010) Comparative performances of DNA barcoding across insect orders. BMC Bioinforma 11:206. doi:10.1186/1471-2105-11-206

Virgilio M, Jordaens K, Breman FC, Backeljau T, De Meyer M (2012) Identifying insects with incomplete DNA barcode libraries, African fruit flies (Diptera: Tephritidae) as a test case. PLoS One 7, e31581. doi:10.1371/journal.pone.0031581

Vogler AP, Monaghan MT (2007) Recent advances in DNA taxonomy. J Zool Syst Evol Res 45:1–10. doi:10.1111/j.1439-0469.2006.00384.x

Wallace LJ, Boilard SMAL, Eagle SHC, Spall JL, Shokralla S, Hajibabaei M (2012) DNA barcodes for everyday life: routine authentication of natural health products. Food Res Int 49:446–452. doi:10.1016/j.foodres.2012.07.048

Wang N, Jacques FMB, Milne RI, Zhang CQ, Yang JB (2012) DNA barcoding of Nyssaceae (Cornales) and taxonomic issues. Bot Stud 53:265–274

Ward J, Peakall R, Gilmore SR, Robertson J (2005) A molecular identification system for grasses: a novel technology for forensic botany. Forensic Sci Int 152:121–131

Ward J, Gilmore SR, Robertson J, Peakall R (2009) A grass molecular identification system for forensic botany: a critical evaluation of the strengths and limitations. J Forensic Sci 54(6):1254–1260. doi:10.1111/j.1556-4029.2009.01196

Whitlock BA, Hale AM, Groff PA (2010) Intraspecific inversions pose a challenge for the trnH-psbA plant DNA barcode. PLoS One 5:e11533. doi:10.1371/journal.pone.0011533

Whitworth TL, Dawson RD, Magalon H, Baudry E (2007) DNA barcoding cannot reliably identify species of the blowfly genus *Protocalliphora* (Diptera: Calliphoridae). Proc Biol Sci 274:1731–1739. doi:10.1098/rspb.2007.0062

Will KW, Rubinoff D (2004) Myth of the molecule: DNA barcodes for species cannot replace morphology for identification and classification. Cladistics 20:47–55. doi:10.1111/j.1096-0031.2003.00008.x
Will KW, Mishler BD, Wheeler QD (2005) The perils of DNA barcoding and the need for integrative taxonomy. Syst Biol 54:844–851. doi:10.1080/10635150500354878
Wiriyakarun S, Yodpetch W, Komatsu K, Zhu S, Ruangrungsi N, Sukrong S (2013) Discrimination of the Thai rejuvenating herbs *Pueraria candollei* (white Kwao Khruea), *Butea superba* (red Kwao Khruea) and *Mucuna collettii* (black Kwao Khruea) using PCR-RFLP. J Nat Med 67:562–570. doi:10.1007/s11418-012-0716-1
Wolf M, Friedrich J, Dandekar T, Müller T (2005) CBCAnalyzer: inferring phylogenies based on compensatory base changes in RNA secondary structures. In Silico Biol 5:291–294
Wu Y, Xu L, Chen L, Wang B, Zhao R (2015) DNA molecular identification of Datura medicinal plants using ITS2 barcode sequence. Zhong Yao Cai 38:1852–1857
Xiang XG, Zhang JB, Lu AM, Li RQ (2011) Molecular identification of species in Juglandaceae: a tiered method. J Syst Evol 49:252–260
Xiao WL, Motley TJ, Unachukwu UJ, Lau CBS, Jiang B, Hong F, Leung PC, Wang QF, Livingston PO, Cassileth BR, Kennelly EJ (2011) Chemical and genetic assessment of variability in commercial Radix Astragali (Astragalus spp) by ion trap LC-MS and nuclear ribosomal DNA barcoding sequence analyses. J Agric Food Chem 59:1548–1556. doi:10.1021/jf1028174
Xu G, Wang X, Liu C, Li W, Wei S, Liu Y, Cheng X, Liu J, Donini P (2013) Authentication of official Da-huang by sequencing and multiplex allele-specific PCR of a short maturase K gene. Genome 56:109–113. doi:10.1139/gen-2012-0182
Yan LJ, Liu J, Möller M, Zhang L, Zhang XM, Li DZ, Gao LM (2015) DNA barcoding of Rhododendron (Ericaceae), the largest Chinese plant genus in biodiversity hotspots of the Himalaya-Hengduan Mountains. Mol Ecol Resour 15:932–944. doi:10.1111/1755-0998.12353
Yao H, Song JY, Ma XY, Liu C, Li Y, Xu HX, Han JP, Duan LS, Chen SL (2009) Identification of *Dendrobium* species by a candidate DNA barcode sequence: the chloroplast *psbA-trnH* intergenic region. Planta Med 75:667–669. doi:10.1055/s-0029-1185385
Yao H, Song J, Liu C, Luo K, Han J, Li Y, Pang X, Xu H, Zhu Y, Xiao P, Chen S (2010) Use of ITS2 region as the universal DNA barcode for plants and animals. PLoS One 510:e13102
Yuan QJ, Zhang B, Jiang D, Zhang WJ, Lin TY, Wang NH, Chiou SJ, Huang LQ (2015) Identification of species and *materia medica* within *Angelica* L. (Umbelliferae) based on phylogeny inferred from DNA barcodes. Mol Ecol Resour 15:358–371. doi:10.1111/1755-0998.12296
Zhang DX, Hewitt GM (1996) Highly conserved nuclear copies of the mitochondrial control region in the desert locust *Schistocerca gregaria*: some implications for population studies. Mol Ecol 5:295–300
Zhang AB, Sikes DS, Muster C, Li SQ (2008) Inferring species membership using DNA sequences with back-propagation neural networks. Syst Biol 57:202–215. doi:10.1080/10635150802032982
Zhang A, Feng J, Ward RD, Wan P, Gao Q, Wu J, Zhao W (2012a) A new method for species identification via protein-coding and non-coding DNA barcodes by combining machine learning with bioinformatic methods. PLoS One 7, e30986. doi:10.1371/journal.pone.0030986
Zhang AB, Muster C, Liang HB, Zhu CD, Crozier R, Wan P, Feng J, Ward RD (2012b) A fuzzy-set-theory-based approach to analyse species membership in DNA barcoding. Mol Ecol 21:1848–1863. doi:10.1111/j.1365-294X.2011.05235.x
Zhang W, Fan X, Zhu S, Zhao H, Fu L (2013) Species-specific identification from incomplete sampling: applying DNA barcodes to monitoring invasive Solanum plants. PLoS One 8:e55927. doi:10.1371/journal.pone.0055927
Zhang J, Chen M, Dong X, Lin R, Fan J, Chen Z (2015a) Evaluation of four commonly used DNA barcoding loci for Chinese medicinal plants of the family Schisandraceae. PLoS One 10, e0125574. doi:10.1371/journal.pone.0125574

Zhang JQ, Meng SY, Wen J, Rao GY (2015b) DNA barcoding of *Rhodiola* (Crassulaceae): a case study on a group of recently diversified medicinal plants from the Qinghai-Tibetan Plateau. PLoS One 10, e0119921. doi:10.1371/journal.pone.0119921

Zhang ZL, Song MF, Guan YH, Li HT, Niu YF, Zhang LX, Ma XJ (2015c) DNA barcoding in medicinal plants: testing the potential of a proposed barcoding marker for identification of Uncaria species from China. Biochem Syst Ecol 60:8–14. doi:10.1016/j.bse.2015.02.017

Zhao S, Chen X, Song J, Pang X, Chen S (2015) Internal transcribed spacer 2 barcode: a good tool for identifying Acanthopanacis cortex. Front Plant Sci 6:840. doi:10.3389/fpls.2015.00840

Zheng S, Liu D, Ren W, Fu J, Huang L, Chen S (2014) Integrated analysis for identifying *Radix Astragali* and its adulterants based on DNA barcoding. Evid Based Complem Alternat Med 2014:1–11. doi:10.1155/2014/843923

Zhu Y, Chen S, Yao H, Tan R, Song J, Luo K, Lu J (2010) DNA barcoding the medicinal plants of the genus Paris. Yao XueXueBao 45:376–382

Chapter 5
Deciphering the Biosynthetic Pathways of Bioactive Compounds *In Planta* Using Omics Approaches

Hsiao-Hang Chung*, Yi-Chang Sung*, and Lie-Fen Shyur

Abstract "Plant bioactive compounds" are secondary metabolites mainly composed of terpenoids, alkaloids, and phenolics that are observed to have varied pharmacological effects in preventing or intervening in human disorders. Due to the great potential of plant secondary metabolites for use in the nutraceutical and pharmaceutical industries, genetic engineering techniques and bioreactor approaches may be efficient ways to mass produce bioactive compounds for industrial use. However, before mass production strategies can be implemented to fulfill industrial needs, a full understanding of the biosynthetic pathways and underlying regulation mechanisms must be gained. The conventional "one enzyme, one gene" study approach is insufficient for elucidating the biosynthetic pathways of most bioactive compounds, especially those of non-model medicinal plants due to lack of available whole genome information. It is foreseen that the emerging "omics" technologies will be useful platforms to provide more global genomics, transcriptomics, proteomics, and metabolomics information to help uncover the biosynthesis pathways of secondary metabolites, especially in medicinal plants. The aim of this review is to summarize the current research progress and knowledge in deciphering the biosynthesis pathways of plant bioactive secondary metabolites selected from the three major compound types using omics approaches.

Keywords Bioactive compound • Biosynthetic pathway • Metabolomics • Omics • Proteomics • Transcriptomics

---

*Author contributed equally with all other contributors.

H.-H. Chung
Department of Horticulture, National Ilan University, Yilan 26047, Taiwan

Y.-C. Sung • L.-F. Shyur (✉)
Agricultural Biotechnology Research Center, Academia Sinica,
No. 128, Sec. 2, Academia Road, Nankang, Taipei 11529, Taiwan
e-mail: lfshyur@ccvax.sinica.edu.tw

H.-S. Tsay et al. (eds.), *Medicinal Plants - Recent Advances in Research and Development*, DOI 10.1007/978-981-10-1085-9_5

Abbreviations

3′OMT	3′-*O*-Methyltransferase
4′OMT	3′-Hydroxyl-*N*-methylcoclaurine 4′-*O*-methyltransferase
6OMT	Norcoclaurine 6-*O*-methyltransferase
7DLGT	7-Deoxyloganetic acid glucosyltransferase
8HGO	8-Hydroxygeraniol oxidoreductase
16OMT	*O*-Methyltransferase
β-AS	β-Amyrin synthase
ANR	Anthocyanidin reductase
ANS	Anthocyanidin synthase
AVLBS	Anhydrovinblastine synthase
BBE	Berberine bridge enzyme
BIA	Benzylisoquinoline alkaloid
BIS1	bHLH iridoid synthase 1
CHI	Chalcone isomerase
CHS	Chalcone synthase
CNMT	Coclaurine *N*-methyltransferase
CODM	Codeine *O*-demethylase
COR	Codeinone reductase
CPP	Copalyl diphosphate
CPR	Cytochrome P450 reductase
CPS	Copalyl diphosphate synthase
CYP	Cytochrome P450
CYP82	*N*-Methylcanadine 1-hydroxylase
D4H	Desacetoxyvindoline-4-hydroxylase
DAT	Deacetylvindoline-4-*O*-acetyltransferase
DDS	Dammarenediol synthase
DFR	Dihydroflavonol reductase
DL7H	7-Deoxyloganic acid 7-hydroxylase
DMAPP	Dimethylallyl diphosphate
DXS	1-Deoxy-D-xylulose-5-phosphate synthase
EST	Expressed sequence tag
F3′5′H	Flavonoid 3′,5′-hydroxylase
F3′H	Flavonoid 3′-hydroxylase
FLS	Flavone synthase
FPP	Farnesyl diphosphate
FPKM	Fragments per kilobase of transcript per million mapped reads
FPPS	Farnesyl diphosphate synthase
G8O	Geraniol-8-oxidase
G10H	Geranyl 10-hydroxylase
GGPP	Geranylgeranyl diphosphate
GGPPS	GGPP synthase
GT	Glycosyltransferase

IFS	Isoflavone synthase
IO	Iridoid oxidase
IPP	Isopentenyl diphosphate
IPPI	IPP isomerase
IS	Iridoid synthase
JA	Jasmonic acid
KSL	Kaurene synthase-like diphosphate synthase
LAMT	Loganic acid methyltransferase
LAR	Leucocyanidin 4-reductase
LDOX	Leucocyanidin dioxygenase
MEP	2-C-Methyl-D-erythritol 4-phosphate
MeJA	Methyl jasmonate
MVA	Mevalonate
N7OMT	Norreticuline 7-*O*-methyltransferase
NCS	Norcoclaurine synthase
NGS	Next generation sequencing
NMCH	(*S*)-*N*-methylcoclaurine 3′-hydroxylase
NMT	*N*-Methyltransferase
OAS	Oleanolic acid synthase
ORCA3	Octadecanoid-responsive *Catharanthus* APETALA2-domain 3
P5βR	Progesterone-5β-reductase
PPD	Protopanaxadiol
PPDS	Protopanaxadiol synthase
PPT	Protopanaxatriol
PPTS	Protopanaxatriol synthase
RPKM	Reads per kilobase per million mapped reads
SM	Secondary metabolite
SalAT	Salutaridinol 7-*O*-acetyltransferase
SalR	Salutaridine reductase
SalSyn	Salutaridine synthase
SE	Squalene epoxidase
SGD	Strictosidine β-D-glucosidase
SLS	Secologanin synthase
SQ	Squalene
SS	Squalene synthase
STOX	(*S*)-Tetrahydroprotoberberine oxidase
STR	Strictosidine synthase
STS	Stilbene synthase
T16H	Tabersonine 16-hydroxylase
T6ODM	Thebaine 6-*O*-demethylase
TDC	Tryptophan decarboxylase
THAS	Tetrahydroalstonine synthase
TF	Transcription factors
TIA	Terpenoid indole alkaloid

TNMT	Tetrahydroprotoberberine *N*-methyltransferase
TPS	Terpene synthases
UGT	UDP-dependent glycosyltransferase
VIGS	Virus-induced gene silencing

5.1 Introduction

Accumulating evidences show that plant secondary metabolites (SMs) play significant roles as signaling molecules between plants or between plants and herbivores or other organisms for plant survival and/or adaptation (Hartmann 2007). Distinct from essential nutrients, plant SMs are recognized as extra-nutritional or sensing molecules that occur in relatively small amounts, but some are known to influence human health. Bioactive compounds in plants can be defined as SMs that have pharmacological or toxicological activities in humans and animals (Bernhoft 2010). Research into bioactive compounds from medicinal plants was started in the early 1800s when morphine was isolated from *Papaver somniferum* (Schmitz 1985). Since then intensive research focused on isolation and structure elucidation of novel bioactive plant SMs has greatly affected both scientific research and the pharmaceutical industry. To date, more than 200,000 SMs have been identified from various plant species (Hartmann 2007). Based on their structural characteristics and biosynthetic pathways, three major classes of plant SMs have been identified: terpenoids (over 36,000 compounds) (Buckingham 2007), alkaloids (around 12,000 compounds) (Croteau et al. 2000), and phenolics (nearly 10,000 compounds) (Taiz and Zeiger 2006).

Traditional medicinal plants are still extensively used in many parts of the world such as Asia, Africa, and South America (Brusotti et al. 2014), and ethnomedical systems which utilize medicinal plants are used by 60 % of the world population (Busia and Kasilo 2010). Many pure compounds isolated from particular medicinal plants have been proven to have therapeutic functions in humans. Morphine from *Papaver somniferum* is an important and well-known plant-derived drug used as a painkiller, which is still irreplaceable in the modern medical system. Another famous example from ethnomedicine is quinine, a critical antimalarial drug isolated from *Cinchona* species by Caventou and Pelletier in 1820. Further examples include khellin from *Ammi visnaga*, which led to the development of sodium chromoglycate as a bischromone to prevent the release of the antiallergenic substances responsible for an asthmatic attack; colchicine from *Colchicum autumnale* that remains a standard treatment for gout; and diosgenin from *Dioscorea* species that is used for the commercial synthesis of contraceptive pills. Vinblastine and vincristine from the Madagascar periwinkle *Catharanthus roseus* are now well established in the treatment of acute leukemia and Hodgkin's disease. Paclitaxel from the bark of Pacific yew is a plant-derived anticancer drug that is used in the treatment of ovarian, breast, lung, pancreatic, and other cancers. Over the last few decades, traditional Chinese medicine (TCM) has become an attractive source from which to discover and explore new bioactive compounds. A large number of bioactive compounds from TCM have been isolated and studied, such as artemisinin, tanshinones, and ginsenosides, etc. A web-based database, TCM Database@Taiwan, has also been established (Chen 2011) which as of July 2013 contained 61,000 TCM compounds. In addition to the discovery of new natural products, mass production of these bioactive compounds for medicinal use is still a challenge for scientists and the pharmaceutical industry. Many strategies have been extensively explored using in vitro propagation systems, such as cell, hairy root, and other organ cultures, in combination with media modification, precursor feeding, elicitation studies, and so on (Bourgaud et al. 2001; Namdeo 2007). However, only a few cases have led to commercial success in enhancing the production of bioactive SMs from medicinal plants using biotechnological approaches. Due to the complex biosynthesis and regulation of SMs, most bioactive compounds are still mainly isolated from wild or cultivated plants. Therefore, to achieve mass production of important or high-value bioactive compounds, it is necessary to conduct basic research in the biochemical and genetic regulation of the SM biosynthetic pathways *in planta*.

The study of SM or bioactive compound biosynthesis in plants has advanced considerably after the development of the radioactive labeling technique in the 1950s. Before that, the formation of plant SMs was predicted according to analogy of molecular structures and organic chemistry reactions, with no direct evidence. Based on feeding experiments with ^{13}C and ^{3}H-labeled compounds, the biosynthesis profiles of different SM classes were proposed in the following decade. At the protein level, breakthroughs in the enzymology of plant SM study were conducted using two techniques developed in the 1970s. The first was application of column chromatography including size exclusion, ion exchange, and affinity methods for protein separation, and the second was the development of in vitro plant tissue culture systems that made the study system easy to handle, especially suspension cell culture (Hahlbrock 1981). Many enzymes were characterized in this period. At the nucleotide level, isolation of genes involved in plant SM biosynthesis and elucidation of their functional expression were achieved in the mid-1980s (Ryder et al. 1984; Kutchan et al. 1988). Later on, more and more genes, encoding enzymes, and transcription factors involved in SM biosynthesis pathways were revealed (Verpoorte and Memelink 2002). To date we know that there are four major biosynthetic pathways involved in the formation of plant SMs, namely, (1) the mevalonic acid (MVA) pathway, (2) the methylerythritol phosphate (MEP) pathway, (3) the shikimate pathway, and (4) the acetate/malonate pathway (Taiz and Zeiger 2006). Terpenoids are produced via the MVA and MEP pathways, which form their basic C_5 unit, isopentenyl diphosphate (IPP) (Ashour et al. 2010). Alkaloids are formed from aromatic and aliphatic amino acids which are synthesized through the shikimic acid pathway and Krebs cycle, respectively. Certain types of alkaloids need to combine monoterpene or diterpene in their backbones, in which the MEP pathway is also involved (Roberts et al. 2010). Phenolic compounds are derived from the acetate/malonate, shikimate, and phenylpropanoid pathways (Crozier et al. 2006).

Current knowledge of SM biosynthesis mainly centers around the upstream pathways which form the backbone skeletons in each class. A number of genetic engineering experiments have shown the potential to improve SM content in several medicinal plants (Oksman-Caldentey and Inzé 2004); however, there are still many missing pieces in knowledge of the regulatory and holistic biosynthetic pathways, especially in the downstream or final modifications. The rapid development of the emerging "omics" technologies, including genomics, transcriptomics, proteomics, and metabolomics, has provided us a global view of system biology in plants at different molecular levels and helps to address how plant SMs are synthesized. In recent years, advancements in technologies, such as next generation sequencing (NGS), and improved analytical technologies, such as hyphenated mass spectrometry (MS) or NMR spectrometry, have extended the width and depth of omics techniques available for deciphering SM biosynthesis. Omics studies have been utilized for the functional identification of genes, proteins, and enzymes involved in plant SM synthesis or metabolism. In this review, we summarize the current development in "omics" approaches in the investigation of selected pharmacologically bioactive compounds from three major classes of SMs, terpenoids, alkaloids, and phenolics, which are being used as drugs in the prevention or therapy of human diseases.

5.2 Terpenoids

Terpenoids or terpenes are the largest group of plant SMs. The name is derived from turpentine which is obtained from the resin of pines (Breitmaier 2006). The basic structure of the terpenoids is a branched five-carbon unit called an isoprene unit (or isoprenoid) synthesized from acetyl-CoA or 3-phosphoglycerate. Repetitive fusion of the five-carbon unit of isoprene produces numerous different types of terpenoid molecules, such as hemiterpene (C_5), monoterpene (C_{10}), sesquiterpene (C_{15}), diterpene (C_{20}), triterpene (C_{30}), and teraterpene (C_{40}).

The biosynthetic pathway of terpenoids can generally be divided into four major steps: (1) formation of isopentenyl diphosphate (IPP) which represents a basic C_5 isoprene unit; (2) repetitive assembly of basic C_5 unit, IPP, to form a range of prenyl diphosphates as the central intermediates for the different classes of terpenoids; (3) formation of parent carbon skeletons by specific terpenoid synthases; and (4) secondary modification of terpenoid skeletons to yield a great variety of compounds in this family (Croteau et al. 2000). There are two distinct routes, i.e., the MVA and MEP pathways, involved in the first step of terpenoid biosynthesis. The classic route, the MVA pathway, which takes place in the cytoplasm commences with three acetyl-CoA molecules (C_2) as a precursor, is used in the formation of sesquiterpenes (C_{15}) and triterpenes (C_{30}). The alternative route, the MEP pathway, which is restricted to the plastids, starts from the condensation of two precursors, pyruvate and glyceraldehyde 3-phosphate (GA-3P), and is used in the biosynthesis of monoterpenes (C_{10}), diterpenes (C_{20}), and tetraterpenes (C_{40}). The second step is to synthesize different sizes of terpenoid intermediates by fusion of basic C_5 units, IPP and its more active allylic isomer, dimethylallyl diphosphate (DMAPP). Farnesyl diphosphate synthase (FPPS) and squalene synthase (SS) catalyze the formation of C_{15} and C_{30} prenyl diphosphates in the cytosol. Geranyl diphosphate synthase (GPPS) and geranylgeranyl diphosphate synthase (GGPPS) produce C_{10} and C_{20} prenyl diphosphates, respectively, in plastids. All the intermediates and enzymes involved in the first two steps of terpenoid biosynthesis have been well investigated (Ashour et al. 2010).

Terpene synthases (TPSs) are responsible for the third step in terpenoid biosynthesis, which catalyzes the formation of various terpene skeletal types. So far one hundred TPS genes have been characterized (Tholl 2006), and some of their protein structures have also been elucidated (Gao et al. 2012). Based on the protein domain structures and enzymatic reaction mechanisms, plant TPSs can be divided into classes I and II. The class I TPSs, known as ionization-initiated TPSs, can catalyze the ionization of the diphosphate ester bond of prenyl diphosphates to form carbocation intermediates. The cyclizations, hydride shifts, or other further rearrangements of carbocation intermediates can lead to formation of a variety of products. For example, Chen et al. (2004) reported that a class I monoterpene synthase from *Arabidopsis thaliana* converted geranyl diphosphates to ten types of volatile monoterpenes including 1,8-cineole, α-thujene, α-pinene, sabinene, β-pinene, myrcene, limonene, β-ocimene, terpinolene, and α-terpineol. In contrast, the class II TPSs belonging to protonation-initiated TPSs provide a proton donor that triggers initial carbocation formation (Christianson 2006). Addition of a proton to a carbon-carbon bond or epoxide generally leads to form a multicyclic structure. Like class I enzymes, diverse products can also be found by single class II TPSs. Furthermore, the cyclic terpene skeletons formed by TPSs are subjected to further enzymatic modification such as reduction, oxidation, peroxidation, methylation, acylation, and conjugation in the last step in plants, which result in a wide array of terpenoid end products and change their physical properties and/or biological activity.

It is well known that terpenoids play many roles in plants, including in growth, development, and defense (Ashour et al. 2010). The essential plant hormones, gibberellins, abscisic acid, and brassinosteroids, are terpene derivatives, involved in stem elongation, germination, flowering, dormancy, senescence, and other functions (Thomas et al. 2005). Other derivatives, e.g., azadirachtin, pulegone, pyrethrin, and atractyloside, function as toxins or deterrents to herbivores for plant defense (Aerts and Mordue 1997; Davies et al. 2007). Carotenoids, a famous class of tetraterpene plant pigments, have a protective function in photosynthetic tissues to avoid photooxidation and have been found to exhibit anticancer activities (Sharoni et al. 2004). Terpenoids have also been used extensively throughout human history, like camphor, a monoterpenoid used as a pest deterrent and preservative or antimicrobial and essential oils composed primarily of terpenes and terpenoids that are used widely as fragrances in perfumery or aromatherapy. A number of terpenoids such as sesquiterpene artemisinin and diterpene paclitaxel have also been developed into therapeutics for treating human diseases. Artemisinin, an effective drug against malaria, is extracted from the Chinese medicinal plant, *Artemisia annua*, and Youyou Tu was awarded the 2015 Nobel Prize in Medicine for the work of discovery of this bioactive plant natural product. Paclitaxel (also known as Taxol) originally isolated from Pacific yew (*Taxus brevifolia*) has been used in the intervention for several forms of cancer, including ovarian cancer, breast cancer, lung cancer, and prostate cancer (Lee et al. 2012).

Since the terpenoids make up such a large SM group in plants, the complicated biosynthetic steps and numerous genes, proteins, and enzymes involved still require concerted research efforts. In this section, we selected two pharmaceutical terpenoid compounds, tanshinones and ginsenosides, from the well-known TCMs danshen and ginseng, respectively, to illustrate the use of advanced omics approaches to update and increase information about their biosynthesis.

5.2.1 *Tanshinones*

Tanshinones are diterpenoids that are a major type of bioactive compound in *Salvia miltiorrhiza* (called "danshen" in TCM), present particularly in the root and rhizome. To date, more than 40 tanshinones including tanshinone I, tanshinone IIA, tanshinone IIB, and cryptotanshinone have been isolated (Ma et al. 2015). Their functional properties have been widely examined and show a range of activities, e.g., antioxidation, anti-inflammation, antiaging and anticancer activities (Robertson et al. 2014; Zhang et al. 2012). Due to the significant medicinal value of tanshinones, their biosynthetic pathways and regulatory mechanisms for further metabolic manipulation have been actively studied recently. The first step of tanshinone biosynthesis is formation of the basic C_5 unit which is proposed to be synthesized mainly through MEP pathway (Fig. 5.1A). Most of the genes involved in this pathway have been identified by deep sequencing (Ma et al. 2012). Five 1-deoxy-*D*-xylulose-5-phosphate synthase (DXS) genes encoding the key enzymes in the MEP pathway have been revealed from the *S. miltiorrhiza* genome. The expression of one DXS gene, *SmDXS2*, at the RNA level demonstrated a similar trend to tanshinone accumulation which might indicate their close relationship. For formation of C_{20} prenyl diphosphates (GGPPs) in the second biosynthesis step, although both *GPPS* and *GGPPS*, in general, are known to be involved in diterpene biosynthesis, recent transcriptomics studies of *S. miltiorrhiza* illustrated only *SmGGPPS* expression corresponding to tanshinone content in different plant tissues (Yang et al. 2013) or under elicitor induction (Gao et al. 2014). GGPP as the universal precursor of all diterpenes can be catalyzed by copalyl diphosphate synthase (SmCPS) and kaurene synthase-like (SmKSL) to form the diterpene skeletons (Fig. 5.1A). Using NGS technology, five *SmCPS* and two *SmKSL* gene homologues were revealed in the *S. miltiorrhiza* genome (Ma et al. 2012). *SmCPS1* has been functionally verified using an RNAi knockdown method in hairy root cultures which resulted in a decrease in dihydrotanshinone I and cryptotanshinone production (Cheng et al. 2014). Based on the structure of the tanshinones (Fig. 5.1A), a series of hydroxylation, dehydrogenation, reduction, and other modification reactions should be involved in the downstream pathway, which were proposed to be catalyzed by cytochrome P450 (CYP) family (Ma et al. 2015; Yang et al. 2013). A comparative transcriptomics approach was utilized for selecting candidate CYP genes from inducible hairy root system of *S. miltiorrhiza* (Guo et al. 2013). According to the RNA-seq profiles and qPCR analysis, finally six CYP candidates (from more than 300 CYP isotigs) were subcloned for further biochemical assay. CYP76AH1 was identified as a ferruginol synthase, adding a hydroxyl group on miltiradiene to produce ferruginol both in vitro and in vivo.

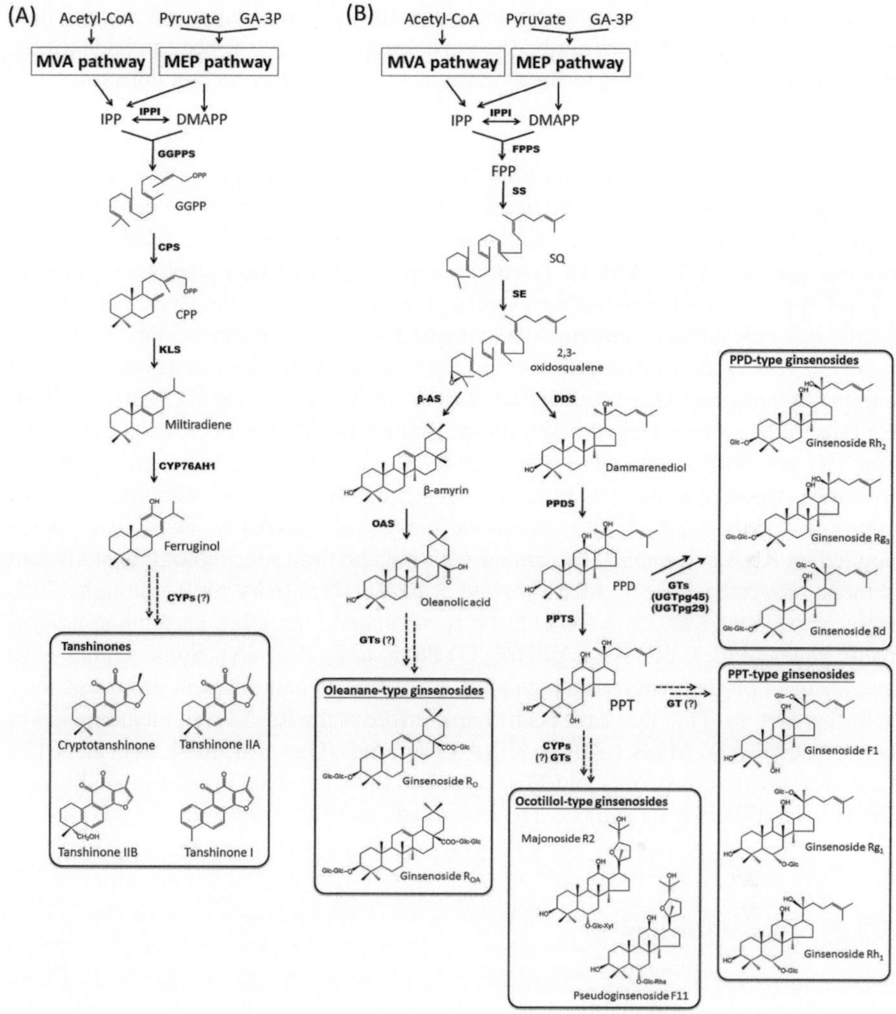

Fig. 5.1 The proposed biosynthetic pathways of bioactive terpenoids tanshinones (**A**) and ginsenosides (**B**) by current knowledge. *Solid arrows* represent known steps, *dash arrows* are unclear steps, and *double arrows* represent multiple steps. *Abbreviations* of compounds: *GA-3P* glyceraldehyde 3-phosphate, *MVA* mevalonate, *MEP* 2-C-methyl-D-erythritol 4-phosphate, *IPP* isopentenyl diphosphate, *DMAPP* dimethylallyl diphosphate, *GGPP* geranylgeranyl diphosphate, *CPP* copalyl diphosphate, *FPP* farnesyl diphosphate, *SQ* squalene, *PPD* protopanaxadiol, *PPT* protopanaxatriol. Abbreviations of enzymes: *IPPI* IPP isomerase, *GGPPS* GGPP synthase, *CPS* copalyl diphosphate synthase, *KSL* kaurene synthase-like diphosphate synthase, *CYP* cytochrome P450, *FPPS* farnesyl diphosphate synthase, *SS* squalene synthase, *SE* squalene epoxidase, *β-AS* β-amyrin synthase, *OAS* oleanolic acid synthase, *GT* glycosyltransferase, *DDS* dammarenediol synthase, *PPDS* protopanaxadiol synthase, *PPTS* protopanaxatriol synthase

The upstream pathway of tanshinone biosynthesis is relatively clear, but the final modification steps, which might be the most complex part, still need to be clarified. Recently, more studies have focused on the post-modification step of converting ferruginol to various tanshinones, and a few CPY genes have been proposed. By comparing the transcriptome profiles of *S. miltiorrhiza* leaf and root tissues, Yang et al. (2013) showed a total of 2863 unigenes with high levels of expression in roots including those genes known to be involved in tanshinone biosynthesis, such as copalyl diphosphate synthase (CPS1), kaurene synthase-like diphosphate synthase (KSL), and ferruginol synthase (CYP76AH1). Twenty, out of twenty-one CYP candidates selected for qPCR analysis, showed higher levels of expression in roots; however, their roles in tanshinone biosynthesis remain to be addressed. Other approaches using methyl jasmonate (MeJA)-treated leaves for transcriptome analysis have narrowed down the number of candidate genes for functional characterization (Luo et al. 2014), and three CYP candidates were picked based on the correlation of their expression trend with *SmCPS1* and *SmKSL* and were suggested to be involved in tanshinone biosynthesis. Recently, transcriptomics combined with metabolomics have become a major approach for the study of the tanshinone biosynthetic pathway. Guo et al. (2013) applied an RNA-seq method in parallel with metabolomics technology to study tanshinone biosynthesis using MeJA-treated *S. miltiorrhiza* hairy roots, through which eight new CYP genes were found to be co-regulated with other tanshinone biosynthetic genes (e.g., CYP71B, CYP76C, CYP81F, and CYP89A). Some studies have also focused on the underlying regulatory mechanisms and different groups of transcription factors (TFs) that have been characterized at the RNA level, including members of the WRKY, MYB, bZIP, and ERF TF families (Gao et al. 2014; Li et al. 2015). However, most of the CYPs and TFs suggested to be involved in tanshinone biosynthesis have still not been subjected to functional assays to validate their exact roles.

5.2.2 *Ginsenosides*

The *Panax* species, commonly known as "ginsengs," have been used as highly valued herbs in traditional Chinese medicine for over 4000 years (Hemmerly 1977). The major bioactive constituents in the root of ginseng plants are called ginsenosides and belong to the class of triterpene saponins. Ginsenosides contain a spectrum of pharmacological effects, including immune-modulating, anticancer, and antidiabetic activities (Kang et al. 2011; Yang et al. 2010). To date, more than 150 ginsenosides have been identified from *Panax* plants and are classified into two major types, dammarane and oleanane, based on their skeleton of aglycones (Fig. 5.1B) (Kim et al. 2015). Dammarane-type ginsenosides can be further divided into three subgroups: protopanaxadiol (PPD), protopanaxatriol (PPT), and ocotillol. A variety of ginsenosides are formed by distinct glycosylation activities on different positions of the chemical skeletons. Due to the importance of their pharmaceutical properties, their biosynthetic pathway has been widely studied for further in vivo or in vitro mass production purposes. As triterpenes, ginsenosides are considered to be mainly derived from the MVA pathway which was supported by $^{13}CO_2$ pulse-chase experiments in field-grown *P. ginseng* (Schramek et al. 2014). However, Zhao et al. (2014) demonstrated that both the MVA and MEP pathways could compensate each other in the *P. ginseng* hairy root system as demonstrated by two chemical inhibitors, mevinolin and fosmidomycin, which suppress the 3-hydroxy-3-methyl-glutaryl-CoA reductase (HMGR) and DXR activities in the MVA and MEP pathways, respectively. Since the genes and enzymes involved in the MVA and MEP pathways are well studied, of more concern is the downstream enzyme genes in the biosynthetic pathway. For instance, FPPS gene (to form a C_{15} unit FPP) and SS gene (to yield a C_{30} squalene) were isolated and functionally characterized in *Panax* (Kim et al. 2014; Lee et al. 2004), and two squalene epoxidases (SE) (*PgSQE1* and *PgSQE2*) that convert squalene to 2,3-oxido-squalene, an important intermediate for all ginsenosides and plant sterols, have been characterized. *PgSQE1*, corresponding to ginsenoside biosynthesis, has been suggested to be a rate-limiting enzyme (Fig. 5.1B) (Han et al. 2010). As a branch point of synthesis of different ginsenosides, 2,3-oxido-squalene can be cyclized by β-amyrin synthase (β-AS) or dammarenediol synthase (DDS) to form the skeleton of oleanane- or dammarane-type ginsenosides, respectively (Han et al. 2006; Zhao et al. 2015).

After the production of these basic structures, sequential hydroxylation by the cytochrome P450 (CYP) family and glycosylation by UDP-dependent glycosyltransferase (UGT) convert them to various ginsenosides. In the oleanane-type pathway, CYP716A52v2 (β-amyrin 28-oxidase), also called oleanolic acid synthase (OAS), catalyzes the oxidation reaction on the 28 position of β-amyrin to produce oleanolic acid which is the precursor of oleanane-type saponin (Han et al. 2013). Regarding to the dammarane-type ginsenosides, the backbones of three subgroups, i.e., PPD type, PPT type, and ocotillol type, are catalyzed by a series of CYPs. CYP716A47, known as PPD synthase (PPDS), converts dammarenediol to the PPD type in *P. ginseng* (Han et al. 2011), and a counterpart PPDS gene, *PqD12H*, was subcloned from *P. quinquefolius* and functionally characterized using yeast expression and a RNAi transgenic hairy root system (Sun et al. 2013). The PPT type was formed by further modification of PPD by CYP716A53v2 which acts as the PPD 6-hydroxylase (known as PPT synthase; PPTS) which has also been examined using in vitro yeast expression (Han et al. 2012). The ocotillol type has been proposed to be formed by one more enzymatic step after PPT which could be an epoxidase (Zhang et al. 2015).

After the establishment of different types of ginsenoside skeleton, deciphering the subsequent post-modification steps is still a big challenge. Two major groups, CYPs and glycosyltransferases (GTs), are proposed to be responsible for the final modifications. GTs which catalyze glycosylations are highly divergent and can be classified into 98 families based on amino acid sequences collected in the CAZy website (http://www.cazy.org/). Khorolragchaa et al. (2014) selected 12 UGTs from a ginseng expressed sequence tag (EST) library, and two of them, *PgUGT1* and *PgUGT2*, exhibited high RNA expression level in the MeJA inducibility experiment of *P. ginseng* roots. By application of a yeast expression system, one recent study reported two UGTs, namely, UGTPg45 and UGTPg29, that can catalyze the formation of ginsenosides Rh2 and Rg3, respectively (Wang et al. 2015). Due to the diversity of ginsenosides with different numbers of glycosyl moieties at different positions, a range of GTs can be assumed to be involved in ginsenoside biosynthesis. Recently, more research using NGS techniques for studying *Panax* species have revealed a series of enzyme genes including 15 CYP genes and 17 UGT genes that were proposed to be involved in the triterpene saponin biosynthetic pathway by in silico phylogenetic analysis (Zhang et al. 2015). Other transcriptome studies identified a large number of candidate *CYP450*s and *UGT*s; however, whether they also play a role in ginsenoside biosynthesis is not yet validated (Chen et al. 2011; Jayakodi et al. 2015). Furthermore, the regulatory mechanisms such as TFs or microRNAs that are involved in ginsenoside biosynthesis need further intensive investigation (Li et al. 2013).

5.3 Alkaloids

Alkaloids are a structurally diverse group of natural chemical compounds containing nitrogen atoms derived from amino acids, which are produced in many organisms like bacteria, fungi, animals, and mostly plants. Plant-derived alkaloids with biological activity can be found in some plant families including the Papaveraceae (e.g., codeine and sanguinarine), Berberidaceae (e.g., berberine), Erythroxylaceae (e.g., cocaine), Apocynaceae (e.g., vinblastine and vincristine), and Asteraceae (e.g., pyrrolizidine alkaloids) (Crozier 2008; Roberts et al. 2010). Alkaloids have great potential in pharmacological and clinical applications. Studies on the biosynthesis pathway of plant alkaloids began with morphine, which was isolated in 1800s followed by subsequent biochemical, chemical, and molecular studies of its metabolism (Roberts et al. 2010). Although many plants and microorganisms can produce alkaloids, certain pharmacologically bioactive alkaloids can only be found in limited plant species with low abundance and tissue specificity (Glenn et al. 2013). Most alkaloids are synthesized from an amino acid as their precursor, such as tyrosine (to produce morphine, aporphine, colchicine, and benzophenanthridine), tryptophan (to form ergot alkaloids and β-carboline alkaloids), arginine (to yield pyrrolizidine alkaloids and nicotiana alkaloids), lysine (to synthesize quinolizidine alkaloids, anabasine, and sedamine), and so on. Some other types of alkaloids also combine a monoterpene or diterpene in their backbone; these types are named quinoline alkaloids, such as strychnine, ajmaline, and camptothecin. The alkaloid biosynthetic pathways are also mainly deduced based on radiolabeling studies; however, there is still a long way to go to clarify the whole picture of alkaloid biosynthesis and regulation.

To date metabolic engineering studies of alkaloids are not comprehensive due to limited understanding of metabolic regulatory mechanisms. Insufficient genetic information about non-model medicinal plants is the main obstacle faced over the past decades (Crozier 2008). In addition, limited genetic basis and information for metabolite variants and their metabolism has become a major challenge in engineering the novel bioactive metabolites (Verpoorte and Memelink 2002). The integrative bioinformatics and functional omics study platforms can be useful to help elucidating alkaloid biosynthesis pathways in non-model plants. Metabolic engineering of nitrogen-containing alkaloids in plants through these "omics" approaches makes it possible to characterize key components at the transcriptional and posttranscriptional levels (Glenn et al. 2013; Martinez-Esteso et al. 2015). In this section, we focus on the biosynthesis mechanisms, enzymatic activity, and transcriptional regulation (Glenn et al. 2013) of plant terpenoid indole alkaloid (TIA) and benzylisoquinoline alkaloid (BIA), which have been developed and used in cancer therapy and treatment of other diseases. We will highlight the current progress achieved by using omics approaches and also the challenges faced in this field.

5.3.1 Benzylisoquinoline Alkaloids (BIAs)

More than 2,500 structures have been elucidated for BIAs (Zulak et al. 2006). Most benzylisoquinolines are present in basal angiosperms including the Ranunculaceae, Papaveraceae, Berberidaceae, and Magnoliaceae (Roberts et al. 2010). BIAs are considered to be a group of alkaloids with pharmacological properties, including codeine and morphine which are used as analgesic and antitussive drugs, berberine and sanguinarine which have antimicrobial activity, noscapine for cough and cancer treatment, and (+)-tubocurarine and papaverine which is used as a muscle relaxant (Hagel and Facchini 2013). The opium poppy (*Papaver somniferum*) (Papaveraceae) contains a great amount of analgesic morphine, codeine, and other related alkaloids in the laticifer cytoplasm (latex). Although BIAs accumulate in the latex tissue of the opium poppy, the related biosynthesis genes and enzymes were observed localized in the proximal sieve elements and companion cells, indicating that there is some missing link or information about the specific BIA metabolism regulation and transportation (Lee et al. 2013). Current knowledge shows that BIA shares a common biosynthetic origin intermediate (*S*)-norcoclaurine from condensation of dopamine and 4-hydroxyphenylacetaldehyde (4-HPAA) with the enzyme norcoclaurine synthase (NCS). (*S*)-Norcoclaurine is the central precursor to all BIAs which is converted to (*S*)-reticuline by an *N*-methyltransferase, a 6-*O*-methyltransferase, and a P450 hydroxylase, 4′-*O*-methyltransferase. (*S*)-Reticuline is the branch point intermediate known to produce different types of BIAs with various additional functionalities such as aromatic ring hydroxylation, *N*-methylation, *O*-methylation, *O*-acetylation, or methylenedioxy bridge formation. Genes involved in these modification steps have been identified, such as berberine bridge enzyme (*BBE*), which is involved in (*S*)-scoulerine formation in the berberine pathway, one specific methyltransferase, along with two P450-dependent enzymes in the sanguinarine pathway, and acyl coenzyme A: salutaridinol-7-*O*-acetyltransferase (*SAT*) and NADPH-dependent enzyme codeinone reductase (*COR*), which are two critical enzymes to yield morphine (Hagel and Facchini 2013; Roberts et al. 2010; Zulak et al. 2006) (Fig. 5.2A). Although some components and committed intermediates of the BIA backbone in biosynthesis pathway have been deciphered, understanding of the metabolic mechanisms and the metabolite diversity is still incomplete (Yang et al. 2014).

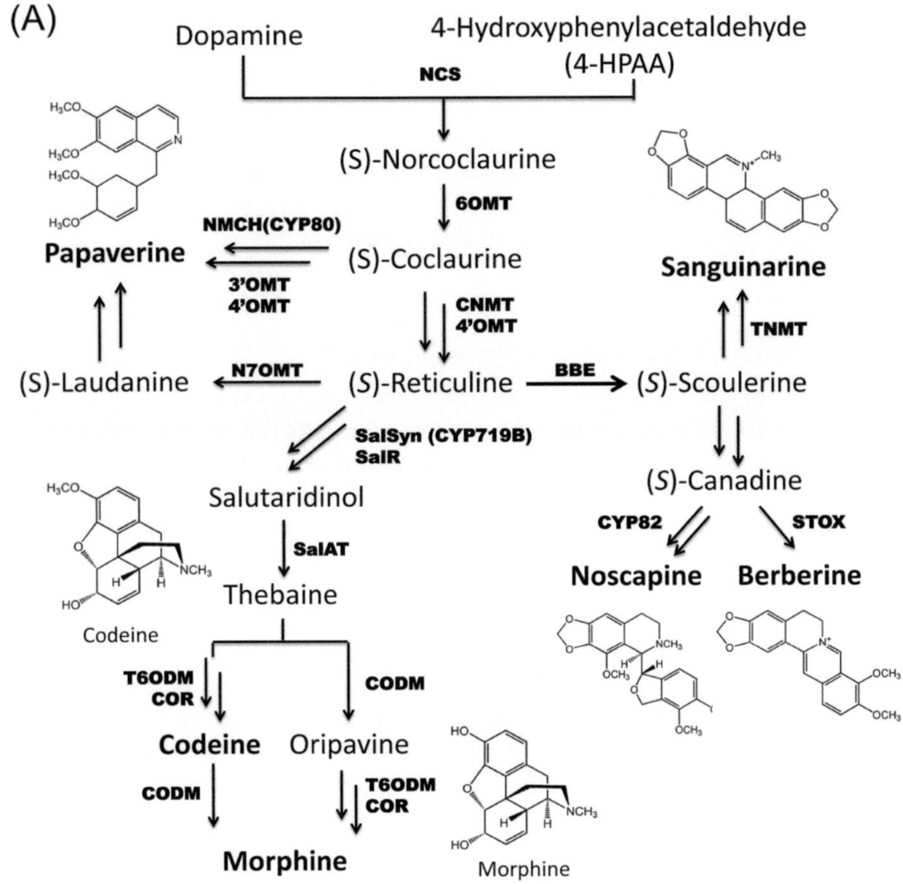

Fig. 5.2 The biosynthetic pathways of selected plant (**A**) benzylisoquinoline alkaloids (BIAs) and (**B**) terpenoid indole alkaloids (TIAs). *Solid arrows* represent known steps and *double arrows* represent multiple steps. The underlined enzymes in **B** are regulated by the circled transcription factors. *Abbreviations*: *NCS* norcoclaurine synthase, *6OMT* norcoclaurine 6-*O*-methyltransferase, *CNMT* coclaurine *N*-methyltransferase, *4'OMT* 3′-hydroxyl-*N*-methylcoclaurine 4′-*O*-methyltransferase, *3'OMT* 3′-*O*-methyltransferase, *NMCH* (*S*)-*N*-methylcoclaurine 3′-hydroxylase, *N7OMT* norreticuline 7-*O*-methyltransferase, *BBE* berberine bridge enzyme, *SalSyn* salutaridine synthase, *SalR* salutaridine reductase, *SalAT* salutaridinol 7-*O*-acetyltransferase, *T6ODM* thebaine 6-*O*-demethylase, *COR* codeinone reductase, *CODM* codeine *O*-demethylase, *TNMT* tetrahydroprotoberberine *N*-methyltransferase, *CYP82 N*-methylcanadine 1-hydroxylase, *STOX* (*S*)-tetrahydroprotoberberine oxidase, *G10H* geranyl 10-hydroxylase, *G8O* geraniol-8-oxidase, *8HGO* 8-hydroxygeraniol oxidoreductase, *IS* iridoid synthase, *P5βR* progesterone-5β-reductase, *CPR* cytochrome P450 reductase, *IO* iridoid oxidase, *7DLGT* 7-deoxyloganetic acid glucosyltransferase, *DL7H* 7-deoxyloganic acid 7-hydroxylase, *LAMT* loganic acid methyltransferase, *SLS* secologanin synthase, *TDC* tryptophan decarboxylase, *STR* strictosidine synthase, *SGD* strictosidine β-D-glucosidase, *T16H* tabersonine 16-hydroxylase, *16OMT* *O*-methyltransferase, *NMT N*-methyltransferase, *D4H* desacetoxyvindoline-4-hydroxylase, *DAT* deacetylvindoline-4-*O*-acetyltransferase, *ORCA3* octadecanoid-responsive *Catharanthus* APETALA2-domain 3, *BIS1* bHLH iridoid synthase 1, *AVLBS* anhydrovinblastine synthase, *THAS* tetrahydroalstonine synthase

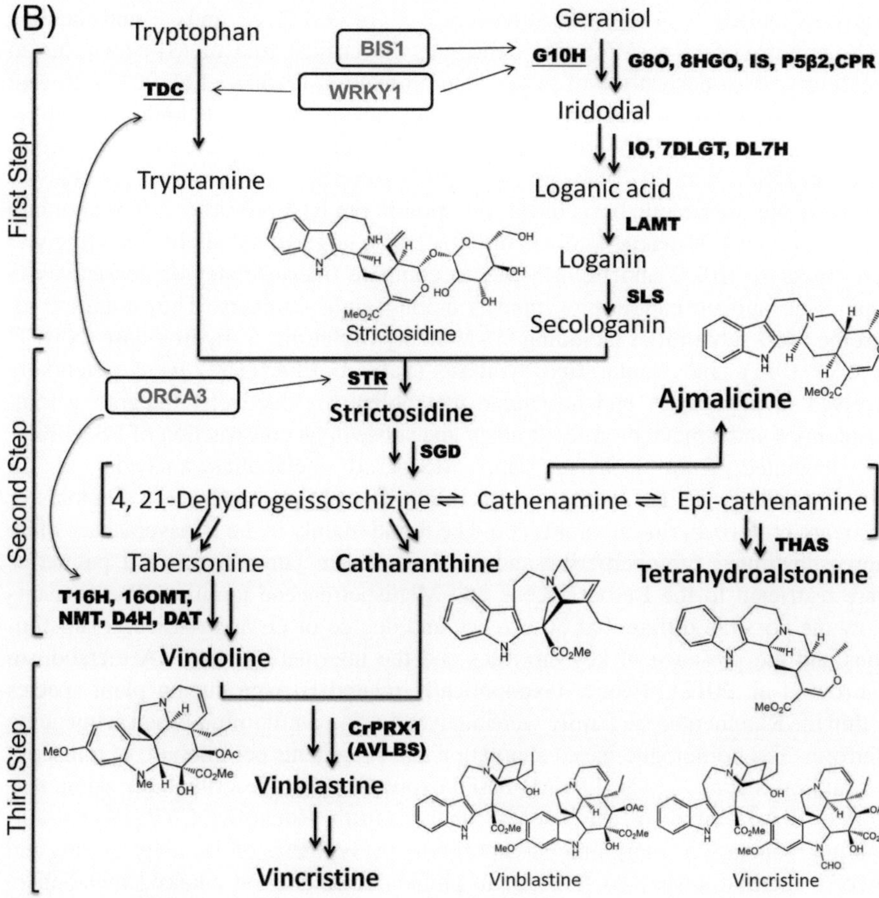

Fig. 5.2 (continued)

Some genes and enzymes involved in different types of BIA metabolism have been identified in opium poppy species, for example, Pictet-Spengler condensation enzymes [(*S*)-norcoclaurine synthase, NCS], SAM-dependent *O*-/*N*-methyltransferases (6OMT, 4OMT, CNMT, and TNMT), P450-dependent monooxygenase (NMCH or called CYP80B1), and FAD-linked oxidoreductase (BBE) (Hagel and Facchini 2013). Recently, transcriptomics tools were introduced into this field to uncover the steps including cytochrome P450-mediated hydroxylation, acetylation, and carboxylesteration in sanguinarine and noscapine metabolism in opium poppy by homology sequence-based comparison integrated with targeted metabolite analysis (Beaudoin and Facchini 2013; Dang et al. 2015; Dang and Facchini 2014). Transcriptomics comparison of *high papaverine mutant* (*pap1*, with higher papaverine accumulated) and normal cultivar revealed the enhanced expression of methyltransferase families like (*S*)-norcoclaurine-6-*O*-methyltransferase (*6OMT*), (*S*)-3-hydroxy-*N*-

methylcoclaurine 4′-*O*-methyltransferase (4′OMT), and norreticuline 7-*O*-methyltransferase (*N7OMT*), indicating the critical role of (*S*)-coclaurine in deciding various derivatives of papaverine production (Pathak et al. 2013). Recent studies on BIA biosynthesis have focused on biosynthetic elucidation of other phylogenetically related BIA-producing plants through comparative transcriptomics strategies (Xiao et al. 2013). Desgagne-Penix's group applied transcriptome analysis to investigate the genetic basis of the variation in the BIA profile in different opium poppy cultivars (Desgagne-Penix et al. 2012). The targeted metabolite profiling was performed by HPLC and LC-MS/MS to compare intermediates or derivatives in papaverine and noscapine biosynthesis among eight cultivars. They found cytochrome P450 subfamilies including (*S*)-*N*-methylcoclaurine 3′-hydroxylase (NMCH or CYP80B1), and salutaridine synthase (SalSyn, EF451150) were potentially involved in papaverine and noscapine metabolism, respectively. Integrative transcriptomics and a metabolomics strategy and subsequent construction of EST libraries generated from RNA-seq and MS-based metabolite analysis on 18 BIA-accumulating plant species revealed that benzo[*c*]phenanthridine alkaloids or upstream protoberberine alkaloids could be found mainly in the Papaveraceae, while more substituted protoberberine and its downstream jatrorrhizine and palmatine were restricted to the Berberidaceae and Menispermaceae families. These results show the possible differential occurrence and degree of *O*- and *N*-methyl substitutions and the presence of key enzymes and the intermediates in BIA metabolism (Farrow et al. 2012). Twenty taxonomically related BIA-producing plant species within the Ranunculaceae family were analyzed using an Illumina-based sequencing platform. The homologue-based annotation and Fragments per kilobase of transcript per million mapped reads (FPKM)-based expression data described a possible network of BIA metabolism. Phylogenetic analysis further identified CYP719 and salutaridine reductase in morphine and aporphine biosynthesis of *Papaver* species and novel *N*-methyltransferases (NMTs) in phthalideisoquinoline alkaloid metabolism (Hagel et al. 2015b). Winzer's group compared metabolite profiling in stem and capsule tissues of three opium poppy varieties and performed Roche 454 pyrosequencing to describe a 10-gene cluster putatively involved in the noscapine biosynthesis. This is the first gene cluster discovered for the BIA pathway. The same research group also used virus-induced gene silencing (VIGS) to correlate specific gene function to noscapine synthesis (Winzer et al. 2012).

In the early 2000s, proteomics studies using two-dimensional gel electrophoresis to analyze alkaloid-containing tissues in opium poppies revealed novel enzymes such as codeinone reductase in morphine biosynthesis (Decker et al. 2000). Further LC-MS/MS and related MS-based high-throughput proteomics analysis deciphered large-scale proteins involved in compound metabolism (Decker et al. 2000). With metabolomics tools, novel metabolic pathways and networks can be predicted by linking genes or enzymes to unknown or described compounds (Yang et al. 2014). One-dimensional NMR coupled with direct flow injection-mass spectrometry (DFI/LC-MS/MS) analysis in the plant order Ranunculales uncovered hundreds of metabolites including lipid derivatives, phenolic compounds, and organic acids in a tissue-specific manner, which could correlate with the variation in content of alkaloids among the tested plant cultivars (Hagel et al. 2015a). The comprehensive metabolite profiling study also suggested that alkaloids can function as defense metabolites against environmental factors or in response to biotic stress. The metabolomics results provide a reference for transcriptomics data to detect biosynthetic pathways or involved intermediates that might be ignored by elucidating individual compounds (Yang et al. 2014; Hagel et al. 2015a, b).

Overexpression (Inui et al. 2012), mutagenesis (Millgate et al. 2004), and knockdown techniques including RNA interference and antisense RNA (Kempe et al. 2009; Allen et al. 2008) have been employed for studying BIA metabolism *in planta*. The VIGS technique has also become a popular reverse genetics tool to study gene function in a rapid and straightforward way (Hagel and Facchini 2013). VIGS has been applied to opium poppy (Hileman et al. 2005) and was used to reveal three *O*-methyltransferases in noscapine metabolism and six enzymes in morphine biosynthesis (Wijekoon and Facchini 2012). Systematic silencing of six known genes in the papaverine biosynthesis pathway, whose corresponding enzyme function was only characterized in vitro, not only confirms their roles but also identified (*S*)-reticuline as a central branch point intermediate (Desgagne-Penix and Facchini 2012). Roles of *N*-methylcanadine 1-hydroxylase (*CYP82Y1* gene) in noscapine biosynthesis and thebaine 6-*O*-demethylase (*T6ODM* gene) and codeine *O*-demethylase (*CODM* gene) in morphine metabolism have also been resolved by VIGS study, which has provided further information for deciphering new steps or enzymes involved (Dang and Facchini 2014; Farrow and Facchini 2013).

Systems biology that integrates transcriptomics, metabolomics, and proteomics approaches has been proposed and widely used in the study of BIA metabolism (Dang et al. 2012), as the traditional concept of "one gene, one enzyme" for resolving metabolic pathways in a linear reaction sequence is oversimplified (Hagel and Facchini 2013). Conducting this study in a more comprehensive and systemic way allows unknown genes, enzymes, intermediates, or steps in biosynthesis to be deciphered, although information like ESTs, translated proteins, and genomics databases for medicinal plants still need to be improved to fully optimize this approach (Martinez-Esteso et al. 2015).

5.3.2 Terpenoid Indole Alkaloids (TIAs)

TIAs are a group of alkaloids consisting of indole and terpenoid moieties mainly found in the Apocynaceae, Loganiaceae, and Rubiaceae families (Zulak et al. 2006). The indole moiety of alkaloids is provided by tryptamine from tryptophan catalyzed via tryptophan decarboxylase (TDC), and then the alkaloids are transported to the vacuole for subsequent condensation with secologanin (O'Connor and Maresh 2006). The terpenoid moiety of TIAs is built from secologanin produced through the secoiridoid pathway, which starts with geraniol and is converted into loganic acid by geraniol 10-hydroxylase (G10H) and cytochrome P450 reductase (CPR). The loganic acid methyltransferase (LAMT) is responsible for the conversion of loganic acid to loganin by methylation, which forms secologanin by secologanin synthase (SLS) (Pan et al. 2016). Strictosidine synthase (STR) condenses tryptamine and secologanin to produce strictosidine, which is then converted into 4,21-dehydrogeissoschizine through strictosidine β-D-glucosidase (SGD)-mediated deglucosylation. The removal of the glycol moiety of strictosidine caused an unstable reactive aglycon which was then converted into 4,21-dehydrogeissoschizine (El-Sayed and Verpoorte 2007). Therefore, it is believed that strictosidine is the precursor of many important TIA compounds, such as the antihypertensive drug ajmalicine; the anticancer drugs vinblastine, vincristine, and camptothecin; and the anti-malaria drug quinine (Zulak et al. 2006; Pan et al. 2016) (Fig. 5.2B). The biosynthesis of the terpenoid indole alkaloids has been vigorously studied for more than a decade due to their pharmacological importance. Madagascar periwinkle (*Catharanthus roseus*), belonging to the Apocynaceae family, is of high economical and pharmacological value and is able to produce more than 130 TIAs. However, the TIAs are accumulated in very low amounts in plants and the complex biosynthesis pathway *in planta* is far beyond current understanding (Ziegler and Facchini 2008). The big challenge is those TIA-producing plants that are non-model medicinal plants for which genome information is scarce or unavailable. As with other pathways, recently, omics tools have been adopted to accelerate the deciphering of unknown parts of the TIA biosynthesis pathway (Pan et al. 2016).

TIA biosynthesis can be divided into three major steps: formation of tryptamine and secologanin, formation of monomeric alkaloids, and formation of bisindole alkaloids (Zhu et al. 2015). In the first step, iridoid, a cyclopenta[*c*] pyran monoterpenoid, is the key precursor in secologanin formation. Transcriptomics data from various TIA-producing medicinal plants were compared to screen out putative homologue genes involved in the iridoid pathway (Facchini et al. 2012). They were also compared to non-secologanin-producing plants to filter out unrelated genes and confirm that 7-deoxyloganic acid 7-hydroxylase (*CrDL7H*, also called *CYP72A224*), homologous to secologanin synthase-like gene in *Camptotheca acuminate*, is required in iridoid metabolism by VIGS study (Salim et al. 2013). The key cyclization step to produce cyclic terpenes to form iridoid was identified by using transcriptomics data, focusing on those genes showing co-expression patterns with the geraniol 10-hydroxylase in NAD(P)H family. Twenty co-regulated genes were selected and two high-ranking genes annotated to be progesterone-5β-reductase (P5βR) were subsequently validated by heterologous expression enzyme assay analyzed by thin-layer chromatography and GC/MS. VIGS has been established in *C. roseus* and is an efficient tool to study gene functions in many physiological responses (Liscombe and O'Connor 2011; Sung et al. 2014). P5βR-silencing plants with lower cyclase levels and TIA accumulation further confirmed the importance of this gene in the iridoid pathway (De Luca et al. 2014; Geu-Flores et al. 2012).

To boost the TIA production *in planta*, TF regulation has become the target, coupled with transcriptomics and metabolomics profiling analyses. In the iridoid pathway, the promoter region of geraniol-8-oxidase (*G8O*) has been analyzed to screen TFs involved in regulation in response to jasmonic acid (JA) treatment

(Suttipanta et al. 2007). Expression patterns of TFs similar to *G8O* in the RNA-seq data were selected, and one TF belonging to the basic helix-loop-helix (bHLH) family with higher transactivation activity was designated as bHLH iridoid synthase 1 gene (*BIS1*), which is capable of inducing upstream genes in the iridoid pathway (Van Moerkercke et al. 2015). Functional validation of metabolite profiling by silencing and/or overexpression revealed that BIS1 controls the production of iridoid, and the final TIA-type products exhibiting bioactivity can be viewed as targets for genetic engineering (Van Moerkercke et al. 2015). One well-studied example is JA-responsive TF APETALA2-domain 3 (ORCA3, octadecanoid-responsive *Catharanthus* APETALA2-domain), whose overexpression could upregulate expression of strictosidine synthase gene (*STR* gene), tryptophan decarboxylase gene (*TDC* gene), and desacetoxyvindoline 4-hydroxylase gene (*D4H* gene) and further increase TIA content in *Catharanthus* (Pan et al. 2012); through protein homologue comparison to *Arabidopsis*, several WRKY transcription factors in *C. roseus* derived from transcriptome sequences were identified and proved to be regulated by JA and participate in early- and mid-stage TIA metabolism regulation by controlling *G10H*, *TDC*, and *STR* expression (Schluttenhofer et al. 2014). These examples demonstrate that understanding of the functions of transcription factors may provide insights into their roles in phytohormone-mediated regulation of TIA compound production.

In the formation of strictosidine steps, loganin is catalyzed by CYP450 enzyme secologanin synthase (SLS, CYP72A1) to open the ring structure and form secologanin. Strictosidine synthase (STR) is responsible for the assembly of secologanin and tryptamine to produce strictosidine. SGD-mediated formation strictosidine aglycone further converts to the intermediates 4,21-dehydrogeissoschizine, cathenamine, and epi-cathenamine. Cathenamine is converted into catharanthine and tabersonine through several steps, but the enzymes involved are still unknown. Biosynthesis of vindoline from tabersonine requires 6 steps including tabersonine 16-hydroxylase (T16H), *O*-methyltransferase (OMT), an unidentified hydroxylase, NMT, D4H, and the last step deacetylvindoline-4-*O*-acetyltransferase (DAT) (Besseau et al. 2013; El-Sayed and Verpoorte 2007). Isoforms of SLS (SLS1 and SLS2) and T16H (T16H1 and T16H2) were found with different expression profiles or in specific tissue compartments, and both SLSs were able to oxidize secologanin into secoxyloganin. However, currently available transcriptome data were unable to provide accurate sequences or expression profiles for these isoforms (de Bernonville et al. 2015). Hence, transcriptome information about *C. roseus* has been reconstructed and reassembled from currently available databases, such as the Medicinal Plant Genomic Resources (Gongora-Castillo et al. 2012), PhytoMetaSyn (Xiao et al. 2013), and CathaCyc/Orcae (Van Moerkercke et al. 2013). This optimized data set can reduce redundancy contigs and differentiate SLS and T16H isoforms which were further confirmed by the expression profiles from FPKM and qPCR results (de Bernonville et al. 2015). This work provides a more complete and optimized transcriptomics database capable of dealing with potential isoform problems and facilitates uncovering of homologues or uncharacterized genes in TIA metabolism.

In *C. roseus*, some enzymes participating in TIA biosynthesis were restricted to certain tissues. For example, *TDC* and *STR* expression can be found in stems, leaves, flowers, and apical meristem of root tips, while *D4H* and *DAT* that are involved in the last steps in vindoline production are only present in the aerial part rather than underground parts (El-Sayed and Verpoorte 2007; St-Pierre et al. 1999). Therefore, to address the differential expression pattern and regulatory factors of leaf-specific vindoline, leaf and root tissues of *C. roseus* were then subjected to do RNA-seq analysis and generated 155 ESTs for homology-based functional categorization. Coupled with semiquantitative RT-PCR expression profiling, they discovered a known *DAT* and one novel *SGD*, which could only be detected in the aerial part of *C. roseus*. This tissue-specific gene expression corresponded to that of vindoline and was restricted to the green tissue of the plant (Shukla et al. 2006).

Champagne's group used a comparative proteomic strategy using 2D difference gel electrophoresis (2D-DIGE) to find novel proteins in two independent *C. roseus* cell lines with various TIA profiling and content. Through MALDI-MS/MS analysis and NCBInr protein database search, 22 enzymes were identified in the key steps of TIA biosynthesis including G10H, two 10HGO isoforms, SLS, tabersonine 16-hydroxylase (CYP71D12), NMT, D4H, and DAT, while 16 proteins predicted as ABC transports were also found to be involved in SM transport. They also found that the differential abundance of 10HGO can correspond to TIA content variation between the two cell lines, indicating the proteomics data can refer to metabolic profiling for functional analysis of the TIA pathway (Champagne et al. 2012). Another example of the use of integrating genomics tools is to describe missing genes in the secologanin-strictosidine pathway. Using RNA-seq and gene expression profile data identical to *LAMT* and *SLS* genes in the early TIA pathway, oxidoreductase, cytochrome P450 monooxygenase, and UGT were identified. Coupled with proteomics, those toolbox enzymes were identified along with GC/MS-based biochemical characterization. The missing enzymes were proposed which were 8-hydroxygeraniol oxidoreductase (8-HGO), iridoid oxidase (CYP76A26), 7-deoxyloganetic acid glucosyl transferase (UGT709C2), and 7-deoxyloganic acid hydroxylase (CYP72A224). Those enzymes were subsequently heterogeneous expressed in *Nicotiana benthamiana* to reconstitute the TIA pathway (Miettinen et al. 2014). The study demonstrated a platform using a synthetic biology approach and application in pharmacologically important bioactive compound production.

In the later bisindole alkaloid formation steps, monomeric alkaloids catharanthine and vindoline are coupled by anhydrovinblastine synthase (AVLBS, known as CrPRX1), the class III basic peroxidases localized in vacuoles (Costa et al. 2008), to synthesize α-3′,4′-anhydrovinblastine, which is further converted into vinblastine and vincristine (Zhu et al. 2015). Nevertheless, the enzymes involved in the last two steps still remain to be uncovered (Pan et al. 2016). To reveal these steps, "CathaCyc" was established by integrating *C. roseus* RNA-seq data sets containing 390 metabolic pathways and 1347 enzymes, including the TIA pathway. The information obtained from the *C. roseus* data sets were annotated to be included in the "MetaCyc," a metabolic pathway database derived from many organisms, and "AraCyc" and "PlantCyc" are derived from *Arabidopsis* studies (Van Moerkercke et al. 2013). Leaf, root, and flower tissues of *C. roseus* were subjected to RNA-seq using the Illumina platform, and the follow-up gene ontology (GO), BLAST, and CathaCyc analyses revealed that the expression of genes and TFs correlated to TIA accumulation was tissue specific. For example, tabersonine 16-hydroxylase and 4-*O*-acetyltransferase are two genes whose expression is restricted to the aerial parts as vindoline and vinblastine are only detected in plant leaves, while other upstream TIA pathway genes exist in both leaf and root. The CathaCyc system has great potential in expression profiling analysis, pathway elucidation, and discovering novel genes (Verma et al. 2014). A transcriptomics study compared three different TIA-producing plant species *Camptotheca acuminate*, *Catharanthus roseus*, and *Rauvolfia serpentina* for camptothecin, vinblastine, and ajmaline production, respectively (Gongora-Castillo et al. 2012). RNA-seq was introduced to analyze diverse sets of plants through their developmental tissues, cultured cells, or elicitor-treated plants. It provides rich sequence information near to full-length sequences of genes and also identified clusters of orthologous and paralogous gene families in those plants with camptothecin, vinblastine, and ajmaline biosynthesis (Gongora-Castillo et al. 2012). Based on the transcriptome database compiled by Gongora-Castillo et al. (2012), one tetrahydroalstonine synthase (*THAS*) gene, a nuclear-localized alcohol dehydrogenase homologue, upregulated by MeJA, was found to convert strictosidine aglycone to tetrahydroalstonine as validated by NMR spectrum, LC-MS, and VIGS study (Stavrinides et al. 2015).

5.4 Phenolics

Phenolics, also known as phenols, are SMs that are extensively distributed in the plant kingdom but are more uncommon in bacteria, fungi, and algae. They are a class of compounds possessing an aromatic ring with hydroxyl groups. Quideau et al. (2011) proposed a general rule for the term "plant phenolics" which should be restricted within the SM group and synthesized through the shikimate/phenylpropanoid pathway and/or the acetate/malonate pathway. This definition covers a very broad range of phenolic compounds from monomeric to polymeric structures. A large number of small phenolic compounds are found ubiquitously in plants such as gallic acid and catechin (Ow and Stupans 2003). Recently, they have been attracting attention as sunscreens to protect plants from UV irradiation, attractants for pollinators and seed-dispersing animals, protectants to defend or avoid microbial infection, allelopathic agents, and signal molecules in mutualistic symbiosis of *Rhizobium* and legumes, etc. (Kliebenstein 2004). Currently, one of the most interesting topics in

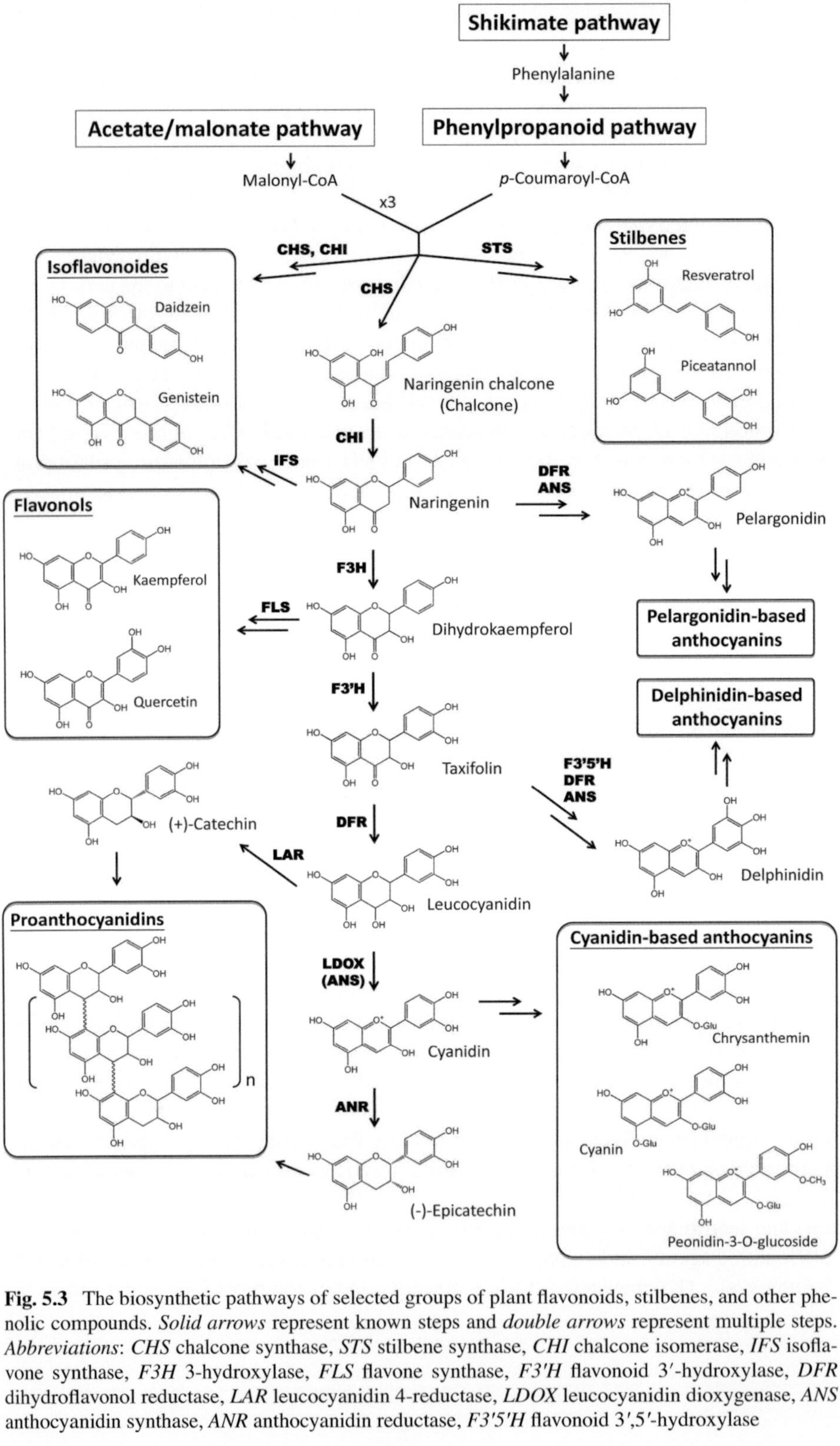

Fig. 5.3 The biosynthetic pathways of selected groups of plant flavonoids, stilbenes, and other phenolic compounds. *Solid arrows* represent known steps and *double arrows* represent multiple steps. *Abbreviations*: *CHS* chalcone synthase, *STS* stilbene synthase, *CHI* chalcone isomerase, *IFS* isoflavone synthase, *F3H* 3-hydroxylase, *FLS* flavone synthase, *F3'H* flavonoid 3′-hydroxylase, *DFR* dihydroflavonol reductase, *LAR* leucocyanidin 4-reductase, *LDOX* leucocyanidin dioxygenase, *ANS* anthocyanidin synthase, *ANR* anthocyanidin reductase, *F3'5'H* flavonoid 3′,5′-hydroxylase

plant phenolics is their beneficial effect on human health. A large volume of biological activities of plant phenolics have been demonstrated via in vitro studies, showing potential in reducing the risk of cardiovascular diseases, cataracts, cancers, etc., and part of their pharmacological activities might be due to their powerful antioxidant and free radical-scavenging properties (Kandaswami and Middleton 1994).

Plant phenolics can be classified into several categories based on carbon number and basic skeleton, such as C_6 (simple phenolics), C_6-C_1 (phenolic acids), C_6-C_2 (acetophenones), C_6-C_3 (hydroxycinnamic acids), C_6-C_4 (naphthoquinones), C_6-C_1-C_6 (xanthones), C_6-C_2-C_6 (stilbenes), C_6-C_3-C_6 (flavonoids), $(C_6\text{-}C_3)_2$ (lignans), $(C_6\text{-}C_3\text{-}C_6)_2$ (biflavonoids), $(C_6)_n$ (catechol melanins), $(C_6\text{-}C_3)_n$ (lignins), and $(C_6\text{-}C_3\text{-}C_6)_n$ (condensed tannins) (Lattanzio et al. 2008). Besides polyphenols that possess more than 2 phenolic cycles, several single aromatic ring-containing phenolics can be found (C_6, C_6-C_1, C_6-C_2, C_6-C_3 and C_6-C_4), some of which are well known in our daily diet. In some articles, phenolics are divided into two groups: flavonoids and non-flavonoids. The flavonoid group comprising two aromatic rings connected by a three-carbon bridge (C_6-C_3-C_6) contains the most natural phenolics and can be found throughout the plant kingdom (Tsao 2010). Flavonoids can be further classified into several subgroups: flavones, flavonols, isoflavones, flavanones, anthocyanidins, etc. The rest of the phenolics belong to the non-flavonoid group. Hydrolysable tannins including gallotannins and ellagitannins are formed by condensation of gallic acids, simple phenolics (C_6) which are derived from 3-dehydro-shikimate in the shikimate pathway. Most phenolic compounds are synthesized through the phenylpropanoid pathway. *p*-Coumaric acid is the precursor of phenolic acids, and *p*-coumaroyl-CoA can be further catalyzed and condensed to produce lignans and lignins. Other types of phenolics, such as stilbenes, flavonoids, and condensed tannins, require another precursor, malonyl-CoA, from the acetate/malonate pathway to yield their basic skeletons. Selective phenolic compounds and their biosynthesis cascades along with the corresponding enzymes in different steps are shown in Fig. 5.3. An overall profile of phenolic biosynthesis has been proposed; however, there are still many vague areas, for example, the regulatory mechanisms and post-modifications.

5.4.1 *Flavonoids and Stilbenes*

Flavonoids with the C_6-C_3-C_6 structure are highly conserved throughout the higher plant kingdom and are produced in response to UV stimulation and biotic and abiotic stresses. Stilbenes contain a C_6-C_2-C_6 basic skeleton and are phytoalexins produced by plants in response to pathogen invasion (Crozier 2008). Flavonoids exist in the human daily diet. Flavonoids contained in food sources include flavaonol (e.g., (+)-catechin and (−)-epicatechin), abundant in tea (Song and Chun 2008), flavonol (e.g., kaempferol) in tomato and red wine (Lopez et al. 2001; Stewart and Steenkamp 2000), and flavanones (e.g., naringin and naringenin) and anthocyanidin (e.g., apigenidin and cyaniding) which are abundant in grapefruits, cherries, and raspberries (Del Rio et al. 2010). The biological activities of flavonoids include antioxidant, hapatoprotective, antibacterial, anti-inflammatory, and anticancer activities (Kumar and Pandey 2013). Some reports have also suggested that consumption of flavonoid-rich foods can inhibit the development of Alzheimer's disease (Williams and Spencer 2012). In the stilbene group, resveratrol and its derivative *trans*-piceid are abundant in grapevines and legumes. They have been reported to possess anticarcinogenic and antitumor properties and are used as herbal remedies for heart diseases (Burns et al. 2002). These examples indicate the importance and application of flavonoids in the human diet and their possible pharmacological usage, and therefore research has put much emphasis on their metabolism.

The biosynthesis pathway of flavonoids is a complex network involving the formation of malonyl-CoA and *p*-coumaroyl-CoA from the upstream malonate pathway and the shikimic acid pathway (Crozier et al. 2006). The enzyme chalcone synthase (CHS) condenses the two early precursors to form naringenin chalcone (Ayabe et al. 1988), while stilbene synthase (STS) with identical function to CHS forms a stilbene backbone from three malonyl-CoA with one *p*-coumaroyl-CoA (Wiese et al. 1994). Naringenin chalcone is then stereospecifically converted to naringenin (flavanone) by chalcone isomerase (CHI), which is a central intermediate and the branch point from which diverse modification produces different types of flavonoids (Boland and Wong 1975). The microsomal cytochrome P450 enzyme isoflavone synthase (IFS) is responsible for the first step in the isoflavonoid biosynthesis branch. This gene has been cloned and the enzyme function has been identified in legumes through homologous-based ETS collection identification (Jung et al. 2000).

Flavonol metabolism starts with flavanone 3-hydroxylase (F3H)-mediated conversion from flavanones to dihydrokaempferol, and then flavone synthase (FLS) characterized in citrus catalyzes the formation of the flavonol kaempferol (Wellmann et al. 2002). For the following anthocyanidin synthesis, dihydrokaempferol is converted to dihydroflavonol by flavonoid 3′-hydroxylase (F3′H) and then dihydroflavonol reductase (DRF) to catalyze the formation of leucocyanidin, which is a key intermediate in anthocyanidin formation (Fischer et al. 2003). The leucocyanidin intermediate is involved in two distinct processes; leucocyanidin 4-reductase (LAR) converts it to flavan-3-ol (+)-catechin, while leucocyanidin dioxygenase (LDOX) catalyzes the synthesis of cyanidin (anthocyanidin) from leucocyanidin and then anthocyanidin reductase (ANR) mediates cyanidin to form (−)-epicatechin (Xie et al. 2003; Tanner et al. 2003). The two monomers, (+)-catechin and (−)-epicatechin, derived from leucocyanidin constitute the extension units for polymeric proanthocyanidin formation (Tanner et al. 2003). Although some key steps in the flavonoid pathways have been reported, most of the knowledge of flavonoid metabolism is derived from *Arabidopsis*, maize, or grapes (Bogs et al. 2006; Saito et al. 2013). The genome information on medicinal plants is limited and may be restricted to only a few hundred ESTs available in a public database (Kalra et al. 2013), making studies on non-model medicinal plants difficult in terms of bioactive compound biosynthesis. Hence, using multiomics strategy systemic biology is useful to identify novel molecular mechanisms related to flavonoid metabolism in medicinal plants (Suzuki et al. 2015; Zamboni et al. 2010).

5.4.2 Berries

Berries are being recognized as some of the healthiest foods on the planet because they have been identified to contain several antioxidants such as vitamins C, E, ellagic acid, anthocyanins, flavonoids, etc. which have been shown to reduce heart disease, slow the aging process, reduce cell damage, and act as anticancer agents (Del Rio et al. 2010). Therefore, a great number of berry fruit transcriptome studies have been actively pursued in recent years, aiming to elucidate the bioactive flavonoid compound biosynthesis. RNA-seq of two developmental stages of cranberry (*Vaccinium macrocarpon*) fruit was annotated based on public databases NCBI nr, NT, GO, and KEGG in which candidate genes involved in flavonoid biosynthesis including *CHS*, *CHI*, *F3H*, *F3′H*, and *LDOX* were observed (Sun et al. 2015). Compared to transcriptomics data for blueberry (*Vaccinium* sp.), cranberry has relatively more abundant flavonoid biosynthesis enzymes such as UDP-glucose flavonoid 3-*O*-glucosyl transferase (UFGT), which correspond to more diverse types of flavonoids existing in cranberry. Some known flavonoid transporters and regulatory transcription factors were also found in this transcriptome resource such as ABC transporters, glutathione *S*-transferases (GST), and WD40, bHLH, and WRKY transcription factors. This study provides comprehensive information on gene IDs and their expression during flavonoid metabolism (Sun et al. 2015). The Korean black raspberry (*Rubus coreanus*) is also rich in anthocyanin, and metabolomics and transcriptomics analyses have been performed to investigate the metabolite changes in the fruit-ripening process. MS-based spectrometry revealed that the content of the flavonoid and anthocyanin groups increased while the amount of sucrose and most carbon-nitrogen compounds, such as amino acids, organic acids, and fatty acids, decreased (Hyun et al. 2014). Transcriptomics data has demonstrated that 28 annotated unigenes are involved in flavonoid metabolism including key genes like *CHI*, *CHS*, and *F3H*. Because the *CHI* gene has been viewed as the major gene deciding the flavonoid pathway and anthocyanin production in many higher plants or crops like rice, tomato, and petunia (Bovy et al. 2007; Hong et al. 2012; Muir et al. 2001), Hyun and co-workers screened for the CHI enzyme family in the Korean black raspberry transcriptome database and confirmed the CHI enzyme functions by complementary tests in *Arabidopsis transparent testa 5-1* (*tt5-1*) mutant which lacks CHI activity. One *Rubus coreanus* Miquel chalcone flavanone isomerase 2 (*RcMCHI2*) gene was identified to encode a CHI enzyme function that may be used as the target for genetic manipulation of flavonoid synthesis in the Korean black raspberry. Profiling of *DFR4* and *LDOX1* gene expression and cyaniding derivative content can be seen as positively correlated during the fruit-ripening process in *Rubus coreanus*, which provides one model system describing the gene-metabolite relationship in non-model plants (Hyun et al. 2014).

5.4.3 Anthocyanin

The flavonoid anthocyanin possesses anti-inflammatory and antioxidation properties and also contributes to flower color (Del Rio et al. 2010). Anthocyanin accumulation is associated with variation in flower color (Weiss 2000) and has been used to elucidate genes involved in flavonoid biosynthesis in *Magnolia sprengeri*. Comparative transcriptomics of red and white flowers from two distinct strains of *Magnolia sprengeri* revealed some key enzymes in flavonoid biosynthesis, such as phenylalanine ammonia-lyase (PAL), cinnamate-4-hydroxylase (C4H), F3H, F3′H, and CHS. Three TF families, MYB, bHLH, and WD40, were also identified in regulating anthocyanidin metabolism. FPKM-based analysis showed that eight of these TF genes exhibit higher (eightfold increase) expression level in red petals than in white petals which are also associated with metabolic processes, gene transcript abundance, and flower color (Shi et al. 2014). The molecular insights regarding flavonoid glycosylation mechanism were also investigated by transcriptome analysis in *Chlorophytum borivilianum* using Illumina's HiSeq 2000 sequencing platform. *C. borivilianum* is an important traditional herbal medicine used as an adaptogen, for antiaging and for general health promotion; however, this plant is now endangered due to overexploitation (Kaushik 2005). Simple sequence repeats (SSR) derived from RNA-seq results were used to identify unigenes from the EST database of the plant. Three groups of genes involved in glycosylation of flavonoids were found including 16 anthocyanidin 3-*O*-glucosyltransferase genes, 14 anthocyanidin 2-*O*-β-glucosyltransferase genes, and 46 flavonol-3-*O*-glucosyltransferase genes. Reads per kilobase per million mapped reads (RPKM)-based quantification of transcripts also helped to differentiate genes involved in different metabolic pathway and provide functional genomics information in *Chlorophytum borivilianum* (Kalra et al. 2013). Moreover, two flavones with certain glycosylation patterns baicalin and wogonoside are rich in root tissues of a Chinese herbal medicine *Scutellaria baicalensis* (SB) with anti-inflammatory and antitumor activities (Takahashi et al. 2011). Investigation of their metabolism was conducted by Illumina/Solexa deep sequencing to reveal organ-specific gene and metabolite profiling (Liu et al. 2015a). The RNA-seq data found 54 unigenes encoding 12 key enzymes in flavonoid biogenesis. Through RPKM-based differential expression profiling in roots, leaves, stems, and flowers followed by qRT-PCR validation, one baicalinase and three baicalein 7-*O*-glucuronosyltransferases were identified to be involved in the transformation of baicalin/wogonoside and baicalein/wogonin, and four candidate 6-hydroxylases and one 8-*O*-methyltransferase were involved in baicalein/baicalin and wogonin/wogonoside biosynthesis. This work provides more insight into the metabolism of these bioactive flavonoids and their relationships to gene expression, metabolite profiling, and organ specificity (Liu et al. 2015a).

5.4.4 Resveratrol

Stilbenes, in particular resveratrol (3,5,4′-trihydroxystilbene), are rich in grapefruit (*Vitis vinifera*) and red wine and show antioxidant or radical-scavenging properties (Goldberg and Soleas 2003). The genome of grape has been sequenced and is available online (http://www.genoscope.cns.fr/spip) (Jaillon et al. 2007; Velasco et al. 2007). With advances in RNA sequencing techniques, the expression profile of some key enzymes and transcription factors has been identified on a global scale. Different stages of grapefruit ripening, known as veraison, were subjected to transcriptome analysis, and the expression profiling of genes was clustered by RPKM-based K-means grouping (Grimplet et al. 2009). Some key genes involved in stilbene biosynthesis including 12 *PAL* genes, 3 *C4H* genes, and 38 *STS* genes have been identified and most of them were found to have low levels of expression in the young and veraison stage but were specifically highly upregulated in ripened fruits. This indicates that the timing of the berry harvest determines the stilbene content in fruits or red wine through transcriptional regulation (Sweetman et al. 2012). Vannozzi et al. studied the *stilbene synthase* (*STS*) family genes and phylogenetically related chalcone synthase genes in determining the formation of stilbene and flavonoid derivatives (Tropf et al. 1994).

Whole transcriptome sequencing of *Vitis vinifera* in response to biotic and abiotic stresses revealed 33 full-length *STS* sequences which were divided into 3 families depending on expression profiles validated by microarray and qRT-PCR analysis. This analysis provides differential expression profiling of *VvSTS* and *VvCHS* showing that they responded to stresses with different timing, indicating the evolutionary relationship between the two gene families and the carbon flow of the two metabolic pathways is under transcriptional control (Vannozzi et al. 2012). Multiomics involving transcriptomics, metabolomics, and genome-wide microarray analysis revealed 238 genes and 2012 metabolites upregulated by UV-C irradiation in *V. vinifera*. Based on gene annotation and GO analysis, the trihydroxystilbene synthase family including *STS* genes was activated, while liquid chromatography-quadrupole time-of-flight mass spectrometry (LC-QTOF/MS) analysis also detected a high intensity of resveratrol induced by UV-C. Genes encoding PAL, 4-coumarate-CoA ligase, and cinnamate 4-hydroxylase in the early stage (the coumaroyl-CoA formation steps) of resveratrol biogenesis were also upregulated, whereas expression in the branch pathway of phenolic compounds like CHS and CHI was not induced by UV-C treatment. The results from the multiomics study updated knowledge of SM metabolism and supported the notion that transcription factor-mediated regulation is responsible for regulating stilbene biosynthesis (Suzuki et al. 2015).

In 2013, some key transcription factors affecting *STS* gene expression in grapevine were deciphered using RNA-seq data. Two genes in the R2R3-MYB TF family, *MYB14* and *MYB15*, displayed a similar expression pattern to that of *STS* genes. Further confirmations were conducted by qRT-PCR that showed a strong co-induction of *MYB14/15* and *STS* genes in response to wounding, UV irradiation, pathogen invasion, and resveratrol metabolite synthesis. Integrated with HPLC and LC-QTOF-MS analysis, the accumulation of glycosylated resveratrol and *trans*-piceid, major resveratrol derivatives in grapevine, was shown to correspond to *MYB14/15* and *STS* isoform gene expression during fruit development, further demonstrating that stilbene biosynthesis regulation is transcription factor dependent (Holl et al. 2013). Understanding the relationship between critical genes/enzymes and their regulatory transcription factors provides direction for genetic engineering of transcription factors to increase yields of pharmacologically valuable SMs in plants (Liu et al. 2015b).

5.5 Future Prospects

Plant SMs, such as aglycones or the glycosides of terpenoids, flavonoids, and alkaloids, have a rich chemical ecology that has been exploited for medicinal purposes over thousands of years. The ability of plants to produce such structurally diverse bioactive compounds has been studied for more than a century. Despite being commonly used in modern pharmaceutical or nutritional science for their therapeutic or disease preventive properties, relatively little is known about the biosynthesis, regulation, and transport of these bioactive molecules. The employment of synthetic biology strategies to sustainably manufacture of bioactive compounds is an attractive option to conserve medicinal plants, crops, and other precious woody plants that produce novel bioactive compounds. However, synthetic production will only become possible if the biosynthetic pathways and the underlying regulatory mechanisms of these SMs are elucidated. As an alternative to relying on laborious single gene or enzyme identification and functional characterization, the technological and methodological advances in systems biology in plants using omics approaches are efficiently enhancing our ability to identify numerous novel genes, enzymes, and/or transcription factors involved in specific bioactive compound synthesis or metabolism *in planta*. Whole genome information is lacking for most non-model medicinal or woody plants and genetic transformation techniques remain to be established. It is therefore suggested that scientists with multidisciplinary expertise in plant molecular biology, genetic engineering and transformation, natural product chemistry, enzymology, protein engineering, omics technology, and bioinformatics contribute research to the important and challenging field of uncovering the biosynthesis of novel bioactive compounds. Furthermore, it is not only necessary to elucidate the entire pathways at the biochemical and genetic levels but also to understand the enzyme interactions, metabolic regulation, and compartmentalization in plants to achieve successful synthetic production of pharmacologically and economically important bioactive compounds.

References

Aerts RJ, Mordue AJ (1997) Feeding deterrence and toxicity of neem triterpenoids. J Chem Ecol 23(9):2117–2132

Allen RS, Miller JA, Chitty JA, Fist AJ, Gerlach WL, Larkin PJ (2008) Metabolic engineering of morphinan alkaloids by over-expression and RNAi suppression of salutaridinol 7-*O*-acetyltransferase in opium poppy. Plant Biotechnol J 6(1):22–30

Ashour M, Wink M, Gershenzon J (2010) Biochemistry of terpenoids: monoterpenes, sesquiterpenes and diterpenes. In: Wink M (ed) Biochemistry of plant secondary metabolism. Wiley, New York, pp 258–303

Ayabe S, Udagawa A, Furuya T (1988) NAD(P)H-dependent 6′-deoxychalcone synthase activity in *Glycyrrhiza echinata* cells induced by yeast extract. Arch Biochem 261(2):458–462

Beaudoin GA, Facchini PJ (2013) Isolation and characterization of a cDNA encoding (*S*)-cis-*N*-methylstylopine 14-hydroxylase from opium poppy, a key enzyme in sanguinarine biosynthesis. Biochem Biophys Res Commun 431(3):597–603

Bernhoft A (2010) A brief review on bioactive compounds in plants. Bioactive compounds in plants-benefits and risks for man and animals. The Norwegian Academy of Science and Letters, Novus forlag, Oslo, pp 11–18

Besseau S, Kellner F, Lanoue A, Thamm AM, Salim V, Schneider B, Geu-Flores F, Hofer R, Guirimand G, Guihur A, Oudin A, Glevarec G, Foureau E, Papon N, Clastre M, Giglioli-Guivarc'h N, St-Pierre B, Werck-Reichhart D, Burlat V, De Luca V, O'Connor SE, Courdavault V (2013) A pair of tabersonine 16-hydroxylases initiates the synthesis of vindoline in an organ-dependent manner in *Catharanthus roseus*. Plant Physiol 163(4):1792–1803

Bogs J, Ebadi A, McDavid D, Robinson SP (2006) Identification of the flavonoid hydroxylases from grapevine and their regulation during fruit development. Plant Physiol 140(1):279–291

Boland MJ, Wong E (1975) Purification and kinetic properties of chalcone-flavanone isomerase from soya bean. Eur J Biochem 50(2):383–389

Bourgaud F, Gravot A, Milesi S, Gontier E (2001) Production of plant secondary metabolites: a historical perspective. Plant Sci 161(5):839–851

Bovy A, Schijlen E, Hall RD (2007) Metabolic engineering of flavonoids in tomato: the potential for metabolomics. Metabolomics 3:399–412

Breitmaier E (2006) Terpenes: flavors, fragrances, pharmaca, pheromones. Wiley-VCH Verlag GmbH & Co. KGaA, Weinheim

Brusotti G, Cesari I, Dentamaro A, Caccialanza G, Massolini G (2014) Isolation and characterization of bioactive compounds from plant resources: the role of analysis in the ethnopharmacological approach. J Pharm Biomed 87:218–228

Buckingham J (2007) Dictionary of natural products. Chapman and Hall CRC, London

Burns J, Yokota T, Ashihara H, Lean ME, Crozier A (2002) Plant foods and herbal sources of resveratrol. J Agric Food Chem 50(11):3337–3340

Busia K, Kasilo OM (2010) Overview of traditional medicine in Ecowas Member States. Afr Health Monit 13:16–24

Champagne A, Rischer H, Oksman-Caldentey KM, Boutry M (2012) In-depth proteome mining of cultured *Catharanthus roseus* cells identifies candidate proteins involved in the synthesis and transport of secondary metabolites. Proteomics 12(23–24):3536–3547

Chen CYC (2011) TCM Database@Taiwan: the world's largest traditional chinese medicine database for drug screening *in silico*. PLos One 6(1), e15939

Chen F, Ro DK, Petri J, Gershenzon J, Bohlmann J, Pichersky E, Tholl D (2004) Characterization of a root-specific *Arabidopsis* terpene synthase responsible for the formation of the volatile monoterpene 1,8-cineole. Plant Physiol 135(4):1956–1966

Chen S, Luo H, Li Y, Sun Y, Wu Q, Niu Y, Song J, Lv A, Zhu Y, Sun C, Steinmetz A, Qian Z (2011) 454 EST analysis detects genes putatively involved in ginsenoside biosynthesis in *Panax ginseng*. Plant Cell Rep 30(9):1593–1601

Cheng QQ, Su P, Hu YT, He YF, Gao W, Huang LQ (2014) RNA interference-mediated repression of *SmCPS* (copalyldiphosphate synthase) expression in hairy roots of *Salvia miltiorrhiza* causes a decrease of tanshinones and sheds light on the functional role of SmCPS. Biotechnol Lett 36(2):363–369

Christianson DW (2006) Structural biology and chemistry of the terpenoid cyclases. Chem Rev 106(8):3412–3442

Costa MM, Hilliou F, Duarte P, Pereira LG, Almeida I, Leech M, Memelink J, Barcelo AR, Sottomayor M (2008) Molecular cloning and characterization of a vacuolar class III peroxidase involved in the metabolism of anticancer alkaloids in *Catharanthus roseus*. Plant Physiol 146(2):403–417

Croteau R, Kutchan TM, Lewis NG (2000) Natural products (secondary metabolites). Biochemistry and molecular biology of plants, 2nd edn. Wiley, Blackwell

Crozier A (2008) Plant secondary metabolites: their role in the human diet. Planta Med 74(3):315–315

Crozier A, Jaganath IB, Clifford MN (2006) Phenols, polyphenols and tannins: an overview. In: Plant secondary metabolites: occurrence, structure and role in the human diet. Wiley, Blackwell

Dang TT, Facchini PJ (2014) CYP82Y1 is *N*-methylcanadine 1-hydroxylase, a key noscapine biosynthetic enzyme in opium poppy. J Biol Chem 289(4):2013–2026

Dang TTT, Onoyovwi A, Farrow SC, Facchini PJ (2012) Biochemical genomics for gene discovery in benzylisoquinoline alkaloid biosynthesis in opium poppy and related species. Methods Enzymol 515:231–266

Dang TT, Chen X, Facchini PJ (2015) Acetylation serves as a protective group in noscapine biosynthesis in opium poppy. Nat Chem Biol 11(2):104–106

Davies TGE, Field LM, Usherwood PNR, Williamson MS (2007) DDT, pyrethrins, pyrethroids and insect sodium channels. Iubmb Life 59(3):151–162

de Bernonville TD, Foureau E, Parage C, Lanoue A, Clastre M, Londono MA, Oudin A, Houille B, Papon N, Besseau S, Glevarec G, Atehortua L, Giglioli-Guivarc'h N, St-Pierre B, De Luca V, O'Connor SE, Courdavault V (2015) Characterization of a second secologanin synthase isoform producing both secologanin and secoxyloganin allows enhanced de novo assembly of a *Catharanthus roseus* transcriptome. BMC Genomics 16:619–637

De Luca V, Salim V, Thamm A, Masada SA, Yu F (2014) Making iridoids/secoiridoids and monoterpenoid indole alkaloids: progress on pathway elucidation. Curr Opin Plant Biol 19:35–42

Decker G, Wanner G, Zenk MH, Lottspeich F (2000) Characterization of proteins in latex of the opium poppy (*Papaver somniferum*) using two-dimensional gel electrophoresis and microsequencing. Electrophoresis 21(16):3500–3516

Del Rio D, Borges G, Crozier A (2010) Berry flavonoids and phenolics: bioavailability and evidence of protective effects. Brit J Nutr 104(S3):S67–S90

Desgagne-Penix I, Facchini PJ (2012) Systematic silencing of benzylisoquinoline alkaloid biosynthetic genes reveals the major route to papaverine in opium poppy. Plant J 72(2):331–344

Desgagne-Penix I, Farrow SC, Cram D, Nowak J, Facchini PJ (2012) Integration of deep transcript and targeted metabolite profiles for eight cultivars of opium poppy. Plant Mol Biol 79(3):295–313

El-Sayed M, Verpoorte R (2007) *Catharanthus* terpenoid indole alkaloids: biosynthesis and regulation. Phytochem Rev 6(2–3):277–305

Facchini PJ, Bohlmann J, Covello PS, De Luca V, Mahadevan R, Page JE, Ro DK, Sensen CW, Storms R, Martin VJ (2012) Synthetic biosystems for the production of high-value plant metabolites. Trends Biotechnol 30(3):127–131

Farrow SC, Facchini PJ (2013) Dioxygenases catalyze O-demethylation and O, O-demethylenation with widespread roles in benzylisoquinoline alkaloid metabolism in opium poppy. J Biol Chem 288(40):28997–29012

Farrow SC, Hagel JM, Facchini PJ (2012) Transcript and metabolite profiling in cell cultures of 18 plant species that produce benzylisoquinoline alkaloids. Phytochemistry 77:79–88

Fischer TC, Halbwirth H, Meisel B, Stich K, Forkmann G (2003) Molecular cloning, substrate specificity of the functionally expressed dihydroflavonol 4-reductases from *Malus domestica* and *Pyrus communis* cultivars and the consequences for flavonoid metabolism. Arch Biochem Biophys 412(2):223–230

Gao Y, Honzatko RB, Peters RJ (2012) Terpenoid synthase structures: a so far incomplete view of complex catalysis. Nat Prod Rep 29(10):1153–1175

Gao W, Sun HX, Xiao HB, Cui GH, Hillwig ML, Jackson A, Wang X, Shen Y, Zhao N, Zhang LX, Wang XJ, Peters RJ, Huang LQ (2014) Combining metabolomics and transcriptomics to characterize tanshinone biosynthesis in *Salvia miltiorrhiza*. BMC Genomics 15:73–86

Geu-Flores F, Sherden NH, Courdavault V, Burlat V, Glenn WS, Wu C, Nims E, Cui Y, O'Connor SE (2012) An alternative route to cyclic terpenes by reductive cyclization in iridoid biosynthesis. Nature 492(7427):138–142

Glenn WS, Runguphan W, O'Connor SE (2013) Recent progress in the metabolic engineering of alkaloids in plant systems. Curr Opin Biotech 24(2):354–365

Goldberg D, Soleas G (2003) Resveratrol: biochemistry, cell biology and the potential role in disease prevention. In: Wine: a scientific exploration. Taylor and Francis, London, pp 160–198

Gongora-Castillo E, McKnight TD, O'Connor S, Childs KL, Buell C, Fedewa G, Hamilton JP, Liscombe DK, Magallanes-Lundback M, Mandadi KK, Nims E, Runguphan W, Vaillancourt B, Varbanova-Herde M, DellaPenna D (2012) Development of transcriptomic resources for interrogating the biosynthesis of monoterpene indole alkaloids in medicinal plant species. PLos One 7(12), e52506

Grimplet J, Cramer GR, Dickerson JA, Mathiason K, Van Hemert J, Fennell AY (2009) VitisNet: "Omics" integration through grapevine molecular networks. PLoS One 4(12), e8365

Guo J, Zhou YJJ, Hillwigc ML, Shen Y, Yang L, Wang YJ, Zhang XA, Liu WJ, Peters RJ, Chen XY, Zhao ZBK, Huang LQ (2013) CYP76AH1 catalyzes turnover of miltiradiene in tanshinones biosynthesis and enables heterologous production of ferruginol in yeasts. Proc Natl Acad Sci Biol 110(29):12108–12113

Hagel JM, Facchini PJ (2013) Benzylisoquinoline alkaloid metabolism: a century of discovery and a brave new world. Plant Cell Physiol 54(5):647–672

Hagel JM, Mandal R, Han B, Han J, Dinsmore DR, Borchers CH, Wishart DS, Facchini PJ (2015a) Metabolome analysis of 20 taxonomically related benzylisoquinoline alkaloid-producing plants. BMC Plant Biol 15:220–236

Hagel JM, Morris JS, Lee EJ, Desgagne-Penix I, Bross CD, Chang L, Chen X, Farrow SC, Zhang Y, Soh J, Sensen CW, Facchini PJ (2015b) Transcriptome analysis of 20 taxonomically related benzylisoquinoline alkaloid-producing plants. BMC Plant Biol 15:227–242

Hahlbrock K (1981) Flavonoids. Biochem Plants 7:425–456

Han JY, Kwon YS, Yang DC, Jung YR, Choi YE (2006) Expression and RNA interference-induced silencing of the dammarenediol synthase gene in *Panax ginseng*. Plant Cell Physiol 47(12):1653–1662

Han JY, In JG, Kwon YS, Choi YE (2010) Regulation of ginsenoside and phytosterol biosynthesis by RNA interferences of squalene epoxidase gene in *Panax ginseng*. Phytochemistry 71(1):36–46

Han JY, Kim HJ, Kwon YS, Choi YE (2011) The Cyt P450 enzyme CYP716A47 catalyzes the formation of protopanaxadiol from dammarenediol-II during ginsenoside biosynthesis in *Panax ginseng*. Plant Cell Physiol 52(12):2062–2073

Han JY, Hwang HS, Choi SW, Kim HJ, Choi YE (2012) Cytochrome P450 CYP716A53v2 catalyzes the formation of protopanaxatriol from protopanaxadiol during ginsenoside biosynthesis in *Panax ginseng*. Plant Cell Physiol 53(9):1535–1545

Han JY, Kim MJ, Ban YW, Hwang HS, Choi YE (2013) The involvement of beta-amyrin 28-oxidase (CYP716A52v2) in oleanane-type ginsenoside biosynthesis in *Panax ginseng*. Plant Cell Physiol 54(12):2034–2046

Hartmann T (2007) From waste products to ecochemicals: fifty years research of plant secondary metabolism. Phytochemistry 68(22–24):2831–2846

Hemmerly TE (1977) A ginseng farm in Lawrence County, Tennessee. Econ Bot 31(2):160–162

Hileman LC, Drea S, Martino G, Litt A, Irish VF (2005) Virus-induced gene silencing is an effective tool for assaying gene function in the basal eudicot species *Papaver somniferum* (opium poppy). Plant J 44(2):334–341

Holl J, Vannozzi A, Czemmel S, D'Onofrio C, Walker AR, Rausch T, Lucchin M, Boss PK, Dry IB, Bogs J (2013) The R2R3-MYB transcription factors MYB14 and MYB15 regulate stilbene biosynthesis in *Vitis vinifera*. Plant Cell 25(10):4135–4149

Hong L, Qian Q, Tang D, Wang K, Li M, Cheng Z (2012) A mutation in the rice chalcone isomerase gene causes the golden hull and internode 1 phenotype. Planta 236(1):141–151

Hyun TK, Lee S, Rim Y, Kumar R, Han X, Lee SY, Lee CH, Kim JY (2014) De-novo RNA sequencing and metabolite profiling to identify genes involved in anthocyanin biosynthesis in Korean black raspberry (*Rubus coreanus* Miquel). PLoS One 9(2), e88292

Inui T, Kawano N, Shitan N, Yazaki K, Kiuchi F, Kawahara N, Sato F, Yoshimatsu K (2012) Improvement of benzylisoquinoline alkaloid productivity by overexpression of 3′-hydroxy-*N*-methylcoclaurine 4′-*O*-methyltransferase in transgenic *Coptis japonica* plants. Biol Pharm Bull 35(5):650–659

Jaillon O, Aury JM, Noel B, Policriti A, Clepet C, Casagrande A, Choisne N, Aubourg S, Vitulo N, Jubin C, Vezzi A, Legeai F, Hugueney P, Dasilva C, Horner D, Mica E, Jublot D, Poulain J, Bruyere C, Billault A, Segurens B, Gouyvenoux M, Ugarte E, Cattonaro F, Anthouard V, Vico V, Del Fabbro C, Alaux M, Di Gaspero G, Dumas V, Felice N, Paillard S, Juman I, Moroldo M, Scalabrin S, Canaguier A, Le Clainche I, Malacrida G, Durand E, Pesole G, Laucou V, Chatelet P, Merdinoglu D, Delledonne M, Pezzotti M, Lecharny A, Scarpelli C, Artiguenave F, Pe ME, Valle G, Morgante M, Caboche M, Adam-Blondon AF, Weissenbach J, Quetier F, Wincker P, French-Italian Public Consortium for Grapevine Genome C (2007) The grapevine genome sequence suggests ancestral hexaploidization in major angiosperm phyla. Nature 449(7161):463–467

Jayakodi M, Lee SC, Lee YS, Park HS, Kim NH, Jang W, Lee HO, Joh HJ, Yang TJ (2015) Comprehensive analysis of *Panax ginseng* root transcriptomes. BMC Plant Biol 15:138–149

Jung W, Yu O, Lau SM, O'Keefe DP, Odell J, Fader G, McGonigle B (2000) Identification and expression of isoflavone synthase, the key enzyme for biosynthesis of isoflavones in legumes. Nat Biotechnol 18(2):208–212

Kalra S, Puniya BL, Kulshreshtha D, Kumar S, Kaur J, Ramachandran S, Singh K (2013) De novo transcriptome sequencing reveals important molecular networks and metabolic pathways of the plant, *Chlorophytum borivilianum*. PLoS One 8(12), e83336

Kandaswami C, Middleton E Jr (1994) Free radical scavenging and antioxidant activity of plant flavonoids. In: Armstrong D (ed) Free radicals in diagnostic medicine. Springer, New York, pp 351–376

Kang JH, Song KH, Woo JK, Park MH, Rhee MH, Choi C, Oh SH (2011) Ginsenoside Rp1 from *Panax ginseng* exhibits anti-cancer activity by down-regulation of the IGF-1R/Akt pathway in breast cancer cells. Plant Food Hum Nutr 66(3):298–305

Kaushik N (2005) Saponins of *Chlorophytum* species. Phytochem Rev 4(2–3):191–196

Kempe K, Higashi Y, Frick S, Sabarna K, Kutchan TM (2009) RNAi suppression of the morphine biosynthetic gene *salAT* and evidence of association of pathway enzymes. Phytochemistry 70(5):579–589

Khorolragchaa A, Kim YJ, Rahimi S, Sukweenadhi J, Jang MG, Yang DC (2014) Grouping and characterization of putative glycosyltransferase genes from *Panax ginseng* Meyer. Gene 536(1):186–192

Kim YK, Kim YB, Uddin MR, Lee S, Kim SU, Park SU (2014) Enhanced triterpene accumulation in *Panax ginseng* hairy roots overexpressing mevalonate-5-pyrophosphate decarboxylase and farnesyl pyrophosphate synthase. ACS Synth Biol 3(10):773–779

Kim YJ, Zhang DB, Yang DC (2015) Biosynthesis and biotechnological production of ginsenosides. Biotechnol Adv 33(6):717–735

Kliebenstein DJ (2004) Secondary metabolites and plant/environment interactions: a view through *Arabidopsis thaliana* tinged glasses. Plant Cell Environ 27(6):675–684

Kumar S, Pandey AK (2013) Chemistry and biological activities of flavonoids: an overview. Scientific World Journal 2013:162750

Kutchan TM, Hampp N, Lottspeich F, Beyreuther K, Zenk MH (1988) The cDNA clone for strictosidine synthase from *Rauvolfia serpentina* DNA sequence determination and expression in *Escherichia coli*. FEBS Lett 237(1–2):40–44

Lattanzio V, Kroon PA, Quideau S, Treutter D (2008) Plant phenolics—secondary metabolites with diverse functions. Rec Adv Polyphen Res 1:1–35

Lee MH, Jeong JH, Seo JW, Shin CG, Kim YS, In JG, Yang DC, Yi JS, Choi YE (2004) Enhanced triterpene and phytosterol biosynthesis in *Panax ginseng* overexpressing squalene synthase gene. Plant Cell Physiol 45(8):976–984

Lee WL, Shiau JY, Shyur LF (2012) Taxol, camptothecin and beyond for cancer therapy. Adv Bot Res 62:133–178

Lee EJ, Hagel JM, Facchini PJ (2013) Role of the phloem in the biochemistry and ecophysiology of benzylisoquinoline alkaloid metabolism. Front Plant Sci 4:182–188

Li C, Zhu Y, Guo X, Sun C, Luo H, Song J, Li Y, Wang L, Qian J, Chen S (2013) Transcriptome analysis reveals ginsenosides biosynthetic genes, microRNAs and simple sequence repeats in *Panax ginseng* CA Meyer. BMC Genomics 14:245–255

Li C, Li D, Shao F, Lu S (2015) Molecular cloning and expression analysis of WRKY transcription factor genes in *Salvia miltiorrhiza*. BMC Genomics 16:200–220

Liscombe DK, O'Connor SE (2011) A virus-induced gene silencing approach to understanding alkaloid metabolism in *Catharanthus roseus*. Phytochemistry 72(16):1969–1977

Liu J, Hou J, Jiang C, Li G, Lu H, Meng F, Shi L (2015a) Deep sequencing of the *Scutellaria baicalensis* Georgi transcriptome reveals flavonoid biosynthetic profiling and organ-specific gene expression. PLoS One 10(8), e0136397

Liu J, Osbourn A, Ma P (2015b) MYB Transcription factors as regulators of phenylpropanoid metabolism in plants. Mol Plant 8(5):689–708

Lopez M, Martinez F, Del Valle C, Orte C, Miro M (2001) Analysis of phenolic constituents of biological interest in red wines by high-performance liquid chromatography. J Chromatogr A 922(1–2):359–363

Luo H, Zhu Y, Song J, Xu L, Sun C, Zhang X, Xu Y, He L, Sun X, Xu H, Wang B, Li X, Li C, Liu J, Chen S (2014) Transcriptional data mining of *Salvia miltiorrhiza* in response to methyl jasmonate to examine the mechanism of bioactive compound biosynthesis and regulation. Physiol Plant 152(2):241–255

Ma Y, Yuan L, Wu B, Li X, Chen S, Lu S (2012) Genome-wide identification and characterization of novel genes involved in terpenoid biosynthesis in *Salvia miltiorrhiza*. J Exp Bot 63:2809–2823

Ma XH, Ma Y, Tang JF, He YL, Liu YC, Ma XJ, Shen Y, Cui GH, Lin HX, Rong QX, Guo J, Huang LQ (2015) The biosynthetic pathways of tanshinones and phenolic acids in *Salvia miltiorrhiza*. Molecules 20:16235–16254

Martinez-Esteso MJ, Martinez-Marquez A, Selles-Marchart S, Morante-Carriel JA, Bru-Martinez R (2015) The role of proteomics in progressing insights into plant secondary metabolism. Front Plant Sci 6:504–511

Miettinen K, Dong L, Navrot N, Schneider T, Burlat V, Pollier J, Woittiez L, van der Krol S, Lugan R, Ilc T, Verpoorte R, Oksman-Caldentey KM, Martinoia E, Bouwmeester H, Goossens A, Memelink J, Werck-Reichhart D (2014) The seco-iridoid pathway from *Catharanthus roseus*. Nat Commun 5:3606

Millgate AG, Pogson BJ, Wilson IW, Kutchan TM, Zenk MH, Gerlach WL, Fist AJ, Larkin PJ (2004) Analgesia: morphine-pathway block in *top1* poppies. Nature 431(7007):413–414

Muir SR, Collins GJ, Robinson S, Hughes S, Bovy A, Ric De Vos CH, van Tunen AJ, Verhoeyen ME (2001) Overexpression of petunia chalcone isomerase in tomato results in fruit containing increased levels of flavonols. Nat Biotechnol 19(5):470–474

Namdeo A (2007) Plant cell elicitation for production of secondary metabolites: a review. Pharmacogn Rev 1(1):69–79
O'Connor SE, Maresh JJ (2006) Chemistry and biology of monoterpene indole alkaloid biosynthesis. Nat Prod Rep 23(4):532–547
Oksman-Caldentey KM, Inzé D (2004) Plant cell factories in the post-genomic era: new ways to produce designer secondary metabolites. Trends Plant Sci 9(9):433–440
Ow YY, Stupans I (2003) Gallic acid and gallic acid derivatives: effects on drug metabolizing enzymes. Curr Drug Metab 4(3):241–248
Pan Q, Wang Q, Yuan F, Xing S, Zhao J, Choi YH, Verpoorte R, Tian Y, Wang G, Tang K (2012) Overexpression of ORCA3 and G10H in *Catharanthus roseus* plants regulated alkaloid biosynthesis and metabolism revealed by NMR-metabolomics. PLos One 7(8), e43038
Pan Q, Mustafa N, Tang K, Choi Y, Verpoorte R (2016) Monoterpenoid indole alkaloids biosynthesis and its regulation in *Catharanthus roseus*: a literature review from genes to metabolites. Phytochem Rev 15(2):221–250
Pathak S, Lakhwani D, Gupta P, Mishra BK, Shukla S, Asif MH, Trivedi PK (2013) Comparative transcriptome analysis using high papaverine mutant of *Papaver somniferum* reveals pathway and uncharacterized steps of papaverine biosynthesis. PLos One 8(5), e65622
Quideau S, Deffieux D, Douat-Casassus C, Pouységu L (2011) Plant polyphenols: chemical properties, biological activities, and synthesis. Angew Chem Int Ed 50(3):586–621
Roberts MF, Strack D, Wink M (2010) Biosynthesis of alkaloids and betalains. In: Annual plant reviews, vol 40: Biochemistry of plant secondary metabolism, Second Edition. Wiley, Blackwell, pp 20–91
Robertson AL, Holmes GR, Bojarczuk AN, Burgon J, Loynes CA, Chimen M, Sawtell AK, Hamza B, Willson J, Walmsley SR, Anderson SR, Coles MC, Farrow SN, Solari R, Jones S, Prince LR, Irimia D, Rainger GE, Kadirkamanathan V, Whyte MK, Renshaw SA (2014) A zebrafish compound screen reveals modulation of neutrophil reverse migration as an anti-inflammatory mechanism. Sci Trans Med 26(6):29
Ryder TB, Cramer CL, Bell JN, Robbins MP, Dixon RA, Lamb CJ (1984) Elicitor rapidly induces chalcone synthase mRNA in Phaseolus vulgaris cells at the onset of the phytoalexin defense response. Proc Natl Acad Sci U S A 81(18):5724–5728
Saito K, Yonekura-Sakakibara K, Nakabayashi R, Higashi Y, Yamazaki M, Tohge T, Fernie AR (2013) The flavonoid biosynthetic pathway in *Arabidopsis*: structural and genetic diversity. Plant Physiol Biochem 72:21–34
Salim V, Yu F, Altarejos J, De Luca V (2013) Virus-induced gene silencing identifies *Catharanthus roseus* 7-deoxyloganic acid-7-hydroxylase, a step in iridoid and monoterpene indole alkaloid biosynthesis. Plant J 76(5):754–765
Schluttenhofer C, Pattanaik S, Patra B, Yuan L (2014) Analyses of *Catharanthus roseus* and *Arabidopsis thaliana* WRKY transcription factors reveal involvement in jasmonate signaling. BMC Genomics 15:502–521
Schmitz R (1985) Friedrich Wilhelm Sertürner and the discovery of morphine. Pharm Hist 27(2):61–74
Schramek N, Huber C, Schmidt S, Dvorski SE, Kmispel N, Ostrozhenkova E, Pena-Rodriguez LM, Cusido RM, Wischmann G, Eisenreich W (2014) Biosynthesis of ginsenosides in field-grown *Panax ginseng*. JSM Biotechnol Bioeng 2(1):1033
Sharoni Y, Stahl W, Danilenko M, Levy J (2004) Anticancer activity of carotenoids: from human studies to cellular processes and gene regulation. In: Carotenoids in health and disease. CRC Press, Boca Raton, pp 165–196
Shi SG, Yang M, Zhang M, Wang P, Kang YX, Liu JJ (2014) Genome-wide transcriptome analysis of genes involved in flavonoid biosynthesis between red and white strains of *Magnolia sprengeri* pamp. BMC Genomics 15:706–776
Shukla AK, Shasany AK, Gupta MM, Khanuja SPS (2006) Transcriptome analysis in *Catharanthus roseus* leaves and roots for comparative terpenoid indole alkaloid profiles. J Exp Bot 57(14):3921–3932

Song WO, Chun OK (2008) Tea is the major source of flavan-3-ol and flavonol in the U.S. diet. J Nutr 138(8):1543S–1547S

Stavrinides A, Tatsis EC, Foureau E, Caputi L, Kellner F, Courdavault V, O'Connor SE (2015) Unlocking the diversity of alkaloids in *Catharanthus roseus*: nuclear localization suggests metabolic channeling in secondary metabolism. Chem Biol 22(3):336–341

Stewart MJ, Steenkamp V (2000) The biochemistry and toxicity of atractyloside: a review. Ther Drug Monit 22:641–649

St-Pierre B, Vazquez-Flota FA, De Luca V (1999) Multicellular compartmentation of *Catharanthus roseus* alkaloid biosynthesis predicts intercellular translocation of a pathway intermediate. Plant Cell 11(5):887–900

Sun Y, Zhao SJ, Liang YL, Cao HJ (2013) Regulation and differential expression of protopanaxadiol synthase in Asian and American ginseng ginsenoside biosynthesis by RNA interferences. Plant Growth Regul 71(3):207–217

Sun H, Liu Y, Gai Y, Geng J, Chen L, Liu H, Kang L, Tian Y, Li Y (2015) De novo sequencing and analysis of the cranberry fruit transcriptome to identify putative genes involved in flavonoid biosynthesis, transport and regulation. BMC Genomics 16:652–669

Sung YC, Lin CP, Chen JC (2014) Optimization of virus-induced gene silencing in *Catharanthus roseus*. Plant Pathol 63(5):1159–1167

Suttipanta N, Pattanaik S, Gunjan S, Xie CH, Littleton J, Yuan L (2007) Promoter analysis of the *Catharanthus roseus* geraniol 10-hydroxylase gene involved in terpenoid indole alkaloid biosynthesis. Biochim Biophys Acta 1769(2):139–148

Suzuki M, Nakabayashi R, Ogata Y, Sakurai N, Tokimatsu T, Goto S, Suzuki M, Jasinski M, Martinoia E, Otagaki S, Matsumoto S, Saito K, Shiratake K (2015) Multiomics in grape berry skin revealed specific induction of the stilbene synthetic pathway by ultraviolet-C irradiation. Plant Physiol 168(1):47–59

Sweetman C, Wong DC, Ford CM, Drew DP (2012) Transcriptome analysis at four developmental stages of grape berry (*Vitis vinifera* cv. Shiraz) provides insights into regulated and coordinated gene expression. BMC Genomics 13:691–715

Taiz L, Zeiger E (2006) Secondary metabolites and plant defense. In: Plant physiology, 5th edn. Sinauer Associates, Sunderland, pp 369–400

Takahashi H, Chen MC, Pham H, Angst E, King JC, Park J, Brovman EY, Ishiguro H, Harris DM, Reber HA, Hines OJ, Gukovskaya AS, Go VL, Eibl G (2011) Baicalein, a component of *Scutellaria baicalensis*, induces apoptosis by *Mcl-1* down-regulation in human pancreatic cancer cells. Biochim Biophys Acta 1813(8):1465–1474

Tanner GJ, Francki KT, Abrahams S, Watson JM, Larkin PJ, Ashton AR (2003) Proanthocyanidin biosynthesis in plants. Purification of legume leucoanthocyanidin reductase and molecular cloning of its cDNA. J Biol Chem 278(34):31647–31656

Tholl D (2006) Terpene synthases and the regulation, diversity and biological roles of terpene metabolism. Curr Opin Plant Biol 9(3):297–304

Thomas SG, Rien I, Steber CM (2005) Gibberellin metabolism and signaling. Vitam Horm 72:289–338

Tropf S, Lanz T, Rensing SA, Schroder J, Schroder G (1994) Evidence that stilbene synthases have developed from chalcone synthases several times in the course of evolution. J Mol Evol 38(6):610–618

Tsao R (2010) Chemistry and biochemistry of dietary polyphenols. Nutrients 2(12):1231–1246

Van Moerkercke A, Fabris M, Pollier J, Baart GJE, Rombauts S, Hasnain G, Rischer H, Memelink J, Oksman-Caldentey KM, Goossens A (2013) CathaCyc, a metabolic pathway database built from *Catharanthus roseus* RNA-Seq data. Plant Cell Physiol 54(5):673–685

Van Moerkercke A, Steensma P, Schweizer F, Pollier J, Gariboldi I, Payne R, Bossche RV, Miettinen K, Espoz J, Purnama PC, Kellner F, Seppanen-Laakso T, O'Connor SE, Rischer H, Memelink J, Goossens A (2015) The bHLH transcription factor BIS1 controls the iridoid branch of the monoterpenoid indole alkaloid pathway in *Catharanthus roseus*. Proc Natl Acad Sci Biol 112(26):8130–8135

Vannozzi A, Dry IB, Fasoli M, Zenoni S, Lucchin M (2012) Genome-wide analysis of the grapevine stilbene synthase multigenic family: genomic organization and expression profiles upon biotic and abiotic stresses. BMC Plant Biol 12:130–151

Velasco R, Zharkikh A, Troggio M, Cartwright DA, Cestaro A, Pruss D, Pindo M, Fitzgerald LM, Vezzulli S, Reid J, Malacarne G, Iliev D, Coppola G, Wardell B, Micheletti D, Macalma T, Facci M, Mitchell JT, Perazzolli M, Eldredge G, Gatto P, Oyzerski R, Moretto M, Gutin N, Stefanini M, Chen Y, Segala C, Davenport C, Dematte L, Mraz A, Battilana J, Stormo K, Costa F, Tao Q, Si-Ammour A, Harkins T, Lackey A, Perbost C, Taillon B, Stella A, Solovyev V, Fawcett JA, Sterck L, Vandepoele K, Grando SM, Toppo S, Moser C, Lanchbury J, Bogden R, Skolnick M, Sgaramella V, Bhatnagar SK, Fontana P, Gutin A, Van de Peer Y, Salamini F, Viola R (2007) A high quality draft consensus sequence of the genome of a heterozygous grapevine variety. PLoS One 2(12), e132

Verma M, Ghangal R, Sharma R, Sinha AK, Jain M (2014) Transcriptome analysis of *Catharanthus roseus* for gene discovery and expression profiling. PLos One 9(7), e103583

Verpoorte R, Memelink J (2002) Engineering secondary metabolite production in plants. Curr Opin Biotechnol 13(2):181–187

Wang P, Wei Y, Fan Y, Liu Q, Wei W, Yang C, Zhang L, Zhao G, Yue J, Yan X, Zhou Z (2015) Production of bioactive ginsenosides Rh2 and Rg3 by metabolically engineered yeasts. Metab Eng 29:97–105

Weiss D (2000) Regulation of flower pigmentation and growth: multiple signaling pathways control anthocyanin synthesis in expanding petals. Physiol Plant 110(2):152–157

Wellmann F, Lukacin R, Moriguchi T, Britsch L, Schiltz E, Matern U (2002) Functional expression and mutational analysis of flavonol synthase from *Citrus unshiu*. Eur J Biochem 269(16):4134–4142

Wiese W, Vornam B, Krause E, Kindl H (1994) Structural organization and differential expression of three stilbene synthase genes located on a 13 kb grapevine DNA fragment. Plant Mol Biol 26(2):667–677

Wijekoon CP, Facchini PJ (2012) Systematic knockdown of morphine pathway enzymes in opium poppy using virus-induced gene silencing. Plant J 69(6):1052–1063

Williams RJ, Spencer JP (2012) Flavonoids, cognition, and dementia: actions, mechanisms, and potential therapeutic utility for Alzheimer disease. Free Radic Biol Med 52(1):35–45

Winzer T, Gazda V, He Z, Kaminski F, Kern M, Larson TR, Li Y, Meade F, Teodor R, Vaistij FE, Walker C, Bowser TA, Graham IA (2012) A *Papaver somniferum* 10-gene cluster for synthesis of the anticancer alkaloid noscapine. Science 336(6089):1704–1708

Xiao M, Zhang Y, Chen X, Lee EJ, Barber CJS, Chakrabarty R, Desgagne-Penix I, Haslam TM, Kim Y-B, Liu E, MacNevin G, Masada-Atsumi S, Reed DW, Stout JM, Zerbe P, Zhang Y, Bohlmann J, Covello PS, De Luca V, Page JE, Ro D-K, Martin VJJ, Facchini PJ, Sensen CW (2013) Transcriptome analysis based on next-generation sequencing of non-model plants producing specialized metabolites of biotechnological interest. J Biotechnol 166(3):122–134

Xie DY, Sharma SB, Paiva NL, Ferreira D, Dixon RA (2003) Role of anthocyanidin reductase, encoded by BANYULS in plant flavonoid biosynthesis. Science 299(5605):396–399

Yang CY, Wang J, Zhao Y, Shen L, Jiang X, Xie ZG, Liang N, Zhang L, Chen ZH (2010) Anti-diabetic effects of Panax notoginseng saponins and its major anti-hyperglycemic components. J Ethnopharmacol 130(2):231–236

Yang L, Ding G, Lin H, Cheng H, Kong Y, Wei Y, Fang X, Liu R, Wang L, Chen X, Yang C (2013) Correction: transcriptome analysis of medicinal plant *Salvia miltiorrhiza* and identification of genes related to tanshinone biosynthesis. PLoS One 8(12), e80464

Yang D, Du X, Yang Z, Liang Z, Guo Z, Liu Y (2014) Transcriptomics, proteomics, and metabolomics to reveal mechanisms underlying plant secondary metabolism. Eng Life Sci 14(5):456–466

Zamboni A, Di Carli M, Guzzo F, Stocchero M, Zenoni S, Ferrarini A, Tononi P, Toffali K, Desiderio A, Lilley KS, Pe ME, Benvenuto E, Delledonne M, Pezzotti M (2010) Identification of putative stage-specific grapevine berry biomarkers and omics data integration into networks. Plant Physiol 154(3):1439–1459

Zhang Y, Jiang P, Ye M, Kim SH, Jiang C, Lu J (2012) Tanshinones: sources, pharmacokinetics and anti-cancer activities. Int J Mol Sci 13:13621–13666

Zhang GH, Ma CH, Zhang JJ, Chen JW, Tang QY, He MH, Xu XZ, Jiang NH, Yang SC (2015) Transcriptome analysis of *Panax vietnamensis* var. *fuscidicus* discovers putative ocotillol-type ginsenosides biosynthesis genes and genetic markers. BMC Genomics 16(1):159

Zhao S, Wang L, Liu L, Liang Y, Sun Y, Wu J (2014) Both the mevalonate and the non-mevalonate pathways are involved in ginsenoside biosynthesis. Plant Cell Rep 33:393–400

Zhao C, Xu T, Liang Y, Zhao S, Ren L, Wang Q, Dou B (2015) Functional analysis of β-amyrin synthase gene in ginsenoside biosynthesis by RNA interference. Plant Cell Rep 34(8):1307–1315

Zhu J, Wang M, Wen W, Yu R (2015) Biosynthesis and regulation of terpenoid indole alkaloids in *Catharanthus roseus*. Pharmacogn Rev 9(17):24–28

Ziegler J, Facchini PJ (2008) Alkaloid biosynthesis: metabolism and trafficking. Annu Rev Plant Biol 59(1):735–769

Zulak KG, Liscombe DK, Ashihara H, Facchini PJ (2006) Alkaloids. In: Crozier A (ed) Plant secondary metabolites: occurrence, structure and role in the human diet. Wiley, Blackwell, pp 102–136

Chapter 6
Memory Booster Plant *Bacopa monniera* (Brahmi): Biotechnology and Molecular Aspects of Bacoside Biosynthesis

Rishi K. Vishwakarma, Uma Kumari, and Bashir M. Khan

Abstract *Bacopa monniera* (Brahmi) has been used in the Ayurvedic system of medicine as a brain tonic, memory enhancer, antianxiety, cardiotonic, anticancer, anti-inflammatory, analgesic, and anticonvulsant agent since ancient times. These pharmacological properties are mainly attributed to the triterpenoid saponins present in the extracts of the plant. Biosynthesis of triterpenoid saponins starts from the isoprenoid pathway through farnesyl pyrophosphate by cyclization of 2,3-oxidosqualene, resulting in the formation of triterpenoid backbones. The plant produces relatively smaller amounts of bacosides, and to overcome this shortage, a large amount of biomass is used in the pharmaceutical preparations. Despite a wealth of medicinal importance, the molecular characterization and pathway engineering of bacoside biosynthesis in *Bacopa* remain unexplored. In this chapter, we have briefly discussed the research findings on in vitro plant regeneration, genetic transformation, and molecular characterization of some of the genes involved in the biosynthesis of bacosides.

Keywords Bacosides • *Bacopa monniera* • Genetic transformation • Micropropagation • Terpenoids biosynthesis • Triterpenoids

R.K. Vishwakarma (✉)
Plant Tissue Culture Division, CSIR-National Chemical Laboratory, Dr. Homi Bhabha Road, Pune 411008, India

Le CBS-Centre de Biochimie Structurale-CNRS, UMR 5048-UM-INSERM U 1054, 29 rue de Navacelles, 34090 Montpellier, France
e-mail: rishivish@gmail.com

U. Kumari • B.M. Khan
Plant Tissue Culture Division, CSIR-National Chemical Laboratory, Dr. Homi Bhabha Road, Pune 411008, Maharashtra, India

H.-S. Tsay et al. (eds.), *Medicinal Plants - Recent Advances in Research and Development*, DOI 10.1007/978-981-10-1085-9_6

Abbreviations

AD	Alzheimer's disease
BA	6-Benzyladenine
CAT	Catalase
DAHPS	3-Deoxy-D-arabino-heptulosonate-7-phosphate synthase
FPS	Farnesyl diphosphate synthase
GSH-Px	Glutathione peroxide
GTs	Glycosyltransferases
HMGR	3-Hydroxy-3-methylglutaryl-CoA reductase
IAA	Indole 3-acetic acid
MCI	Mild cognitive impairment
MK	Mevalonate kinase
MVD	Mevalonate-5-pyrophosphate decarboxylase
OSC	Oxidosqualene cyclase
PR1	Pathogenesis-related protein 1
SOD	Superoxide dismutase
SQS	Squalene synthase
WHO	World Health Organization

6.1 Introduction

Increasing age in human is a key risk factor for mild cognitive impairment (MCI) that indicate dementia, isolated memory impairment, including Alzheimer's disease (AD), and other predominant neurodegenerative disorders (Yankner et al. 2008). The responsible causes for dementia and brain aging are very complex and not yet fully understood. Since the last many years, various laboratories across the world started to gear up research to develop promising health and lifestyle interventions so that elderly people can remain both physically and cognitively healthy (Lowsky et al. 2014). Over the past decades, herbal medicines have gained a wide acceptance due to the growing faith in them. However, a continuous supply of the source materials often becomes a major constraint due to the overexploitation of wild plants, destruction of natural habitats of medicinal plants owing to human development, and lack of alternative cultivation practices.

The Indian medicinal herb, *Bacopa monniera* (Brahmi), may be a great source to be utilized as a dietary antioxidant, to shield the aging brain against oxidative damage and cognitive debility. A number of researchers have shown that the standardized *Bacopa* extract improved cognitive functions predominantly in the aged group (Stough et al. 2001, 2008; Nathan et al. 2004). The antioxidant properties of *Bacopa* have also been elucidated in various animal and in vitro research using the standardized plant extract (Bhattacharya et al. 2000; Russo and Borrelli 2005; Kapoor et al. 2009; Singh 2013). The medicinal properties of *B. monniera* are mainly contributed by the presence of triterpenoid saponins called bacosides. In spite of having a significant medicinal value, molecular studies on *B. monniera*, especially bacoside biosynthesis, have not been explored much. The main purpose of this chapter is to summarize the research outcomes on *Bacopa* as a great potential therapeutic antioxidant to reduce oxidative stress and dementia in the aging brain and the active metabolites which are responsible for its medicinal properties. Also, we have discussed biotechnological approaches to improve *B. monniera* regarding medicinal properties and molecular aspects of bacoside biosynthetic pathway.

6.2 *Bacopa monniera*

Bacopa monniera (synonyms: *Bacopa monnieri*, *Herpestis monniera*, etc., commonly known as Brahmi), a member of the family Scrophulariaceae (figwort or snapdragon family), is a small herb with light purple flowers. This plant grows in the wet swamp and sandy areas. The plant is a prostrate or creeping, tender, annual herb, rooting at all nodes with various climbing branches (Fig. 6.1). Leaves are simple, opposite, sessile, entire, fleshy, and obscurely veined. Flowers and fruits appear in summer. Dispersal and mass propagation are by seeds and stem fragments.

Fig. 6.1 (**A**) *Bacopa* plants. (**B**) *Bacopa* leaves. (**C**) *Bacopa* flower

Crushed leaves of Brahmi have a unique "lemon" scent. The plant is distributed in the major part of the plains of India, Pakistan, Afghanistan, Nepal, Sri Lanka, Africa, and Australia. *B. monniera* is commercially cultivated for medicinal purposes, and the annual production is around 40,000–50,000 kg per hectare (Russo and Borrelli 2005). The whole plant is medicinally useful (Bone 1996).

The medicinal herb has been marked in Ayurvedic literature since about 800 BC and documented as a treatment for the mental ailments in the *Charaka Samhita* (Singh and Dhawan 1997). The Export-Import Bank of India made a regional study and placed *B. monniera* on the second rank in the priority list of most important medicinal plants, assessed by their medicinal prominence, commercial worth, and the potential for further research and development. According to an estimate, the annual requirement of the plant was projected to be about 12.700 tons of dry material, valued at approximately Rs 15 billion, and it comes exclusively from a natural source which is leading to their continuing exhaustion (Ahmad 1993).

6.3 Uses of *B. monniera*

B. monniera, also known as "the thinking person's herb," is broadly used in traditional Indian Ayurvedic system of medicine as an effective nervine boost to improve memory, cerebral, and thought functions (Rastogi et al. 1994). It is also used in the treatment of asthma, leprosy, hoarseness, renal disease, water retention, and blood cleaning (Singh and Dhawan 1982). In Ayurveda, the use of *Bacopa* is not limited to as a brain tonic but also to treat catarrhal complaints, gastrointestinal disturbances (due to excessive tobacco use), habitual abortions and high blood sugar (due to anxiety), hysteria, epilepsy, etc. (Chopra et al. 1956; Nadkarni 1976).

A medicine called "Memory Plus" which contains the standardized formulation of Brahmi plant extract has been marketed in India. According to the definition of herbal drugs in the guidelines of herbal medicines stipulated by the World Health Organization (WHO) in 1991, an herbal medicinal product that has been used conventionally without evident side effects does not require strict regulatory action unless evidence demands a revised risk/benefit analysis. Subsequently, like Memory Plus, several formulations containing *B. monniera* extracts standardized for bacoside content have been marketed globally. The assessment of one such formulation, BacoMind™ in healthy adult volunteers, has shown that the given oral dose of 300 mg once a day for an initial 15 days and 450 mg once a day for the next 15 days is safe and acceptable. However, some minor side effects like gastrointestinal ailments were reported in 3 out of 23 volunteers, without affecting general physical, systemic, hematological, biochemical, and electrocardiographic parameters (Pravina et al. 2007). Also, *Bacopa* can be used as an antioxidant and anticarcinogenic agent primarily against those cancerous compounds that facilitate their effects by means of free radical oxygen formation (Deb et al. 2008).

Because of its medicinal properties, *B. monniera* has been used for the treatment of a number of disorders, principally those having symptoms of anxiety, intellect, and poor memory (Singh and Dhawan 1997). Also, it has been used in epilepsy, insanity, and retardation (Mathur et al. 2002) and to neutralize the effects of mental stress and neurosis and renaissance of sensory organs (Sivarajan and Balachandran 1994). The renowned nootropic plant has been reported to possess sedative (Malhotra and Das 1959), cardiotonic (Mathur et al. 2002), cognitive enhancer (Nathan et al. 2001; Roodenrys et al. 2002), broncho-vasodilator (Channa et al. 2003), hepatoprotective (Sumathy and Nongbri 2008; Sumathy et al. 2001), antidepressant (Sairam et al. 2002), calcium antagonistic (Dar and Channa 1999), smooth muscle relaxant (Dar and Channa 1997), neuropharmacological (Russo and Borrelli 2005), cell-stabilizing (Samiulla et al. 2001), and antiulcer (Sairam et al. 2001) properties. Studies have shown that *B. monniera* plant possesses anti-stress effect through the modulation of Hsp 70 expression, superoxide dismutase, and cytochrome p450 inhibitory activities in the rat brain (Chowdhuri et al. 2002) and antioxidant activity (Tripathi et al. 1996; Simpson et al. 2015). The studies of Anbarasi et al. (2005a, b) demonstrated the protective effects of bacoside A (a triterpenoid) on brain damage induced by cigarette smoking.

B. monniera has been shown to improve the rate of learning in a brightness discrimination task and a restricted avoidance task and improves retention. It has been demonstrated by savings in relearning and attenuates amnesia induced by immobilization, electroconvulsive shock, and scopolamine (Singh and Dhawan 1997). Their findings also suggest that the administration of extracted bacoside A and B influenced the cholinergic system. In a separate study, Singh and coworkers reported that the stem extract of *B. monniera* caused inhibition of corrosion of aluminum in 0.5 M NaOH solution and weight loss measurements. The findings showed that *Bacopa* stem extract was an effective inhibitor and the inhibition competencies obtained from polarization and weight loss experiments were in good agreement (Singh et al. 2012).

6.4 Medicinally Active Compounds in *B. monniera*

Given the significance of this plant in the aboriginal system of remedy, several laboratories have performed chemical analyses of this plant. The first detailed investigations were documented in 1931 when Bose and Bose described the isolation of the alkaloid "brahmine" from *B. monniera* (Bose and Bose 1931). Much later, other alkaloids like nicotine and herpestine were also reported (Chopra et al. 1956). Several glycosides such as asiaticoside and thanakunicide, flavonoids (apigenin and luteolin), phytochemicals such as betulinic acid, wogonin, oroxindin, betulic acid, stigmasterol, β-sitosterol, as well as brahamoside, brahminoside, brahmic acid, isobrahmic acid, vallerine, pectic acid, fatty acids, tannin, volatile oil, ascorbic acid, thanakunic acid, and asiatic acid have also been reported (Mathew et al. 2010).

Bacoside A, the principal active constituent responsible for the memory-assisting action of *B. monniera*, was recognized as 3-(α-L-arabinopyranosyl)-O-β-D-glucopyranoside-10, 20-dihydroxy-16-keto-dammar-24-ene (Chatterji et al. 1965).

The compound usually exists together with its optical isomer bacoside B and probably an artifact produced during the extraction process of bacoside A (Rastogi 1990) as shown in Fig. 6.2. The chemical and physical degradation studies established the chemical composition of bacosides present in the polar fraction of the plant extract. On acid hydrolysis, bacosides yield a mixture of aglycones, bacogenin A1, A2, and A3 which are by-products (Kulshreshtha and Rastogi 1973), and two confirmed sapogenins called jujubogenin and pseudojujubogenin (Rastogi et al. 1994).

Later, another compound from *Bacopa* extract was identified as ebelin lactone pseudojujubogenin, referred as bacogenin A4 (Rastogi et al. 1994). Sequentially, some other components like bacoside A1 were isolated and characterized as 3-O-[α-L-arabinofuranosyl(1-3)-β-L-arabinopyranosyl] jujubogenin (Garai et al. 1996a; Rastogi et al. 1994). Garai et al. (1996b), by using spectroscopic and chemical transformation methods, isolated three other new dammarane-type triterpenoid saponins of biological importance, namely, bacopasaponins A, B, and C which are identified as 3-O-α-L-arabinopyranosyl-20-O-α-L-arabinopyranosyl-jujubogenin, 3-O-[α-L-arabinofuranosyl(1-2)α-L-arabinopyranosyl] pseudojujubogenin, and 3-O-[β-D-glucopyranosyl(1-3) {α-L-arabinofuranosyl(1-2)}α-L-arabinopyranosyl] pseudojujubogenin. By using similar methods, Garai et al. (1996a) isolated a new dammarane-type pseudojujubogenin glycoside, bacopasaponin D, defined as 3-O-[α-L-arabinofuranosyl(1-2)β-D-glucopyranosyl] pseudojujubogenin.

In addition, Hou et al. (2002) isolated a new saponin, 3-O-[α-1-arabinofuranosyl-(1-2)]-α-L-arabinopyranosyl jujubogenin, named bacopasaponin G; a new matsutake alcohol derivative, (3R)-1-octan-3yl-(6-O-sulfonyl)-β-D-glucopyranoside; a new phenylethanoid glycoside, 3,4-dihydroxyphenyl ethyl alcohol (2-O-feruloyl)-β-D-glucopyranoside; and a new glycoside, phenylethyl alcohol [5-O-p-hydroxybenzoyl-β-D-apiofuranosyl-(1-2)]-β-D-glucopyranoside. Also, two new constituents from *Bacopa*, bacopaside N1 and bacopaside N2, have been reported (Sivaramakrishna et al. 2005).

6.5 Mechanism of Action

The therapeutic potential of *Bacopa* plant in treatment and prevention of various neurological disorders is due to its ability to reduce NO-induced cellular adaptations (Russo et al. 2003a). It was also reported that *Bacopa* extract rich in saponins was able to induce a dose-dependent increase in several enzymes including superoxide dismutase (SOD), catalase (CAT), and glutathione peroxide (GSH-Px) activities in rat frontal cortex, striatum, and hippocampus (Bhattacharya et al. 2000). The triterpenoid saponins are responsible for *Bacopa's* ability to enhance nerve impulse transmission. The bacosides support the restoration of impaired neurons by enhancing kinase activity, neuronal synthesis, and restoration of synaptic activity and eventually nerve impulse transmission and boosting of the synthesis of new proteins in the brain (Singh and Dhawan 1997).

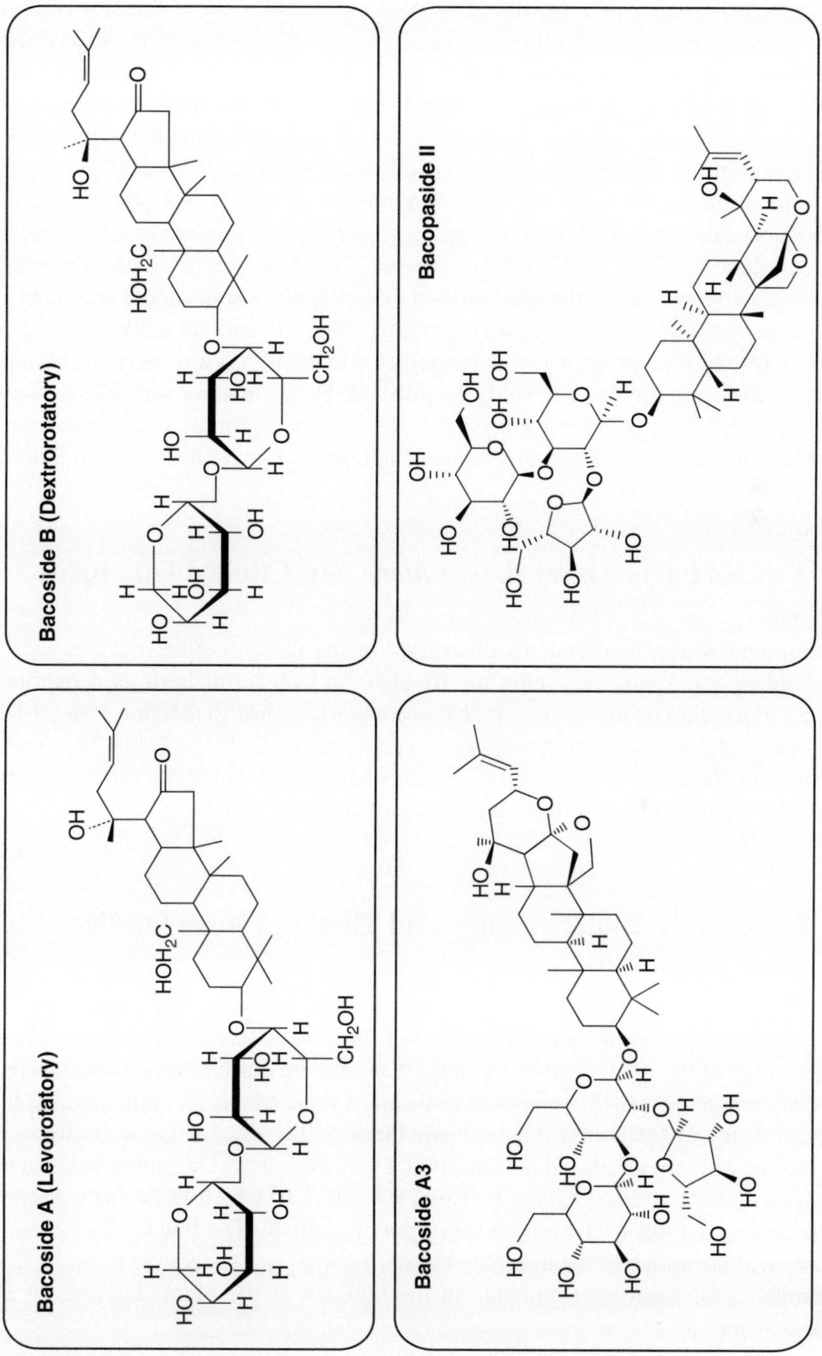

Fig. 6.2 Chemical structures of four different bacosides from *Bacopa monniera*

Research on animals has shown that *Bacopa* extracts control the expression of certain enzymes responsible for the generation and scavenging of reactive oxygen species in the brain (Chowdhuri et al. 2002). In vitro research has also shown that *Bacopa* exerts a shielding effect against DNA damage in astrocytes and human fibroblasts (Russo et al. 2003a, b). *Bacopa* also has a relaxant effect on pulmonary arteries, aorta, trachea, and bronchial tissue, which is possibly mediated by inhibition of calcium ion influx into plasma membranes (Channa et al. 2003). *Bacopa* appears to stabilize mast cells in vitro (Samiulla et al. 2001) and possesses anti-inflammatory activity via inhibition of prostaglandin synthesis and lysosomal membrane stabilization (Jain et al. 1994). The possible mechanism of an anticancer effect of *Bacopa* extracts is the inhibition of DNA replication in cancer cell lines as demonstrated in vitro research (Elangovan et al. 1995). In another study, the neuroprotective role of *Bacopa* extract was investigated in the hippocampus of the temporal lobe of the epileptic rat. The study established that the extract was able to show the therapeutic effect by moving back the changes in glutamate receptor binding and NMDA R1 gene expression that happened during epilepsy (Paulose et al. 2008).

6.6 Known Facts About *B. monniera* and Clinical Highlights

As mentioned above, *B. monniera* plant used in Indian Ayurvedic medicine has a long-standing reputation for being an effective and powerful herb as a memory booster and fighting against stress. It also promotes liver health (Menon et al. 2010) and protects against neonatal hypoglycemia (Thomas et al. 2013). It has been studied clinically for its effect on various diseases and disorders mostly on rat models and in vitro. Some clinical studies are summarized in Table 6.1.

6.7 Updates on Biotechnology and Tissue Culture Studies on *B. monniera*

In this section, the recent updates on *Bacopa* tissue culture are highlighted. For a detailed study of the genetic diversity in the *B. monniera* germ plasm, 24 accessions from different geographical regions of India and 1 from Malaysia maintained at the Central Institute of Medicinal and Aromatic Plants (CIMAP), Lucknow, India, were analyzed for random amplified polymorphic DNA variation. This study has shown the similarity between accessions between 0.8 and 1.0, which indicates a narrow genetic base and a low to medium level of polymorphism. The low level of genetic variation was attributed to the interplay between sexual and vegetative reproduction and similarity of local environments in the habitats of the plant (Darokar et al. 2001).

Table 6.1 Some clinical studies on *B. monniera* extracts

Conditions/capability	Treatment/dosages	Effects	References
Cognitive effects	300 mg/day (standardized to 55 % combined bacoside A and B)	No significant effects on processes of memory performance were found with acute administration of *Bacopa* extracts	Vohora et al. (2000) and Nathan et al. (2001)
		Improved memory attainment and retention in elderly persons	Morgan and Stevens (2010)
Anxiety and depression	12 g/day	Decreased anxiety, improvement in concentration and memory span, no side effects	Singh and Singh (1980)
	Methanolic extract of *B. monniera* (100 and 200 mg/kg)	Dose-dependent antidepressant effects and increased locomotion	Mannan et al. (2015)
Epilepsy	High dose (close to 50 % of LD50)	Anticonvulsant activity was observed	Martis et al. (1992)
Bronchitis and asthma	0.1–0.7 mg/mL ethanolic extract	Vasorelaxant action inhibited hypotension and bradycardia	Dar and Channa (1997)
Gastrointestinal disorder	5 g/day	Direct spasmolytic activity on intestinal smooth muscle through inhibition of calcium ion transport across cell membrane	Dar and Channa (1999)
Gastric ulcers in rats	20 mg/kg	Anti-ulcerogenic reinforced the mucosal obstacle and reduced mucosal exfoliation	Rao et al. (2000)
Castor oil-induced diarrhea in mice	500 mg/kg	It improved mean latent period and reduced frequency of defecation	Ajalus et al. (2013)
Cardiovascular effects	100 mg/mL	Vasodilatory effect on calcium chloride-induced tissue contraction	Rashid et al. (1990)
Hypothyroidism	High dose (200–400 mg/day)	By stimulating synthesis or release of T4 but not T3	Kar et al. (2002)
Cancer	–	By having cytotoxic effect on sarcoma-180 cells	Elangovan et al. (1995)

(continued)

Table 6.1 (continued)

Conditions/capability	Treatment/dosages	Effects	References
N-Nitrosodiethylamine (DEN)-induced damage in rat liver	15 mg/kg body weight/day of bacoside A orally	Maintained the antioxidant system and protected the rats from DEN-induced hepatotoxicity	Janani et al. (2009)
	50 micromol/L	Significantly inhibited human breast cancer cell line MDA-MB-231 adhesion, migration, and Matrigel invasion in vitro	Peng et al. (2010)
Drug toxicity (morphine)	1 mg	Significantly reduced the naloxone-induced withdrawal effect	Sumathi et al. (2002)
Cold stress-induced neurodegeneration in rats	40 mg/kg	When extracts administered orally, produced a neuroprotective effect in cold stress-induced hippocampal neurodegeneration of rats	Kumar et al. (2015)
Scopolamine-induced cognitive impairments in rat brain	10, 20, and 40 mg/kg	It enhanced cognition and neuromodulator tendency through controlled expression of AChE, BDNF, MUS-1, and CREB and also by changing the levels of neurotransmitters in the hippocampus of rat brain	Pandareesh et al. (2015)
Aluminum-induced oxidative stress and hippocampus damage in rats	40 mg/kg	Structural derangement in the hippocampus by aluminum is directly proportionate with increased lipid peroxidation by *Bacopa* extracts	Tripathi et al. (2011) and Nannepaga et al. (2014)
H_2O_2-induced oxidative stress	40 mg/kg	Free radical scavenging activity by scavenging 2,2-diphenyl-1-picrylhydrazyl, 2,2′-azino-bis(3-ethylbenzothiazoline-6-sulfonic acid), superoxide radical, and nitric oxide radicals	Pandareesh et al. (2016)

(continued)

Table 6.1 (continued)

Conditions/capability	Treatment/dosages	Effects	References
Renal oxido-lipidemic stress in hypercholesterolemic rats	40 mg/kg	Acts as renoprotective agent by abating the renal oxido-lipidemic stress through NOS level regulation and thus protects nephrons in hypercholesterolemic rats	Kamesh and Sumathi (2014)
Contextual fear	80 mg/kg Standardized extract of *Bacopa monniera* (CDRI-08)	Advances hippocampus-dependent contextual memory by differentially regulating histone acetylation and protein phosphatases in the hippocampus	Preethi et al. (2014)
Lead-induced oxidative stress (in rat)	10 mg/kg	It can alleviate the lead-induced oxidative stress tissue precisely by pharmacologic interventions which involved both chelation and antioxidant roles	Velaga et al. (2014)
Ischemia-reperfusion injury (in rat heart)	75 mg/kg	A restored antioxidant network of the myocardium and reduced myocardial apoptosis, caspase 3, and Bax protein expression	Mohanty et al. (2010)
Morphine-induced liver and kidney toxicity (Rats)	40 mg/kg	Histopathological changes in the liver and kidneys	Sumathi and Niranjali (2009)
Antimicrobial effects	1 mg/mL	Ether extract of *B. monniera* showed antimicrobial activity against four different bacteria and one fungus, namely, *Salmonella typhi*, *Pseudomonas aeruginosa*, *Staphylococcus aureus*, *Vibrio cholera*, and *Candida albicans*	Azad et al. (2012)
Anti-inflammatory and analgesic effects	100 mg/kg	Through modulation of pro-inflammatory mediator release	Mathur et al. (2010)
Anticonvulsant	50 and 55 mg/kg	Significant anticonvulsant activity in rats with action mechanism akin to benzodiazepines (GABA agonist)	Kaushik et al. (2009)

6.7.1 *Development of a Plant Regeneration System in* B. monniera

The plant has very high morphogenic potential, and hence explants readily respond to the application of plant growth regulators (PGRs). From different laboratories, several studies on in vitro propagation of *B. monniera* have been published (Praveen et al. 2009; Ceasar et al. 2010; Jain et al. 2012; Kumari et al. 2015). In the majority of the tissue, culture studies on *B. monniera*, 6-benzyl aminopurine (BAP), were found to be the most effective PGR for inducing adventitious shoots from different explants (Singh and Dhawan 1982), induction of somatic embryogenesis (Tiwari et al. 1998), and development of a successful protocol for the mass propagation of the plant (Tiwari et al. 2001). Most recently, we have developed a highly efficient regeneration method of *B. monniera* at CSIR-National Chemical Laboratory, Pune, India (Kumari et al. 2015). In vitro shoots developed in nodal segments cultured on semisolid MS medium (Murashige and Skoog 1962) containing 0.5 mg l^{-1} BA. Further multiple shoot regeneration occurred in leaves derived from in vitro shoots. Longitudinal cuts were made in leaves and then inoculated on MS basal medium supplemented with 1–2 mg l^{-1} BA and 0–0.2 mg l^{-1} IAA. After 2 weeks of incubation on the above media, leaf explants inoculated on MS basal medium with different combinations of BA and IAA concentrations showed the highest frequency of shoot regeneration (98.33 %) on MS basal medium having 2 mg l^{-1} BA and 0.2 mg l^{-1} IAA. The multiple shoots were induced at the cut regions of the lamina. After 3 weeks of growth, multiple shoots produced from leaves were shifted to another medium for root induction (Fig. 6.3). Similar to our study, there are several

Fig. 6.3 Multiple shoots regeneration in leaf explant of *B. monniera*. (**A**) A 3-week-old leaf explants on shoot induction media. (**B, C**) Multiple shoots showing induction of roots

other reports on shoot induction in *B. monniera*, where leaf explant was more responsive compared to other explants (Shrivastava and Rajani 1999; Tiwari et al. 1998, 2000, 2001; Ceasar et al. 2010; Mahender et al. 2012).

6.7.2 Production of Bacosides by Suspension Cultures of **Bacopa monniera**

Cell suspension cultures for *B. monniera* were established from leaf explants obtained from in vitro grown plants. Two cell lines showed a five- to sixfold increase in fresh and dry weight in approximately 6 weeks of culture in MS medium supplemented with 1 mg/L α-naphthalene acetic acid and 0.5 mg/L kinetin. In these two cell lines, bacosides A and B started appearing from the tenth day of culture and accumulated progressively up to the 40th day. Overall bacoside A content was found to be higher than bacoside B throughout the culture period, with a maximum of 1 % dry weight basis in 40-day cultures, while bacoside B accumulation was limited to 0.25–0.37 % of dry weight in 40-day cultures (Rahman et al. 2002).

6.7.3 Genetic Transformation of **Bacopa monniera**

A regeneration system for *Bacopa* has already been established in our previous work (Kumari et al. 2015). To exploit this protocol to produce transformed *Bacopa* plants for altered bacoside content, various terpenoid pathway genes (HMG-CoA reductase, farnesyl diphosphate synthase, squalene synthase, etc.) were cloned and used for further transformation experiments. Before starting with the specific gene overexpression in *Bacopa* plant, we optimized the transformation protocol using blank pCAMBIA 1301 vector. We compared two methods of gene transfer named *Agrobacterium*-mediated (Kumari et al. 2015) and the particle bombardment method (unpublished). Standard method was used for *Agrobacterium*-mediated transformation. For particle bombardment-mediated gene transfer method, leaf explants were aseptically excised from in vitro shoots and placed as a monolayer in Petri plates containing shoot regeneration medium and pre-cultured for 24 h. These leaves were bombarded with 1 μm gold particles coated with pCAMBIA1301 plasmid DNA using the protocol described by Sanford et al. (1993). Every shot was comprised of a mixture of 1 mg of gold particles and 1 μg of plasmid DNA. BioRad PDS 1000/He was used for leaf bombardment at helium gas pressure of 1300 psi with 6 cm target distance. The bombarded leaves were incubated on shoot regeneration medium for another 48 h. Transformed explants were selected on hygromycin B (10 mg/mL), and cefotaxime (200 mg/mL) was used to eliminate the access growth of the bacteria. After 4 weeks on the selection medium, the shoots survived in the presence of hygromycin B from both *A. tumefaciens* and particle

Fig. 6.4 Putative transgenic *B. monniera* plants obtained from *Agrobacterium*-mediated transformation. (**A**) Shoots survived on hygromycin B after 2 weeks of transformation. (**B**) Transgenic plants on selection medium. (**C**) Transgenic plants in the pot. (**D**) Transgenic plants in the greenhouse. (**E**) Histochemical GUS assay (*I*), control leaf (*II*), leaf and shoot of transgenic *B. monniera* displaying expression of *gus* gene. (**F**) PCR amplification analysis of putative transgenic lines using *hpt*II gene-specific primers and genomic DNA as a template. Lane M, low-range DNA ruler; Lane 1–10, putatively transformed plants; Lane 11, non-transformed plant (negative control); Lane 12, plasmid pCAMBIA1301 (positive control). (**G**) Southern blot analysis. DNA blot hybridization of two transgenic *Bacopa* plants' genomic DNA using *hpt*II gene fragment as a hybridizing probe (Kumari et al. 2015)

bombardment and were transferred to the shoot proliferation medium. The GUS assay was performed to confirm the transformation event in the leaves of treated shoots as described by Jefferson et al. (1987). The rooted shoots were shifted to the soil-sand-peat mixture for further establishment (Fig. 6.4) (Kumari et al. 2015). Putative transformed plants were screened by PCR for the presence of the *hpt*II gene in the genome using genomic DNA as template. To avoid the false positive due to bacterial contamination during *A. tumefaciens*-mediated transformation of the plants, PCR amplification was carried out using 16S rRNA primers (Aggarwal et al. 2011) which is specific to bacterial 16S rRNA.

There are several reports on the successful *Agrobacterium*-mediated transformation of *B. monniera* (Mahender et al. 2012; Ramesh et al. 2011; Nisha et al. 2003). Nisha et al. (2003) reported 60 % transformation efficiency using callus, while later it was found to be 68.8 % by Ramesh et al. (2011) where nodal explants were used for genetic transformation. Further in another report, the transformation efficiency achieved by Mahender et al. (2012) was 70.6 %, with an average of 10.4 ± 0.15 transgenic plantlets per leaf explants. In our most recent study, the transformation efficiency of *Agrobacterium*-mediated transformation was much higher (82.50 %) compared to the particle bombardment method (26.67 %). It indicates that for the efficient production of transgenic *B. monniera* plants with predictable transgene expression, the *Agrobacterium*-mediated method offers considerable advantages over particle bombardment.

6.8 Overexpression of Endogenous *HMGR* (*BmHMGR*), *FPS* (*BmFPS*), and *SQS* (*BmSQS*) in *Bacopa monniera*

Pharmaceutically important compounds from medicinal herb *B. monniera* are mainly triterpenoid saponins collectively called as bacosides, and they are present in very low amounts. Therefore, there is an urgent need to develop designer *Bacopa* plants with altered triterpenoid contents. In our laboratory at CSIR-National Chemical Laboratory, Pune, India, we have isolated and characterized various genes involved in bacoside biosynthesis in *Bacopa* (Table 6.2). We further initiated work

Table 6.2 Genes isolated and characterized from *Bacopa monniera*

Name of genes	Accession no.	Annotations	References
Acetyl-CoA C-acetyltransferase	FJ947159	*BmAACT*	Vishwakarma et al. (2013a)
3-hydroxy-3-methylglutaryl-CoA reductase (HMGR)	HM222606	*BmHMGR 1*	Unpublished
	HM222607	*BmHMGR 2*	
Mevalonate kinase (MK)	JQ670899	*BmMK*	Kumari et al. (2014)
Mevalonate-5-pyrophosphate decarboxylase (MVD)	JN116821	*BmMVD*	Abbassi et al. (2015, 2016)
Farnesyl diphosphate (FPP) synthase	GU385740	*BmFPS*	Vishwakarma et al. (2012)
Squalene synthase (SQS)	GU734711	*BmSQS*	Vishwakarma et al. (2015)
Oxidosqualene cyclase (OSC)	HM769762	*BmOSC*	Vishwakarma et al. (2013b)
Glycosyltransferases (GTs)	FJ586244	*BmGT*	Sharma et al. (2011) and Ruby et al. (2014)
		UGT74W1	
3-deoxy-D-arabino-heptulosonate--7-phosphate (DAHP) synthase	HQ687488	*BmDAHP*	Unpublished
Pathogenesis-related protein 1 (PR-1)	JN642525	*BmPR1*	Unpublished

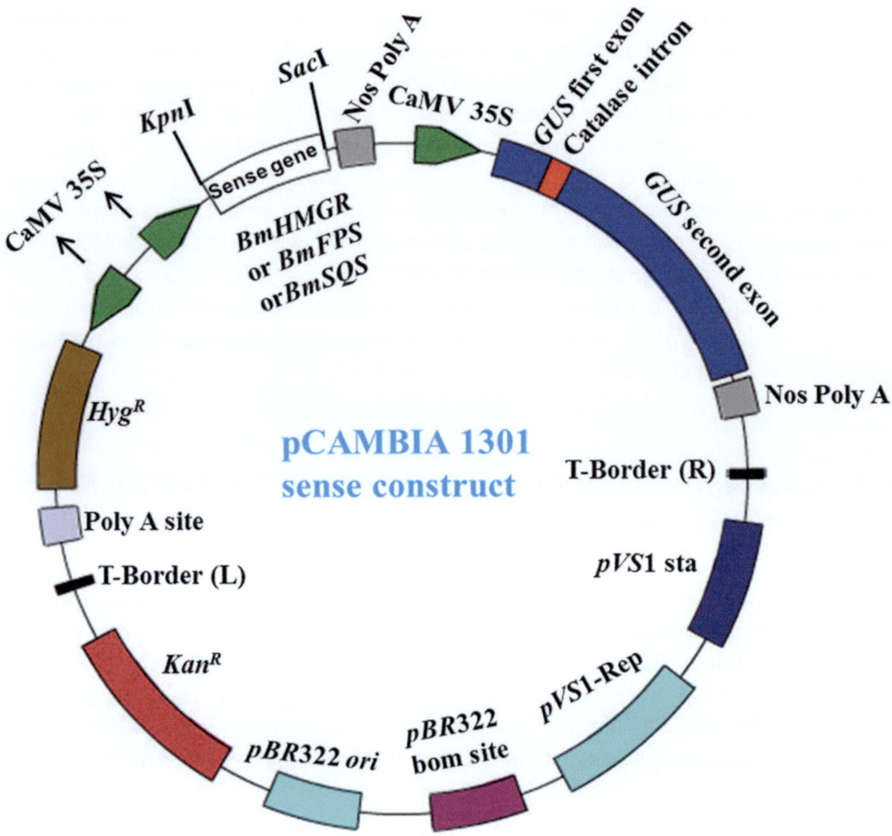

Fig. 6.5 Map of recombinant pCAMBIA 1301 harboring target gene

on overexpression of HMG-CoA reductase, FPP synthase, and squalene synthase *in planta*. The full-length mRNA of *Bacopa HMGR*, *FPS*, and *SQS* was cloned in sense orientation in pCAMBIA 1301 binary vector and used for plant transformation (Fig. 6.5).

Agrobacterium-mediated transformation method was adopted to transform the *Bacopa* plant, as mentioned above (Kumari et al. 2015). Leaf explants were cocultivated with *Agrobacterium* (GV2260) harboring recombinant pCAMBIA 1301 with respective sense constructs. Transformants were analyzed by *hpt*II gene-specific PCR, GUS assay, quantitative real-time PCR, and HPLC. The transformation efficiency and overexpression effect are summarized in Table 6.3.

B. monniera plants, overexpressing *BmHMGR*, *BmFPS*, and *BmSQS*, were successfully generated, and the transformation efficiencies were found to be more than 80 %. *BmHMGR*-transformed *Bacopa* plants showed two- to four fold higher *HMGR* transcript accumulation as compared to control, whereas *BmFPS*- and *BmSQS*-transformed lines showed 3.7- to 5.4-fold and 1.4- to 4.08-fold higher tran-

Table 6.3 Characterization of *Bacopa* plant overexpressing *BmHMGR*, *BmFPS*, and *BmSQS*

Transformed events	No. of transformed plants analyzed	Transformation efficiency (%) by GUS assay	Overexpression of transcript (mean-fold)	Fold increase in bacoside content
BmFPS	27	66.66	2–4-fold	2.7–6.2-fold bacopasaponin C and 3.7–5.9-fold bacoside A
BmHMGR	15	60.00	4–6-fold	To be analyzed
BmSQS	21	71.42	2–5-fold	To be analyzed

script accumulation, respectively. These transformed lines were also analyzed for the effect of overexpression of a particular gene on the expression of other genes of the same pathway. *BmHMGR*-transgenic lines showed 8.2- to 39.1-fold BmFPS, 2- to 92.4-fold *BmSQS*, and 5.8- to 16.4-fold *BmOSC* (*Bacopa monniera* oxidosqualene cyclase) mRNA expression as compared to the control. On the other hand, *BmFPS*- and *BmSQS*-transformed lines showed slight changes in *BmHMGR* transcripts level, whereas *BmSQS* and *BmOSC* expressions were up to 37.79- and 20.68-fold higher, respectively, in *BmFPS*-transformed lines. In the case of *BmSQS* overexpressing events, the expression level of *BmFPS* and *BmOSC* was up to 7.4- and 94.35-fold higher than control plants, respectively. These data suggested that overexpression of one pathway gene induces the expression of other genes within the pathway (unpublished data). HPLC analysis of *BmFPS*-transformed lines showed 2.7- to 6.2-fold higher level of bacopasaponin C and 3.7- to 5.9-fold higher level of bacoside A as compared to control plants. These results indicate that the development of elite lines of *Bacopa* overexpressing pathway genes may provide insight into the regulatory mechanism of bacoside biosynthesis.

6.9 Conclusions

In *Bacopa* plants, bacosides are present in very low quantities. Hence, there is a need to develop designer *Bacopa* plants by genetic engineering of the biosynthetic pathway to increase the yield of such compounds. The biosynthesis of terpenoids takes place via isoprenoid pathway. In view of this, attempts were made to improve bacoside yielding properties of *Bacopa* through genetic engineering. In this context, we could isolate and characterize different genes including 3-hydroxy-3-methylglutaryl-coenzyme A reductase (HMGR), farnesyl pyrophosphate synthase (FPS), and squalene synthase (SQS) involved in terpenoid biosynthesis in *B. monniera*. We have established an efficient plant regeneration and stable transformation system and generated three transformed lines overexpressing endogenous terpenoid biosynthetic pathway genes. Our findings could pave the way for further investigations and understanding of bacoside biosynthesis.

References

Abbassi SJ, Vishwakarma RK, Patel P, Kumari U, Khan BM (2015) *Bacopa monniera* recombinant mevalonate diphosphate decarboxylase: biochemical characterization. Int J Biol Macromol 79:661–668. doi:10.1016/j.ijbiomac.2015.05.041

Abbassi S, Patel K, Khan B, Bhosale S, Gaikwad S (2016) Functional and conformational transitions of mevalonate diphosphate decarboxylase from *Bacopa monniera*. Int J Biol Macromol 83:160–170. doi:10.1016/j.ijbiomac.2015.11.067

Aggarwal D, Kumar A, Reddy MS (2011) *Agrobacterium tumefaciens* mediated transformation of selected elite clones(s) of *Eucalyptus tereticornis*. Acta Physiol Plant 33:41603–41611. doi:10.1007/s11738-010-0695-3

Ahmad RU (1993) Medicinal plants used in ISM- their procurement, cultivation, regeneration and import/export aspects: a report. In: Govil JN, Singh VK, Hashmi S (eds) Medicinal plants: new vistas of research (Part-1). Today and Tomorrow Printers and Publisher, New Delhi, pp 221–225

Ajalus SMD, Chakma N, Rahman M, Salahuddin M, Kumar SS (2013) Assessment of analgesic, antidiarrheal and cytotoxic activity of ethanolic extract of the whole plant of *Bacopa monnieri* Linn. Int Res J Pharm 3(10):98–101. doi:10.13140/RG.2.1.3744.9127

Anbarasi K, Vani G, Balakrishna K, Devi CS (2005a) Effects of bacoside-A on membrane-bound ATPases in the brain of rats exposed to cigarette smoke. J Biochem Mol Toxicol 19:59–65. doi:10.1002/jbt.20050

Anbarasi K, Vani G, Devi CS (2005b) Protective effects of bacoside-A on cigarette smoking-induced brain mitochondrial dysfunction in rats. J Environ Pathol Toxicol Oncol 24:225–234. doi:10.1615/JEnvPathToxOncol.v24.i3.80

Azad AK, Awang M, Rahman MM (2012) Phytochemical and microbiological evaluation of a local medicinal plant *Bacopa monnieri* (l.) Penn. Int J Curr Pharm Rev Res 3(3):66–78

Bhattacharya SK, Bhattacharya A, Kumar A, Ghosal S (2000) Antioxidant activity of *Bacopa monniera* in rat frontal cortex, striatum, and hippocampus. Phytother Res 3(14):174–179. doi:10.1002/(SICI)1099-1573(200005), 14:3<174::AID-PTR624>3.0.CO;2-O

Bone K (1996) Clinical applications of Ayurvedic and Chinese herbs: monographs for the western herbal practitioner. Physiotherapy press, Warwick

Bose KC, Bose NK (1931) Observations on the actions and uses of *Herpestis monniera*. J Indian Med Assoc 1:60

Ceasar SA, Maxwell SL, Prasad BK, Karthigan M, Ignacimuthu S (2010) Highly efficient shoot regeneration of *Bacopa monnieri* (L.) using a two-stage culture procedure and assessment of genetic integrity of micropropagated plants by RAPD. Acta Physiol Plant 32:443–452. doi:10.1007/s11738-009-0419-8

Channa S, Dar A, Yaqoob M, Anjum S, Sultani Z, Attaur R (2003) Broncho-vasodilatory activity of fractions and pure constituents isolated from *Bacopa monniera*. J Ethnopharmacol 86:27–35. doi:10.1016/S0378-8741(03)00013-8

Chatterji N, Rastogi RP, Dhar ML (1965) Chemical examination of *Bacopa monniera* Wettst: part II. The constitution of Bacoside A. Indian J Chem 3:24–30

Chopra RN, Nayar L, Chopra IC (1956) Glossary of Indian medicinal plants. Council of Scientific and Industrial Research, New Delhi, p 32

Chowdhuri DK, Parmar D, Kakkar P, Shukla R, Seth PK, Srimal RC (2002) Antistress effects of bacosides of *Bacopa monnieri*: modulation of Hsp70 expression, superoxide dismutase and cytochrome P450 activity in rat brain. Phytother Res 16:639–645. doi:10.1002/ptr.1023

Dar A, Channa S (1997) Relaxant effect of ethanol extract of *Bacopa monniera* on trachea, pulmonary artery and aorta from rabbit and guinea-pig. Phytother Res 11:323–325. doi:10.1002/(SICI)1099-1573(199706)11:4<323::AID-PTR93>3.0.CO;2-H

Dar A, Channa S (1999) Calcium antagonistic activity of *Bacopa monniera* on vascular and intestinal smooth muscles of rabbit and guinea-pig. J Ethnopharmacol 66:167–174. doi:10.1016/S0378-8741(98)00240-2

Darokar MP, Suman PSK, Shasany AK, Kumar S (2001) Low levels of genetic diversity detected by RAPD analysis in geographically distinct accessions of *Bacopa monnieri*. Genet Resour Crop Evol 48:555–558. doi:10.1023/A:1013800101604

Deb DD, Kapoor P, Dighe RP, Padmaja R, Anand MS, D'Souza P, Deepak M, Mural B, Agrawal A (2008) *In vitro* safety evaluation and anticlastogenic effects of BacoMindTM on human lymphocytes. Biomed Environ Sci 21(1):7–23. doi:10.1016/S0895-3988(08)60002-1

Elangovan V, Govindasamy S, Ramamoorthy N, Balasubramanian K (1995) In vitro studies on the anticancer activity of *Bacopa monnieri*. Fitoterapia 66:211–215

Garai S, Mahato SB, Ohtani K, Yamasaki K (1996a) Bacopasaponin D – a pseudojujubogenin glycoside from *Bacopa monniera*. Phytochemistry 43:447–449. doi:10.1016/0031-9422(96)00250-6

Garai S, Mahato SB, Ohtani K, Yamasaki K (1996b) Dammarane-type triterpenoid saponins from *Bacopa monniera*. Phytochemistry 42:815–820. doi:10.1016/0031-9422(95)00936-1

Hou CC, Lin SJ, Cheng JT, Hsu FL (2002) Bacopaside III, bacopasaponin G, and bacopasides A, B, and C from *Bacopa monniera*. J Nat Prod 65:1759–1763. doi:10.1021/np020238w

Jain P, Khanna NK, Trehan N, Pendse VK, Godhwani JL (1994) Antiinflammatory effects of an Ayurvedic preparation, Brahmi Rasayan, in rodents. Indian J Exp Biol 32:633–636

Jain N, Sharma V, Ramawat KG (2012) Shoot culture of *Bacopa monnieri*: standardization of explant, vessels, and bioreactor for growth and antioxidant capacity. Physiol Mol Biol Plants 18(2):185–190. doi:10.1007/s12298-012-0103-0

Janani P, Sivakumari K, Parthasarathy C (2009) Hepatoprotective activity of bacoside A against N-nitrosodiethylamine-induced liver toxicity in adult rats. Cell Biol Toxicol 25(5):425–434. doi:10.1007/s10565-008-9096-4

Jefferson RA, Kavanagh TA, Bevan MW (1987) GUS fusion: β-glucuronidase as a sensitive and versatile gene fusion marker in higher plants. EMBO J 6:3901–3907

Kamesh V, Sumathi T (2014) Nephroprotective potential of *Bacopa monniera* on hypercholesterolemia induced nephropathy via the NO signaling pathway. Pharm Biol 52(10):1327–1334. doi:10.3109/13880209.2014.891142

Kapoor R, Srivastava S, Kakkar P (2009) *Bacopa monnieri* modulates antioxidant responses in brain and kidney of diabetic rats. Environ Toxicol Pharmacol 27(1):62–69. doi:10.1016/j.etap.2008.08.007

Kar A, Panda S, Bharti S (2002) Relative efficacy of three medicinal plant extracts in the alteration of thyroid hormone concentrations in male mice. J Ethnopharmacol 81:281–285. doi:10.1016/S0378-8741(02)00048-X

Kaushik D, Tripathi A, Tripathi R, Ganachari M, Khan SA (2009) Anticonvulsant activity of *Bacopa monniera* in rodents. Braz J Pharm Sci 45:643–649. doi:10.1590/S1984-82502009000400006

Kulshreshtha DK, Rastogi RP (1973) Identification of ebelin lactone from Bacoside-A and the nature of its genuine sapogenin. Phytochemistry 12:2074–2081. doi:10.1016/S0031-9422(00)91552-8

Kumar SS, Saraswathi P, Vijayaraghavan R (2015) Effect of *Bacopa monniera* on cold stress induced neurodegeneration in hippocampus of wistar rats: a histomorphometric study. J Clin Diagn Res 9(1):AF05–AF07. doi:10.7860/JCDR/2015/10199.5423

Kumari U, Vishwakarma RK, Sonawane P, Abbassi S, Khan BM (2014) Biochemical characterization of recombinant mevalonate kinase from *Bacopa monniera*. Int J Biol Macromol 72:776–783. doi:10.1016/j.ijbiomac.2014.09.030

Kumari U, Vishwakarma RK, Gupta N, Ruby, Shirgurkar MV, Khan BM (2015) Efficient shoots regeneration and genetic transformation of *Bacopa monniera*. Physiol Mol Biol Plants 21(2):261–267. doi:10.1007/s12298-015-0290-6

Lowsky DJ, Olshansky SJ, Bhattacharya J, Goldman DP (2014) Heterogeneity in healthy aging. J Gerontol Biol Sci 6(69):640–649. doi:10.1093/gerona/glt162

Mahender A, Mallesham B, Srinivas K, Kumar GK, Rao KV, Rajesh Y, Zhang P, Sadanandam A (2012) A rapid and efficient method for in vitro shoot organogenesis and production of transgenic *Bacopa monnieri* L. mediated by *Agrobacterium tumefaciens*. In Vitro Cell Dev Biol Plant 48:153–159. doi:10.1007/s11627-011-9421-0

Malhotra CL, Das PK (1959) Pharmacological studies of *Herpestis monniera* Linn. (Brahmi). Indian J Med Res 47:294–305

Mannan A, Abir AB, Rahman R (2015) Antidepressant-like effects of methanolic extract of *Bacopa monniera* in mice. BMC Complement Altern Med 15:337. doi:10.1186/s12906-015-0866-2

Martis G, Rao A, Karanth KS (1992) Neuropharmacological activity of *Herpestis monniera*. Fitoterapia 63:399–404

Mathew J, Paul J, Nandhu MS, Paulose CS (2010) *Bacopa monnieri* and Bacoside-A for ameliorating epilepsy associated behavioral deficits. Fitoterapia 81:315–322. doi:10.1016/j.fitote.2009.11.005

Mathur S, Gupta MM, Ram M, Sharma S, Kumar S (2002) Herb yield and bacoside A content of field-grown *Bacopa monnieri* accessions. J Herbs Spices Med Plants 9:11–18. doi:10.1300/J044v09n01_03

Mathur A, Verma SK, Purohit R, Singh SK, Mathur D, Prasad G, Dua VK (2010) Pharmacological investigation of *Bacopa monnieri* on the basis of antioxidant, antimicrobial and anti-inflammatory properties. J Chem Pharm Res 2(6):191–198

Menon BR, Rathi MA, Thirumoorthi L, Gopalakrishnan VK (2010) Potential effect of *Bacopa monnieri* on nitrobenzene-induced liver damage in rats. Indian J Clin Biochem 25(4):401–404. doi:10.1007/s12291-010-0048-4

Mohanty IR, Maheshwari U, Joseph D, Deshmukh Y (2010) *Bacopa monniera* protects rat heart against ischaemia-reperfusion injury: role of key apoptotic regulatory proteins and enzymes. J Pharm Pharmacol 62(9):1175–1184. doi:10.1111/j.2042-7158.2010.01155.x

Morgan A, Stevens J (2010) Does *Bacopa monnieri* improve memory performance in older persons? Results of a randomized, placebo-controlled, double-blind trial. J Altern Complement Med 16(7):753–759. doi:10.1089/acm.2009.0342

Murashige T, Skoog F (1962) A revised medium for rapid growth and bioassay with tobacco tissue cultures. Physiol Plant 15:473–497

Nadkarni KM (1976) Indian Materia Medica. Popular Prakashan Private, Bombay, pp 624–625

Nannepaga JS, Korivi M, Tirumanyam M, Bommavaram M, Kuo CH (2014) Neuroprotective effects of *Bacopa monniera* whole-plant extract against aluminum-induced hippocampus damage in rats: evidence from electron microscopic images. Chin J Physiol 57(5):279–285. doi:10.4077/CJP.2014.BAC221

Nathan PJ, Clarke J, Lloyd J, Hutchison CW, Downey L, Stough C (2001) The acute effects of an extract of *Bacopa monniera* (Brahmi) on cognitive function in healthy normal subjects. Hum Psychopharmacol 16:345–351. doi:10.1002/hup.306

Nathan PJ, Tanner S, Lloyd J, Harrison B, Curran L, Oliver C, Stough C (2004) Effects of a combined extract of *Ginkgo biloba* and *Bacopa monniera* on cognitive function in healthy humans. Hum Psychopharmacol 2(19):91–96. doi:10.1002/hup.544

Nisha KK, Seetha K, Rajmohan K, Purushothama MG (2003) *Agrobacterium tumefaciens* mediated transformation of Brahmi [*Bacopa monnieri* (L.)Wettst.], a popular medicinal herb of India. Curr Sci 85:85–89

Pandareesh MD, Anand T, Khanum F (2015) Cognition enhancing and neuromodulatory propensity of *Bacopa monniera* extract against scopolamine induced cognitive impairments in rat hippocampus. Neurochem Res. doi:10.1007/s11064-015-1780-1

Pandareesh MD, Anand T, Bhat PV (2016) Cytoprotective propensity of *Bacopa monniera* against hydrogen peroxide induced oxidative damage in neuronal and lung epithelial cells. Cytotechnology 68(1):157–172. doi:10.1007/s10616-014-9767-3

Paulose CS, Chathu F, Khan SR, Krishnakumar A (2008) Neuroprotective role of *Bacopa monnieri* extract in epilepsy and effect of glucose supplementation during hypoxia: glutamate receptor gene expression. Neurochem Res 33:1663–1671. doi:10.1007/s11064-007-9513-8

Peng L, Zhou Y, de Kong Y, Zhang WD (2010) Antitumor activities of dammarane triterpene saponins from *Bacopa monniera*. Phytother Res 24(6):864–868. doi:10.1002/ptr.3034

Praveen N, Naik PM, Manohar SH, Nayeem A, Murthy HN (2009) In vitro regeneration of Brahmi shoots using semisolid and liquid cultures and quantitative analysis of bacoside A. Acta Physiol Plant 31:723–728. doi:10.1007/s11738-009-0284-5

Pravina K, Ravindra KR, Goudar KS, Vinod DR, Joshua AJ, Wasim P, Venkateshwarlu K, Saxena VS, Amit A (2007) Safety evaluation of BacoMindTM in healthy volunteers: a phase I study. Phytomedicine 14:301–308. doi:10.1016/j.phymed.2007.03.010

Preethi J, Singh HK, Venkataraman JS, Rajan KE (2014) Standardized extract of *Bacopa monniera* (CDRI-08) improves contextual fear memory by differentially regulating the activity of histone acetylation and protein phosphatases (PP1α, PP2A) in hippocampus. Cell Mol Neurobiol 34(4):577–589. doi:10.1007/s10571-014-0042-0

Rahman LU, Verma PC, Singh D, Gupta MM, Banerjee S (2002) Bacosides production by suspension cultures of *Bacopa monnieri* (L). Biotechnol Lett 24:1427–1429. doi:10.1023/A:1019815018436

Ramesh M, Karthikeyan A, Vijayakumar K, Largia MJV, Pandian SK (2011) *Agrobacterium*-mediated transformation of pharmaceutically important Indian medicinal herb *Bacopa monnieri* (L.). J Med Plant Res 5:2316–2321

Rao CHV, Sairam K, Goel RK (2000) Experimental evaluation of *Bacopa monniera* on rat gastric ulceration and secretion. Indian J Physiol Pharmacol 44:435–441

Rashid S, Firyal L, Ahmad M, Usmanghani K (1990) Cardiovascular effects of *Bacopa monnieri* (L.) Pennel extract in rabbits. Pak J Pharm Sci 3:57–62

Rastogi RP (1990) Compendium of Indian medicinal plants, vol 1. CSIR, New Delhi, pp 118–122

Rastogi S, Pal R, Kulshreshtha DK (1994) Bacoside A3 – a triterpenoid saponin from *Bacopa monniera*. Phytochemistry 36:133–137. doi:10.1016/S0031-9422(00)97026-2

Roodenrys S, Booth D, Bulzoni S, Phipps A, Micallef C, Smoker J (2002) Chronic effects of Brahmi (*Bacopa monnieri*) on human memory. Neuropsychopharmacology 27:279–281. doi:10.1016/S0893-133X(01)00419-5

Ruby, Santosh Kumar RJ, Vishwakarma RK, Singh S, Khan BM (2014) Molecular cloning and characterization of genistein 4′-O-glucoside specific glycosyltransferase from *Bacopa monniera*. Mol Biol Rep 41(7):4675–4688. doi:10.1007/s11033-014-3338-8

Russo A, Borrelli F (2005) *Bacopa monniera*, a reputed nootropic plant: an overview. Phytomedicine 12:305–317. doi:10.1016/j.phymed.2003.12.008

Russo A, Borrelli F, Campisi A, Acquaviva R, Raciti G, Vanella A (2003a) Nitric oxide-related toxicity in cultured astrocytes: effect of *Bacopa monniera*. Life Sci 73:1517–1526. doi:10.1016/S0024-3205(03)00476-4

Russo A, Izzo AA, Borrelli F, Renis M, Vanella A (2003b) Free radical scavenging capacity and protective effect of *Bacopa monniera* L. on DNA damage. Phytother Res 17:870–875. doi:10.1002/ptr.1061

Sairam K, Rao CV, Babu MD, Goel RK (2001) Prophylactic and curative effects of *Bacopa monniera* in gastric ulcer models. Phytomedicine 8:423–430. doi:10.1078/S0944-7113(04)70060-4

Sairam K, Dorababu M, Goel RK, Bhattacharya SK (2002) Antidepressant activity of standardized extract of *Bacopa monniera* in experimental models of depression in rats. Phytomedicine 9:207–211. doi:10.1078/0944-7113-00116

Samiulla DS, Prashanth D, Amit A (2001) Mast cell stabilizing activity of *Bacopa monniera*. Fitoterapia 72:284–285. doi:10.1016/S0367-326X(00)00309-9

Sanford JC, Smith FD, Russell JA (1993) Optimizing the biolistic process for different biological applications. Methods Enzymol 217:483–509. doi:10.1016/0076-6879(93)17086-K

Sharma R, Ruby Z, Khan BM, Suresh CG (2011) In silico substrate specificity in bmgt and bmgt2 genes of *Bacopa monniera* glycosyltransferases. Online J Bioinforma 12:413–433

Shrivastava N, Rajani M (1999) Multiple shoot regeneration and tissue culture studies on *Bacopa monnieri* (L.). Pennell. Plant Cell Rep 18:919–923. doi:10.1007/s002990050684

Simpson T, Pase M, Stough C (2015) *Bacopa monnieri* as an antioxidant therapy to reduce oxidative stress in the aging brain. Evid Based Complement Alternat Med 2015:615384. doi:10.1155/2015/615384

Singh HK (2013) Brain enhancing ingredients from Ayurvedic medicine: quintessential example of *Bacopa monniera*, a narrative review. Nutrients 2(5):478–497. doi:10.3390/nu5020478

Singh HK, Dhawan BN (1982) Effect of *Bacopa monniera* Linn. (Brahmi) extract on avoidance responses in rat. J Ethnopharmacol 5:205–214. doi:10.1016/0378-8741(82)90044-7

Singh HK, Dhawan BN (1997) Neuropsychopharmacological effects of the Ayurvedic nootropic *Bacopa monniera* Linn (Brahmi). Indian J Pharmacol 29:S359–S365

Singh RH, Singh L (1980) Studies on the antianxiety effect of the medha rasayana drug Brahmi (*Bacopa monniera* Wettst.). J Res Ayurveda Siddha 1:133–148

Singh A, Ebenso EE, Quraishi MA (2012) Stem extract of Brahmi (*Bacopa monnieri*) as green corrosion inhibitor for aluminum in NaOH solution. Int J Electrochem Sci 7:3409–3419

Sivarajan VV, Balachandran I (1994) Ayurvedic drugs and their plant sources. Oxford and IBH Publishing Co, New Delhi, pp 97–99

Sivaramakrishna C, Rao CV, Trimurtulu G, Vanisree M, Subbaraju GV (2005) Triterpenoid glycosides from *Bacopa monnieri*. Phytochemistry 66:2719–2728. doi:10.1016/j.phytochem.2005.09.016

Stough C, Lloyd J, Clarke J, Downey L, Hutchison C, Rodgers T, Nathan PJ (2001) The chronic effects of an extract of *Bacopa monniera* (Brahmi) on cognitive function in healthy human subjects. Psychopharmacology (Berl) 4(156):481–484. doi:10.1007/s002130100815

Stough C, Downey LA, Lloyd J, Silber B, Redman S, Hutchison C, Wesnes K, Nathan PJ (2008) Examining the nootropic effects of a special extract of *Bacopa monniera* on human cognitive functioning: 90-day double-blind placebo-controlled randomized trial. Phytother Res 12(22):1629–1634. doi:10.1002/ptr.2537

Sumathi T, Niranjali DS (2009) Effect of *Bacopa monniera* on liver and kidney toxicity in chronic use of opioids. Phytomedicine 16(10):897–903. doi:10.1016/j.phymed.2009.03.005

Sumathy T, Nongbri A (2008) Hepatoprotective effects of Bacoside-A, a major constituent of *Bacopa monniera* Linn. Phytomedicine 15:801–905. doi:10.1016/j.phymed.2007.11.020

Sumathy T, Subramanian S, Govindasamy S, Balakrishna K, Veluchamy G (2001) Protective role of *Bacopa monniera* on morphine induced hepatotoxicity in rats. Phytother Res 15:643–645. doi:10.1002/ptr.1007

Sumathy T, Govindasamy S, Balakrishna K, Veluchamy G (2002) Protective role of *Bacopa monniera* on morphine-induced brain mitochondrial enzyme activity in rats. Fitoterapia 73:381–385. doi:10.1016/S0367-326X(02)00114-4

Thomas RB, Joy S, Ajayan MS, Paulose CS (2013) Neuroprotective potential of *Bacopa monnieri* and Bacoside A against dopamine receptor dysfunction in the cerebral cortex of neonatal hypoglycemic rats. Cell Mol Neurobiol 33(8):1065–1074. doi:10.1007/s10571-013-9973-0

Tiwari V, Singh BD, Tiwari KN (1998) Shoot regeneration and somatic embryogenesis from different explants of Brahmi [*Bacopa monniera* (L.) Wettst.]. Plant Cell Rep 17:538–543. doi:10.1007/s002990050438

Tiwari V, Tiwari KN, Singh BD (2000) Suitability of liquid culture for in vitro multiplication of *Bacopa monniera* (L.) Wettst. Phytomorphology 50:337–342

Tiwari V, Tiwari KN, Singh BD (2001) Comparative studies of cytokinins on *in vitro* propagation of *Bacopa monnieri*. Plant Cell Tissue Organ 66:9–16. doi:10.1023/A:1010652006417

Tripathi YB, Chaurasia S, Tripathi E, Upadhyay A, Dubey GP (1996) *Bacopa monniera* Linn. as an antioxidant: mechanism of action. Indian J Exp Biol 34:523–526

Tripathi S, Mahdi AA, Hasan M, Mitra K, Mahdi F (2011) Protective potential of *Bacopa monniera* (Brahmi) extract on aluminum induced cerebellar toxicity and associated neuromuscular status in aged rats. Cell Mol Biol 57(1):3–15. doi:10.1170/T896

Velaga MK, Basuri CK, Robinson Taylor KS, Yallapragada PR, Rajanna S, Rajanna B (2014) Ameliorative effects of *Bacopa monniera* on lead-induced oxidative stress in different regions of rat brain. Drug Chem Toxicol 37(3):357–364. doi:10.3109/01480545.2013.866137

Vishwakarma RK, Patel KA, Sonawane P, Singh S, Ruby, Kumari U, Agrawal DC, Khan BM (2012) Molecular characterization of farnesyl pyrophosphate synthase from *Bacopa monniera* by comparative modeling and docking studies. Bioinformation 8(22):1075–1081. doi:10.6026/97320630081075

Vishwakarma RK, Singh S, Ruby, Sonawane P, Srivastava S, Kumari U, Santosh Kumar RJ, Khan BM (2013a) Molecular cloning, biochemical characterization, and differential expression of an Acetyl-CoA C-Acetyltransferase gene (AACT) of Brahmi (*Bacopa monniera*). Plant Mol Biol Report 31:547–557. doi:10.1007/s11105-012-0523-6

Vishwakarma RK, Sonawane P, Singh S, Kumari U, Ruby, Khan BM (2013b) Molecular characterization and differential expression studies of an oxidosqualene cyclase (OSC) gene of Brahmi (*Bacopa monniera*). Physiol Mol Biol Plants 19(4):547–553. doi:10.1007/s12298-013-0195-1

Vishwakarma RK, Patel K, Sonawane P, Kumari U, Singh S, Ruby, Abbassi S, Agrawal DC, Tsay HS, Khan BM (2015) Squalene synthase gene from medicinal herb *Bacopa monniera*: molecular characterization, differential expression, comparative modeling, and docking studies. Plant Mol Biol Report 6(33):1675–1685. doi:10.1007/s11105-015-0864-z

Vohora D, Pal SN, Pillai KK (2000) Protection from phenytoin-induced cognitive deficit by *Bacopa monniera*, a reputed Indian nootropic plant. J Ethnopharmacol 71:383–390. doi:10.1016/S0378-8741(99)00213-5

Yankner BA, Lu T, Loerch P (2008) The aging brain. Annu Rev Pathol-Mech 3:41–66. doi:10.1146/annurev.pathmechdis.2.010506.092044

Chapter 7
Metabolic Engineering: Achieving New Insights to Ameliorate Metabolic Profiles in *Withania somnifera*

Neha Patel, Parth Patel, and Bashir M. Khan

Abstract *Withania somnifera*, commonly known as Indian ginseng, has been used for centuries in Ayurvedic system of medicine for its antitumor, antioxidant, antiaging, antiserotogenic, and antistress activities. The various medicinal properties of the plant are accredited to the steroidal lactones (withanolides) present in the plant. Withanolides are synthesized by diverting the metabolite flux away from the isoprenoid pathway by the reductive condensation of farnesyl diphosphate to squalene through the activity of the enzyme squalene synthase. This enzyme squalene synthase is a major branch point involved in the regulation of withanolides. Owing to low concentrations of these bioactive compounds in plant, large biomass is utilized for the preparation of medicinal formulations in pharmaceutical industries to fulfill the growing commercial demand. To protect *Withania* spp. from becoming an endangered species, the activity of squalene synthase has been well exploited. This chapter is focused on the engineering of isoprenoid biosynthetic pathway in *W. somnifera* by the introduction of squalene synthase gene to improve the yield of desired product.

Keywords *Agrobacterium tumefaciens* • Metabolic engineering • Squalene synthase • Triterpenoids • *Withania somnifera* • Withanolides

N. Patel (✉)
Division of Plant Sciences, Research School of Biology, College of Medicine, Biology and Environment, The Australian National University, Canberra ACT 0200, Australia
e-mail: neha.patel@anu.edu.au

P. Patel
Molecular Microbial Biochemistry Laboratory, Department of Biochemistry, The Maharaja Sayajirao University of Baroda, Vadodara 390002, Gujarat, India

B.M. Khan
Plant Tissue Culture Division, CSIR-National Chemical Laboratory, Dr. Homi Bhabha Road, Pune 411008, Maharashtra, India

H.-S. Tsay et al. (eds.), *Medicinal Plants - Recent Advances in Research and Development*, DOI 10.1007/978-981-10-1085-9_7

Abbreviations

BAP	6-Benzylaminopurine
bp	Base pair
cDNA	Complementary DNA
CoA	Coenzyme A
CR	Callus-like root
DMAPP	3,3-Dimethylallyl pyrophosphate
EC	Enzyme Commission
FPP	Farnesyl pyrophosphate
GABA	γ-Aminobutyric acid
GC	Gas chromatography
GGPP	Geranylgeranyl pyrophosphate
GPP	Geranyl pyrophosphate
GUS	β-Glucuronidase
HMG	3-Hydroxy-3-methylglutaryl
HR	Hairy root
IBA	Indole-3-butyric acid
IPP	Isopentenyl-5-pyrophosphate
kDa	Kilodalton
MEP	2-C-methyl-D-erythritol 4-phosphate
Mg	Magnesium
mRNA	Messenger RNA
MS	Mass spectrometry
MS media	Murashige and Skoog media
MVA	Mevalonic acid
NADPH	Nicotinamide adenine dinucleotide phosphate
PGR	Plant growth regulators
PPi	Inorganic phosphate
PSPP	Presqualene diphosphate
qRT-PCR	Quantitative real-time PCR
SE	Standard error
spp	Species
SQS	Squalene synthase
WsSQS	*Withania somnifera* squalene synthase

7.1 Introduction

The isoprenoid biosynthetic pathway is an important cellular metabolic process present in all higher eukaryotes where the central intermediate, isopentenyl-5-pyrophosphate (IPP), serves as the basis of biosynthesis of many molecules used by the organisms in various processes. Triterpenoids (C_{30} isoprenoids) are one of the diverse groups of natural organic products and include various steroidal sapogenins, sterols, steroidal lactones, and alkaloids. The occurrence of these compounds is very common in members of the Solanaceae family. *Withania somnifera* (L.) Dunal, a member of the Solanaceae family, has been widely studied for its anticancer and adaptogenic activities (Marie Winters 2006), and Ayurveda has acknowledged this plant for its medicinal importance. The therapeutic efficiencies of the plant are accredited to the characteristic secondary metabolites called withanolides. These chemical markers in plants are steroidal lactones characterized by C_{28} ergostane-type steroids where lactone ring is a characteristic feature, and their structural variation accounts for a broad range of the plant efficacy. These compounds are found to be localized mainly in leaves and roots depending on the type of withanolide, and their concentration usually ranges from 0.001 to 0.5 % of dry weight (Mirjalili et al. 2009a, b). Such a low abundance of withanolides in plants is incapable of meeting the present demand for medicinal formulations. Earlier, the chemical synthesis of withanolides had been attempted (Kovganko and Kashkan 1997; Gamoh et al. 1984; Jana et al. 2011), but it required extensive experimentation due to the structural complexities and variations in the compounds, thus resulting in the low yields. Alternatively, tissue culture techniques had been carried out to enhance withanolide content in the plant; however, no success has been achieved so far to produce these metabolites at the commercial scale (Ray and Jha 2001). Well-documented reports are available, detailing plant cell cultures and hairy root cultures developed for the extraction of important metabolites from *Withania* extracts (Murthy et al. 2008; Vitali et al. 1996); however, synthesis of withanolides by these in vitro culture techniques is still below the levels required for optimal economical use. To overcome this limitation, researchers have employed genetic engineering to modify the critical steps of the biosynthetic pathways to improve the production of secondary metabolites. In this chapter, we describe metabolic engineering of the isoprenoid pathway using squalane synthase as a tool to enhance secondary metabolite contents in *W. somnifera*.

7.2 *Withania somnifera* (L.) Dunal (Ashwagandha)

Withania, usually found in India, Congo, Spain, Canary Islands, South Africa, Egypt, Morocco, Jordan, Palestine, Pakistan, and Afghanistan, is cultivated as an annual shrub and grows well in sandy loam or light red soil, having pH of 7.5–8.0 with good drainage, and at temperatures between 20 and 35 °C. The plant has an erect branching, ovate leaves with sparse hair above and dense beneath, clothed with minutely stellate hairy tomentum. The flowers are bisexual with a long pedicel, greenish or lurid yellow in axillary fascicles, which occurs during spring and fall. Fruits of *W. somnifera* are globose berries enclosed in a persistent calyx carrying seeds (Fig. 7.1).

W. somnifera is comprehensively studied for the chemistry of its constituent chemical composition. Steroidal lactones (withanolides), alkaloids

Fig. 7.1 *Withania somnifera* (**a**) seeds, (**b**) flowers, (**c**) fruits, and (**d**) plant

(isopelletierineanaferine), tannin, and flavonoids have been identified, extracted, and isolated (Atta-ur-Rahman et al. 1991, 1993). Currently, *Withania* roots, berries, and aerial parts are reported to contain more than 40 withanolides, 12 alkaloids, and numerous sitoindosides (withanolides with glucose at C_{27}) (Choudhary et al. 1996).

W. somnifera is a well-recognized plant for its medicinal value; and the Vedas describe its use for treatment of several diseases, including as herbal tonic and health food (Dhuley 2000). Traditionally, *W. somnifera* is used to enhance endurance, strength, vigor, health, and increased synthesis of blood, vital fluids, semen, lymph, and muscle fats. The plant has found its special place for antitumor, antiaging, antistress, antioxidant, and anti-inflammatory properties. It is known to counteract dehydration, bone weakness, chronic fatigue, loose teeth, debility, muscle tension, and emaciation. Apart from claims as a potent aphrodisiac, sedative, and rejuvenative properties, its immunomodulator and life-prolonging effects have also been demonstrated (Agrawal et al. 1999). It has often been called Indian ginseng due to its similar restorative properties to ginseng roots (Singh and Kumar 1998). Various parts of the plant have shown beneficial effects in treating arthritis and geriatric problems and reported to possess antiserotogenic and anabolic properties (Prakash et al. 2001), as well as being known for its abortifacient, astringent, deobstruent, and narcotic properties (Mirjalili et al. 2009b). The fruits and seeds of the plant are diuretic and hypnotic and used for curdling of plant milk to prepare vegetable cheese. They can be used as substitutes of soap due to their high saponin content (Saritha and Naidu 2007). These saponins have found applications in many herbal medicines due to their allelopathic, anticholesterolemic, antimicrobial, hemolytic, and adjuvant properties (Madina et al. 2007).

Kumar and Kushwaha (2006) evaluated the potential of various parts of *W. somnifera* to cure several fatal diseases. Ayurveda, Siddha, and Unani medicines report over 200 formulations from its roots, treating various physiological disorders (Asthana and Raina 1989; Singh and Kumar 1998), as well as treating constipation, rheumatism, memory loss, and spermatorrhea (Watt 1972; Mirjalili et al. 2009b). It's bitter leaves are used as anthelmintic; bruised leaves and fruits are locally applied to tumors and tubercular glands, carbuncles, and ulcers (Kapoor 2001). The wild and cultivated plants of *W. somnifera* contain the same alkaloids but exhibit different morphologies and therapeutic actions (Kaul 1957).

7.2.1 Anti-inflammatory Properties

The effectiveness of *W. somnifera* in a variety of rheumatologic conditions may be due in part to its anti-inflammatory properties. Rats given powdered roots orally 1 h before being given injections of Freund's complete adjuvant over a period of 3 days showed considerable reduction in inflammation (Begum and Sadique 1988).

7.2.2 Antitumor Properties

The use of *W. somnifera* on animal cell cultures caused a decrease in level of the nuclear factor kappa B, suppression of intercellular tumor necrosis factor, and a potential apoptotic signaling in cancerous cell lines (Ichikawa et al. 2006). *W. somnifera* has been reported to reduce the size of tumor (Prakash et al. 2002; Jayaprakasam et al. 2003). Antitumor and radiosensitizing effects of Withaferin were also seen in mouse Ehrlich ascites carcinoma in vivo (Sharada et al. 1996).

7.2.3 Antistress Effect

W. somnifera has traditionally been used to stabilize mood in patients with behavioral disturbances. Research has revealed that the herb produces antidepressant and antianxiety effects in rodents comparable to the antidepressant drug imipramine and the antianxiety drug lorazepam (Archana and Namasivayam 1999).

7.2.4 Antioxidant Effect

Sitoindosides VII-X and Withaferin A found in *W. somnifera* are powerful antioxidants which were tested for their antioxidant activity using the major free radical scavenging enzymes, superoxide dismutase (SOD), catalase (CAT), and glutathione peroxidase (GPX) in the rat brain frontal cortex and striatum (Dhuley 2000).

7.2.5 Immunomodulatory Properties

In a study using mice, administration of powdered root extract was found to enhance total white blood cell count. Also, this extract inhibited delayed-type hypersensitivity reactions and enhanced phagocytic activity of macrophages when compared to the control group (Davis and Kuttan 2002).

7.2.6 Hematopoietic Effect

Administration of *W. somnifera* extract was found to reduce significantly leukopenia induced by cyclophosphamide (CTX) treatment in Swiss albino mice. The major activity of *Withania* may be due to the stimulation of stem cell proliferation (Davis and Kuttan 1998).

7.2.7 Effects on the Nervous System

Total alkaloid extract (ashwagandholine, AG) of *Withania* roots has been studied for its effects on the central nervous system (Malhotra et al. 1965). Effects of sitoindosides VII-X and Withaferin isolated from aqueous methanol extract of roots of cultivated varieties of *Withania* were studied on brain cholinergic, glutamatergic, and GABAergic receptors in male Wistar rats (Schliebs et al. 1997).

7.3 Withanolides

Withanolides are the major chemical constituents of *W. somnifera* (Bandyopadhyay et al. 2007). To date, about 400 withanolides or closely related congeners have been discovered in altogether 58 solanaceous species belonging to 22 genera. Withanolides possess properties like tumor inhibiting, hepatoprotective, antifungal, antibacterial, antifeedant, and insecticidal, as well as phytotoxicity and antifertility effects and leishmanicidal and trypanocidal activities (Veleiro et al. 2005).

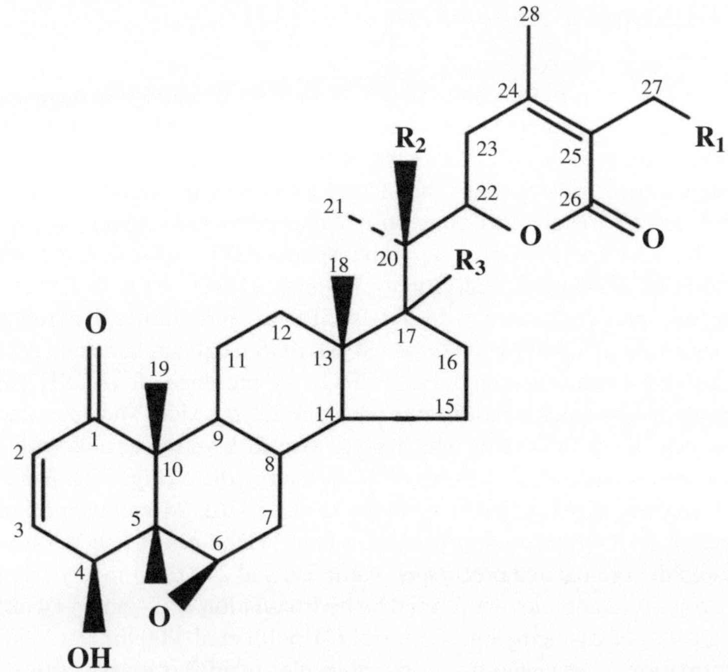

Fig. 7.2 Basic skeleton of some withanolides (Alfonso and Kapetanidis 1994). Withaferin A R_1=OH, R_2=H, R_3=H; Withanolide D R_1=H, R_2=OH, R_3=H; 27-Deoxywithaferin A R_1=H, R_2=H, R_3=H; 27-Hydroxywithanolide D R_1=OH, R_2=OH, R_3=H; Dihydrodeoxywithaferin A 2,3-diH, R_1=H, R_2=H, R_3=H; Dihydrowithaferin A 2,3-diH, R_1=H, R_2=OH, R_3=; 17-Hydroxywithaferin A, R_1=R_3=OH, R_2=H

Withanolides consist of C_{28} steroidal lactones occurring naturally, which are built on a rearranged or otherwise intact ergostane backbone, containing a six-membered lactone ring formed by C_{22}-C_{26} oxidization (Glotter 1991) (Fig. 7.2). "Withanolide" represents the term for 22-hydroxyergostan-26-oic acid-22,26-olide (Lavie et al. 1965a, b), and their structural diversity arises from the modifications of the carbocyclic skeleton or the side chain (Mirjalili et al. 2009b). The genera *Datura*, *Lycium*, *Dunalia*, and *Withania* comprise a group of metabolites consisting of withanolides having an unmodified β-side chain, e.g., Withaferin A. Unmodified α-side chains are quite unusual, occurring in *Jaborosa* and *Withania*, e.g., jaborosalactol N. Withaferin, the first withanolide, was discovered from the leaves of *W. somnifera* (Yarden and Lavie 1962). It was characterized as a new class of steroids containing an α, β-unsaturated lactone linked to C_{17} of the sterane skeleton (Lavie et al. 1965a) which later turned out to be 2,3-dihydro-3-methoxywithaferin A, co-occurring with Withaferin A (Lavie et al. 1965b). Withaferin A occurs in a very low abundance, up to 0.2–0.3 % of dry weight (Abraham et al. 1968). Indian *W. somnifera* roots contain 0.13–0.31 % total alkaloid content, as well as starch, hentriacontane, reducing sugars, dulcitol, glycosides, withanicil, and different peroxidases.

7.3.1 Withanolide Biosynthesis

The biogenetic aspects of the chemical constituents of *W. somnifera* including withanolides are not well explored (Kirson et al. 1977; Nittala and Lavie 1981; Ray and Gupta 1994). The pathway starts with the activation of acetate to form acetyl coenzyme A which then fuses with another acetyl-CoA to form mevalonic acid. Living system synthesizes metabolites using only *R*-mevalonic acid, which later loses one carbon moiety to form IPP. Geranyl pyrophosphate (GPP) is formed by head-to-tail condensation of 3,3-dimethyl allyl pyrophosphate (DMAPP) with IPP. A second IPP molecule then condenses with *Trans* GPP to form farnesyl pyrophosphate (FPP), a substrate of squalene synthase (SQS). SQS catalyzes squalene production by a head-to-head condensation of two FPPs in the presence of NADPH as a coenzyme, which is then quickly oxidized to squalene 2,3-epoxide. The latter undergoes ring closure to form lanosterol which gives rise to several steroidal triterpenoid backbones via 24-methylenecholesterol as an intermediate (Fig. 7.3). Based on *W. somnifera* feeding experiments by Lockley et al. (1976), 24-methylenecholesterol was proposed as a withanolide precursor, where Withaferin A and Withanolide D incorporated the radioactive precursors, while labeled 24-(*R, S*)-methyl-cholesterol failed to do so. Withanolides are formed by hydroxylation of C_{22} and δ-lactonization between C_{22}-C_{26} of 24-methylenecholestrol (Mirjalili et al. 2009b).

The formation of squalene from two molecules of FPP is a very critical step to regulate the metabolic flow toward withanolide biosynthesis rather than the usual isoprenoid pathway. This is catalyzed by squalene synthase, a 47 kDa transmembrane dual function enzyme (Gupta et al. 2012), which forms the basis of the study of this chapter. FPP can be directed by overexpression of squalene synthase to the production of squalene, a precursor of triterpenoids. It has been suggested that active biosynthesis of withanolides occurs when sterol formation is sacrificed (Kamisako et al. 1984; Flores-Sánchez et al. 2002).

7.3.2 Biosynthesis of Triterpenoids

Triterpenoids are the widespread group of naturally occurring isoprenoids distributed throughout the plant kingdom. These compounds have drawn much attention in research due to their active involvement in primary and secondary metabolism in plants and their diverse biological activities which include their applications in human health and nutrition. Genetic engineering has emerged as a powerful technology for manipulation of triterpenoid biosynthetic pathway to improve the recovery of desired end products. To achieve this, a better understanding of triterpenoid biosynthesis is required.

The C_5 isopentenyl-5-pyrophosphate, the building block of isoprenoids, is generated by two different pathways along with its isomer, DMAPP, the mevalonic acid (MVA) pathway localized in cytoplasm, and the 2-C-methyl-D-erythritol

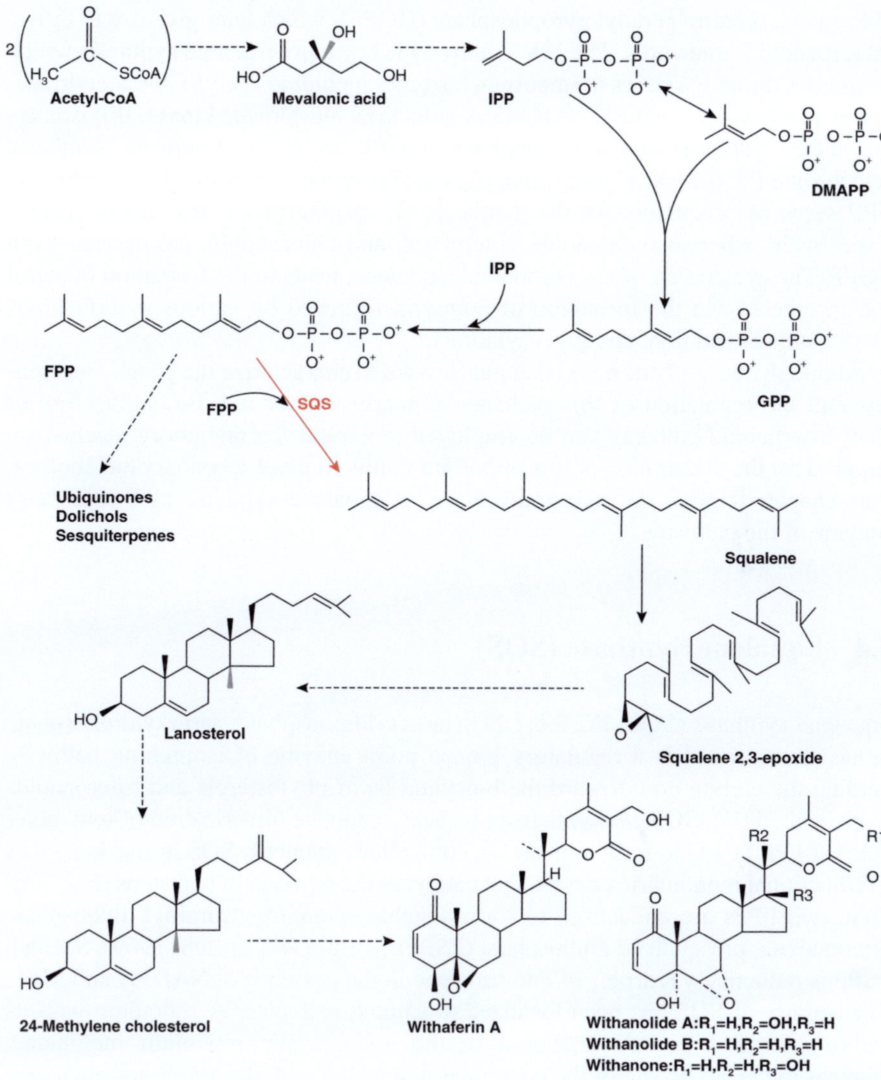

Fig. 7.3 Schematic representation of withanolide biosynthetic pathway. Abbreviations: *IPP* isopentenyl-5-pyrophosphate, *DMAPP* dimethylallyl pyrophosphate, *GPP* geranyl pyrophosphate, and *FPP* farnesyl pyrophosphate. Multiple steps involved in the reaction are shown by dashed arrows (Patel et al. 2015)

4-phosphate (MEP) pathway occurring in plastids (Eisenreich et al. 1998; Lichtenthaler 1999; Rohmer 1999). Monoterpenoids, diterpenoids, and tetraterpenoids are biosynthesized via MEP pathway, whereas sterols, triterpenoids, and sesquiterpenoids are derived via MVA pathway. The combination of C_5 isoprene units leads to the formation of other intermediates of the pathway such as C_{10} GPP, C_{15}

FPP, and C_{20} geranylgeranyl pyrophosphate (GGPP), which later give rise to different terpenoid frameworks. The MVA pathway is key to triterpenoid synthesis which progresses through a series of important enzymes including acetyl-CoA acetyltransferase, HMG-CoA synthase, HMG-CoA reductase, mevalonate kinase, IPP isomerase, FPP synthase, squalene synthase, GGPP synthase, β-amyrin synthase, cytochrome P450s, and glycosyltransferases (Spurgeon and Porter 1981). FPP and GPP serve as precursors for the synthesis of sesquiterpenes and monoterpenes, respectively, whereas carotenoids, diterpenes, and chlorophylls are derived from GGPP. The cyclization of C_{30} compound, squalene, leads to the formation of sterol and triterpenes via the formation of β-amyrin followed by various modifications (oxidation, substitution, and glycosylation).

Although many efforts have been put forward to characterize the genetic machinery, still the regulation of this pathway is poorly understood. Bioengineering of plant triterpenoid pathway can be employed to exploit the regulatory mechanism required for the elaboration of this important family of plant secondary metabolites. This chapter focuses on taking advantage of squalene synthase as a regulatory enzyme of the pathway.

7.4 Squalene Synthase (SQS)

Squalene synthase (SQS, EC 2.5.1.21) (farnesyldiphosphate: farnesyldiphosphate farnesyltransferase) is a regulatory branch point enzyme of isoprenoid pathway, guiding the carbon pool toward the biosynthesis of phytosterols and triterpenoids (Abe et al. 1993). SQS catalyzes head-to-head reductive dimerization of two molecules of FPP (1′–1) to form a linear C_{30} compound, squalene. SQS is also known as a bifunctional monomeric enzyme that catalyzes the reaction in two steps (Fig. 7.4). First, two FPPs are condensed to form a stable cyclopropylcarbinyl diphosphate intermediate, presqualene diphosphate (PSPP) (Rilling and Epstein 1969). Second, PSPP is reductively rearranged into squalene in the presence of NADPH and Mg^{2+}. The activity of SQS has been localized to smooth endoplasmic reticulum with its carboxy-terminal portion anchored to the endoplasmic reticulum membrane, whereas the catalytic site of the enzyme is associated with the amino-terminal portion of the protein found on the cytoplasmic face of the endoplasmic reticulum (Robinson et al. 1993).

SQS cDNA clones have been isolated and characterized from various plant species, such as rice, maize, soybean (Hata et al. 1997), tobacco (Devarenne et al. 1998; Hanley et al. 1996), *Arabidopsis thaliana* (Kribii et al. 1997; Nakashima et al. 1995), *Panax ginseng* (Kim et al. 2011a), *Diospyros kaki* (Zhou et al. 2012), *Centella asiatica* (Kim et al. 2005), *Lotus japonicus* (Akamine et al. 2003), *Capsicum annuum* (Lee et al. 2002), *Glycyrrhiza glabra* (Hayashi et al. 1999), *W. somnifera* (Gupta et al. 2012), and other plants, as well as mammals (Inoue et al. 1995; Robinson et al. 1993) and yeast (Jennings et al. 1991; Merkulov et al. 2000).

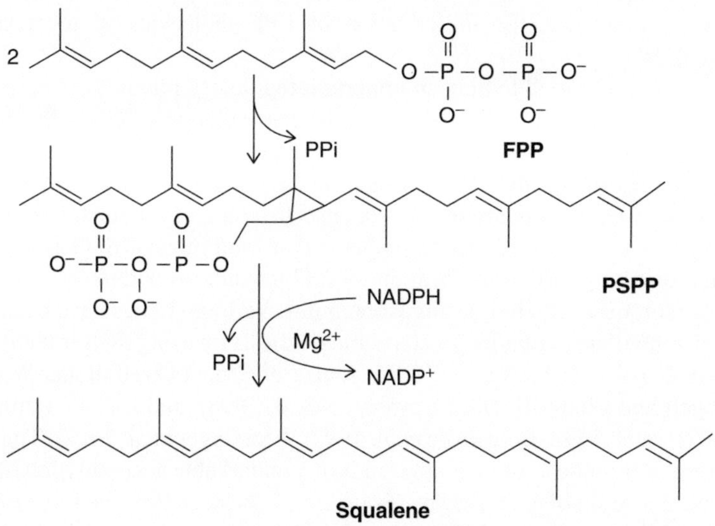

Fig. 7.4 Condensation of two FPP molecules to form squalene, catalyzed by squalene synthase

With SQS being a transmembrane protein, several biochemical investigations were carried out to obtain soluble and active recombinant SQS. SQS amino acid sequence analysis provided insights for engineering more soluble variants. Jennings et al. (1991) cloned the yeast *SQS* and suggested that the enzyme consisted of a large cytosolic domain anchored to the endoplasmic reticulum by a single C-terminal transmembrane helix. Subsequently, a soluble and fully active version of recombinant yeast SQS was constructed by deletion of a C-terminal hydrophobic region from the enzyme (LoGrasso et al. 1993; Yoshioka et al. 1999). Similar C-terminal hydrophobic domains were also found in the *Schizosaccharomyces pombe* (Robinson et al. 1993) and *Homo sapiens* (Jiang et al. 1993). Membrane-bound SQS has been purified to homogeneity from microsomal membranes of *Saccharomyces cerevisiae* (Sasiak and Rilling 1988) and in a truncated soluble form from rat liver (McKenzie et al. 1992). In plants, the enzyme has been solubilized and partially purified from daffodil microsomal membranes (Belingheri et al. 1991) and from tobacco cell suspension cultures (Hanley and Chappell 1992). C-terminal truncation was also carried out in *C. annuum* by removing the last 24 amino acids, and fully active SQS was purified from recombinant *Escherichia coli* (Lee et al. 2002).

Recently, two isoforms of *SQS* were identified from *W. somnifera* that contained an open reading frame of 1236 and 1242 bp encoding polypeptides of 412 and 414 amino acids, respectively (Gupta et al. 2012). Both isoforms shared 99 % similarity and identity with each other. When compared with other plant species, they showed maximum similarity and identity with *C. annuum* followed by *Solanum tuberosum* and *Nicotiana tabacum*. For recombinant production of the enzymes, 24 hydrophobic amino acids were removed from the carboxy terminus, and proteins were expressed in *E. coli*. The heterologously expressed truncated proteins were found to be active in catalyzing the dimerization of FPP molecules to form squalene (Gupta et al. 2012).

SQS has been reported to play an important regulatory role in the triterpene and steroid biosynthetic pathway. Because of its particular position at the interface between hydrophilic and hydrophobic intermediates, SQS constitutes a major control point for regulating the sterol branch in directing FPP molecules into either sterols or non-sterol isoprenoids in response to changing cellular requirements (Wentzinger et al. 2002). Evidences support that inhibition of SQS is a potential means of redirecting FPP away from the sterol biosynthesis and toward the synthesis of other commercially interesting isoprenoids. Various studies have been carried out to understand the regulation of sterol biosynthesis by using SQS mutants (Karst and Lacroute 1977; Tozawa et al. 1999), fungal elicitors (Threlfall and Whitehead 1988; Vogeli and Chappell 1988; Devarenne et al. 1998), and specific inhibitors of SQS (Baxter et al. 1992; Bergstrom et al. 1993; Wentzinger et al. 2002). The disruption of sterol biosynthesis at SQS step leads to a remarkable accumulation of FPP in an erg9 mutant strain of *S. cerevisiae* (Song 2003). Besides FPP, the increase of IPP and GGPP was also observed when the rat liver cells were treated with zaragozic acid A, a potent inhibitor of SQS (Keller 1996), and similar effects have been observed in plants (Fulton et al. 1995). In another study on *S. cerevisiae*, the quantity of amorphadiene increased to fivefold by the downregulation of squalene synthase, while the production of ergosterol decreased due to the decrease in squalene (Paradise et al. 2008). The similar studies have been carried out in many plants, and the results of overexpression of squalene synthase are summarized in Table 7.1.

Table 7.1 Studies of *SQS* overexpression in plants

S. No.	Transformed plant	Introduced gene	Results	References
1.	*P. ginseng*	*PgSS1*	1.6-t o 3-fold higher ginsenoside, increased phytosterol and triterpene saponins	Lee et al. (2004)
2.	*Eleutherococcus senticosus*	*PgSS1*	2- to 2.5-fold increase in phytosterols and triterpene saponins	Seo et al. (2005)
3.	*Glycyrrhiza uralensis*	*GuSQS1*	2.6-fold higher glycyrrhizin	Lu et al. (2008)
4.	*Withania coagulans*	*AtSS1*	Increased phytosterols and withanolides	Mirjalili et al. (2011)
5.	*Bupleurum falcatum*	*BfSS1*	Increased phytosterols and saikosaponins	Kim et al. (2011b)
6.	*Artemisia annua*	Antisense *AaSQS*	23.2 % increase in artemisinin	Wang et al. (2012)
7.	*W. somnifera*	*WsSS*	2.5-fold increase in Withanolide A	Grover et al. (2012)
		WsSQS	Twofold increase in total withanolides	Patel et al. (2015)

7.5 Tissue Culture Studies on *W. somnifera*

W. somnifera is in great demand owing to the presence of medicinally important secondary metabolites. The plant lacks the natural ability for regenerative propagation, and hence it is difficult to fix the variability generated by sexual recombination (Jayanthi and Sharma 1991). In vitro propagation and tissue culture have provided a means for the continuous production of disease-free healthy plants and their metabolites. Micropropagation in *Withania* has been studied through direct multiple shoot formations using different explants such as shoot tip (Jayanthi and Sharma 1991), leaf (Sharma et al. 2010), leaf disks (Abhyankar and Chinchanikar 1996), hypocotyls and axillary leaf (Rani and Grover 1999), node (Siddique et al. 2004), and internode (Valizadeh and Valizadeh 2009), while indirect regeneration has also been achieved through callus (Rani et al. 2003).

Recently, an efficient method of multiple shoot regeneration has been developed by using 1 cm long meristematic (apical and nodal) segments as explants excised from laboratory grown *W. somnifera* plants (Patel et al. 2014). Initially, *W. somnifera* seeds were inoculated on semisolid ½ MS media where radical emergence was observed after 15 days. After 4 weeks, the germinated seeds were transferred to ½ MS liquid medium onto the filter paper support to grow seedlings. The nodal and apical segments were excised from these seedlings and introduced on MS medium containing kinetin (0.1 mg/L) and BAP (0.2 mg/L) to support highest shoot multiplication and elongation (Fig. 7.5a–c). The regenerated shoots were separated and transferred to rooting medium containing 2 mg/L IBA (Fig. 7.5d, e). The completely grown plants were transferred to pots and acclimatized to a glass house (Fig. 7.5f, g), which were further utilized for obtaining explants for *Agrobacterium tumefaciens*-mediated plant transformation with squalene synthase gene.

Multiple shoot cultures of *W. somnifera* from shoot tip explants grown on MS medium supplemented with 1 mg/L BAP accumulated 0.04 % Withaferin A and 0.06 % Withanolide D. The addition of 4 % sucrose to the medium enhanced accumulation of both withanolides to 0.16 % and 0.08 %, respectively (Ray and Jha 2001).

Studies using nodal segments as explants for *W. somnifera* shoot cultures have been reported for biosynthesis of Withanolide A. Significant variation in Withanolide A production was observed in these cultures, depending upon the composition of plant growth regulators (PGRs) in the culture media and genotype of explant. A particular concentration of BAP (1.0 ppm) and kinetin (0.5 ppm) resulted in the highest concentration of Withanolide A in in vitro green shoots of *W. somnifera* (Sangwan et al. 2007).

Several studies have been carried out to investigate the potential of hairy root cultures transformed with *Agrobacterium rhizogenes* for the production of secondary metabolites (Mano et al. 1986; Payne et al. 1987; Shanks and Morgan 1999). In a previous report, three different *A. rhizogenes* strains were used to establish

Fig. 7.5 Tissue culture of *Withania somnifera*. (**a**) Explant transferred to proliferation medium, (**b**) 10-day-old explant, (**c**) 5-week-old plant with only shoots, (**d**) plant transferred to rooting medium, (**e**) rooted plantlet, (**f**) plant transferred in pot, and (**g**) successfully hardened plant in glass house (Patel et al. 2015)

W. somnifera hairy roots, and their ability to produce withanolides was analyzed especially for Withaferin A (Banerjee et al. 1994). Hairy root growth as well as transformation ability of strain A4 was significantly high followed by LBA 9402 strain, while no transformation events were reported by LBA 9360. Maximum Withaferin A synthesis was observed in 10-week-old hairy roots. Another report showed two morphologies of transformed roots in *W. coagulans* where leaf sections were inoculated with *A. rhizogenes* C58C1 (pRiA4): callus-like roots (CR) showing higher withanolide production and typical hairy roots (HR) showing fast growth but lower production of withanolides (Mirjalili et al. 2009a).

Vitali et al. (1996) reported production of withanolides from *A. rhizogenes*-transformed *W. somnifera* cultures grown in vitro, either without PGRs or on BAP or 2,4-D supplemented MS media. The hairy roots showed no withanolides, while shoot and callus cultures showed limited synthesis. Hairy root cultures of *W. somnifera* were established by transformation with *A. rhizogenes* strain R1601 which showed 2.7-fold higher accumulation of Withanolide A as compared to non-transformed roots (Murthy et al. 2008). Several other hairy root and plant cell culture studies on *Withania* using *A. rhizogenes* did not show significant improvement in accumulation of withanolides but could only produce some specific secondary metabolites (Bandyopadhyay et al. 2007; Kumar et al. 2005). However, a serious limitation of hairy root cultures is that these can produce metabolites specific to root tissues only.

7.6 Genetic Engineering of Isoprenoid Pathway in *W. somnifera* Using Squalene Synthase as a Target Gene

Squalene synthase is identified as a potential target for metabolic engineering of plant secondary metabolite biosynthetic pathway in *W. somnifera*, as it regulates at a branch point of the pathway diverting the carbon flux toward the withanolide biosynthesis. Earlier, in several studies, genes that express the enzymes of isoprenoid pathway have been manipulated to enhance the carbon flow into specific downstream products in different plant species (Re et al. 1995; Hey et al. 2006; Chen et al. 2000). The genetic manipulations suggested that, in most cases, synthesis of the upstream isoprenoid precursors, IPP and DMAPP, should be maximized to boost the production of downstream metabolites (Kumar et al. 2012). The regulatory role of SQS in triterpene sterol biosynthesis has been demonstrated in various plants (Vogeli and Chappell 1988; Devarenne et al. 1998, 2002; Wentzinger et al. 2002).

In *A. annua*, the antisense squalene synthase resulted in the increased production of artemisinin by diverting the carbon flux away from squalene (Wang et al. 2012). Lee et al. (2004) demonstrated the role of squalene synthase (*PgSS1*) gene in the biosynthesis of phytosterols and triterpene saponins in *P. ginseng*. The overexpression of *PgSS1* in adventitious roots of transgenic *P. ginseng* resulted in upregulation of all the downstream genes, thus causing the remarkable increase of phytosterols as well as ginsenoside contents. A similar study on *E. senticosus* showed that overexpression of *SQS* in transformed plants leads to higher triterpene saponin content (Seo et al. 2005). *SQS* derived from *P. ginseng* (*PgSS1*) was introduced in *E. senticosus* through *Agrobacterium*-mediated transformation, and the transgenic plants were tested for the level of SQS enzyme which increased up to threefold as compared to wild-type plants, while phytosterols and triterpene saponins increased by 2- to 2.5-fold.

A positive correlation was found between the expression levels of SQS and the amount of triterpenes produced in *Ganoderma lucidum* (Zhao et al. 2007). Analysis of *SQS* levels by qRT-PCR and SQS protein by Western blotting in fungal mycelia and mushroom primordia revealed that gene and protein expression corroborated with the accumulation of triterpenes with the developmental stages. It was concluded that the lower enzyme expression level in mycelia was directly correlated with low triterpene content, whereas the higher enzyme level in fruiting bodies accounted for the increased triterpene synthesis. Hairy roots were established in *G. uralensis* through *A. rhizogenes* carrying *SQS*, which synthesized 2.5-fold higher glycyrrhizin than control (Lu et al. 2008).

Similarly, *B. falcatum* transgenic roots overexpressing squalene synthase (*BfSS1*) in the sense orientation resulted in the accumulation of downstream genes such as squalene epoxidase and cycloartenol synthase but unexpectedly decreased the mRNA level of β-amyrin synthase (β-AS), a triterpene synthase mRNA. The treatment of wild-type roots with methyl jasmonate stimulated β-AS mRNA accumulation and saikosaponin production but suppressed phytosterol production. Methyl jasmonate treatment on transgenic roots did not stimulate β-AS mRNA accumulation but still showed enhanced saikosaponin and phytosterol production. This concluded that overexpression of *SQS* is more effective in regulating downstream genes in triterpene and phytosterol biosynthesis than elicitor treatment (Kim et al. 2011b). In *W. coagulans*, the overexpression of *SQS* resulted in the increased production of phytosterols, withanolides, and triterpenoids (Mirjalili et al. 2011). *SQS* of *A. thaliana* (*SS1*) was introduced in *W. coagulans* via *A. rhizogenes* A4, and the engineered hairy roots were studied for metabolite accumulation. It was observed that the transgenic hairy roots biosynthesized more phytosterols and withanolides as compared to the control roots. In another study on *W. somnifera*, overexpression of *SQS* resulted in fourfold higher SQS activity and 2.5-fold higher Withanolide A contents. The cell suspension cultures also produced Withaferin A which was absent in non-transformed cell cultures (Grover et al. 2012).

Recently, genetic engineering of isoprenoid pathway using *SQS* was performed in *W. somnifera* where a systematic approach of plant regeneration and transformation was applied (Patel et al. 2015). The apical and nodal segments excised from in vitro grown plants were used as explants for *A. tumefaciens*-mediated transformation. A combination of growth hormones (0.1 mg/L kinetin and 0.2 mg/L BAP) in the medium supported maximum multiple shoot proliferation. Later, the positive transformants were selected by GUS assay (Fig. 7.6a, b) and hygromycin sensitivity, followed by the confirmation through PCR. The qRT-PCR analysis of the transformants showed that *SQS* mRNA levels increased by two- to fivefold as compared to wild-type plants (Fig. 7.6c). SQS enzyme levels and activity were also determined using GC-MS which showed a 2- to 3.5-fold increase. The transformed plants were also investigated for the tissue-specific withanolide content which showed a two fold increase in total withanolides in the plant. The withanolide increase of about 1.9-, 1.8-, and 1.5-fold was achieved in leaf, root, and stem, respectively (Fig. 7.6d). The leaf tissue showed increased level of withanolides; however, the maximum withanolides accumulated in the transformed roots. The metabolite profile of four different withanolides was studied: Withanolide A, Withanolide B, Withaferin A, and Withanone. Withanolide A and Withaferin A increased up to 2- to 2.25- and 4- to 4.5-fold in leaf tissue and roots, respectively. These studies demonstrate that pathway engineering of withanolide biosynthetic pathway could be utilized as an unconventional approach for the production of these compounds. The detailed description of wide range of therapeutic properties of *W. somnifera* and withanolides, and elaboration of secondary metabolic pathway and their engineering are beyond the scope of this chapter; however, one can refer to the reviews (Mirjalili et al. 2009b; Mishra et al. 2000) for a better understanding.

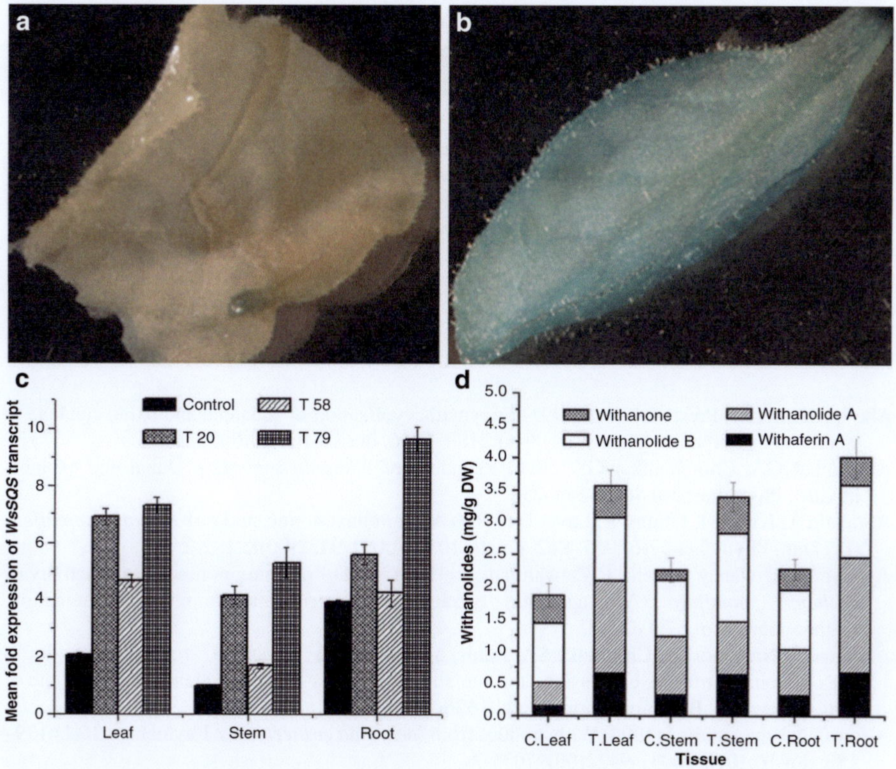

Fig. 7.6 Analysis of transformed *W. somnifera* plants by histochemical Gus assay showing (**a**) control leaf, (**b**) leaf of transformed plant, and (**c**) *WsSQS* transcript analysis in different tissues of transformed *W. somnifera* lines by qRT-PCR. T20, T58, and T79 are the three best performing transformed events. Means of three biological replicates were considered and (**d**) improved production of withanolides in transformed tissues overexpressing *WsSQS*. Vertical bars indicate the mean values±SE from three independent experiments. *C* control, *T* transformed (Patel et al. 2015)

7.7 Conclusions

Pathway engineering is an efficient technology leading to the precious end products and has been performed successfully with various higher plants. The approach provides a better understanding of the biosynthetic pathways via incorporation of regulatory genes and enzymes that are often uncovered. Since these genes and enzymes are already characterized, it is now critical to look into their regulation in the pathway. This chapter demonstrates metabolic pathway engineering in *W. somnifera* leading to the increased production of withanolides and also provides understanding of the metabolic networks for the improvement of the pharmacological components of various medicinally valuable plants. Squalene synthase is actively involved in the regulation of withanolide biosynthesis, and metabolic engineering of *Withania* for the overexpression of *SQS* will provide insight about the regulation mechanism of the pathway for further advancement of the process.

Acknowledgments The authors thank the Council of Scientific and Industrial Research (CSIR), New Delhi, and CSIR-National Chemical Laboratory, Pune, India, for the financial support. NP and PP acknowledge the University Grants Commission (UGC), New Delhi, India, for providing Senior Research Fellowship and DSK Postdoctoral Fellowship, respectively.

References

Abe I, Rohmer M, Prestwich GD (1993) Enzymatic cyclization of squalene and oxidosqualene to sterols and triterpenes. Chem Rev 93(6):2189–2206. doi:10.1021/cr00022a009

Abhyankar GA, Chinchanikar GS (1996) Response of *Withania somnifera* Dunal leaf explants in vitro. Phytomorphol 46(3):249–252

Abraham A, Kirson I, Glotter E, Lavie D (1968) A chemotaxonomic study of *Withania somnifera* (L.) Dun. Phytochem 7(6):957–962. doi:10.1016/S0031-9422(00)82182-2

Agarwal R, Diwanay S, Patki P, Patwardhan B (1999) Studies on immunomodulatory activity of *Withania somnifera* (Ashwagandha) extracts in experimental immune inflammation. J Ethnopharmacol 67(1):27–35

Akamine S, Nakamori K, Chechetka SA, Banba M, Umehara Y, Kouchi H, Izui K, Hata S (2003) cDNA cloning, mRNA expression, and mutational analysis of the squalene synthase gene of *Lotus japonicus*. Biochim Biophys Acta 1626(1–3):97–101

Alfonso D, Kapetanidis I (1994) Withanolides from *Iochroma gesnerioides*. Phytochem 36(1):179–183. doi:10.1016/S0031-9422(00)97035-3

Archana R, Namasivayam A (1999) Antistressor effect of *Withania somnifera*. J Ethnopharmacol 64(1):91–3

Asthana R, Raina MK (1989) Pharmacology of *W. somnifera* (L.) Dunal – a review. Indian Drugs 26:199–205

Atta-ur-Rahman, Jamal AS, Choudary MI, Asif I (1991) Two withanolides from *Withania somnifera*. Phytochem 30(11):3824–3825. doi:10.1016/0031-9422(91)80125-K

Atta-ur-Rahman, Abbas S, Dur-e-Shawar, Jamal AS, Choudhary MI (1993) New withanolides from *Withania* spp. J Nat Prod 56(7):1000–1006. doi:10.1021/np50097a003

Bandyopadhyay M, Jha S, Tepfer D (2007) Changes in morphological phenotypes and withanolide composition of Ri-transformed roots of *Withania somnifera*. Plant Cell Rep 26(5):599–609. doi:10.1007/s00299-006-0260-0

Banerjee S, Naqvi AA, Mandal S, Ahuja PS (1994) Transformation of *Withania somnifera* (L.) Dunal by *Agrobacterium rhizogenes*: Infectivity and phytochemical studies. Phytother Res 8(8):452–455. doi:10.1002/ptr.2650080803

Baxter A, Fitzgerald BJ, Hutson JL, McCarthy AD, Motteram JM, Ross BC, Sapra M, Snowden MA, Watson NS, Williams RJ et al (1992) Squalestatin 1, a potent inhibitor of squalene synthase, which lowers serum cholesterol in vivo. J Biol Chem 267(17):11705–11708

Begum VH, Sadique J (1988) Long-term effect of herbal drug *Withania somnifera* on adjuvant-induced arthritis in rats. Indian J Exp Biol 26(11):877–882

Belingheri L, Beyer P, Kleinig H, Gleizes M (1991) Solubilization and partial purification of squalene synthase from daffodil microsomal membranes. FEBS Lett 292(1–2):34–36. doi:10.1016/0014-5793(91)80827-P

Bergstrom JD, Kurtz MM, Rew DJ, Amend AM, Karkas JD, Bostedor RG, Bansal VS, Dufresne C, Van Middlesworth FL, Hensens OD et al (1993) Zaragozic acids: a family of fungal metabolites that are picomolar competitive inhibitors of squalene synthase. Proc Natl Acad Sci U S A 90(1):80–84

Chen D, Ye H, Li G (2000) Expression of a chimeric farnesyl diphosphate synthase gene in *Artemisia annua* L. transgenic plants via *Agrobacterium tumefaciens*-mediated transformation. Plant Sci 155(2):179–185. doi:10.1016/S0168-9452(00)00217-X

Choudary MI, Abbas S, Jamal AS, Atta-ur-Rahman (1996) *Withania somnifera* – a source of exotic withanolides. Heterocycles 42(2):555–563. doi:10.3987/COM-94-6935

Davis L, Kuttan G (1998) Suppressive effect of cyclophosphamide-induced toxicity by *Withania somnifera* extract in mice. J Ethnopharmacol 62(3):209–214. doi:10.1016/S0378-8741(98)00039-7

Davis L, Kuttan G (2002) Effect of *Withania somnifera* on CTL activity. J Exp Clin Cancer Res 21(1):115–118

Devarenne TP, Shin DH, Back K, Yin S, Chappell J (1998) Molecular characterization of tobacco squalene synthase and regulation in response to fungal elicitor. Arch Biochem Biophys 349(2):205–215. doi:10.1006/abbi.1997.0463

Devarenne TP, Ghosh A, Chappell J (2002) Regulation of squalene synthase, a key enzyme of sterol biosynthesis, in tobacco. Plant Physiol 129(3):1095–1106. doi:10.1104/pp.001438

Dhuley JN (2000) Adaptogenic and cardioprotective action of ashwagandha in rats and frogs. J Ethnopharmacol 70(1):57–63. doi:10.1016/S0378-8741(99)00177-4

Eisenreich W, Schwarz M, Cartayrade A, Arigoni D, Zenk MH, Bacher A (1998) The deoxyxylulose phosphate pathway of terpenoid biosynthesis in plants and microorganisms. Chem Biol 5(9):221–233. doi:10.1016/S1074-5521(98)90002-3

Flores-Sánchez IJ, Ortega-López J, Montes-Horcasitas MC, Ramos-Valdivia AC (2002) Biosynthesis of sterols and triterpenes in cell suspension cultures of *Uncaria tomentosa*. Plant Cell Physiol 43(12):1502–1509. doi:10.1093/pcp/pcf181

Fulton DC, Tait M, Threlfall DR (1995) Comparative study of the inhibition of rat and tobacco squalene synthase by squalestatins. Phytochemistry 38(5):1137–1141. doi:10.1016/0031-9422(94)00798-X

Gamoh K, Hirayama M, lkekawa N (1984) Stereocontrolled synthesis of withanolide D and related compounds. J Chem Soc Perkin Trans 1:449–454. doi:10.1039/p19840000449

Glotter E (1991) Withanolides and related ergostane-type steroids. Nat Prod Rep 8(4):415–440. doi:10.1039/NP9910800415

Grover A, Samuel G, Bisaria VS, Sundar D (2012) Enhanced withanolide production by overexpression of squalene synthase in *Withania somnifera*. J Biosci Bioeng 115(6):680–685. doi:10.1016/j.jbiosc.2012.12.011

Gupta N, Sharma P, Santosh Kumar RJ, Vishwakarma RK, Khan BM (2012) Functional characterization and differential expression studies of squalene synthase from *Withania somnifera*. Mol Biol Rep 39(9):8803–8812. doi:10.1007/s11033-012-1743-4

Hanley K, Chappell J (1992) Solubilization, partial purification, and immunodetection of squalene synthetase from tobacco cell suspension cultures. Plant Physiol 98(1):215–220. doi:10.1104/pp.98.1.215

Hanley KM, Nicolas O, Donaldson TB, Smith-Monroy C, Robinson GW, Hellmann GM (1996) Molecular cloning, in vitro expression and characterization of a plant squalene synthetase cDNA. Plant Mol Biol 30(6):1139–1151. doi:10.1007/BF00019548

Hata S, Sanmiya K, Kouchi H, Matsuoka M, Yamamoto N, Izui K (1997) cDNA cloning of squalene synthase genes from Mono – and dicotyledonous plants, and expression of the gene in rice. Plant Cell Physiol 38(12):1409–1413

Hayashi H, Hirota A, Hiraoka N, Ikeshiro Y (1999) Molecular cloning and characterization of two cDNAs for *Glycyrrhiza glabra* squalene synthase. Biol Pharm Bull 22(9):947–950. doi:10.1248/bpb.19.1387

Hey SJ, Powers SJ, Beale MH, Hawkins ND, Ward JL, Halford NG (2006) Enhanced seed phytosterol accumulation through expression of a modified HMG-CoA reductase. Plant Biotechnol 4(2):219–229

Ichikawa H, Takada Y, Shishodia S, Jayaprakasam B, Nair MG, Aggarwal BB (2006) Withanolides potentiate apoptosis, inhibit invasion, and abolish osteoclastogenesis through suppression of nuclear factor-kappa B (NF-kappa B) activation and NF-kappaB-regulated gene expression. Mol Cancer Ther 5(6):1434–1445. doi:10.1158/1535-7163.MCT-06-0096

Inoue T, Osumi T, Hata S (1995) Molecular cloning and functional expression of a cDNA for mouse squalene synthase. Biochim Biophys Acta 1260(1):49–54. doi:10.1016/0167-4781(94)00178-6

Jana CK, Hoecker J, Woods TM, Jessen HJ, Neuburger M, Gademann K (2011) Synthesis of withanolide A, biological evaluation of its neuritogenic properties, and studies on secretase inhibition. Angew Chem Int Ed Engl 50(36):8407–8411. doi:10.1002/anie.201101869

Jayanti S, Sharma AK (1991) Micropropagation of *Withania somnifera* from germinating seeds and shoot tips. Plant Cell Tiss Organ Cult 26(2):71–73. doi:10.1007/BF00036108

Jayaprakasam B, Zhang Y, Seeram N, Nair M (2003) Growth inhibition of tumor cell lines by withanolides from *Withania somnifera* leaves. Life Sci 74(1):125–132. doi:10.1016/j.lfs.2003.07.007

Jennings SM, Tsay YH, Fisch TM, Robinson GW (1991) Molecular cloning and characterization of the yeast gene for squalene synthetase. Proc Natl Acad Sci U S A 88(14):6038–6042

Jiang G, McKenzie TL, Conrad DG, Shechter I (1993) Transcriptional regulation by lovastatin and 25-hydroxycholesterol in HepG2 cells and molecular cloning and expression of the cDNA for the human hepatic squalene synthase. J Biol Chem 268(17):12818–12824

Kamisako W, Morimoto K, Makino I, Isoi K (1984) Changes in triterpenoid content during the growth cycle of cultured plant cells. Plant Cell Physiol 25(8):1571–1574

Kapoor LD (2001) Handbook of Ayurvedic medicinal plants. CRC Press, London, pp 337–338

Karst F, Lacroute F (1977) Ertosterol biosynthesis in *Saccharomyces cerevisiae*: mutants deficient in the early steps of the pathway. Mol Gen Genet 154(3):269–277

Kaul KN (1957) On the origin, distribution and cultivation of Ashwagandha, the so called *Withania somnifera* Dunal, of Indian literature on materia medica. Pharmaceut and Drug Committee, symposium on the utilization of the Indian medicinal plants, Lucknow, pp 7–8

Keller RK (1996) Squalene synthase inhibition alters metabolism of nonsterols in rat liver. Biochim Biophys Acta 1303(3):169–179

Kim O-T, Seong N-S, Kim M-Y, Hwang B (2005) Isolation and characterization of squalene synthase cDNA from *Centella asiatica* (L.) urban. J Plant Biol 48(3):263–269. doi:10.1007/BF03030521

Kim T-D, Han J-Y, Huh GH, Choi Y-E (2011a) Expression and functional characterization of three squalene synthase genes associated with saponin biosynthesis in *Panax ginseng*. Plant Cell Physiol 52(1):125–137. doi:10.1093/pcp/pcq179

Kim YS, Cho JH, Park S, Han JY, Back K, Choi YE (2011b) Gene regulation patterns in triterpene biosynthetic pathway driven by overexpression of squalene synthase and methyl jasmonate elicitation in *Bupleurum falcatum*. Planta 233(2):343–355. doi:10.1007/s00425-010-1292-9

Kirson I, Abraham A, Lavie D (1977) Chemical analysis of hybrids of *Withania somnifera* L. (Dun.) Chemotypes III (Israel) by Indian I (Delhi). Israel J Chem 16(1):22–24. doi:10.1002/ijch.197700007

Kovganko NV, Kashkan ZN (1997) Advances in the chemical synthesis of withanolides. Chem Nat Compd 33(2):133–145. doi:10.1007/BF02291529

Kribii R, Arro M, Del Arco A, Gonzalez V, Balcells L, Delourme D, Ferrer A, Karst F, Boronat A (1997) Cloning and characterization of the *Arabidopsis thaliana* SQS1 gene encoding squalene synthase – involvement of the C-terminal region of the enzyme in the channeling of squalene through the sterol pathway. Eur J Biochem 249(1):61–69. doi:10.1111/j.1432-1033.1997.00061.x

Kumar P, Kushwaha RA (2006) Medicinal evaluation of *Withania somnifera* (L.) Dunal (Ashwagandha). Asian J Chem 18(2):1401–1404

Kumar V, Murthy KNC, Bhamidi S, Sudha CG, Ravishankar GA (2005) Genetically modified hairy roots of *Withania somnifera* dunal: a potent source of rejuvenating principles. Rejuv Res 8(1):37–45. doi:10.1089/rej.2005.8.37

Kumar S, Hahn FM, Baidoo E, Kahlon TS, Wood DF, McMahan CM, Cornish K, Keasling JD, Daniell H, Whalen MC (2012) Remodeling the isoprenoid pathway in tobacco by expressing the cytoplasmic mevalonate pathway in chloroplasts. Metab Eng 14(1):19–28. doi:10.1016/j.ymben.2011.11.005

Lavie D, Glotter E, Shvo Y (1965a) Constituents of *Withania somnifera*. III. The side chain of Withaferin A. J Org Chem 30(6):1774–1778. doi:10.1021/jo01017a015

Lavie D, Glotter E, Shvo Y (1965b) Constituents of *Withania somnifera*. IV. The structure of withaferin A. J Chem Soc 1965:7517–7531. doi:10.1039/JR9650007517

Lee JH, Yoon YH, Kim HY, Shin DH, Kim DU, Lee IJ, Kim KU (2002) Cloning and expression of squalene synthase cDNA from hot pepper (*Capsicum annuum*). Mol Cells 13(3):436–443

Lee MH, Jeong JH, Seo JW, Shin CG, Kim YS, In JG, Yang DC, Yi JS, Choi YE (2004) Enhanced triterpene and phytosterol biosynthesis in *Panax ginseng* overexpressing squalene synthase gene. Plant Cell Physiol 45(8):976–984. doi:10.1093/pcp/pch126

Lichtenthaler HK (1999) The 1-Deoxy-D-Xylulose-5-phosphate pathway of isoprenoid biosynthesis in plants. Annu Rev Plant Physiol Plant Mol Biol 50:47–65. doi:10.1146/annurev.arplant.50.1.47

Lockley WJS, Rees HH, Goodwin TW (1976) Biosynthesis of steroidal withanolides in *Withania somnifera*. Phytochemistry 15(6):937–939. doi:10.1016/S0031-9422(00)84374-5

LoGrasso PV, Soltis DA, Boettcher BR (1993) Overexpression, purification, and kinetic characterization of a carboxyl-terminal-truncated yeast squalene synthetase. Arch Biochem Biophys 307(1):193–199. doi:10.1006/abbi.1993.1578

Lu HY, Liu JM, Zhang HC, Yin T, Gao SL (2008) Ri-mediated Transformation of *Glycyrrhiza uralensis* with a Squalene Synthase Gene (*GuSQS1*) for Production of Glycyrrhizin. Plant Mol Biol Rep 26(1):1–11. doi:10.1007/s11105-008-0018-7

Madina BR, Sharma LK, Chaturvedi P, Sangwan RS, Tuli R (2007) Purification and characterization of a novel glucosyltransferase specific to 27 β-hydroxy steroidal lactones from *Withania somnifera* and its role in stress responses. Biochim Biophys Acta 1774(9):1199–1207. doi:10.1016/j.bbapap.2007.06.015

Malhotra CL, Mehta VL, Das PK, Dhalla NS (1965) Studies on *Withania*-ashwagandha, Kaul. IV. The effect of total alkaloids (ashwagandholine) on the central nervous system. Indian J Physiol Pharmacol 9(3):127–136

Mano Y, Nabeshima S, Matsui C, Ohkawa H (1986) Production of tropane alkaloids by hairy root cultures of *Scopolia japonica*. Agr Biol Chem 50(11):2715–2722. doi:10.1080/00021369.1986.10867820

Marie Winters ND (2006) Ancient medicine, modern use: *Withania somnifera* and its potent role in integrative oncology. Altern Med Rev 11(4):269–277

McKenzie TL, Jiang G, Straubhaar JR, Conrad DG, Shechter I (1992) Molecular cloning, expression, and characterization of the cDNA for the rat hepatic squalene synthase. J Biol Chem 267(30):21368–21374

Merkulov S, Assema FV, Springer J, Carmen AFD, Mooibroek H (2000) Cloning and characterization of the *Yarrowia lipolytica* squalene synthase (*SQS1*) gene and functional complementation of the *Saccharomyces cerevisiae* erg9 mutation. Yeast 16(3):197–206. doi:10.1002/(SICI)1097-0061(200002)16:3<197::AID-YEA513>3.0.CO;2-L

Mirjalili MH, Fakhr-Tabatabaei SM, Bonfill M, Alizadeh H, Cusido RM, Ghassempour A, Palazón (2009a) Morphology and withanolide production of *Withania coagulans* hairy root cultures. Eng Life Sci 9(3):197–204. doi:10.1002/elsc.200800081

Mirjalili MH, Moyano E, Bonfill M, Cusido RM, Palazón J (2009b) Steroidal lactones from *Withania somnifera*, an ancient plant for novel medicine. Molecules 14(7):2373–2393. doi:10.3390/molecules14072373

Mirjalili MH, Moyano E, Bonfill M, Cusido RM, Palazon J (2011) Overexpression of the *Arabidopsis thaliana* squalene synthase gene in *Withania coagulans* hairy root cultures. Biol Plant 55(2):357–360. doi:10.1007/s10535-011-0054-2

Mishra L-C, Singh BB, Dagenais S (2000) Scientific basis for the therapeutic use of *Withania somnifera* (Ashwagandha): a review. Altern Med Rev 5(4):334–346

Murthy HN, Dijkstra C, Anthony P, White DA, Davey MR, Power JB, Hahn EJ, Paek KY (2008) Establishment of *Withania somnifera* hairy root cultures for the production of withanolide A. J Integr Plant Biol 50(8):975–981. doi:10.1111/j.1744-7909.2008.00680.x

Nakashima T, Inoue T, Oka A, Nishino T, Osumi T, Hata S (1995) Cloning, expression, and characterization of cDNAs encoding *Arabidopsis thaliana* squalene synthase. P Natl Acad Sci USA 92(6):2328–2332

Nittala SS, Lavie D (1981) Chemistry and genetics of withanolides in *Withania somnifera* hybrids. Phytochemistry 20(12):2741–2748. doi:10.1016/0031-9422(81)85278-8

Paradise EM, Kirby J, Chan R, Keasling JD (2008) Redirection of flux through the FPP branch-point in *Saccharomyces cerevisiae* by down-regulating squalene synthase. Biotechnol Bioeng 100(2):371–378. doi:10.1002/bit.21766

Patel N, Patel P, Kumari U, Kendurkar SV, Khan BM (2014) Microprojectile bombardment assisted agroinfection increases transformation efficiency of *Withania somnifera* (L.). Res Biotechnol 5(4):13–24

Patel N, Patel P, Kendurkar SV, Thulasiram HV, Khan BM (2015) Overexpression of squalene synthase in *Withania somnifera* leads to enhanced withanolide biosynthesis. Plant Cell Tiss Organ Cult 122(2):409–420. doi:10.1007/s11240-015-0778-3

Payne J, Hamill JD, Robins RJ, Rhodes MJC (1987) Production of hyoscyamine by 'Hairy Root' cultures of *Datura stramonium*. Planta Med 53(5):474–478. doi:10.1055/s-2006-962776

Prakash J, Gupta SK, Kochupillai V, Gupta YK, Joshi S (2001) Chemopreventive activity of *Withania somnifera* in experimentally induced fibrosarcoma tumors in Swiss albino mice. Phytother Res 15(3):240–244. doi:10.1002/ptr.779

Prakash J, Gupta SK, Dinda AK (2002) *Withania somnifera* root extract prevents DMBA-induced squamous cell carcinoma of skin in Swiss albino mice. Nutr Cancer 42:91–97. doi:10.1207/S15327914NC421_12

Rani G, Grover IS (1999) In vitro callus induction and regeneration studies in *Withania somnifera*. Plant Cell Tiss Org Cult 57(1):23–27. doi:10.1023/A:1006329532561

Rani G, Virk GS, Nagpal A (2003) Callus induction and plantlet regeneration in *Withania somnifera* (L.) dunal. In Vitro Cell Dev Biol Plant 39(5):468–474. doi:10.1079/IVP2003449

Ray AB, Gupta M (1994) Withasteroids, a growing group of naturally occurring steroidal lactones. In: Zechmeister L, Herz W, Kirby GW, Moore RE, Steglich W, Tamm C (eds) Progress in the chemistry of organic natural products, vol 63, pp 1–106. doi:10.1007/978-3-7091-9281-8_1

Ray S, Jha S (2001) Production of withaferin A in shoot cultures of *Withania somnifera*. Planta Med 67(5):432–436. doi:10.1055/s-2001-15811

Re EB, Jones D, Learned RM (1995) Co-expression of native and introduced genes reveals cryptic regulation of HMG CoA reductase expression in *Arabidopsis*. Plant J 7(5):771–784. doi:10.1046/j.1365-313X.1995.07050771.x

Rilling HC, Epstein WW (1969) Mechanism of squalene biosynthesis. Presqualene, a pyrophosphorylated precursor to squalene. J Am Chem Soc 91(4):1041–1042. doi:10.1021/ja01032a052

Robinson GW, Tsay YH, Kienzle BK, Smith-Monroy CA, Bishop RW (1993) Conservation between human and fungal squalene synthetases: similarities in structure, function, and regulation. Mol Cell Biol 13(5):2706–2717. doi:10.1128/MCB.13.5.2706

Rohmer M (1999) The discovery of a mevalonate-independent pathway for isoprenoid biosynthesis in bacteria, algae and higher plants. Nat Prod Rep 16:565–574. doi:10.1039/A709175C

Sangwan RS, Chaurasiya ND, Lal P, Misra L, Uniyal GC, Tuli R, Sangwan NS (2007) Withanolide A biogeneration in in vitro shoot cultures of ashwagandha (*Withania somnifera* Dunal), a main medicinal plant in Ayurveda. Chem Pharm Bull 55(9):1371–1375. doi:10.1248/cpb.55.1371

Saritha KV, Naidu CV (2007) In vitro flowering of *Withania somnifera* Dunal. An important antitumor medicinal plant. Plant Sci 172(4):847–851. doi:10.1016/j.plantsci.2006.12.016

Sasiak K, Rilling HC (1988) Purification to homogeneity and some properties of squalene synthetase. Arch Biochem Biophys 260(2):622–627. doi:10.1016/0003-9861(88)90490-0

Schliebs R, Liebmann A, Bhattacharya SK, Kumar A, Ghosal S, Biql V (1997) Systemic administration of defined extracts from *Withania somnifera* (Indian Ginseng) and Shilajit differentially affects cholinergic but not glutamatergic and GABAergic markers in rat brain. Neurochem Int 30(2):181–190. doi:10.1016/S0197-0186(96)00025-3

Seo JW, Jeong JH, Shin CG, Lo SC, Han SS, Yu KW, Harada E, Han JY, Choi YE (2005) Overexpression of squalene synthase in *Eleutherococcus senticosus* increases phytosterol and triterpene accumulation. Phytochemistry 66(8):869–877. doi:10.1016/j.phytochem.2005.02.016

Shanks JV, Morgan J (1999) Plant 'hairy root' culture. Curr Opin Biotechnol 10(2):151–155. doi:10.1016/S0958-1669(99)80026-3

Sharada AC, Solomon FE, Devi PU (1996) Antitumor and radiosensitizing effects of withaferin A on mouse Ehrlich ascites carcinoma *in vivo*. Acta Oncol 35(1):95–100. doi:10.3109/02841869609098486

Sharma MM, Ali DJ, Batra A (2010) Plant regeneration through in vitro somatic embryogenesis in Ashwaganda (*Withania somnifera* (L.) Dunal). Researcher 2(3):1–6

Siddique NA, Bari MA, Shahnewaz S, Rahman MH, Khan MSI, Islam MS (2004) Plant regeneration of *Withania somnifera* (L.) Dunal (Ashwagandha) from nodal segments derived callus an endangered medicinal plant in Bangladesh. J Biol Sci 4(2):219–223. doi:10.3923/jbs.2004.219.223

Singh S, Kumar S (1998) *Withania somnifera*: the Indian ginseng ashwagandha. Central Institute of Medicinal and Aromatic Plants, Lucknow

Song L (2003) Detection of farnesyl diphosphate accumulation in yeast ERG9 mutants. Anal Biochem 317(2):180–185. doi:10.1016/S0003-2697(03)00138-6

Spurgeon SL, Porter JW (1981) Biosynthesis of isoprenoid compounds, vol 1. Wiley, New York, pp 1–46

Threlfall DR, Whitehead IM (1988) Coordinated inhibition of squalene synthetase and induction of enzymes of sesquiterpenoids phytoalexin biosynthesis in cultures of *Nicotiana tabacum*. Phytochemistry 27(8):2567–2580. doi:10.1016/0031-9422(88)87028-6

Tozawa R, Ishibashi S, Osuga J, Yagyu H, Oka T, Chen Z, Ohashi K, Perrey S, Shionoiri F, Yahagi N, Harada K, Gotoda T, Yazaki Y, Yamada N (1999) Embryonic lethality and defective neural tube closure in mice lacking squalene synthase. J Biol Chem 274(43):30843–30848. doi:10.1074/jbc.274.43.30843

Valizadeh J, Valizadeh M (2009) In vitro callus induction and plant regeneration from *Withania coagulans*: a valuable medicinal Plant. Pak J Biol Sci 12(21):1415–1419. doi:10.3923/pjbs.2009.1415.1419

Veleiro AS, Oberti JC, Burton G (2005) Chemistry and bioactivity of withanolides from South American Solanaceae. In: Atta-ur-Rahman (ed) Studies in natural products chemistry 32 (part L). Elsevier, Amsterdam, pp 1019–1052

Vitali G, Conte L, Nicoletti M (1996) Withanolide composition and in vitro culture of Italian *Withania somnifera*. Planta Med 62(3):287–288. doi:10.1055/s-2006-957884

Vogeli U, Chappell J (1988) Induction of sesquiterpene cyclase and suppression of squalene synthetase activities in plant cell cultures treated with fungal elicitor. Plant Physiol 88(4):1291–1296. doi:10.1104/pp.88.4.1291

Wang H, Song Y, Shen H, Liu Y, Li Z, Wang H, Chen J, Liu B, Ye H (2012) Effect of antisense squalene synthase gene expression on the increase of artemisinin content in *Artemisia anuua* . Transgenic plants – advances and limitations. PhD. Yelda Ozden Çiftçi (Ed.), ISBN: 978-953-51-0181-9, InTech, pp 397–406. Available from: http://www.intechopen.com/books

Watt GA (1972) Dictionary of the economic products of India, vol 6. Cosmo Publication, Delhi, p 309

Wentzinger LF, Bach TJ, Hartmann MA (2002) Inhibition of squalene synthase and squalene epoxidase in tobacco cells triggers an up-regulation of 3-hydroxy-3-methylglutaryl coenzyme a reductase. Plant Physiol 130(1):334–346. doi:10.1104/pp.004655

Yarden A, Lavie D (1962) Constituents of *Withania somnifera*. I. Functional groups of withaferin. J Chem Soc 567:2925–2927. doi:10.1039/JR9620002925

Yoshioka H, Yamada N, Doke N (1999) cDNA cloning of sesquiterpene cyclase and squalene synthase, and expression of the genes in potato tuber infected with *Phytophthora infestans*. Plant Cell Physiol 40(9):993–998

Zhao MW, Liang WQ, Zhang DB, Wang N, Wang CG, Pan YJ (2007) Cloning and characterization of squalene synthase (*SQS*) gene from *Ganoderma lucidum*. J Microbiol Biotechnol 17(7):1106–1112

Zhou C, Zhao D, Sheng Y, Liang G, Tao J (2012) Molecular cloning and expression of squalene synthase and 2,3-oxidosqualene cyclase genes in persimmon (*Diospyros kaki L.*) fruits. Mol Biol Rep 39:1125–1132. doi:10.1007/s11033-011-0841-z

Chapter 8
Salvia miltiorrhiza: A Medicinal Herb from Metabolites to Pathway Engineering

Sushim Kumar Gupta, Poonam Sharma, Dinesh Chandra Agrawal, and Hsin-Sheng Tsay

Abstract The dried, red-colored roots of *Salvia miltiorrhiza* (Lamiaceae) also called "danshen" in colloquial language in China have been used in traditional Chinese medicine for thousands of years to treat hypertension and other cardiovascular ailments. The main constituents of danshen are hydrophilic phenolic acids and lipophilic tanshinones. The various uses of danshen in traditional as well as modern medicines have motivated an intensive research on compounds in *S. miltiorrhiza*. In recent years, more than 110 compounds have been isolated from *S. miltiorrhiza* and their structure was identified. Tanshinones and their derivatives have been demonstrated to possess properties of slowing down or curing various ailments related to cardiovascular, cerebrovascular, respiratory, liver, nervous system, cancer, Alzheimer's, and Parkinson's diseases. With the increasing demand of this herb, an unrestricted collection to supply raw materials and the extraction of its constituents have severely threatened the natural habitats of *S miltiorrhiza*. This has prompted the researchers to develop alternative strategies for metabolite production. Several in vitro methodologies have been established to generate callus, cell suspension culture, hairy roots, and plant regeneration. Different regulators and elicitors for plant growth have been employed to enhance levels of different constituents. The advent of sequencing technologies, whole genome, and expression data has helped to provide insights and identification of pathway genes involved in the biosynthesis. This book chapter gives a brief description of in vitro methodologies, use of different elicitors, gene functions, genetic modifications, expression profiling for a better understanding, and enhancement of the constituents in *S. miltiorrhiza*.

S.K. Gupta (✉)
Bacterial Epidemiology and Antimicrobial Resistance Research Unit, USDA, ARS, Athens, GA 30605, USA

Department of Applied Chemistry, Chaoyang University of Technology, Taichung, Taiwan
e-mail: sushimgupta@gmail.com

P. Sharma
Division of Biochemical Sciences, CSIR-National Chemical Laboratory, Pune, India

D.C. Agrawal • H.-S. Tsay (✉)
Department of Applied Chemistry, Chaoyang University of Technology, Taichung, Taiwan
e-mail: hstsay@cyut.edu.tw

H.-S. Tsay et al. (eds.), *Medicinal Plants - Recent Advances in Research and Development*, DOI 10.1007/978-981-10-1085-9_8

Keywords Callus • Danshen • Hairy roots • Rosmarinic acid • *Salvia miltiorrhiza* • Tanshinones

Abbreviations

2,4-D	2,4-Dichlorophenoxyacetic acid
4CL	4-Coumaroyl:CoA ligase
ABA	Abscisic acid
AOPP	L-a-aminooxy-beta-phenylpropionic acid
ATM	Activation tagging mutagenesis
ATMT	*A. tumefaciens*-mediated transformation
BA	N6-benzyladenine
BABA	β-Aminobutyric acid
C_4H	Cinnamic acid 4-hydroxylase
cDNA	Complimentary DNA
COG	Clusters of orthologous groups
CPS	Copalyl diphosphate synthase
DMAPP	Dimethylallyl diphosphate
DNA	Deoxyribonucleic acid
DW	Distilled water
DXS	1-Deoxy-D-xylulose-5-phosphate synthase
FPP	Farnesyl diphosphate
FPPS	Farnesyl diphosphate synthase
GA_3	Gibberellic acid
GC	Gas chromatography
GC-MS	Gas chromatography-mass spectrometry
g-DNA	Genomic deoxyribonucleic acid
GGPP	Geranylgeranyl diphosphate
GGPPS	Geranylgeranyl diphosphate synthase
GPP	Geranyl diphosphate
GPPS	Geranyl diphosphate synthase
HDR	1-Hydroxy-2-methyl-2-(E)-butenyl-4-diphosphate reductase
HMGR	3-Hydroxy-3-methylglutaryl CoA reductase
HPLC	High-performance liquid chromatography
HPPD	4-Hydroxyphenylpyruvate dioxygenase
HPPR	4-Hydroxyphenylpyruvate reductase
IDSs	Isoprenyl diphosphate synthases
IPP	Isopentenyl diphosphate
KS	Kaurene synthase
LAB	Lithospermic acid B
MeJA	Methyl jasmonate
MEP	2-C-Methyl-D-erythritol 4-phosphate
MS	Murashige and Skoog
MVA	Mevalonate

NAA	1-Naphthaleneacetic acid
NR	NCBI nonredundant
PAL	Phenylalanine ammonia-lyase
PGRs	Plant growth regulators
RA	Rosmarinic acid
RACE	Rapid amplification of cDNA ends
RAS	Rosmarinic acid synthase
RDRs	RNA-dependent RNA polymerases
ri	Root inducing
RNA	Ribonucleic acid
RNAi	RNA interference
SA	Salicylic acid
SEM	*S*canning electron microscope
TAT	Tyrosine aminotransferase
T-DNA	Transfer DNA
TDZ	Thidiazuron
ti	Tumor inducing
TLC	Thin-layer chromatography
TPS	Terpene synthases
UV-B	Ultraviolet-B radiation
WPM	Woody plant medium
YE	Yeast extract

8.1 Introduction

Salvia miltiorrhiza, a perennial flowering plant, belongs to the family Lamiaceae (Fig. 8.1A) and mainly distributed in China, Japan, and other Southeast Asian countries. It grows on grassy lands in forests and hillsides and along stream banks at an elevation of 90–1200 m (Standley and Williams 1973). *S. miltiorrhiza* dried roots

Fig. 8.1 (**A**) *Salvia miltiorrhiza* plant, (**B**) *S. miltiorrhiza* roots, (**C**) dried roots of *S. miltiorrhiza*

(Fig. 8.1B, C) commonly known as "danshen" in Chinese have been used in age-old traditional Chinese medicine for thousands of years (Cheng 2007). Danshen contains hydrophilic phenolics comprising caffeic acid, danshensu, protocatechuic acid, protocatechuic aldehyde, rosmarinic acid (RA), and salvianolic acids A and B as well as hydrophobic diterpene quinones such as cryptotanshinone, tanshinone I and tanshinone IIA, etc. (Liu et al. 2007; Wang 2010; Luo et al. 2013). Many of these colored compounds impart a reddish appearance to the roots. Also, *S. miltiorrhiza* leaves are a rich source of health-promoting phenolics and natural antioxidants. Dried leaves of *S. miltiorrhiza* have been used to obtain essential oils (hexadecanoic acid, germacrene D, phytol, β-caryophyllene, and methyl linolenate) by hydro-distillation and have been analyzed by gas chromatography (GC) and gas chromatography-mass spectrometry (GC-MS) (Li and Wang 2009). The different organic acid (acetone or methanol) extracts of *S. miltiorrhiza* leaves contain several phenolic compounds such as salvianolic acid B and RA, and these organic constituents play an important role as antioxidants. These compounds are implicated as a probable source of phenolic antioxidants for food, pharmaceutical, cosmetics, and nutraceutical industries (Toth et al. 2003; Zhang et al. 2010b). Phenolic components (salvianolic acid B and RA) analyzed at different growth stages in the leaves of *S. miltiorrhiza* revealed that the localization of these components in leaves changes with the plant growth stage. HPLC was used to analyze the main active components (tanshinoate B, danshensu, isotanshinone IIA, and cryptotanshinone) of *S. miltiorrhiza* in different commercially available crude drugs, and preparations of danshen and variations were observed in them. Among the four components measured in the crude drugs, tanshinone B content was in the highest amount. However, isotanshinone IIA and cryptotanshinone contents have been found in varying amounts (Zhang et al. 2002; Wan et al. 2009). Among the organic solvents, methanol has been found to be the best solvent used in the extraction of cryptotanshinone, tanshinone I, and tanshinone IIA from *S miltiorrhiza*. However, two other methodologies, ionic liquid-modified silica sorbents (Tian et al. 2009) and task-specific nonionic surfactant-assisted extraction (Bi et al. 2011), also have been used for the extraction of different tanshinones in *S. miltiorrhiza*.

Different compounds (>110) of *S. miltiorrhiza* have been already isolated and structurally identified (Su et al. 2015). Chemical structures of some of the important compounds in *S. miltiorrhiza* have been shown in Fig. 8.2 (Xing et al. 2015). Due to its enormous medicinal importance, numerous studies were carried out on tanshinones and their derivatives, demonstrating either a slowdown or cure of various ailments including cardiovascular diseases (Morse and Choi 2002; Mao et al. 2009; Liu et al. 2010; Zhou et al. 2005; Lee et al. 2012; Hung et al. 2009), cerebrovascular diseases (Zhou et al. 2011), respiratory diseases (Li et al. 2011), liver diseases (Yin et al. 2014; Kang et al. 2004; Ahn et al. 2010), nervous system diseases (Isacchi et al. 2011; Tan et al. 2009), cancer (Kim et al. 2011; Chen et al. 2010; Yang et al. 2011; Shan et al. 2009; Dai et al. 2012), Alzheimer's disease (Wong et al. 2010; Zhou et al. 2011; Mei et al. 2012), and Parkinson's disease (Zhang et al. 2010a; Tian et al. 2008). Since the last two decades, researchers have been motivated to study

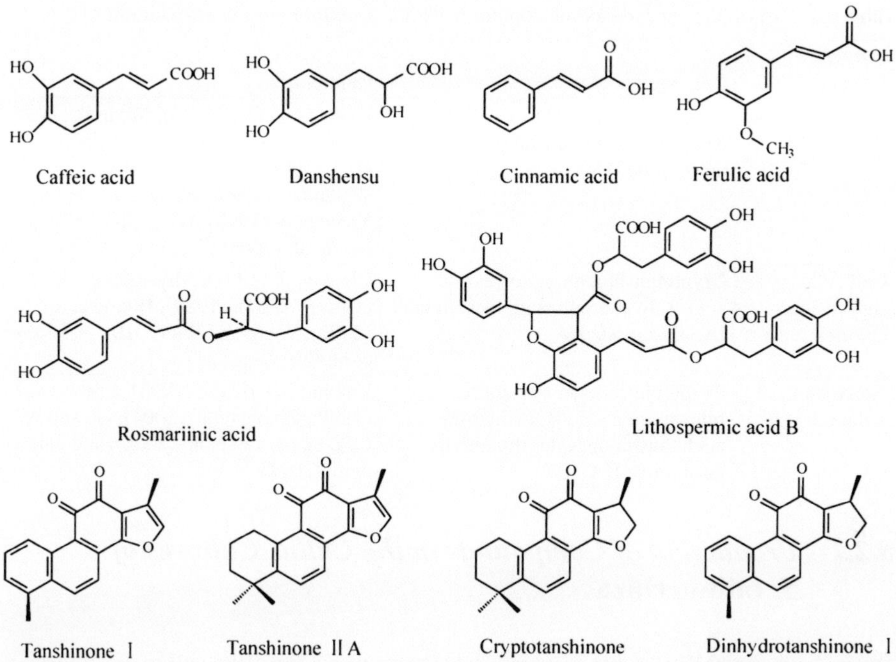

Fig. 8.2 Chemical structures of some of the important compounds in *S. miltiorrhiza* (Xing et al. 2015)

the biosynthesis and biotechnological production of tanshinones mainly due to the vital role of danshen in Chinese traditional and modern medicine. Hence, researchers have tried to develop methodologies to enhance the metabolite production by in vitro techniques, through metabolic engineering or synthetic biology. The present review encompasses a concise description of various in vitro techniques, use of different elicitors, gene functions, genetic modification, expression profiling, and pathway engineering to enhance the constituents in *S. miltiorrhiza*.

8.2 In Vitro Methodologies to Enhance Metabolite Production in *S. miltiorrhiza*

Studies on various in vitro culture methodologies such as callus, cell suspension, hairy root cultures, and their end products in *S. miltiorrhiza* have been summarized in Table 8.1.

Table 8.1 List of plant secondary metabolites from *Salvia miltiorrhiza* in vitro cultures

In vitro cultures	Compounds produced	References
Callus cultures	Cryptotanshinone, ferruginol, rosmarinic acid, tanshinone IA, tanshinone IIA	Nakanishi et al. (1983), Miyasaka et al. (1989), Hu et al. (1992), Morimoto et al. (1994), Gao et al. (1996), Waldemar (1996), Wang et al. (1998), Yuan et al. (1990), Wu et al. (2003), and Lee et al. (2008)
Cell suspension cultures	Cryptotanshinone, lithospermic acid B, rosmarinic acid, tanshinone IA, tanshinone IIA	Chen et al. (1997), Miyasaka et al. (1985), (1987), (1989), Huang et al. (2000), Wang and Wu (2010), and Jiao et al. (2012)
Hairy root cultures	Cryptotanshinone, ferruginol, lithospermic acid B, rosmarinic acid, tanshinones, tanshinone IA, tanshinone IIA	Zhi and Alfermann (1993), Chen et al. (1999), Zhang et al. (2004), Ge and Wu (2005a, b), Yan et al. (2006), and Gupta et al. (2011)

8.2.1 Production of Compounds in the Callus Cultures of S. miltiorrhiza

Plants have the ability to generate disorganized cell masses, like callus or tumors, in response to various, biotic and abiotic stimuli (wounding or infection caused by a pathogen). This nature of plant was exploited under in vitro conditions earlier by Gautheret (1939). A defined ratio of auxin and cytokinin has been commonly used to generate callus. In some species, abscisic acid (ABA) also induced callus and may replace auxin or cytokinin in the formation of callus (Goren et al. 1979; Hu et al. 2000). Genes can be inserted into callus cells using biolistic bombardment or by *Agrobacterium tumefaciens*-mediated transformation (Chen et al. 1997; Song and Wang 2011). In earlier studies on *S. miltiorrhiza* in Japan, Nakanishi et al. (1983) induced various undifferentiated cell lines from seedlings raised on MS (Murashige and Skoog 1962) medium and isolated a line producing cryptotanshinone and ferruginol. It was observed that cryptotanshinone was produced in traces, decreasing slowly in successive subcultures. Different plant parts were successfully used to generate callus in *S. miltiorrhiza*, e.g., epicotyls, hypocotyls (Waldemar 1996; Miyasaka et al. 1989), shoot tips (Morimoto et al. 1994), leaf (Wu et al. 2003), seeds (Waldemar 1996), and plantlets from in vitro-germinated seeds (Miyasaka et al. 1989; Gao et al. 1996). Different plant growth regulators (PGRs) were supplemented to semisolid MS medium to either produce plant metabolites or regenerate whole plants. Wu et al. (2003) established a valuable method for induction and proliferation of *S. miltiorrhiza* callus and achieved cryptotanshinone production under the influence of PGRs. Leaf explants of *S. miltiorrhiza* induced callus upon culturing on semisolid MS basal medium containing 1 mg/l 2,4-dichlorophenoxyacetic acid (2,4-D) and 3 % sucrose and incubation at 25 ± 1 °C in the dark for a month. The induced first-generation callus was subcultured (200 mg/test tube) thrice at

20-day intervals on MS medium containing 1.0 mg/l 2,4-D and 0.5 mg/l N6-benzyladenine (BA). The growing callus (200 mg) was re-cultured in test tubes containing MS basal medium with 0.1, 0.2, 0.5, 1.0, and 2.0 mg/l of BA, 3 % sucrose, and 1 % Difco Bacto agar. The calli grown on the MS basal medium containing 0.2 mg/l BA were cultured for 8, 16, 24, 30, and 60 days in darkness. Later, the methanol extract of callus was subjected to in vacuo to dryness followed by HPLC quantitative analysis for cryptotanshinone. There was no cryptotanshinone production in the control callus (grown on medium without PGRs); however, low amount of cryptotanshinone was detected when PGRs were added together (0.26 mg/g dry weight). The highest yield of cryptotanshinone (4.59 mg/g dry weight) was seen in the calli cultured for 60 days (Wu et al. 2003). The calli grown on solid and liquid media including a bioreactor system have been described (Krajewska-Patan et al. 2007). The calli abstained from synthesizing the tanshinones, however, produced (2.85–5.72 %) phenolic compounds. A further supplement of yeast extract (YE) in the medium caused a marginal rise in the diterpene production but reduced amount of RA in the calli.

A time course study of ferruginol and cryptotanshinone production in the immobilized *S. miltiorrhiza* cell cultures showed continuous secretion of ferruginol and cryptotanshinone in the medium. At day 25, it was observed that 74 % of cryptotanshinone and 38 % of ferruginol released into the medium. The immobilized cells produced 39 % of ferruginol and 61 % of cryptotanshinone, in comparison to cell suspension cultures. The secondary metabolite production declined with the increase in time, due to a negative feedback affected by accumulation of lipophilic secondary metabolites, rather than the death of immobilized cells (Miyasaka et al. 1985). In a report on the effect of various nutrient components in cell suspension cultures of *S. miltiorrhiza* for ferruginol and cryptotanshinone production, it was observed that the production of ferruginol and cryptotanshinone depended on sugar, nitrogen sources, and thiamine, while kinetin, $MnSO_4$, and phosphates exhibited marginal beneficial effects (Miyasaka et al. 1987). They found that other constituents of MS medium were not very critical. Thus, they formulated a much simpler medium for cryptotanshinone production and proposed a viable mechanism for the synthesis of cryptotanshinone in cell suspension culture (Miyasaka et al. 1987).

Yuan et al. (1990) developed a methodology to get a continuous supply of metabolites from immobilized callus. Immobilized callus was cultured in MS liquid medium supplemented with 0.1 mg/l kinetin and 1 mg/l NAA and incubation at room temperature for 1 month. After that, medium analyzed with HPLC and TLC showed that immobilized callus secreted tanshinone IIA and cryptotanshinone consistently in the medium. Further comparative analysis of compounds was carried out between callus suspension culture and immobilized callus. It was noticed that in contrast to suspension cultures, immobilized callus was stable for a longer period along with the high content of the product and suitability for downstream processing (Yuan et al. 1990). In another study, Hu et al. (1992) analyzed tanshinone IA, tanshinone IIA, and cryptotanshinone constituents in the calli obtained from the root, stem, leaf, and petiole explants of *S. miltiorrhiza* and found that the highest amounts of these constituents were in calli derived from the root (Hu et al. 1992).

8.2.2 Production of Compounds in the Cell Suspension Cultures of S. miltiorrhiza

Tanshinone production in cell suspension cultures of *S. miltiorrhiza* has increased significantly due to various strategies including optimization of culture media, stimulation by elicitors, and nutrient feeding operations (Wang and Wu 2010). Chen et al. (1997) were successful in establishing transformed cell cultures by infecting *S. miltiorrhiza* plantlets with *A. tumefaciens* strain C58. The transformed cell suspension cultures produced high tanshinone in 6,7-V medium (Veliky and Martin 1970). Though B5 (Gamborg et al. 1968) medium supplemented with 30 g/l sucrose supported the maximum growth of cells, however, it reduced the accumulation of tanshinone. This was overcome by transferring the cell aggregates to medium containing yeast extract. Further, when these cells were grown under illumination, they inhibited the production of tanshinone, and the cells turned green (Chen et al. 1997).

The influence of elicitor salicylic acid (SA) on the production of RA and related enzyme synthesis in *S. miltiorrhiza* cell suspension cultures was reported by Jiao et al. (2012). The cells were treated with SA and AOPP (L-a-aminooxy-beta-phenylpropionic acid), which are the competitive inhibitors of *tyrosine aminotransferase* (*TAT*) and activities of related enzymes, such as *phenylalanine ammonia-lyase* (*PAL*). The treated cells were assayed for *TAT* and *PAL* activities, and its RA contents were analyzed. It was observed that in the cells treated with 6.25 mg/l SA at the sixth day of suspension culture, *PAL* activity maximized at 4 h, which was 124 % higher than the control and the RA content maximized (5.914 mg/g dw) at 8 h. The treatment of cells with 0.1 μmol/l AOPP demonstrated that the effect of AOPPs was negligible on the *TAT* activity, while the *PAL* activity was reduced by 44 % compared to the control at 6 h. Also, the RA (4.709 mg/g dry weight) content reduced with the decreased *PAL* activity. The cells accumulated more RA when treated with AOPP and SA. This indicated that SA leads to the accumulation of RA in cell suspension cultures of *S. miltiorrhiza* with higher *PAL* activity, and the rate-limiting effect of *PAL* was higher than the *TAT* (Jiao et al. 2012). The role of biologically active secondary metabolites using a simpler system comprising of a yeast elicitor and a Ti C58-transformed *S. miltiorrhiza* cell line in the defense responses of plants was reported by Chen and Chen (2000). The outcome of yeast elicitor on the accumulation of RA and cryptotanshinone was carried out through dosage and time course manner. The yeast elicitor reduced RA content and enhanced the level of cryptotanshinone significantly. The results here suggested the potential role of RA and cryptotanshinone in the passive and active defense responses against the attack of pathogens (Chen and Chen 2000).

A report on the effect of different PGRs on tanshinone accumulation was carried out in the in vitro regenerated callus, and it was observed that the media supplemented with 2,4-D inhibited the synthesis of hydrophilic compounds, while kinetin promoted the synthesis of tanshinone (Wang et al. 1998). Contrary to this in another study, it was observed that 2,4-D promoted cell growth in suspension culture, but inhibited lithospermic acid B synthesis, while it had no influence on RA synthesis. Gibberellic acids (GA_3) promoted lithospermic acid B and RA synthesis but inhibited cell growth (Huang et al. 2000).

8.2.3 Production of Compounds Through Hairy Root Cultures of S. miltiorrhiza

Agrobacterium rhizogenes-mediated hairy root generation has become a matter of intensive research interest in the last two decades due to its ability to produce and accumulate secondary metabolites. The hairy root system is a preferred method due to a rapid growth of roots in a culture medium free of PGRs, negative geotropism, genetic stability, and robust lateral branching. The accumulation of secondary metabolites in hairy roots may be higher or similar as synthesized in intact plant root from which hairy roots are generated (Sevon and Oksman-Caldentey 2002). Hairy roots are generated from different plant parts, and usually, wounded plant parts are infected with *A. rhizogenes*. During the *A. rhizogenes*-mediated transformation, a fragment of T-DNA located in the root-inducing Ri plasmid is transferred to plant cells, and this fragment gets associated with the genome and expressed the genes contained in this fragment similar to endogenous genes of plant cells and induced hairy roots. The next step is to select a stable clone with healthy and rapidly growing roots for establishing hairy root cultures, followed by checking their ability to produce various compounds of interest under the influence of different carbon source, ionic concentrations, and pH on the medium.

In our laboratory at the Chaoyang University of Technology, Taiwan, we have analyzed the influence of different PGRs on the root growth and accumulation of three tanshinone constituents in hairy roots of *S. miltiorrhiza* (Gupta et al. 2011). *A. rhizogenes* strain BCRC15010 was used to infect leaves of *S. miltiorrhiza*, and about 78 % of explants were responded to infections. Among all the tested media, the B5 liquid medium was most suitable for the growth of hairy roots. A robust hairy root line was established on this medium under darkness. The transfer of Ti plasmid fragments in the transformed hairy root lines was confirmed by PCR using "*rol*" B and C gene-specific primers. The root growth was optimum in B5 liquid medium, and an increase in dry weight was observed over a few consecutive weeks. The maximum dry weight increase (34-fold) was recorded after 12 weeks on the medium containing TDZ. HPLC analysis revealed an increased accumulation of cryptotanshinone, tanshinone I, and tanshinone IIA over the following weeks. An elevation in the production of cryptotanshinone, tanshinone I, and tanshinone IIA was observed under the influence of different PGRs (auxins, cytokinins, and ABA). An increase of five- and 7.5-fold tanshinone I and cryptotanshinone accumulation was observed with the addition of 1 mg/l ABA and TDZ, respectively, compared to the greenhouse-grown roots. The presence of TDZ, ABA, and BA in the medium also enhanced cryptotanshinone accumulation, and it increased 6.3-, 5.0-, and 3.75-fold, respectively, in hairy roots compared to a commercial herbal sample. This protocol can be implemented for the upscale production of tanshinone in *S. miltiorrhiza*.

In another study, Chen et al. (1999) established hairy root cultures by inoculating *A. rhizogenes* ATCC 15834 along with the sterile plantlets of *S. miltiorrhiza*, and the transgenics were confirmed by PCR to detect the inserted T-DNA. The presence of lithospermic acid B (LAB), RA, other related phenolic compounds, and the water-soluble active components in hairy roots was analyzed by HPLC, which confirmed that hairy roots had the ability to produce them. The effect of different basal media, MS, MS-NH (MS without ammonium nitrate), B5, WPM, and 6,7-V on hairy root growth and phenolic compound accumulation was analyzed, and it was observed that the root growth and phenolic compound accumulation was favored by MS-NH and 6,7-V media. Further, MS-NH medium was also tested for the time course of biomass production and phenolic compound accumulation, and it was observed that the RA content was stable throughout the study (approximately 0.48 % of dry weight), while accumulation of LAB varied between 0.73 % and 1.61 % of dry weight and further declined at the stationary phase of growth (Chen et al. 1999). The RA accumulation in the hairy root culture of *S. miltiorrhiza* in response to yeast extracts (YE) was achieved by Yan et al. (2006).

Hairy root cultures of *S. miltiorrhiza* were established by infecting the plantlets with different *A. rhizogenes* strains LBA 9402, ATCC 15834, TR 105, R 1601, and A 4 1027, and these strains displayed different abilities to induce hairy roots. Hairy root initiation was promoted by the use of acetosyringone. The transgenic hairy root lines were positive for agropine and mannopine, when assayed for opine by TLC. Different metabolites (tanshinone I, tanshinone IIA, tanshinone IIB, tanshinone V, dihydrotanshinone I, cryptotanshinone, tanshinone VI, and the colorless diterpene ferruginol) were quantified by HPLC in hairy roots as well as in the culture medium, and tanshinones were observed as the main product in the hairy root cultures. The effect of modified ammonium nitrate level was evident in the medium when the effect of culture condition on hairy root growth and the production of diterpene was studied (Zhi and Alfermann 1993).

In a study by Zhang et al. (2004), secondary metabolite accumulation in the hairy roots of *S. miltiorrhiza* was induced by Ag^+, which was employed as an abiotic elicitor. In the culture, a twofold increase in the yield of the three diterpenoid tanshinones (tanshinone I, tanshinone IIA, and cryptotanshinone) was noticed between 12 and 22 days post-inoculation with 15–40 μM $Ag_2S_2O_3$. The defined dose of Ag^+ affected the total tanshinone yields, and 30 μM Ag^+ was most suitable. However, Ag^+ dose and the day when Ag^+ was added to the culture affected the tanshinone stimulation. The growth of hairy roots was also controlled in a dose-dependent manner using Ag^+. The increase in biomass and tanshinone was noticed when sucrose feeding or medium renewal was employed before the addition of Ag^+ to the culture. About 6.6-fold increase in tanshinone yield in contrast to the control (55.7 mg/l versus 7.3 mg/l) was noticed when combined medium renewal and Ag^+ treatment was employed. The function of Ag^+ as an elicitor in the hairy root culture was verified by its capability to induce the characteristic elicitor responses of plants, the elevated level of cross-cell membrane ion fluxes, and the production of reactive oxygen species (Zhang et al. 2004).

In a similar study, Ge and Wu (2005b) assessed the role of biotic elicitor (carbohydrate fraction of yeast extract {YE, 100 mg/mL}) and an abiotic elicitor (Ag+ {30 mM}) on the accumulation of diterpenoid tanshinones and related secondary metabolism pathways in hairy root cultures of *S. miltiorrhiza*. It was observed that the Ag+ stimulated the activity of *3-hydroxy-3-methylglutaryl CoA reductase* (*HMGR*), while both Ag+ and YE stimulated the activity of *1-deoxy-D-xylulose 5-phosphate synthase* (*DXS*), and YE was a better stimulator for *DXS* enzyme. Further, they found that the tanshinone accumulation induced by Ag+ elicitors was inhibited by both MVA (inhibitor mevinolin) and non-MVA (inhibitor fosmidomycin) pathway inhibitors. However, tanshinone accumulation induced by YE elicitors was inhibited by the non-MVA pathway inhibitor fosmidomycin only. Hence, the non-MVA pathway (*DXS* activity) plays a major role in tanshinone accumulation induced by the two elicitors, and it may also depend on the cross talk between the MVA and non-MVA pathways. Furthermore, an increase in tanshinone production and *DXS* activity was observed with YE, when hairy roots were pretreated with Ag+ for 24–48 h.

The effect of β-aminobutyric acid (BABA), a nonprotein amino acid, either alone or in combination with YE, was examined on the production of diterpenoid tanshinones in *S. miltiorrhiza* hairy root cultures (Ge and Wu 2005a). Different concentrations (0.1, 1, and 2 mM) of BABA in a dose-dependent manner enhanced the total yield of three major tanshinones (cryptotanshinone, tanshinone I, and tanshinone IIA) accumulation, which increased by 4.5-folds. The tanshinone production further increased in YE-induced roots which were pretreated with BABA for a few days. The pretreatment for 3 days with 1 mM BABA and further treatment with YE ended up with a 9.4-fold increase in tanshinone content in roots and 6.3-fold increase in volumetric tanshinone yield in culture. It was also noticed that the BABA was more effective in enhancing tanshinone production compared to methyl jasmonate. These observations suggested that BABA might strongly initiate elicitor-induced secondary metabolite production in hairy roots (Ge and Wu 2005a).

Yan et al. (2006) examined the role of biotic elicitor (YE) and an abiotic elicitor (Ag^+) on RA accumulation and activities of pathway enzyme in hairy roots of *S. miltiorrhiza*. Both elicitors increased the RA accumulation and total phenolic contents in the hairy roots; however, accumulation was greatly enhanced by YE treatment. Further variations in the enzyme activities were observed when cells were induced with elicitors; an elevated activity for *tyrosine aminotransferase* (*TAT*) and a sharp decline in the *phenylalanine ammonia-lyase* (*PAL*) activity were observed in elicitor-induced hairy roots. These results showed a clear correlation between elicitor-induced biosynthesis of RA and phenolic compounds with the *TAT* activity in *S. miltiorrhiza* hairy roots (Yan et al. 2006).

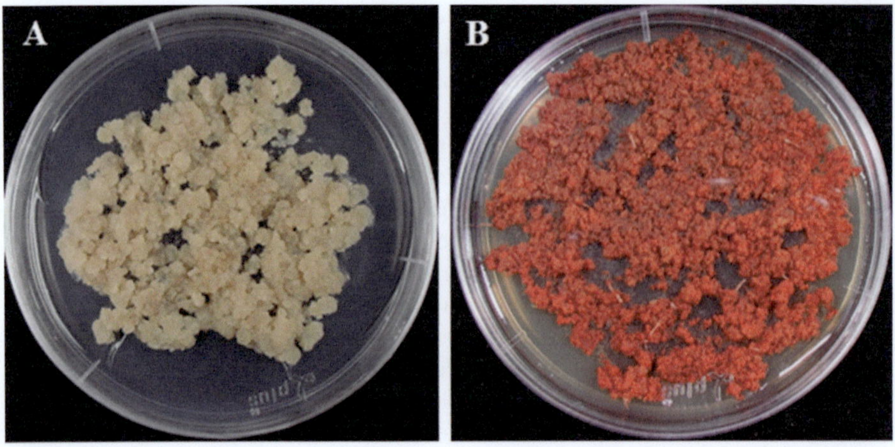

Fig. 8.3 (**A**) Non-transformed (control) callus of *S. miltiorrhiza*, (**B**) ATMT callus line of *S. miltiorrhiza* showing *red*-colored callus

8.2.4 T-DNA Activation Tagging-Mediated Enhanced Production of Compounds in S. miltiorrhiza

T-DNA activation method utilizes tandem copies of the enhancer sequence or constitutive promoter present on T-DNA. This fragment increases the expression of neighboring genes present on any side of the randomly integrated T-DNA tag, resulting in the upregulation of genes. This strategy has been implemented to target transcriptional factors specific to pathway genes to obtain the target products at a faster rate. There are two reports on activation tagging in *S. miltiorrhiza* from our laboratory in Taiwan (Lee et al. 2008; Ho et al. 2013). Lee et al. (2008) generated an activation tagging mutagenesis (ATM) population of *S. miltiorrhiza* by *A. tumefaciens*-mediated transformation (ATMT) system. The conditions for *Agrobacterium*-mediated transformation of *S. miltiorrhiza callus* were optimized by using *A. tumefaciens* harboring plasmid pCAMBIA1302. The construct pTAG-8 used in the ATMT consisted of eight copies of the A1 domain of the CaMV 35S promoter, and it was assumed that the 35S enhancer element might upregulate the expression of nearby gene expression without changing the original expression pattern. Different transgenic callus lines were confirmed by GUS assay followed by g-DNA isolation and Southern blot analysis. Non-transformed callus of *S. miltiorrhiza* was off-white in color (Fig. 8.3A), while out of 1,435 ATMT lines screened, only six (T1–T6) transgenic lines exhibited a bright red color on a 4.5 μM 2–4 D containing medium (Fig. 8.3B). Thus, calli of six ATMT lines growing on medium with producing tanshinone were easily distinguishable phenotypically by its red color. This allowed easy screening of tanshinone producing ATM population. The six (T1–T6) transgenic lines were compared with non-transgenic lines to check the growth rate and tanshinone content (Lee et al. 2008).

In another study, Ho et al. (2013) successfully generated an activation-tagged transgenic *S. miltiorrhiza* plant (SH41) having dissimilar leaf morphology and high diterpene content in its roots. PCR (using *hptII* primers) and Southern blotting were used to validate the genetic transformation of the SH41 plant. The ATM transgenic SH41 plant had broader leaves in contrast to non-transformed plants. Also, SEM showed that the leaf blades of SH41 plants exhibited thickening of the palisade and spongy tissues, high numbers of trichomes, woody, and bigger guard cells. High-performance liquid chromatography (HPLC) analysis of SH41 roots showed a significant increase in tanshinone I (3.7-fold) and tanshinone IIA (twofold) contents in comparison to the wild plants (Ho et al. 2013).

8.3 Genes, Expression Analysis, and Metabolites

Metabolic engineering is used for the modification of genetic as well as regulatory processes in the cells to increase the production of targeted substance (Lessard 1996). This technique has been studied extensively in various commercially important plants including *S. miltiorrhiza*, to understand the role of individual genes in biochemical pathways which help to pick suitable genes to upregulate or downregulate to provide commercial, agronomic, and/or postharvest processing features (Kinney 1998). However, it is necessary to have a proper knowledge of gene function and how it impacts the biosynthetic pathway to accumulate specific products. Metabolic engineering encompasses several strategies to improve productivity through either downregulation or overexpression of biosynthetic pathway genes. The following sections comprise a brief description of pathway genes and their roles (Table 8.2).

8.3.1 Functional Prediction of Metabolically Important Genes

Metabolic engineering or pathway engineering can be implemented to improve the targeted product biosynthesis by enhancing the rate-limiting steps or blocking competitive pathways. The biosynthesis and storage of secondary metabolites are mostly tissue specific, and consequently, genes involved in the synthesis and their regulators are also expressed specifically in a particular organ or tissue. The tanshinone biosynthesis pathway is divided into three steps: the first step is the formation of precursors for all the terpenoids, which involve *IPP* (*isopentenyl diphosphate*) and its isomer *DMAPP* (*dimethylallyl diphosphate*) through the *MEP* (*2-C-methyl-D-erythritol 4-phosphate*) pathway in plastid and/or the MVA (mevalonate) pathway in the cytoplasm. The second step is the construction of skeletons of tanshinones in which the intermediate diphosphate precursors such as *GPP* (*geranyl diphosphate*), *FPP* (*farnesyl diphosphate*), and *GGPP* (*geranylgeranyl diphosphate*) are synthesized under the catalysis of *IDSs* (*isoprenyl diphosphate synthases*), *GPPS*

Table 8.2 Biosynthetic pathway genes and their role in the synthesis of metabolites in *Salvia miltiorrhiza*

Genes	Roles	Involvement	References
4CL	Activation of cinnamic acid and its derivatives to their corresponding thioesters	Water-soluble phenolic compound biosynthesis (rosmarinic acid, salvianolic acid B, and lithospermic acid B)	Zhao et al. (2006) and Jin et al. (2012)
C4H	Conversion of cinnamic acid into 4-hydroxy-cinnamate	Rosmarinic acid and lithospermic acid B biosynthesis	Xiao et al. (2011)
TAT	4-Hydroxyphenyllactate synthesis	Rosmarinic acid and lithospermic acid B biosynthesis	Huang et al. (2008) and Xiao et al. (2011)
RAS	4-Coumaroyl-3′4′-dihydroxyphenyllactic acid synthesis	Rosmarinic acid	Ma et al. (2012)
HPPR	3,4-Dihydroxyphenyllactic acid synthesis	Rosmarinic acid and lithospermic acid B biosynthesis	Xiao et al. (2011)
PAL	Cinnamic acid synthesis	Water-soluble phenolic compound biosynthesis	Song and Wang (2011)
CPS	Copalyl diphosphate synthesis	Tanshinone biosynthesis	Cheng et al. (2014) and Cui et al. (2015)
GGPPS	Geranylgeranyl diphosphate synthesis	Tanshinone biosynthesis	Kai et al. (2010), (2011). and Ma et al. 2012
HMGR	Mevalonic acid synthesis	Tanshinone biosynthesis	Kai et al. (2011) and Ma et al. (2012)
DXS	1-Deoxy-D-xylulose 5-phosphate synthesis	Tanshinone biosynthesis	Kai et al. (2011) and Ma et al. (2012)
HDR	Plastidial terpenoid precursor synthesis	Diterpenoid tanshinone biosynthesis	Ma et al. (2012) and Hao et al. (2013)
IPPS	Isoprene precursor isopentenyl diphosphate synthesis	Tanshinone biosynthesis	Ma et al. (2012)
DMAPPS	Dimethylallyl diphosphate synthesis	Tanshinone biosynthesis	Ma et al. (2012)
GPPS	Geranyl diphosphate synthesis	Tanshinone biosynthesis	Ma et al. (2012)
FPPS	Farnesyl diphosphate synthesis	Tanshinone biosynthesis	Ma et al. (2012)
IDS	Prenyl diphosphate synthesis	Tanshinone biosynthesis	Ma et al. (2012)

(*including geranyl diphosphate synthase*), *FPPS* (*farnesyl diphosphate synthase*), and *GGPPS* (*geranylgeranyl diphosphate synthase*), and the third step is the post-modification of the skeleton that involves the formation of diverse terpenoids under the catalysis of *TPSs* (*terpene synthases/cyclases*), such as *CPS* (*copalyl diphosphate synthase*) and *KS* (*kaurene synthase*), and various terpenoid-modifying enzymes. However, the biosynthesis of phenolic compounds in *S. miltiorrhiza* involves two biosynthetic pathways: (1) phenylpropanoid pathway and (2) tyrosine-derived pathway. The major enzymes involved in the pathways are P*AL* (*phenylalanine ammonia-lyase*), C_4H (*cinnamic acid 4-hydroxylase*), *4CL* (*4-coumaroyl:CoA ligase*), *TAT* (*tyrosine aminotransferase*), *HPPR* (*4-hydroxyphenylpyruvate reductase*), *and RAS* (*rosmarinic acid synthase*) (Ma et al. 2015).

The role of *4CL* genes in the biosynthesis of water-soluble phenolics in *S. miltiorrhiza* was investigated by Zhao et al. (2006). And to do so, they performed nested RT-PCR with degenerate primers followed by 5′/3′ rapid amplification of cDNA ends (RACE) and cloned two cDNAs (*Sm4CL1* and *Sm4CL2*) encoding divergent *4CL* members. The gene was cloned into a pRSET expression vector, and 4CL activity was checked with purified recombinant proteins. The substrate specificity for *Sm4CL1* was distinct from that of *Sm4CL2*, and the Km values of *Sm4CL1* and *Sm4CL2* for the substrate 4-coumaric acid were (72.20 ± 4.10) and (6.50 ± 1.45) μmol/l, respectively. These results along with Northern blotting and other information suggested that *Sm4CL2* is involved in the biosynthesis of water-soluble phenolic compounds, whereas the involvement of *Sm4CL1* was not so significant in phenolic biosynthesis. A further single copy of these genes was predicted by Southern blot (Zhao et al. 2006).

Jin et al. (2012) analyzed the expression of *4CL2* and *4CL3* genes and predicted their possible role in the phenolic acid biosynthesis (Jin et al. 2012). The biosynthesis of RA and LAB was increased through genetic manipulation, which provided an effective approach to upscale the production of these compounds in hairy root culture systems using bioreactors. Xiao et al. (2011) have successfully engineered RA biosynthesis pathway for the production of RA and LAB in *S. miltiorrhiza* hairy root cultures. Overexpression of *C4H*, *TAT*, and *HPPR* or coexpression of *TAT and HPPR* and suppression of *4-hydroxyphenylpyruvate dioxygenase* (*hppd*) resulted in the higher production of RA and LAB in the hairy roots compared to wild type (control).

Huang and colleagues (2008) characterized a novel *TAT* gene involved the in RA biosynthesis from *S miltiorrhiza*, which shared significant homology with other known *TAT* genes. This gene was constitutively expressed in different plant parts (stem, root, and leaf); however, under normal conditions, it was overexpressed in the stem. The gene was overexpressed when the plants were exposed to the signaling components of defense/stress pathways such as methyl jasmonate (MeJA), ABA, SA, and ultraviolet-B radiation (UV-B) (Huang et al. 2008). Kai and colleagues (2011) have reported the overexpression of tanshinone biosynthetic pathway genes *HMGR* (*3-hydroxy-3-methylglutaryl CoA reductase*), *DXS* (*1-deoxy-d-xylulose-5-phosphate synthase*), and *GGPPS* in hairy root lines, which led to higher levels of tanshinone accumulations. Among the three genes studied, tanshinone production was significantly higher with *DXS gene* compared to *HMGR gene*. However, G*GPPS gene* played a crucial role in inducing the accumulation of tanshinone compared to upstream enzyme *HMGR* or *SmDXS*, while coexpression of *HMGR* and *GGPPS* genes led to a *4.74-fold* increase in tanshinone production compared to control (Kai et al. 2011).

In another study carried out by Kai et al. (2010) a full-length cDNA encoding *GGPPS* was isolated from *S. miltiorrhiza* using RACE. *GGPPS* catalyzes the biosynthesis of GGPP, a key precursor for diterpene synthesis including tanshinone. The addition of SA upregulated the expression of *SmGGPPS* in leaves. However, the addition of MeJA inhibited the expression of *SmGGPPS*, indicated the elicitor-responsive nature of *SmGGPPS*, and helped to predict the role of *SmGGPPS* in the tanshinone biosynthesis and genetic manipulation to increase tanshinone production in *S. miltiorrhiza* (Kai et al. 2010).

8.3.2 Expression Profiling

Recent advances in sequencing and comparative transcriptome analysis have improved our understanding of the role of different genes in biosynthetic pathway. Cui et al. (2011) used cDNA microarray to analyze genes involved in tanshinone biosynthesis and to identify variations in the gene expression profiles in *S. miltiorrhiza* at different stages of hairy root development. Out of 4,354 cDNA clones, about 203 genes were singled out on the microarray, and a total of 114 unique differentially expressed cDNA clones were detected: six and 96 genes were differentially expressed in 45-day- and 60-day-old hairy root, respectively, compared with 30-day-old hairy root, and 12 genes were expressed at different growth stages. Among the 96 genes expressed in 60-day-old hairy root, about 57 genes were upregulated, and 26 genes represented 29 metabolism-related enzymes. A 6.63-fold expression of *CPS* gene was observed in the hairy root, and differential expression pattern was observed for genes involved in tanshinone biosynthesis and P450 genes. These analyses helped further identification of enzymes involved in the tanshinone biosynthesis (Cui et al. 2011).

Wenping et al. (2011) identified a set of novel genes involved in the secondary metabolite biosynthetic pathways related to the synthesis of terpenoid-derived tanshinones and salvianolic acids using Solexa deep RNA sequencing (RNA-seq) to evaluate the transcriptome of *S. miltiorrhiza*. RNA-seq data was generated using Solexa sequencing, and reads were assembled using SOAP de novo assembler. As a part of five major secondary metabolite biosynthetic pathways covering almost all nodes in the phenylpropanoid and terpenoid pathways, 1,539 unigenes were identified. RNA-seq data provided a better understanding of the functional characteristics such as more than 70 novel transcripts linked to phenylpropanoid and terpenoid pathways and spatiotemporal expression patterns of ten novel transcripts related to terpenoid and phenolic acid biosynthesis. These preliminary data provide useful information for manipulating the pathways of secondary metabolite to produce health-promoting compounds in *S. miltiorrhiza* (Wenping et al. 2011). TGICL software was used to cluster these unigenes to acquire nonredundant unigenes as long as possible followed by BLASTX aligned protein databases of NR (NCBI nonredundant protein sequences), Swiss-Prot (high-quality annotated and nonredundant protein sequence database), KEGG (Kyoto Encyclopedia of Genes and Genomes – a reference resource for gene and protein annotation), and COG (clusters of orthologous groups). A total of 56.774 unigenes of length ≥200 bp was generated of which 34,340 were annotated, and 2,545 unigenes were associated with specific pathways (Pertea et al. 2003). Shao and Lu (2014) studied the RNA-dependent RNA polymerases (RDRs) in *S. miltiorrhiza.* RDRs act as important constituents of the small RNA biogenesis pathways and play a key role in posttranscriptional gene silencing (PTGS) and antiviral defense mechanisms. Five full-length *S. miltiorrhiza RDR* genes termed *SmRDR1–SmRDR5* were identified through genome-wide prediction and subsequent molecular cloning, and further study suggested their role in the development and response to abiotic and biotic stresses (Shao and Lu 2014).

Ma et al. (2012) have studied the genome-wide identification and characterization of putative genes involved in the terpenoid biosynthesis in *S. miltiorrhiza.* Draft genome sequence of *S. miltiorrhiza* was looked up for the homologs of terpenoid biosynthesis-related proteins from various plant species using BLASTX, and 40 terpenoid biosynthesis-related genes were detected, of which 27 were novel. These 27 genes were grouped into 19 families comprising ten single and nine multigene families. Some of the enzymes, such as *DXS*, *HDR*, *HMGR*, and *GGPPS*, are encoded by multigene with variable expression profiles and subcellular localizations, indicating the complexity of terpenoid biosynthesis in *S. miltiorrhiza* (Ma et al. 2012).

8.3.3 Gene Functions and Pathway Engineering

In the post-genomic era, studies on loss of gene functions are crucial to determining its role in the biosynthetic pathway and the synthesis of secondary metabolites. RNA interference is one of the profound methodologies applied to generate knockout plants and to gain insight into the functions of different genes involved in biosynthetic pathways. RNA silencing is a novel gene regulatory mechanism that limits the transcript level either by lowering mRNA synthesis or by sequence-specific RNA degradation process (Bernstein et al. 2001). It was demonstrated that RNAi-mediated gene silencing had been used to silence many endogenous target genes, resulting in metabolically engineered plants with improved storage capacity, virus resistance, oil content, and health benefits (Mansoor et al. 2006).

The efficient method to generate hairy root in *S. miltiorrhiza* has presented a simple system for studying the biochemical properties and gene functions (Gupta et al. 2011). Thus, the easily regenerated hairy roots have been used as a model system to elucidate the role of various pathway genes as well as to alter pathway gene functions. It was evident in the work of Li et al. (2008) that in *S. miltiorrhiza*, tanshinones are mostly accumulated in roots, with minimal amounts in the aerial parts (Li et al. 2008). Song and Wang (2011) studied the RNAi-mediated downregulation of the *phenylalanine ammonia-lyase* (*PAL*) gene which caused abnormal phenotypes and reduced RA synthesis in *S. miltiorrhiza. PAL* is the first enzyme involved in the phenylpropanoid biosynthesis and involved in the accumulation of RA and its derivatives. To construct RNAi, a 217-bp gene fragment was selected from the 3′ end of the *PAL1 gene. The generated RNAi construct was introduced into S. miltiorrhiza* through *A. tumefaciens*-mediated transformation. The RNAi-introduced transgenics showed different phenotypes such as stunted growth, delayed root formation, altered leaves, and reduced lignin deposition. A low level of *PAL* activity and reduced phenolic acids content (20–70 %) were recorded in transgenic lines. The expression patterns of other related genes of phenylpropanoid pathways such as *C4H*, *4CL2*, and *TAT were also affected in generated transgenic lines. These* are related genes in the RA pathway. Metabolic analysis of the transgenic lines clearly indicated a reduced pattern of accumulations of many phenolic acids such as RA and salvianolic acid B. Thus, this study clearly elucidated the role of *PAL* in the synthesis of major water-soluble *phenolics* (Song and Wang 2011). Cheng et al. (2014) have reported RNA interference-mediated repression of *CPS gene (*the first key enzyme in tanshinone biosynthesis) expression in hairy roots of *S. miltiorrhiza. To examine the expression of CPS and its effect on tanshinone accumulation, transformed RNAi constructs in A. rhizogenes were used to generate hairy roots in S. miltiorrhiza. The generated hairy roots with RNAi resulted in the low level of CPS expression (26 %). The generated hairy roots were examined for the dihydrotanshinone I and cryptotanshinone levels, which lessened by 53 % and 38 %, respectively. This confirms* that *CPS* is an important enzyme for tanshinone biosynthesis in *S. miltiorrhiza (*Cheng et al. 2014).

Hao et al. (2013) have reported the cloning, molecular characterization, and functional analysis of *1-hydroxy-2-methyl-2-(E)-butenyl-4-diphosphate reductase* (*HDR*), an enzyme in the plastid MEP pathway, which produced isoprenoid precursors for diterpenoid tanshinone biosynthesis in *S. miltiorrhiza.* An elevated level of *HDR1* gene expression was detected in leaves, while the gene expression in stem and roots was low. The introduction of 0.1 mM MeJA and SA induced the expression of HDR1. However, no induction was observed with 0.1 mM ABA, in the hairy roots of *S. miltiorrhiza*. It was introduced into *Escherichia coli HDR* mutant MG1655 to determine the function of *HDRI*. The functional color assay in *E. coli* displayed that HDR1 accelerated β-carotene biosynthesis, revealing that *HDR1* encodes a functional protein. An elevated level of expression of *HDR1* resulted in higher yield of tanshinones in cultured *S. miltiorrhiza* hairy roots. Thus, *HDR1* is a new and key enzyme involved in diterpenoid tanshinone biosynthesis in *S. miltiorrhiza* (Hao et al. 2013). Cui et al. (2015) have reported the characterization of *CPS* and kaurene synthase-like cyclase enzymatic families from *S. miltiorrhiza*, which has paved the way to identify unique pathways and elucidated the role of separate *CPS* *s* in tanshinone production in roots versus aerial tissues (*CPS1* and *CPS2*, respectively) and also the distinct production of *ent*-13-epi-manoyl oxide by *CPS4* and *S. miltiorrhiza* kaurene synthase-like2 in floral sepals. *CPS5* plays a role in gibberellin plant hormone biosynthesis. The expression of *CPS1* was suppressed by RNA interference, which resulted in significant reduction of tanshinones, and metabolomic analysis indicated 21 potential intermediates, revealing an intricated network of tanshinone metabolism defined by important biosynthetic steps (Cui et al. 2015).

8.4 Conclusions

For almost 20 years, more emphasis was given to the production of secondary metabolites from in vitro cultures of *S. miltiorrhiza*. The use of elicitors and generation of transgenics have been employed to elevate metabolite production to meet the increasing demand. The advent of next-generation sequencing technology and the latest methods to analyze transcriptome data has provided efficient tools for biosynthetic pathway analysis. Now it has become easy to trace and isolate multiple pathway genes involved in the metabolite biosynthesis. With the help of RNA sequencing data, it becomes easy to predict the gene expression pattern while particular elicitor or stress employed, which further enabled to engineer the biosynthetic pathway genes to get the product of interest at the higher end.

Acknowledgment The authors gratefully acknowledged a research grant (NSC96-2313-B324-001 and NSC98-2811-B324-001) from the National Science Council, Taiwan.

References

Ahn SY, Kim SK, Lee SH, Ahn SY, Kang SW, Chung JH, Kim SD, Lee BC (2010) Renoprotective effect of tanshinone IIA, an active component of *Salvia miltiorrhiza*, on rats with chronic kidney disease. Phytother Res 24(12):1886–1892. doi:10.1002/ptr.3347

Bernstein E, Caudy AA, Hammond SM, Hannon GJ (2001) Role for a bidentate ribonuclease in the initiation step of RNA interference. Nature 409(6818):363–366. doi:10.1038/35053110

Bi W, Tian M, Row KH (2011) Extraction and concentration of tanshinones in *Salvia miltiorrhiza* Bunge by task-specific non-ionic surfactant assistance. Food Chem 126(4):1985–1990. doi:10.1016/j.foodchem.2010.12.059

Chen H, Chen F (2000) Effect of yeast elicitor on the secondary metabolism of Ti-transformed *Salvia miltiorrhiza* cell suspension cultures. Plant Cell Rep 19(7):710–717. doi:10.1007/s002999900166

Chen H, Yuan JP, Chen F, Zhang YL, Song JY (1997) Tanshinone production in Ti-transformed *Salvia miltiorrhiza* cell suspension cultures. J Biotechnol 58(3):147–156, doi:S0168165697001442

Chen H, Chen F, Zhang YL, Song JY (1999) Production of lithospermic acid B and rosmarinic acid in hairy root cultures of *Salvia miltiorrhiza*. J Ind Microbiol Biotechnol 122(3):133–138. doi:10.1038/sj.jim.2900624

Chen W, Luo Y, Liu L, Zhou H, Xu B, Han X, Shen T, Liu Z, Lu Y, Huang S (2010) Cryptotanshinone inhibits cancer cell proliferation by suppressing mammalian target of rapamycin-mediated cyclin D1 expression and Rb phosphorylation. Cancer Prev Res (Phila) 3(8):1015–1025, doi:1940-6207.CAPR-10-0020

Cheng TO (2007) Cardiovascular effects of danshen. Int J Cardiol 121:9–22. doi:10.1016/j.ijcard.2007.01.004

Cheng Q, Su P, Hu Y, He Y, Gao W, Huang L (2014) RNA interference-mediated repression of SmCPS (copalyldiphosphate synthase) expression in hairy roots of *Salvia miltiorrhiza* causes a decrease of tanshinones and sheds light on the functional role of *SmCPS*. Biotechnol Lett 36(2):363–369. doi:10.1007/s10529-013-1358-4

Cui G, Huang L, Tang X, Zhao J (2011) Candidate genes involved in tanshinone biosynthesis in hairy roots of *Salvia miltiorrhiza* revealed by cDNA microarray. Mol Biol Rep 38(4):2471–2478. doi:10.1007/s11033-010-0383-9

Cui G, Duan L, Jin B, Qian J, Xue Z, Shen G, Snyder JH, Song J, Chen S, Huang L, Peters RJ, Qi X (2015) Functional divergence of diterpene syntheses in the medicinal plant *Salvia miltiorrhiza*. Plant Physiol 169(3):1607–1618. doi:pp.15.00695

Dai ZK, Qin JK, Huang JE, Luo Y, Xu Q, Zhao HL (2012) Tanshinone IIA activates calcium-dependent apoptosis signaling pathway in human hepatoma cells. J Nat Med 66(1):192–201. doi:10.1007/s11418-011-0576-0

Gamborg OL, Miller RA, Ojima K (1968) Nutrient requirement of suspensions cultures of soybean root cells. Exp Cell Res 50:151–158. doi:10.1016/0014-4827(68)90403-5

Gao SL, Zhu DN, Cai ZH, Xu DR (1996) Autotetraploid plants from colchicine-treated bud culture of *Salvia miltiorrhiza* Bge. Plant Cell Tissue Org 47:73–77. doi:10.1007/BF02318968

Gautheret R (1939) Sur la possibilité de réaliser la culture indéfinie des tissues de tubercules de carotte. C R Soc Biol Paris 208:118–120

Ge X, Wu J (2005a) Tanshinone production and isoprenoid pathways in *Salvia miltiorrhiza* hairy roots induced by Ag+ and yeast elicitor. Plant Sci 168(2):487–491. doi:10.1016/j.plantsci.2004.09.012

Ge X, Wu J (2005b) Induction and potentiation of diterpenoid tanshinone accumulation in *Salvia miltiorrhiza* by hairy roots by â-aminobutyric acid. Appl Microbiol Biotechnol 68(2):183–188. doi:10.1007/s00253-004-1873-2

Goren R, Altman A, Giladi I (1979) Role of ethylene in abscisic acid-induced callus formation in Citrus Bud cultures. Plant Physiol 63(2):280–282

Gupta SK, Liu RB, Liaw SY, Chan HS, Tsay HS (2011) Enhanced tanshinone production in hairy roots of "*Salvia miltiorrhiza* Bunge" under the influence of plant growth regulators in liquid culture. Bot Stud 52(4):435–443

Hao G, Shi R, Tao R, Fang Q, Jiang X, Ji H, Feng L, Huang L (2013) Cloning, molecular characterization and functional analysis of 1-hydroxy-2-methyl-2-(E)-butenyl-4-diphosphate reductase (HDR) gene for diterpenoid tanshinone biosynthesis in *Salvia miltiorrhiza* Bge. f. alba. Plant Physiol Biochem 70:21–32, doi:S0981-9428(13)00175-7

Ho HS, Vishwakarma RK, Chen ECF, Tsay HS (2013) Activation tagging in *Salvia miltiorrhiza* can cause increased leaf size and accumulation of tanshinone I and IIA in its roots. Bot Stud 54(1):1–8

Hu Y, Zhang Z, Hu Z, Wu Y, Shen X, Zhang G, Zhou X (1992) Callus culture and bioactive ingredients of *Salvia miltiorrhiza*. Plant Physiol Commun 28(6):424–425

Hu Y, Bao F, Li J (2000) Promotive effect of brassinosteroids on cell division involves a distinct CycD3-induction pathway in *Arabidopsis*. Plant J 24(5):693–701. doi:10.1046/j.1365-313x.2000.00915.x

Huang L, Liu D, Hu Z (2000) Effects of phytohormones on growth and content of depsides in *Salvia miltiorrhiza* suspension cells. Zhong Yao Cai 23(1):1–4

Huang B, Yi B, Duan Y, Sun L, Yu X, Guo J, Chen W (2008) Characterization and expression profiling of tyrosine aminotransferase gene from *Salvia miltiorrhiza* (Dan-shen) in rosmarinic acid biosynthesis pathway. Mol Biol Rep 35(4):601–612. doi:10.1007/s11033-007-9130-2

Hung YC, Wang PW, Pan TL, Bazylak G, Leu YL (2009) Proteomic screening of antioxidant effects exhibited by radix *Salvia miltiorrhiza* aqueous extract in cultured rat aortic smooth muscle cells under homocysteine treatment. J Ethnopharmacol 124(3):463–474, doi:S0378-8741(09)00309-2

Isacchi B, Fabbri V, Galeotti N, Bergonzi MC, Karioti A, Ghelardini C, Vannucchi MG, Bilia AR (2011) Salvianolic acid B and its liposomal formulations: anti-hyperalgesic activity in the treatment of neuropathic pain. Eur J Pharm Sci 44(4):552–558, doi:S0928-0987(11)00338-1

Jiao M, Cao R, Chen H, Hao W, Dong J (2012) Effects of salicylic acid on the synthesis of rosmarinic acid and related enzymes in the suspension cultures of *Salvia miltiorrhiza*. Shengwu Gongcheng Xuebao/Chinese. J Biotechnol 28(3):320–328

Jin XQ, Chen ZW, Tan RH, Zhao SJ, Hu ZB (2012) Isolation and functional analysis of 4-coumarate: coenzyme A ligase gene promoters from *Salvia miltiorrhiza*. Biol Plant 56(2):261–268

Kai G, Liao P, Zhang T, Zhou W, Wang J, Xu H, Liu Y, Zhang L (2010) Characterization, expression profiling, and functional identification of a gene encoding geranylgeranyl diphosphate synthase from *Salvia miltiorrhiza*. Biotechnol Bioproc Eng 15(2):236–245. doi:10.1007/s12257-009-0123-y

Kai G, Xu H, Zhou C, Liao P, Xiao J, Luo X, You L, Zhang L (2011) Metabolic engineering tanshinone biosynthetic pathway in *Salvia miltiorrhiza* hairy root cultures. Metab Eng 13(3):319–327. doi:10.1016/j.ymben.2011.02.003, doi:S1096-7176(11)00015-2 [pii]

Kang DG, Oh H, Sohn EJ, Hur TY, Lee KC, Kim KJ, Kim TY, Lee HS (2004) Lithospermic acid B isolated from *Salvia miltiorrhiza* ameliorates ischemia/reperfusion-induced renal injury in rats. Life Sci 75(15):1801–1816. doi:10.1016/j.lfs.2004.02.034

Kim JH, Jeong SJ, Kwon TR, Yun SM, Jung JH, Kim M, Lee HJ, Lee MH, Ko SG, Chen CY, Kim SH (2011) Cryptotanshinone enhances TNF-a-induced apoptosis in chronic myeloid leukemia KBM-5 cells. Apoptosis 16(7):696–706. doi:10.1007/s10495-011-0605-1

Kinney AJ (1998) Manipulating flux through plant metabolic pathways. Curr Opin Plant Biol 1(2):173–178, doi:S1369-5266(98)80021-6

Krajewska-Patan A, Dreger M, Gorska-Paukszta M, Moecisz A, Mielcarek S, Baraniak M, Buchwald W, Marecik R, Grajek W, Mrozikiewicz PM (2007) *Salvia miltiorrhiza* Bunge *in vitro* cultivation. Herb Pol 53(4):88–96

Lee CY, Agrawal DC, Wang CS, Yu SM, Chen JJ, Tsay HS (2008) T-DNA activation tagging as a tool to isolate *Salvia miltiorrhiza* transgenic lines for higher yields of tanshinones. Planta Med 74(7):780–786. doi:10.1055/s-2008-1074527

Lee SE, Jeong SI, Yang H, Jeong SH, Jang YP, Park CS, Kim J, Park YS (2012) Extract of *Salvia miltiorrhiza* (danshen) induces Nrf2-mediated heme oxygenase-1 expression as a cytoprotective action in RAW264.7 macrophages. J Ethnopharmacol 139(2):541–548. doi:10.1016/j.jep.2011.11.046

Lessard P (1996) Metabolic engineering: the concept coalesces. Nat Biotechnol 14(13):1654–1655. doi:10.1038/nbt1296-1654

Li X, Wang ZZ (2009) Chemical composition, antimicrobial and antioxidant activities of the essential oil in leaves of *Salvia miltiorrhiza* Bunge. J Essent Oil Res 21(5):476–480. doi:10.1080/10412905.2009.9700222

Li TJ, Dong JE, Liang ZS, Shu ZM, Wan GW (2008) Distributional difference of fat-soluble compounds in the roots, stems, and leaves of four Salvia plants. Fen Zi Xi Bao Sheng Wu Xue Bao 41(1):44–52

Li J, Xu M, Fan Q, Xie X, Zhang Y, Mu D, Zhao P, Zhang B, Cao F, Wang Y, Jin F, Li Z (2011) Tanshinone IIA ameliorates seawater exposure-induced lung injury by inhibiting aquaporins (AQP) 1 and AQP5 expression in lung. Respir Physiol Neurobiol 176(1–2):39–49, doi:S1569-9048(11)00006-1

Liu AH, Guo H, Ye M, Lin YH, Sun JH, Xu M, Guo DA (2007) Detection, characterization and identification of phenolic acids in danshen using high-performance liquid chromatography with diode array detection and electrospray ionization mass spectrometry. J Chromatogr A 1161(1–2):170–182. doi:10.1016/j.chroma.2007.05.081, doi:S0021-9673(07)00986-7 [pii]

Liu HB, Xu J, Peng Y, Zhou JJ, Xiao PG (2010) Targets of danshen's active components for activating blood circulation activities. Acta Phys -Chim Sin 26(1):199–205. doi:10.3866/PKU.WHXB20091222

Luo X, Li K, Kai G, Chen S (2013) Biosynthesis of tanshinones and metabolic regulation in Danshen (*Salvia miltiorrhiza*). Asian J Chem 25(18):10028–10032. doi:10.14233/ajchem.2013.15295

Ma Y, Yuan L, Wu B, Li X, Chen S, Lu S (2012) Genome-wide identification and characterization of novel genes involved in terpenoid biosynthesis in *Salvia miltiorrhiza*. J Exp Bot 63(7):2809–2823. doi:10.1093/jxb/err466

Ma XH, Ma Y, Tang JF, He YL, Liu YC, Ma XJ, Shen Y, Cui GH, Lin HX, Rong QX, Guo J, Huang LQ (2015) The biosynthetic pathways of tanshinones and phenolic acids in *Salvia miltiorrhiza*. Molecules 20(9):16235–16254, doi:molecules200916235

Mansoor S, Amin I, Hussain M, Zafar Y, Briddon RW (2006) Engineering novel traits in plants through RNA interference. Trends Plant Sci 11(11):559–565, doi:S1360-1385(06)00255-X

Mao H, Zhang H, Wang H, Wang Y, Zhao F, Hu L, Yanagihara N, Gao X (2009) Dual effects of a lipophilic extract of *Salvia miltiorrhiza* (danshen) on catecholamine secretion in cultured bovine adrenal medullary cells. J Ethnopharmacol 125(1):59–67, doi:S0378-8741(09)00376-6

Mei Z, Yan P, Situ B, Mou Y, Liu P (2012) Cryptotanshinione inhibits beta-amyloid aggregation and protects damage from beta-amyloid in SH-SY5Y cells. Neurochem Res 37(3):622–628. doi:10.1007/s11064-011-0652-6

Miyasaka H, Nasu M, Yamamoto T, Yoneda K (1985) Production of ferruginol by cell suspension cultures of *Salvia miltiorrhiza*. Phytochemistry 24(9):1931–1933. doi:10.1016/S0031-9422(00)83094-0

Miyasaka H, Nasu M, Yamamoto T, Shiomi Y, Ohno H, Endo Y, Yoneda K (1987) Effect of nutritional factors on cryptotanshinone and ferruginol production by cell suspension cultures of *Salvia miltiorrhiza*. Phytochemistry 26(5):1421–1424. doi:10.1016/S0031-9422(00)81826-9

Miyasaka H, Nasu M, Yoneda K (1989) *Salvia miltiorrhiza*: *in vitro* production of cryptotanshinone and feruginol. In: Bajaj YPS (ed) Biotechnology in agriculture and forestry medicinal and aromatic plants, IIth edn. Springer Verlag, Berlin, pp 417–430

Morimoto S, Goto Y, Shoyama Y (1994) Production of lithospermic acid B and rosmarinic acid in callus tissue and regenerated plantlets of *Salvia miltiorrhiza*. J Nat Prod 57(6):817–823. doi:10.1021/np50108a020

Morse D, Choi AM (2002) Heme oxygenase-1: the "emerging molecule" has arrived. Am J Respir Cell Mol Biol 27(1):8–16. doi:10.1165/ajrcmb.27.1.4862

Murashige T, Skoog F (1962) A revised medium for rapid growth and bioassay with tobacco tissue cultures. Physiol Planta 15:473–497. doi:10.1111/j.1399-3054.1962.tb08052.x

Nakanishi T, Miyasaka H, Nasu M, Hashimoto H, Yoneda K (1983) Production of cryptotanshinone and ferruginol in cultured cells of *Salvia miltiorrhiza*. Phytochemistry 22:721–722. doi:10.1016/S0031-9422(00)86969-1

Pertea G, Huang X, Liang F, Antonescu V, Sultana R, Karamycheva S, Lee Y, White J, Cheung F, Parvizi B, Tsai J, Quackenbush J (2003) TIGR Gene Indices clustering tools (TGICL): a software system for fast clustering of large EST datasets. Bioinformatics 19(5):651–652. doi:10.1093/bioinformatics/btg034

Sevon N, Oksman-Caldentey KM (2002) Agrobacterium rhizogenes-mediated transformation: root cultures as a source of alkaloids. Planta Med 68(10):859–868. doi:10.1055/s-2002-34924

Shan X, Shen YF, Xie YK, Chen JC, Shi HQ, Yu ZP, Song QT, Zhou MT, Zhang QY (2009) Inhibitory effects of tanshinone II-A on invasion and metastasis of human colon carcinoma cells. Acta Pharmacol Sin 30(11):1537–1542, doi:aps2009139

Shao F, Lu S (2014) Identification, molecular cloning and expression analysis of five RNA-dependent RNA polymerase genes in *Salvia miltiorrhiza*. PLoS ONE 9(4), e95117. doi:10.1371/journal.pone.0095117

Song J, Wang Z (2011) RNAi-mediated suppression of the phenylalanine ammonia-lyase gene in *Salvia miltiorrhiza* causes abnormal phenotypes and a reduction in rosmarinic acid biosynthesis. J Plant Res 124(1):183–192. doi:10.1007/s10265-010-0350-5

Standley P, Williams L (1973) Flora of Guatemala. Fieldiana Bot 24:237–317

Su CY, Ming QL, Rahman K, Han T, Qin LP (2015) *Salvia miltiorrhiza*: traditional medicinal uses, chemistry, and pharmacology. Chin J Nat Med 13(3):163–182, doi:S1875-5364(15)30002-9

Tan EC, Bahrami S, Kozlov AV, Kurvers HA, Ter-Laak HJ, Nohl H, Redl H, Goris RJ (2009) The oxidative response in the chronic constriction injury model of neuropathic pain. J Surg Res 152(1):84–88, doi:S0022-4804(08)00239-4

Tian LL, Wang XJ, Sun YN, Li CR, Xing YL, Zhao HB, Duan M, Zhou Z, Wang SQ (2008) Salvianolic acid B, an antioxidant from *Salvia miltiorrhiza*, prevents 6-hydroxydopamine induced apoptosis in SH-SY5Y cells. Int J Biochem Cell Biol 40(3):409–422, doi:S1357-2725(07)00261-0

Tian ML, Yan HY, Row KH (2009) Solid-phase extraction of tanshinones from *Salvia miltiorrhiza* Bunge using ionic liquid-modified silica sorbents. J Chromatogr B 877:738–742. doi:10.1016/j.jchromb.2009.02.012

Toth J, Mrlianova M, Tekelova D, Korenova M (2003) Rosmarinic acid- an important phenolic active compound of Lemon balm (*Melissa officinalis* L.). Acta Fac Pharm Univ Comenianae 50(139):146

Veliky IA, Martin SM (1970) A fermenter for plant cell suspension cultures. Can J Microbiol 16:223–226

Waldemar B (1996) Micropropagation and phenolic acids production of *Salvia miltiorrhiza* Buzge. In: Z. B (ed) Eucarpia series. Zuchtungsforschung, pp 327–329

Wan XL, Wang YE, Row KH (2009) Separation of tanshinone I, tanshinone IIA, and cryptotanshinone from *Salvia miltiorrhiza* Bunge by normal phase HPLC. J Liq Chromatogr R T 32(4):544–552. doi:10.1080/10826070802671507

Wang BQ (2010) *Salvia miltiorrhiza*: chemical and pharmaceutical review of a medicinal plant. J Med Plants Res 4(25):2813–2820, ISSN 1996-0875 ©2010

Wang JW, Wu JY (2010) Tanshinone biosynthesis in *Salvia miltiorrhiza* and production in plant tissue cultures. Appl Microbiol Biotechnol 88(2):437–449. doi:10.1007/s00253-010-2797-7

Wang K, Luo Q, Cheng H (1998) Production of allochtonic formations in callus cells of *Salvia miltiorrhiza* Bge. Zhongguo Zhong Yao Za Zhi 23(10):592–594

Wenping H, Yuan Z, Jie S, Lijun Z, Zhezhi W (2011) De novo transcriptome sequencing in *Salvia miltiorrhiza* to identify genes involved in the biosynthesis of active ingredients. Genomics 98(4):272–279, doi:S0888-7543(11)00080-2

Wong KK, Ho MT, Lin HQ, Lau KF, Rudd JA, Chung RC, Fung KP, Shaw PC, Wan DC (2010) Cryptotanshinone, an acetylcholinesterase inhibitor from *Salvia miltiorrhiza*, ameliorates scopolamine-induced amnesia in Morris water maze task. Planta Med 76(3):228–234. doi:10.1055/s-0029-1186084

Wu CT, Mulabagal V, Nalawade SM, Chen CL, Yang TF, Tsay HS (2003) Isolation and quantitative analysis of cryptotanshinone, an active quinoid diterpene formed in callus of *Salvia miltiorrhiza* BUNGE. Biol Pharm Bull 26(6):845–848. doi:10.1248/bpb.26.845

Xiao Y, Zhang L, Gao S, Saechao S, Di P, Chen J, Chen W (2011) The c4h, tat, hppr and hppd genes prompted engineering of the rosmarinic acid biosynthetic pathway in *Salvia miltiorrhiza* hairy root cultures. PLoS ONE 6(12), e29713. doi:10.1371/journal.pone.0029713

Xing BC, Yang DF, Guo W, Liang ZS, Yan XJ, Zhu YH, Liu Y (2015) Ag+ as a more effective elicitor for the production of tanshinones than phenolic acids in *Salvia miltiorrhiza* hairy roots. Molecules 20(1):309–324. doi:10.3390/molecules20010309

Yan Q, Shi M, Ng J, Wu JY (2006) Elicitor-induced rosmarinic acid accumulation and secondary metabolism enzyme activities in *Salvia miltiorrhiza* hairy roots. Plant Sci 170(4):853–858. doi:10.1016/j.plantsci.2005.12.004

Yang Y, Ge PJ, Jiang L, Li FL, Zhu QY (2011) Modulation of growth and angiogenic potential of oral squamous carcinoma cells *in vitro* using salvianolic acid B. BMC Complement Altern Med 11(1):54–61. doi:10.1186/1472-6882-11-54

Yin D, Yin J, Yang Y, Chen S, Gao X (2014) Renoprotection of danshen injection on streptozotocin-induced diabetic rats, associated with tubular function and structure. J Ethnol Pharmacol 151(1):667–674, doi:S0378-8741(13)00819-2

Yuan JM, Tao LL, Xu JT (1990) Immobilization of callus tissue cells of *Salvia miltiorrhiza* and the characteristics of their products. Chin J Biotechnol 6(3):199–205

Zhang H, Yu C, Jia JY, Leung SW, Siow YL, Man RY, Zhu DY (2002) Contents of four active components in different commercial crude drugs and preparations of danshen (*Salvia miltiorrhiza*). Acta Pharmacol Sin 23(12):1163–1168, ISSN 16 71-4 083

Zhang C, Yan Q, Cheuk W, Wu J (2004) Enhancement of tanshinone production in *Salvia miltiorrhiza* hairy root cultures by Ag+ elicitation and nutrient feeding. Planta Med 70(2):147–151. doi:10.1055/s-2004-815492

Zhang HN, An CN, Zhang HN, Pu XP (2010a) Protocatechuic acid inhibits neurotoxicity induced by MPTP in vivo. Neurosci Lett 474(2):99–103, doi:S0304-3940(10)00289-2

Zhang Y, Li X, Wang Z (2010b) Antioxidant activities of leaf extract of *Salvia miltiorrhiza* Bunge and related phenolic constituents. Food Chem Toxicol 48(10):2656–2662. doi:10.1016/j.fct.2010.06.036

Zhao SJ, Hu ZB, Liu D, Leung FCC (2006) Two divergent members of 4-coumarate: coenzyme A ligase from *Salvia miltiorrhiza* Bunge: cDNA cloning and functional study. J Integr Plant Biol 48(11):1355–1364. doi:10.1111/j.1744-7909.2006.00302.x

Zhi BH, Alfermann AW (1993) Diterpenoid production in hairy root cultures of *Salvia miltiorrhiza*. Phytochemistry 32(3):699–703. doi:10.1016/S0031-9422(00)95156-2

Zhou L, Zuo Z, Chow MS (2005) Danshen: an overview of its chemistry, pharmacology, pharmacokinetics, and clinical use. J Clin Pharmacol 45(12):1345–1359. doi:10.1177/0091270005282630

Zhou Y, Li W, Xu L, Chen L (2011) In *Salvia miltiorrhiza*, phenolic acids possess protective properties against amyloid ß-induced cytotoxicity, and tanshinones act as acetylcholinesterase inhibitors. Environ Toxicol Pharmacol 31(3):443–452. doi:10.1016/j.etap.2011.02.006

Chapter 9
Biotechnology of Medicinal Plants in Taiwan: Studies on In Vitro Propagation and Influence of Ventilation Closures on Hyperhydricity in Cultures

Dinesh Chandra Agrawal, Hung-Chi Chang, Chia-Chen Chen, Chao-Lin Kuo, Emily Chin-Fun Chen, and Hsin-Sheng Tsay

Abstract Medicinal plants have played a significant role in maintaining human health and improving the quality of life for thousands of years. In the past few decades, there has been an exponential rise in demand for herbal medicines, which in turn has resulted in an overexploitation and dwindling supply of medicinal plants in the wild. Therefore, it is imperative to explore all possible modes of plant propagation and large-scale cultivation. Taiwan is the home of many highly valued medicinal herbs used in the traditional Chinese medicines. In our laboratories, for more than two decades, tissue culture techniques have been used successfully for the propagation of several medicinally important plant species. The present article reviews the studies carried out on in vitro propagation of selected medicinal plants in Taiwan (*Glossogyne tenuifolia*, *Saussurea involucrata*, *Gentiana scabra*, and *Drynaria fortunei*). The article also includes the studies on induction of somatic embryogenesis in medicinal herb *Peucedanum japonicum* and influence of different

D.C. Agrawal (✉) • E.C.-F. Chen
Department of Applied Chemistry, Chaoyang University of Technology, Taichung, Taiwan
e-mail: dc.agrawal99@gmail.com

H.-C. Chang
Department of Applied Chemistry, Chaoyang University of Technology, Taichung, Taiwan

Department of Golden-Ager Industry Management, Chaoyang University of Technology, Taichung, Taiwan

C.-C. Chen • C.-L. Kuo
Department of Chinese Pharmaceutical Sciences and Chinese Medicine Resources, China Medical University, Taichung, Taiwan

H.-S. Tsay (✉)
Department of Applied Chemistry, Chaoyang University of Technology, Taichung, Taiwan

Department of Agronomy, National Chung Hsing University, Taichung, Taiwan
e-mail: hstsay@cyut.edu.tw

H.-S. Tsay et al. (eds.), *Medicinal Plants - Recent Advances in Research and Development*, DOI 10.1007/978-981-10-1085-9_9

ventilation closures on hyperhydricity (vitrification) in cultures of *Scrophularia yoshimurae*, *Bupleurum kaoi*, *Gentiana scabra*, and *Glossogyne tenuifolia*.

Keywords Hyperhydricity • In vitro propagation • Medicinal plants • Somatic embryogenesis • Tissue culture • Vitrification

Abbreviations

2,4-D	2,4-Dichlorophenoxyacetic acid
ABA	Abscisic acid
AF	Aluminum foil
BA	6-Benzyladenine
CYUT	Chaoyang University of Technology
DP	Dispense paper
DPPH	2,2-Diphenyl-1-picrylhydrazyl
EC50	Effective concentration (half maximal)
IAA	Indole-3-acetic acid
IBA	Indole-3-butyric acid
LED	Light-emitting diode
MS	Murashige and Skoog's medium
NAA	α-Naphthaleneacetic acid
SEM	Scanning electron microscopy
TARI	Taiwan Agricultural Research Institute
TEAC	Trolox equivalent antioxidant capacity
TFTF	Tree fern trunk fiber

9.1 Introduction

Plants have been a prime natural source of traditional medicines all over the world since prehistoric times. The old tradition of medicinal plant application has turned into a highly profitable business in the global market, resulting in the release of a large number of herbal products. In modern times, several distinct chemicals derived from plants are used as important drugs. However, management of medicinal plant resources has become a matter of urgency due to a rapidly increasing number of threatened plant species. Taiwan, an island situated in the West Pacific Ocean, has a diverse weather ranging from tropical to oceanic in the south to semitropical or temperate at mountain altitudes in the north. The central mountain range of Taiwan is the home of many highly valued medicinal herbs (Tsay and Agrawal 2005). Like other parts of the world, medicinal herbs in Taiwan are indiscriminately collected in large quantities from the wild to meet an ever-increasing demand for traditional crude drugs. To cope up with the alarming situation, biotechnological tools are being increasingly applied for mass propagation, production of bioactive compounds, and genetic improvement of medicinal plants. Tissue culture is useful for multiplication and conservation of endangered or rare species, which are difficult to propagate by conventional techniques, and save them from extinction. Micropropagation has several advantages over conventional methods of plant propagation because of its high multiplication rate. Also, most of the plants raised from seeds are highly heterozygous and show great variations in growth, habit, and yield and may not be suitable for commercial release due their poor quality of products (Yadav et al. 2012). Likewise, a majority of plants are not amenable to vegetative propagation through cutting and grafting. Moreover, many plants propagated by conventional vegetative means might contain systemic bacteria, fungi, and viruses (Murch et al. 2000), while medicinal plants propagated in vitro are mostly true to type (Chaturvedi et al. 2007).

There are many native medicinal plants in Taiwan with the therapeutic value which have been neglected for years. Since 1988, the National Science Council (NSC) under the Ministry of Science and Technology (MOST), Taiwan, has been promoting research on traditional Chinese medicinal plants (Tsay and Agrawal 2005). Biotechnology has played an important role in the mass propagation of economically important medicinal herbs and in the production of a number of bioactive compounds which are used as important drugs. Hence, several strategies including tissue cultures have been adopted across the world to improve the production of plant metabolites.

In our laboratories located at the Taiwan Agricultural Research Institute (TARI) and Chaoyang University of Technology (CYUT), significant progress has been made in the development of tissue culture protocols and isolation of bioactive compounds of a large number of traditional Chinese medicinal plant species. *Saussurea involucrata* and *Glossogyne tenuifolia* through shoot morphogenesis and *Angelica sinensis*, *Corydalis yanhusuo*, and *Peucedanum japonicum* through somatic embryogenesis have been developed. In addition, cell suspension cultures of *Taxus mairei*, *Angelica dahurica*, *Angelica sinensis*, *Dioscorea doryophora*, *Gentiana davidii*, and *Bupleurum falcatum* have been established. These cell culture systems have been successfully used to produce pharmaceutically important compounds like imperatorin from *Angelica dahurica*, corydaline and tetrahydropalmatine from in vitro-grown tubers of *Corydalis yanhusuo*, alkyl ferulates from *Dendrobium tosaense*, diosgenin from *Dioscorea doryophora*, gentiopicroside and swertiamarin from *Gentiana davidii*, anthraquinones from *Polygonum multiflorum*, cryptotanshinone from *Salvia miltiorrhiza*, harpagoside from *Scrophularia yoshimurae*, anthocyanins from *Solanum melongena*, syringin and rutin from *Saussurea involucrata*, oleanolic acid and luteolin from *Glossogyne tenuifolia*, chlorogenic acid and rutin from *Peucedanum japonicum*, and paclitaxel from *Taxus mairei*. The work has been reviewed earlier in several articles (Tsay 1999; Nalawade et al. 2003; Mulabagal et al. 2004; Nalawade and Tsay 2004; Tsay and Agrawal 2005; Mulabagal and Tsay 2007; Gupta et al. 2012; Tsay et al. 2015). The present article reviews the work carried out on in vitro propagation, somatic embryogenesis, and control of hyperhydricity (vitrification) in cultures of selected medicinal plant species in Taiwan.

9.2 In Vitro Propagation

In vitro culture constitutes the most efficient method of plant propagation of medicinal plant species world over, where the wild plant population is at a critical stage. Different plant parts such as the leaf, stem, node, internode, shoot tips, roots, and tubers are used as the starting material (explants) in culture on a synthetic medium, in sterile conditions, in a controlled environment, and in a reduced space (culture containers). Plant regeneration could occur either from explants with preexisting meristems or through callus (undifferentiated mass of cells) via shoot organogenesis or somatic embryogenesis. In our laboratory, in vitro propagation of several medicinal plants has been achieved (Table 9.1). In this section, we will briefly describe in vitro propagation results of four medicinal plant species: *Glossogyne tenuifolia, Saussurea involucrata, Gentiana scabra, and Drynaria fortunei.* The developed protocols have application in micropropagation, germplasm conservation, and commercial cultivation of these medicinal plants.

9.2.1 In Vitro Propagation of **Glossogyne tenuifolia** *Cassini*

Glossogyne tenuifolia Cassini (Asteraceae), a perennial herb, is a native species to Taiwan which originated in the Penghu Islands where for a long time it has been used to make a traditional healthy food and herbal tea. In the traditional Chinese medicine system, it has been used as an antipyretic and hepatoprotective source. *G. tenuifolia* has several pharmacological properties like cytotoxicity to several human cancer cells (Hsu et al. 2005), immunomodulation (Ha et al. 2006), and antioxidation (Yang et al. 2006). In a more recent study, *G. tenuifolia* extract suppressed the survival of mature osteoclasts by inhibiting osteoclast survival signaling pathways including NF-B, JNK, p38, and Akt, indicating its potential in the osteoclast-related diseases such as osteoporosis (Wang et al. 2014). The supply of natural plant material of *G. tenuifolia* is seasonal and restricted to only a few months. To overcome this constraint, we successfully developed a micropropagation protocol of *G. tenuifolia* (Chen et al. 2014) (Fig. 9.1a–d).

Table 9.1 Development of in vitro propagation protocols of medicinal plant species in Taiwan

Plant species/family	Methodologies	Explants	Reference
Adenophora triphylla (Campanulaceae)	In vitro culture	Internodes	Chen et al. (2001)
Angelica sinensis (Umbelliferae)	Somatic embryogenesis	Immature zygotic embryos	Huang et al. (1997) and Tsay and Huang (1998)
Anoectochilus formosanus (Orchidaceae)	In vitro culture	Axillary buds, shoot tips, seeds	Liu et al. (1987) and Shiau et al. (2002)
Bupleurum falcatum (Umbelliferae)	In vitro culture	Terminal and lateral buds	Hsu et al. (1993)
Bupleurum kaoi (Apiaceae)	In vitro culture	Axillary buds	Chen et al. (2004)
Corydalis yanhusuo (Fumariaceae)	Somatic embryogenesis	Tubers	Sagare et al. (2000), Lee et al. (2001), and Kuo et al. (2002)
Dendrobium linawianum (Orchidaceae)	In vitro culture	Lateral buds	Chen et al. (1995)
Dendrobium tosaense (Orchidaceae)	In vitro *culture*	Seed	Lo et al. (2004)
Drynaria fortunei (Polypodiaceae)	In vitro culture	Spores	Chang et al. (2007a)
Fritillaria hupehensis (Liliaceae)	In vitro culture	Bulb scales	Chen et al. (2000) and Shiau et al. (2000)
Gentiana davidii var. formosana (Gentianaceae)	In vitro culture	Stem nodes	Chueh et al. (2000, 2001)
Gentiana scabra (Gentianaceae)	In vitro culture	Shoot tips	Huang et al. (2014)
Glossogyne tenuifolia (Asteraceae)	In vitro culture	Shoot tips	Chen et al. (2014)
Limonium wrightii (Plumbaginaceae)	In vitro culture	Shoot tips	Huang et al. (2000)
Peucedanum japonicum (Umbelliferae)	Somatic embryogenesis	Leaf, stem, petiole, root	Chen et al. (2016)
Polygonum multiflorum (Polygonaceae)	In vitro culture	Nodal segments	Lin et al. (2003)
Saussurea involucrata (Asteraceae)	In vitro culture	Shoot base	Kuo et al. (2015)
Scrophularia yoshimurae (Scrophulariaceae)	In vitro culture	Shoot tips, stem, nodes, internodes	Lin et al. (1998) and Sagare et al. (2001)
Zingiber zerumbet (Zingiberaceae)	In vitro culture	Shoot tips	Hsu et al. (1991)

Fig. 9.1 In vitro propagation of *Glossogyne tenuifolia*. (**a**) Wild plants of *G. tenuifolia*, (**b**) multiple shoots, (**c**) proliferation of multiple shoots, (**d**) tissue culture plant in greenhouse. Bar: **b** = 1.5 cm; **c** = 1.4 cm; **d** = 3 cm

9.2.2 In Vitro Propagation of **Saussurea involucrata** *(Kar. et Kir.)*

S. involucrata Kar. et Kir. commonly known as "snow lotus" belongs to the family Asteraceae. The species mainly grows in the high rocky mountains in Xinjiang province of China (Fu 1992). In the traditional Chinese medicine system, the plant has been used for the treatment of several diseases such as a cough and cold, stomachache, rheumatoid arthritis, dysmenorrhea, and altitude sickness. In modern times, several pharmacological properties such as anticancer, cardiotonic, anti-inflammatory, antifatigue, and abortifacient have been reported (Jia et al. 2005). In a recent study, Byambaragchaa and coworkers have reported the potential of *S. involucrata* as an antitumor agent. It was observed that *S. involucrata* inhibited the growth of metastatic cells in a dose- and time-dependent manner (Byambaragchaa et al. 2013).

Market demand for *S. involucrata* plants is ever increasing; however, there is a severe shortage of plant material owing to the excessive collection of wild plants. The ecological destruction of the natural habits and the slow growth rates are other factors for the shortage of plants. The wild plant population has diminished to the extent that *S. involucrata* is now considered as an endangered species. Hence, it was imperative to develop alternative means of propagation. Recently, in our laboratory, Kuo and coworkers have reported an in vitro culture protocol for *S. involucrata* using response surface methodology (RSM) (Kuo et al. 2015). In vitro raised seedlings of *S. involucrata* were used as a source of explants to carry out further experiments. Shoot base explants induced multiple shoots when cultured on 3/4 strength of MS basal medium (Murashige and Skoog 1962) containing BA (1.0 mg/L) and NAA (1.5 mg/L) (Fig. 9.2a) or on 1/4X MS basal medium with BA (2.0 mg/L) and NAA (0.5 mg/L) (Fig. 9.2b). Elongation of shoots was achieved on MS basal medium with a reduced concentration of BA (0.5 mg/L) (Fig. 9.2c). Cent percent

Fig. 9.2 In vitro propagation of *S. involucrata*. (**a**) Multiple shoots induced on ¾ MS, BA (1 mg/L) + NAA (1.5 mg/L); (**b**) multiple shoots induced on ¼ MS, BA (2 mg/L) + NAA (0.5 mg/L); (**c**) elongation of multiple shoots on ½ MS, BA (0.5 mg/L); (**d**) tissue culture plants in greenhouse. Medium in **a**–**c** contained sucrose 3 %, agar 0.9 %. *Bar*: **a**–**c** = 1 cm

shoots were rooted in vitro on 1/2X MS basal medium containing indole-3-butyric acid (IAA) (1.0 mg/L) for 1 week and then transferred to a medium devoid of IAA. Ex vitro acclimation of plantlets was achieved by the "Sachet technique," and tissue culture plants could be established successfully in the university greenhouse (Kuo et al. 2015) (Fig. 9.2d).

9.2.3 *In Vitro Propagation of* Gentiana scabra *Bunge*

Gentiana scabra Bunge belonging to the family Gentianaceae is a perennial herb distributed in the northern parts of China (Han and Wang 1993). In traditional Chinese medicines, dry roots of *G. scabra* have been used in the treatment of indigestion, gastric infections, and inflammation (Tang and Eisenbrand 1992). Recent studies have confirmed the antirheumatic, antipyretic, analgesic, anti-inflammatory, diuretic, and hypoglycemic properties of *Gentiana* spp. (Chen et al. 2008; Wani et al. 2011). Due to the excessive collection from natural habitats, wild plant population has significantly declined. Also, natural regeneration is hampered owing to shorter dormancy period and poor seed germination. Therefore, it became necessary

Fig. 9.3 In vitro propagation of *Gentiana scabra*. (**a**) Multiple shoots, (**b**) rooted shoots, (**c**) tissue culture plant in the greenhouse. *Bar*: **a**–**c** = 2 cm

to optimize a propagation method by tissue culture. Recently, Huang and coworkers have reported an efficient micropropagation protocol of *G. scabra* (Huang et al. 2014). Multiple shoots (Fig. 9.3a) induced on the culture of shoot apices on half-strength Murashige and Skoog's basal medium containing 2.0 mg/L BA, 3 % sucrose, and 0.9 % Difco agar. Rooting in shoot cultures (Fig. 9.3b) was achieved on half-strength MS basal medium containing 0.1 mg/L NAA, 3 % sucrose, and 0.3 % gelrite. Rooted shoots established successfully and resulted in the maximum survival of 96 % by following a two-stage ventilation closure procedure during in vitro culture, and a transparent sachet technique during the hardening process. Tissue culture plants flowered after 5 months of transfer to pots (Fig. 9.3c) (Huang et al. 2014).

9.2.4 In Vitro Propagation of Drynaria fortunei *(Kunze)*

Drynaria fortunei (Kunze) J. Sm belonging to the family Polypodiaceae is a fern species growing mainly on tree trunks (Fig. 9.4a). In traditional Chinese medicines, it is known as "Gu-Sui-Bu." *D. fortunei* is mainly distributed in Taiwan, China, Vietnam, and Thailand. The fern is commonly used to treat bone injuries and has been shown to be effective in the treatment of inflammation, hyperlipidemia, and arteriosclerosis (Anonymous 2005). Jeong and coworkers in their in vitro studies demonstrated that "Gu-Sui-Bu" prevented osteoporosis by antiresorptive action in bone cells (Jeong et al. 2005). Also, in a separate study, Wong and Rabie found that "Gu-Sui-Bu" extract taken orally increased bone density (Wong and Rabie 2006). There is no commercial cultivation of *D. fortunei* at present. Hence, due to over collection, this fern is rapidly disappearing from the natural habits. The problem is compounded by the fact that it is not the frond but a finger-thick fleshy rhizome of

Fig. 9.4 In vitro propagation of *Drynaria fortunei*, (**a**) *D. fortune* plants in wild, (**b**) germinating spore, (**c**) gametophytes on different basal media, (**d**) gametophytes in culture flasks, (**e**) sporophytes,(**f**) sporophytes growing in pots in the greenhouse. *Bar*: **b** = 30 μm, **c** = 1.3 cm, **d** = 1.8 cm, **e** = 1.4 cm, **f** = 2 cm

D. fortunei which is used as the medicine. A severe shortage of *D. fortunei* prompted us to develop an in vitro propagation protocol which could help in the large-scale cultivation of this fern (Chang et al. 2007a). We could successfully achieve spore germination (Fig. 9.4b), gametophyte development (Fig. 9.4c–e), and change in the reproductive phase of *D. fortunei* in response to pH and LED light spectra. Gametophytes when transferred to plastic tray cells containing a potting mix of tree fern trunk fiber mix (TFTF mix) and peat moss developed sporophytes (Fig. 9.4f) or normal fern plants in the University greenhouse (Chang et al. 2007a).

In another report, ethanol and aqueous extracts of six folk medicinal ferns called "Gu-Sui-Bu" or "Shibu" in Taiwan were characterized for their antioxidant, scavenging activities, reducing power, total polyphenols, flavonols, flavonoids, condensed tannins, and proanthocyanidin contents (Chang et al. 2007b). These six ferns were rhizomes of (1) *Drynaria fortunei* and (2) *Pseudodrynaria coronans* (both from Polypodiaceae) and (3) *Davallia divaricata*, (4) *Davallia mariesii*, (5) *Davallia solida*, and (6) *Humata griffithiana* (all from Davalliaceae). In the commercial crude drug market, these six "Gu-Sui-Bu" have been claimed to cure body ache, inflammation, cancer, aging, blood stasis, and bone injuries. Investigations revealed that in contrast to the ethanol extracts, aqueous extracts of most samples showed higher antioxidant properties and polyphenol contents. Thus, it appears that the aqueous formulation of "Gu-Sui-Bu" is more potent than the ethanol one (Chang et al. 2007b).

9.3 In Vitro Plant Regeneration via Somatic Embryogenesis in *Peucedanum japonicum* Thunb

Earlier we have published two reports on induction of somatic embryogenesis and plant regeneration in two medicinal plant species *Angelica sinensis* (Tsay and Huang 1998) and *Corydalis yanhusuo* (Sagare et al. 2000; Kuo et al. 2002). More recently, we have reported an in vitro induction of somatic embryogenesis plant regeneration protocol in another medicinally important plant *P. japonicum* Thunb (Chen et al. 2016). *P. japonicum*, a perennial herb belonging to the family Umbelliferae, grows in Japan, the Philippines, China, Taiwan, and Korea. In Okinawa, Japan, leaves of this herb have been used traditionally for the treatment of cough. In Taiwan, roots of *P. japonicum* have been used as a folk medicine for cold and neuralgic diseases (Chen et al. 1996). Several coumarins isolated from roots and whole plant possess pharmacological properties such as antiplatelet aggregation (Chen et al. 1996), antioxidant (Hisamoto et al. 2003), anti-inflammatory, and antibacterial (Yang et al. 2009), antidiabetic (Nukitrangsan et al. 2012), and antiobesity (Nukitrangsan et al. 2012; Nugara et al. 2014). Induction of somatic embryogenesis in *P. japonicum* was achieved by both direct and indirect methods (Fig. 9.5a–j). Calli derived from in vitro roots developed somatic embryos when after three subcultures these were transferred from MS basal medium containing 2,4-dichlorophenoxyacetic acid (2,4-D) (0.1–5 mg/L) to the medium with abscisic acid (ABA) (0.5–4 mg/L) and exposed to eight different light spectra provided by light-emitting diode (LED) sources. ABA concentrations and LED light spectra had an influence on proliferation of callus and the number of somatic embryos induced. Conversion of somatic embryos to plantlets and secondary somatic embryogenesis were observed on a medium with ABA or exposure of somatic embryos to red or blue lights in a specially designed incubation chamber. Tissue culture plants (4 months old) derived from somatic embryos had significantly higher levels of chlorogenic acid (10.5 mg/g dw) compared to the controls (0.55 mg/g dw) (a commercial product sold in the Japanese market) (Chen et al. 2016).

9.4 Influence of Ventilation Closures on Hyperhydricity in Tissue Cultures

In vitro shoot cultures often exhibit hyperhydricity or vitrification which is a morphological, anatomical, and physiological abnormality (Kevers et al. 1984; Debergh et al. 1992). The type of ventilation closure affects the microenvironment inside a culture vessel and results in an unbalanced gaseous composition which in turn leads to the hyperhydric condition of in vitro shoots and plantlets. The hyperhydricity in cultures can be controlled by the use of more air-permeable materials as container closures (Zobayed et al. 2001; Lai et al. 2005). It is necessary to seal or cover culture vessels by a ventilation closure to avoid contamination by microorganisms.

Fig. 9.5 Plant regeneration in *Peucedanum japonicum* by somatic embryogenesis. (**a**) Wild plants (Inset: seed), (**b**) embryogenic callus, (**c**) somatic embryos (globular stage), (**d**) somatic embryos induced in callus, (**e**) embryos induced in the cut end of petiole, (**f**) embryos induced in leaf explant, (**g**) secondary somatic embryos, (**h**) somatic embryos conversion to plantlets, (**i**) somatic embryos derived in vitro plantlets, (**j**) potted tissue culture plants in the greenhouse. *Bar*: **a** = 6 cm; **b** = 1.6 cm; **c–f** = 1.5 mm; **g** = 5 mm; **h**= mm; **i** = 1.3 cm; **j** = 1.6 cm

However, plant species respond differently to the type of ventilation closures. Therefore, in a micropropagation protocol, it is important to optimize a closure type most suitable for a particular plant species. Various types of ventilation closure materials like micropore, parafilm, aluminum, and dispense paper (Chen et al. 2006a; 2006b) and polyvinyl chloride for eggplants (Ribeiro et al. 2009) and plastic films for in vitro cultures of neem (*Azadirachta indica*) (Rodrigues et al. 2012) have been used. These ventilation closures sometimes cause restriction of gaseous exchange between the container atmosphere and outside environment (Buddendorf-Joosten and Woltering 1994) and result in the poor aeration and hyperhydric condition of cultures. Zobayed and coworkers observed that that type of ventilation closures used to cover culture containers affected anatomical features of leaves in tobacco and cauliflower plants and that these effects were linked to hyperhydricity (Zobayed et al. 1999a, b, 2001). In an earlier study, Walker and coworkers have reported that physical and chemical microenvironments of culture containers had effects on the growth rate and other morphological/physiological characteristics of

Fig. 9.6 In vitro rooting of shoots in *S. yoshimurae* influenced by gelling agent and ventilation closure of vessel: (**a**) medium with agar, aluminum foil (AF) as ventilation closure; (**b**) medium with gelrite, AF; (**c**) roots in agar medium, DP (DP) as ventilation closure; (**d**) medium with agar, dispense paper; (**e**) medium with gelrite, DP.; (**f**) roots in gelrite medium, DP (*Bar* = 1.7 cm)

plants (Walker et al. 1988). In this section, we have briefly reviewed studies carried out in our laboratories on the influence of different ventilation closures on hyperhydricity in cultures of four medicinal plant species: *Scrophularia yoshimurae*, *Bupleurum kaoi*, *Gentiana scabra*, and *Glossogyne tenuifolia*.

9.4.1 Influence of Ventilation Closures on In Vitro Cultures of Scrophularia yoshimurae *Yamazaki*

Scrophularia yoshimurae Yamazaki (Scrophulariaceae) is a perennial herb native to Taiwan. Locally, it is known as "Xuanshen" in traditional Chinese medicines and is a substitute for *Scrophularia ningpoensis* (Chiu and Chang 1998). The plant is used for the treatment of constipation, inflammation, tonsillitis, and laryngitis. It is used for lowering blood sugar levels and blood pressure and has antioxidant and antibacterial properties. It has reported being cardiotonic (Reid 1996). Wild plants of *S. yoshimurae* have restricted distribution in the central mountain range of Taiwan (Liu 1998). Therefore, it became imperative to develop an alternative method of propagation by tissue culture. In our laboratory, though, in vitro plant regeneration method of *S. yoshimurae* was standardized (Lin et al. 1998; Sagare et al. 2001). However, high frequency of hyperhydric shoots was a major problem. Hence, we further evaluated the influence of different ventilation closures on the growth parameters of leaf, root, and plant survival rate and also carried out a scanning electron microscopy (SEM) study of the leaf surfaces of *S. yoshimurae* plants derived from these different ventilation closures. The results of the present study should be of immense help in understanding the underlying scientific principles behind different rates of acclimation achieved with various container closures and also boost the commercial production and conservation of *Scrophularia yoshimurae*.

Different ventilation closures like aluminum foil (AF) and dispense paper (DP) had significant effects on leaf growth parameters, number and quality of roots (Fig. 9.6a–f), and acclimation rates. Also, it was observed that ventilation closures had an influence on the anatomical features of the leaf surface. Density and size of epidermal cell and stomata, the size of guard cells, and stomata aperture were affected by the type of container closure. Leaf surfaces investigated by scanning electron microscope (SEM) revealed that leaves derived from AF treatment had higher densities of epidermal cell and stomata than the DP treatments. Changes in leaf anatomical feature in turn affected rates of plant survival. Thus, well-ventilated container closure by DP improved the survival rates of *S. yoshimurae* in the greenhouse conditions (Chen et al. 2006a).

9.4.2 *Influence of Ventilation Closures on Ex Vitro Acclimation Rates in* **Bupleurum kaoi** *Liu*

Genus *Bupleurum* Liu (Apiaceae) represents one of the most successful and widely used herbal drugs for the treatment of many diseases over the past 2,000 years in the traditional Chinese system of medicine. Roots of *Bupleurum* species commonly known as "Chai-hu" contain pharmaceutically active compounds known for anti-inflammatory, analgesic, antiallergic, and hepatoprotective properties (Kan 1985; Hiraoka 1989; Tang and Eisenbrand 1992; Lin and Yen 1999).

B. kaoi Liu is a native species to Taiwan. Though Chen and coworkers developed an in vitro propagation protocol in *B. kaoi*, however, a high percentage of hyperhydric shoots was a serious problem (Chen et al. 2004). Experiments were carried out to control the hyperhydricity in shoot cultures and improve the micropropagation protocol. A two-stage ventilation closure treatment was found the most suitable. In this treatment, culture vessels were first closed with two layers of aluminum foil (2AF) and incubated for 4, 3, 2, or 1 week. After that, aluminum foil was replaced by three layers of dispense paper (3DP) followed by incubation for 2, 3, 4, and 5 weeks, respectively. Further scanning electron microscope (SEM) examination of adaxial and abaxial surfaces in vitro leaves was carried to observe the densities of epidermal cells and stomata (Chen et al. 2006a).

In another study on *B. kaoi*, it was observed that certain anatomical features in leaves were affected by the type of ventilation closures used. The leaves from culture vessels closed with two layers of aluminum foil (AF) and 6 weeks of incubation lacked epicuticular wax and possessed larger stomata, higher stomata density, and fewer functional stomata compared to those of plants treated with two layers of aluminum foil for 2 weeks and then AF replaced by three layers of dispense paper

(DP) for 4 weeks. Also, this combination of AL and DP improved the ex vitro survival rates of in vitro regenerated plants of *B. kaoi* (Chen et al. 2006b).

9.4.3 *Influence of Ventilation Closures on In Vitro Cultures of* **Gentiana scabra** *Bunge*

In our laboratory, though induction and multiplication of shoot cultures in *Gentiana scabra* could be achieved, however, it was observed that in some treatments, shoots were hyperhydric which affected the hardening process and ex vitro survival percentage. Therefore, we carried out experiments with different combinations of ventilation closures to control hyperhydricity in cultures of *G. scabra* and to improve ex vitro acclimation rates (Huang et al. 2014). Two layers of aluminum foil (AF) and two to four layers of dispense papers (DP) as ventilation closures in different combinations were tested. For all the treatments, the medium composition was kept constant as half-strength MS basal medium containing 0.1 mg/L–1 NAA, 3 % sucrose, and 0.3 % gelrite. It was found that aluminum foil, a less air-permeable material for the first 4 weeks, followed by dispense papers (more air permeable) for the next 4 weeks, was an adequate ventilation treatment for optimum root/shoot growth and subsequent survival of *G. scabra* plantlets. The highest survival percentage of 96 was recorded with this two-stage ventilation closure treatment (Huang et al. 2014).

9.5 Conclusions

Biotechnology has played an important role in the propagation of important medicinal herbs in Taiwan, especially whose wild populations have become critical. For more than two decades, our research groups at the Chaoyang University of Technology (CYUT) and the Taiwan Agricultural Research Institute (TARI) have been successful in the development of protocols for in vitro propagation of a large number of medicinal herbs used in the traditional Chinese medicines. Our studies demonstrated that ventilation closures had an influence on anatomical features of in vitro cultures and ex vitro acclimation rates of tissue culture plants. Developed protocols have application in rapid mass propagation, germplasm conservation, and commercial cultivation of these medicinal plant species.

Acknowledgments Financial support in the form of research grants (NSC-91-2317-B-324-001, NSC-91-2313-B-324-002, and NSC-94-2317-B-324-001) from the National Science Council (94 Agri Sci-5.2.1-Sci-al) and from the Council of Agriculture, Taiwan, is gratefully acknowledged.

Disclosure The authors declare that all the experiments undertaken in the case of *Saussurea involucrata* comply with the current laws of Taiwan.

References

Anonymous (2005) Pharmacopoeia Commission of People Republic of China (ChPC), vol 1. Chemical Industry Press, Beijing, pp 179–180

Buddendorf-Joosten JMC, Woltering EJ (1994) Components of the gaseous environment and their effects on plant growth and development *in vitro*. Plant Growth Regul 15:1–16

Byambaragchaa M, Cruz de la J, Yang SH, Hwang SG (2013) Anti-metastatic potential of ethanol extract of *Saussurea involucrata* against hepatic cancer *in vitro*. Asian Pac J Cancer Prev 14(9):5397–5402. doi:http://dx.doi.org/10.7314/ APJCP.2013.14.9.5397

Chang HC, Agrawal DC, Kuo CL, Wen JL, Chen CC, Tsay HS (2007a) *In vitro* culture of D*rynaria fortunei*, a fern species source of Chinese medicine 'Gu-sui-bu'. In Vitro Cell Dev Biol 43:133–139. doi:10.1007/s11627-007-9037-6

Chang HC, Huang GJ, Agrawal DC, Kuo CL, Wu CR, Tsay HS (2007b) Antioxidant activities and polyphenol contents of six folk medicinal ferns used as "Gusuibu". Bot Stud 48:397–406

Chaturvedi HC, Jain M, Kidwai NR (2007) Cloning of medicinal plants through tissue culture – a review. Indian J Exp Biol 45(11):937–948

Chen CC, Na C, Hsu CM, Tsay HS (1995) Studies on the clonal propagation of *Dendrobium linawianum*. Formos Sci 48:69–78

Chen IS, Chang CT, Sheen WS, Teng CM, Tsai IL, Duh CY, Ko FN (1996) Coumarins and antiplatelet aggregation constituents from Formosan *Peucedanum japonicum*. Phytochemistry 41:525–530

Chen UC, Tai CD, Chen CC, Tsay HS (2000) Studies on the tissue culture of *Fritillaria hupehensis* Hsiao et K. C. Hsia. II. Influence of medium component and light treatment on in vitro rooting, and acclimatization of bulblet. J Agric Res China 49:48–55

Chen CC, Chen SJ, Sagare AP, Tsay HS (2001) Adventitious shoot regeneration from stem internode explants of *Adenophora triphylla* (Thunb.) A. DC. (Campanulaceae) – an important medicinal herb. Bot Bull Acad Sin 42:1–7

Chen UC, Yeh MS, Tsay HS (2004) Studies on *in vitro* shoot multiplication of native medicinal herbs – *Bupleurum kaoi* Liu, Chao et Chuang by axillary bud culture. J Agric Res China 53:27–38

Chen UC, Hsia CN, Agrawal DC, Tsay HS (2006a) Influence of ventilation closures on plant growth parameters, acclimation and anatomy of leaf surface in *Scrophularia yoshimurae* Yamazaki – a medicinal plant native to Taiwan. Bot Stud 47:259–266

Chen UC, Hsia CN, Yeh MS, Agrawal DC, Tsay HS (2006b) *In vitro* micropropagation and *ex vitro* acclimation of *Bupleurum kaoi* – an endangered medicinal plant native to Taiwan. In Vitro Cell Dev Biol 42:128–133

Chen L, Wang HB, Sun XL, Sun W (2008) Study on the analgesic and anti-inflammatory activities of Gentiopicroside. Nat Prod Res Dev 20:903–906

Chen CC, Chang HC, Kuo CL, Agrawal DC, Wu CR, Tsay HS (2014) *In vitro* propagation and analysis of secondary metabolites in *Glossogyne tenuifolia* (Hsiang-Ju) – a medicinal plant native to Taiwan. Bot Stud 55:45. doi:10.1186/s40529-014-0045-7

Chen CC, Agrawal DC, Lee MR, Lee RJ, Kuo CL, Wu CR, Tsay HS, Chang HC (2016) Somatic embryogenesis and LC-MS analysis of chlorogenic acid and rutin in a medicinal herb *Peucedanum japonicum* Thunb. Bot Stud 57:9. doi:10.1186/s40529-016-0124-z

Chiu NY, Chang KH (1998) *Scrophularia yoshimurae*. In: The illustrated medicinal plants of Taiwan, vol 5. SMC Pub. Inc., Taipei, p 194 (Chinese)

Chueh FS, Chen CC, Tsay HS (2000) Studies on factors affecting the establishment of *Gentiana davidii* var. *formosana* (Hayata) T. N. Ho cell suspension cultures. J Food Drug Anal 5:297–303

Chueh FS, Chen CC, Sagare AP, Tsay HS (2001) Quantitative determination of secoiridoid glucosides in *in vitro* propagated plants of *Gentian a davidii* var. *formosana* by high-performance liquid chromatography. Planta Med 67:70–73

Debergh P, Aitken-Christie J, Cohen D, Grout B, von Arnold S, Zimmerman R, Ziv M (1992) Reconsideration of the term "Vitrification" as used in micropropagation. Plant Cell Tissue Organ Cult 30:135–140

Fu LG (1992) China plant red data book – rare and endangered plants, vol 1. Chinese Science Press, Beijing

Gupta SK, Kuo CL, Chang HC, Chan HS, Chen ECF, Chueh FS, Tsay HS (2012) In: Shyur LF, Lau ASY (eds) Recent trends in medicinal plants research. Chapter 2. *In vitro* propagation and approaches for metabolites production in medicinal plants. Adv Bot Res 62: 35–55. doi:10.1016/B978-0-12-394591-4.00002-7

Ha CL, Weng CY, Wang L, Lian TW, Wu MJ (2006) Immunomodulatory effect of *Glossogyne tenuifolia* in murine peritoneal macrophages and splenocytes. J Ethnopharmacol 107:116–125

Han QZ, Wang ZX (1993) Flora of Dalian district, vol II. Dalian University of Science and Technology Press, Dalian, p 588p (In Chinese)

Hiraoka N (1989) *Bupleurum falcatum* L. embryogenesis and the production of saikosaponins. In: Bajaj YPS (ed) Biotechnology in agriculture and forestry, vol 7. Springer, Berlin, pp 69–81

Hisamoto M, Kikuzaki H, Ohigashi H, Nakatani N (2003) Antioxidant compounds from the leaves of *Peucedanum japonicum* thunb. J Agric Food Chem 51:5255–5261. doi:10.1021/jf0262458

Hsu JY, Yeh CC, Tsay HS (1991) *In vitro* propagation of *Zingiber zerumbet* Smith. J Agric Res China 40(2):171–177

Hsu JY, Liu SY, Tsay HS (1993) Studies on tissue culture of *Bupleurum falcatum* L. cv. Taichung No. 1. J Agric Res China 42(3):245–252

Hsu HF, Huang JY, Chang CL, Wu CC, Chang FR, Wu YC (2005) Antioxidant activity, cytotoxicity, and DNA information of *Glossogyne tenuifolia*. J Agric Food Chem 53:6117–6125

Huang HL, Chen CC, Chen UC, Tsay HS (1997) Studies on tissue culture of *Angelica sinensis* (Oliv.) Diels. (II) Influence of culture methods on somatic embryos formation from culture suspension cells. Chin Agron J 7:87–93

Huang CL, Hsieh MT, Hsieh WC, Sagare AP, Tsay HS (2000) *In vitro* propagation of *Limonium wrightii* (Hance) Ktze. (Plumbaginaceae), an ethnomedicinal plant, from shoot tip, leaf, and inflorescence node explants. In Vitro Cell Dev Biol Plant 36:220–224

Huang SH, Agrawal DC, Wu FS, Tsay HS (2014) *In vitro* propagation of *Gentiana scabra* Bunge – an important medicinal plant in the Chinese system of medicines. Bot Stud 55:56. doi:10.1186/s40529-014-0056-4

Jeong JC, Lee JW, Yoon CH, Lee YC, Chung KH, Kim MG, Kim CH (2005) Stimulative effects of *Drynaria* rhizome extracts on the proliferation and differentiation of osteoblastic MC3T3-E1 cells. J Ethnopharmacol 96:489–495

Jia JM, Wu CF, Liu W, Yu H, Hao Y, Zheng JH, Ji YR (2005) Anti-inflammatory and analgesic activities of the tissue culture of *Saussurea involucrata*. Biol Pharm Bull 28(9):1612–1614

Kan WS (1985) *Bupleurum falcatum* L., *Bupleurum kaoi* Liu, Chao et Chuang In: Manual of medicinal plants in Taiwan. National Research Institute of Chinese Medicine, Taipei, Taiwan, pp 650–651

Kevers C, Coumans M, Coumans-Gilles MF, Gaspar T (1984) Physiological and biochemical events leading to vitrification of plants cultured *in vitro*. Physiol Plant 61:69–74

Kuo CL, Sagare AP, Lo SF, Chen CC, Tsay HS (2002) Abscisic acid (ABA) promotes the development of somatic embryos on converted somatic embryos of *Corydalis yanhusuo* (Fumariaceae). J Plant Physiol 159(4):423–427

Kuo CL, Agrawal DC, Chang HC, Chiu YT, Huang CP, Chen YL, Huang SH, Tsay HS (2015) *In vitro culture and* production of syringin and rutin in *Saussurea involucrata* (Kar. et Kir.) – an endangered medicinal plant. Bot Stud 56:12. doi:10.1186/s40529-015-0092-8

Lai CC, Lin HM, Nalawade SM, Fang W, Tsay HS (2005) Hyperhydricity in shoot cultures of *Scrophularia yoshimurae* can be effectively reduced by ventilation of vessels. J Plant Physiol 162:355–361

Lee YL, Sagare AP, Lee CY, Feng HT, Ko YC, Shaw JF, Tsay HS (2001) Formation of protoberberine type alkaloids by the tubers of somatic embryo-derived plants of *Corydalis yanhusuo*. Planta Med 67:839–842

Lin CC, Yen MH (1999) Development of crude drug resources and evaluation of *Bupleurum kaoi* (in Chinese with English abstract). In: Lin CY, Lu HS, Liu SY, Chen CC, Chang CG (eds) Proceedings of the symposium on development and utilization of medicinal plants. Taiwan Agricultural Research Institute, Taichung, pp 51–72

Lin HM, Chen SM, Chen CC, Tsay HS (1998) Mass propagation of *Scrophularia yoshimurae* Yamazaki by tissue culture. J Agric Assoc China 183:1–12

Lin LC, Nalawade SM, Mulabagal V, Yeh MS, Tsay HS (2003) Micropropagation of *Polygonum multiflorum* Thumb and quantitative analysis of the anthraquinones emodin and physcion formed *in vitro* propagated shoots and plants. Biol Pharm Bull 26(10):1467–1471

Liu HY (1998) Flora of Taiwan. In: Huang TC (ed) vol 4, Department of Botany, National Taiwan University, Taipei, Taiwan, pp 624–625

Liu SY, Tsay HS, Huang HC, Hu MF, Yeh CC (1987) Comparison of growth characteristics and nutrient composition between plants of *Anoectochilus* species from mass vegetative propagation by tissue culture techniques. J Agric Res China 36:357–366

Lo SF, Nalawade SM, Chen CC, Tsay HS (2004) Asymbiotic germination of seeds, plantlet development and *ex-vitro* establishment of plants of *Dendrobium tosaense* Makino – a medicinally important orchid. In Vitro Cell Dev Biol Plant 40:528–535

Mulabagal V, Tsay HS (2007) Plant cell cultures: production of biologically active secondary metabolites from medicinal plants of Taiwan. In: Kayser O, Quax W (eds) Medicinal plant biotechnology, from basic research to industrial applications. WILEY-VCH Verlag GMBH & Co., Weinheim, pp 267–285

Mulabagal V, Lee CY, Lo SF, Nalawade SM, Lin CY, Tsay HS (2004) Studies on the production of some important secondary metabolites from medicinal plants by plant tissue cultures. Bot Bull Acad Sin 45:1–22

Murashige T, Skoog F (1962) A revised medium for rapid growth and bio-assays with tobacco tissue cultures. Physiol Plant 15:473–497

Murch SJ, Krishna RS, Saxena PK (2000) Phyto-pharmaceuticals: mass production, standardization, and conservation. Herb Med 4(2):39–43

Nalawade SM, Tsay HS (2004) *In vitro* propagation of some important Chinese medicinal plants and their sustainable usage. In Vitro Cell Dev Biol Plant 40:143–154

Nalawade SM, Sagare AP, Lee CY, Kuo CL, Tsay HS (2003) Studies on tissue culture of Chinese medicinal plant resources in Taiwan and their sustainable utilization. Bot Bull Acad Sin 44:79–98

Nugara RN, Inafuku M, Iwasaki H, Oku H (2014) Partially purified *Peucedanum japonicum* Thunb extracts exert anti-obesity effects *in vitro*. Nutrition 30:575–583. doi:10.1016/j.nut.2013.09.017

Nukitrangsan N, Okabe T, Toda T, Inafuku M, Iwasaki H, Oku H (2012) Anti-obesity activity of *Peucedanum japonicum* Thunb extract in obese diabetic animal model C57 BL/6 J Ham Slc-ob/ob mice. Int J Life Sci Med Res 2:28–34. doi:10.5963/LSMR0202004

Reid DP (1996) *Adenophora tetraphylla* (Campanulaceae). In: Amsel A, McClellan M, Maitland D (eds) Chinese herbal medicine. Shambhala Publications, Boston, p 98

Ribeiro APO, Picoli EAT, Lani ERG, Vendrame WA, Otoni WC (2009) The influence of flask sealing on *in vitro* morphogenesis of eggplant (*Solanum melongena* L.). In Vitro Cell Dev Plant 45:421–428

Rodrigues M, Costa TF, Festucci-Buselli R, Silva L, Otoni W (2012) Effects of flask sealing and growth regulators on *in vitro* propagation of neem (*Azadirachta indica* A. Juss). In Vitro Cell Dev Plant 48:67–72

Sagare AP, Lee YL, Lin TC, Chen CC, Tsay HS (2000) Cytokinin induced somatic embryogenesis and plant regeneration in *Corydalis yanhusuo* (Fumariaceae) – a medicinal plant. Plant Sci 160:139–147

Sagare AP, Kuo CL, Chueh FS, Tsay HS (2001) *De novo* regeneration of *Scrophularia yoshimurae* Yamazaki (Scrophulariaceae) and quantitative analysis of harpagoside, an iridoid glucoside, formed in aerial and underground parts of in vitro propagated and wild plants by HP LC. Biol Pharm Bull 24:1311–1315

Shiau YJ, Tai CD, Chen CC, Tsay HS (2000) Studies on the tissue culture of *Fritillaria hupehensis* Hsiao et K. C. Hsia. I. The induction of bulb scale and embryogenic callus from different source explants. J Agric Res China 49(4):29–38

Shiau YJ, Sagare AP, Chen UC, Yang SR, Tsay HS (2002) Conservation of *Anoectochilus formosanus* Hayata by artificial cross-pollination and *in vitro* culture of seeds. Bot Bull Acad Sin 43:123–130

Tang W, Eisenbrand G (1992) Chinese drugs of plant origin: chemistry, pharmacology, and use in traditional and modern medicine. Springer, Berlin, pp 549–553

Tsay HS (1999) Tissue culture technology of medicinal herbs and its application of medicinal herbs and its application in Taiwan. In: Chou CH, Waller GR, Reinhardt C (eds) Biodiversity and allelopathy: from organisms to ecosystems in the pacific. Academia Sinica, Taipei, pp 137–144

Tsay HS, Agrawal DC (2005) Tissue culture technology of Chinese medicinal plant resources in Taiwan and their sustainable utilization. Int J Appl Sci Eng 3(3):215–223

Tsay HS, Huang HL (1998) Somatic embryo formation and germination from immature embryo-derived suspension- cultured cells of *Angelica sinensis* (Oliv.) Diels. Plant Cell Rep 17:670–674

Tsay HS, Gupta SH, Huang SH, Kuo CL, Chueh FS, Chang HC (2015) *In vitro* studies and biotechnology of Taiwan native species of the gentianaceae. In: Rybczyński JJ, Davey MR, Mikula A (eds) The gentianaceae – volume 2: biotechnology and applications. Springer, Berlin, p 452. doi:10.1007/978-3-642-54102-5

Walker PN, Heuser CW, Heinemann PH (1988) Micropropagation: studies of gaseous environments. Acta Horticult 230:145–151

Wang SW, Kuoa HC, Hsu HF, Tu YK, Chenge TT, Houng JY (2014) Inhibitory activity on RANKL-mediated osteoclastogenesis of *Glossogyne tenuifolia* extract. J Funct Foods 6:215–223

Wani B, Ramamoorthy D, Ganai B (2011) Preliminary phytochemical screening and evaluation of analgesic activity of methanolic extract of roots of *Gentiana kurroo* Royle in experimental animal models. Int J Pharm Sci 3:164–166

Wong RW, Rabie AB (2006) Systemic effect of crude extract from the rhizome of *Drynaria fortunei* on bone formation in mice. Phytother Res 20:313–315

Yadav K, Singh N, Verma S (2012) Plant tissue culture: a biotechnological tool for solving the problem of propagation of multipurpose endangered medicinal plants in India. J Agric Tech 8(1):305–318

Yang JH, Tsai SY, Han CM, Shih CC, Mau JL (2006) Antioxidant properties of *Glossogyne tenuifolia*. Am J Chin Med 34:707–720

Yang EJ, Kim SS, Oh TH, Song G, Kim KN, Kim JY, Lee NH, Hyun CG (2009) *Peucedanum japonicum* and citrus unshiu essential oils inhibit the growth of antibiotic-resistant skin pathogens. Ann Microbiol 59:623–628. doi:10.1007/BF03175155

Zobayed SMA, Armstrong J, Armstrong W (1999a) Cauliflower-shoot culture: effects of different types of ventilation on growth and physiology. Plant Sci 141:209–217

Zobayed SMA, Armstrong J, Armstrong W (1999b) Evaluation of a closed system, diffusive and humidity induced convective throughflow ventilation on the growth and physiology of cauliflower *in vitro*. Plant Cell Tissue Organ Cult 59:113–123

Zobayed SMA, Armstrong J, Armstrong W (2001) Leaf anatomy of *in vitro* tobacco and cauliflower plantlets as affected by different types of ventilation. Plant Sci 161:537–548

Chapter 10
Propagation and Bioreactor Technology of Medicinal Plants: Case Studies on Paclitaxel, 10-Deacetylbaccatin III, and Camptothecin

Shu-Hwa Chang, Cheng-Kuen Ho, and Fen-Hui Chen

Abstract The article consists of the studies carried out in the Forest Research Institute, Taiwan, for the production of commercially high-value compounds such as paclitaxel, 10-deacetylbaccatin III (10-DAB), and camptothecin (CPT). Two strategies were followed to produce these compounds: by cultivation of plants in a farm and by cell and hairy root cultures in suitable bioreactors. Both strategies essentially need a selection of target compounds and elite materials from natural genetic resources followed by optimization of efficient cultivation practices in a farm and establishment of an efficient in vitro propagation system by application of biotechnological tools including bioreactors. The production of medicinal plants in a farm is easier to farmers in contrast to bioreactors which often require high investment, carry high risk, and pose greater technical barriers. It was found that cell or hairy root cultures of *Taxus*, *Camptotheca*, and *Nothapodytes* grow well in wave bioreactor and thus may be a better approach for small and medium drug enterprises in Taiwan.

Keywords Bioreactors • Camptothecin • 10-Deacetylbaccatin III • Medicinal plants • Propagation systems • Paclitaxel

Abbreviations

2,4-D	2,4-Dichlorophenoxyacetic acid
10-DAB	10-Deacetylbaccatin III
BA	6-Benzyladenine
cGMP	Current good manufacturing practice
CPT	Camptothecin
DBAT	10-Deacetylbaccatin III-10-O-acetyltransferase

S.-H. Chang • C.-K. Ho (✉) • F.-H. Chen
Taiwan Forestry Research Institute, 53 Nanhai Rd, Taipei, Taiwan, Republic of China
e-mail: ckho@tfri.gov.tw

H.-S. Tsay et al. (eds.), *Medicinal Plants - Recent Advances in Research and Development*, DOI 10.1007/978-981-10-1085-9_10

DBH	Diameter at breath height
FDA	Food and Drug Administration
HCl	Hydrogen chloride
IAA	Indole-3-acetic acid
IBA	Indole-3-butyric acid
MJ	Methyl jasmonate
MS medium	Murashige and Skoog medium
NAA	1-Naphthaleneacetic acid
PVP	Polyvinylpyrrolidone
TDZ	Thidiazuron (N-phenyl-N′-1, 2, 3-thiadiazol-5-ylurea)
WPM	Woody plant medium

10.1 Introduction

There are several reviews on both micropropagation of medicinal plants (Máthé et al. 2015; Sharma and Vashistha 2015; Yaadwinder 2010) and production of secondary metabolites by bioreactors (Orhan 2012; Paek et al. 2014). The use of bioreactor to produce valuable drugs, featuring high efficiency, reasonable costs, and eco-friendly, is regarded as a better system than farm production (Georgiev 2014). However, high cost of commercial bioreactor at development stage remains a cause of concern. Murthy and Lee (2014) indicated that the cost of producing adventitious ginseng roots in farm cultivation was $35 USD/kg which is lower than that of $47 USD/kg in a bioreactor. Lehmann et al. (2014) pointed out that disposable bioreactors allowing rapid and cheap development and manufacture of products may be a viable option.

A successful commercial application process includes selection of compounds from genetic resources, propagation for preservation and cultivation, selection and preservation of callus and root lines, evaluation of the performance of cultures transformed with foreign genes, detection of long-term somaclonal variation in cultures, and design of bioreactors suitable for different culture systems. When comparing the efficiency of different production methods, several important considerations such as production level of raw materials, target compounds, cost involved, and time frame of production need to be addressed. In the present article, two production strategies, farming and tissue culture technologies and bioreactors for the production of three high-value compounds, paclitaxel, 10-deacetylbaccatin III (10-DAB), and camptothecin (CPT), have been discussed.

10.2 Strategies for the Production of Valuable Medicinally Important Compounds

Valuable medicinally important compounds such as 10-DAB, paclitaxel, and CPT exist in a great variation in plants. The study begins with the selection of target compounds in plants followed by the establishment of propagation methods both in field and under controlled laboratory conditions. The latter consists of various procedures like cell suspension/callus/hairy root cultures, genetic transformation *in planta* and of biosynthetic pathways, and production in bioreactors. A summary of two approaches for production of target plants and important compounds is shown in Flowchart 10.1. To ensure the production can work at the industrial level instead of remaining at the laboratory stage, we suggest to use the two approaches simultaneously.

In the first step, an elite plant having high yield of target compound and good growth potential is identified, and then further procedures are followed step by step as shown in Flowchart 10.1. The vegetative propagation via cutting is explored first. If the growth of rooted cuttings (stecklings) is very slow, then grafting or tissue culture is explored next. Several propagation methods can usually be carried out simultaneously to preserve plant material, especially if the selected target plants are located in remote areas and are difficult to be collected again. For clonal test, planting of clones at different sites is important to test the genetic stability of plants and the biomass increment. Once the superior clones are selected, mass propagation techniques via cutting and tissue culture are followed. Establishment of scion gardens and cryopreservation of shoot tip in vitro can avoid the risk of loss of germplasm. Callus lines derived from different plant sources are examined according to their biomass increment and concentration of target compounds. Further selection in liquid cultures is carried out by determining the patterns of cell growth cycles, the changes in the concentration of target compound via additives or elicitors, and the

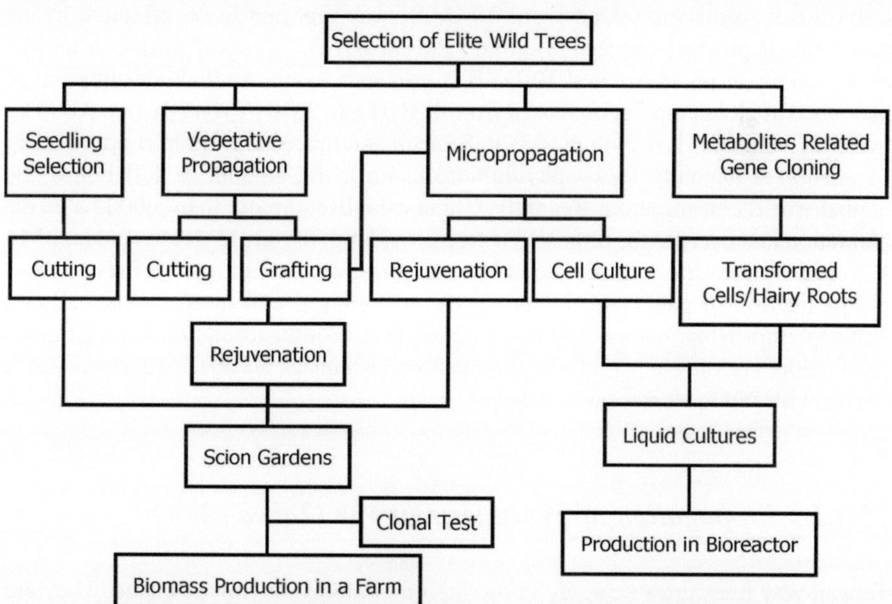

Flowchart 10.1 Strategies for the production of valuable medicinally important compounds by farming and under controlled laboratory conditions

stability of cultures after subculture for a long duration. Once the stable cell lines are confirmed, these are ready for scale-up culture in a bioreactor to evaluate the feasibility of commercial production. Both original clones and callus lines preserved at low temperature in the modified medium may be adapted for long-term storage. The possibility of occurrence of somaclonal variations in liquid culture is high. Therefore, any variation in cell growth and target compound production in each batch culture in the bioreactor has to be checked at regular intervals.

To avoid somaclonal variation, cell or root lines transformed with *Agrobacterium tumefaciens* or *A. rhizogenes* that can grow rapidly and relatively stable in hormone-free medium are desirable. Cultures transformed with specific genes related to biosynthesis usually enhance the production of target compounds.

10.3 Production of Paclitaxel and 10-DAB

In 1992, paclitaxel was found to be effective in treating ovarian cancer, while 10-DAB converted into docetaxel via semisynthesis to treat the ovarian cancer was approved in 1995. Annual sale of taxol was 1.5 billion USD in 1999, which decreased to 1 billion USD in 2011 (Malik et al. 2011). The value of docetaxel was $3.1 billion USD in 2010 (en.wikipedia.org/wiki/Docetaxel). In the years 2005–2013, abraxane, a protein-bound paclitaxel, was approved to treat breast cancer, non-small cell lung cancer, and advanced pancreatic cancer (Kratz 2014). Cabazitaxel, semisynthesized from 10-DAB, was reported to be effective in the treatment of prostate cancer (Tyagi and Prasad 2015). In cancer drug market, the requirement for paclitaxel and 10-DAB is ever increasing. Annual consumption of paclitaxel at global market increased from 1,310 kg in 2010 (Yvon 2013) to 3,000 kg in 2007 (Nhat 2011). Although Phyton Biotech announced that it can supply 500 kg of paclitaxel annually (www.phytonbiotech.com), the amount is still below the annual world consumption. Recently, China established more than 3,000 ha *Taxus* plantation and became the largest yew tree provider in the world (Nosov et al. 2014). Nevertheless, several recent reviews have emphasized the continuity of research on the production of paclitaxel, including the selection of cell lines, callus, and cell cultures, improving method for taxane production, biotransformation strategies, and production via different kinds of bioreactors (Khani et al. 2012; Li et al. 2015; Malik et al. 2011; Nosov et al. 2014).

10.3.1 Propagation of Taxus sumatrana *(Taiwan Yew)*

Taiwan yew trees grow scarcely in the high mountains on Taiwan Island. They are giant trees about 30 m high and 1 m in DBH (diameter at breath height). In our study, shoots were collected from 35 wild trees. Most germplasm of these trees were preserved in vitro and could be traced to their original mother trees in the wild. The age of these trees was estimated to be above 1000 years. Most materials were used for propagation by cutting, grafting, and culturing in vitro. Small parts were ground and extracted using methanol to determine their taxane concentrations (Ho et al. 1997). The variation of 10-DAB and taxol contents among different rooted clones was significant. Taiwan yew trees are rich in 10-DAB. Some clones contain very high contents of these two compounds. Because the cuttings were from very old trees, rooting of shoots was very poor, and rooted cuttings grew slowly. To propagate hard-to-root trees, regrafting scions onto capped seedlings and shoot tip cultures were carried out. To improve rooting rate and quality, the base of leafy shoots was dipped in *Agrobacterium rhizogenes* suspension cultures for 10 s, then packed them in a sealed PE plastic bag, and incubated for 2 days in a growth chamber at 25 °C. Rooting rates, numbers, and length of treated cuttings were found to be two to three times greater than those treated with indole-3-butyric acid (IBA) (Chen et al. 2003).

Tissue culture can rejuvenate aged materials from old trees, while regrafting shoots cannot overcome plagiotropism. Based on the literature, bud explants derived from wild trees should be collected in spring and cultured on 1/2MS (Murashige and Skoog) medium supplemented with 1 g/L activated charcoal and 100 mg/L silver nitrate (Chang et al. 2001). Immature embryos isolated from seeds with green aril were observed to germinate after 7 days in culture on 1/2 MS medium supplemented with PVP (polyvinylpyrrolidone) (Chang and Yang 1996).

10.3.2 Production of Paclitaxel and 10-DAB by Farming

In Taiwan, farming yew trees by harvesting branches annually to obtain compounds paclitaxel and 10-DAB has been carried out for the last 14 years. Two *Taxus* varieties, named as Taiwan Yew No.1 for producing both paclitaxel and 10-DAB and Taiwan Yew No.2 for producing 10-DAB, were selected. Mass propagation was carried out by planting 10-cm long shoot cutting in plug trays after treating with 3 mg/L of IBA (Fig. 10.1A). Rooted cuttings were transferred into 2.5 × 14 cm pots for 1 year (Fig. 10.1B) and then planted in the field at a density of 12,000 stecklings/ha. Branches were harvested using a tea harvester machine at the height of 50 cm above trees after planting for 3 years when trees grew about 1 m tall (Fig. 10.1C). We could harvest about 5 tons dry matter/ha annually for more than 10 years at one farm. On average, 1 ton dry matter of plant material produced 0.3–0.4 kg of paclitaxel and 1.5–2 kg of 10-DAB. The break-in period for the investment is approximately 5 years after planting in a farm. Due to the enormous economic value of these two compounds, licensees are protected under the certification of plant variety right in Taiwan.

Since cultivation can easily suffer from an adverse climate in a large open area, agroforestry system is a viable option. In our experience, annual biomass production was stable under agroforestry system, even after several attacks of typhoon, heavy rain, and drought (Fig. 10.1D).

Fig. 10.1 Farming system of Taiwan yew. (**A**) Mass propagation by cutting in plug trays; (**B**) stecklings raised in a nursery; (**C**) harvesting of branches using a tea harvester machine; (**D**) agroforestry system to resist adverse climate

10.3.3 Production of Paclitaxel and 10-DAB in a Bioreactor

10.3.3.1 Cell Cultures in a Bioreactor

We have selected several callus lines with rapid growth and high taxane content from cultures derived from different tissue explants and genetic resources (Chang et al. 1996). Callus lines were selected further by culturing at different inoculum volumes in flasks containing various combinations of plant growth regulators (Chang et al. 2004). We observed that methyl jasmonate (MJ) treatment increased the production of 10-DAB, baccatin III, and paclitaxel. This is in conformity with the result indicated in the reviews of Malik et al. (2011) and Murthy and Lee (2014). Both paclitaxel and baccatin III changed greatly in different compositions of media when MJ was added on day 14 during the culture period in 25-ml flask cultures. These callus cultures produced higher baccatin III content than the paclitaxel and 10-DAB. More than 67 % of baccatin III was found to be released into cultural medium. The cell cultures grew rapidly when scaled-up in larger flasks, a bubble column, or stirred tank bioreactors. The maximum volume of cell cultures in a 2-L stirred bioreactor reached at the 20th week when 100 ml of cells was used as an

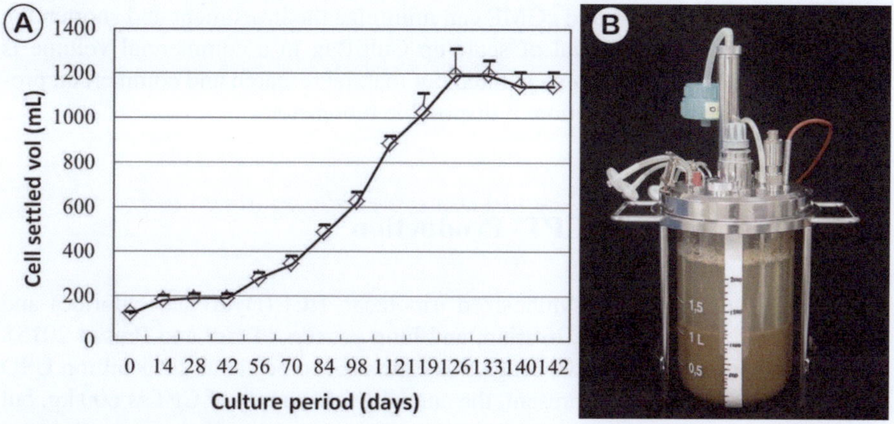

Fig. 10.2 (**A**) The cell growth cycle measured in a cell settled volume in three 2-L stirred bioreactors; (**B**) cell cultures in a bioreactor

initial inoculum (Fig. 10.2A, B) (unpublished data). However, the production of taxanes reduced gradually after continuous subculture for 3 years. It could be due to somaclonal variation, decreasing transcript level of related genes as observed by Li et al. (2009) and/or DNA methylation (Fu et al. 2012). To overcome this problem, the callus and shoots which could induce the best cell lines were cultured in a medium containing ABA or mannitol and incubated in cold at 4 °C and 12 °C, respectively, for more than 1.5 years. These cultures regained the vigor and production of taxanes (Chang et al. 2005a). Another method is to obtain transgenic cells harboring auxin gene or paclitaxel synthesis-related genes with 35S promoter. The cells transformed with auxin gene derived from *A. tumefaciens* could grow in hormone-free medium that can result in high biomass, and taxanes yield for more than 5 years (unpublished data). The concentration of IAA (indole-3-acetic acid) in transgenic callus was found to be 12 times greater than normal ones (unpublished data). Transgenic cell cultures harboring 10-deacetyl baccatin III-10-O-acetyltransferase (DBAT) gene required MJ to enhance the production of taxanes (Ho et al. 2005).

10.3.3.2 Hairy Root Cultures in a Bioreactor

Hairy root cultures are more stable and resistant to stress than cell cultures. Several transgenic root lines transformed with *A. tumefaciens* 281 and *A. rhizogenes* 1600, 1601, and 15834 were obtained. A significant variation in concentrations of taxanes among hairy root lines was observed. Hairy roots grew rapidly in 580-ml and 2-L airlift bioreactors. We observed that biomass increased five times after culturing for 11 weeks (unpublished data). Compared to farming, a bioreactor system requires more time to set up, skilled technicians, more regulations to meet, and higher risks of scale-up failure. To overcome these barriers, disposable bioreactors or "CultiBags" by the regulations of FDA and cGMP can minimize the investment and shorten the production time. The potential of scale-up CultiBag in a commercial volume is high. Lehmann et al. (2014) also pointed out that the research and commercial production are going in the direction of disposable bioreactors.

10.4 Camptothecin (CPT) Production

CPT is a precursor of semisynthesized irinotecan HCl (Hydrogen chloride) and topotecan HCl to treat colon, ovarian, and lung cancers (Tyagi and Prasad 2015). The world market for these two drugs was estimated to be about 2.2 $ billion USD in 2008 (Cui et al. 2015). At present, the annual world supply of CPT is 600 kg, but the demand is estimated to be more than 3,000 kg annually. The source plants of CPT are *Camptotheca acuminata* and *Nothapodytes nimmoniana* (Cui et al. 2015). Because of overharvesting from wild populations, these two tree species are listed as endangered species in China (Li et al. 2014) and in India (Sivakamasundari et al. 2015). Unlike paclitaxel production, CPT production relies mainly on farm cultivation, though there are some successful instances of scale-up bioreactor production.

Li (2014) introduced a strategy to develop *Camptotheca* as pharmaceutical crops in the USA. In Taiwan, between 1994 and 2005, about 200 ha of *N. nimmoniana* was planted by Yakult Co. of Japan to produce CPT. It was reported that the company earned about $2 billion USD annually from the *N. nimmoniana* plantation. In 2003, we began to survey the genetic resources of *N. nimmoniana* and cooperated with local biotech companies in the farming of *Camptotheca*. Also, a comparative study was carried out to evaluate the performance of a farm and bioreactor in the production of compounds from these two species.

10.4.1 Production of CPT from Camptotheca acuminata

10.4.1.1 Propagation of *Camptotheca acuminata*

In Taiwan, *C. acuminata* plants introduced from China in 1949 were planted in the Taipei Botanical Garden. Currently, these trees remain the main source of planting materials of *Camptotheca* cultivation at different sites in Taiwan. In our study, seedlings from four plantations were raised for 1 year at the Taipei Botanical Garden. Then shoots (10 cm long) were collected monthly to analyze CPT contents in them. It was observed that the maximum CPT contents were found in the shoots collected in the month of May to July (unpublished). Out of the four plantations, shoots from L2 plantation showed the highest CPT content.

Commercial propagation of *Camptotheca* uses seedlings since seed collection and growth in nursery are easier. Rooting of cuttings derived from aged *Camptotheca*

Fig. 10.3 Commercial plantation of *Camptotheca*. (**A**) One-year-old plantation; (**B**) manual harvesting of young shoots

trees is hard to achieve. However, cuttings from young potted seedlings easily root as described by Trueman and Richardson (2011). Micropropagation of *Camptotheca* has been reported by Lineberger et al. (1998), Liu and Li (2001), Chang et al. (2005b), and Li and Liu (2005).

In our study (Chang et al. 2005b), in vitro seedlings could be raised by inoculation of disinfected seeds on WPM (woody plant medium) medium and incubation of 6–8 weeks. Shoot tips and stem explants derived from these seedlings induced an average of 6.4 shoot buds per explant when cultured in WPM medium containing 0.01 mg/L NAA (1-naphthaleneacetic acid) and 0.5 mg/L 6-benzyladenine (BA). Elongated shoots induced rooting on WPM medium containing 1 mg/L IBA. To identify a CPT-rich clone, we analyzed the CPT contents in in vitro seedlings derived from three mother trees and their in vitro shoot clones, respectively. It was observed that the CPT contents varied in the range of 0.9–2 mg/g dry weight in the seedlings from the same mother tree. These suggest that there is a possibility of identifying superior clones having higher contents of CPT for further mass propagation by tissue culture.

10.4.1.2 Cultivation of *Camptotheca* in a Farm for the Production of CPT

In a commercial farm of *Camptotheca*, the planting density was about 10,000 plants/ha (Fig. 10.3A). After planting of 1 year, 3-cm long shoots were collected (Fig. 10.3B). CPT yield per ha from shoot dry biomass was estimated to be about 2.58 kg (data from Dr. Yu HM, personal communication). However, CPT content significantly decreases during the process of drying of plant materials; therefore, in a commercial production, time taken in drying process needs to be reduced considerably. That labor cost for harvesting young shoots is very high, and so far, there is no mechanical automation device for harvesting young shoots.

10.4.1.3 Production of CPT in a Bioreactor

10.4.1.3.1 Cell Cultures in a Bioreactor

Callus cultures induced from leaf explants of *Camptotheca* cultured in WPM medium containing 1–3 mg/L 2,4-D (2,4-Dichlorophenoxyacetic acid) or NAA could grow rapidly (Chang et al. 2006). CPT content in callus cultured in medium with NAA (up to 0.307 mg/g dry weight) was greater than in the callus grown in the medium with 2,4-D (up to 0.069 mg/g dry weight). When callus was continuously subcultured for 3 years, CPT content decreased, especially in the medium containing 2,4-D.

Suspension cultures of *C. acuminata* were grown in a medium with 2,4-D, while for nodule cultures medium in containing NAA was better (Chang et al. 2007).

Production of CPT in Hairy Root Cultures of *Camptotheca acuminata*

In a study by Liu and Cui (2007), hairy roots of *Camptotheca acuminata* produced 0.14 mg/g dry weight of CPT in the medium after 30 day of incubation. In another study by Lorence et al. (2004), hairy root cultures of *C. acuminata* were established from tissue transformed with *A. rhizogenes* strains ATCC 15834 and R-1000. Integration of the genes responsible for the hairy root phenotype (rol genes) into the plant genome was verified by DNA gel blot analysis. The hairy roots produce and secrete CPT into the medium. The cultures were able to synthesize the CPT equal to, and sometimes greater than, the roots in planta, i.e., 1.0 mg/g dry weight. Ni et al. (2011) obtained high CPT content (1.12 mg/g dry weight) in hairy root cultures. ORCA3 gene, a jasmonate-responsive APETALA2-domain transcript factor, was introduced into the hairy roots. When cultured in 250-ml flasks for 60 days, the hairy roots produced CPT content 1.5 times higher than control hairy roots (without ORCA3 gene). These studies show that hairy root cultures have potential of producing higher CPT contents than cell cultures. However, so far there is no report on further scale-up of hairy root culture of *Camptotheca acuminata* in a bioreactor.

10.4.2 Production of CPT from Nothapodytes nimmoniana

10.4.2.1 Propagation of *Nothapodytes*

Nothapodytes nimmoniana grows in India, Ceylon, and Cambodia and from Southern China to Ryukyu Islands. It is native to Orchid and Green islands of Taiwan. The planting density used in commercial plantations is about 24,000 seedlings/ha. To meet the demand of seedlings in Taiwan, we established two seed orchards of elite plants with higher contents of compounds. To propagate these elite trees or its seedlings, the techniques of rooted cuttings and micropropagation procedures were established.

Chang et al. (2008) optimized a micropropagation protocol of *Nothapodytes*. Shoot tip explants cultured in MS medium containing 0.25 mg/L TDZ (thidiazuron)

Fig. 10.4 Commercial plantation of *Nothapodytes*. (**A**) 1.5-year-old plants; (**B**) plants pruned to 50 cm height (*arrow*); (**C**) plants with higher number of branches after 1 year of harvest

for 10 weeks induced more than 20 shoots per explant. Elongation of these shoots was achieved in MS medium containing 0.05 mg/L BA (Benzyladenine) and incubation for 1 month. Rooting in these shoots was achieved in the MS medium containing 0.5 mg/L IBA or NAA (Chang et al. 2008; Rajasekharan et al. 2010). Induction of somatic embryogenesis in *Nothapodytes* has been reported (Rajasekharan et al. 2010; Khadke and Kuvalekar 2013). Using MS medium containing 0.2 mg/L TDZ to culture embryos can induce somatic embryos successfully (Rajasekharan et al. 2010). Leaf and stem explants obtained from nursery-grown plants induced somatic embryos in a medium with TDZ concentration of 0.5–3.0 mg/L (Khadke and Kuvalekar 2013). Thus, micropropagation of superior clones is more effective than the propagation by cutting.

10.4.2.2 Production of CPT by Farming

Improved seedlings obtained from seed orchards were used to establish high-density *Nothapodytes* plantation (24,000 plants/ha). Plants grew to 1.5 m tall after 1.5 years of planting (Fig. 10.4A). These plants were pruned to the 50 cm height to collect branches (Fig. 10.4B). The first harvest collected weighed 0.7–4.4 tons/ha of dry biomass and 0.6–6.6 kg/ha of CPT, depending on clones (Ho et al. 2007). In the second harvest, plants had higher number of branches (Fig. 10.4C). It was reflected in the higher dry biomass (13 kg/ha) and a higher CPT content (13.8 kg/ha). The yield of CPT was maintained for four harvests. Thus, it seems that *Nothapodytes* is easier to grow as a medicinal crop than the *Camptotheca*.

10.4.2.3 Production of CPT in Tissue Cultures Including Bioreactor

10.4.1.3.1 Cell Cultures of *Nothapodytes*

Cotyledon, hypocotyl, and stem explants of *Nothapodytes* induced nodular calli when cultured in MS medium containing 2 mg/L NAA and 0.5–1 mg/L BA, while friable calli were observed in medium with 2,4-D (Tsay et al. 2008). A similar result was reported by Fulzele et al. (2001). Nodular calli in both solid and liquid media grew faster than the friable calli. Also, only nodular calli had CPT content in the range of 0.106–0.331 mg/g dry weight (Tsay et al. 2008). In another study, Karwasara and Dixit (2013) devised a modified MS medium containing 0.5 mM phosphate, a nitrogen source feeding ratio of 50/10 mM NH_4^+/NO_3^-, and 3 % sucrose with additional 2 % sucrose feeding (added on day 12 of the cell culture cycle) with 2 mg/L NAA and 0.2 mg/L kinetin. Finally, the selective medium has 1.7- and 2.3-fold higher intracellular and extracellular camptothecin content over the control culture (0.029 and 0.0082 μg/g dry weight), respectively.

10.4.2.3.2 Hairy Root Cultures of *Nothapodytes* in a Bioreactor

Hairy roots of *Nothapodytes* were obtained by infection of stem and leaf explants with *A. rhizogenes* strains. Root lines cultured in MS medium without any plant growth regulators showed rapid growth and high CPT contents of concentration (0.537–1.555 mg/g) (Chang et al. 2014). Liquid cultures in flasks containing the same medium for 30 days produced 9.8 mg/L CPT, in which 0.6 mg/L (6 %) was in roots and remaining 9.2 mg/L (94 %) in the medium. The CPT concentration detected in the medium was about 1.048 mg/g based on dry weight of hairy roots. This CPT yield is comparative to that produced by the hairy root culture of *Ophiorrhiza pumila* (Cui et al. 2015).

The results of three hairy root culture bioreactors, i.e., stirred, airlift, and wave bioreactors, were compared after 5 months of cultivation. Although hairy roots could resist high shear stress of stirred bioreactor (Fig. 10.5A, D), they grew faster in airlift (Fig. 10.5B, E) and wave bioreactor (Fig. 10.5C, F) because of lower stress. The space in the vessels was occupied by high-density roots. The average fresh biomass increment was 2.2 g/L/day in an airlift bioreactor and 4.8 g/L/day in a wave bioreactor (unpublished data). In another study, Sudo et al. (2002) recorded biomass increment of 1.2 g/day/L (from 10 to 218 g for 8 weeks) in *O. pumila* root cultured in a 3-L stirred bioreactor. CPT content in hairy roots in a 10-L airlift bioreactor was 2.135 mg/g, indicating its potential of commercial application.

10.5 Conclusions

This article describes the production of three high-value compounds paclitaxel, 10-deacetylbaccatin III (10-DAB), and camptothecin (CPT) in the Forest Research Institute in Taiwan. Two strategies like cultivation of plants in farms and biotechnological tools like cell and tissue cultures including bioreactor studies have been briefly described. Studies carried out by other researchers to propagate these plants

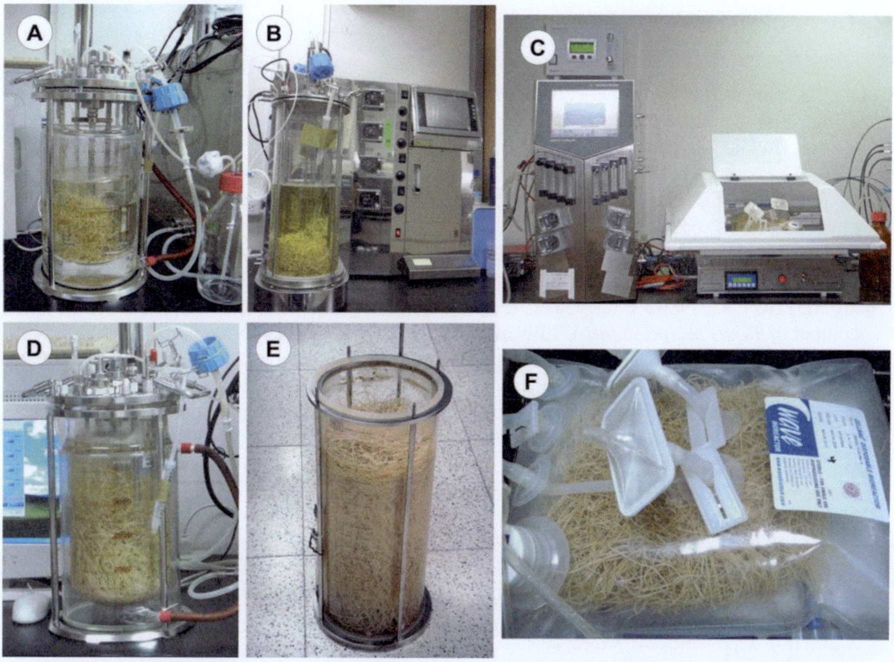

Fig. 10.5 Different bioreactors showing initial and final biomass of hairy roots after 5 months. (**A**) 5-L stirred bioreactor (initial biomass 5 g); (**B**) 10-L airlift bioreactor (initial biomass 10 g); (**C**) 1-L wave bioreactor (initial biomass 5 g); (**D**) 5-L stirred bioreactor after 5 months (final biomass 675 g); (**E**) 10-L airlift bioreactor after 5 months (final biomass 3,284 g); (**F**) 1-L wave bioreactor after 5 months (final biomass 732 g)

and to produce these three compounds also have been reviewed. Both strategies, cultivation in a farm and biomass production in culture systems including bioreactors, have its share of merits and demerits. No single strategy works effectively for all the plants and the production of every medicinally important compound. In our studies, we found that cell or hairy root cultures of *Taxus*, *Camptotheca*, and *Nothapodytes* grow well in wave bioreactor, which is suitable for small and medium drug enterprises in Taiwan. Although bioreactor production has several advantages, however, the production of paclitaxel by farming of yew crops and CPT from farming of *Nothapodytes* is still the major source for these two high-value compounds in Taiwan at present. Farm and bioreactor production systems may coexist just like the production of paclitaxel and ginseng. Finally, as the global demand of herbal products and compounds is increasing, so is the innovation in technologies. Combinations of technologies will open new avenues of production of plants and pharmaceutically high-value compounds.

References

Chang SH, Yang JC (1996) Enhancement of plant formation from embryo cultures of *Taxus mairei* using suitable culture medium and PVP. Bot Bull Acad Sin 37(1):35–40

Chang SH, Ho CK, Tsay JY (1996) Callus initiation, growth, and taxol production of *Taxus mairei*. Taiwan J For Sci 11(4):445–453

Chang SH, Tsay JY, Chen ZZ, Ho CK (2001) Micropropagation of *Taxus mairei* from mature trees. Plant Cell Rep 20:491–502

Chang SH, Ho CK, Tsay JY (2004) Cell cultures and taxane production of *Taxus mairei*. Taiwan J For Sci 19(1):11–20

Chang SH, Ho CK, Tsay JY (2005a) Effect of cold storage on the survival, growth, and taxane content of *Taxus mairei* in vitro. Taiwan J For Sci 21(1):49–59

Chang SH, Tsay JY, Ho CK (2005b) Micropropagation and evaluation of the camptothecin content in *Camptotheca acuminata*. Taiwan J For Sci 20(4):331–340

Chang SH, Tsay JY, Ho CK, Huang CY (2006) Callus culture and camptothecin production in *Camptotheca acuminata*. Taiwan J For Sci 21(4):513–521

Chang SH, Ho CK, Tsay JY, Chen KF, Huang CY (2007) Crown gall culture and camptothecin production of *Camptotheca acuminata*. Taiwan J For Sci 22(4):395–404. doi:10.1007/s002990100362

Chang SH, Tsay JY, Yeh YC, Ck H (2008) Micropropagation of *Nothapodytes nimmoniana*. Q J Chin For 41(1):17–27

Chang SH, Chen FH, Tsay JY, Chen J, Huang CY, Lu WL, Ho CK (2014) Establishment of hairy root cultures of *Nothapodytes nimmoniana* to produce camptothecin. Taiwan J For Sci 29(3):193–204

Chen KP, Ho CK, Kuo SR (2003) Enhancement of *Agrobacterium rhizogenes* on root formation of *Taxus mairei* cuttings of mature trees. Taiwan J For Sci 18(3):213–223

Cui L, Ni X, Ji Q, Ten X, Yang Y, Wu C, Zekria D, Zhang D, Kai G (2015) Co-overexpression of geraniol-10-hydroxylase and strictosidine synthase improves anti-cancer drug camptothecin accumulation in *Ophiorrhiza pumila*. Sci Rep 5:8227. doi:10.1038/srep08227

Fu C, Li L, Wu W, Li M, Yu X, Yu L (2012) Assessment of genetic and epigenetic variation during long-term *Taxus* cell culture. Plant Cell Rep 31:1321–1331. doi:10.1007/s00299-012-1251-y

Fulzele DP, Satdive RK, Pol BB (2001) Growth and production of camptothecin by cell suspension cultures of *Nothapodytes foetida*. Planta Med 67:150–152. doi:10.1055/s-2001-11519

Georgiev M (2014) Design of bioreactors for plant cell and organ cultures. In: Paek KY, Murthy HN, Zhong JJ (eds) Production of biomass and bioactive compounds using bioreactor technology. Springer, New York, pp 3–17. doi:10.1007/978-94-017-9223-3_1

Ho CK, Chang SH, Chen ZZ (1997) Content variation of taxanes in needles and stems of *Taxus mairei* trees naturally distributed in Taiwan. Taiwan J For Sci 12(1):23–37

Ho CK, Chang SH, Lung JH, Tsai CJ, Chen KP (2005) The strategies to increase taxol production by using *Taxus mairei* cells transformed with TS and DBAT Genes. Int J Appl Sci Eng 3(3):179–185

Ho CK, Chang SH, Tsai, Chen KP, Wu JT, Hung JI (2007) Camptothecin production from superior clones of *Nothapodytes nimmoniana*, Series of Books in Taiwan Forestry 180. Taiwan Forestry Research Institute, Taipei, pp 141–148

Karwasara VS, Dixit VK (2013) Culture medium optimization for camptothecin production in cell suspension cultures of *Nothapodytes nimmoniana* (J. Grah.) Mabberley. Plant Biotechnol Rep 7:357–369. doi:10.1007/s11816-012-0270-z

Khadke S, Kuvalekar A (2013) Direct somatic embryogenesis and plant regeneration from leaf and stem explants of *Nothapodytes foetida*: a critically and endangered plant species. Int J Plant Animal Environ Sci 3:257–264. doi:10.1016/j.scienta.2010.06.016

Khani S, Barar J, Movafeghi A, Omidi Y (2012) Production of anticancer secondary metabolites impact of bioprocess engineering. In: Orhan IE (ed) Biotechnological production of plant secondary metabolites. Bentham Science Publishers, Bentham e Books, Dubai, pp 215–240. doi:10.2174/978160805114411201010215

Kratz F (2014) A clinical update of using albumin as a drug vehicle—a commentary. J Control Release 190:331–336. doi:10.1016/j.jconrel.2014.03.013

Lehmann N, Dittler I, Lämsä M, Ritala A, Rischer H, Eibl D, Oksman-Caldentey KM, Eibl R (2014) Disposable bioreactors for cultivation of plant cell cultures. In: Paek KY, Murthy HN, Zhong JJ (eds) Production of biomass and bioactive compounds using bioreactor technology. Springer, New York, pp 17–44. doi:10.1007/978-94-017-9223-3_2

Li S (2014) Development of *Camptotheca decaisne* as pharmaceutical crops. Pharm Crops 5:85–88. doi:10.2174/2210290601405010085

Li Z, Liu Z (2005) Plant regeneration from leaf petioles in *Camptotheca acuminata*. In Vitro Cell Dev Bio-Plant 41(3):262–265. doi:10.1007/s11627-001-0016-z

Li LQ, Fu CH, Zhao CF, Xia J, Wu WJ, Yu LJ (2009) Efficient extraction of RNA and analysis of gene expression in a long-term *Taxus* cell culture using real-time RT-PCR. Z Naturforsch 64:125–130. doi:10.1515/znc-2009-1-220

Li S, Zhang W, Northrup K, Zhang D (2014) Distribution of *Camptotheca* Decaisne: endangered status. Pharm Crops 5:135–139. doi:10.2174/2210290601405010135

Li Y, Zhang G, Pfeifer BA (2015) Current and emerging options for taxol production. Adv Biochem Eng Biotechnol 148:405–425. doi:10.1007/10_2014_292

Lineberger D, Reed D, Rumpho M (1998) Micropropagation of *Camptothecin acuminata*. Hort Sci 33(4):604

Liu Z, Cui Y (2007) Study on hairy root cultures of *Camptotheca acuminata* (Nyssaceae) and the camptothecin (CPT) from medium. Ecol Environ 4:1266–1270

Liu Z, Li Z (2001) Micropropagation of *Camptotheca acuminata* decaisne from axillary buds, shoot tips, and seed embryos in a tissue culture system. In Vitro Cell Develop Bio-Plant 37(1):84–88. doi:10.1007/s11627-001-0016-z

Lorence A, Medina F, Nessler CL (2004) Camptothecin and 10-hydroxy- camptothecin from *Camptotheca acuminata* hairy root. Plant Cell Rep 22:437–441. doi:10.1007/s00299-003-0708-4

Malik S, Cusidó RM, Mirajalili MH, Moyano E, Palazón J, Bonfill M (2011) Production of the anticancer dry taxol in *Taxus baccata* suspension cultures: a review. Process Biochem 46:23–34. doi:10.1016/j.procbio.2010.09.004

Máthé Á, Hassan F, Kader AA (2015) In vitro micropropagation of medicinal and aromatic plants. In: Máthé Á (ed) Medicinal and aromatic plants of the World. Springer, Dordrecht, pp 305–336. doi:10.1007/978-94-017-9810-5_15

Murthy HN, Lee EJ (2014) Production of secondary metabolites from cell and organ cultures: strategies and approaches for biomass improvement and metabolite accumulation. Plant Cell Tiss Organ Cult 118:1–16. doi:10.1007/s11240-014-0467-7

Nhat HN (2011) NMR studies in nanoparticle drug delivery systems: paclitaxel loaded nanocarriers. In: Atta-ur-Rahman, Choudhary I (eds) Structure-activity relationship studies in drug development by NMR spectroscopy. Bentham Science Publishers, Bentham e Books, Dubai, pp 184–213. doi:10.2174/978160805164910184

Ni X, Wen S, Wang W, Wang X, Xu H, Kai G (2011) Enhancement of camptothecin production in *Camptotheca acuminate* hairy roots by overexpressing ORCA3 gene. J Appl Pharm Sci 1(8):85–88

Nosov AM, Popova EV, Kochkin DV (2014) Isoprenoid production via plant cell cultures: biosynthesis, accumulation and scallion-up to bioreactors. In: Murthy HN, Zhong JJ, Paek KY (eds) Production of biomass and bioactive compounds using bioreactor technology. Springer, Dordrecht, pp 563–624. doi:10.1007/978-94-017-9223-3_23

Orhan IE (2012) Biotechnological production of plant secondary metabolites. Bentham Science Publishers, Bentham e Books, Dubai. doi:10.2174/978160805114411201010036

Paek KY, Hosakatte NM, Zhong JJ (2014) Production of biomass and bioactive compounds using bioreactor technology. Springer, New York. doi:10.1007/978-94-017-9223-3

Rajasekharan PE, Abdul Kareem VK, Vasantha KT (2010) Optimizations of protocol for in vitro multiplication and conservation of *Nothapodytes nimmoniana*, an endangered medicinal plant. Acta Hort (ISHS) 865:53–58. doi:10.17660/ActaHortic.2010.865.5

Sharma H, Vashistha BD (2015) Plant tissue culture: a biological tool for solving the problem of propagation of medicinally important woody plants- a review. Int J Adv Res 3:402–411

Sivakamasundari, Karuppusamy S, Parthipan R (2015) Survey on the RET-listed medicinal plants in Thadagamalai Range of Kanyakumari District, Tamilnadu. Biodivers Endanger Species 3:1–4. doi:10.4172/2332-2543.1000151

Sudo H, Yamakawa T, Yamazaki M, Aimi N, Saio K (2002) Bioreactor production of camptothecin by hairy root cultures of *Ophiorrhiza pumila*. Biotechnol Lett 24:359–363

Trueman SJ, Richardson DM (2011) Propagation and chlorophyll fluorescence of *Camptotheca acuminata* cuttings. J Med Plant Res 5:1–6

Tsay JY, Chang SH, Ho CK, Huang JU (2008) Induction, growth and CPT analysis of callus for *Nothapodytes nimmoniana*. In: Ho CK, Cheng CD, Liao SN (eds) Symposium of tree breeding and silviculture. Taiwan Forestry Research Institute, Taipei, pp 321–329

Tyagi AK, Prasad S (2015) Drug discovery inspired by mother nature for cancer therapy. Biochem Physiol 4, e128. doi:10.4172/2168-9652.1000e128

Yaadwinder S (2010) In vitro micropropagation of medicinal plants by tissue culture. Plymouth Stud Sci 4(1):432–449

Yvon D (2013) Paclitaxel and docetaxel leading the antineoplastic market. Health Articles. Available via http://www.articlesfactory.com/articles/health/paclitaxel-and-docetaxel-leading-the-antineoplastic-market.html. Accessed 16 May 2016

Chapter 11
Pharmacological Applications of Lucidone: A Naturally Occurring Cyclopentenedione

K.J. Senthil Kumar and Sheng-Yang Wang

Abstract *Lindera erythrocarpa*, commonly known as spicewood or spicebush, has a long history of use as a traditional remedy and culinary spice. Recent studies have reported that lucidone, an active constituent of fruits and leaves of *Lindera erythrocarpa*, has various beneficial properties, such as antioxidant, anti-inflammatory, hepatoprotective, dermatoprotective, hypolipidemic, and skin-whitening effects. The pleiotropic activities of lucidone derive from its unique chemistry as well as its ability to modulate multiple signaling pathways, such as inflammatory signaling pathways regulated by NF-κB and MAPKs; cytoprotective pathway that depends on Nrf2 activation and inhibition of apoptosis; hypolipidemic pathway regulated by PPRγ and C/EBPα; and anti-melanogenic pathway modulated by MITF. Also, lucidone is remarkably low cytotoxic and exhibits limited bioavailability. These findings suggest that lucidone is a promising agent for the treatment of inflammatory and oxidative diseases.

Keywords Anti-inflammation • Anti-melanogenesis • Dermatoprotection • Hepatoprotection • Hypolipidemic • *Lindera erythrocarpa* • Lucidone

Abbreviations

AAPH	2,2′-Azobis(2-amidinopropane) dihydrochloride
ALT	Alanine aminotransferase
AP-1	Activator protein-1
ARE	Antioxidant response element
AST	Aspartate aminotransferase

K.J. Senthil Kumar
Department of Forestry, National Chung Hsing University, Taichung, Taiwan

S.-Y. Wang (✉)
Department of Forestry, National Chung Hsing University, Taichung, Taiwan

Agricultural Biotechnology Research Center, Academia Sinica, Taipei, Taiwan
e-mail: taiwanfir@dragon.nchu.edu.tw

H.-S. Tsay et al. (eds.), *Medicinal Plants - Recent Advances in Research and Development*, DOI 10.1007/978-981-10-1085-9_11

ATF-2	Activating transcription factor-2
cAMP	Cyclic adenosine monophosphate
COX-2	Cyclooxygenase-2
EMSA	Electrophoretic mobility shift assay
ERK	Extracellular signal-regulated kinase
GSH	Glutathione
HF	High fat
HO-1	Heme oxygenase-1
IκB	Inhibitor of nuclear factor kappa-B
IKK	IκB kinase
iNOS	Inducible nitric oxide synthase
JNK	c-JUN N-terminal kinase
LDH	Lactate dehydrogenase
LPS	Lipopolysaccharide
MAPK	Mitogen-activated protein kinase
MDA	Malondialdehyde
MITF	Microphthalmia-associated transcription factor
MKK	Mitogen kinase kinase
NAFLD	Nonalcoholic fatty liver diseases
NF-κB	Nuclear factor kappa-B
NO	Nitric oxide
Nrf2	Nuclear factor E2-related factor-2
PGE_2	Prostaglandin-E_2
PPARγ	Peroxisome proliferator-activated receptor-γ
ROS	Reactive oxygen species
SAPK	Stress-activated protein kinase
TGF-β	Transforming growth factor-β
TNF-α	Tumor necrosis factor alpha
TRP	Tyrosinase-related proteins
UVA	Ultraviolet A
α-MSH	Alpha-melanocyte-stimulating hormone

11.1 Introduction

Phytochemicals, nonnutritive components existing in fruits, vegetables, edible macrofungus, algae, and bacteria, are increasingly gaining popularity over conventional synthetic drugs, primarily because they act via multiple molecular targets, which synergize to prevent efficiently or treat chronic disorders. Phytochemicals are safe with nontoxic or minimal toxic side effects with better bioavailability. It is known that the metabolism of plants is divided into two major types: primary and secondary. The substances that are common to living things and essential to cell maintenance such as lipids, proteins, carbohydrates, and nucleic acids are originated from the primary metabolism. Moreover, substances originated from several biosynthetic pathways that are restricted to determined groups of organisms are results of the secondary metabolism including, polyphenols, alkaloids, terpenes, steroids, saponins, flavonoids, lignans, tannins, cyclopentene diones, polysaccharides, fatty acids, and organic acids. Given the great structural diversity of phytochemicals, it is not feasible to define the structure-activity relationship to deduce their underlying molecular mechanisms. A better approach is to elucidate their medicinal properties by analyzing modulations in signal transduction pathways.

Cyclopentenediones (CPDs) are secondary metabolites found in higher plants, fungi, algae, and cyanobacteria. A common denominator of CPDs is the cyclopent-4-ene-1,3-dione skeleton with various functional groups. Most of the CPDs were primarily isolated from plants or macrofungus, although synthetic analogs with new biological properties and greater pharmacological efficacies were subsequently studied (Sevcikova et al. 2014). There are several pharmacologically important CPDs that have been investigated. Coruscanones A and B isolated from *Piper coruscans* exhibit potent antifungal properties (Babu et al. 2006). Asterredione isolated from *Aspergillus terreus* occurring in the rhizosphere of *Opuntia versicolor*, using bioassay-guided fractionation and the cytotoxic evaluation with a panel of three sentinel cancer cell lines, NCI-H460 (non-small cell lung cancer), MCF-7 (breast cancer), and SF-268 (CNS glioma), showed a moderate cytotoxic effects (Wijeratne et al. 2003). Involutone was isolated from the mushroom *Paxillus involutus* using ethanol, methanol, or n-butanol extracts (Antkowiak et al. 2003). Nostotrebin 6, a bis-cyclopentenedione isolated from a methanol extract of the cyanobacterial strain *Nostoc* sp. str. Lukesova 27/97, significantly inhibited acetylcholinesterase and butyrylcholinesterase activities with an IC_{50} value of 5.5 μM and 6.1 μM, respectively (Zelik et al. 2010). Linderone, methyllinderone, lucidone, and methyllucidone isolated from ethyl acetate extracts of fruits of *Lindera erythrocarpa* and lucidone exhibited potent anti-inflammatory effect against lipopolysaccharide-induced inflammation in murine macrophages RAW264.7 cells in vitro (Wang et al. 2008). Subsequent investigations with lucidone showed antioxidant (Kumar et al. 2013), anti-inflammatory (Senthil Kumar et al. 2010; Senthil Kumar and Wang 2009), anti-melanogenic (Kumar et al. 2010), hepatoprotective (Chen et al. 2013; Senthil Kumar et al. 2012), and hypolipidemic (Hsieh and Wang 2013) effects. In addition, methyllucidone showed a significant inhibition of farnesyl protein transferase and antitumor activity in various human cancer cell lines in vitro (Oh et al. 2005). Fermentation extracts of *Streptomyces* strain, K93-0711, produced madindolines A and B which displayed a dose-dependent inhibition of MH60 cells, an interleukin-6 (IL-6)-dependent cell line, whereas these compounds did not show any antimicrobial activities (Yang et al. 2013). Moreover, synthetic analogs of cyclopentenedione-derived TX-1123 and TX-1925 showed antitumor activity through the inhibition of protein tyrosine kinase activities (Surh et al. 2000). G2201-C, a cyclopentenedione antibiotic produced by *Streptomyces cattleya*, was found to be moderately active in vitro against Gram-positive bacteria, weakly active against Gram-negative bacteria, and inactive against fungi. Also, G2201-C was toxic to mice (Noble et al. 1978). Chrysotriones A and B, two 2-acylcyclopentene-

1,3-dione derivatives isolated from the fruiting bodies of Basidiomycete *Hygrophorus chrysodon*, showed antifungal activity against *Fusarium verticillioides* (Tichotova et al. 2011). A large number of CPDs have been shown to possess antibacterial and antifungal properties, despite recent studies indicating that CPDs are potent anti-inflammatory, hepatoprotective, and neuroprotective agents (Sevcikova et al. 2014). Among the known CPDs, lucidone is the most extensively studied. This chapter provides comprehensive information on the biological effects and pharmacological importance of lucidone.

11.2 Source, Isolation, Chemical Properties, and Synthesis of Lucidone

11.2.1 Source of Lucidone

Many species in the *Lauraceae* family have been used in folk medicine and culinary purposes. The genus *Lindera* belonging to Lauraceae is one of the economically important genera commonly known as spicewood, spicebush, and Benjamin bush. The Latin name *Lindera* commemorates the Swedish botanist Johan Linder (1676–1724). *Lindera* species widespread in Eastern Asia and few in Northern America have several important medicinal plants including *L. lucida*, *L. erythrocarpa*, *L. aggregata*, *L. glauca*, *L. obtusiloba*, *L. reflexa*, *L. akoensis*, *L. oxyphylla*, *L. umbellata*, *L. melissifolia*, *L. pulcherrima*, *L. communis*, *L. neesiana*, *L. fruticose*, *L. angustifolia*, *L. chunni*, and *L. strychnifolia. L. erythrocarpa* Makino (Fig. 11.1A) is an important species distributed mainly in Eastern Asia including Taiwan, Japan, Korea, and China (Oh et al. 2005). The fruits of *L. erythrocarpa* are used as a folk medicine for analgesic, digestive, diuretic, antidote, and antibacterial activities (Wang et al. 2008).

A

B

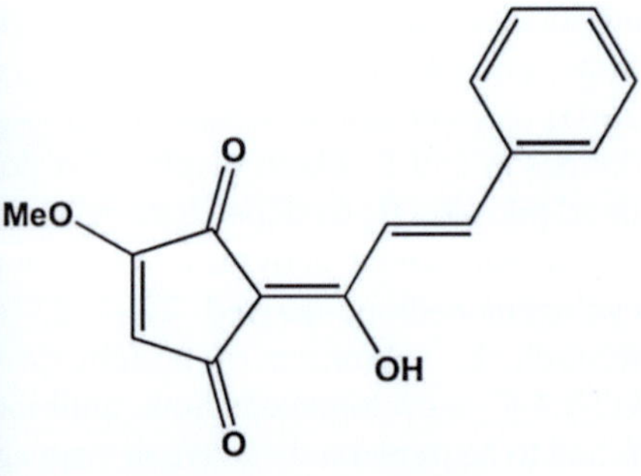

Fig. 11.1 (**A**) Fruits of *Lindera erythrocarpa* Makino. (**B**) Chemical structure of lucidone

The genus *Lindera* is one of the major sources of naturally occurring cyclopentenedione such as linderone, methyllinderone, lucidone, and methyllucidone. Lucidone (Fig. 11.1B) was first isolated in 1968 from the fruits of *Lindera lucida* (Syn. *L. malaccensis* or *L. selangorensis*) using chloroform extracts (Comai et al. 2010). The obtained yellow crystalline compound was immediately identified as lucidone, and the structure indicated that lucidone exhibits cyclopent-4-ene-1,3-dione tautomerism. The detailed study on the structure of lucidone was carried out by Ng et al. (1990). They reported that external tautomers exist in lucidone through strong intramolecular hydrogen bonding (Ng et al. 1990). After over three decades, lucidone was subsequently isolated from other relative species such as *Lindera erythrocarpa* Makino (Wang et al. 2008).

11.2.2 Isolation and Chemical Properties of Lucidone

The dried fruits of *L. erythrocarpa* Makino (2 kg) were extracted with EtOH. The total crude extract was concentrated under vacuum to yield a residue (124.3 g). One hundred grams of EtOH crude extract was suspended in H_2O in the ratio of 1:1 and successively partitioned with n-hexane (n-hex) and ethyl acetate (EA), yielding n-hex soluble fraction (16.0 %), EA-soluble fraction (45.6 %), and EA-insoluble fraction (34.3 %). The EA-soluble fraction (15 g) was chromatographed on a silica gel column, eluted with a gradient of n-hex/EA (95/5–100/0) to give a total of 12 subfractions (EA-1 to EA-12). When EA subfraction-5 (EA-5) was further separated by semi-preparative HPLC using Cosmogel column, eluted with n-hex/dichloromethane/EA solvent system, it resulted into four major compounds: linderone, methyllinderone, lucidone, and methyllucidone (Wang et al. 2008). The structures of these four compounds were confirmed by spectroscopic analyses. The amount of lucidone in the EtOH extract was further analyzed by HPLC. It was found that the total content of lucidone was 6.50 % in the EtOH extract of *L. erythrocarpa* Makino fruits (Wang et al. 2008).

Lucidone((2Z)-2-[(2E)-1-hydroxy-3-phenylprop-2-en-1-ylidene]-4-methoxycyclopent-4-ene-1,3-dione) is a yellow powder, soluble in organic solvents such as dimethyl sulfoxide (DMSO), EtOH, MeOH, or acetone and has a melting point of 166.5–168.5 °C. Its molecular formula is $C_{15}H_{12}O_4$, and the molecular weight is 256.0752 g/mol. Accurate mass measured that was analyzed by high-resolution electron impact mass spectrometry (HREIMS) resulted in [M] + m/z 256.0752. Its chemical formula is $C_{15}H_{12}O_4$ (calcd m/z 256.0735). ^{1}H NMR ($CDCl_3$): δ 3.96 (s. 3H), 7.40 (m, 3H), 7.63 (m, 3H), and 7.71 (d, 1H, J = 18 Hz) (Oh et al. 2005).

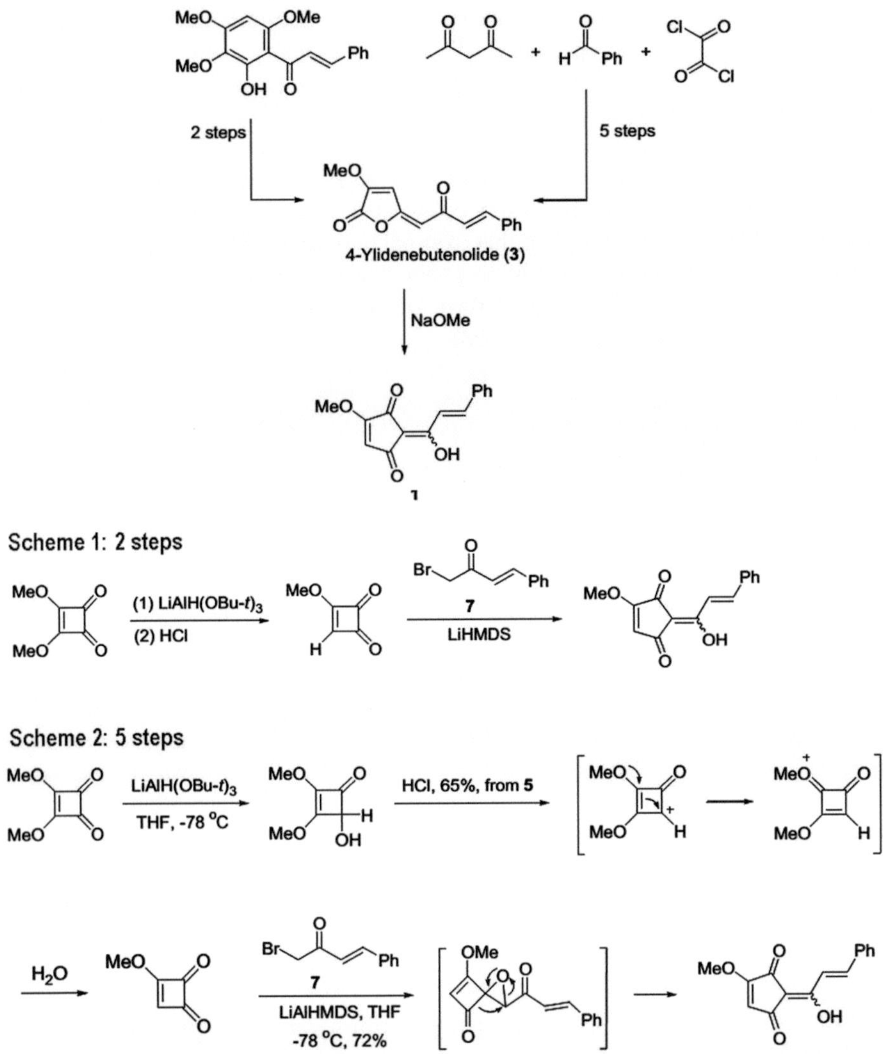

Fig. 11.2 Concise synthesis of lucidone (Adapted from Wu et al. 2013)

11.2.3 Synthesis of Lucidone

As lucidone exhibits various pharmacological properties, it was chemically synthesized by various research groups. Wu et al. (2013), utilized "one-pot" reduction/rearrangement of dimethyl squarate and Darzens/ring expansion of the monomethoxy cyclobutenedione. The synthesis of lucidone was accomplished in two steps and obtained 46 % total yield (Zhao et al. 2006). A concise synthesis of lucidone is depicted in Fig. 11.2.

11.3 Pharmacological Properties of Lucidone

Accumulating evidence suggests that lucidone has a diverse range of pharmacological properties such as anti-inflammatory, antioxidant, hepatoprotective, neuroprotective, dermatoprotective, and skin-whitening effects through the modulations in molecular targets, supporting the notion that it acts upon various biochemical and molecular cascades. Lucidone modulates various targets either through direct interaction or via modulation of gene expression. Various molecular targets modulated by lucidone include transcription factors, growth factors, and their receptors, cytokines, enzymes, and genes regulating inflammation and oxidative stress.

11.3.1 Anti-inflammatory Activities

Inflammation is a complicated and crucial physiological response to many pathological conditions including tissue injury and microbial invasion, which is manifested with redness, swelling, and pain. Macrophages play a functional role in coordinating the immune response to invading pathogens through phagocytosis and cytokine secretion. Activation of macrophages by endotoxins or pathogenic microorganisms produces a vast amount of pro-inflammatory molecules including nitric oxide (NO), prostaglandin E_2 (PGE_2), tumor necrosis factor-α (TNF-α), and interleukin-1β/6 (IL-1β/6) (Joshi and Mathela 2012). The constant production of these molecules involves a variety of inflammatory disorders, such as rheumatoid arthritis, atherosclerosis, asthma, hepatitis, pulmonary fibrosis, and cancer. Therefore, inhibition of these pro-inflammatory molecules represents an ideal target for minimizing the burden of inflammatory diseases (Joshi and Mathela 2012).

The production of pro-inflammatory molecules was mediated by a variety of soluble factors and signaling events. For example, NF-κB-dependent gene expression plays a crucial role in inflammatory responses and increases the expression of genes encoding inducible nitric oxide synthase (iNOS) and cyclooxygenase-2 (COX-2). TNF-α and IL-1β are enzymes involved in the synthesis of NO, PGE_2, TNF-α, and IL-1β, respectively. Under the normal physiological condition, NF-kB is sequestered in the cytoplasm by complex with its negative regulator, inhibitor κBs (IκBα, IκBβ, and IκBγ). Upon activation of IκBs by their upstream kinase, inhibitor κB kinase (IKKα), NF-κB is disassociated and exported to the nucleus, where its target genes may be activated (Joshi and Mathela 2012).

Clinically used anti-inflammatory drugs exhibit several side effects on humans and a high cost of treatment, as in the case of biologics. Natural products derived from anti-inflammatory agents offer promising options for the development of drugs for treating inflammatory diseases. Although, for centuries, herbal products have been utilized to treat or minimize the inflammatory disorders, the most successful example is curcumin from the tuber of *Curcuma longa* (turmeric). The anti-inflammatory activity of *L. erythrocarpa* was evaluated in vitro through measuring NO production from LPS-stimulated macrophages (Wang et al. 2008; Senthil Kumar and Wang 2009). Our findings showed that EA-soluble fraction derived from EtOH extracts of *L. erythrocarpa* fruits exhibited potent NO inhibitory effect. Also, four cyclopentene diones, namely, linderone, methyllinderone, lucidone, and cis/trans-methyllucidone, were identified as major compounds of this active fraction. In further analysis with the bioactivity-guided fraction procedure, lucidone showed a strong NO inhibitory activity with an EC_{50} value of 4.22 μg/mL (Wang et al. 2008). Also, lucidone significantly inhibited the secretion of PGE_2 and TNF-α in LPS-induced murine macrophage RAW264.7 cells. Further analysis with immunoblotting and Q-PCR revealed that the inhibition of pro-inflammatory molecules occurred through the downregulation of their corresponding mediator genes, iNOS and COX-2. Electrophoretic mobility shift assay (EMSA) and Western blotting suggested that the inhibition of pro-inflammatory genes by lucidone is caused by the suppression of nuclear export and transcriptional activation of the redox-sensitive transcription factor, NF-κB. Also, lucidone increased the protein stability of the IκBα, an endogenous repressor of NF-κB, through the inhibition of its phosphorylation and proteasomal degradation. It was found that lucidone significantly blocked the LPS-induced IKKα, an upstream kinase of the IκBα (Senthil Kumar and Wang 2009). These reports strongly suggest that lucidone inhibits production of pro-inflammatory molecules through the suppression of the redox-sensitive NF-κB signaling pathway (Fig. 11.3).

Though the role of NF-κB signaling pathway in inflammation is well characterized, however, only a few studies have reported the alternative inflammatory signaling pathway such as activator protein 1 (AP-1) pathway. AP-1, an early transcription factor, regulates iNOS and COX-2 expression in macrophage cells, either alone or in association with NF-kB. Lee et al. (2007) reported that mitogen-activated protein kinases (MAPKs) including p38 MAPK, JNK/SAPK, and ERK1/2 trigger transcriptional activation of AP-1. The LPS-induced activation of MAPKs, particularly p38 MAPK and JNK/SAPK, was significantly prevented by lucidone in a dose-dependent manner. Furthermore, lucidone treatment inhibited the nuclear translocation and transcriptional activity of ATF-2 (a member of AP-1 family in LPS-induced macrophages) in vitro (Senthil Kumar and Wang 2009) (Fig. 11.3).

Subsequent in vivo studies have shown that lucidone protects mice from acute inflammation (Senthil Kumar et al. 2010). Acute systemic inflammation in male ICR mice was induced by injection of LPS (5 μg/kg), and the protective effect of lucidone (50–200 mg/kg) was determined by measuring the production of pro-inflammatory molecules. Lucidone treatment strongly reduced NO, PGE_2, and TNF-α levels in mice blood serum. Protein and mRNA analyses of the liver samples confirmed that lucidone inhibited the production of pro-inflammatory molecules through the downregulation of their corresponding mediators, iNOS and COX-2, followed by the suppression of transcriptional activation of NF-κB and AP-1 signaling pathways (Senthil Kumar et al. 2010). In another study, lucidone was proved to be effective against inflammation in a rodent model. A topical application of

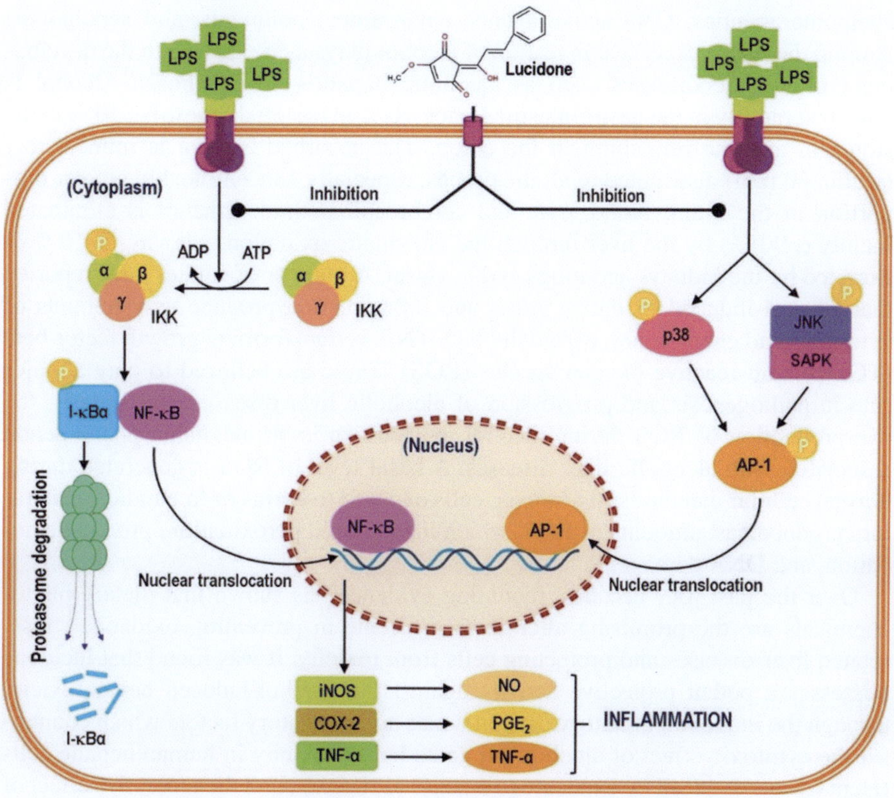

Fig. 11.3 Schematic depiction of the anti-inflammatory mechanism of lucidone

lucidone at a dose of 0.5 and 1 mg/ear significantly reduced croton-oil-induced ear edema in mice. The percentage of edema reduction in treated mice was 44 % and 25 %, respectively (Wang et al. 2008). These reports concluded that lucidone exerted anti-inflammatory effects by inhibiting the expression of pro-inflammatory factors and their corresponding transcriptional factors (Fig. 11.3).

11.3.2 Hepatoprotective Activities

The liver is the prime organ which regulates many metabolic functions and plays an important role in the maintenance of an internal environment of the body through its multiple and diverse functions. The chronic liver diseases represent a global concern. There is a progressive increase in the incidence of hepatic damage mainly due to viral infection, hepatotoxic chemicals (alcohol), toxin in food (especially aflatoxins), peroxides (particularly peroxidized edible oil), pharmaceutical antibiotics, chemotherapeutics, CNS active agents, environment pollutants, and xenobiotics. Among these, the excessive ingestion of alcohol plays a crucial role in the development of hepatic diseases such as alcoholic hepatitis and cirrhosis. Alcohol is absorbed rapidly in the gastrointestinal tract: 70 % in the small intestine, 20 % in the stomach, and the remaining in the colon. The absorbed alcohol is immediately (within 60 min) distributed to all the tissues, especially concentrated in greater proportion in the brain, blood, eye, and cerebrospinal fluid. Ethanol is eliminated mainly (>90 %) by the liver through the enzymatic oxidation pathway; 5–10 % is excreted by the kidneys and lungs and in sweat. A number of studies have reported that ethanol-induced oxidative stress and inflammation produce vast amounts of cytokines and chemokines, especially NO, TNF-α, transforming growth factor-beta (TGF-β), and reactive oxygen species (ROS). These are believed to play a major role in pathogenesis and progression of alcoholic liver diseases. Particularly, the overproduction of ROS during alcohol metabolism is an inevitable phenomenon associated with alcoholic liver diseases. A basal level of ROS is generated during normal cellular metabolism; however, cells exposed to toxins or free radical generators produce vast amounts of ROS, which induce lipid peroxidation, protein degradation, and DNA damage.

Over the past four decades, mounting evidence has shown that dietary phytochemicals are the promising alternative medicine in preventing oxidative stress-related liver diseases and protecting cells from toxicity. It was found that lucidone possesses a potent protective activity against the alcohol-induced hepatotoxicity through the induction of antioxidant enzymes and regulatory factors which counteract the cytotoxic effect of alcohol. An acute hepatotoxicity in human hepatic cells (HepG2) was induced by exposure to ethanol (100 mM), and the protective effect of lucidone was determined by pretreatment of cells with increasing concentrations of lucidone (1–10 μg/mL) for 2 h. The notable signs of ethanol-intoxicated hepatic injury are leakage of hepatic transaminases and cytokines into the circulatory system. Serum biochemical analysis shows that the ethanol-induced increase in the production of alanine aminotransferase (ALT) and aspirate aminotransferase (AST) was significantly inhibited by lucidone in a dose-dependent manner without appreciable cytotoxic effects. The crucial role of pro-inflammatory cytokines and chemokines such as NO, TNF-α, and IL-1β in the development of alcoholic liver diseases has been well demonstrated (Zhao et al. 2015). Especially, TNF-α is of great relevance to liver pathology, as increased enzyme levels and protein expression were observed in many forms of liver diseases, such as alcoholic liver diseases. Also, it mediated hepatocyte apoptosis (Fernandez-Checa et al. 2005). Fernández Checa et al. (2002) reported that mice deficient in TNF-a receptor (TNFR1) failed to develop alcoholic liver diseases except steatosis which indicates the crucial role of TNF-α in the alcoholic liver diseases. Therefore, controlled production of TNF-α primarily is implied to prevent alcoholic liver diseases in humans. It is noteworthy that pretreatment of hepatic cells with lucidone significantly prevented alcohol-induced TNF-α production. This observation was concomitant with a previous report that lucidone inhibits LPS-induced TNF-α secretion in macrophage cells. In addition to the pro-inflammatory cytokines, NO, a chemokine, has been demonstrated as another key player of liver diseases (Chen et al. 2015). It was demonstrated that the exposure of human astrocytoma cells to ethanol increased NO production in vitro (Davis et al. 2002). Also, a mixture of pro-inflammatory cytokines induced NO production in human hepatocyte HepG2 cells (Majano et al. 2004). An in vivo study showed that acute or chronic alcohol exposure increased NO production in rat circulatory system (Deng and Deitrich 2007). However, in our study, for the first time, we reported an ethanol-induced NO production in HepG2 cells (Senthil Kumar et al. 2012), though pretreatment with lucidone significantly blocked the increase of NO production (Senthil Kumar et al. 2012). This data is in conformity with our previous report that lucidone inhibited LPS-induced NO production in murine macrophage cells (Senthil Kumar and Wang 2009). Reduced glutathione content is often used for the evaluation of oxidative stress in biological systems. Augmentation of GSH/GSSG ratio has been demonstrated to protect the liver from oxidative stress. In our earlier study, it has been reported that exposure of hepatic cells to ethanol significantly increased GSH depletion, whereas pretreatment with lucidone prevents such reduction in GSH protein levels (Senthil Kumar et al. 2012). During the alcohol metabolism, alcohol is oxidized into aldehydes, especially acetaldehydes (ADAs) and malonaldehydes (MDAs). Elevated MDA and decreased antioxidant capacities have been used as biomarkers of oxidative stress. Interestingly, pretreatment with lucidone significantly blocked ethanol-induced lipid peroxidation as evidenced by decreased malonaldehyde (MDA) level in HepG2 cells (Senthil Kumar et al. 2012).

Ethanol-mediated ROS generation plays a critical role in the development of alcoholic liver diseases and in limiting the expression of cytoprotective genes. Therefore, the removal of ROS accumulation through cellular antioxidant defense system could maintain the intracellular redox homeostasis. Lucidone pretreatment significantly prevented ethanol-induced intracellular ROS accumulation in cultured HepG2 cells (Senthil Kumar et al. 2012). Further cell-free antioxidant analysis such as DPPH and iron-chelating assays revealed that lucidone does not have the ability to directly scavenge free radicals. In such conditions, most of the eukaryotic cells are fortified with primary and secondary defense against oxidative stresses. Particularly, phase II enzymes such as heme oxygenase-1 (HO-1), NAD(P)H:quinone oxidoreductase 1 (NQO1), and glutathione-*S*-transferase (GST) get rapidly activated by an endogenous mechanism through which oxidative toxicants are removed before they could damage the DNA. Conversely, excessive or chronic oxidative stress increases weakened defense and reduces endogenous antioxidants. In these conditions, induction of antioxidant defense by external factors is an important component of the cellular stress response. Many natural products have been reported to have beneficial effects on alcoholic liver diseases: polyphenols, flavonoids, terpenoids, carotenoids, vitamins, silymarin, curcumin, *N*-acetylcysteine, and antroquinonol are well known for their high antioxidant contents. These components not only act as free radical scavengers but also modulate signal transduction pathways and gene expression patterns. Particularly, HO-1 is a rate-limiting enzyme that catalyzes heme into biliverdin, free iron, and carbon monoxide. Induction of HO-1 has been reported to minimize cellular injuries, oxidative stress, pro-

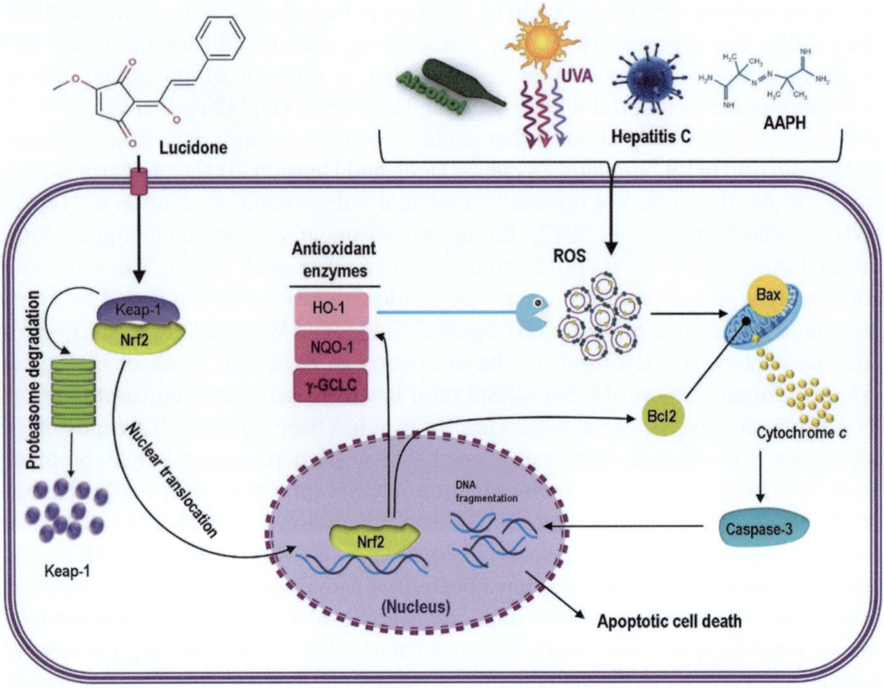

Fig. 11.4 Schematic diagram of the antioxidant defense mechanism or cytoprotective effects of lucidone

inflammatory cytokine production, and proapoptotic inducer activation (Braudeau et al. 2003). A recent study has reported that ethanol exposure prominently reduced the endogenous HO-1 level in hepatocytes (Bao et al. 2010). In our study, it was demonstrated that the ethanol-induced decline in HO-1 level was significantly prevented by lucidone in hepatic cells, while Surh (2003) reported that lucidone treatment significantly induced the transcriptional activation of Nrf2, a binding of redox-sensitive transcription activator to antioxidant response element (ARE) in the upstream promoter region of many antioxidant genes including HO-1 (Surh 2003). Previous studies also indicated that the increased Nrf2 activity in hepatic tissues is highly hepatoprotective during chemical- or ethanol-induced oxidative stress (Farombi et al. 2008; Yao et al. 2007). It is also found that lucidone-mediated induction of antioxidant genes, such as HO-1, is associated with the alcohol-induced oxidative stress (Fig. 11.4). These results suggest that lucidone could be a potential lead compound for the treatment of alcoholic liver diseases.

Hepatitis C virus (HCV) is one of the five known hepatitis viruses: A, B, C, D, and E. HCV is a leading causative agent of hepatocellular carcinoma (HCC) disease in developed countries, and it is estimated that 3 % of the world population is infected with this virus (Stauber and Stadlbauer 2006). To determine the potential effects of lucidone on HCV replication, Ava5 cells, a parent Huh-7-derived cell line harboring an HCV sub-genomic RNA replicon, were treated with increasing concentration of lucidone (5–50 μM) for 4 days or a single dose (50 μM) for various time points (24–96 h). Results of Western blotting analysis showed that lucidone markedly decreased the HCV NS5B protein levels in a concentration- and time-dependent manner (Chen et al. 2013). Also, lucidone significantly suppressed HCV RNA levels with an EC_{50} value of 15 ± 0.5 μM. Next, the cell viability assay (MTS assay) revealed that lucidone is not cytotoxic at effective antiviral concentrations as indicated by 50 % cytotoxic concentration (CC_{50}) value of 620 ± 5 μM (Chen et al. 2013). In addition, HCV JFH-1 infectious assay confirmed the inhibitory effect of lucidone on viral RNA replication, with 50 % effective concentration (EC_{50}) of 20 ± 1.1 μM, which is an acceptable selectivity index (SI; CC_{50}/EC_{50}) of ~31. Also, treatment with lucidone significantly induced HO-1 expression and led to the increase of its product biliverdin for induction of antiviral interferon (INF) response and inhibition of HCV NS3/4A protease activity. Conversely, the anti-HCV activity of lucidone was barely observed in HO-1 or Nrf2 silenced cells, indicating that the anti-HCV property of lucidone was due to the induction of Nrf2-mediated HO-1 expression (Fig. 11.4). Moreover, co-treatment of lucidone with alpha interferon, the protease inhibitor telaprevir, the NS5A inhibitor BMS-790052, or the NS5B polymerase inhibitor PSI-7977 synergistically suppressed HCV RNA replication. These findings suggest that lucidone could be a potential lead or a supplement for the development of new anti-HCV agents (Chen et al. 2013).

11.3.3 Hypolipidemic Activity

Obesity is a complex disorder involving an accumulation of excessive body fat. Its incidence is dramatically increasing in developed and developing countries due to diet and lifestyle changes. Overweight is defined as a body mass index (BMI) of 25–30 and obesity as a BMI >30, with the latter being classified into class I (BMI of 30–35), class II (BMI of 35–40), and class III (BMI >40). Obesity can also be differentiated between peripheral and central obesity, with the latter having more implication in the metabolism (Watt and Charlton 2010). According to the World Health Organization (WHO), the worldwide prevalence of obesity has nearly doubled between 1980 and 2008, and more than 10 % of the adults aged 20 and over were obese in 2008. Obesity increases the risk of developing various pathological conditions including insulin resistance, type 2 diabetes, hypertension, hyperlipidemia, atherosclerosis, and nonalcoholic fatty liver diseases (NAFLD). Risk factors for the above are smoking, hypertension, serum cholesterol, genetic factors, physical activity, hormones, alcohol, and diseases related to the thyroid, kidney, and liver. Adipose tissue has been recognized as an active endocrine organ and a main energy store of the body. Extensive adipocyte remodeling by adipocyte hyperplasia (increased number of adipocytes), adipocyte hypertrophy (increased size of adipocytes), and angiogenesis (neovasculature) is the crucial factor involving obesity.

Excess fat accumulation from free energy intake such as high-fat diet promotes the release of free fatty acids into the circulation from adipocytes, and hyperplasia results from the complex interplay between proliferation and differentiation of preadipocytes. Increasing evidence suggests that natural products have the potential to suppress preadipocyte differentiation and adipogenesis in 3T3-L1 cells and prevent obesity in animal models. For example, curcumin, an active ingredient of turmeric (*Curcuma longa*), inhibited preadipocyte differentiation and blocked body weight gain in diet-induced obesity mice through the downregulation of lipogenesis in the liver (Ferguson et al. 2016). Berberine, an isoquinoline alkaloid isolated from the roots of *Berberis aristata*, inhibits adipocyte differentiation by suppressing PPARγ and reducing the secretion of adipogenic enzymes. Also, berberine reduced serum glucose level and body weight gain in high-fat-diet-induced mice and decreased lipid levels in the circulation of both obese and normal SD rats (Pang et al. 2015). In a similar way, epigallocatechin gallate inhibits preadipocyte differentiation through the suppression of PPARγ pathway via activating AMPK (Moon et al. 2007) and also attenuates fatty liver formation in diet-induced obese mice. These studies strongly suggest that natural products have the potential to inhibit preadipocyte differentiation in vitro and hypolipidemic effects in vivo. We recently reported that lucidone inhibited 3T3-L1 adipocyte differentiation by suppressing the transcription of master regulators of adipogenesis including PPARγ and C/EBPα (Hsieh and Wang 2013). Also, lucidone downregulates the expression levels of genes involved in lipogenesis including *LXR-α*, *LPL*, *aP2*, *adiponectin*, and *GLUT4* (Hsieh and Wang 2013). The dietary intake of lucidone significantly reduced high-fat-diet-induced body weight gain and epididymal and perirenal fat accumulation presumably resulting from a reduction in adipocyte diameter. Mice fed a HF diet with lucidone improved hyperglycemia, hyperinsulinemia, dyslipidemia, and hepatomegaly without kidney lesion (Hsieh and Wang 2013). This study suggests that lucidone as a nutraceutical supplement prevents obesity and associated metabolic disorders (Fig. 11.5).

11.3.4 Dermatoprotective Effect

In humans, the skin is the largest organ of the integumentary system. It protects the body from microbes and noxious substances (toxic chemicals and ultraviolet radiation), which result in skin aging, inflammation, and cancer. Several phytochemicals have been reported as potent skin-protecting agents. For example, sauchinone, a lignan from *Saururus chinensis*, protects human skin keratinocytes against ultraviolet B-induced photoaging by regulating HO-1-mediated antioxidant defense mechanism. Sauchinone also inhibits UVB-induced matrix metalloproteinase-1 (MMP-1) and reduction in type-1 collagen in skin keratinocytes (Park et al. 2013). A recent study showed that lucidone protects human skin keratinocytes (HaCaT cells) from free radical inducer 2,2′-Azobis(2-amidinopropane) dihydrochloride (AAPH)-induced oxidative damage and inflammation. Lucidone pretreatment

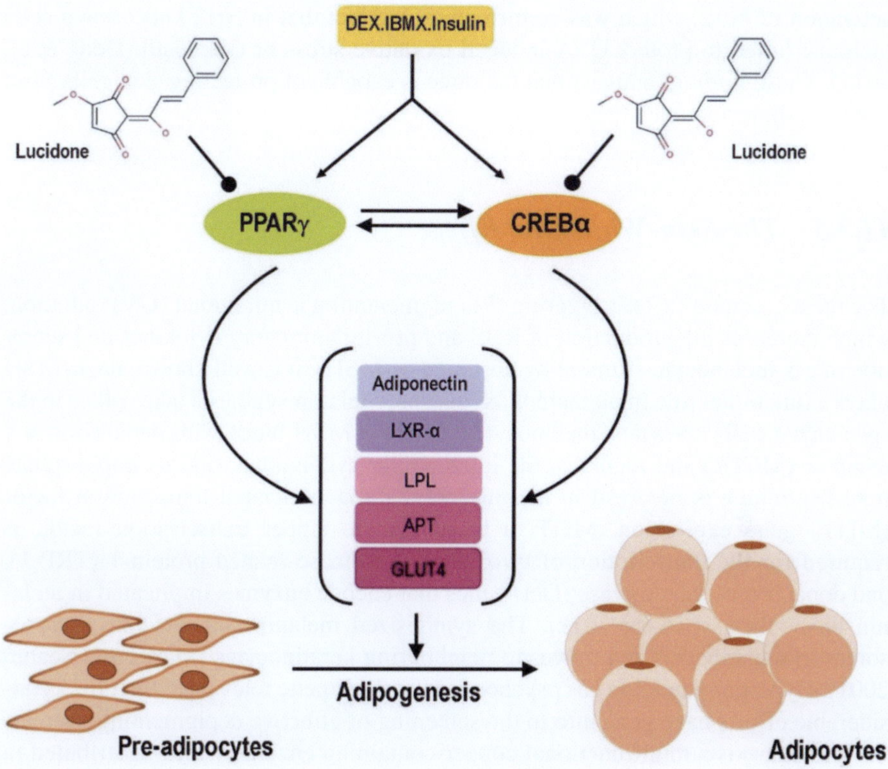

Fig. 11.5 Schematic representation of hypolipidemic effects of lucidone (3T3-L1 preadipocytes were stimulated with a mixture of 1 μM dexamethasone (DEX), 0.52 mM isobutylmethylxanthine (IBMX), and 0.17 μM insulin for 2 days)

(0.5–10 μg/mL) markedly increased HaCaT cell viability and inhibited AAPH-induced intracellular ROS generation, lipid peroxidation, and DNA damage (Kumar et al. 2013). The protective and preventive mechanisms of lucidone are mediated by the induction of an antioxidant gene HO-1 gene through the transcriptional activation of Nrf2 (Fig. 11.4). The study also reported that pretreatment with lucidone significantly inhibited AAPH-induced inflammatory chemokine prostaglandin E_2 (PGE_2) production and the expression of cyclooxygenase-2 (COX-2) in HaCaT cells. Additionally, lucidone protects human keratinocytes against AAPH-induced inflammation through the suppression of NF-κB and MAPKs signaling pathways. Another study revealed that pretreatment with lucidone (1–4 μM) significantly protected keratinocytes from UVA (15 J/cm^2)-induced cell death, excessive ROS generation, LDH release, lipid peroxidation, and DNA damage (Deng et al. 2011). In addition, lucidone inhibited the UVA-induced apoptosis of HaCaT cells (Deng et al. 2011). The antioxidant potential of lucidone was directly correlated with the induction of antioxidant genes, including HO-1, NQO-1, and γ-GCLC by transcriptional activation of Nrf2, which was confirmed by the fact that in Nrf2 knockdown cells lucidone failed to protect UVA-induced oxidative stress or cell death (Deng et al. 2011). These findings suggest that lucidone is capable of protecting skin cells from UVA-irradiated damage (Fig. 11.4).

11.3.5 The Skin-Whitening Effect

The most common factor triggering skin pigmentation is ultraviolet (UV) radiation, which increases the production of ROS and pro-inflammatory cytokines and secretion of a α-melanocyte-stimulating hormone (α-MSH). It is well known that α-MSH plays a functional role in melanin biosynthesis. Melanin synthesis takes place in the specialized cells known as melanocytes, where α-MSH binds with melanocortin 1 receptor (MC1R) and regulates the intracellular cyclic adenosine monophosphate (cAMP), which is involved in the microphthalmia-associated transcription factor (MITF) gene expression. MITF, a basic leucine zipper transcription factor, is required for the transcription of tyrosinase, tyrosinase-related protein-1 (TRP-1), and dopachrome tautomerase (Dct) genes that encode enzymes implicated in melanin biosynthesis (Chang 2012). The synthesized melanin converts into melanosomes in melanocytes and moves to neighboring keratinocytes (Marks and Seabra 2001). Hyperpigmentation has psychosocial and cosmetic relevance; therefore, considerable efforts have gone into to the screening of effective depigmenting agents.

Tyrosinase is a multifunctional copper-containing enzyme widely distributed in plants and animals. It plays a key role in melanin biosynthesis. Tyrosinase is majorly involved in the first two steps of melanin biosynthesis pathway: hydroxylation of monophenol to o-diphenol (monophenolase or cresolase activity) and oxidation of diphenol to o-quinones (diphenolase or catecholase activity). These both steps use molecular oxygen followed by a series of nonenzymatic steps. Therefore, tyrosinase inhibitors can be clinically useful for the treatment of some dermatological disorders associated with hyperpigmentation. Many researchers are working on to isolate tyrosinase inhibitors from natural sources such as arbutin, kojic acid, gallic acid, ascorbic acid, and hydroquinones (Kim and Uyama 2005) (Fig. 11.6).

A previous study by Kang et al. (2008) reported that EtOH extracts and subfractions of leaves of *Lindera erythrocarpa* exerted antioxidant and anti-melanogenic effects in vitro. The tyrosinase inhibitory effect of ethanol extract was higher than hexane fraction. In contrast, CH_2Cl_2 fraction showed a higher inhibitory effect on α-MSH-stimulated melanin biosynthesis in melanoma B16F10 cell. Bioactive fraction-guided investigations led to the isolation of two compounds, lucidone and methyl linderone, as characterized by spectroscopic techniques including 1D, 2D NMR, and HR-MS. Lucidone and methyl linderone compounds were acting as a potent tyrosinase inhibitors compared to positive control (arbutin). These results suggest that extract of *Lindera erythrocarpa* could be used as a functional biomaterial in developing a skin-whitening agent having the antioxidant activity (Lin et al. 2007). However, this study barely explained the molecular mechanism involved in

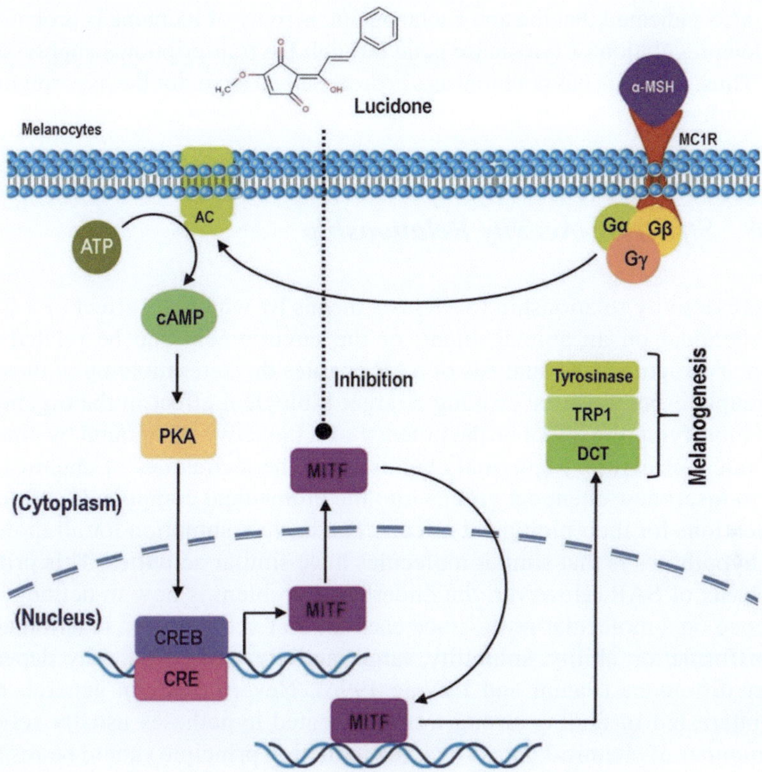

Fig. 11.6 Schematic representation of the anti-melanogenic effect of lucidone

the anti-melanogenic effects. After 2 years, we found that lucidone, isolated from the fruits of *Lindera erythrocarpa*, strongly inhibits mushroom tyrosinase activity (Kumar et al. 2010). The effects of lucidone on tyrosinase were further examined in α-MSH-induced B16 melanoma cells. Lucidone significantly inhibited tyrosinase activity and led to decreased melanin content in cultured B16 melanoma cells. Further molecular analysis showed that lucidone significantly attenuates the expression of tyrosinase and MITF proteins in a dose-dependent manner (Kumar et al. 2010). The reduction in tyrosinase protein expression by lucidone is extended to its transcriptional levels as lucidone significantly inhibited α-MSH-induced tyrosinase mRNA expression (Kumar et al. 2010). A previous report indicates that α-MSH-induced MITF activation is negatively regulated by ERK1/2 (Chang 2012). However, according to our results, lucidone did not play a major role in the induction of ERK activation. It is likely that the hydroxyl group in lucidone plays a key role in the direct tyrosinase inhibition. Most of the tyrosinase inhibitors are polyphenol derivatives such as flavonoids, resveratrol, 3,5-dihydroxyphenyl decanoate, and 5-(hydroxymethyl)-2-furfural that are rich in hydroxyl and methoxy groups. Our results also indicated that the anti-melanogenic activity of lucidone is probably due to its downregulation of tyrosinase gene through the transcriptional suppression of MITF. Thus, lucidone has potential as a cosmeceutical agent for the hyperpigmented skin disorders.

11.3.6 Structure-Activity Relationship

Structure-activity relationship (SAR) is a means by which the effect of a drug or toxic chemical on an animal, plant, or the environment can be related to its molecular structure. The analysis of SAR enables the determination of the chemical groups responsible for evoking a target biological effect in the organism. It allows modifying the effect or the potency of a bioactive compound by changing its chemical structure. Medicinal chemists use the techniques of chemical synthesis to insert new chemical groups into the biomedical compounds and test the modifications for their biological effects. The basic assumption for all molecule-based hypotheses is that similar molecules have similar activities. This principle is the basis of SAR. However, the underlying problem is how to define a small difference on a molecular level, since each kind of activity, e.g., reaction ability, biotransformation ability, solubility, target activity, and so on, may depend on another difference (Patani and LaVoie 1996). Nevertheless, in general, one is more interested in finding strong trends. Created hypotheses usually rely on a finite number of chemical data. Thus, the induction principle should be respected to avoid over fitted hypotheses and driving over fitted and useless interpretations on structural/molecular data.

α,β-Unsaturated ketones have displayed diverse biological activities such as antimicrobial, antitumor, and plant growth regulation (Altalbawy 2013). The structure-activity relationship of these compounds has pointed out that the biological activity is attributed to the presence of α,β-unsaturated carbonyl group (Bag et al. 2013). Structurally, α,β-unsaturated ketones can be considered as a Michael acceptor, an active moiety often employed in the design of enzyme inhibitors.

Surh et al. (2000) studied anti-inflammatory activity of curcumin in TPA-induced mice dorsal skin. Curcumin potentially suppressed NF-κB and AP-1 activity (Surh et al. 2000). It is already known that curcumin has two α,β-unsaturated ketone moieties (Fig. 11.7). Thus, this compound may covalently interact with nucleophilic sites of the above transcription factors through Michael addition, thereby hampering DNA binding capability. Surh et al. also demonstrated the antioxidant-mediated hepatoprotective efficacy of curcumin (Surh et al. 2000). He proposed that curcumin has α,β-unsaturated ketone moiety and can, therefore, act as Michael reaction acceptors that can modify cysteine thiols located in Keap1 protein.

A previous study has shown that lucidone is chemically architectured with a cyclopentenedione ring, enol hydroxyl functionality, and styryl moiety (Ng et al.

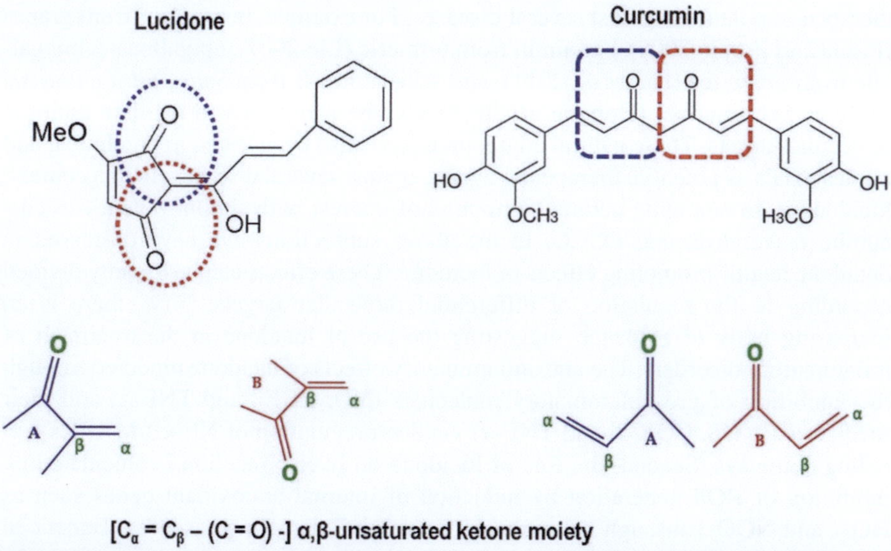

Fig. 11.7 A comparative structure-activity relationship of lucidone and curcumin

1990). The cyclopentenedione ring is made up of two α,β-unsaturated ketone moiety functional components (Fig. 11.7). We hypothesized that lucidone might have anti-inflammatory and hepatoprotective effects. Thereby, in our study, we designed to evaluate the anti-inflammatory and hepatoprotective activity of lucidone compared with curcumin.

11.4 Conclusion and Future Perspectives

Over the past five decades, changes in lifestyle patterns including diet and physical activity and the incidences of chronic and metabolic diseases such as liver diseases, lung diseases, cardiovascular diseases, diabetes, and certain types of cancers are increasing worldwide and hence are a public health concern with major economic impacts. However, the pathogenesis of these diseases is different and regulated by one or more molecular candidates that are commonly up- or downregulated leading to the notion that these could be common molecular targets in the prevention or therapeutic interventions of diseases. Since human civilization, ethnomedicine in the form of herbs and food has been contributing to disease prevention and therapy; however, rigorous experimental-based evidence in support of ethnomedicine-derived notions would lead to products relevant to the drug development. Several phytocompounds such as those in edible plants or spices are known to target multiple molecular signaling pathways, thus providing a promising preventive or therapeutic potential against several diseases. For example, resveratrol from grapes (Pirola and Frojdo 2008), curcumin from turmeric (Lin 2007), epigallocatechin gallate from green tea (Pan et al. 2011), and schisandrin B from *Schisandra chinensis* (Fructus Schisandrae) (Hong et al. 2015) have been reported to regulate multiple molecular targets. These natural compounds have also been tested in preclinical and clinical trials as potential therapeutic agents against several diseases. In this context, lucidone is an emerging natural compound of interest with similar potency as curcumin, resveratrol, and EGCG. In the above subsections, we have discussed in detail the health-promoting effects of lucidone. These effects can be broadly divided according to the regulation of differential molecular targets. First, there is an increasing body of evidence suggesting the use of lucidone in the treatment of inflammatory disorders. The anti-inflammatory effects of lucidone rendered through the inhibition of pro-inflammatory molecules (NO, PGE_2, and TNF-α) and their mediators (iNOS, COX-2, and TNF-α) via downregulation of NF-κB/MAPKs signaling pathways. Second, the role of lucidone on liver protection is elucidated by inhibition of ROS generation by induction of internal antioxidant genes such as HO-1 and NQO-1 through the upregulation of Nrf2 signaling cascades. Beneficial health effects of lucidone are further extended to its potential role to treat other ailments such hepatitis C infections as well as liver diseases through the induction of Nrf2-mediated antioxidant defense mechanism. Third, in vitro and in vivo experimental evidence suggests the use of lucidone in the treatment of metabolic diseases. Much of these effects rendered through lucidone's efficacy to inhibit preadipocyte differentiation in vitro and hypolipidemic effects in vivo. Lucidone inhibited 3T3-L1 adipocyte differentiation by suppressing PPARγ and C/EBPα transcription factors. Lucidone also downregulates the expression levels of *LXR-α*, *LPL*, *aP2*, *adiponectin*, and *GLUT4* in vitro. A dietary intake of lucidone significantly reduced high-fat-diet-induced body weight gain and epididymal and perirenal fat accumulation presumably resulting from a reduction in adipocyte diameter. Mice fed an HF diet with lucidone improved hyperglycemia, hyperinsulinemia, dyslipidemia, and hepatomegaly without kidney lesion. Fourth, lucidone treatment protects skin keratinocytes from free radical-induced oxidative damage and inflammation and UVA-induced apoptotic cell death in vitro. The molecular mechanisms involved in the induction of antioxidant genes HO-1, NQO-1, and γ-GCLC through the transcriptional activation of Nrf2. In addition, lucidone inhibits free radical-induced inflammation in keratinocytes by inhibiting pro-inflammatory chemokine PGE_2 and its corresponding mediator COX-2 through the transcriptional suppression of NF-κB/MAPKs. Lucidone treatment provokes α-MSH-induced hyperpigmentation in melanocytes through the inhibition of tyrosinase enzymes and their transcription factor MITF. Moreover, cell-free analysis confirms that lucidone can inhibit tyrosinase enzyme activity directly. However, there is little information on the bioavailability, pharmacokinetics, and pharmacodynamics of lucidone about its beneficial health effects. The scientific knowledge in this area is limited, and hence extensive preclinical and clinical research needs to be carried out before advocating the safe and efficacious use of lucidone and lucidone-rich plant extracts against the prevention and control of diseases. Furthermore, such research may assist in the development of evidence-based regulation of lucidone and lucidone-containing products as they become increasingly popular and enter the market.

References

Altalbawy FMA (2013) Synthesis and antimicrobial evaluation of some novel bis-α, β-unsaturated ketones, nicotinonitrile, 1,2-dihydropyridine-3-carbonitrile, fused thieno[2,3-b]pyridine and pyrazolo[3,4-b]pyridine derivatives. Int J Mol Sci 14(2):2967–2979. doi:10.1139/v02-194

Antkowiak R, Antkowiak WZ, Banczyk I et al (2003) A new phenolic metabolite, involutone, isolated from the mushroom *Paxillus involutus*. Can J Chem 81(1):118–124. doi:10.1139/v02-194

Babu KS, Li XC, Jacob MR et al (2006) Synthesis, antifungal activity, and structure-activity relationships of coruscanone A analogues. J Med Chem 49(26):7877–7886. doi:10.1021/jm061123i

Bag S, Ghosh S, Tulsan R et al (2013) Design, synthesis and biological activity of multifunctional α, β- unsaturated carbonyl scaffolds for Alzheimer's disease. Bioorg Med Chem Lett 23(9):2614–2618. doi:10.1016/j.bmcl.2013.02.103

Bao W, Li K, Rong S et al (2010) Curcumin alleviates ethanol-induced hepatocytes oxidative damage involving heme oxygenase-1 induction. J Ethnopharmacol 128(2):549–553. doi:10.1016/j.jep.2010.01.029

Braudeau C, Bouchet D, Toquet C et al (2003) Generation of heme oxygenase-1-transgenic rats. Exp Biol Med (Maywood) 228(5):466–471

Chang TS (2012) Natural melanogenesis inhibitors acting through the down-regulation of tyrosinase activity. Materials 5(9):1661

Chen WC, Wang SY, Chiu CC et al (2013) Lucidone suppresses hepatitis C virus replication by Nrf2-mediated heme oxygenase-1 induction. Antimicrob Agents Ch 57(3):1180–1191. doi:10.1128/AAC.02053-12

Chen S, Wang L, Zhang W et al (2015) Secondary metabolites from the root of *Lindera reflexa* Hemsl. Fitoterapia 105:222–227. doi:10.1016/j.fitote.2015.07.005

Comai S, Dall'Acqua S, Grillo A et al (2010) Essential oil of *Lindera neesiana* fruit: chemical analysis and its potential use in topical applications. Fitoterapia 81(1):11–16. doi:10.1016/j.fitote.2009.06.017

Davis RL, Dertien J, Syapin PJ (2002) Ethanol-induced modulation of inducible nitric-oxide synthase activity in human A172 astrocytoma cells. Alcohol Clin Exp Res 26(9):1404–1411. doi:10.1097/01.ALC.0000030841.92766.80

Deng XS, Deitrich RA (2007) Ethanol metabolism and effects: nitric oxide and its interaction. Curr Clin Pharmacol 2(2):145–153

Deng Z, Zhong H, Cui S et al (2011) Cytotoxic sesquiterpenoids from the fruits of *Lindera communis*. Fitoterapia 82(7):1044–1046. doi:10.1016/j.fitote.2011.06.012

Farombi EO, Shrotriya S, Na HK et al (2008) Curcumin attenuates dimethylnitrosamine-induced liver injury in rats through Nrf2-mediated induction of heme oxygenase-1. Food Chem Toxicol 46(4):1279–1287. doi:10.1016/j.fct.2007.09.095

Ferguson BS, Nam H, Morrison RF (2016) Curcumin inhibits 3T3-L1 preadipocyte proliferation by mechanisms involving post-transcriptional p27 regulation. Biochem Biophys Rep 5:16–21. doi:10.1016/j.bbrep.2015.11.014

Fernández Checa JC, Bellentani S, Tiribelli C (2002) Alcohol-induced liver disease: from molecular damage to treatment. Rev Med Chil 130:681–690

Fernandez-Checa JC, Colell A, Mari M et al (2005) Ceramide, tumor necrosis factor and alcohol-induced liver disease. Alcohol Clin Exp Res 29(11):151S–157S

Hong M, Li S, Tan HY (2015) Current status of herbal medicines in chronic liver disease therapy: the biological effects, molecular targets and future prospects. Int J Mol Sci 16(12):28705–28745. doi:10.3390/ijms161226126

Hsieh YH, Wang SY (2013) Lucidone from *Lindera erythrocarpa* Makino fruits suppresses adipogenesis in 3T3-L1 cells and attenuates obesity and consequent metabolic disorders in high-fat diet C57BL/6 mice. Phytomedicine 20(5):394–400. doi:10.1016/j.phymed.2012.11.007

Joshi SC, Mathela CS (2012) Antioxidant and antibacterial activities of the leaf essential oil and its constituents furanodienone and curzerenone from *Lindera pulcherrima* (Nees.) Benth. ex Hook. f. Phcog Res 4(2):80–84. doi:10.4103/0974-8490.94721

Kang MC, Ko RK, Lee JY, et al (2008) Antioxidant activity and whitening effect of *Lindera erythrocarpa* extract. www.earticle.net/article.aspx?sn=99232

Kim YJ, Uyama H (2005) Tyrosinase inhibitors from natural and synthetic sources: structure, inhibition mechanism and perspective for the future. Cell Mol Life Sci 62(15):1707–1723. doi:10.1007/s00018-005-5054-y

Kumar KJS, Yang JC, Chu FH et al (2010) Lucidone, a novel melanin inhibitor from the fruit of *Lindera erythrocarpa* Makino. Phytother Res 24(8):1158–1165. doi:10.1002/ptr.3018

Kumar KJS, Yang HL, Tsai YC et al (2013) Lucidone protects human skin keratinocytes against free radical-induced oxidative damage and inflammation through the up-regulation of HO-1/Nrf2 antioxidant genes and down-regulation of NF-kappaB signaling pathway. Food Chem Toxicol 59:55–66. doi:10.1016/j.fct.2013.04.055

Lee JY, Park HR, Oh YK et al (2007) Effects of transcription factor activator protein-1 on interleukin-8 expression and enteritis in response to *Clostridium difficile* toxin A. J Mol Med 85(12):1393–1404. doi:10.1007/s00109-007-0237-7

Lin JK (2007) Molecular targets of curcumin. Adv Exp Med Biol 595:227–243. doi:10.1007/978-0-387-46401-5_10

Lin CT, Chu FH, Chang ST et al (2007) Secoaggregatalactone-A from *Lindera aggregata* induces apoptosis in human hepatoma hep G2 cells. Planta Med 73(15):1548–1553. doi:10.1055/s-2007-993739

Majano PL, Medina J, Zubia I et al (2004) *N*-Acetyl-cysteine modulates inducible nitric oxide synthase gene expression in human hepatocytes. J Hepatol 40(4):632–637. doi:10.1016/j.jhep.2003.12.009

Marks MS, Seabra MC (2001) The melanosome: membrane dynamics in black and white. Nat Rev Mol Cell Biol 2(10):738–748

Moon HS, Chung CS, Lee HG et al (2007) Inhibitory effect of (-)-epigallocatechin-3-gallate on lipid accumulation of 3T3-L1 cells. Obesity 15(11):2571–2582. doi:10.1038/oby.2007.309

Ng S, Lee HH, Bennett GJ (1990) 13C NMR study on linderones and lucidones. Magn Reson Chem 28(4):337–342. doi:10.1002/mrc.1260280411

Noble M, Noble D, Fletton RA (1978) G2201-C, a new cyclopentenedione antibiotic, isolated from the fermentation broth of *Streptomyces cattleya*. J Antibiot 31(1):15–18

Oh HM, Choi SK, Lee JM et al (2005) Cyclopentenediones, inhibitors of farnesyl protein transferase and anti-tumor compounds, isolated from the fruit of *Lindera erythrocarpa* Makino. Bioorg Med Chem 13(22):6182–6187. doi:10.1016/j.bmc.2005.06.029

Pan MH, Chiou YS, Wang YJ et al (2011) Multistage carcinogenesis process as molecular targets in cancer chemoprevention by epicatechin-3-gallate. Food Funct 2(2):101–110. doi:10.1039/c0fo00174k

Pang B, Zhao LH, Zhou Q et al (2015) Application of berberine on treating type 2 diabetes mellitus. Int J Entocrinol 2015:905749. doi:10.1155/2015/905749

Park G, Kim HG, Sim Y et al (2013) Sauchinone, a lignan from *Saururus chinensis*, protects human skin keratinocytes against ultraviolet B-induced photoaging by regulating the oxidative defense system. Biol Pharm Bull 36(7):1134–1139

Patani GA, LaVoie EJ (1996) Bioisosterism: a rational approach in drug design. Chem Rev 96(8):3147–3176

Pirola L, Frojdo S (2008) Resveratrol: one molecule, many targets. IUBMB Life 60(5):323–332. doi:10.1002/iub.47

Senthil Kumar KJ, Wang SY (2009) Lucidone inhibits iNOS and COX-2 expression in LPS-induced RAW 264.7 murine macrophage cells *via* NF-kappaB and MAPKs signaling pathways. Planta Med 75(5):494–500. doi:10.1055/s-0029-1185309

Senthil Kumar KJ, Hsieh HW, Wang SY (2010) Anti-inflammatory effect of lucidone in mice *via* inhibition of NF-kappaB/MAP kinase pathway. Int Immunopharmacol 10(4):385–392. doi:10.1016/j.intimp.2009.12.013

Senthil Kumar KJ, Liao JW, Xiao JH et al (2012) Hepatoprotective effect of lucidone against alcohol-induced oxidative stress in human hepatic HepG2 cells through the up-regulation of HO-1/Nrf-2 antioxidant genes. Toxicol *in Vitro* 26(5):700–708. doi:10.1016/j.tiv.2012.03.012

Sevcikova Z, Pour M, Novak D et al (2014) Chemical properties and biological activities of cyclopentenediones: a review. Mini Rev Med Chem 14(4):322–331

Stauber RE, Stadlbauer V (2006) Novel approaches for therapy of chronic hepatitis C. J Clin Virol 36(2):87–94. doi:10.1016/j.jcv.2006.02.003

Surh YJ (2003) Cancer chemoprevention with dietary phytochemicals. Nat Rev Cancer 3(10):768–780. doi:10.1038/nrc1189

Surh YJ, Han S, Keum YS et al (2000) Inhibitory effects of curcumin and capsaicin on phorbol ester-induced activation of eukaryotic transcription factors, NF-kappaB and AP-1. Biofactors 12(1–4):107–112

Tichotova L, Matousova E, Spulak M et al (2011) Synthesis and biological activity of desmethoxy analogues of coruscanone A. Bioorg Med Chem Lett 21(20):6062–6066. doi:10.1016/j.bmcl.2011.08.059

Wang SY, Lan XY, Xiao JH et al (2008) Antiinflammatory activity of *Lindera erythrocarpa* fruits. Phytother Res 22(2):213–216. doi:10.1002/ptr.2289

Watt KD, Charlton MR (2010) Metabolic syndrome and liver transplantation: a review and guide to management. J Hepatol 53(1):199–206. doi:10.1016/j.jhep.2010.01.040

Wijeratne EM, Turbyville TJ, Zhang Z et al (2003) Cytotoxic constituents of *Aspergillus terreus* from the rhizosphere of *Opuntia versicolor* of the Sonoran Desert. J Nat Prod 66(12):1567–1573. doi:10.1021/np030266u

Wu L, Xiao F, Hu X (2013) Concise synthesis of lucidone and methyl lucidone. Chin J Org Chem 33(7):1587

Yang CP, Huang GJ, Huang HC et al (2013) The effect of the aerial part of *Lindera akoensis* on lipopolysaccharides (LPS)-induced nitric oxide production in RAW264.7 cells. Int J Mol Sci 14(5):9168–9181. doi:10.3390/ijms14059168

Yao P, Nussler A, Liu L et al (2007) Quercetin protects human hepatocytes from ethanol-derived oxidative stress by inducing heme oxygenase-1 *via* the MAPK/Nrf2 pathways. J Hepatol 47(2):253–261. doi:10.1016/j.jhep.2007.02.008

Zelik P, Lukesova A, Cejka J et al (2010) Nostotrebin 6, a bis(cyclopentenedione) with cholinesterase inhibitory activity isolated from Nostoc sp. str. Lukesova 27/97. J Enzym Inhib Med Chem 25(3):414–420. doi:10.3109/14756360903213481

Zhao Q, Zhao Y, Wang K (2006) Antinociceptive and free radical scavenging activities of alkaloids isolated from *Lindera angustifolia* Chen. J Ethnopharmacol 106(3):408–413. doi:10.1016/j.jep.2006.01.019

Zhao YY, Xiao M, Zhang CL et al (2015) Associations between the tumor necrosis factor-alpha gene and interleukin-10 gene polymorphisms and risk of alcoholic liver disease: a meta-analysis. Clin Res Hepatol Gastroenterol. doi:10.1016/j.clinre.2015.10.007

Chapter 12
Pharmacokinetics of Phytopharmaceuticals: A Peek into Contingencies and Impediments in Herbal Drug Development

Ranjeet Prasad Dash and Manish Nivsarkar

Abstract Plant secondary metabolites have been extensively used in the treatment of many diseases and have served as compounds of interest, both in their natural form and as templates for synthetic modification. With the help of the traditional knowledge, indigenous people try to derive therapeutic materials from thousands of plants; however, their safety and efficacy remain a vital concern. Natural product-based drug discovery involves the identification of new chemical entities (NCEs) of potential therapeutic interest through chemical synthesis or isolation from natural sources. Although some of these drugs have entered the international pharmacopeia through the study of ethnopharmacology and traditional medicine, they are very small in number. It is because of limitations with the availability of proper guidelines for standardization, manufacture, and quality control, which are required for herbal medicinal products. Data regarding the safety and efficacy needs to be generated from preclinical and clinical pharmacokinetic and pharmacodynamic studies. A better understanding of pharmacokinetics and bioavailability of phytopharmaceuticals can be of immense help in designing the rational dosage regimens. Based on the preclinical pharmacokinetic data, suitable formulations may be developed to ensure optimum efficacy and safety. In this article, the authors would like to share their research experiences about various aspects of pharmacokinetics, which need to be addressed to generate reliable data on safety and efficacy of herbal drugs. This information would be helpful in designing rationalized preclinical pharmacokinetic studies.

R.P. Dash
Department of Drug Discovery and Development, Harrison School of Pharmacy, Auburn University, Auburn, AL 36849, USA

Department of Pharmacology and Toxicology, B. V. Patel Pharmaceutical Education and Research Development (PERD) Centre, S. G. Highway, Thaltej, Ahmedabad 380054, Gujarat, India

M. Nivsarkar (✉)
Department of Pharmacology and Toxicology, B. V. Patel Pharmaceutical Education and Research Development (PERD) Centre, S. G. Highway, Thaltej, Ahmedabad 380054, Gujarat, India
e-mail: manishnivsarkar@gmail.com; perd@perdcentre.com

H.-S. Tsay et al. (eds.), *Medicinal Plants - Recent Advances in Research and Development*, DOI 10.1007/978-981-10-1085-9_12

Keywords Herbal drugs • Pharmacokinetics • Phytopharmaceuticals

Abbreviations

AIDS	Acquired immune deficiency syndrome
CE	Capillary electrophoresis
CE-MS	Capillary electrophoresis coupled with mass spectrophotometer
CYP_{450}	Cytochrome P_{450}
GC	Gas chromatography
GC-MS	Gas chromatography coupled with mass spectrophotometer
GCP	Good clinical practices
GLP	Good laboratory practices
HIV	Human immunodeficiency virus
HPLC	High-performance liquid chromatography
IND	Investigational new drug
LC	Liquid chromatography
LC-MS	Liquid chromatography coupled with mass spectrophotometer
NCEs	New chemical entities
NDA	New drug application
NMR	Nuclear magnetic resonance
USFDA	United States Food and Drug Administration
WHO	World Health Organization

12.1 Introduction

Substances derived from animals, plants, and microbes of both terrestrial and marine origin have been used as medicine to prevent, treat, and mitigate many diseases, since antiquity. Researchers have tried to unravel the hidden treasures in nature using the available traditional knowledge and ethnomedical claim. More than 80 % of the complementary and alternative medicines practiced are herbs (Harvey 2008). The diversity and complexity of chemical structures of natural products provide a valuable and rich pool for the discovery of new templates and drug candidates, viz., aspirin, digitoxin, morphine, quinine, and pilocarpine, along with the antibiotics, penicillin, streptomycin, chloramphenicol, chlortetracycline, cephalosporin, erythromycin, and vancomycin (Butler 2004). In the last decade, the rate of new chemical entities (NCEs) approved under "N" (natural origin) category has increased from 35 to 50 %. Half of the 20 approved NCEs and majority of the antitumor agents fall under the "N" category (David and Gordan 2012). The primary share among these natural medicines is held by the medicinal plants. Moreover, the advancement of new promising techniques, viz., bioassay-guided fractionation, improved screening format, reagent production, data management, and mechanism-based screening, has hastened the drug development procedure from natural sources. It has led to further exploration of the traditional herbal literature of different countries, especially Indian and Chinese, to get NCEs for the existing and new therapeutic targets (Cragg and Newman 2013).

The World Health Organization (WHO) estimated that herbal medicines provide primary health care for approximately 3.5–4 billion people worldwide (Farnsworth 1988). This increased demand of the herbal medicinal products is because of its believed safety and efficacy over the synthetic alternatives. However, with this increased demand of the herbal medicinal products, both as a single entity and extracts and fractions, the guidelines for the industries have become more stringent. USFDA has published draft guidance for industry on the quality of botanical drug products, which establishes a clear understanding on the marketing and use of herbal products as over-the-counter products and as new drug candidates. The USFDA is expected to introduce regulations for access to the market of medicinal products (Mahady 2001).

The prescription rate of herbal medicines has gone up to 4 % in the United States. Herbal medicine, folk medicine, traditional medicine, nutraceuticals, holistic medicine, natural medicine, and phytopharmaceuticals are the synonyms for the therapeutics derived from plant sources for treatment of various ailments (Obodozie 2012). The use of herbal medicines has increased nowadays as a complimentary therapy among the patients suffering from chronic, endemic, and social ailments. It has been observed that herbal medicines are being used by more than 80 % of the Asian and African population, and 65 % of the American population (Eisenberg et al. 1998). Furthermore, the use of herbal medicines is being supported by World Health Organization.

The ethnobotanical claims reported in the different traditional literature on herbs and their preparations have been utilized in the discovery of biologically active NCEs. This approach is based on the use; e.g., *Andrographis paniculata* was used for dysentery in ethnomedicine, and the compound responsible for the activity was isolated as andrographolide. Morphine from *Papaver somniferum*, berberine from *Berberis aristata*, and picroside from *Picrorhiza kurroa* are some examples of this approach (Katiyar et al. 2012). However, the presence of a large number of chemical entities makes the whole process of extraction, isolation, and identification of the pharmacologically active compound highly complex and cumbersome.

Moreover, these pharmacologically active chemical entities also have some disadvantages. They rarely have the same degree of activity as the unrefined extract at comparable concentrations or dose of the active component (Wagner and Ulrich-Merzenich 2009). It has been attributed to the absence of interacting substances present in the extract. These interacting substances are chemical entities having different physical and chemical properties. They have the potential to modify the bioavailability of a bioactive compound, thus resulting in reduced response. Inside the body, these components may behave differently and show some unpredictable interactions. One of our previously reported studies on oral bioavailability of vasicine and vasicinone demonstrated that the chief chemical compounds of *Adhatoda vasica*, viz., vasicine and vasicinone, are more bioavailable when administered as "vasa swaras" (aqueous extract of the dried leaves of *Adhatoda vasica* obtained after macerating the leaves in water for 12 h), rather than a single chemical entity at similar doses in rats (Dash et al. 2010). This study gave an insight that the pharmacokinetic basis of herbal extracts and fractions and their comparison with the single chemical entity present therein needs to be revisited. Therefore, we attempted to develop a capsule formulation of "vasa swaras" after lyophilization with an aim to enhance the stability and bioavailability of vasicine. The results of our study showed that the oral bioavailability of vasicine was higher from lyophilized "vasa swaras" as compared to vasicine as a single compound (Vyas et al. 2011). However, in the case of berberine, we observed it is more bioavailable orally when administered as a single compound rather than the herbal formulation named as "rasont," which is an aqueous decoction of the roots and stem barks of *Berberis aristata* (Patil et al. 2012). In another study, we observed that picrosides I and II were more bioavailable orally when administered as a crude extract of *Picrorhiza kurroa* as compared to an enriched extract "kutkin," which is a mixture of picrosides I and II (Upadhyay et al. 2013). Thus, it can be inferred that the pharmacokinetic and pharmacodynamic profile of an active chemical compound obtained from the herbal medicinal preparation is governed by the presence of other chemical moieties within the preparation, which may enhance or reduce the bioavailability of the chief chemical compound responsible for its activity on the therapeutic targets. Furthermore, there is a need for a justified approach in using these herbs for the treatment of diseases, as they were used in the traditional literature, or to formulate the preparation in the form of an equivalent alternative. Research on these aspects will lead to the development of a proper rationale for the efficacy and safety of the herbal medicinal products. Most of the published review articles on pharmacokinetic studies of herbal drugs have restricted their discussion on the pharmacokinetic behavior of various compounds and extracts of herbal origin and their interactions with other chemical constituents or physiological systems. However, we realized the need of presenting experimental design scaffold based on our practical experiences which may be helpful to other researchers for designing their experiments on the pharmacokinetics of herbal drugs. In this article, we would be discussing our unpublished research experience about various aspects of pharmacokinetics, which need to be addressed to generate reliable data on the safety and efficacy of herbal drugs.

12.2 Significance of Pharmacokinetics of Herbal Drugs

Herbal medicines have become popular these days as complementary therapies for various disorders such as metabolic disorders and cancer (Hollander and Mechanick 2008). These are available as single compounds, enriched extracts, or polyherbal preparations. The polyherbal formulations behave very differently as compared to the single compounds due to the interaction of different chemical constituents, present therein, with multiple receptors/targets in the biological system (Chan 1995; Plumb et al. 1999). The presence of the multiple chemical constituents is also likely to exhibit pharmacokinetic and metabolic interactions. Thus, it becomes difficult to monitor and predict the exact pharmacokinetic profile of all the chemical constituents present in a polyherbal formulation (He et al. 2011; Xie et al. 2012). As a result, pharmacokinetic studies of polyherbal formulations have been a long-standing hurdle in phytotherapy research. Conventional scientific approach for studying the pharmacological behavior of a multicomponent therapy is to isolate the individual compound and study the effect of these compounds on various targets and enzymes, independently. Based on the available findings, the conclusion is drawn. The same approach is followed for pharmacokinetic studies of polyherbal formulations. However, the data obtained from these types of studies do not reveal the real results (Xue and Roy 2003). The results of such experiments failed to capture the complex pharmacokinetic behavior of polyherbal medicines as the pharmacokinetic profile of a compound in a multicomponent assay may be significantly different from that in a single-compound assay due to drug-drug as well as drug-receptor interactions. Thus, it becomes technically challenging to handle the multicomponent assay due to the presence of a wide array of variables that includes the complexities in dose, routes of exposure, inter- and intra-individual genetic differences, gut microbiota, diet, lifestyle, and environment, biological matrices, physiological responses, as well as other xenobiotics intentionally or unintentionally, present therein (Nassar and Talaat 2004).

There is no regulatory need to study the pharmacokinetic profile of herbal drugs (Zhou et al. 2010). The herbal medicines undergo phase I and/or II metabolism to yield active or inactive metabolites. However, to rationalize the use of herbal medicines, a clear understanding of their pharmacokinetics and metabolism in the physiological system is required. This information will be helpful in establishing the potential dose-effect and dose-concentration relationships which would allow us to conduct proper therapeutic monitoring for herbal remedies and perform real individualized herbal therapy (He et al. 2010).

However, due to lack of appropriate pharmacokinetic data, herbal medicines are mostly used as a supplemental therapy and could not stand parallel regarding physician's reliability and market share on modern allopathic medicines. Thus, a fundamental knowledge of pharmacokinetics of phytopharmaceuticals will be helpful in authenticating their use. Recent advances in chromatographic and spectroscopic techniques such as LC, GC, and CE, as well as hyphenated techniques such as LC-MS, GC-MS, and CE-MS and NMR, have facilitated the simultaneous detection and quantification of various chemical constituents present in very low concentrations (Tolonen et al. 2009). The usefulness and application of these modern analytical techniques have been illustrated in the report by Amagaya et al. (2001). These modern analytical tools and techniques have been found to meet the regulatory standards for the generation of highly consistent pharmacokinetic and pharmacodynamic data in GLP and GCP laboratory set-up (Williamson and Evans 2000).

12.3 Regulatory Aspect of Herbal Drug Pharmacokinetics

Preclinical pharmacokinetic studies are one of the requirements for any investigational new drugs (INDs). The same condition is also applicable to herbal medicinal products, which is still lacking in most of the cases. There arises a need for a classified guideline that can elaborate various steps needed to perform the pharmacokinetics of herbal drugs. The use and launch of herbal medicines can be rationalized only after having a complete knowledge and understanding about their absorption, distribution, metabolism, and excretion as well as pharmacokinetic and pharmacodynamic interactions with the existing conventional/allopathic drugs. The primary challenges for pharmacokinetics of herbal drugs are the presence of multiple analytes, limitations in identifying the biomarkers, and inability to predict the metabolic fate during in vivo studies. Furthermore, there are no clear regulatory guidelines regarding the pharmacokinetic study of herbal drugs (Obodozie 2012).

12.4 Spectrometric and Chromatographic Analysis of Herbal Drugs

The development of an analytical method has always been a difficult task for herbal extracts, due to the presence of multiple components in the extracts. It becomes primarily difficult for an assay using HPLC, where there are always chances of elution of multiple compounds at the same retention time as that of the analyte of interest. Furthermore, there are probabilities of column damage due to the accumulation of various unwanted compounds during analysis. With the advent of mass spectrometry, this problem has been much reduced owing to higher specificity and sensitivity of the instrument. However, the optimization of the method using mass spectrometry is still problematic for herbal extracts, since many compounds may have a similar mass-to-charge ratio, which can enhance or suppress the response of the desired analyte. Furthermore, elution of any interfering compound with the same mass-to-charge ratio and same retention time of the desired analyte could further magnify this problem.

12.5 Purity of the Standards

The purity of the standard or reference materials used is of great importance. The standards used for the analysis need to be of the highest purity to avoid any hurdle in the development of a method. Especially, purity of the standard is important in a study, where we need to develop a method for some herbal extracts and simultaneously monitor concentrations of multiple analytes. Researchers prefer to develop a single method for simultaneous quantification of drug and its metabolite.

The simultaneous method development requires preparation of combined stock and working solutions containing all the analytes to be monitored in their respective concentrations. However, if one analyte contains another analyte as an impurity, it would be difficult to develop a method for simultaneous quantification of these two analytes. For example, if we need to develop a single method for quantifying vasicine and vasicinone, where vasicine standard contains vasicinone as an impurity, then the linearity and lower-quality control samples may fail for vasicinone (if its limit of quantification is equal or the same as that of vasicine) due to contribution of area from the vasicinone which is present as an impurity.

12.6 Design of Pharmacokinetic Experiments

The pharmacokinetic studies for herbal drugs are conducted for both isolated single compounds and extracts depending on their intended use. However, experimental design for both cases is different.

The pharmacokinetic studies of a single isolated compound are carried out like that for any NCE. However, the experimental design is different for herbal extracts. The first step in the pharmacokinetic studies for herbal extract is the collection of plant material for the preparation of herbal extracts or formulations. It is a known fact that the quantities of phytoconstituents vary on several factors such as the source of collection, time, place, altitude, etc. (Bernhoft 2010). Thus, it is very important to maintain consistency with these factors. It is advisable to collect sufficient amount of plant material that may suffice the need of entire study to avoid the variability. These materials may be shade-dried and stored in appropriate conditions (protecting from light and moisture) till use. Consistency in these factors may ease the standardization process and could provide a clue about the quantities of phytoconstituents. The second step involves preparation of the herbal extracts or formulations. For isolated single compound, it is possible to conduct both oral and intravenous pharmacokinetic studies. However, intravenous pharmacokinetic studies are difficult to perform for herbal extracts due to solubility-related issues. The solubility of all the chemical constituents present in an extract or polyherbal formulation is likely to be different in the selected solvent for intravenous dosing. The oral preparations either for isolated single compound or extract may be formulated as a suspension, but the intravenous preparation needs to be a homogenous solution. Subsequent identification of the chemical compound(s) that need to be monitored in the biological matrix has to be done. Based on the number and nature of analyte(s), a method has to be developed. A suitable analytical method has to be developed to quantify the analyte(s) that need to be monitored in the herbal extract or herbal preparation. The result of this experiment will be helpful in deciding the dose for pharmacokinetic studies and establishing the range of calibration curve. Another aspect that needs to be considered is the standardization of the herbal extracts by quantification of the amount of analytes (chief chemical constituent) in the herbal preparation that is to be dosed to animals. The next most important step is the dose calculation. The dose for pharmacokinetic studies need not be the same as that of the efficacy dose. The dose has to be finalized based on the sensitivity of the instrument and should not be too high. It is advisable to perform analysis on a mass spectrophotometer, as it facilitates quantification of even a minute quantity of an analyte. Another reason to perform the pharmacokinetic study at low dose is to avoid saturation-mediated kinetics. If the concentration of the analyte is too high in the formulation, then this may saturate the drug-metabolizing enzymes and inhibit the metabolism of the excess amount of active compound present in the system. This will subsequently lead to false-positive results about high oral bioavailability.

Another aspect which is important for dose selection is the concentration of individual analytes in the herbal extracts, which needs to be monitored in the biological matrix of pharmacokinetic studies. The concentration need not be same for the all the analytes. There arises a situation, where we can quantify one analyte, but others may be below the detection level. In this case, the dose linearity experiment needs to be designed in a small group of animals. The dose may be increased up to that point till we did not get enzyme saturation-mediated concentration and attain sufficient detectability, or we may increase the matrix volume to be analyzed. Thus, it is advisable to perform a pilot study in a small group of animals, which may give an idea regarding the dose and matrix volume to be analyzed. After finalization of dose and sensitivity requirement for analysis, a range of final calibration curve needs to be finalized. It should be followed by final analytical method optimization and validation. The final part of the experiment comprises of pharmacokinetic study and data analysis. Figure 12.1 shows the flowchart of a pharmacokinetic experimental design.

12.7 Significance of Traditional Herbal Formulations

Research pertaining to herbal drugs has been primarily focused on isolation and characterization of chemical compounds present therein for various therapeutic indications. However, one primary aspect which we should mind during our experimental design is the "type of formulation" as it was prescribed and prepared according to the ethnomedical claims. If we revisit the Ayurveda, which is a traditional Indian literature, we would find that most of the preparations were administered as asavas (alcoholic infusions), arishtas (alcoholic decoctions), bhashmas (metallic/mineral preparations), lehya (semisolid topical formulations), gutti (tablets), etc. The herbal therapies were practiced to achieve maximum efficacy with minimum toxicity and to attain optimum pharmacokinetic behavior. For example, most of the herbal preparations containing "brahmi," scientifically known as *Bacopa monnieri*, were administered along with "ghee" (butter made from the milk of a buffalo or cow) in order to improve the lipophilicity and enhance the permeability of its principal constituents, viz., bacosides across the blood-brain barrier (Dash and Sharma 2009). Thus, before the experimental design for herbal drugs pertaining to their activity, toxicity, and pharmacokinetic studies, the nature and properties of the

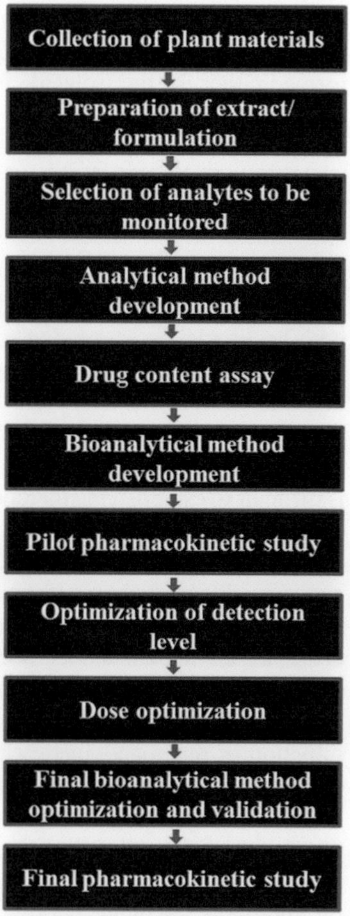

Fig. 12.1 Flowchart depicting a pharmacokinetic experimental design

traditional formulation need to be checked. Another aspect is to establish a comparison between the efficacy and pharmacokinetic behavior of the herbal drugs when administered as traditional preparation or as a single chemical entity. Based on the results, new approaches to formulations could be designed.

12.8 In Vitro Pharmacokinetic Studies of Herbal Drugs

In vitro pharmacokinetic studies have the same importance as that of in vivo studies, since these provide us information about the plasma protein binding, membrane transport, metabolite identification, CYP_{450} enzyme inhibition, metabolic stability, and time-dependent inhibition (Zhang et al. 2012). In vitro experiments can easily be performed with a single isolated compound; however, it is difficult to generate the in vitro study data for the herbal extracts, due to the presence of multiple compounds in the extract which may interfere with the behavior of the analyte of interest. Currently, the in vitro pharmacokinetic studies of herbal drugs are performed either for the isolated chief chemical compounds or the enriched fractions containing the chief chemical compounds.

12.9 Conclusions

Recognition of the medical and economic benefits of herbal medicinal products with health claims is growing worldwide. Herbal preparations are being manufactured and marketed by different pharmaceutical companies regardless of their standardization and clinical data. It results in ambiguities regarding the dose, pharmacokinetics, and safety aspects of these preparations. Without proper data on standardization, as well as pharmacokinetics and pharmacodynamics of herbals, it is difficult to establish the therapeutic potential of herbal preparations, scientifically. A plant-based medicinal product elicits its effect when the pharmacologically active compounds reach and maintain optimum concentration at the sites of action. The concentration at the target sites is primarily controlled by the dose levels and metabolic fate of the active compounds. To ensure the safe and optimal use of the herbal drugs, a clear understanding of the pharmacokinetic profile of the active compounds is required. Therefore, the pharmacokinetic and pharmacodynamic bases of herbal medicinal products need to be established. The assessment of bioavailability, determination of pharmacokinetic characteristics, and use of pharmacokinetic/pharmacodynamic modeling can aid in the most rational use of herbal products.

In summary, along with the search of new chemical entities for different therapeutic categories, there is a need for the establishment of pharmacological basis regarding safety and efficacy of the potent herbal drugs. Research must be focused on improving the bioavailability of the herbal drugs to make use of the potent molecules which fail to show optimum efficacy due to bioavailability-related problems. Amalgamation of new drug delivery system research with phytopharmaceutical research is needed for the development of novel dosage forms of natural products. Along with the formulation, emphasis must be given to the route of drug administration to get the optimum therapeutic benefit. Prospects on formulation from the traditional literature need to be decoded to resolve the problems related to bioavailability. Establishment of the pharmacokinetic parameters may be helpful in designing optimum dosage regimen for the herbal drugs. The synergistic activities of other chemical moieties present in the herbal drugs on the bioavailability of the chief therapeutic moiety need to be decoded. All these aspects may strengthen the Investigational New Drug/New Drug Application (IND/NDA) filing process of botanicals in USFDA and other European countries.

References

Amagaya S, Iizuka A, Makino B, Kubo M, Komatsu Y, Cheng FC, Ruo TI, Itoh T, Terasawa K (2001) General pharmacological properties of Sho-seiryu-to (TJ-19) extracts. Phytomedicine 8:338–347, doi: 0944-7113/01/08/05-338

Bernhoft A (2010) A brief review on bioactive compounds in plants. In: Bernhoft A (ed) Bioactive compounds in plants – benefits and risks for man and animals. The Norwegian Academy of Science and Letters, Norway, pp 11–17

Butler MS (2004) The role of natural product chemistry in drug discovery. J Nat Prod 67:2141–2153. doi:10.1021/np040106y

Chan K (1995) Progress in traditional Chinese medicine. Trends Pharmacol Sci 16:182–187. doi:10.1016/S0165-6147(00)89019-7

Cragg GM, Newman DJ (2013) Natural products: a continuing source of novel drug leads. Biochim Biophys Acta 1830:3670–3695. doi:10.1016/S0165-6147(00)89019-7

Dash B, Sharma RK (2009) Charaka Samhita. Chowkhamba Sanskrit Series Office, p 45, 58, 447, 456

Dash RP, Chauhan BF, Anandjiwala S, Nivsarkar M (2010) Comparative pharmacokinetics profile of Vasa Swaras with vasicine and vasicinone. Chromatographia 71:609–615. doi:10.1365/s10337-010-1517-x

David JN, Gordan MC (2012) Natural products as source of new drugs over the 30 years from 1981 to 2010. J Nat Prod 75:311–335. doi:10.1021/np200906s

Eisenberg DM, Davis RB, Ettner SL, Appel S, Wilkey S, Van Rompay M, Kessler RC (1998) Trends in alternative medicine use in the United States, 1990–1997: results of a follow-up national survey. JAMA 280:1569–1575. doi:10.1001/jama.280.18.1569

Farnsworth NR (1988) Screening plants for new medicines. In: Wilson EO (ed) Biodiversity. National Academy Press, Washington, DC, pp 83–97

Harvey AL (2008) Natural product in drug discovery. Drug Discov Today 13:894–901. doi:10.1016/j.drudis.2008.07.004

He SM, Li CG, Liu JP, Chan E, Duan W, Zhou SF (2010) Disposition pathways and pharmacokinetics of herbal medicines in humans. Curr Med Chem 17:4072–4133. doi:10.2174/092986710793205336

He SM, Chan E, Zhou SF (2011) ADME properties of herbal medicines in humans: evidence, challenges, and strategies. Curr Pharm Des 17:357–407. doi:10.2174/138161211795164194

Hollander JM, Mechanick JI (2008) Complementary and alternative medicine and the management of the metabolic syndrome. J Am Diet Assoc 108:495–509. doi:10.1016/j.jada.2007.12.007

Katiyar C, Gupta A, Kanjilal S, Katiyar S (2012) Drug discovery from plant sources: an integrated approach. Ayu 33:10–19. doi:10.4103/0974-8520.100295

Mahady GB (2001) Global harmonization of herbal health claims. J Nutr 131:1120S–1123S, doi: 0022-3166/0

Nassar AEF, Talaat RE (2004) Strategies for dealing with metabolite elucidation in drug discovery and development. Drug Discov Today 9:317–327. doi:10.1016/S1359-6446(03)03018-6

Obodozie OO (2012) Pharmacokinetics and drug interactions of herbal medicines: a missing critical step in the phytomedicine/drug development process. In: Noreddin A (ed) Readings in advanced pharmacokinetics-theory, methods and applications. InTech, Croatia, pp 127–156

Patil S, Dash RP, Anandjiwala S, Nivsarkar M (2012) Simultaneous quantification of berberine and lysergol by HPLC-UV: evidence that lysergol enhances the oral bioavailability of berberine in rats. Biomed Chromatogr 26:1170–1175. doi:10.1002/bmc.2674

Plumb GW, Garcia-Conesa MT, Kroon PA, Rhodes M, Ridley S, Williamson G (1999) Metabolism of chlorogenic acid by human plasma, liver, intestine and gut microflora. J Sci Food Agric 79:390–392. doi:10.1002/(SICI)1097-0010(19990301)79:3<390::AID-JSFA258>3.0.CO;2-0

Tolonen A, Turpeinen M, Pelkonen O (2009) Liquid chromatography-mass spectrometry in *in vitro* drug metabolite screening. Drug Discov Today 14:120–133. doi:10.1016/j.drudis.2008.11.002

Upadhyay D, Dash RP, Anandjiwala S, Nivsarkar M (2013) Comparative pharmacokinetic profiles of picrosides I and II from kutkin, *Picrorhiza kurroa* extract and its formulation in rats. Fitoterapia 85:76–83. doi:10.1016/j.fitote.2013.01.004

Vyas T, Dash RP, Anandjiwala S, Nivsarkar M (2011) Formulation and pharmacokinetic evaluation of hard gelatin capsule encapsulating lyophilized Vasa Swaras for improved stability and oral bioavailability of vasicine. Fitoterapia 82:446–453. doi:10.1016/j.fitote.2010.12.005

Wagner H, Ulrich-Merzenich G (2009) Synergy research: approaching a new generation of phytopharmaceuticals. Phytomedicine 16:97–110. doi:10.1016/j.phymed.2008.12.018

Williamson EM, Evans FJ (2000) Cannabinoids in clinical practice. Drugs 60:1305–1314. doi:10.2165/00003495-200060060-00005

Xie G, Zhao A, Zhao L, Chen T, Chen H, Qi X, Zheng X, Ni Y, Cheng Y, Lan K, Yao C, Qiu M, Jia W (2012) Metabolic fate of tea polyphenols in humans. J Proteome Res 11:3449–3457. doi:10.1021/pr300318m

Xue T, Roy R (2003) Studying traditional Chinese medicine. Science 300:740–741. doi:10.1126/science.300.5620.740

Zhang D, Luo G, Ding X, Lu C (2012) Preclinical experimental models of drug metabolism and disposition in drug discovery and development. Acta Pharm Sin B 2:549–561. doi:10.1016/j.apsb.2012.10.004

Zhou SF, Zhou ZW, Huang M (2010) Pharmacokinetics and disposition of plant natural products in humans and the clinical and toxicological implications. Curr Org Chem 14:1833–1868. doi:10.2174/138527210792927672

Chapter 13
Green Tea and Its Role in Cancer Prevention and Therapy

Uriel Bachrach and Zohara Yaniv

Abstract Cancer is one of the major causes of mortality worldwide, and despite desperate attempts, many patients still suffer from poor prognosis. Hence, efforts for discovering and developing more potent and effective anticancer agents continue. A growing body of research and experiments indicates the potential of some medicinal plants as a possible source of anticancer agents. In recent years, the health benefits of consuming green tea (derived from the plant *Camellia sinensis*) have been extensively documented. The ailments which can be treated and/or prevented include different types of cancer, heart and liver diseases, and neuroprotective and antioxidant activities. Many of these beneficial effects are related to tea catechins, particularly (-)-epigallocatechin-3-gallate (EGCG) content. Green tea consumption is also linked to the prevention of many types of cancer including breast, prostate, lung, colon, and stomach cancers. Moreover, cancer rates in Asian countries such as Japan and China where green tea is consumed in large quantities are significantly low according to epidemiological studies. These associations are confirmed by experiments with animals as well as cultured cancer cells. The use of EGCG instead of crude green tea extracts permitted studies to elucidate the mode of anticancer of green tea. Clinical studies demonstrating the prevention of cancer by green tea or by EGCG were recently questioned. The use of the nontoxic green tea or EGCG as anticancer agent is highly recommended.

Keywords Apoptosis • Cancer prevention • Cancer therapy • EGCG • Green tea • Oncogene • Signal transduction

U. Bachrach
Department of Molecular Biology, Hebrew University-Hadassah Medical School, Jerusalem, Israel
e-mail: urielb@ekmd.huji.ac.il

Z. Yaniv (✉)
Department of Genetic Resources, Institute of Plant Sciences, Agricultural Research Organization, The Volcani Center, Bet Dagan, Israel
e-mail: zoharayb@netvision.net.il

H.-S. Tsay et al. (eds.), *Medicinal Plants - Recent Advances in Research and Development*, DOI 10.1007/978-981-10-1085-9_13

Fig. 13.1 *Camellia sinensis* branch and flower (Wikipedia)

Abbreviations

EGCG	(-)-Epigallocatechin-3-gallate
MAPKs	Mitogen-activated protein kinases
ODC	Ornithine decarboxylase
PSA	Prostate-specific antigen

13.1 Introduction

Natural products, mainly plants and their constituents, have been used in the cure of diseases from ancient times. In the Bible, we find that Adam and Eve ate the fruits of the "Tree of Wisdom," which grew in Paradise. It resulted in "opening their eyes" and realizing that they were naked. So, it is clear that medicinal plants grew in Paradise. Tea is derived from the leaf of the plant *Camellia sinensis* (Fig. 13.1). Chinese people had become aware, as early as 4000–5000 years ago, that tea could cure and prevent some human diseases (Mair and Hoh 2009). Today, hundreds of millions of people around the world drink tea. It appears that one type of tea, in particular, green tea, has many health benefits. Green tea is processed by steaming fresh tea leaves immediately after harvest. The result is minimal oxidation of the naturally occurring polyphenols in the tea leaves. Black tea is processed by drying tea leaves and by crushing upon harvesting to encourage oxidation, which converts

Fig. 13.2 Various tea leaves (contribution of Zohara Yaniv)

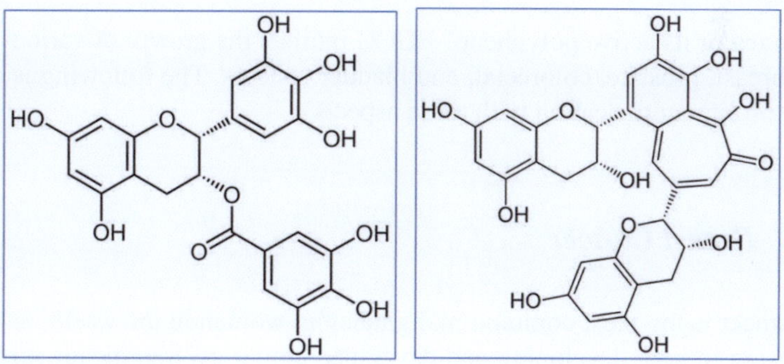

EGCG (green tea) Theaflavin (black tea)

Fig. 13.3 Tea polyphenols

indigenous tea polyphenols to inactive oxidation products. This results in the appearance of the typical red-brown color and aroma of black tea (Fig. 13.2). It is generally accepted that next to water, tea is the most consumed beverage in the world, with a per capita consumption of 120 ml/day (Katiyar and Mukhtar 1996). Of the total amount of tea produced in the world, 78 % is black, and 20 % is green. Black tea is used primarily in the West, whereas green tea is mainly consumed in Asian countries (Mukhtar and Ahmad 2000). Botanical evidence indicates that leaves of green tea contain polyphenols (Fig. 13.3) including (-)-epigallocatechin-3-gallate (EGCG), which amount to 8–12 % of the dried green tea leaves. One cup of green tea will contain approximately 30–40 mg of EGCG, which is claimed to possess many therapeutic activities. These include anticancer activities, cardiovascular therapies, antioxidant activities, and anti-inflammatory behaviors (Rahmani et al. 2015).

13.2 Anticancer Activities

Cancer is a multifunctional disease including genetic and metabolic alterations. This disease is still one of the major causes of mortality worldwide. Therefore, the effort for developing new and better anticancer agents has never been stopped. According to the tradition and experimental studies, a great number of medicinal plants have been reported to have anticancer properties (Bachrach 2012). Green tea extracts showed beneficial effects on cancer-related diseases and quality of life of individuals suffering from different forms of cancer. It has been demonstrated in in vitro and in vivo studies. Clinical trials supported the use of green tea extracts in treating cancer patients, and the results of the clinical studies became so popular that even the BBC stated on December 5, 2005, that:

> Green tea extract may help patients with a form of leukemia. It is exciting that research is now demonstrating that this agent may provide new hopes for chronic lymphocytic leukemia (CLL) patients.

Green tea or its active polyphenol-EGCG inhibits the growth of various cancers such as breast, prostate, colorectal, and bladder cancers. The following is a review of the latest literature dealing with these aspects.

13.2.1 Breast Cancer

Breast cancer is the most common malignancy in women in the world, and its rate is increasing in both developing and developed countries. Significant studies conducted in Asia, where green tea is consumed in large quantities, tend to show a low onset of breast cancer. The rate of breast cancer in China is 28.7 per 100,000 women per year, which is four- to fivefold lower than rates in developed countries (Bray et al. 2004). A case-control study conducted in Southeast China included more than 1000 women with confirmed breast cancer. These patients drank between 249 and 750 g of dried green tea leaves per annum. A significant inverse relationship between the severity of the disease and the amount of green tea consumed was established (Zhang et al. 2007). Epidemiological studies showed a strong negative correlation between consumption of green tea and the onset and frequency of many types of cancer. Thus, the onset of breast cancer in Japan was delayed by 8.7 years after daily drinking ten cups of tea, compared to a delay of 3.0 years after drinking only three cups of tea per day (Fujiki et al. 1998). The inverse association between risk of breast cancer and green tea intake was also observed among Japanese hospitalized patients (Inoue et al. 2001). These findings were not limited to people drinking green tea, as similar results were obtained when EGCG, and no crude green tea was consumed. Studies with Japanese breast cancer patients revealed (Nakachi et al. 1998) that increased consumption of EGCG decreased the recurrence of stage I and II breast cancer. No improvement in prognosis was observed in stage III breast cancer. Suganuma et al. (1999) also reported that high daily consumption of green tea was associated with a lower recurrence rate among stage I and II breast cancer patients.

Kumar et al. (2009) studied the effect of tea consumption on the risk of breast cancer in 5082 women in Wisconsin. They found a 37 % reduction in breast cancer risk among women who consumed at least three cups of green tea per day. Clement (2009) also observed that habitual green tea consumption attenuates the risk associated with breast cancer.

The reduced rate of cancer incidence among people in Asia, who consume green tea, cannot be attributed to genetic factors. Foreign-born Asian Americans who reside in North America and who consume Western food have an increased risk of cancer, based on the years of residence in North America (Shimizu et al. 1991). The rate for breast cancer was substantially higher when migration took place early in life. These findings suggest that lifestyle in the young rather than in later life is important in the etiology of breast cancer (Shimizu et al. 1991). Similarly, cancer rates for Korean American immigrants have increased for prostate, colon, and rectal cancers, based on the years of residence in the United States (Lee et al. 2007). Many authors agreed that green tea reduced the risk of breast cancer. However, recent studies conducted in Japan (Iwasaki et al. 2010a, b; Iwasaki and Tsugane 2011) showed that there was no correlation between breast cancer risk and drinking of green tea. These studies should be repeated and extended.

13.2.2 Prostate Cancer

Every year nearly 200,000 men in the United States are diagnosed with prostate cancer, and the other 29,000 succumb to the disease (Johnson et al. 2010). Genetic backgrounds may contribute to prostate cancer risk. Men who have a first-degree relative (father or brother) with prostate cancer have twice the risk of developing prostate cancer, and those with two first-degree relatives affected have a fivefold greater risk compared with men with no family history. About 99 % of cases occur in those over the age of 50 years (Johnson et al. 2010). Prostate cancer incidence in China, where green tea is consumed in large quantities, is the lowest in the world. Gupta et al. (2001) and Adhami et al. (2004) reported that tea polyphenols could inhibit the development of prostate cancer in animals. The chemopreventive effect of green tea was studied by Henning et al. (2011), who found that green tea extracts were more effective than black tea extracts. Bettuzzi et al. (2006) reported that in humans, a high dose of green tea catechins (200 mg catechin – given three times per day) prevented prostate cancers. The incidence of cancer in men in the treated group was 3 %, while it was 30 % in the control group. It implies that a chemopreventive effect of 90 % was achieved. A larger cohort study involving 49,000 Japanese men who were followed up for 14 years showed a marked reduction in the risk of advanced prostate cancer in those with habitual consumption of green tea (Kurahashi et al. 2008; McLarty et al. 2009) and reported that patients with prostate cancer, who received a daily dose of 800 mg of EGCG, showed a significant reduction of the prostate-specific antigen (PSA), with no elevation of liver enzymes. They concluded that EGCG could play a potential role in the treatment or prevention of prostate cancer. Kurahashi et al. (2008) studied the effect of green tea consumption on 404 cancer patients in Japan and found a dose-dependent decrease in the risk of advanced prostate cancer. In China, the effect of drinking more than three cups of green tea on the development of prostate cancer was also studied (Jian et al. 2004). Again, it was found that prostate cancer risk declined with increasing frequency, duration, and quantity of green tea consumed. Henning et al. (2011) studied the chemopreventive effects of tea on prostate cancer and found that the effect of black tea was much weaker than green tea.

On the other hand, recent studies (Kumar et al. 2015) raised some questions whether EGCG, 200 mg per day, could reduce the likelihood of prostate cancer diagnosis and whether green tea could be used for the prevention of prostate cancer. These studies should be repeated and confirmed.

13.2.3 Other Types of Cancer

Chronic Lymphocytic Leukemia (CLL) In 2013, 42 patients with chronic lymphocytic leukemia were treated with EGCG at a dose of 2000 mg, twice daily for 6 months. This treatment resulted in the reduction of absolute lymphocyte count in 29 (69 %) of the patients (Shanafelt et al. 2013).

Colorectal Cancer This cancer represents the third most common and the second deadliest type of cancer for both men and women in the United States claiming over 50,000 lives in 2014 (Pabla et al. 2015). Yang et al. (2007) evaluated the association between green tea consumption and colorectal cancer risk in a cohort of 67,710 Chinese women. They found an inverse association between green tea drinkers and cancer for both colon and rectal cancers. Kumar et al. (2007) and Larsen and Dashwood (2009) reported that EGCG might be a beneficial therapeutic agent in treating colon cancer. A cohort study in Singapore (Sun et al. 2007) found that, for men, green tea, but not black tea, consumption affected the risk of the advanced stage of colon cancer.

Stomach Cancer It was shown that among 711 green tea drinkers in Shanghai, the incidence of stomach cancer was low (Yu et al. 1995). Wang et al. (2015) studied the effect of green tea extracts on 160 stomach cancer patients. They found that larger amount of consumption, lower temperature, and longer interval were strongly associated with a lower risk of stomach cancer.

Cervical Lesions Ahn et al. (2003) investigated the clinical efficacy of green tea extracts given to 27 patients with cervical lesions caused by human papillomavirus (HPV). These patients received 200 mg of EGCG for 12 weeks. A 69 % response rate was noted in patients treated with green tea extracts compared with a 10 % response rate in untreated controls.

Biliary Tract Cancer Consumption of 120 ml green tea extracts per day decreased the risk of biliary tract cancer (Makiuchi et al. 2016).

Liver Cancer The influence of drinking green tea on 3694 liver cancer patients from China, Japan, and Singapore was tested (Huang et al. 2016). A significant association between highest green tea consumption and reduced risk of liver cancer was reported. No association was observed when patients drank only one cup of green tea per day. The protective effect of green tea consumption on the risk of liver cancer was observed only for the group of Asian women, but not for men.

Gastric Cancer Five gastric cancer cell lines were found to be sensitive to EGCG treatment and induced apoptosis (Onoda et al. 2011).

13.2.4 Brain Functions

The loss of cognitive function due to the structure and function damage of neuronal cells is a common process including Parkinson's and Alzheimer's diseases. A study conducted in the United States demonstrated a decrease in Parkinson's disease in the population who consumed two cups or more of green tea per day (Hu et al. 2007). Similar results were obtained by studying how green tea drinking affected cognition of people in Malta (Caruana and Vassallo 2015). A slower progression of Alzheimer's disease was observed in humans treated with EGCG and conducting voluntary exercise (Walker et al. 2015). Green tea and the EGCG compound, in particular, could boost memory and could even be of benefit in the prevention of various neurodegenerative diseases (Winreb et al. 2008).

13.3 Modifications of the Active Components

13.3.1 Nanoparticles

The possibility of nanotechnology was put forward to improve the bioavailability of the active components of green tea. Siddiqui et al. (2009) proposed to control cancer by using polylactic acid-polyethylene glycol nanoparticles that encapsulated EGCG. Results indicated that the effect of the new formulations on prostate cancer was ten times higher than that of the formulations with non-nanomaterials after 24 h of administration. In inhibition of apoptosis, 3 μg nano-EGCG showed 57 % inhibition, whereas 30 μg ordinary EGCG showed a 35 % inhibition (Siddiqui et al. 2010).

dePace et al. (2013) synthesized EGCG-encapsulated chitosan-coated nanoliposomes. The stability of EGCG was significantly enhanced, and the inhibition of proliferation of breast cancer cells was improved.

13.3.2 Structure Modification

Lambert et al. (2006) improved the activity of EGCG in the human body by structural chemical modification of the drug. They substituted eight OH groups of EGCG by OAc groups. This new molecule was termed "a prodrug." When the new compound entered the cells, a new EGCG drug was formed by the activity of the enzyme esterase. The amount of the new EGCG was increased up to 30 times. When pro-EGCG was orally administered to rats, the anticancer activity was increased by twofold (Landis-Piwowar et al. 2007). Dou et al. (2008) and Dou (2009) also prepared a peracetate-protected EGCG and demonstrated its potential use in cancer prevention and treatment. Two novel fluoro-substituted EGCG analogs increased the anticancer activity of green tea polyphenols (Yang et al. 2010). Gelatin-based 200 nm nanoparticles (Shutava et al. 2009) and colloidal mesoporous silica-encapsulated EGCG greatly promoted the efficacy of EGCG on breast tumors (Ding et al. 2015). Other chemical modifications of EGCG were proposed by Yi et al. (2014). These studies opened new possibilities for improving the biological activities of EGCG.

13.3.3 Safety

Along with the use of green tea as a therapeutic agent, its toxicity was also investigated. Mazzanti et al. (2015) found that liver diseases resulted if high concentrations of green tea extracts were used as antiobesity agents (containing 25 % catechins). If the amounts of daily green tea extracts are controlled (daily dose of 30 g tea polyphenols), no toxicity was observed. However, they recommended that green tea extracts should not be taken on an empty stomach for the safety of the consumers. It was concluded that the consumption of highly concentrated green tea extracts in the empty stomach was more likely to lead to adverse effects (Sarma et al. 2008). Pregnant women should also avoid drinking green tea extracts at high concentrations. Lambert et al. (2010) administered 1500 mg/kg EGCG orally to mice and found that the activity of alanine aminotransferases increased by 138 times and the survival was decreased by 85 %. These studies implied that high dosages of EGCG were toxic to the liver. It has been suggested that the equivalent of seven to eight cups, three times daily, can be taken safely for at least 6 months (Pisters et al. 2001).

13.3.4 *Selective Toxicity*

Selective toxicity means that the drug must be effective against the target but have minimal or no toxicity to humans or normal counterparts. In general, the efficacy of the treatment of cancer is directly proportional to the ability of the drug to selectively target the cancer cell, thus improving the life of the patients. Green tea has been used for many years without observing disturbing side effects. The selective anticancer toxicity of green tea and/or EGCG was also confirmed experimentally.

Chen et al. (1998) studied the effect of EGCG on breast cancer cells and their respective normal counterparts. After exposure to EGCG at 200 μM, for 8 h, more than 50 % of the cancer cells became apoptotic. In contrast, less than 1 % of the normal counterpart was affected. Wang and Bachrach (2002) tested the effect of EGCG on fibroblasts in which transformation by H-*ras* was controlled. It has been demonstrated that EGCG did not affect the growth of normal fibroblast, whereas the transformed cells became apoptotic.

13.4 Clinical Studies

The first clinical trials on the effects of tea polyphenols on humans were conducted at the MD Anderson Cancer Center in collaboration with the Memorial Sloan Kettering Cancer Center. To examine the safety and possible efficacy of consuming the equivalent of >10 cups of green tea per day, 30 cancer patients suffering from advanced solid tumors were treated with green tea for >6 months. The treatment appeared to be beneficial (Mukhtar and Ahmad 2000).

Bettuzzi et al. (2006) tested 60 volunteers with the predominant premalignant lesion of prostate cancer. The patients received 600 mg of green tea extracts daily. One year later, only 3 % of the patients who received the green tea extract developed prostate cancer, compared with 30 % in the placebo group. Choan et al. (2005) treated 19 patients with prostate cancer with green tea extracts (250 mg twice daily) for 2 months. This treatment reduced the levels of prostate-specific antigens (PSAs). A clinical trial with 60 volunteers who received 600 mg of green tea extracts per day for 1 year revealed that in the treated patient, only one developed prostate cancer, while nine cancers were found in the placebo controls Brausi et al. (2008). If the green tea extracts contain caffeine in high concentrations, gastrointestinal effects were observed.

13.5 Mode of Action

After establishing the biological activities of green tea and of EGCG, it was of great importance to find out the mode of their actions. It could provide some information whether toxic side effects could be avoided.

13.5.1 *Tyrosine Kinase*

The scheme of signal transduction pathways is illustrated in Fig. 13.4a. Membrane-associated receptors for tyrosine kinase which appear on the top of the scheme demonstrated an important role in signal transduction and malignant transformation processes. Increased tyrosine phosphorylation of a 130 kD protein has been implicated in cell transformation by c-H-*ras* and by Src family tyrosine kinase (Wang and Bachrach 2002). The phosphorylation of the 130 kD protein in the transformed cells was twofold higher than those of the normal cells. Adding 20 μM EGCG to those transformed cells caused a 50 % inhibition of tyrosine phosphorylation. However, no decrease in phosphorylation was observed in the neutral counterpart kinase (Wang and Bachrach 2002). These findings suggest that EGCG elicited a specific and a preferred anticancer effect. Similar results were obtained by Larsen et al. (2010), Shirakami et al. (2012), and Colomer et al. (2016), who confirmed that the anticancer activity of EGCG could be linked to the inhibition of tyrosine phosphorylation.

13.5.2 *Mitogen-Activated Protein Kinases (MAPKs)*

Growth factors (mitogens) are known to bind to specific receptors which are located on the cellular membrane. The mitogen receptor complexes then trigger a cascade of events, such as the activation of the oncogene Ras. The activation of protein kinases is considered as the next step in signal transduction (Fig. 13.4a). MAPKs are phosphorylated by MAPK/ERK kinases, which in turn are activated by Raf. Human breast cancer cells, exposed to EGCG, revealed a dose-dependent growth inhibition and a decrease in cyclin-dependent kinases (Deguchi et al. 2002). Wang and Bachrach (2002) reported that the concentration of Ras was higher in the fibroblast-transformed cells, as compared to the normal counterpart. The content of Ras in normal cells was hardly affected by EGCG at 5 μM concentration, while the concentration of this protein in the transformed cells was reduced by 35 %. Higher concentrations of EGCG (10 μM) reduced the Ras content of normal cells by 17 % and that of the transformed cells by approximately 50 %. These findings demonstrate that EGCG exerts an inhibitory effect on Ras synthesis. Normal NIH fibroblasts were treated with EGCG for 12 h and then exposed to transforming agents to explore the possibility that EGCG can prevent carcinogenesis. EGCG at 10 μM concentrations inhibited the expression of ERK 1/2 and MAPK (Wang and Bachrach 2002). Similar results were reported by Chen et al. (2008) and Singh et al. (2011). These findings suggest that EGCG can prevent cancer in addition to its therapeutic activity.

Jun is a nuclear proto-oncogene. It constitutes a part of the transcription factor AP-1 and therefore plays an important role in growth processes. The effect of EGCG on the expression of transcription factors was studied in vitro. It was found by Lai

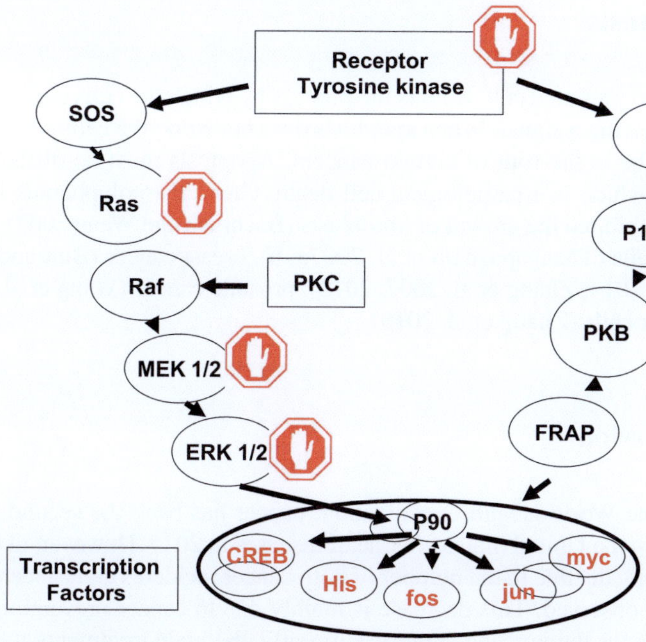

Fig. 13.4a Signal transduction (contribution of the author)

et al. (2004) and by Bachrach and Wang (2002) that green tea EGCG downregulated the expression of c-Fos and c-Jun. The effect of EGCG on the expression of c-Fos and c-Myc was also studied by Chen et al. (1998).

The enzyme ornithine decarboxylase (ODC, EC. 4.1.1.17) catalyzes the conversion of ornithine into putrescine (Fig. 13.4b). It is the most effective rate-limiting step in the biosynthesis of the polyamines: spermidine and spermine. These aliphatic polycations are closely linked with growth processes, and their concentrations increase in cancer cells. Overproduction of ODC is associated with malignant transformation. On the other hand, inhibition of the activity of ODC causes cell apoptosis. It was therefore of interest to find out whether green tea and/or EGCG inhibit ODC activity, prevent growth, and affect carcinogenesis. Wang and Bachrach (2002) showed that the activity of ODC in fibroblast-transformed cells was twofold higher than that in the normal controls. After exposure to 5 μM of EGCG for 12 h, the activity of ODC in the transformed cells was similar to the activity of normal cells (Wang and Bachrach 2002). Similar results were also reported by Stoner and Mukhtar (1995). It can be concluded that EGCG can inhibit the activity of ODC and thus impair growth. If the activity of ODC is inhibited by green tea or by EGCG, then putrescine levels will decline (Fig. 13.4b). It will subsequently lead to the inhibition of the synthesis of spermidine, which in turn will reduce the induction of the oncogenes myc and jun (Fig. 13.4b) and inhibit cellular proliferations. A similar scheme for the effect of green tea on signal transduction processes was also proposed by Rahmani et al. (2015).

13.6 Apoptosis

Apoptosis or programmed cell death is the process by which the cells destroy themselves after receiving a signal. When apoptosis does not work, the cells can multiply freely and can be at the root of certain cancers. Apoptosis must be differentiated from necrosis, which is a pathological cell death. Green tea polyphenols induced apoptosis and inhibited the growth of fibroblasts (Bachrach and Wang 2002), human breast cancer cells (Thangapazham et al. 2007a, b), breast cancer (Butt and Sultan 2009; Hu et al. 2014; Zhang et al. 2007, 2012), prostate cancer (Wang et al. 2014), and liver cancer cells (Zhang et al. 2015).

13.7 Discussion

According to the World Health Organization, cancer has been the second leading cause of death in the United States after heart diseases in 2015. However, in the past few years, a slight decline in the number of both cancer incidents and cancer-related deaths is being observed. This decrease is mainly due to cancer prevention and to new approaches for therapy and surgery. Currently, the main treatments for cancer are chemotherapy, radiotherapy, and surgery. Most of the drugs used in chemotherapy are accompanied by several unwanted side effects. Also, patients with poor prognosis are not cured by the drugs. Therefore, a search for new anticancer agents with fewer side effects and higher efficiency and accuracy received high priority. Natural compounds are good sources for the development of new remedies for different disease. A number of phytochemicals isolated from medicinal plants have been shown to decrease cell proliferation and to induce apoptosis. Epidemiologic observations and laboratory studies have indicated that tea consumption may have beneficial effects in reducing certain types of cancer in some populations. One of the key issues in chemoprevention with phytochemicals is to find out whether the activity is due to a single active compound present in the extract. Numerous studies indicated that (-)-epigallocatechin-3-gallate (EGCG) is the active component in green tea but is not found in black tea. There is evidence from in vitro and animal studies that green tea and EGCG are potent antioxidants, inhibit cancer, prevent atherosclerosis, and control blood sugar in the body. Researchers at the University of Kansas feel that EGCG is at least 100 times more effective than vitamin C and 25 times better than vitamin E in protecting cells and their genetic material, DNA, from damage by free radicals. EGCG carries twice the antioxidant of resveratrol found in red wine. In this review, we focus our attention on anticancer activities of green tea and EGCG. We provide evidence that these phytochemicals inhibit the growth of various types of cancer both in vivo and in vitro studies. These remedies, which cause apoptosis, have been used in different countries without reporting any harmful side effects. It appears that they preferentially and selectively attack cancer cells but do not harm healthy counterparts. Therefore, they seem to be an ideal cancer

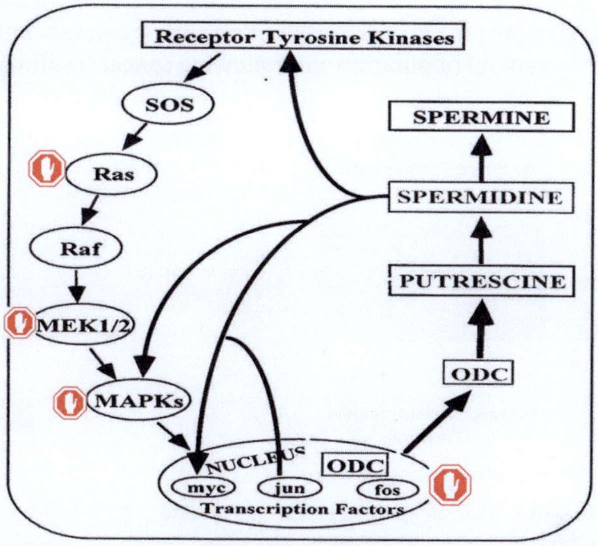

Fig. 13.4b Oncogenes (= inhibition)

agent. Moreover, as EGCG is a single pure and defined reagent, its stability and its mode of action can be monitored easily. Green tea or EGCG can arrest growth of cancer by different mechanisms which include antioxidant action and apoptosis. However, it appears that cell cycle arrest by modulating signal transduction processes can explain the biological activities of green tea and EGCG. Figures 13.4a and 13.4b clearly shows that these phytochemicals interfere with the phosphorylation of tyrosine and protein kinases like ERK1/2 and also suppress the expression of Jun and Fos oncogenes. The synthesis of polyamines, which promote growth, is also inhibited by green tea, due to the reduction of ornithine decarboxylase (ODC) activity (Fig. 13.4b). Therefore, green tea has been shown to exert its chemopreventive effect by targeting various intracellular signaling cascades.

Green tea and EGCG inhibit the proliferation of various cancer cells without affecting normal counterparts. Therefore, green tea can prevent cancer before the onset of the disease or block the progress of the syndrome in sick patients (Fig. 13.5). As EGCG can be obtained from commercial sources, it can serve as an important tool to combat cancer.

13.8 Conclusions

As previously stated, cancer is a major cause of mortality worldwide. Currently, the main treatments for cancer are chemotherapy, radiotherapy, and surgery. Most of the drugs used in chemotherapy are accompanied by several unwanted side effects. Therefore, a search for new anticancer agents with fewer side effects and higher efficiency and accuracy received high priority. Dietary habits influence the risk of

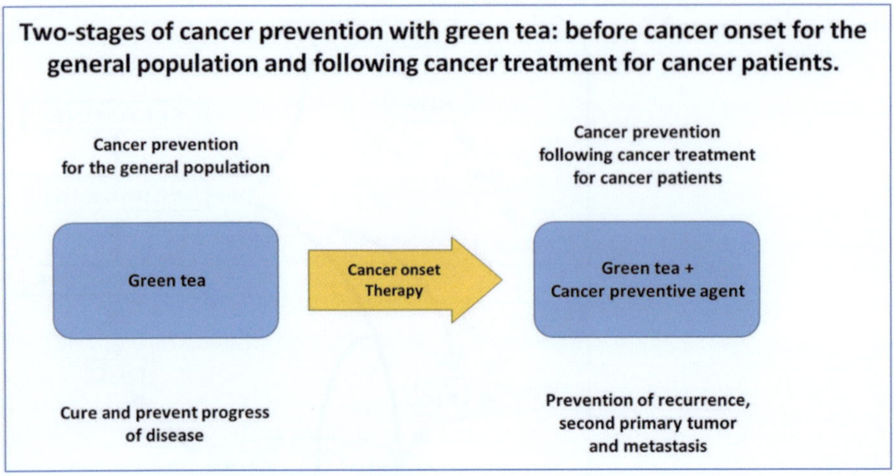

Fig. 13.5 Cure and prevent progress of disease

developing a variety of diseases, especially cancer. Tea derived from the leaf of the plant *Camellia sinensis* is, next to water, the most consumed beverage in the world. When the leaves of *Camellia sinensis* are steamed, green tea is produced. Epidemiologic observations and laboratory studies indicated that green tea consumption might have beneficial effects in reducing certain types of cancers and in preventing its outbreak. Thus, the incidence of cancer is very low in Asian countries like China and Japan, where green tea is consumed in large quantities. Numerous studies suggested that the polyphenol (-)-epigallocatechin-3-gallate (EGCG) is the active component of green tea. The use of EGCG instead of crude green tea extracts permitted studies to elucidate the mode of anticancer of green tea. Green tea or EGCG can arrest the growth of cancer by different mechanisms which include antioxidant actions and apoptosis. They also regulate oncogene activities and modulate signal transduction processes. As green tea and EGCG inhibit the proliferation of cancer cells without affecting normal counterparts, their use as anticancer agents is highly recommended.

References

Adhami VM, Siddiqui IA, Ahmad N, Gupta S, Mukhtar H (2004) Oral consumption of green tea polyphenols inhibits insulin-like growth factor-I-induced signaling in an autochthonous mouse model of prostate cancer. Cancer Res 64(23):8715–8722

Ahn WS, Yoo J, Huh SW, Kim CK, Lee JM, Namkoong SE, Bae SM, Lee IP (2003) Protective effects of green tea extracts (polyphenon E and EGCG) on human cervical lesions. Eur Cancer Prev Organ 12(5):383–390

Bachrach U, Wang YC (2002) Cancer therapy and prevention by green tea: role of ornithine decarboxylase. Amino Acids 22(1):1–13

Bachrach U, Wang YC, Tabib A (2001) Polyamines: new cues in cellular signal transduction. News Physiol Sci 16:106–109

Bettuzzi S, Brausi M, Rizzi F, Castagnetti G, Peracchia G, Corti A (2006) Chemoprevention of human prostate cancer by oral administration of green tea catechins in volunteers with high-grade prostate intraepithelial neoplasia: a preliminary report from a one-year proof-of-principle study. Cancer Res 66(2):1234–1240

Brausi M, Rizzi F, Bettuzzi S (2008) Chemoprevention of human prostate cancer by green tea catechins: two years later. A follow-up update. Eur Urol 54(2):472–473

Bray F, McCarron P, Parkin DM (2004) The changing global patterns of female breast cancer incidence and mortality. Breast Cancer Res 6(6):229–239. doi:10.1186/bcr932

Butt MS, Sultan MT (2009) Green tea: nature's defense against malignancies. Crit Rev Food Sci 49(5):463–473

Caruana M, Vassallo N (2015) Tea polyphenols in Parkinson's disease. Adv Exp Med Biol 863:117–137

Chen D, Milacic V, Chen MS, Wan SB, Lam WH, Huo C, Landis-Piwowar KR, Cui QC, Wali A, Chan TH, Dou QP (2008) Tea polyphenols, their biological effects and potential molecular targets. Histol Histopathol 3(4):487–496

Chen ZP, Schell JB, Ho CT, Chen KY (1998) Green tea epigallocatechin gallate shows a pronounced growth inhibitory effect on cancerous cells but not on their normal counterparts. Cancer Lett 129(2):173–179

Choan E, Segal R, Jonker D, Malone S, Reaume N, Eapen L, Gallant VA (2005) Prospective clinical trial of green tea for hormone refractory prostate cancer: an evaluation of the complementary/alternative therapy approach. Urol Oncol 23(2):108–113

Clement Y (2009) Can green tea do that? A literature review of the clinical evidence. Prev Med 49(2–3):83–87

Colomer R, Sarrats A, Lupu R, Puig T (2016) Natural polyphenols and their synthetic analogs as emerging anticancer agents. Curr Drug Targets PMID 26758667

de Pace RC, Liu X, Sun M, Nie S, Zhang J, Cai Q, Gao W, Pan X, Fan Z, Wang S (2013) Anticancer activities of (-)-epigallocatechin-3-gallate encapsulated nanoliposomes in MCF7 breast cancer cells. J Lipos Res 23(3):187–196

Deguchi H, Fujii T, Nakagawa S, Koga T, Shirouzu K (2002) Analysis of cell growth inhibitory effects of catechin through MAPK in human breast cancer cell line T47D. Int J Oncol 21(6):1301–1305

Ding J, Yao J, Xue J, Li R, Bao B, Jiang L, Zhu JJ, He Z (2015) Tumor-homing cell-penetrating peptide linked to colloidal mesoporous silica encapsulated (-)-epigallocatechin-3-gallate as drug delivery system for breast cancer therapy in vivo. Appl Mater Interfaces 7(32):18145–18155

Dou QP (2009) Molecular mechanisms of green tea polyphenols. Nutr Cancer 61(6):827–835

Dou QP, Landis-Piwowar KR, Chen D, Huo C, Wan SB, Chan TH (2008) Green tea polyphenols as a natural tumor cell proteasome inhibitor. Inflammopharmacology 16(5):208–212

Fujiki H, Suganuma M, Okabe S, Sueoka N, Komori A, Sueoka E, Kozu T, Tada Y, Suga K, Imai K, Nakachi K (1998) Cancer inhibition by green tea. Mutat Res 402(1–2):307–310

Gupta S, Hastak K, Ahmad N, Lewin JS, Mukhtar H (2001) Inhibition of prostate carcinogenesis in TRAMP mice by oral infusion of green tea polyphenols. Proc Natl Acad Sci U S A 98(18):10350–10355

Henning SM, Wang P, Heber D (2011) Chemopreventive effects of tea in prostate cancer: green tea versus black tea. Mol Nutr Food Res 55(6):905–920

Hu G, Bidel S, Jousilahti P, Antikainen R, Tuomilehto J (2007) Coffee and tea consumption and the risk of Parkinson's disease. Mov Disord 22(15):2242–2248

Hu G, Zhang L, Rong Y, Ni X, Sun Y (2014) Downstream carcinogenesis signaling pathways by green tea polyphenols: a translational perspective of chemoprevention and treatment for cancers. Curr Drug Metab 15(1):14–22

Huang YQ, Lu X, Min H, Wu QQ, Shi XT, Bian KQ, Zou XP (2016) Green tea and liver cancer risk: a meta-analysis of prospective cohort studies in Asian populations. Nutrition 32(1):3–8

Inoue M, Tajima K, Mizutani M, Iwata H, Iwase T, Miura S, Hirose K, Hamajima N, Tominaga S (2001) Regular consumption of green tea and the risk of breast cancer recurrence: a follow-up study from the Hospital-based Epidemiologic Research Program at Aichi Cancer Center (HERPACC), Japan. Cancer Lett 167(2):175–182

Iwasaki M, Tsugane S (2011) Risk factors for breast cancer: epidemiological evidence from Japanese studies. Cancer Sci 102(9):1607–1614

Iwasaki M, Inoue M, Sasazuki S, Miura T, Sawada N, Yamaji T, Shimazu T, Willett WC, Tsugane S (2010a) Plasma tea polyphenol levels and subsequent risk of breast cancer among Japanese women: a nested case-control study. Breast Cancer Res Treat 124(3):827–834

Iwasaki M, Inoue M, Sasazuki S, Sawada N, Yamaji T, Shimazu T, Willett WC, Tsugane S (2010b) Green tea drinking and subsequent risk of breast cancer in a population-based cohort of Japanese women. Breast Cancer Res 12(5):R88

Jian L, Xie LP, Lee AH, Binns CW (2004) Protective effect of green tea against prostate cancer: a case-control study in southeast China. Int J Cancer 108(1):130–135

Johnson JJ, Bailey HH, Mukhtar H (2010) Green tea polyphenols for prostate cancer chemoprevention: a translational perspective. Phytomedicine 17(1):3–13

Katiyar S, Mukhtar H (1996) Tea in chemoprevention of cancer. Int J Oncol 8(2):221–238

Kumar N, Shibata D, Helm J, Coppola D, Malafa M (2007) Green tea polyphenols in the prevention of colon cancer. Front Biosci 12:2309–2315

Kumar N, Titus-Ernstoff L, Newcomb PA, Trentham-Dietz A, Anic G, Egan KM (2009) Tea consumption and risk of breast cancer. Cancer Epidemiol Biomarkers Prev 18(1):341–345

Kumar NB, Pow-Sang J, Egan KM, Spiess PE, Dickinson S, Salup R, Helal M, McLarty J, Williams CR, Schreiber F, Parnes HL, Sebti S, Kazi A, Kang L, Quinn G, Smith T, Yue B, Diaz K, Chornokur G, Crocker T, Schell MJ (2015) Randomized, placebo-controlled trial of green tea catechins for prostate cancer prevention. Cancer Prev Res 8(10):879–887

Kurahashi N, Sasazuki S, Iwasaki M, Inoue M, Tsugane S, Group JS (2008) Green tea consumption and prostate cancer risk in Japanese men: a prospective study. Am J Epidemiol 167(1):71–77

Lai HC, Chao WT, Chen YT, Yang VC (2004) Effect of EGCG, a major component of green tea, on the expression of Ets-1, c-Fos, and c-Jun during angiogenesis *in vitro*. Cancer Lett 213(2):181–188

Lambert JD, Sang S, Hong J, Kwon SJ, Lee MJ, Ho CT, Yang CS (2006) Peracetylation as a means of enhancing in vitro bioactivity and bioavailability of epigallocatechin-3-gallate. Drug Metab Dispos 34(12):2111–2116

Lambert JD, Kennett MJ, Sang S, Reuhl KR, Ju J, Yang CS (2010) Hepatotoxicity of high oral dose (-)-epigallocatechin-3-gallate in mice. Food Chem Toxicol 48(1):409–416

Landis-Piwowar KR, Huo C, Chen D, Milacic V, Shi G, Chan TH, Dou QP (2007) A novel prodrug of the green tea polyphenol (-)-epigallocatechin-3-gallate as a potential anticancer agent. Cancer Res 67(9):4303–4310

Larsen CA, Dashwood RH (2009) Suppression of Met activation in human colon cancer cells treated with (-)-epigallocatechin-3-gallate: minor role of hydrogen peroxide. Biochem Biophys Res Commun 389(3):527–530

Larsen CA, Dashwood RH, Bisson WH (2010) Tea catechins as inhibitors of receptor tyrosine kinases: mechanistic insights and human relevance. Pharmacol Res 62(6):457–464

Lee J, Demissie K, Lu SE, Rhoads GG (2007) Cancer incidence among Korean-American immigrants in the United States and native Koreans in South Korea. Cancer Control 14(1):78–85

Mair VH, Hoh E (2009) The true history of tea. Thames & Hudson, New York

Makiuchi T, Sobue T, Kitamura T, Ishihara J, Sawada N, Iwasaki M, Sasazuki S, Yamaji T, Shimazu T, Tsugane S (2016) Association between green tea/coffee consumption and biliary tract cancer: a population-based cohort study in Japan. Cancer Sci 107(1):76–83

Mazzanti G, Di Sotto A, Vitalone A (2015) Hepatotoxicity of green tea: an update. Arch Toxicol 89(8):1175–1191

McLarty J, Bigelow RL, Smith M, Elmajian D, Ankem M, Cardelli JA (2009) Tea polyphenols decrease serum levels of prostate-specific antigen, hepatocyte growth factor, and vascular endothelial growth factor in prostate cancer patients and inhibit production of hepatocyte growth factor and vascular endothelial growth factor in vitro. Cancer Prev Res 2(7):673–682

Mukhtar H, Ahmad N (2000) Tea polyphenols: prevention of cancer and optimizing health. Am J Clin Nutr 71(6 Suppl):1698S–1702S, discussion 1703S

Nakachi K, Suemasu K, Suga K, Takeo T, Imai K, Higashi Y (1998) Influence of drinking green tea on breast cancer malignancy among Japanese patients. Gann 89(3):254–261

Onoda C, Kuribayashi K, Nirasawa S, Tsuji N, Tanaka M, Kobayashi D, Watanabe N (2011) (-)-Epigallocatechin-3-gallate induces apoptosis in gastric cancer cell lines by down-regulating surviving expression. Int J Oncol 38(5):1403–1408

Pabla B, Bissonnette M, Konda VJ (2015) Colon cancer and the epidermal growth factor receptor: current treatment paradigms, the importance of diet, and the role of chemoprevention. World J Clin Oncol 6(5):133–141

Pisters KM, Newman RA, Coldman B, Shin DM, Khuri FR, Hong WK, Glisson BS, Lee JS (2001) Phase I trial of oral green tea extract in adult patients with solid tumors. J Clin Oncol 19(6):1830–1838

Rahmani AH, Al Shabrmi FM, Allemailem KS, Aly SM, Khan MA (2015) Implications of green tea and its constituents in the prevention of cancer via the modulation of cell signalling pathway. Biomed Res Int 2015:925640. doi:10.1155/2015/925640

Sarma DN, Barrett ML, Chavez ML, Gardiner P, Ko R, Mahady GB, Marles RJ, Pellicore LS, Giancaspro GI, Low Dog T (2008) Safety of green tea extracts: a systematic review by the US Pharmacopeia. Drug Saf 31(6):469–484

Shanafelt TD, Call TG, Zent CS, Leis JF, LaPlant B, Bowen DA, Roos M, Laumann K, Ghosh AK, Lesnick C, Lee MJ, Yang CS, Jelinek DF, Erlichman C, Kay NE (2013) Phase 2 trial of daily, oral Polyphenon E in patients with asymptomatic, Rai stage 0 to II chronic lymphocytic leukemia. Cancer 119(2):363–370

Shimizu H, Ross RK, Bernstein L, Yatani R, Henderson BE, Mack TM (1991) Cancers of the prostate and breast among Japanese and white immigrants in Los Angeles County. Br J Cancer 63(6):963–966

Shirakami Y, Shimizu M, Moriwaki H (2012) Cancer chemoprevention with green tea catechins: from bench to bed. Curr Drug Targets 13(14):1842–1857

Shutava TG, Balkundi SS, Vangala P, Steffan JJ, Bigelow RL, Cardelli JA, O'Neal DP, Lvov YM (2009) Layer-by-layer-coated gelatin nanoparticles as a vehicle for delivery of natural polyphenols. Nano 3(7):1877–1885. doi:10.1021/nn900451a

Siddiqui IA, Adhami VM, Bharali DJ, Hafeez BB, Asim M, Khwaja SI, Ahmad N, Cui H, Mousa SA, Mukhtar H (2009) Introducing nano chemoprevention as a novel approach for cancer control: proof of principle with green tea polyphenol epigallocatechin-3-gallate. Cancer Res 69(5):1712–1716. doi:10.1158/0008-5472.CAN-08-3978

Siddiqui IA, Adhami VM, Ahmad N, Mukhtar H (2010) Nanochemoprevention: sustained release of bioactive food components for cancer prevention. Nutr Cancer 62(7):883–890. doi:10.1080/01635581.2010.509537

Singh BN, Shankar S, Srivastava RK (2011) Green tea catechin, epigallocatechin-3-gallate (EGCG): mechanisms, perspectives, and clinical applications. Biochem Pharmacol 82(12):1807–1821. doi:10.1016/j.bcp.2011.07.093

Stoner GD, Mukhtar H (1995) Polyphenols as cancer chemopreventive agents. J Cell Biochem Suppl 22:169–180

Suganuma M, Okabe S, Sueoka N, Sueoka E, Matsuyama S, Imai K, Nakachi K, Fujiki H (1999) Green tea and cancer chemoprevention. Mutat Res 428(1–2):339–344

Sun CL, Yuan JM, Koh WP, Lee HP, Yu MC (2007) Green tea and black tea consumption in relation to colorectal cancer risk: the Singapore Chinese Health Study. Carcinogenesis 28(10):2143–2148

Thangapazham RL, Passi N, Maheshwari RK (2007a) Green tea polyphenol and epigallocatechin gallate induce apoptosis and inhibit invasion in human breast cancer cells. Cancer Biol Ther 6(12):1938–1943

Thangapazham RL, Singh AK, Sharma A, Warren J, Gaddipati JP, Maheshwari RK (2007b) Green tea polyphenols and its constituent epigallocatechin gallate inhibits proliferation of human breast cancer cells in vitro and in vivo. Cancer Lett 245(1–2):232–241, doi:http://dx.doi.org/10.1016/j.canlet.2006.01.027

Walker JM, Klakotskaia D, Ajit D, Weisman GA, Wood WG, Sun GY, Serfozo P, Simonyi A, Schachtman TR (2015) Beneficial effects of dietary EGCG and voluntary exercise on behavior in an Alzheimer's disease mouse model. J Alzheimers Dis 44(2):561–572. doi:10.3233/JAD-140981

Wang YC, Bachrach U (2002) The specific anti-cancer activity of green tea (-)-epigallocatechin-3-gallate (EGCG). Amino Acid 22(2):131–143. doi:10.1007/s007260200002

Wang P, Wang B, Chung S, Wu Y, Henning SM, Vadgama JV (2014) Increased chemopreventive effect by combining arctigenin, green tea polyphenol, and curcumin in prostate and breast cancer cells. RSC Adv 4(66):35242–35250. doi:10.1039/C4RA06616B

Wang Y, Duan H, Yang H (2015) A case-control study of stomach cancer in relation to Camellia sinensis in China. Surg Oncol 24(2):67–70. doi:10.1016/j.suronc.2015.02.002

Winreb O, Amit T, Youdim MB (2008) The application of proteomics for studying the neurorescue activity of the polyphenol (-)-epigallocatechin-3-gallate. Arch Biochem Biophys 476(2):152–160

Yang G, Shu XO, Li H, Chow WH, Ji BT, Zhang X, Gao YT, Zheng W (2007) Prospective cohort study of green tea consumption and colorectal cancer risk in women. Cancer Epidemiol Biomarkers Prev 16(6):1219–1223

Yang H, Sun DK, Chen D, Cui QC, Gu YY, Jiang T, Chen W, Wan SB, Dou QP (2010) Antitumor activity of novel fluoro-substituted (-)-epigallocatechin-3-gallate analogs. Cancer Lett 292(1):48–53

Yaniv Z (2012) Contribution of selected medicinal plants for cancer prevention and therapy. Acta Fac Med Naissensis 29(3):117–123

Yi S, Wang Y, Huang Y, Xia L, Sun L, Lenaghan SC, Zhang M (2014) Tea nanoparticles for immunostimulation and chemo-drug delivery in cancer treatment. J Biomed Nanotechnol 10(6):1016–1029

Yu GP, Hsieh CC, Wang LY, Yu SZ, Li XL, Jin TH (1995) Green-tea consumption and risk of stomach cancer: a population-based case-control study in Shanghai, China. Cancer Causes Control 6(6):532–538

Zhang M, Holman CD, Huang JP, Xie X (2007) Green tea and the prevention of breast cancer: a case-control study in Southeast China. Carcinogenesis 28(5):1074–1078

Zhang G, Wang Y, Zhang Y, Wan X, Li J, Liu K, Wang F, Liu K, Liu Q, Yang C, Yu P, Huang Y, Wang S, Jiang P, Qu Z, Luan J, Duan H, Zhang L, Hou A, Jin S, Hsieh TC, Wu E (2012) Anticancer activities of tea epigallocatechin-3-gallate in breast cancer patients under radiotherapy. Curr Mol Med 12(2):163–176

Zhang Y, Duan W, Owusu L, Wu D, Xin Y (2015) Epigallocatechin-3-gallate induces the apoptosis of hepatocellular carcinoma LM6 cells but not non-cancerous liver cells. Int J Mol Med 35(1):117–124

Chapter 14
Chinese Medicinal Herbs as Source of Rational Anticancer Therapy

Karuppaiya Palaniyandi, Shenghua Wang, and Fang Chen

Abstract Since ancient times, plants have been the source of traditional medicines for the treatment of various diseases throughout the world. Even today, the majority of purified drugs used clinically derived directly or indirectly from plants and other natural sources. Traditional Chinese medicine (TCM), the oldest form of medicine that has originated in ancient China, is rooted in the ancient philosophy of Taoism and dates back to more than 2500 years. As a holistic medicine, TCM has its theory and methods. TCM practitioners use a combination of different herbal medicines and various mind and body practices, such as acupuncture and tai chi, to treat or prevent health problems. Traditional Chinese medicine-based anticancer therapy is an old system but young in the field of modern medical science. In China, TCM herbals are widely used for cancer therapy and have gained increasing recognition worldwide in the recent years. The TCM-based herbal plants are being pursued by pharmaceutical companies as natural and rich sources for drug discovery. TCM has been recognized as a new source of anticancer drugs and new chemotherapy adjuvant to enhance the efficacy of chemotherapy and to ameliorate the side effects of cancer chemotherapies. Though TCM has been used since ancient times for preventing and healing disease, the scientific evidence at a molecular level behind the TCM is still mostly unknown, is rarely discussed, and needs to be known widely. Thus, this chapter reviews TCM oncology theory and its approach toward cancer, therapeutic effects, and various anticancer compounds obtained from TCM herbal plants with the hope of providing a better understanding on the role of drugs in the treatment of cancer.

Keywords Anticancer • Chemotherapy • Drugs • Herbs • Oncology • Phytomedicine

---

K. Palaniyandi • S. Wang • F. Chen (✉)
Department of Botany, Key Laboratory of Bio-Resources and Eco-Environment Ministry of Education, College of Life Sciences, Sichuan University,
Chengdu 610064, People's Republic of China
e-mail: chenfang@scu.edu.cn

H.-S. Tsay et al. (eds.), *Medicinal Plants - Recent Advances in Research and Development*, DOI 10.1007/978-981-10-1085-9_14

Abbreviations

ATP	Adenosine triphosphate
DHA	Dihydroartemisinin
DNA	Deoxyribonucleic acid
HCC	Hepatocellular carcinoma
HPV	Human papillomavirus
RNA	Ribonucleic acid
ROS	Reactive oxygen species
SAHA	Suberoylanilide hydroxamic acid
TCM	Traditional Chinese medicine
TRAIL	TNF-related apoptosis ligand
VPA	Valproic acid

14.1 Introduction

14.1.1 History and Recent Advancement of TCM in Light of Herbal Medicine in Cancer

Traditional Chinese medicine (TCM) (Zhongyi in Chinese) is an important integral part of Chinese culture. It holds an important position in primary healthcare system in rural as well as urban areas in China because of its 5000-year-old tradition (Chen et al. 2012). TCM has its philosophy, diagnosis, and treatment system (Wang et al. 2012) which follows the usage of Chinese herbs against diseases and disorders. The utilization of medicinal herbs as medicine was documented over a thousand years ago which is the combination of simplicity and complexity, facts and fiction, and superstition and folk culture. The development of Chinese herbal medicine, however, is greatly connected with the society and the evolution of civilization. In Chinese history, shamans (Wu) who dominated in health care by incantation in a combination of ashes and herbs. Irrespective of the knowledge of herbs and their usage, Chinese considered them as "divine herbs." People began to notice various functions of the "weeds" they were ready to eradicate from the field at the tilling. These weeds occupied a major position in the health-care system. Mysterious curing power and ancient tradition make the Chinese herbal medicine as a natural healing system.

The commonest medical practice in China is Western medicine. Statistics shows that out of 65,911 hospitals, around 800 were designated for practicing Chinese herbal medicine. Out of 1,244,000 doctors, only 300,000 were traditional Chinese medical doctors in 1980 (Lee 1994). Despite the prevalence of Western medicine in modern China, Chinese herbal medicine is considered as national medicine. It has been an integral part of Chinese culture and medicine system since Emperor

Shennong taught them how to use herbs in diet and therapy (Ergil et al. 2002). Chinese herbal medicine is an interesting but complicated subject. To study the medicinal services in China, one would expect to see the dominance of herbal medicine. Though Chinese herbal medicine is the world's oldest, safest, and most comprehensive medical system, it has not been thoroughly explored in the West. The study has been hampered by the lack of appropriate translations of Chinese medical classics, unavailable herbal sources in the West, and poor information exchange between China and the West concerning the development of herbal medicine. Hence, any attempts to understand this medicine system should take into account the complexity of Chinese philosophy and cosmology and their relationship with natural healing system. In other words, it can be achieved only by examining and gaining thorough knowledge in Taoism, yin-yang, the five elements theory, as well as the culture that has prevailed for a thousand years (Lee 1994). According to TCM, the imbalance of the whole body and mind network leads to the development of cancer (Macek 1984). Therefore, it is believed that cancer can be resolved by improving body strength and rebalancing body and mind network.

The word tumor, "ai" in Chinese, appears in prehistoric cave carving in Northwest China as far back as 1000 BC. The word "tumor" probably referred to the appearance of a lump over some parts of the human body (Bao et al. 2006). Later another word (qi), which carried the same literal meaning, appeared around 1200 AD, which referred to ulcerative lumps that often described together with deformities in the limbs, kyphosis in the spine, and occurrences of sinuses and fistulae in the human body (Li 1996). Cancerous conditions might initiate general states of anemia or debilitation and clues relevant to cancer therapy that might be included in the treatment system of these clinical states. Systematic treatment related to such tumorous conditions neither clear nor specialized. Unclear causative pathology results in mixed clinical presentations with regard to treatment which include policies like those used for infection and inflammation, while the general physiological state of the individual sufferer would provide another basis for treatment.

14.1.2 TCM-Based Herbal Medicine Approach to the Cancer Therapy

Cancer remains as a life-threatening disease and a leading cause of death throughout the world, particularly in developing countries, as its invasion mechanism has been difficult to control. Its incidence and mortality rates continue to rise every year. For cancer treatment, conventional medicine, chemotherapy, surgery, and radiotherapy are main conventional approaches, but its efficacy and safety remain a primary concern due to its toxicity and other side effects (Hsiao and Liu 2010; Tan et al. 2011). Moreover, current anticancer regimens are associated with the significant level of toxicity, side effects, and the emergence of drug resistance. One major challenge to relieve cancer burden is to develop highly effective anticancer drugs with specificity on cancers but little or no side effects on normal mammalian cells (Crespo-Ortiz and Wei 2011). Long history and rich experience of TCM in the treatment of cancer is evident from ancient medicinal books which deal with pediatrics, internal medicine, gynecology, surgery, and related fields. Basic principles and theories of TCM treatment have led to the opening of new avenues in the field of oncology which includes the TCM anticancer therapy, mechanism of action of TCM drug, its principles, tumor etiology, clinical treatment, prognostic evaluation and pathogenesis, diagnosis, and review of the literature on TCM books.

TCM oncology has its own characteristic features which are quantitative, are standardized, and are based on specific regulations. According to syndrome differentiation and concept of holism, the specificity of TCM treatment differs. Like other disorders, cancerous conditions are well known in the TCM treatment which forms a strong base in the field of TCM oncology (Zhou 2007). Literature shows that *Huang Di Nei Jing Di*, a classic medicinal book, described the pathogenesis, appearances, and treatment principles of tumor of the muscle, tendon, and bone carcinomas. However, this book failed to differentiate between malignant and nonmalignant tumors. This difference initially appeared in the ancient Chinese medicinal book *Wei Ji Bao Shu* in the Sung Dynasty (Hsiao and Liu 2010).

Based on the TCM oncology theory, imbalances between endogenous conditions of the human body and exogenous pathogenic factors lead to cancer. The endogenous conditions of the body play a crucial and dominant role in the onset of cancer, that is to say, when the body's immune system fails, factors can induce cancer. The pathogenic factors (such as heat, accumulated toxins, blood stasis) attack when a person is weak in the immune system. Moreover, malfunction communication between the physical condition of the body and mind may also trigger cancer progression (Macek 1984). *The Yellow Emperor's* by Huangdi Neijing Suwen highlights that the "treatments should aim at the disease fundamentals." The doctors should give much more importance to the body instead of tumor alone. Moreover, the doctors should use a methodology of treatment which is rational, adequate, and standardized rather than giving importance to the tumor.

The ancient Chinese healers formed three main principles of treatment to cancerous conditions:

(i) Toxic internal derangement (toxic derangement was balanced with detoxification through the cooling down of heat)
(ii) Circulatory stagnation (circulatory stagnation was removed with the activation of blood movements and resolution of bruises)
(iii) Collapse of defense (collapse of defense was rebuilding with the promotion of internal strength)

Furthermore, the observation by ancient healers suggests that toxic material could have a controlling effect on cancer growth. Therefore, treatment depending on the use of toxic preparations to counteract toxic disorders was also developed (Leung et al. 2007). So, TCM doctor assesses cancer as a systemic disease associated with the state of the whole body or disturbance of the molecular signaling pathways or network. Regarding TCM, "cancer" is defined as the manifestation of failure in the body's ability to handle pathogenic factors, but not a disease in indigenous cells or organs. TCM-based approach and the treatment philosophy highlight complete modulation and body improvement then removing the tumor mass or killing the cancerous cells. TCM treatment strategy is particularly required for a cancer patient at the late stages. The treatment is mainly focused on improving the quality of patient life, thereby extending life expectancy.

The other major principle of TCM cancer treatment is the emphasis on an individual therapy. According to the treatment based on symptom pattern differentiation principle, the same type of cancer in different persons, the diagnosis and treatment schemes could be very different. Alternatively, in TCM treatment, the doctors confirm the diagnosis and prepare the treatment based on the assessment of the pattern of symptoms manifest in each individual. For cancer treatment, several herbs are used together as a whole, but not the purified compounds. Thus, multiple effective compounds from herbs deliver a complete, effective, and integrated treatment of cancer through multiple targets and their associated pathways. This approach is in step with the outlook of TCM that cancer is a systemic disease which requires holistic approach and drugs that can produce therapeutic actions through multiple targets. While this approach differs from that of conventional medicine, the effects of treatment still come down to biochemistry. If treatments are effective, then there must be underlying mechanisms that can be investigated and verified scientifically. Understanding these mechanisms can help us expand the efficacy of both Western and Chinese medicines in a rational way (Hsiao and Liu 2010).

In the eighth century BC, Chinese healers defined health by the concepts of the yin and yang, which formed the theoretical base of TCM. Chinese medicine practitioners (CMP) consider patient's symptoms in the context of an imbalance between yin and yang according to the TCM yin-yang equilibrium theory, which was described in the *Internal Classic of Medicine*. According to yin and yang, if energies of a man are kept in a state of equilibrium, the body will be strong with sound spirit; if energies are dissociated, then the indispensable energy declines and finally exhausted. A healthy man is one whose physique, muscle, blood, and qi are harmonious and appropriate with each other. Clinical studies show that a large number of patients who are in the late stage of cancer are not suitable for operation, chemotherapy, or radiotherapy. Such patients undergo treatment through TCM rather than chemotherapy and radiotherapy. It is because TCM treatment stresses on the support of vital energy and elimination of the cause of tumor, that is, causative agents. It mainly focuses on the restoration of yin and yang balance that can decrease the damage of tumor cells. In other words, TCM treatment deals with raising and lowering the resistant power decisively, which affects the growth of the tumor. This phenomenon explains how TCM enhances the life span of cancerous patients (Wu and Wu 1997).

A large number of patients place their hope on TCM, and few of them obtain a good clinical outcome. It has been proved that TCM delays the growth rate of the tumor and has an inhibitory effect in certain cancer, thereby enhancing the effects of radiotherapy or chemotherapy or decreasing their toxic effects.

14.2 Therapeutic Potential of TCM Herbs on Cancer Treatment

In China, more than 10,000 drugs for TCM and 50,000 compound formulas are available, and nearly 3000 drugs and 300 formulas have been screened and validated for anticancer therapies (Zhou 2007). In TCM, every patient condition is handled individually, and there is no predetermined treatment procedure available. In past decades, studies have been carried out in laboratories and at clinical level to prove the effectiveness of TCM against cancer. On the other hand, treatment through TCM methods has its own bottleneck. Limited sample size, lack of quality assurance of herbal products, and randomized controlled trials are few reasons, which are the limitations of TCM in the field of oncology. Clinical studies published in this field are just trials without rigorous randomization or involved in matched case-control studies (Sagar and Wong 2008). At the same time, there are some contradictory reports against the TCM in the field of oncology.

In TCM preparation, a combination of herbs is used as "formulas," believing that the combinations enhance their benefits and at the same time diminish side effects. The TCM practitioners prepare the herbal formulas to suit individual cancer patients based on the diagnosis. Due to multiple herbs combined in the herbal formulas, interaction of active compounds in the different herbs synergistically exerts its effects in several ways such as (i) no toxicity to noncancerous cells, protect the cell damage from the chemotherapy and radiotherapy, (ii) enhance the potential of chemo-/radiotherapy, (iii) can reduce infection and inflammation around the tumor tissues, (iv) can enhance immune power and energy level in the body, (v) can enhance the condition and quality of life, and (vi) can prolong later-stage cancer patient's life span. The scientific implication of TCM-based herbal medicine has revealed its potential in therapeutics. TCM formulations are composed of many herbs, and these herbs decocted singly or in combination may vary in the types of active ingredients and in the quantity of the contents. Due to multiple active chemicals in the TCM, drugs may have action on multiple receptor sites and targets. Based on the drug's target, the anticancer herbal drugs can be divided into three categories: (i) drugs that inhibit topoisomerases (Topos) and thereby transcription, (ii) drugs that kill tumor cells through apoptotic pathways, and (iii) drugs that alter signaling pathway(s) required for the maintenance of transforming phenotypes of the tumor cells.

Several natural compounds obtained from TCM herbs are already known to possess the anticancer activity. Drugs targeting proapoptotic pathways and topoisomerases I and II are known to trigger the cell death. In China, *Camptotheca acuminata* (xi shu in Chinese) is a well-known TCM herb and has long been used in cancer treatment for many years. The active compound camptothecin extracted from this Chinese herb is known to target type I topoisomerase (Topo I) enzymes which can break the double-stranded DNA, relax the strand, and reanneal the DNA strand. The intrusion of Topo I induces apoptosis and cell cycle perturbations (Hsiang et al.

1985; Wall 1998; Oberlies and Kroll 2004; Zhang et al. 1999). The discovery of active compounds camptothecin and taxol from TCM herbs led to the identification drugs targeting Topo I and Topo II and considered this finding as historical achievements in natural products. Chemotherapeutic agents target cell death through apoptosis, including the natural product-derived drugs (Kaufmann and Earnshaw 2000; Fulda and Debatin 2006). In TCM, systemic herbal treatments and topical herbal applications appear to be effective in treating cancer-related pain. Previous studies have shown that effectiveness in controlling the pain to be over 90 % (Yang et al. 1995). Drugs, herbal preparations (single or multiple herbs), pills (such as *Liu Shen Wan* pill, *Xi Huang* pill, *Pin Xiao* capsule, *Shen Yi* capsule), oral liquid, or injections prepared in TCM (*Ai Di* injection) were found to be effective in the cancer therapy. Traditionally used multiple Chinese herb extracts such as *Anemarrhena asphodeloides*, *Artemisia argyi*, *Commiphora myrrha*, *Duchesnea indica*, *Gleditsia sinensis*, *Ligustrum lucidum*, *Rheum palmatum*, *Rubia cordifolia*, *Salvia chinensis*, *Scutellaria barbata*, *Uncaria rhynchophylla*, and *Vaccaria segetalis* have demonstrated growth inhibitory activity against several cancer cell lines, compared to the limited inhibitory activity against normal cell proliferation (Shoemaker et al. 2005).

In TCM oncology, herbs alone are occasionally associated with tumor regression. For example, a combination of herbs for the sole treatment of pathologically proven squamous cell carcinoma of the lung in a 51-year-old female patient attained complete regression (Liang et al. 2004). TCM herbs, such as *Salvia miltiorrhiza*, inhibited the tumor edema and increased tumor perfusion, oxygenation, and response to radiotherapy (Sagar et al. 1995; Peigen et al. 1996). Some herbs may directly sensitize neoplastic cells to radiotherapy (Huali et al. 1994), and others such as *Panax ginseng* and *Panax quinquefolius* water extract containing Rh2 ginsenoside may radioprotect properties in normal tissues via antioxidative and immunomodulating properties (Lee et al. 2005).

Traditionally, Chinese herbs have been used in combination for the treatment of cancer. For example, in *Coptidis rhizoma* herb, DNA microdata studies evaluated the antiproliferative activity and eight constituent molecules against eight human pancreatic cancer cell lines (Hara et al. 2005), and among 12,600 genes, nearly 27 genes identified showed a strong correlation with the 50 % inhibitory dose (ID50) of *C. rhizoma* after 72-h exposure. Hierarchical cluster analysis with correlation coefficients between expression levels of these 27 genes related to *C. rhizoma* and the ID50 of each constituent molecule classified these test molecules into two clusters, one consisting of *C. rhizoma* and *berberine* and the other consisting of the remaining seven molecules. Therefore, it suggests that one specific phytochemical, *berberine*, can account for the majority of the antiproliferative activity of *C. rhizoma* and that DNA microarray analyses can be used to improve our understanding of the actions of an intact herb. In contrast, there appears to be merit in using combinations of herbs and their derivatives. For example, PHY106 (*Radix scutellariae*, *Paeonia lactiflora pall*, *Fructus ziziphi*, *Radix glycyrrhizae*) is an authenticated combination of herbs that may increase the efficacy and reduce the adverse effects of the cytotoxic drug, capecitabine, which is used in treating colorectal cancer (Farrell 2003).

In TCM oncology, a malignant tumor is a stagnation of *Qi* (energy) and blood. *Qi* is considered as a model for cellular information (both inter and intra) and potential energy transfer. Moreover, the above information correlates with abnormal conditions, signal transduction, cell contact, and electrophysiology of cancerous cells (Coffey 1998; Cuzick et al. 1998; Kang et al. 2000). Literature review shows that in malignant tumors, fluid content is increased and blood supply becomes stagnant (Baxter and Jain 1989; Boucher and Jain 1992; Sagar et al. 1993; Milosevic et al. 1998). TCM reveals that stagnation of blood is associated with a tumor and it is due to poor oxygenation. Finally, the cancer cells that survive in such low-oxygenated condition are found resistant to radiotherapy and chemotherapy (Brizel et al. 1997; Fyles et al. 1998).

The movement of blood and Qi within the malignant tumor is the main role of de-stagnation or detoxification herbs in TCM. It is similar to the use of anticoagulants like heparin and Coumadin, an adjunctive treatment to chemotherapy which prevents the development of blood-borne metastases in lab condition and also improves the survival of cancer patients (Lebeau et al. 1994; Hejna et al. 1999). Moreover, many herbs show positive results for antiangiogenic property (Yance and Sagar 2006). TCM plays a key role in cancer therapy by harmonizing the energy balance between the mind and different organs of the body and thereby triggering the immune system. In Fu Zheng treatment, herbs including *Rx Ginseng*, *Ganoderma*, *Rx Astragalus membranaceus*, *Rx Angelica sinensis*, *Cordyceps sinensis*, and *Fructus lycii* are found effective in enhancing defense mechanism. It is also found that Fu Zheng herbs increased the count of NK and OKT4 (immune-enhancing lymphocyte) cells (Sagar and Wong 2008). In China, *Fu Zheng* herbs have been reported to increase survival rate when combined with radiotherapy for patients with nasopharyngeal cancer and once combined with chemotherapy for patients with stomach and liver cancer (Macek 1984; Wang 1990). In cancer patients, symptoms of fatigue, depression, pain, and specific symptoms such as gastrointestinal side effects and myelosuppression are effectively managed by TCM-based herbal medicines (Sagar and Wong 2008).

14.3 Bioactive Compounds from TCM Herbs with Anticancer Activity

Natural products represent a rich reservoir of potential small-chemical molecules which exert antiproliferation and anticancer activity. Numerous studies have shown that traditional Chinese herbal medicine in combination with chemo- or radiotherapy can be used to enhance the efficacy and diminish the side effects and complications caused by chemo- and radiotherapy in the cancer treatment (Qi et al. 2010). Therefore, an understanding of Chinese herbal medicines is needed by physicians and other health-care providers. Preclinical and clinical studies have shown that Chinese herbal medicines possess great advantages regarding suppressing tumor progression, increasing the sensitivity of chemo- and radiotherapeutics, improving an organism's immune system function, and lessening the damage caused by chemo- and radiotherapeutics. By reducing side effects and complications during the chemo- and radiotherapy, these Chinese herbal medicines have a significant effect on reducing cancer-related fatigue and pain; improving respiratory tract infections and gastrointestinal side effects including diarrhea, nausea, and vomiting; protecting liver function; and even ameliorating the symptoms of cachexia. While it is realistic to look forward to the production of effective anticancer drugs developed from Chinese herbs, which may follow new innovative pathways of cancer control, one could choose to start clinical research using herbs of expected promise immediately. The need to look for an innovative supplement to the conventional cancer therapy is obvious because of its disappointing limitations and the cytotoxic complications. The current state on pharmacological research is aiming at molecular targets with an eventual aim toward drug production. Here we have reviewed 15 active compounds isolated from Chinese herbal plants which exhibited strong anticancer activity.

14.3.1 Artemisinin and Its Derivatives

Artemisinin (Fig. 14.1A), also known as qinghaosu, in Chinese is an active antimalarial sesquiterpene lactone-containing unusual peroxide bond. This endoperoxide compound is extracted from the Chinese herb *Artemisia annua* (qing hao in Chinese), which was used in traditional Chinese herbal medicine over two millennia for the treatment of malaria and fever (Tan et al. 2011). The plant's dry weight contains only 0.01–0.8 % of artemisinin (Jung et al. 2004). Anticancer activity of artemisinin has been reported in human trials (Zhang et al. 2008) and individual clinical cases (Singh and Verma 2002; Berger et al. 2005). Artemisinin has anticancer effects in experimental models of hepatocellular carcinoma cells (Hou et al. 2008; Deng et al. 2013). Interestingly, this peroxide bond is thought to be responsible for the potential anticancer action of artemisinin. The anticancer mechanism of artemisinin is cleavage of the endoperoxide bridge by the relatively high concentrations of iron in cancer cells, resulting in free radicals such as ROS and subsequent oxidative damage as well as iron depletion in the cells (Firestone and Sundar 2009). Artemisinin derivatives such as dihydroartemisinin (DHA) and artesunate (semi-synthesized) exert potential anticancer activities in vitro and in vitro (Tan et al. 2011).

Artemisinin and its derivatives exhibit potential anticancer activity in a variety of human cancer cell line including those of leukemia and other cancer cells of the breast, ovary, liver, lung, pancreas, and colon (Lu et al. 2009; Ba et al. 2012; Lai et al. 2013). Artemisinin exhibited differential cytotoxicity effect in P815 (murine mastocytoma) and BSR (kidney adenocarcinoma of a hamster) cell lines. Also, *in vivo* studies demonstrated that oral administration of artemisinin inhibited solid tumor development (Tilaoui et al. 2014). Many studies demonstrated artemisinin cytotoxicity with blocking cytokines, inhibiting tumor invasion, migration, and metastasis (Das 2015). Artemisinin and its derivatives exert synergistic effects combined with other anticancer drugs/chemotherapy drugs. The combination of DHA with TRAIL significantly enhances the cell death in prostate cancer cells mediated

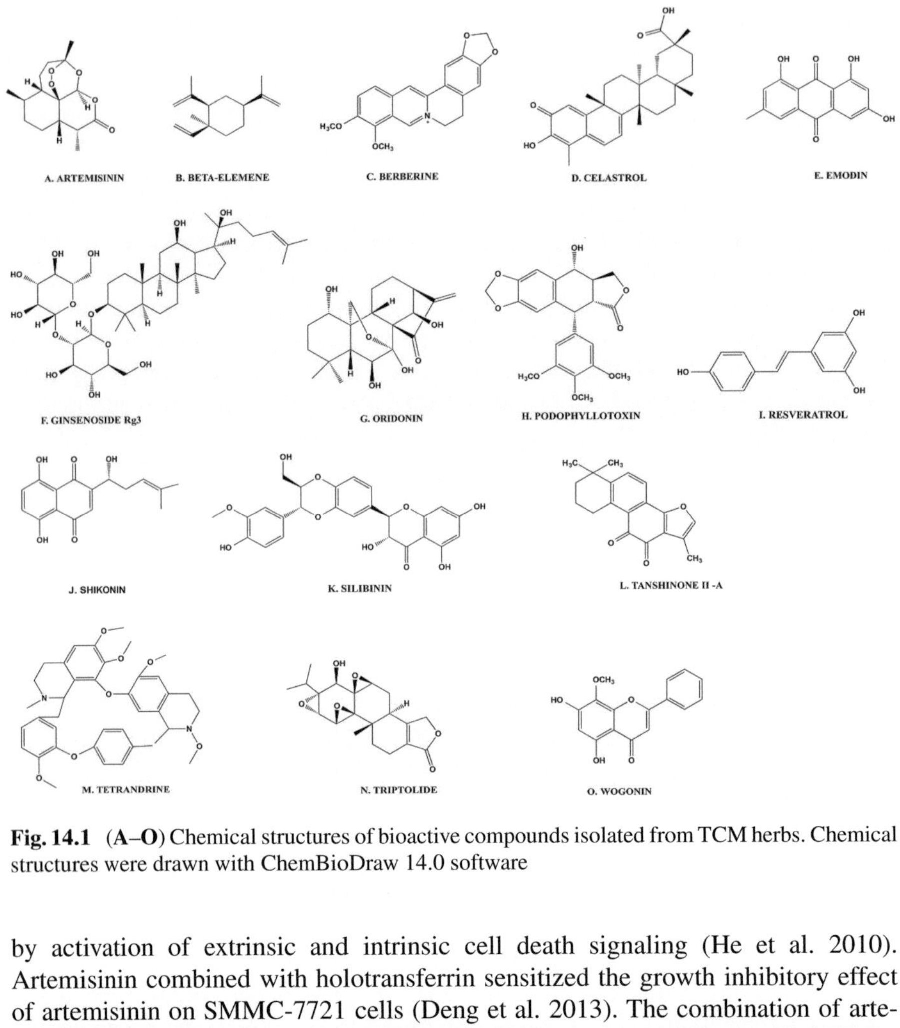

Fig. 14.1 (**A–O**) Chemical structures of bioactive compounds isolated from TCM herbs. Chemical structures were drawn with ChemBioDraw 14.0 software

by activation of extrinsic and intrinsic cell death signaling (He et al. 2010). Artemisinin combined with holotransferrin sensitized the growth inhibitory effect of artemisinin on SMMC-7721 cells (Deng et al. 2013). The combination of artemisinin with resveratrol significantly increased the apoptosis and necrosis in HeLa and HepG2 cells rather than using alone. Further, this combination effectively inhibited the proliferation of cancer cells and enhanced migration, apoptosis, necrosis, and reactive oxygen species (ROS) levels (Li et al. 2014b). Recent research findings demonstrate that artemisinin derivatives (dimeric and trimeric) exert higher antitumor activity than their monomeric counterparts (Das 2015).

14.3.2 Berberine

Berberine (Fig. 14.1B) is a quaternary ammonium from protoberberine group of isoquinoline alkaloid from Chinese medicinal herbs such as *Coptis chinensis* Franch, *Coptis deltoidea*, or *Coptis teeta* (huang lian in Chinese). It has been used in traditional Chinese herbal medicine for more than 3000 years for the treatment of diarrhea, relieving fidgetiness, detoxifying, women's abdomen ailments, and eye inflammation (Sun et al. 2009). Research studies have shown that berberine exerts anticancer activity in vivo and in vitro (Sun et al. 2009). Berberine suppresses the growth of a wide variety of tumor cells including breast cancer (Kim et al. 2009), leukemia, melanoma (Serafim et al. 2008), epidermoid carcinoma, hepatoma, pancreatic cancer (Pinto-Garcia et al. 2010), oral carcinoma, tongue carcinoma (Ho et al. 2009), glioblastoma, prostate carcinoma, and gastric carcinoma (Auyeung and Ko 2009; Tang et al. 2009).

Berberine has drawn extensive attention due to its antineoplastic activities against various cancer cells including skin, liver, lung, esophagus, stomach, colon, and brain cancer cells (Sun et al. 2009; Tang et al. 2009). Its antineoplastic activities are also involved in the cell cycle arrest, cell migration and invasion mediated by regulation of multiple pathways, and induction of apoptosis (Tan et al. 2011). Berberine also induces apoptosis via an intrinsic (mitochondrial) pathway and by arresting the cell cycle at the G_1/S phase in Huh-7 liver cancer cells (Yip and Ho 2013), human breast cancer MCF-7 cells (Patil et al. 2010), endometrial cancer cells (Kuo et al. 2015), human glioblastoma T98G cells, and human tongue squamous carcinoma (Ho et al. 2009; Eom et al. 2008). Berberine can also induce autophagic cell death in HepG2 cells and MHCC97-L cells by inhibition of rapamycin signaling pathway and cell death inhibitor 3-methyladenine through beclin-1 activation (Wang et al. 2010a). It was reported that berberine modified LC3 in human lung cancer A549 cells indicating that autophagy plays a vital role in berberine-induced cancer cell death (Peng et al. 2008). Berberine combined with evodiamine increased the inhibition rate of SMMC-7721 cells when compared to berberine and evodiamine treatment individually (Wang et al. 2008, 2012). Berberine, combined with cisplatin or As_2O_3 (chemotherapy drug), exerts significant cytotoxicity in HeLa and SH-SY5Y cells when compared to monotherapy (Kim et al. 2007; Youn et al. 2008). The combination of berberine with gamma-radiation significantly enhanced the apoptotic effect in HepG2 cells through the p38 MAPK pathway and ROS generation (Hur et al. 2009).

Future research focuses on berberine drug delivery in HCC cell lines using sophisticated methodologies such as novel drug delivery which may be helpful in developing efficient strategies and providing a possible methodology for clinical use (Tillhon et al. 2012). Berberine exerted a profound inhibitory effect on the expression of Id-1, which is a key regulator of HCC development and metastasis (Tsang et al. 2015). In spite of having exhibited natural anticancer drug, the molecular mechanism of berberine on cancer cells remains unclear (Hu et al. 2013). A recent study finds that berberine inhibits the RET gene transcription by specifically targeting the promoter region of this gene, thereby reducing the RET protein expression in MTC-TT cells (Sun and Kumarasamy 2015). Recent research studies showed that natural berberine isolated from plant source has good safety profile with low toxicity (Yi et al. 2013). In xenografts model, it inhibited the tumor growth mediated through the induction of senescence and downregulation of EGFR. Berberine can also be useful in the treatment of glioblastoma due its inhibition capability of EGFR (Liu et al. 2015b). Thus, further development of berberine for cancer prevention and treatment may be beneficial.

14.3.3 *β-Elemene*

β-Elemene (Fig. 14.1C) is an anticancer drug, which was extracted from Chinese herbal plants such as *Curcuma wenyujin* (wenyujin in Chinese). Elemene is mainly composed of β-, δ-, and α-elemene. Experimental studies of β-elemene exhibit its anticancer activities in various cancer cells including lung, colon, prostate, ovary, laryngeal, brain, breast, and cervical carcinomas (Li et al. 2005, 2010; Wang et al. 2005a). An investigation showed that β-elemene can pass through the blood-brain barrier and suggested that β-elemene is effective in treating a cerebral malignancy (Wu et al. 2009). β-elemene also induces mitochondrial-mediated apoptosis in prostate cancer and non-small cell lung carcinoma (NSCLC) cells (Wang et al. 2005; Li et al. 2010b).

β-Elemene is effective in inducing mitochondrial apoptosis of RA-FLS, which is mediated through induction of ROS formation and p38 MAPK activation (Zou et al. 2016). In a mouse model of intraocular melanoma, β-elemene inhibits tumor growth by downregulating the expression of uPA, uPAR, MMP-2, and MMP-9 (Shi et al. 2015). A recent study showed that β-elemene induced cell apoptosis in human glioblastoma cells (Zhu et al. 2015). β-Elemene inhibits the growth of human non-small cell lung carcinoma (NSCLC) cell via both extracellular signal-regulated kinase 1/2 (ERK1/ERK2) and AMP-activated protein kinase alpha (AMPKα)-mediated inhibition of transcription factor Sp1. Moreover, β-elemene inhibited expression of DNA methyltransferase 1 (DNMT1) by activation of ERK1/ERK2 and AMPKα signaling pathways (Zhao et al. 2015). A recent study suggests that β-elemene triggers cell cycle arrest by activating the p38 MAPK pathway (Zhang et al. 2015). The chemotherapeutic drug β-elemene suppressed the proliferation of esophageal carcinoma ECA-109 cells by regulating the inhibition of hTERT expression by lncRNA CDKN2B-AS1 (Hu et al. 2015).

β-Elemene with some chemotherapeutic drugs synergistically exerts anticancer activities (Tan et al. 2011). The combination of β-elemene with etoposide phosphate (VP-16) induced apoptosis compared with VP-16 alone in NSCLC cells (Zhang et al. 2011). Also, combined with taxanes, it induced G_2/M arrest in human non-small cell lung cancer and led to the blockade of the cell cycle by modulating the G_2 cell cycle checkpoint (Zhao et al. 2007). Furthermore, the combination of β-elemene and paclitaxel significantly inhibited the proliferation of MB-468 cells by decreasing the expression of cyclin B1 and increasing the expression of p27 (kip1) when compared with paclitaxel injection alone (Cai et al. 2013). China Food and Drug Administration has approved β-elemene (second-class drug) as an anticancer drug for tumor therapies (Tan et al. 2011).

14.3.4 Celastrol

Celastrol (Fig. 14.1D) or tripterine, a quinone methide triterpenoid compound, is abundantly found in the root skin and bark of Chinese herb *Tripterygium wilfordii* (viz., thunder god vine). This Chinese medicinal herb has promising therapeutic potential including relieving pain, arthritis, and autoimmune and inflammatory conditions and removing dampness. Celastrol has gained much attention due to its potential anticancer activity (He et al. 2009) and anti-inflammatory properties (Jung et al. 2007). Celastrol attenuated prostate carcinoma 3 (PC-3) cell proliferation via downregulation of IL-6 gene expression through the NF-kB-dependent pathway (Chiang et al. 2014). Celastrol induces unfolded protein response (UPR)-dependent cell death in head and neck cancer (Fribley et al. 2015). Studies have shown that formation of celastrol liposomes, to avoid the use of toxic solubilizing agents, exhibit antitumor efficacy in a glioma (Huang et al. 2012a, b), lung carcinoma model (Song et al. 2011), and prostate cancer cell (Wolfram et al. 2014). Celastrol also induced cell apoptosis with ROS accumulation, loss of mitochondrial membrane potentials, cleavage of PARP, and caspases in a dose- and time-dependent manner in A549 parental and A549/DDP cisplatin-resistant cells (Shi 2013). Further, a study reported that celastrol-induced apoptosis in TNBC cells mediated through PI3K/Akt signaling pathway, mitochondrial dysfunction, and oxidative stress (Shrivastava et al. 2015). Celastrol and ABT-737 synergistically exert anticancer activity in HCC cells mediated through suppressed proliferation and induced apoptosis (Zhu et al. 2012). Similarly, combinatorial treatment with SAHA significantly enhanced the anticancer activity of human cancer cells in vitro and in vivo mediated by E-cadherin upregulation and inhibition of NF-kB (Zheng et al. 2014).

14.3.5 Emodin

Emodin (Fig. 14.1E) is an anthraquinone extracted from *Rheum palmatum* L. (zhangyedahuang in Chinese). This herb has been used in TCM for the treatment of gallstones, inflammation, hepatitis, and osteomyelitis and also a known vasorelaxant and diuretic (Teng et al. 2007). Emodin has been reported to possess antibacterial, anti-inflammatory, antiviral, anti-ulcerogenic, anticancer, immunosuppressive, and chemopreventive effects (Hsu and Chung 2012). Emodin induces ROS generation and the activation of the ATM-p53-Bax-dependent signaling pathway in human lung adenocarcinoma A549 cells (Lai et al. 2009). It has been reported that emodin exerts potential anticancer effects in pancreatic cancer cells by downregulating the expression of survivin and β-catenin (Guo et al. 2009). Emodin breaks DNA double strand by stabilizing topoisomerase II-DNA cleavage complexes and inhibiting ATP hydrolysis of topoisomerase II (Li et al. 2010c).

Emodin exerts antitumor activity in many human cancer cells including breast, liver, gallbladder, colon, pancreatic, prostate, cervical, ovarian cancer, human tongue squamous, gastric, and esophageal cancer (Wei et al. 2013). Emodin triggers cell apoptosis in various cancer cell lines including liver cancer cell line through a multifaceted complex cascade of events (Yu et al. 2013) such as orthotopic HCC model by blocking STAT3 activation (Subramaniam et al. 2013a), human hepatocellular carcinoma cells through the induction of death receptors and downregulation of cell survival proteins (Subramaniam et al. 2013b), human lung adenocarcinoma A549 cells through modifying the extrinsic pathways, and the induction of cell cycle arrest (Li et al. 2014a, b) and human colorectal cancer cells by activating ROS/p38/p53/Puma signaling (Liu et al. 2015a).

Emodin and AZT synergistically enhanced the growth suppression in K562/ADM cells via altered cell cycle distribution and led to an accumulation of cells in S phase. At the same time, it decreases the expression of MDR1 mRNA/p-gp protein (Chen et al. 2013a). Emodin combined with curcumin synergistically inhibits proliferation, survival, and invasion of breast cancer cells. Moreover, emodin with curcumin upregulated miR-34a, and this microRNA helps mediate anticancer efficacy in breast cancer by downregulating Bcl-2 and Bmi-1 (Guo et al. 2013). Emodin synergistically enhances the anticancer activity of cisplatin in human gastric cancer (SNU-5) cells by inducing apoptosis as well as cell cycle arrest (Huang et al. 2015). Emodin combined with curcumin effectively downregulates TGF-β signaling pathway in human cervical cancer cells (Thacker and Karunagaran 2015).

14.3.6 Ginsenoside Rg3

Ginsenoside Rg3 (Fig. 14.1F) is a biologically active compound extracted from Chinese herb *Panax ginseng* (ren shen, Chinese ginseng). Ginsenoside Rg3 from *Panax* has been considered as an important compound of traditional Chinese medicine over a century (Bensky and Gamble 1993). Shen Yi capsule containing ginsenoside Rg3 has been approved by China Food and Drug Administration to be used as a drug for cancer treatment in China (Yue et al. 2006). Ginsenoside Rg3 exerts an antitumor effect on human hepatocellular carcinoma cell lines via the intrinsic apoptotic pathway (Jiang et al. 2011; Zhang et al. 2012; Park et al. 2012). The antiproliferative activity of ginsenoside Rg3 is associated with the inactivation of NF-kB (Kim et al. 2010a, b), the modulation of MAPKs (Kim et al. 2004), the downregulation of Wnt/β-catenin signaling (He et al. 2011), and the decreased HDAC3 and increased acetylation of p53 both in vitro and in vivo conditions (Shan et al. 2014). Ginsenoside Rg3 promotes cell apoptosis in cancer cell including human hepatocellular carcinoma cell (Zhang et al. 2012) and human gastric cancer via the endogenous mitochondrial-mediated caspase-dependent apoptotic pathway (Park et al. 2014).

Ginsenoside Rg_3 combined with chemotherapy enhanced the efficacy of cancer treatment. Synergetic treatment of ginsenoside Rg3 with docetaxel enhanced the efficacy tumor inhibition on colon cancer cells through the inhibition of the constitutively activated NF-kB (Kim et al. 2009). Combined treatment with gemcitabine enhanced the tumor growth suppression, prolongation of patient life, decreased VEGF expression, and microvessel density in tumors (Liu et al. 2009). When combined with docetaxel, it affects apoptosis and G_1 cell cycle arrest in prostate cancer cells accompanied by the inhibition of NF-kB activity (Kim et al. 2010b). In another study, ginsenoside Rg3 combined with cisplatin inhibited the bladder cancer cells proliferation via the activation of the intrinsic apoptotic pathway and the enhancement of cell cycle alterations (Lee et al. 2014).

14.3.7 Oridonin

Oridonin (Fig. 14.1G) is a naturally occurring bioactive diterpenoid produced from *Rabdosia rubescens* (Hemsl.) Hara (donglingcao in Chinese). Native residents of Henan province have long been using donglingcao to treat a sore throat, an esophageal cancer, and a tonsillitis. In 1977, donglingcao was included in the book of *Chinese Pharmacopoeia.* Oridonin is a main active compound of donglingcao, which has multiple biological activities such as anti-inflammatory, antibacterial, and antitumor effects (Tan et al. 2011). Oridonin is known to inhibit DNA, RNA, and protein synthesis (Wang et al. 1987), downregulate human telomerase reverse transcriptase mRNA expression, and decrease telomerase (Liu et al. 2004).

Oridonin can potentiate the effects of gemcitabine in pancreatic cancer through the mitogen-activated protein kinase (MAPK)-p38 signaling pathway, which is dependent on p53 activation (Bu et al. 2012). It can also inhibit the tumor cell invasion and metastasis in vitro mediated by regulation of integrinβ1/FAK pathway and by decreasing the expression of MMPs (Wang et al. 2013a, b). Moreover, oridonin can also affect the anticancer activity by inducing apoptosis and by inhibiting cell proliferation in multiple myeloma cell line U266 (Owona and Schluesener 2015). Oridonin inhibited SGC-7901 cell proliferation, cell cycling arrest in the G_2/M phase, and also decreased the protein expression of cyclin-B1 and CDK1 (Gao et al. 2013b). Oridonin exerts antitumor actions through induction of apoptotic response by inhibition of mTORC1 function (Wang et al. 2014). Oridonin triggers cell apoptosis mainly through mitochondrial-mediated pathways. In a study, oridonin upregulated BIM-S by inhibiting the expression of miR-17 and miR-20a, leading to mitochondria-dependent apoptosis. Besides, inhibiting miR-17 or miR-20a also augmented the proapoptotic activity of oridonin (Weng et al. 2014). Oridonin exhibits apoptosis effect on highly metastatic breast cancer cells MDA-MB-231 and might be associated with DNA damage and activation of intrinsic and extrinsic pathways (Wang et al. 2014). The combinatorial treatment of oridonin with VPA also induced apoptosis via MAPK signaling pathway (Shi et al.

2016). Oridonin and arsenic trioxide (AS_2O_3) synergistically exert tumor growth suppression in murine HCC model compared to single treatment in vivo (Chen et al. 2012). When oridonin was combined with gemcitabine, it inhibited the proliferation of the pancreatic cancer cell line, BxPC-3, potentiated the apoptosis induced by gemcitabine, induced G_1 cell cycle arrest, and activated p38 and p53 (Bu et al. 2012). The lower concentration of both oridonin and VPA in combination induced apoptosis via intrinsic and extrinsic apoptosis pathways and also inhibited the proliferation of HL-60 cells (Shi et al. 2016).

14.3.8 Podophyllotoxin and Its Derivatives

Podophyllotoxin (Fig. 14.1H) is a naturally occurring aryltetralin lignan extracted from root and rhizome of Chinese may apple, also known as bajiaolian in Chinese (*Dysosma pleiantha*, *Dysosma versipellis*). It has long been used in TCM for medicinal purpose as an anti-wart and the treatment of cancer (Karuppaiya 2013; Karuppaiya and Tsay 2015). This natural lignan has been known for approximately 1000 years since its first application in folk medicines to its recent podophyllotoxin-derived antitumor drugs (Liu et al. 2015c). Interest in podophyllotoxin was initiated by Kaplan (Kaplan 1942) who demonstrated its curative effect against tumor growth and subsequently by King and Sullivan (King and Sullivan 1946; Sullivan and Wechsler 1947) who found its antiproliferative effect to be similar to that of colchicine at the cellular level. In fact, podophyllotoxin is included in many pharmacopeias and used as an antiviral agent in the treatment of condyloma acuminatum caused by human papillomavirus (HPV) (Syed et al. 1995) and other venereal and perianal warts (Perez-Figarado and Baden 1976). Podophyllotoxin is well known for its antitumor activity. However, the clinical application of podophyllotoxin and its analogs in the treatment of cancer has been limited by severe toxic side effects during administration of the drugs (Gordaliza 2000). Podophyllotoxin functions as a radiosensitizer by promoting apoptosis via activation of an ROS/p38/caspase pathway and suppression of ERK (Choi et al. 2015). Derivatives of podophyllotoxin known as etoposide and teniposide have improved pharmacological activity over the parent compound and are being used in the treatment of different types of cancer including leukemia, lymphoma, glioblastoma, and lung and testicular cancers (Bohlin 1996).

14.3.9 Resveratrol

Resveratrol (Fig. 14.1I) is a stilbenoid, a type of natural polyphenol compound extracted from Chinese medicinal herb *Polygonum cuspidatum* (Hu et al. 2014). Recent studies demonstrated that resveratrol can inhibit or slow down the progression of a wide variety of tumors (Han et al. 2015). Resveratrol was also reported to induce apoptosis mediated by reactive oxygen species (ROS), ERK1/ERK2, and p53-dependent pathway (Lin et al. 2011; Miki et al. 2012). Also, resveratrol significantly inhibited cell migration and invasion leading to suppression of metastasis (Yeh et al. 2013; Wang et al. 2013a, b). Resveratrol treatment in Huh-7 cells inhibited cell migration and invasion mediated by inhibiting phosphorylation JNK 1/2 and SP-1 DNA activities. Also, it inhibited metastasis by downregulation of uPA expression (Yeh et al. 2013). An in vitro and in vivo investigation showed that resveratrol effectively inhibits the growth of lung cancer in a dose-dependent manner in nude mice (Yin et al. 2013). Molecular mechanism investigations revealed that resveratrol could inhibit the activity of phosphoinositide 3-kinase/AKT and p38 mitogen-activated protein kinase signaling pathways in chondrosarcoma cells, which is important in the regulation of proliferation, apoptosis, and invasion in various cancer cell types (Dai et al. 2015). The combined use of resveratrol with oxaliplatin inhibited cell growth in Caco-2 cells mediated by the induction of cell death, and altered cytokine profile of cocultured macrophages and combinatorial treatment has no cytotoxicity in human foreskin fibroblasts and platelets (Kaminski et al. 2014). Further, resveratrol and paclitaxel co-encapsulated micelle combination synergistically exerts enhanced antitumor activity in human lung adenocarcinoma epithelial (A549/T) cell line and mice sarcoma 180 (S180) cells by inducing ROS-dependent apoptosis and inhibiting autophagy. Moreover, it exhibits very limited toxicity toward the normal human hepatic (L02) cell strain and normal human kidney (HK-2) cell line which could be from manipulating ROS levels by redox modulation and reducing protective autophagy (Hu et al. 2014). A recent investigation on resveratrol derivatives showed a significantly increased antitumor efficacy in cervical cancer HeLa cells than resveratrol (Jin et al. 2015). These findings suggest that resveratrol might be a promising phytomedicine for cancer therapy, and further efforts are needed to explore this potential therapeutic strategy.

14.3.10 Shikonin

Shikonin (Fig. 14.1J) is a major natural anthraquinone compound extracted from Chinese herbal therapeutic plant *Lithospermum erythrorhizon* (zicao in Chinese). Shikonin possesses numerous pharmacological activities such as anti-inflammatory, antitumor, wound healing ability (Chen et al. 2003), and inhibition of type 1 HIV-induced cytopathology (Ueba et al. 1993; Yamasaki et al. 1993; Chen et al. 2003). The clinical study revealed that shikonin was effective in treating late-stage lung cancer patients who are not suitable for operation, radiotherapy, and chemotherapy (Guo et al. 1991). Also, shikonin induces apoptosis in human prostate cancer, osteosarcoma cells, chronic myelogenous leukemia (CML) cells and cervical, bladder, and melanoma cancer cells through the various mechanism such as endoplasmic reticulum stress, ROS, intrinsic pathway, extracellular signal-regulated kinase pathway, and classic caspase-dependent pathway (Chang et al. 2010; Mao et al. 2008; Wu et al. 2004a, b; Yeh et al. 2007; Yoon et al. 1999; Gara et al. 2015). Shikonin inhibits tumor invasion in human high-metastatic adenoid cystic carcinoma cells via the NF-kB signaling pathway (Min et al. 2011). In H22 allografts and PC-3 xenografts, it decreased proteasomal activities by inhibiting tumor growth (Yang et al. 2009a, b). Shikonin derivatives also exert potent antitumor activity and inhibit topoisomerase II (Yang et al. 2006). Shikonin induces necroptosis in caspase-3-negative MCF-7 cells, HL-60, and K562 cells (Han et al. 2007, 2009). Shikonin inhibits the proliferation and induces apoptosis in Tca-8113 human oral cancer cells (Min et al. 2008; Nie et al. 2010). There are several different mechanisms involved in the anticancer activities of shikonin, and these may vary with cancer cell type and treatment methodology. Therefore, shikonin may directly or indirectly inhibit or modulate disease-related cellular targets in cancer.

14.3.11 Silibinin

Silibinin (Fig. 14.1K), a flavonoid derived from *Silybum marianum* (shuifeiji), has long been used in traditional Chinese medicine to clear heat and relieve toxic material, to soothe the liver, and to promote bile blow (Wang et al. 2014a, b). Silibinin has been shown to possess strong anticancer activity against many cancer cell lines, such as skin, prostate, colon, bladder, kidney, and lung cancers (Mohan et al. 2004; Li et al. 2008; Rajamanickam et al. 2010; Singh et al. 2009), particularly cell migration, invasion, and metastasis (Singh et al. 2008).

Silibinin exerts anticancer activity in large cell carcinoma cells (H1299 and H460) and a bronchoalveolar carcinoma cell line (H322) by inhibited cell growth and targeted cell cycle progressing causing a prominent G1 arrest (Mateen et al. 2010). Silibinin also exerts antiproliferative effects on several human cancer cells via multiple cellular proliferative pathways including receptor tyrosine kinases (RTKs), androgen receptors, signal transducers and activators of transcription (STATs), and NF-kB (Li et al. 2010d). Further, silibinin induced apoptosis in human bladder transitional cell papilloma RT4 cells (Tyagi et al. 2006), human breast cancer MCF-7 cells (Klampfer 2006), SCLC (SHP-77 cells), and NSCLC (A549 cells) (Sharma et al. 2003) via extrinsic and intrinsic apoptotic pathways. Synergetic effects of silibinin in combination with paclitaxel exhibit beneficial chemotherapeutic strategy, especially in patients with tumor refractory to paclitaxel alone (Zhou et al. 2008). The synergetic effect of silibinin with 1,25-dihydroxyvitamin D3 enhances the expression of differentiation-promoting and inhibits genes in acute myelogenous leukemia cells, and it can be neutralized by a specific pharmacological inhibitor (Wang et al. 2010a, b). A study reported that silibinin combined with chitosan enhanced the antitumor activity in mice (Yousra et al. 2013). Moreover, its combination with anticancer drugs curcumin, piperine, and quercetin showed enhanced anticancer activity on Ovkar-3, HL60, HEPG2, Colo205, and A549 cancer cell lines (Moorthi et al. 2014). These research findings suggest the therapeutic potential of silibinin as an anticancer drug.

14.3.12 Tanshinone IIA

Tanshinone IIA (Fig. 14.1L) is abundantly accumulated in the root of TCM therapeutic herb *Salvia miltiorrhiza* Bunge, a Chinese herb (danshen in Chinese), which exerts various pharmacological activities, inducing anticancer (Wu et al. 1991), antioxidative, anti-inflammatory (Cao et al. 1996), and myocardial infarction and protecting or preventing angina pectoris (Zhao et al. 1996). The literature shows that it could inhibit growth and proliferation of SMMC-7721 cells by arresting the cell cycle in G_0/G_1 phase and induce apoptosis and proliferation of human hepatocellular carcinoma (Yuan et al. 2004).

Tanshinone IIA exhibited cytotoxicity against human endothelial cells through activation of NQO1, which induces a calcium imbalance and mitochondrial dysfunction, thus stimulating caspase activity (Yang et al. 2005). Tanshinone IIA exerts potential anticancer activity on both ER-positive and ER-negative breast cancers. It could be attributed in part to its inhibition of proliferation and apoptosis induction of cancer cells through upregulation and downregulation of multiple genes involved in cell cycle regulation, cell proliferation, apoptosis, signal transduction, transcriptional regulation, angiogenesis, and invasive potential and metastatic potential of cancer cells (Wang et al. 2005b). Tanshinone IIA effectively inhibits in vitro and *in vivo* invasion and metastasis in human hepatocellular carcinoma cells by reducing the expression of the metalloproteinases MMP2 and MMP9 and by blocking NF-kB activation (Yuxian et al. 2008). Tan IIA in combination with epirubicin increased apoptosis by the decreasing phosphorylation of Akt in BT-20 cells (Chan et al. 2011). Tanshinone IIA induced apoptosis via mitochondria-dependent caspase pathway and also induced growth inhibition in human oral cancer KB cells (Tseng et al. 2014). In a J5 xenograft animal model, tanshinone IIA inhibited tumor growth by increasing Bax and caspase-3 and decreasing CD31 expression in vivo (Chien et al. 2012). Tanshinone IIA exhibited potential anticancer effects on BEL-7402 cells via activation of calcium-dependent apoptosis pathway and upregulation of MT 1A expression and G_0/G_1 arrest. Tanshinone IIA exhibited its potential for the treatment of HCC due to its nontoxicity to human amniotic mesenchymal stem cells (HAMCs) (Dai et al. 2012). Since Tan IIA is hard to be absorbed through the intestinal pathway, sodium tanshinone IIA sulfonate (STS) was developed to raise the bioavailability (Shang et al. 2012).

14.3.13 Tetrandrine

Tetrandrine (Fig. 14.1M) is a bisbenzylisoquinoline alkaloid extracted from the roots of *Stephania tetrandra* S. Moore. It is an ancient ingredient in traditional Chinese herbal medicine and is broadly used in China to treat patients with arthritis, hypertension, inflammation, and even silicosis (Cao 1996). Tetrandrine exerts its anticancer activity in the micromolar concentrations. It has been reported to exert a broad range of pharmacological activities including immunomodulating, anti-hepatofibrogenetic, anti-inflammatory, antiarrhythmic, anti-portal hypertension, anticancer, and neuroprotective activities (Ji 2011; Lu et al. 2012). Many research studies have demonstrated that tetrandrine exerts anticancer activity in vivo (Li et al. 2012; Qin et al. 2013; Chen et al. 2014; Xiao et al. 2015). Tetrandrine inducing cell cycle arrest depends on cancer types (Lu et al. 2012). Tetrandrine induces apoptosis in the U-2OS and MG-63 osteosarcoma cell lines (Tao et al. 2014) and human gastric cancer BGC-823 cells (Qin et al. 2013) through the mitochondrial pathway. Tetrandrine promotes apoptosis in human renal cell carcinoma 786-O, 769-P, and ACHN in vitro via caspase cascade activation and upregulation of p21 and p27 (Chen et al. 2014). In human oral cancer HSC-3 cells, tetrandrine induced apoptosis as well as autophagy mediated by PARP and caspase/beclin-1/LC3-I/II signaling pathways (Yu et al. 2014). Tetrandrine with chemotherapy drug doxorubicin synergistically exerts anticancer metastatic and antiangiogenic activities in 4T1 tumor-bearing BALB/c mice model than doxorubicin-alone treatment (Gao et al. 2013a, b). Synergistic antitumor activity of tetrandrine in combination with sorafenib induced apoptosis in both in vitro cell culture and in vivo xenograft model via intrinsic apoptosis pathway through ROS/Akt signaling (Wan et al. 2013). Coadministration of tetrandrine with chloroquine induced ROS accumulation and cell apoptosis and decreased tumor growth in tumor xenograft model in mice (Mei et al. 2015).

14.3.14 Triptolide

Triptolide (Fig. 14.1N) a diterpenoid triepoxide is the principal active ingredient of Chinese herb *Tripterygium wilfordii* Hook. f. (leigongteng). Leigongteng root has long been used as natural medicine in China for many years to treat inflammation and autoimmune diseases such as rheumatoid arthritis, chronic nephritis, chronic hepatitis, and lupus erythematosus (Jia 1985). Triptolide exhibits multiple effects on immunosuppressive, anti-inflammatory, antifertility, apoptotic, angiogenic, and antineoplastic activities, metastasis, and drug resistance (Li et al. 2010a; Zhao et al. 2010a; Zhou et al. 2007, 2011; Zhu et al. 2009). Triptolide induced apoptosis in various cell lines via different pathways such as mitochondrial-mediated pathway and overexpression of cytomembrane death receptor in a caspase-8-dependent manner (Zhou et al. 2007; Clawson et al. 2010; Carter et al. 2006; Wan et al. 2006).

Triptolide manifested the potent antiangiogenic activity against vessel formation in zebra fish embryo model (He et al. 2009) and colon cancer cells (Johnson et al. 2011). Triptolide also inhibits cell migration and is involved in cancer metastasis (Zhang et al. 2006; Johnson et al. 2011). Further, triptolide inhibits human breast cancer cells (Liang and Fu 2008) and also inhibits metastasis of melanoma cells to the lungs and spleen of mice (Yang et al. 2003). Also, clinical trials in China revealed that triptolide could achieve a total remission rate of 71 % in mononucleocytic leukemia and 87 % in granulocytic leukemia, which was more effective than any other chemotherapeutic agent and is currently available (Jiang et al. 2001; Lu et al. 1992). Triptolide significantly enhanced anticancer activity in combination with anticancer drugs or therapy such as 5-fluorouracil (Tang et al. 2007), hydroxycamptothecin (Yang et al. 2011), chemotherapy (idarubicin, AraC) (Pigneux et al. 2008), TRAIL (Carter et al. 2008), and ionizing radiation (Wang et al. 2007). These results revealed that triptolide from leigongteng is the potential anticancer compound in treating cancer.

14.3.15 Wogonin

Wogonin (Fig. 14.1O) is an O-methylated flavone, a flavonoid-like compound isolated from Chinese herb *Scutellaria baicalensis* Georgi (huang qin), and it has been recognized as an anticancer drug with minimal toxicity (Li-Weber 2009). *S. baicalensis* is one of the most popular and multipurpose herbs used extensively in TCM medicine and modern herbal preparations (Murch et al. 2004). It was officially included in *Chinese Pharmacopoeia* against a wide range of diseases such as cancer, hepatitis, cirrhosis, jaundice, hepatoma, leukemia, hyperlipidemia, arteriosclerosis, and inflammation (Chang et al. 2002; Li et al. 2013). Wogonin is the main active medicinal constituents in huang qin and was shown to inhibit the proliferation of various human hepatoma cell lines (Chang et al. 2002; Okamura et al. 1999). Wogonin inhibited the proliferation of bladder cancer cell lines EJ-1, KU-1, MBT-2 (Ikemoto et al. 2000). The whole huang qin herb extracts were effective in reducing prostate cancer (Hsieh et al. 2002); alcoholic extracts inhibited liver fibrosis (Nan et al. 2002) and delayed apoptosis in neuronal cells (Suk et al. 2003). The individual isolated compounds reduced the symptoms of type 1 allergic reactions (Lim 2003). Also, wogonin has been shown to exert cytostatic and proapoptotic effects on several tumor cell lines (Ikemoto et al. 2000; Chen et al. 2002; Lee et al. 2002).

Wogonin has apoptosis-inducing effect in malignant T cells through the intrinsic (mitochondrial) pathway via two important mechanisms and suppresses the growth of human T-cell leukemia xenografts in vivo. Moreover, it may also induce mitochondrial damage by enhancing Ca^{2+} uptake (Baumann et al. 2008). Wogonin showed minimally or no toxicity on normal peripheral T Cells (Baumann et al. 2008), TIG-1 cells (Himeji et al. 2007), and human prostate epithelial cell (Lee et al. 2008). Furthermore, wogonin induced cell cycle inhibition at G_1 phase in human cervical carcinoma HeLa cells. It was related to an inhibition of Cdk4/cyclin D1 and the upregulation of $p21^{Cip1}$ and p53, both at mRNA and protein levels (Yang et al. 2009b) and at the G_2/M phase in THP-1 cells (Himeji et al. 2007). Also, wogonin suppressed the growth and tumor angiogenesis of human gastric carcinoma in nude mice. Further, it inhibited tube formation in HUVEC by inhibiting VEGF receptor 2 (VEGFR2) instead of VEGFR1 phosphorylation and VEGF-stimulated migration (Lu et al. 2008).

The anticancer activity of wogonin and its induced apoptosis in a wide spectrum of human cancer cells have been reported in human breast cancer MCF-7 cells (Huang et al. 2012a), myeloma cell RPMI 8226 (Zhang et al. 2013), nasopharyngeal carcinoma cells (Chow et al. 2011), hepatocellular carcinoma SK-HEP-1 (Chen et al. 2002), and human glioma cancer cells (Tsai et al. 2012). The anticancer mechanism of wogonin is mediated by the induction of apoptosis and cell differentiation and is regulated by various genes and proteins (Chen et al. 2013b). In combination with etoposide, the synergistic effects significantly improve apoptosis in cancer cells similar to verapamil and cyclosporine A (Lee et al. 2007, 2009; Enomoto et al. 2011). Similar results were obtained when wogonin was combined with 5-FU in human gastric MGC-803-transplanted nude mice and MGC-803 cells (Zhao et al. 2010a, b). A study showed that combination of wogonin with ABT-263 promotes *in vivo* tumor regression in a human T-cell leukemia xenograft mouse model and decreasing the risk of adverse side effects. Further, this study suggests that wogonin may be used as an adjuvant for ABT-263-based anticancer therapy (Polier et al. 2015). However, the molecular mechanism of its potent anticancer activity remains unclear and needs further investigations. Wogonin exerts multiple anticancer effects in various types of tumor cells due to its toxicity. Also, it has low or no toxicity to normal cells when combined with other anticancer drugs/chemotherapy agents.

14.4 Conclusions

Traditional Chinese medicine has a long and rich history and practical experiences in the treatment of cancer. For thousands of years, TCM has been proven by clinical practice that it has a rich scientific connotation and has developed a school of its own in global traditional medicine for its exact curative effects and integrated system information. During the last two decades, accumulating research reports have shown that many bioactive compounds have been isolated from Chinese herbal medicine plants, and these compounds are attractive candidates for the development of new anticancer drugs. The majority of the TCM herb-derived bioactive compounds discussed above are the same as conventional anticancer drugs which not only induce apoptosis but also are involved in various molecular pathways. Although fewer side effects have been observed in natural product-based anticancer therapy, anticancer drug from a natural product is safer than chemically synthesized drugs. Recently, the potential of bioactive natural products from medicinal herbs used in TCM has been recognized in the Western medicine. However, molecular mechanism, pharmacological analysis, toxicological analysis, preclinical analysis, and phases I–III clinical trials with cutting-edge technologies need further investigations to develop natural products derived from TCM herbs as rational anticancer drugs.

References

Auyeung KK, Ko JK (2009) *Coptis chinensis* inhibits hepatocellular carcinoma cell growth through nonsteroidal anti-inflammatory drug-activated gene activation. Inter J Mol Med 24:571–577. doi:10.3892/ijmm_00000267

Ba Q, Zhou N, Duan J, Chen T, Hao M, Yang X, Li J, Yin J, Chu R, Wang H (2012) Dihydroartemisinin exerts its anticancer activity through depleting cellular iron via transferrin receptor-1. PLoS One 7:e42703. doi:10.1371/journal.pone.0042703

Bao YX, Wong CK, Leung SF, Chan AT, Li PW, Wong EL, Leung PC, Fung KP, Yin YB, Lam CW (2006) Clinical studies of immunomodulatory activities of Yunzhi-Danshen in patients with nasopharyngeal carcinoma. J Altern Complement Med 12:771–776. doi:10.1089/acm.2006.12.771

Baumann S, Fas SC, Giaisi M, Müller WW, Merling A, Gülow K, Edler L, Krammer PH, Li-Weber M (2008) Wogonin preferentially kills malignant lymphocytes and suppresses T-cell tumor growth by inducing PLCγ1- and Ca2+ dependent apoptosis. Blood 111:2354–2363. doi:10.1182/blood-2007-06-096198

Baxter LT, Jain RK (1989) Transport of fluid and macromolecules in tumors. 1. Role of interstitial pressure and convection. Microvasc Res 37:77–104

Bensky D, Gamble A (1993) Chinese herbal medicine materia medica. Eastland Press, Seattle

Berger TG, Dieckmann D, Efferth T (2005) Artesunate in the treatment of metastatic uveal melanoma–first experiences. Oncol Rep 14:1599–1603. doi:10.3892/or.14.6.1599

Bohlin L, Rosen B (1996) Podophyllotoxin derivatives: drug discovery and development. Drug Discov Today 1:343–351

Boucher Y, Jain RK (1992) Microvascular pressure is the principle driving force for interstitial hypertension in solid tumors: implications for vascular collapse. Cancer Res 52:5110–5114

Brizel DM, Sibley GS, Prosnitz LR, Scher RL, Dewhirst MW (1997) Tumor hypoxia adversely affects the prognosis of carcinoma of the head and neck. Int J Radiat Oncol Biol Phys 38:285–289

Bu H, Luo J, Chen H, Zhang J, Li H, Guo H, Wang Z, Lin S (2012) Oridonin enhances antitumor activity of gemcitabine in pancreatic cancer through MAPK-p38 signaling pathway. Int J Oncol 41:949–958. doi:10.3892/ijo.2012.1519

Cai DY, Gao X, Wu XH, Hong TT (2013) Synergistic effect of beta-elemene injection combined paclitaxel injection on human breast cancer MB-468 cells: an *in vitro* study. Zhongguo Zhong Xi Yi Jie He Za Zhi 33:978–982

Cao ZF (1996) Scavenging effect of tetrandrine of active oxygen radicals. Planta Med 62:413–414

Cao EH, Liu XQ, Wang JJ, Xu NF (1996) Effect of natural antioxidant tanshinone II-A on DNA damage by lipid peroxidation in liver cells. Free Radical Bio Med 20:801–806

Carter BZ, Mak DH, Schober WD, McQueen T, Harris D, Estrov Z, Evans RL, Andreeff M (2006) Triptolide induces caspase-dependent cell death mediated via the mitochondrial pathway in leukemic cells. Blood 108:630–637. doi:10.1182/blood-2005-09-3898

Carter BZ, Mak DH, Schober WD, Dietrich MF, Pinilla C, Vassilev LT, Reed JC, Andreeff M (2008) Triptolide sensitizes AML cells to TRAIL-induced apoptosis via decrease of XIAP and p53-mediated increase of DR5. Blood 111:3742–3750. doi:10.1182/blood-2007-05-091504

Chan SE, Lai HW, Su CC, Kuo SJ, Chien SY, Lin HY, Chen DR (2011) Effect of supplementation of Tanshinone IIA and Sodium Tanshinone IIA sulfonate on the anticancer effect of epirubicin: an *in vitro* study. Evid Based Complement Alternat Med 2011. doi:10.1155/2011/841564

Chang WH, Chem CH, Lu FJ (2002) Different effects of baicalein, baicalin and wogonin on mitochondrial function, glutathione content and cell cycle progression in human hepatoma cell lines. Planta Med 68:128–132. doi:10.1055/s-2002-20246

Chang IC, Huang YJ, Chiang TI, Yeh CW, Hsu LS (2010) Shikonin induces apoptosis through reactive oxygen species/extracellular signal-regulated kinase pathway in osteosarcoma cells. Biol Pharm Bull 33:816–824. doi:10.1248/bpb.33.816

Chen YC, Shen SC, Lee WR, Lin HY, Ko CH, Shih CM, Yang LL (2002) Wogonin and fisetin induction of apoptosis through activation of caspase 3 cascade and alternative expression of p21 protein in hepatocellular carcinoma cells SK-HEP-1. Arch Toxicol 76:351–359. doi:10.1007/s00204-002-0346-6

Chen X, Yang L, Zhang N, Turpin JA, Buckheit RW, Osterling C, Oppenheim JJ, Howard OZ (2003) Shikonin, a component of Chinese herbal medicine, inhibits chemokine receptor function and suppresses human immunodeficiency virus type 1. Antimicrob Agent Chemother 47:2810–2816. doi:10.1128/AAC.47.9.2810-2816.2003

Chen G, Wang K, Yang B, Tang B, Chen J, Hua Z (2012) Synergistic antitumor activity of oridonin and arsenic trioxide on hepatocellular carcinoma cells. Int J Oncol 40:139–147. doi:10.3892/ijo.2011.1210

Chen P, Liu Y, Sun Y, Chen C, Qi Y, Zhang Y (2013a) AZT and emodin exhibit synergistic growth-inhibitory effects on K562/ADM cells by inducing S phase cell cycle arrest and suppressing MDR1 mRNA/p-gp protein expression. Pharm Biol 51:1586–1591. doi:10.3109/13880209.2013.803257

Chen XM, Bai Y, Zhong YJ, Xie XL, Long HW, Yang YY, Wu SG, Jia Q, Wang XH (2013b) Wogonin has multiple anti-cancer effects by regulating c-Myc/SKP2/Fbw7α and HDAC1/HDAC2 pathways and inducing apoptosis in human lung adenocarcinoma cell line A549. PLoS One 8(11):e79201. doi:10.1371/journal.pone.0079201

Chen T, Ji B, Chen Y (2014) Tetrandrine triggers apoptosis and cell cycle arrest in human renal cell carcinoma cells. J Natural Med 68:46–52. doi:10.1007/s11418-013-0765-0

Chiang KC, Tsui KH, Chung LC, Yeh CN, Chen WT, Chang PL, Juang HH (2014) Celastrol blocks interleukin-6 gene expression via downregulation of NF-kB in prostate carcinoma cells. PLoS One 9:e93151. doi:10.1371/journal.pone.0093151

Chien SY, Kuo SJ, Chen YL, Chen DR, Cheng CY, Su CC (2012) Tanshinone IIA inhibits human hepatocellular carcinoma J5 cell growth by increasing Bax and caspase 3 and decreasing CD31 expression *in vivo*. Mol Med Rep 5:282–286. doi:10.3892/mmr.2011.631

Choi JY, Cho HJ, Hwang SG, Kim WJ, Kim JI, Um HD, Park JK (2015) Podophyllotoxin acetate enhances γ-ionizing radiation-induced apoptotic cell death by stimulating the ROS/p38/caspase pathway. Biomed Pharmacother 70:111–118. doi:10.1016/j.biopha.2014.12.038

Chow SE, Chang YL, Chuang SF, Wang JS (2011) Wogonin induced apoptosis in human nasopharyngeal carcinoma cells by targeting GSK-3b and DNp63. Cancer Chemother Pharmacol 68:835–845. doi:10.1007/s00280-010-1552-1

Clawson KA, Borja-Cacho D, Antonoff MB, Saluja AK, Vickers SM (2010) Triptolide and TRAIL combination enhances apoptosis in cholangiocarcinoma. J Surg Res 163:244–249

Coffey DS (1998) Self-organization, complexity and chaos: the new biology for medicine. Nat Med 4:882–885. doi:10.1016/j.jss.2010.03.067

Crespo-Ortiz MP, Wei MQ (2011) Antitumor activity of artemisinin and its derivatives: from a well-known antimalarial agent to a potential anticancer drug. BioMed Res Inter 22:2012. doi:10.1155/2012/247597

Cuzick J, Holland R, Barth V, Davies R, Faupel M, Fentiman I, Frischbier HJ, LaMarque JL, Merson M, Sacchini V, Vanel D (1998) Electropotential measurements as a new diagnostic modality for breast cancer. Lancet 352:359–363

Dai ZK, Qin JK, Huang JE, Luo Y, Xu Q, Zhao HL (2012) Tanshinone IIA activates calcium-dependent apoptosis signaling pathway in human hepatoma cells. J Nat Med 66:192–201. doi:10.1007/s11418-011-0576-0

Dai Z, Lei P, Xie J, Hu Y (2015) Antitumor effect of resveratrol on chondrosarcoma cells via phosphoinositide 3-kinase/AKT and p38 mitogen-activated protein kinase pathways. Mol Med Rep 12:3151–3155. doi:10.3892/mmr.2015.3683

Das AK (2015) Anticancer effect of antimalarial artemisinin compounds. Ann Med Health Sci Res 5:93–102. doi:10.4103/2141-9248.153609

Deng XR, Liu ZX, Liu F, Pan L, Yu HP, Jiang JP, Zhang JJ, Liu L, Yu J (2013) Holotransferrin enhances selective anticancer activity of artemisinin against human hepatocellular carcinoma cells. J Huazhong Univ Sci Tec [Med Sci] 33:862–865

Enomoto R, Koshiba C, Suzuki C, Lee E (2011) Wogonin potentiates the antitumor action of etoposide and ameliorates its adverse effects. Cancer Chemother Pharmacol 67:1063–1072. doi:10.1007/s00280-010-1396-8

Eom KS, Hong JM, Youn MJ, So HS, Park R, Kim JM, Kim TY (2008) Berberine induces G1 arrest and apoptosis in human glioblastoma T98G cells through mitochondrial/caspases pathway. Biol Pharmaceut Bullet 31:558–562. doi:10.1248/bpb.31.558

Ergil KV, Kramer EJ, Ng AT (2002) Chinese herbal medicines. West J Med 176:275

Farrell MP (2003) Ph I/IIA randomized study of PHY906 as a modulator of chemotherapy in patients with advanced colorectal cancer. Clin Colorectal Cancer 2:253–256

Firestone GL, Sundar SN (2009) Anticancer activities of artemisinin and its bioactive derivatives. Expert Rev Mol Med 11:1–15. doi:10.1017/S1462399409001239

Fribley AM, Miller JR, Brownell AL, Garshott DM, Zeng Q, Reist TE, Narula N, Cai P, Xi Y, Callaghan MU, Kodali V (2015) Celastrol induces unfolded protein response-dependent cell death in head and neck cancer. Exp Cell Res 330:412–422. doi:10.1016/j.yexcr.2014.08.014

Fulda S, Debatin KM (2006) Extrinsic versus intrinsic apoptosis pathways in anticancer chemotherapy. Oncogene 25:4798–4811

Fyles AW, Milosevic M, Wong R, Kavanagh MC, Pintilie M, Sun A, Chapman W, Levin W, Manchul L, Keane TJ, Hill RP (1998) Oxygenation predicts radiation response and survival in patients with cervix cancer. Radiother Oncol 48:149–156

Gao JL, Ji X, He TC, Zhang Q, He K, Zhao Y, Chen SH, Lv GY (2013a) Tetrandrine suppresses cancer angiogenesis and metastasis in 4T1 tumor bearing mice. Evid Based Complement Alter Med 2013:12. doi:10.1155/2013/265061

Gao SY, Li J, Qu XY, Zhu N, Ji YB (2013b) Downregulation of Cdk1 and cyclinB1 expression contributes to oridonin-induced cell cycle arrest at G2/M phase and growth inhibition in SGC-7901 gastric cancer cells. Asian Pacific J Cancer Prev 15:6437–6441

Gara RK, Srivastava VK, Duggal S, Bagga JK, Bhatt ML, Sanyal S, Mishra DP (2015) Shikonin selectively induces apoptosis in human prostate cancer cells through the endoplasmic reticulum stress and mitochondrial apoptotic pathway. J Biomed Sci 22:26. doi:10.1186/s12929-015-0127-1

Gordaliza M, Castro MA, Miguel del Corral JM, Feliciano AS (2000) Antitumor properties of podophyllotoxin and related compounds. Curr Pharm Des 6:1811–1839

Guo XP, Zhang XY, Zhang SD (1991) Clinical trial on the effects of shikonin mixture on later stage lung cancer. Zhong Xi Yi Jie He Za Zhi 11:598–599

Guo Q, Chen Y, Zhang B, Kang M, Xie Q, Wu Y (2009) Potentiation of the effect of gemcitabine by emodin in pancreatic cancer is associated with survivin inhibition. Biochem Pharmacol 77:1674–1683. doi:10.1016/j.bcp.2009.02.021

Guo J, Li W, Shi H, Xie X, Li L, Tang H, Wu M, Kong Y, Yang L, Gao J, Liu P (2013) Synergistic effects of curcumin with emodin against the proliferation and invasion of breast cancer cells through upregulation of miR-34a. Mol Cell Biochem 382:103–111. doi:10.1007/s11010-013-1723-6

Han W, Li L, Qiu S, Lu Q, Pan Q, Gu Y, Luo J, Hu X (2007) Shikonin circumvents cancer drug resistance by induction of a necroptotic death. Mol Cancer Ther 6:1641–1649

Han W, Xie J, Li L, Liu Z, Hu X (2009) Necrostatin-1 reverts shikonin-induced necroptosis to apoptosis. Apoptosis 14:674–686. doi:10.1007/s10495-009-0334-x

Han G, Xia J, Gao J, Inagaki Y, Tang W, Kokudo N (2015) Anti-tumor effects and cellular mechanisms of resveratrol. Drug Disc Ther 9:1–2. doi:10.5582/ddt.2015.01007

Hara A, Iizuka N, Hamamoto Y, Uchimura S, Miyamoto T, Tsunedomi R, Miyamoto K, Hazama S, Okita K, Oka M (2005) Molecular dissection of a medicinal herb with anti-tumor activity by oligonucleotide microarray. Life Sci 77:991–1002

He MF, Liu L, Ge W, Shaw PC, Jiang R, Wu LW, But PP (2009) Antiangiogenic activity of *Tripterygium wilfordii* and its terpenoids. J Ethnopharmacol 121:61–68. doi:10.1016/j.jep.2008.09.033

He Q, Shi J, Shen XL, An J, Sun H, Wang L, Hu YJ, Sun Q, Fu LC, Sheikh MS, Huang Y (2010) Dihydroartemisinin upregulates death receptor 5 expression and cooperates with TRAIL to induce apoptosis in human prostate cancer cells. Cancer Biol Ther 9:819–824. doi:10.4161/cbt.9.10.11552

He BC, Gao JL, Luo X, Luo J, Shen J, Wang L, Zhou Q, Wang YT, Luu HH, Haydon RC, Wang CZ, Du W, Yuan CS, He TC, Zhang BQ (2011) Ginsenoside Rg3 inhibits colorectal tumor growth through the down-regulation of Wnt/ß- catenin signaling. Int J Oncol 38:437–445. doi:10.3892/ijo.2010.858

Hejna M, Raderer M, Zielinski CC (1999) Inhibition of metastases by anticoagulants. J Natl Cancer Inst 91:22–36

Himeji M, Ohtsuki T, Fukazawa H, Tanaka M, Yazaki SI, Ui S, Nishio K, Yamamoto H, Tasaka K, Mimura A (2007) Difference of growth-inhibitory effect of Scutellaria baicalensis-producing flavonoid wogonin among human cancer cells and normal diploid cell. Cancer Lett 245:269–274

Ho YT, Lu CC, Yang JS, Chiang JH, Li TC, Ip SW, Hsia TC, Liao CL, Lin JG, Wood WG, Chung JG (2009) Berberine induced apoptosis via promoting the expression of caspase-8, -9 and -3, apoptosis-inducing factor and endonuclease G in SCC-4 human tongue squamous carcinoma cancer cells. Anticancer Res 29:4063–4070

Hou J, Wang D, Zhang R, Wang H (2008) Experimental therapy of hepatoma with artemisinin and its derivatives: *in vitro* and *in vivo* activity, chemosensitization, and mechanisms of action. Clin Cancer Res 14:5519–5530. doi:10.1158/1078-0432.CCR-08-0197

Hsiang YH, Hertzberg R, Hecht S, Liu LF (1985) Camptothecin induces protein- linked DNA breaks via mammalian DNA topoisomerase I. J Biol Chem 260:14873–14878

Hsiao WLW, Liu L (2010) The role of traditional Chinese herbal medicines in cancer therapy from TCM Theory to mechanistic insights. Planta Med 76:1118–1131. doi:10.1055/s-0030-1250186

Hsu SC, Chung JG (2012) Anticancer potential of emodin. BioMedicine 2:108–116. doi:10.1016/j.biomed.2012.03.003

Hsieh TC, Lu X, Chea J, Wu JM (2002) Prevention and management of prostate cancer using PC-SPES: a scientific perspective. J Nutr 132:3513–3517

Hu Y, Wang S, Wu X, Zhang J, Chen R, Chen M, Wang Y (2013) Chinese herbal medicine derived compounds for cancer therapy: a focus on hepatocellular carcinoma. J Ethnopharmacol 149:601–612. doi:10.1016/j.jep.2013.07.030

Hu MY, Zhu JF, Qiu L (2014) Polymer micelle-based combination therapy of paclitaxel and resveratrol with enhanced and selective antitumor activity. RSC Adv 4:64151–64161. doi:10.1039/c4ra09761k

Hu Z, Wu H, Li Y, Hou Q, Wang Y, Li S, Xia B, Wu S (2015) β-Elemene inhibits the proliferation of esophageal squamous cell carcinoma by regulating long noncoding RNA-mediated inhibitionofhTERTexpression.Anti-CancerDrug26:531–539.doi:10.1097/CAD.0000000000000216

Huali S, Shaojin D, Guiqing Y (1994) Free radical mechanism in enhancement of radiosensitization by SRSBR. J Tradit Chin Med 14:51–55

Huang KF, Huang YQ, Diao Y (2012a) Wogonin induces apoptosis and downregulates survivin in human breast cancer MCF-7 cells by modulating PI3K–AKT pathway. Inter Immunopharmacol 12:334–341. doi:10.1016/j.intimp.2011.12.004

Huang Y, Zhou D, Hang T, Wu Z, Liu J, Xu Q, Xie X, Zuo J, Wang Z, Zhou Y (2012b) Preparation, characterization, and assessment of the antiglioma effects of liposomal celastrol. Anti-Cancer Drug 23:515–524. doi:10.1097/CAD.0b013e3283514b68

Huang L, Wang XB, Yu QM, Luo QY, Zhang ZZ (2015) Synergistic cancer growth-inhibitory effect of emodin and low-dose cisplatin on gastric cancer cells *in vitro*. Trop J Pharmaceut Res 14:1427–1434. doi:10.4314/tjpr.v14i8.15

Hur JM, Hyun MS, Lim SY, Lee WY, Kim D (2009) The combination of berberine and irradiation enhances anti-cancer effects via activation of p38 MAPK pathway and ROS generation in human hepatoma cells. J Cell Biochem 107:955–964. doi:10.1002/jcb.22198

Ikemoto S, Sugimura K, Yoshida N, Yasumoto R, Wada S, Yamamoto K, Kishimoto T (2000) Antitumor effects of *Scutellariae radix* and its components baicalein, baicalin, and wogonin on bladder cancer cell lines. Urology 55:951–955

Ji YB (2011) Active ingredients of traditional Chinese medicine: pharmacology and application, People's Medical Publishing House Cp., LTD

Jia L (1985) Chemistry and pharmacology and clinical application of the plants of *Tripterygium* family. Yao Xue Tong Bao 20:1001–1005

Jiang XH, Wong BC, Lin MC, Zhu GH, Kung HF, Jiang SH, Yang D, Lam SK (2001) Functional p53 is required for triptolide-induced apoptosis and AP-1 and nuclear factor-kB activation in gastric cancer cells. Oncogene 20:8009–8018

Jiang JW, Chen XM, Chen XH, Zheng SS (2011) Ginsenoside Rg3 inhibit hepatocellular carcinoma growth via intrinsic apoptotic pathway. World J Gastroenterol 17:3605–3613. doi:10.3748/wjg.v17.i31.3605

Jin L, Ren YJ, Du C (2015) Synthesis and antitumor activities of resveratrol derivatives on cervical cancer Hela cells. Chem Nat Compound 51:652–655. doi:10.1007/s10600-015-1377-6

Johnson SM, Wang X, Mark Evers B (2011) Triptolide inhibits proliferation and migration of colon cancer cells by inhibition of cell cycle regulators and cytokine receptors. J Surg Res 168:197–205. doi:10.1016/j.jss.2009.07.002

Jung M, Lee K, Kim H, Park M (2004) Recent advances in artemisinin and its derivatives as antimalarial and antitumor agents. Curr Med Chem 11:1265–1284. doi:10.2174/0929867043365233

Jung HW, Chung YS, Kim YS, Park YK (2007) Celastrol inhibits production of nitric oxide and proinflammatory cytokines through MAPK signal transduction and NF-kB in LPS-stimulated BV-2 microglial cells. Exp Mol Med 39:715–721. doi:10.1038/emm.2007.78

Kaminski BM, Weigert A, Scherzberg MC, Ley S, Gilbert B, Brecht K, Brüne B, Steinhilber D, Stein J, Ulrich-Rückert S (2014) Resveratrol-induced potentiation of the antitumor effects of oxaliplatin is accompanied by an altered cytokine profile of human monocyte-derived macrophages. Apoptosis 19:1136–1147. doi:10.1007/s10495-014-0988-x

Kang K, Kang B, Lee B, Che J, Li G, Trosko JE, Lee Y (2000) Preventive effect of epicatechin and ginsenoside Rb(2) on the inhibition of gap junctional intercellular communication by TP and H(2)O(2). Cancer Lett 152:97–106. doi:10.1016/S0304-3835(99)00438-3

Kaplan IW (1942) Condylomata acuminata. N Orleans Med Surg J 94:388–390

Karuppaiya P (2013) *In vitro* production of anti-cancer compounds, expression of lignin biosynthetic pathway gene (s) and anti-cancer activity of biogenic metal nanoparticles synthesized from *Dysosma pleiantha*. Dissertation. Chaoyang University of Technology, Taiwan

Karuppaiya P, Tsay HS (2015) Therapeutic values, chemical constituents and toxicity of Taiwanese Dysosma pleiantha-A review. Toxicol Lett 236:90–97. doi:10.1016/j.toxlet.2015.05.004

Kaufmann SH, Earnshaw WC (2000) Induction of apoptosis by cancer chemotherapy. Exp Cell Res 256:42–49

Kim HS, Lee EH, Ko SR, Choi KJ, Park JH, Im DS (2004) Effects of ginsenoside Rg3 and Rh2 on the proliferation of prostate cancer cells. Arch Pharm Res 27:429–435

Kim DW, Ahan SH, Kim TY (2007) Enhancement of arsenic trioxide (As(2)O(3))- mediated apoptosis using berberine in human neuroblastoma SH-SY5Y cells. J Korean Neurosurg Soc 42:392–399. doi:10.3340/jkns.2007.42.5.392

Kim SM, Lee SY, Yuk DY, Moon DC, Choi SS, Kim Y, Han SB, Oh KW, Hong JT (2009) Inhibition of NF-kappaB by ginsenoside Rg3 enhances the susceptibility of colon cancer cells to docetaxel. Arch Pharm Res 32:755–765. doi:10.1007/s12272-009-1515-4

Kim JB, Yu JH, Ko E, Lee KW, Song AK, Park SY, Shin I, Han W, Noh DY (2010a) The alkaloid Berberine inhibits the growth of Anoikis-resistant MCF-7 and MDA-MB-231 breast cancer cell lines by inducing cell cycle arrest. Phytomedicine 17:436–440. doi:10.1016/j.phymed.2009.08.012

Kim SM, Lee SY, Cho JS, Son SM, Choi SS, Yun YP, Yoo HS, do Yoon Y, Oh KW, Han SB, Hong JT (2010b) Combination of ginsenoside Rg3 with docetaxel enhances the susceptibility of prostate cancer cells via inhibition of NF-kappaB. Eur J Pharmacol 631:1–9. doi:10.1016/j.ejphar.2009.12.018

King ML, Sullivan MM (1946) The similarity of the effect of podophyllin and colchicine and their use in the treatment of *Condylomata acuminata*. Science 104:244–245

Klampfer L (2006) Signal transducers and activators of transcription (STATs): novel targets of chemopreventive and chemotherapeutic drugs. Curr Cancer Drug Targets 6:107–121

Kuo HP, Lee YJ, Hsu CY, Lee SL, Hsu SC, Chuang TC, Liu JY, Kuo CL, Ho CT, Kao MC (2015) Growth-suppressive effect of berberine on endometrial carcinoma cells: Role of mitochondrial and PI3K/Akt pathway. J Funct Food 17:600–609. doi:10.1016/j.jff.2015.06.006

Lai JM, Chang JT, Wen CL, Hsu SL (2009) Emodin induces a reactive oxygen species-dependent and ATM-p53-Bax mediated cytotoxicity in lung cancer cells. Eur J Pharmacol 623:1–9. doi:10.1016/j.ejphar.2009.08.031

Lai HC, Singh NP, Sasaki T (2013) Development of artemisinin compounds for cancer treatment. Invest New Drugs 31:230–246. doi:10.1007/s10637-012-9873-z

Lebeau B, Chastang C, Brechot JM, Capron F, Dautzenberg B, Delaisements C, Mornet M, Brun J, Hurdebourcq JP, Lemarie E (1994) Subcutaneous heparin treatment increases survival in small cell lung cancer. Cancer 74:38–45

Lee Hsiao Hung (1994) Chinese Herbal Medicine In Handbook of Chinese popular culture edited by Wu Dingbo and Patrick D. Murphy. Greenwood Press, Westport, Connecticut, London

Lee WR, Shen SC, Lin HY, Hou WC, Yang LL, Chen YC (2002) Wogonin and fisetin induce apoptosis in human promyeloleukemic cells, accompanied by a decrease of reactive oxygen species, and activation of caspase 3 and Ca (2+)-dependent endonuclease. Biochem Pharmacol 63:225–236. doi:10.1016/S0006-2952(01)00876-0

Lee KT, Johnke RM, Allison RR, O'Brien KF, Dobbs LJ (2005) Radioprotective potential of ginseng. Mutagenesis 20:237–243. doi:10.1093/mutage/gei041

Lee E, Enomoto R, Suzuki C, Ohno M, Ohashi T, Miyauchi A, Tanimoto E, Maeda K, Hirano H, Yokoi T, Sugahara C (2007) Wogonin, a plant flavone, potentiates etoposide-induced apoptosis in cancer cells. Ann N Y Acad Sci 1095:521–526. doi:10.1196/annals.1397.056

Lee DH, Kim C, Zhang L, Lee YJ (2008) Role of p53, PUMA, and Bax in wogonin-induced apoptosis in human cancer cells. Biochem Pharmacol 75:2020–2033. doi:10.1016/j.bcp.2008.02.023

Lee SY, Kim GT, Roh SH, Song JS, Kim HJ, Hong SS, Kwon SW, Park JH (2009) Proteomic analysis of the anti-cancer effect of 20S-ginsenoside Rg3 in human colon cancer cell lines. Biosci Biotechnol Biochem 73:811–816. doi:10.1271/bbb.80637

Lee YJ, Lee S, Ho JN, Byun SS, Hong SK, Lee SE, Lee E (2014) Synergistic antitumor effect of ginsenoside Rg3 and cisplatin in cisplatin resistant bladder tumor cell line. Oncol Rep 32:1803–1808. doi:10.3892/or.2014.3452

Leung PC, Ooi V, Wong EL, Au WC, Wong CK, Lam WK, Leung SF, Mok TS (2007) Chinese medicine and cancer treatment in Hong Kong: a general review. In: Leung PC, Fong H (eds) Alternative treatment for cancer. World Scientific Publishing Co. Pte. Ltd., Singapore, pp 65–74

Lim BO (2003) Effects of wogonin, wogonoside, and 3,5,7,2′,6′-pentahydroxyflavone on chemical mediator production in peritoneal exduate cells and immunoglobulin E of rat mesenteric lymph node lymphocytes. J Ethnopharmacol 84:23–29

Li-Weber M (2009) New therapeutic aspects of flavones: the anticancer properties of *Scutellaria* and its main active constituents Wogonin, Baicalein and Baicalin. Cancer Treat Rev 35:57–68. doi.org/10.1016/j.ctrv.2008.09.005

Li PW (1996) Clinical oncology for Chinese and western practitioners. Chinese Medicine Publisher, Beijing

Li X, Wang G, Zhao J, Ding H, Cunningham C, Chen F, Flynn DC, Reed E, Li QQ (2005) Antiproliferative effect of beta-elemene in chemoresistant ovarian carcinoma cells is mediated through arrest of the cell cycle at the G2-M phase. Cell Mol Life Sci 62:894–904. doi:10.1007/s00018-005-5027-1

Li L, Gao Y, Zhang L, Zeng J, He D, Sun Y (2008) Silibinin inhibits cell growth and induces apoptosis by caspase activation, down-regulating survivin and blocking EGFR–ERK activation in renal cell carcinoma. Cancer Lett 272:61–69. doi:10.1016/j.canlet.2008.06.033

Li H, Takai N, Yuge A, Furukawa Y, Tsuno A, Tsukamoto Y, Kong S, Moriyama M, Narahara H (2010a) Novel target genes responsive to the anti- growth activity of triptolide in endometrial and ovarian cancer cells. Cancer Lett 297:198–206. doi:10.1016/j.canlet.2010.05.012

Li QQ, Wang GD, Huang FR, Banda M, Reed E (2010b) Antineoplastic effect of beta-elemene on prostate cancer cells and other types of solid tumor cells. J Pharm Pharmacol 62:1018–1027. doi:10.1111/j.2042-7158.2010.01135.x

Li Y, Luan Y, Qi X, Li M, Gong L, Xue X, Wu X, Wu Y, Chen M, Xing G, Yao J, Ren J (2010c) Emodin triggers DNA double-strand breaks by stabilizing topoisomerase II-DNA cleavage complexes and by inhibiting ATP hydrolysis of topoisomerase II. Toxicol Sci 118:435–443. doi:10.1093/toxsci/kfq282

Li L, Zeng J, Gao Y, He D (2010d) Targeting silibinin in the antiproliferative pathway. Expert Opin Investig Drugs 19:243–255. doi:10.1517/13543780903533631

Li X, Hua XD, Zhu Z, Liu B, Lu X (2012) Enhanced *in vitro* and in *vivo* therapeutic efficacy of codrug-loaded nanoparticles against liver cancer. Inter J Nanomed 7:5183. doi:10.2147/IJN.S34886

Li CR, Zhang GW, Niu YB, Pan YL, Zhai YK, Mei QB (2013) Antiosteoporosis effect of Radix Scutellariae extract on density and microstructure of long bones in tail-suspended Sprague-Dawley rats. Evid Based Complement Alter Med 2013:8. doi:10.1155/2013/753703

Li WY, Ng YF, Zhang H, Guo ZD, Guo DJ, Kwan YW, Leung GPH, Lee SMY, Yu PHF, Chan SW (2014a) Emodin elicits cytotoxicity in human lung adenocarcinoma A549 cells through inducing apoptosis. Inflammopharmacology 22:127–134. doi:10.1007/s10787-013-0186-4

Li P, Yang S, Dou M, Chen Y, Zhang J, Zhao X (2014b) Synergic effects of artemisinin and resveratrol in cancer cells. J Cancer Res Clin Oncol 140(12):2065–2075. doi:10.1007/s00432-014-1771-7

Liang M, Fu J (2008) Triptolide inhibits interferon-gamma-induced programmed death-1-ligand 1 surface expression in breast cancer cells. Cancer Lett 2008(270):337–341. doi:10.1016/j.canlet.2008.05.025

Liang HL, Xue CC, Li CG (2004) Regression of squamous cell carcinoma of the lung by Chinese herbal medicine: a case with an 8-year follow-up. Lung Cancer 43:355–360. doi:10.1016/j.lungcan.2003.08.035

Lin HY, Tang HY, Davis FB, Davis PJ (2011) Resveratrol and apoptosis. Ann N Y Acad Sci 1215:79–88. doi:10.1111/j.1749-6632.2010.05846.x

Liu JJ, Wu XY, Peng J, Pan XL, Lu HL (2004) Antiproliferation effects of oridonin on HL-60 cells. Ann Hematol 83:691–695. doi:10.1007/s00277-004-0919-y

Liu TG, Huang Y, Cui DD, Huang XB, Mao SH, Ji LL, Song HB, Yi C (2009) Inhibitory effect of ginsenoside Rg3 combined with gemcitabine on angiogenesis and growth of lung cancer in mice. BMC Cancer 9:250. doi:10.1186/1471-2407-9-250

Liu B, Yuan B, Zhang L, Mu W, Wang C (2015a) ROS/p38/p53/Puma signaling pathway is involved in emodin-induced apoptosis of human colorectal cancer cells. Int J Clin Exp Med 8:15413

Liu Q, Xu X, Zhao M, Wei Z, Li X, Zhang X, Liu Z, Gong Y, Shao C (2015b) Berberine induces senescence of human glioblastoma cells by downregulating the EGFR–MEK–ERK signaling pathway. Mol Cancer Ther 14:355–363. doi:10.1158/1535-7163.MCT-14-0634

Liu YQ, Tian J, Qian K, Zhao XB, Morris-Natschke SL, Yang L, Nan X, Tian X, Lee KH (2015c) Recent progress on C-4-modified podophyllotoxin analogs as potent antitumor agents. Med Res Rev 35:1–62. doi:10.1002/med.21319

Lu LH, Lian YY, He GY, Lin SP, Huan SH, Chen ZZ, Dend HX, Zheng YL (1992) Clinical study of triptolide in treatment of acute leukemia. Clin Exp Investig Hematol 3:1–3

Lu N, Gao Y, Ling Y, Chen Y, Yang Y, Gu HY, Qi Q, Liu W, Wang XT, You QD, Guo QL (2008) Wogonin suppresses tumor growth *in vivo* and VEGF- induced angiogenesis through inhibiting tyrosine phosphorylation of VEGFR2. Life Sci 82:956–963. doi:10.1016/j.lfs.2008.02.013

Lu YY, Chen TS, Qu JL, Pan WL, Sun L, Wei XB (2009) Dihydroartemisinin (DHA) induces caspase-3-dependent apoptosis in human lung adenocarcinoma ASTC-a-1 cells. J Biomed Sci 16:16. doi:10.1186/1423-0127-16-16

Lu JJ, Bao JL, Chen XP, Huang M (2012) Wang YT (2012) Alkaloids isolated from natural herbs as the anticancer agents. Evid Based Complement Alter Med. doi:10.1155/2012/485042

Macek C (1984) East meets west to balance immunologic yin and yang. JAMA 251:433–441

Mao X, Yu CR, Li WH, Li WX (2008) Induction of apoptosis by shikonin through a ROS/JNK-mediated process in Bcr/Abl-positive chronic myelogenous leukemia (CML) cells. Cell Res 18:879–888. doi:10.1038/cr.2008.86

Mateen S, Tyagi A, Agarwal C, Singh RP, Agarwal R (2010) Silibinin inhibits human nonsmall cell lung cancer cell growth through cell cycle arrest by modulating expression and function of key cell-cycle regulators. Mol Carcinog 49:247–258. doi:10.1002/mc.20595

Mei L, Chen Y, Wang Z, Wang J, Wan J, Yu C, Liu X, Li W (2015) Synergistic antitumour effects of tetrandrine and chloroquine combination therapy in human cancer: a potential antagonistic role for p21. Br J Pharmacol 172:2232–2245. doi:10.1111/bph.13045

Miki H, Uehara N, Kimura A, Sasaki T, Yuri T, Yoshizawa K, Tsubura A (2012) Resveratrol induces apoptosis via ROS-triggered autophagy in human colon cancer cells. Int J Oncol 40:1020–1028. doi:10.3892/ijo.2012.1325

Milosevic MF, Fyles AW, Wong R, Pintilie M, Kavanagh MC, Levin W, Manchul LA, Keane TJ, Hill RP (1998) Interstitial fluid pressure in cervical carcinoma. Within tumor heterogeneity, and relation to oxygen tension. Cancer 82:2418–2426

Min R, Tong J, Wenjun Y, Wenhu D, Xiaojian Z, Jiacai H, Jian Z, Wantao C, Chenping Z (2008) Growth inhibition and induction of apoptosis in human oral squamous cell carcinoma Tca-8113 cell lines by Shikonin was partly through the inactivation of NF-kappaB pathway. Phytother Res 22:407–415. doi:10.1002/ptr.2340

Min R, Zun Z, Min Y, Wenhu D, Wenjun Y, Chenping Z (2011) Shikonin inhibits tumor invasion via down-regulation of NF-kappaB-mediated MMP-9 expression in human ACC-M cells. Oral Dis 17:362–369. doi:10.1111/j.1601-0825.2010.01758.x

Mohan S, Dhanalakshmi S, Mallikarjuna GU, Singh RP, Agarwal R (2004) Silibinin modulates UVB-induced apoptosis via mitochondrial proteins, caspases activation, and mitogen-activated protein kinase signaling in human epidermoid carcinoma A431 cells. Biochem Biophys Res Commun 320:183–189. doi:10.1016/j.bbrc.2004.05.153

Moorthi C, Senthil Kumar C, Kathiresan K (2014) Synergistic anti-cancer activity of curcumin and bio-enhancers combination against various cancer cell lines. Int J Pharm Pharm Sci 6:901–903

Murch SJ, Rupasinghe HV, Goodenowe D, Saxena PK (2004) A metabolomic analysis of medicinal diversity in Huang-qin (*Scutellaria baicalensis* Georgi) genotypes: discovery of novel compounds. Plant Cell Rep 23:419–425. doi:10.1007/s00299-004-0862-3

Nan JX, Park EJ, Kim YC, Ko G, Sohn DH (2002) *Scutellaria baicalensis* inhibits liver fibrosis induced by bile duct ligation or carbon tetrachloride in rats. J Pharm Pharmacol 54:555–5633

Nie YK, Zhu LS, Yu HM (2010) Shikonin inhibits the proliferation and induces the apoptosis of human HepG2 cells. Can J Physiol Pharm 88:1138–1146. doi:10.1139/Y10-085

Oberlies NH, Kroll DJ (2004) Camptothecin and taxol: historic achievements in natural products research. J Nat Prod 67:129–135. doi:10.1021/np030498t

Okamura N, Miki H, Ishida S, Ono H, Yano A, Tanaka T, Ono Y, Yagi A (1999) Simultaneous determination of baicalin, wogonoside, baicalein, wogonin, berberine, coptisine, palmatine, jateorrhizine and glycyrrhizin in Kampo medicines by ion-pair high-performance liquid chromatography. Biol Pharm Bull 22:1015–1021

Owona BA, Schluesener HJ (2015) Molecular insight in the multifunctional effects of oridonin. Drugs R&D 15:233–244. doi:10.1007/s40268-015-0102-z

Park HM, Kim SJ, Kim JS, Kang HS (2012) Reactive oxygen species mediated ginsenoside Rg3- and Rh2-induced apoptosis in hepatoma cells through mitochondrial signaling pathways. Food Chem Toxicol 50:2736–2741. doi:10.1016/j.fct.2012.05.027

Park EH, Kim YJ, Yamabe N, Park SH, Kim HK, Jang HJ, Kim JH, Cheon GJ, Ham J, Kang KS (2014) Stereospecific anticancer effects of ginsenoside Rg3 epimers isolated from heat-processed American ginseng on human gastric cancer cell. J Ginseng Res 38:22–27. doi:10.1016/j.jgr.2013.11.007

Patil JB, Kim J, Jayaprakasha GK (2010) Berberine induces apoptosis in breast cancer cells (MCF-7) through mitochondrial-dependent pathway. Eur J Pharmacol 645:70–78. doi:10.1016/j.ejphar.2010.07.037

Peigen K, Yi T, Yaping T (1996) Radix *Salvia miltiorrhizae* treatment results in decreased lipid peroxidation in reperfusion injury. J Tradit Chin Med 16:138–142

Peng PL, Kuo WH, Tseng HC, Chou FP (2008) Synergistic tumor-killing effect of radiation and berberine combined treatment in lung cancer: the contribution of autophagic cell death. Int J Radiat Oncol 70:529–542. doi:10.1016/j.ijrobp.2007.08.034

Pérez-Figarado RA, Baden HP (1976) Dermatology 10:1

Pigneux A, Mahon FX, Uhalde M, Jeanneteau M, Lacombe F, Milpied N, Reiffers J, Belloc F (2008) Triptolide cooperates with chemotherapy to induce apoptosis in acute myeloid leukemia cells. Exp Hematol 36:1648–1659. doi:10.1016/j.exphem.2008.08.002

Pinto-Garcia L, Efferth T, Torres A, Hoheisel JD, Youns M (2010) Berberine inhibits cell growth and mediates caspase-independent cell death in human pancreatic cancer cells. Planta Med 76:1155–1161. doi:10.1055/s-0030-1249931

Polier G, Giaisi M, Köhler R, Müller WW, Lutz C, Buss EC, Krammer PH, Li-Weber M (2015) Targeting CDK9 by wogonin and related natural flavones potentiates the anti-cancer efficacy of the Bcl-2 family inhibitor ABT-263. Int J Cancer 136:688–698. doi:10.1002/ijc.29009

Qi F, Li A, Inagaki Y, Gao J, Li J, Kokudo N, Li XK, Tang W (2010) Chinese herbal medicines as adjuvant treatment during chemo or radiotherapy for cancer. Biosci Trends 4:297–307

Qin R, Shen H, Cao Y, Fang Y, Li H, Chen Q, Xu W (2013) Tetrandrine induces mitochondria-mediated apoptosis in human gastric cancer BGC-823 cells. PLoS One 8(10):e76486. doi:10.1371/journal.pone.0076486

Rajamanickam S, Velmurugan B, Kaur M, Singh RP, Agarwal R (2010) Chemoprevention of intestinal tumorigenesis in APCmin/+ mice by silibinin. Cancer Res 70:2368–2378. doi:10.1158/0008-5472.CAN-09-3249

Sagar SM, Wong RK (2008) Chinese medicine and biomodulation in cancer patients-part one. Curr Oncol 15:42. doi:10.3747/co.v15i1.237

Sagar SM, Klassen GA, Barclay KD, Aldrich JE (1993) Tumour blood flow: measurement and manipulation for therapeutic gain. Cancer Treat Rev 19:299–349

Sagar SM, Singh G, Hodson DI, Whitton AC (1995) Nitric oxide and anticancer therapy. Cancer Treat Rev 21:159–181

Serafim TL, Oliveira PJ, Sardao VA, Perkins E, Parke D, Holy J (2008) Different concentrations of berberine result in distinct cellular localization patterns and cell cycle effects in a melanoma cell line. Cancer Chemoth Pharm 61:1007–1018. doi:10.1007/s00280-007-0558-9

Shan X, Fu Y-S, Aziz F, Wang X-Q, Yan Q, Liu JW (2014) Ginsenoside Rg3 Inhibits Melanoma cell Proliferation through down-Regulation of histone deacetylase 3 (HDAC3) and increase of p53 acetylation. PLoS One 9:e115401. doi:10.1371/journal.pone.0115401

Shang Q, Xu H, Huang L (2012) Tanshinone IIA: a promising natural cardioprotective agent. Evid Based Complement Alter Med. doi:10.1155/2012/716459

Sharma G, Singh RP, Chan DC, Agarwal R (2003) Silibinin induces growth inhibition and apoptotic cell death in human lung carcinoma cells. Anticancer Res 23:2649–2655

Shi Z (2013) Overcoming cisplatin resistance by celastrol in cancer cells. Cancer Res 73:2245. doi:10.1158/1538-7445.am2013-2245

Shi H, Liu L, Liu LM, Geng J, Chen L (2015) Inhibition of tumor growth by β-elemene through downregulation of the expression of uPA, uPAR, MMP-2, and MMP-9 in a murine intraocular melanoma model. Melanoma Res 25:15–21. doi:10.1097/CMR.0000000000000124

Shi M, Ren X, Wang X, Wang H, Liu G, Yuan X, Zheng S, Yu L, Pan S, Song G, Guo Q (2016) A novel combination of oridonin and valproic acid in enhancement of apoptosis induction of HL-60 leukemia cells. Int J Oncol 48:734–746. doi:10.3892/ijo.2015.3294

Shoemaker M, Hamilton B, Dairkee SH, Cohen I, Campbell MJ (2005) *In vitro* anticancer activity of twelve Chinese medicinal herbs. Phytother Res 19:649–651. doi:10.1002/ptr.1702

Shrivastava S, Jeengar MK, Reddy VS, Reddy GB, Naidu VG (2015) Anticancer effect of celastrol on human triple negative breast cancer: possible involvement of oxidative stress, mitochondrial dysfunction, apoptosis and PI3K/Akt pathways. Exp Mol Pathol 98:313–327. doi:10.1016/j.yexmp.2015.03.031

Singh NP, Verma KB (2002) Case report of a laryngeal squamous cell carcinoma treated with artesunate. Arch Oncol 10:279–280. doi:10.2298/aoo0204279s

Singh RP, Raina K, Sharma G, Agarwal R (2008) Silibinin inhibits established prostate tumor growth, progression, invasion, and metastasis and suppresses tumor angiogenesis and epithelial-mesenchymal transition in transgenic adenocarcinoma of the mouse prostate model mice. Clin Cancer 14:7773–7780. doi:10.1158/1078-0432.CCR-08-1309

Singh RP, Raina K, Deep G, Chan D, Agarwal R (2009) Silibinin suppresses growth of human prostate carcinoma PC-3 orthotopic xenograft via activation of extracellular signal-regulated kinase 1/2 and inhibition of signal transducers and activators of transcription signaling. Clin Cancer Res 15:613–621. doi:10.1158/1078-0432.CCR-08-1846

Song J, Shi F, Zhang Z, Zhu F, Xue J, Tan X, Zhang L, Jia X (2011) Formulation and evaluation of celastrol-loaded liposomes. Molecules 16:7880–7892. doi:10.3390/molecules16097880

Suk K, Lee H, Kang SS, Cho GJ, Choi WS (2003) Flavonoid baicalein attenuates activation-induced cell death of brain microglia. J Pharmacol Exp Ther 305:638–645

Subramaniam A, Loo SY, Rajendran P, Manu KA, Perumal E, Li F, Shanmugam MK, Siveen KS, Park JI, Ahn KS, Hui KM (2013a) An anthraquinone derivative, emodin sensitizes hepatocellular carcinoma cells to TRAIL induced apoptosis through the induction of death receptors and downregulation of cell survival proteins. Apoptosis 18:1175–1187. doi:10.1007/s10495-013-0851-5

Subramaniam A, Shanmugam MK, Ong TH, Li F, Perumal E, Chen L, Vali S, Abbasi T, Kapoor S, Ahn KS, Kumar AP (2013b) Emodin inhibits growth and induces apoptosis in an orthotopic hepatocellular carcinoma model by blocking activation of STAT3. Br J Pharmacol 170:807–821. doi:10.1111/bph.12302

Sullivan BJ, Wechsler HJ (1947) The cytological effects of podophyllin. Science 105:433

Sun D, Kumarasamy V (2015) A natural alkaloid, berberine, as a novel transcriptional repressor of RET proto-oncogene in medullary thyroid carcinoma. Cancer Res 75:5546. doi:10.1158/1538-7445.am2015-5546

Sun Y, Xun K, Wang Y, Chen X (2009) A systematic review of the anticancer properties of berberine, a natural product from Chinese herbs. Anticancer Drug 20:757–769. doi:10.1097/CAD.0b013e328330d95b

Syed TA, Cheema KM, Khayyami M, Ahmad SA, Ahmad SH, Ahmad S (1995) Dermatology 191:129

Tan W, Lu J, Huang M, Li Y, Chen M, Wu G, Gong J, Zhong Z, Xu Z, Dang Y, Guo J (2011) Anti-cancer natural products isolated from Chinese medicinal herbs. Chin Med 6:27. doi:10.1186/1749-8546-6-27

Tang XY, Zhu YQ, Tao WH, Wei B, Lin XL (2007) Synergistic effect of triptolide combined with 5-fluorouracil on colon carcinoma. Postgrad Med J 83:338–343. doi:10.1136/pgmj.2006.055426

Tang J, Feng Y, Tsao S, Wang N, Curtain R, Wang Y (2009) Berberine and *Coptidis Rhizoma* as novel antineoplastic agents: a review of traditional use and biomedical investigations. J Ethnopharmacol 126:5–17. doi:10.1016/j.jep.2009.08.009

Tao LJ, Zhou XD, Shen CC, Liang CZ, Liu B, Tao Y, Tao HM (2014) Tetrandrine induces apoptosis and triggers a caspase cascade in U2-OS and MG-63 cells through the intrinsic and extrinsic pathways. Mol Med Rep 9:345–349. doi:10.3892/mmr.2013.1761

Teng ZH, Zhou SY, Ran YH, Liu XY, Yang RT, Yang X, Yuan CJ, Mei QB (2007) Cellular absorption of anthraquinones emodin and chrysophanol in human intestinal Caco-2 cells. Biosci Biotechnol Biochem 71:1636–1643. doi:10.1016/s1875-5364(09)60025-x

Thacker PC, Karunagaran D (2015) Curcumin and emodin down-regulate TGF-β signaling pathway in human cervical cancer cells. PLoS One 10:e0120045. doi:10.1371/journal.pone.0120045

Tilaoui M, Mouse HA, Jaafari A, Zyad A (2014) Differential effect of artemisinin against cancer cell lines. Nat Med Bioprospect 4:189–196. doi:10.1007/s13659-014-0024-4

Tillhon M, Ortiz LM, Lombardi P, Scovassi AI (2012) Berberine: new perspectives for old remedies. Biochem Pharmacol 84:1260–1267. doi:10.1016/j.bcp.2012.07.018

Tsai CF, Yeh WL, Huang SM, Tan TW, Lu DY (2012) Wogonin induces reactive oxygen species production and cell apoptosis in human glioma cancer cells. Int J Mol 8:9877–9892. doi:10.3390/ijms13089877

Tsang CM, Cheung KC, Cheung YC, Man K, Lui VW, Tsao SW, Feng Y (2015) Berberine suppresses Id-1 expression and inhibits the growth and development of lung metastases in hepatocellular carcinoma. Biochim Biophys Acta 1852:541–551. doi:10.1016/j.bbadis.2014

Tseng PY, Lu WC, Hsieh MJ, Chien SY, Chen MK (2014) Tanshinone IIA induces apoptosis in human oral cancer KB cells through a mitochondria-dependent pathway. BioMed Res Inter 13:2014. doi:10.1155/2014/540516

Tyagi A, Singh RP, Agarwal C, Agarwal R (2006) Silibinin activates p53-caspase 2 pathway and causes caspase-mediated cleavage of Cip1/p21 in apoptosis induction in bladder transitional cell papilloma RT4 cells: evidence for a regulatory loop between p53 and caspase 2. Carcinogenesis 27:2269–2280. doi:10.1093/carcin/bgl098

Ueba N, Otake T, Yamazaki K, Mori H, Morimoto M (1993) Anti-HIV-1 activities of crude drug, shikon (Lithospermi radix), *in vitro*. Nippon Rinsho 51:207–212

Wall ME (1998) Camptothecin and taxol: discovery to clinic. Med Res Rev 18:299–314

Wan CK, Wang C, Cheung HY, Yang M, Fong WF (2006) Triptolide induces Bcl-2 cleavage and mitochondria dependent apoptosis in p53-deficient HL-60 cells. Cancer Lett 241:31–41. doi:10.1016/j.canlet.2005.10.001

Wan J, Liu T, Mei L, Li J, Gong K, Yu C, Li W (2013) Synergistic antitumor activity of sorafenib in combination with tetrandrine is mediated by reactive oxygen species (ROS)/Akt signaling. Br J Cancer 109:342–350. doi:10.1038/bjc.2013.334

Wang GT (1990) Treatment of operated late gastric carcinoma with prescription of strengthening the patient's resistance and dispelling the invading evil in combination with chemotherapy: follow-up study of 158 patients and experimental study in animals. Chung Hsi I Chieh Ho Tsa Chih 10:712–716

Wang MY, Lin C, Zhang TM (1987) Effects of oridonin on DNA, RNA and protein syntheses of leukemia L 1210 cells studied by autoradiography. Zhongguo Yao Li Xue Bao 8:164–165

Wang G, Li X, Huang F, Zhao J, Ding H, Cunningham C, Coad JE, Flynn DC, Reed E, Li QQ (2005a) Antitumor effect of beta-elemene in non-small-cell lung cancer cells is mediated via induction of cell cycle arrest and apoptotic cell death. Cell Mol Life Sci 62:881–893. doi:10.1007/s00018-005-5017-3

Wang X, Wei Y, Yuan S, Liu G, Lu Y, Zhang J, Wang W (2005b) Potential anticancer activity of tanshinone IIA against human breast cancer. Int J Cancer 116:799–807. doi:10.1002/ijc.20880

Wang W, Yang S, Su Y, Xiao Z, Wang C, Li X, Lin L, Fenton BM, Paoni SF, Ding I, Keng P, Okunieff P, Zhang L (2007) Enhanced antitumor effect of combined triptolide and ionizing radiation. Clin Cancer Res 13:4891–4899. doi:10.1158/1078-0432.CCR-07-0416

Wang XN, Han X, Xu LN, Yin LH, Xu YW, Qi Y, Peng JY (2008) Enhancement of apoptosis of human hepatocellular carcinoma SMMC-7721 cells through synergy of berberine and evodiamine. Phytomedicine 15:1062–1068. doi:10.1016/j.phymed.2008.05.002

Wang N, Feng Y, Zhu M, Tsang CM, Man K, Tong Y, Tsao SW (2010a) Berberine induces autophagic cell death and mitochondrial apoptosis in liver cancer cells: the cellular mechanism. J Cell Biochem 111:1426–1436. doi:10.1002/jcb.22869

Wang X, Gocek E, Novik V, Harrison JS, Danilenko M, Studzinski GP (2010b) Inhibition of Cot1/Tlp2 oncogene in AML cells reduces ERK5 activation and up-regulates p27Kip1 concomitant with enhancement of differentiation and cell cycle arrest induced by silibinin and 1,25- dihydroxyvitamin D(3). Cell Cycle 9:4542–4551. doi:10.4161/cc.9.22.13790

Wang S, Wu X, Tan M, Gong J, Tan W, Bian B, Chen M, Wang Y (2012) Fighting fire with fire: poisonous Chinese herbal medicine for cancer therapy. J Ethnopharmacol 140:33–45. doi:10.1016/j.jep.2011.12.041

Wang H, Zhang H, Tang L, Chen H, Wu C, Zhao M, Yang Y, Chen X, Liu G (2013a) Resveratrol inhibits TGF-β1-induced epithelial-to-mesenchymal transition and suppresses lung cancer invasion and metastasis. Toxicology 303:139–146. doi:10.1016/j.tox.2012.09.017

Wang S, Zhong Z, Wan J, Tan W, Wu G, Chen M, Wang Y (2013b) Oridonin induces apoptosis, inhibits migration and invasion on highly metastatic human breast cancer cells. Am J Chin Med 41:177–196. doi:10.1142/S0192415X13500134

Wang L, Waltenberger B, Pferschy-Wenzig EM, Blunder M, Liu X, Malainer C, Blazevic T, Schwaiger S, Rollinger JM, Heiss EH, Schuster D (2014a) Natural product agonists of peroxisome proliferator-activated receptor gamma (PPARγ): a review. Biochem Pharmacol 92:73–89. doi:10.1016/j.bcp.2014.07.018

Wang YY, Lv YF, Lu L, Cai L (2014b) Oridonin inhibits mTOR signaling and the growth of lung cancer tumors. Anti-Cancer Drug 25:1192–1200. doi:10.1097/CAD.0000000000000154

Wei W, Lin S, Liu D, Wang Z (2013) The distinct mechanisms of the antitumor activity of emodin in different types of cancer (Review). Oncol Rep 30:2555–2562. doi:10.3892/or.2013.2741

Weng H, Huang H, Dong B, Zhao P, Zhou H, Qu L (2014) Inhibition of miR-17 and miR-20a by oridonin triggers apoptosis and reverses chemoresistance by derepressing BIM-S. Cancer Res 74:4409–4419. doi:10.1158/0008-5472.CAN-13-1748

Wolfram J, Suri K, Huang Y, Molinaro R, Borsoi C, Scott B, Boom K, Paolino D, Fresta M, Wang J, Ferrari M (2014) Evaluation of anticancer activity of celastrol liposomes in prostate cancer cells. J Microencapsul 31:501–517. doi:10.3109/02652048.2013.879932

Wu LN, Wu QA (1997) Yellow Emperor's canon of internal medicine. Zhongguo ke xue ji shu chu ban she, Beijing

Wu WL, Chang WL, Chen CF (1991) Cytotoxic activities of tanshinones against human carcinoma cell lines. Am J Chin Med 19:207–216

Wu Z, Wu L, Li L, Tashiro S, Onodera S, Ikejima T (2004a) p53-mediated cell cycle arrest and apoptosis induced by shikonin via a caspase-9-dependent mechanism in human malignant melanoma A375-S2 cells. J Pharmacol Sci 94:166–176. doi:10.1254/jphs.94.166

Wu Z, Wu LJ, Li LH, Tashiro S, Onodera S, Ikejima T (2004b) Shikonin regulates HeLa cell death via caspase-3 activation and blockage of DNA synthesis. J Asian Nat Prod Res 6:155–166. doi:10.1080/1028602032000169622

Wu XS, Xie T, Lin J, Fan HZ, Huang-Fu HJ, Ni LF, Yan HF (2009) An investigation of the ability of elemene to pass through the blood–brain barrier and its effect on brain carcinomas. J Pharm Pharmacol 61:1653–1656. doi:10.1211/jpp.61.12.0010

Xiao W, Jiang Y, Men Q, Yuan L, Huang Z, Liu T, Li W, Liu X (2015) Tetrandrine induces G1/S cell cycle arrest through the ROS/Akt pathway in EOMA cells and inhibits angiogenesis *in vivo*. Int J Oncol 46:360–368. doi:10.3892/ijo.2014.2735

Yamasaki K, Otake T, Mori H, Morimoto M, Ueba N, Kurokawa N, Shiota K, Yuge T (1993) Screening test of crude drug extract on anti-HIV activity. Yakugaku Zasshi 113:818–824

Yance DR, Sagar SM (2006) Targeting angiogenesis with integrative cancer therapies. Integrat Cancer Ther 5:9–29. doi:10.1177/1534735405285562

Yang G, Liu J, Xie J, Jiang S, Song M, Li H, Qi F, Liu Y, Huang T, He X (1995) Controlling cancerous pain with analgesic powders for cancers. J Tradit Chin Med 15:174–177

Yang S, Chen J, Guo Z, Xu XM, Wang L, Pei XF, Yang J, Underhill CB, Zhang L (2003) Triptolide inhibits the growth and metastasis of solid tumors. Mol Cancer Ther 2:65–72

Yang LJ, Jeng CJ, Kung HN, Chang CC, Wang AG, Chau GY, Don MJ, Chau YP (2005) Tanshinone IIA isolated from Salvia miltiorrhiza elicits the cell death of human endothelial cells. J Biomed Sci 12:347–361. doi:10.1007/s11373-005-0973-z

Yang F, Chen Y, Duan W, Zhang C, Zhu H, Ding J (2006) SH-7, a newly synthesized shikonin derivative, exerting its potent antitumor activities as a topoisomerase inhibitor. Int J Cancer 119:1184–1193. doi:10.1002/ijc.21943

Yang H, Zhou P, Huang H, Chen D, Ma N, Cui QC, Shen S, Dong W, Zhang X, Lian W, Wang X, Dou QP, Liu J (2009a) Shikonin exerts antitumor activity via proteasome inhibition and cell death induction *in vitro* and *in vivo*. Int J Cancer 124:2450–2459. doi:10.1002/ijc.24195

Yang L, Zhang HW, Hu R, Yang Y, Qi Q, Lu N, Liu W, Chu YY, You QD, Guo QL (2009b) Wogonin induces G1 phase arrest through inhibiting Cdk4 and cyclin D1 concomitant with an elevation in p21Cip1 in human cervical carcinoma HeLa cells. Biochem Cell Biol 87:933–942. doi:10.1139/O09-060

Yang SW, Wang W, Xie XY, Zhu WP, Li FQ (2011) *In vitro* synergistic cytotoxic effect of triptolide combined with hydroxycamptothecin on pancreatic cancer cells. Am J Chin Med 39:121–134. doi:10.1142/S0192415X11008695

Yeh CC, Kuo HM, Li TM, Lin JP, Yu FS, Lu HF, Chung JG, Yang JS (2007) Shikonin- induced apoptosis involves caspase-3 activity in a human bladder cancer cell line (T24). *In Vivo* 21:1011–1019

Yeh CB, Hsieh MJ, Lin CW, Chiou HL, Lin PY, Chen TY, Yang SF (2013) The antimetastatic effects of resveratrol on hepatocellular carcinoma through the downregulation of a metastasis-associated protease by SP-1 modulation. PLoS One 8:e56661. doi:10.1371/journal.pone.0056661

Yi J, Ye X, Wang D, He K, Yang Y, Liu X, Li X (2013) Safety evaluation of main alkaloids from Rhizoma Coptidis. J Ethnopharmacol 145:303–310. doi:10.1016/j.jep.2012.10.062

Yin HT, Tian QZ, Guan L, Zhou Y, Huang XE, Zhang H (2013) *In vitro* and *in vivo* evaluation of the antitumor efficiency of resveratrol against lung cancer. Asia Pacific J Cancer Prev 14:1703–1706. doi:10.7314/APJCP.2013.14.3.1703

Yip NK, Ho WS (2013) Berberine induces apoptosis via the mitochondrial pathway in liver cancer cells. Oncol Rep 30:1107–1112. doi:10.3892/or.2013.2543

Yoon Y, Kim YO, Lim NY, Jeon WK, Sung HJ (1999) Shikonin, an ingredient of *Lithospermum erythrorhizon* induced apoptosis in HL60 human promyelocytic leukemia cell line. Planta Med 65:532–535. doi:10.1055/s-1999-14010

Youn MJ, So HS, Cho HJ, Kim HJ, Kim Y, Lee JH, Sohn JS, Kim YK, Chung SY, Park R (2008) Berberine, a natural product, combined with cisplatin enhanced apoptosis through a mitochondria/caspase-mediated pathway in HeLa cells. Biol Pharmaceut Bullet 31:789–795. doi:10.1248/bpb.31.789

Yousra M, El Galil KH, Gabr MM, Eissa LA, El-Shishtawy MM (2013) Antitumor effect of Chitosan and Silibinin and their combination in mice bearing Ehrlich ascites tumors: impact of p53 and p21. Curr Topic Biochem Res 15:29–49

Yu JQ, Bao W, Lei JC (2013) Emodin regulates apoptotic pathway in human liver cancer cells. Phytother Res 27:251–257. doi:10.1002/ptr.4703

Yu FS, Yu CS, Chen JC, Yang JL, Lu HF, Chang SJ, Lin MW, Chung JG (2014) Tetrandrine induces apoptosis via caspase 8, 9 and 3 and poly (ADP ribose) polymerase dependent pathways and autophagy through beclin1/LC3I, II signaling pathways in human oral cancer HSC 3 cells. Environ Toxicol 31:395–406. doi:10.1002/tox.22053

Yuan SL, Wei YQ, Wang XJ, Xiao F, Li SF, Zhang J (2004) Growth inhibition and apoptosis induction of tanshinone II-A on human hepatocellular carcinoma cells. World J Gastroenterol 10:2024–2028. doi:10.3748/wjg.v10.i14.2024

Yue PY, Wong DY, Wu PK, Leung PY, Mak NK, Yeung HW, Liu L, Cai Z, Jiang ZH, Fan TP, Wong RN (2006) The angiosuppressive effects of 20 (R)-ginsenoside Rg 3. Biochem Pharmacol 72:437–445. doi:10.1016/j.bcp.2006.04.034

Yuxian X, Feng T, Ren L, Zhengcai L (2008) Tanshinone II-A inhibits invasion and metastasis of human hepatocellular carcinoma cells *in vitro* and *in vivo*. Tumori 95:789–795

Zhang XW, Qing C, Xu B (1999) Apoptosis induction and cell cycle perturbation in human hepatoma hep G2 cells by 10-hydroxycamptothecin. Anticancer Drug 10:569–576

Zhang C, Cui GH, Liu F, Wu QL, Chen Y (2006) Inhibitory effect of triptolide on lymph node metastasis in patients with non-Hodgkin lymphoma by regulating SDF-1/CXCR4 axis *in vitro*. Acta Pharmacol Sin 27:1438–1446. doi:10.1111/j.1745-7254.2006.00415.x

Zhang ZY, Yu SQ, Miao LY, Huang XY, Zhang XP, Zhu YP, Xia XH, Li DQ (2008) Artesunate combined with vinorelbine plus cisplatin in treatment of advanced non-small cell lung cancer: a randomized controlled trial. J Chinese Integr Med 6:134–138. doi:10.3736/jcim20080206

Zhang F, Xu L, Qu X, Zhao M, Jin B, Kang J, Liu Y, Hu X (2011) Synergistic antitumor effect of β-elemene and etoposide is mediated via induction of cell apoptosis and cell cycle arrest in non-small cell lung carcinoma cells. Mol Med Rep 4:1189–1193. doi:10.3892/mmr.2011.537

Zhang C, Liu L, Yu Y, Chen B, Tang C, Li X (2012) Antitumor effects of ginsenoside Rg3 on human hepatocellular carcinoma cells. Mol Med Rep 5:1295–1298. doi:10.3892/mmr.2012.808

Zhang M, Liu LP, Chen Y, Tian XY, Qin J, Wang D, Li Z, Mo SL (2013) Wogonin induces apoptosis in RPMI 8226, a human myeloma cell line, by downregulating phospho-Akt and overexpressing Bax. Life Sci 92:55–62. doi:10.1016/j.lfs.2012.10.023

Zhang GN, Ashby CR, Zhang YK, Chen ZS, Guo H (2015) The reversal of antineoplastic drug resistance in cancer cells by β-elemene. Chin J Cancer 34:45. doi:10.1186/s40880-015-0048-0

Zhao BL, Jiang W, Zhao Y, Hou JW, Xin WJ (1996) Scavenging effects of salvia miltiorrhiza on free radicals and its protection for myocardial mitochondrial membranes from ischemia-reperfusion injury. Biochem Mol Biol Inter 38:1171–1182

Zhao J, Li QQ, Zou B, Wang G, Li X, Kim JE, Cuff CF, Huang L, Reed E, Gardner K (2007) *In vitro* combination characterization of the new anticancer plant drug β-elemene with taxanes against human lung carcinoma. Inter J Oncol 31:241–252. doi:10.3892/ijo.31.2.241

Zhao F, Chen Y, Zeng L, Li R, Zeng R, Wen L, Liu Y, Zhang C (2010a) Role of triptolide in cell proliferation, cell cycle arrest, apoptosis and histone methylation in multiple myeloma U266 cells. Eur J Pharmacol 646:1–11. doi:10.1016/j.ejphar.2010.05.034

Zhao Q, Wang J, Zou MJ, Hu R, Zhao L, Qiang L, Rong JJ, You QD, Guo QL (2010b) Wogonin potentiates the antitumor effects of low dose 5-fluorouracil against gastric cancer through induction of apoptosis by down- regulation of NF-kappaB and regulation of its metabolism. Toxicol Lett 197:201–210. doi:10.1016/j.toxlet.2010.05.019

Zhao S, Wu J, Zheng F, Tang Q, Yang L, Li L, Wu W, Hann SS (2015) β-elemene inhibited expression of DNA methyltransferase 1 through activation of ERK1/2 and AMPKα signaling pathways in human lung cancer cells: the role of Sp1. J Cell Mol Med 19:630–641. doi:10.1111/jcmm.12476

Zheng L, Fu Y, Zhuang L, Gai R, Ma J, Lou J, Zhu H, He Q, Yang B (2014) Simultaneous NF-kB inhibition and E-cadherin upregulation mediate mutually synergistic anticancer activity of celastrol and SAHA *in vitro* and *in vivo*. Int J Cancer 135:1721–1732. doi:10.1002/ijc.28810

Zhou DH (2007) Recent status and outlook of traditional Chinese medicine in cancer treatment. In: Leung PC, Fong H (eds) Alternative treatment for cancer medicine. World Scientific Publishing Co. Pte. Ltd, Singapore, pp 55–64

Zhou GX, Ding XL, Huang JF, Zhang H, Wu SB (2007) Suppression of 5- lipoxygenase gene is involved in triptolide-induced apoptosis in pancreatic tumor cell lines. Biochim Biophys Acta 1770:1021–1027. doi:10.1016/j.bbagen.2007.03.002

Zhou L, Liu P, Chen B, Wang Y, Wang X, Internati MC, Wachtel MS, Frezza EE (2008) Silibinin restores paclitaxel sensitivity to paclitaxel-resistant human ovarian carcinoma cells. Anticancer Res 28:1119–1127

Zhou GS, Hu Z, Fang HT, Zhang FX, Pan XF, Chen XQ, Hu AM, Xu L, Zhou GB (2011) Biologic activity of triptolide in t(8;21) acute myeloid leukemia cells. Leuk Res 35:214–218. doi:10.1016/j.leukres.2010.07.013

Zhu W, Hu H, Qiu P, Yan G (2009) Triptolide induces apoptosis in human anaplastic thyroid carcinoma cells by a p53-independent but NF- kappaB- related mechanism. Oncol Rep 22:1397–1401. doi:10.3892/or_00000580

Zhu H, Yang W, He LJ, Ding WJ, Zheng L, Huang P, Lu W, He QJ, Yang B (2012) Upregulating Noxa by ER stress, celastrol exerts synergistic anti-cancer activity in combination with ABT-737 in human hepatocellular carcinoma cells. PLoS One 7:1–12. doi:10.1371/journal.pone.0052333

Zhu T, Li X, Luo L, Wang X, Li Z, Xie P, Gao X, Song Z, Su J, Liang G (2015) Reversion of malignant phenotypes of human glioblastoma cells by β-elemene through β-catenin-mediated regulation of stemness-, differentiation-and epithelial-to-mesenchymal transition-related molecules. J Tranlation Med 13:356. doi:10.1186/s12967-015-0727-2

Zou S, Wang C, Cui Z, Guo P, Meng Q, Shi X, Gao Y, Yang G, Han Z (2016) β-Elemene induces apoptosis of human rheumatoid arthritis fibroblast-like synoviocytes via reactive oxygen species-dependent activation of p38 mitogen-activated protein kinase. Pharmacol Rep 68:7–11. doi:10.1016/j.pharep.2015.06.004

Chapter 15
An Update on Antitumor Activity of *Angelica* Species

Amira Mira and Kuniyoshi Shimizu

Abstract Genus *Angelica* (Family: Apiaceae) comprises more than 60 species of medicinal plants. Many of these species have long been used in ancient traditional medicine systems, especially in the Eastern countries. For many years, several species of *Angelica*, e.g., *A. archangelica*, *A. dahurica*, *A. keiskei, A. gigas*, *A. pubescens, A. sinensis, A. shikokiana* have been used traditionally as anti-inflammatory, diuretic, expectorant, and diaphoretic. Also, these have been used as remedies for hepatitis, arthritis, indigestion, coughs, chronic bronchitis, pleurisy, fever, rheumatism, bacterial and fungal infections, and diseases of the urinary organs. Several classes of compounds such as coumarins, acetylenic compounds, chalcones, sesquiterpenes, and polysaccharides have been reported from these plants. This chapter outlines the results of various scientific studies on *Angelica* species that were reported to have anticancer and antitumor activities.

Keywords *Angelica* • Antitumor • Chalcones • Coumarins

Abbreviations

ALT	Alanine transaminase
AME	*A. shikokiana* methanol extract
Chk2	Checkpoint kinase 2
DDY	Deutschland, Denken, and Yoken mice
DMBA	7,12-Dimethylbenz[a]anthracene
GBM	Glioblastoma multiforme
GGT	Gamma-glutamyl transpeptidase

A. Mira
Department of Pharmacognosy, Faculty of Pharmacy, Mansoura University, Mansoura 35516, Egypt

K. Shimizu (✉)
Division of Systematic Forest and Forest Products Sciences, Department of Agro-Environmental Sciences, Faculty of Agriculture, Graduate School of Kyushu University, Fukuoka 812-8581, Japan
e-mail: shimizu@agr.kyushu-u.ac.jp

H.-S. Tsay et al. (eds.), *Medicinal Plants - Recent Advances in Research and Development*, DOI 10.1007/978-981-10-1085-9_15

HCC	Hepatocellular carcinoma
HDAC8	Histone deacetylase 8
iNOS	Inducible nitric oxide synthase
MMP-9	Matrix metalloproteinase 9
NO	Nitric oxide
PANC-1	Human pancreas cancer cell line
PKC	Protein kinase C
RES	Reticuloendothelial system
TAA	Thioacetamide
TNF	Tumor necrosis factor
TPA	12-O-Tetradecanoylphorbol-13-acetate
VEGF-C	Vascular endothelial growth factor C

15.1 Introduction

The genus *Angelica* L. belongs to the family Apiaceae (alt. Umbelliferae). It comprises more than 60 species of medicinally important biennial or perennial herbs (Martin and Hutchins 1980). Plants of this genus are widely distributed in Asia, Europe, and North America. *Angelica* is one of the most important genera of medicinal plants that are still in use in both traditional medicine systems of the Far East and the Western countries (Sarker and Nahar 2004). To date, several phytochemical and pharmacological studies resulted in the discovery of a number of active compounds from certain species of *Angelica*. The scientific studies reported the antitumor effectiveness of some of the isolated compounds either singled or in combination with known antitumor compounds (Jung et al. 2014). A previous review article summarized the use of different species of *Angelica* in herbal medicine (Sarker and Nahar 2004), with a little mention of the antitumor activities. Therefore, this chapter shed the light on the reported studies on certain *Angelica* species that were investigated to have antitumor activities.

15.2 *Angelica radix*

Angelica radix as a crude Chinese herbal drug was investigated in vivo for its antitumor and tumor necrosis factor (TNF) production in DDY mice. It was observed that treatments with *A. radix* showed a good survival rate and sometimes a complete cure. The antitumor mechanism was proved to be related to the stimulation of reticuloendothelial system (RES) and the high-level production of TNF (Haranaka et al. 1985).

15.3 *Angelica archangelica*

Angelica archangelica Linn. is known in European countries by the name of garden *angelica* or *Angelica officinalis*. It is occasionally found native in cold and moist places in Scotland but is more abundant in countries further north, as in Lapland and Iceland (Bhat et al. 2011). The tincture of fruits of *A. archangelica* showed an antiproliferative effect against the human pancreas cancer cell line, PANC-1, with an EC_{50} value of 28.6 μg/mL (Sigurdsson et al. 2004). Imperatorin and xanthotoxin (Fig. 15.1a, b), two abundant furanocoumarins, were proved to be responsible for the cytotoxic activity with EC_{50} values equivalent to 10 and 17 μM, respectively (Sigurdsson et al. 2004). On the other side, terpenoid compounds showed no cytotoxic activities (Sigurdsson et al. 2004). In addition to the furanocoumarins, essential oils of the fruits *A. archangelica* growing in Iceland showed significant

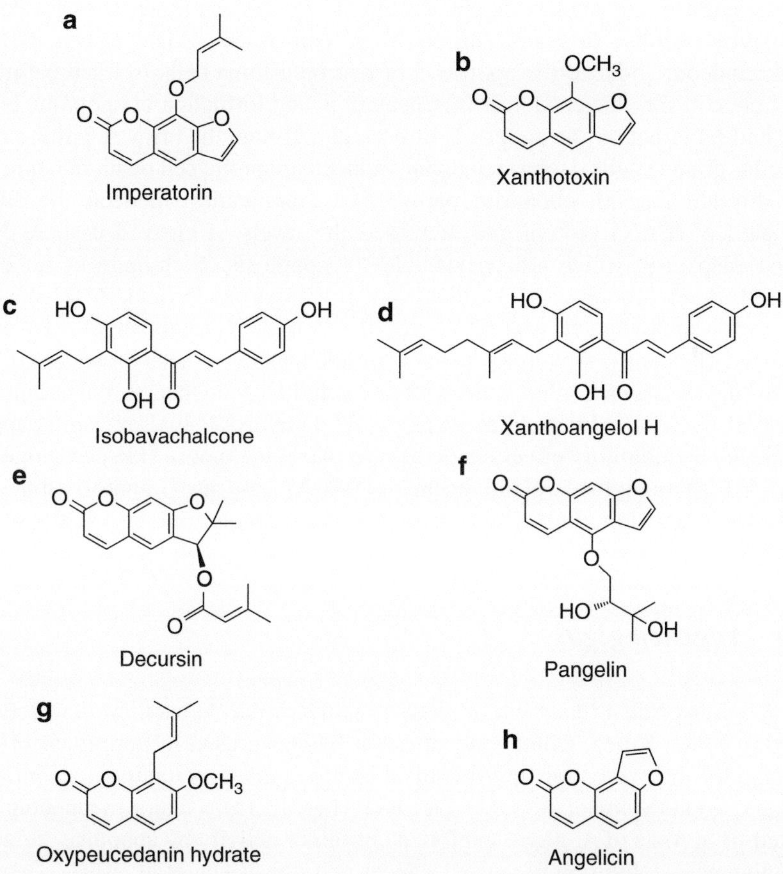

Fig. 15.1 (**a–h**) Chemical structures of compounds of *Angelica* species having antitumor activities

antiproliferative activities against PANC-1 and Crl mouse breast cancer cells. β-Phellandrene was the most active component of the essential oil (Sigurdsson et al. 2005a). The leaf extract of *A. archangelica* showed moderate in vitro antiproliferative activity on the Crl cells with an EC_{50} of 87.6 μg/mL and significant reduction of tumor size in mice (Sigurdsson et al. 2005b).

15.4 *Angelica keiskei*

Angelica keiskei Koidzumi (Japanese common name "Ashitaba") is a perennial herb growing along the Pacific coast of Japan and has been used traditionally in the treatment of several ailments such as constipation, hypertension, and neuro-diseases. Chalcones isolated from the stem extracts reportedly showed inhibitory effects against tumor promoter activity (Ogawa et al. 2003). The most cytotoxic chalcones, isobavachalcone and xanthoangelol H (Fig. 15.1c, d), exhibited strong selective cytotoxicity against human neuroblastoma cell lines (IMR-32 and NB-39). Isobavachalcone induced the apoptosis of neuroblastoma cells by chromatin condensation, cell shrinkage, nuclear fragmentation, and formation of apoptotic bodies while had no effect on the apoptosis of normal cell line, the primary culture of rat cerebellar granule cells. Isobavachalcone induced apoptotic cell death in neuroblastoma through the mitochondrial pathway. It significantly induced the BCL2-associated X (BAX) protein and increased the levels of cleaved caspase-3 and cleaved caspase-9, which are responsible for apoptosis (Nishimura et al. 2007). Biotransformed products (2″,3″-dihydro-4″,3″-dihydroxyderricin, 6″,7″-dihydro-7″-hydroxyxanthoangelol, and 6″,7″-dihydro-7″-hydroxyxanthoangelol F) of the prenylated chalcones, 4-hydroxyderricin, xanthoangelol, and xanthoangelol F, isolated from *A. keiskei* recorded more cytotoxic activities than their parent compounds against HL60 cells (Akihisa et al. 2012). 6″,7″-Dihydro-7″-hydroxyxanthoangelol F displayed an inhibitory effect on the in vivo two-stage mouse skin carcinogenesis using 7,12-dimethylbenz[a]anthracene (DMBA) as an initiator and with 12-O-tetradecanoylphorbol-13-acetate (TPA) as a promoter (Akihisa et al. 2012).

15.5 *Angelica gigas*

Angelica gigas Nakai (dang gui or giant *Angelica* or purple parsnip) is widely dispersed in South Korea, China, and Japan. It has been used in traditional oriental medicine for the prevention and treatment of blood diseases including anemia and as a tonic agent (Moon et al. 2011). Decursin (Fig. 15.1e), a major pyranocoumarin isolated from roots of *A. gigas*, exhibited antitumor activity by apoptosis induction or angiogenesis inhibition in various cancer cell lines (Ahn et al. 1997). Decursin revealed a potential value in restricting breast cancer metastasis. It showed an inhibitory activity on TPA-induced matrix metalloproteinase 9 (MMP-9) expression and MCF-7 cell invasion through the suppression of NF-kB. Also, decursin suppressed the TPA-induced phosphorylation of p38 MAPK and translocation of PKCα from the cytosol to the cell membrane (Kim et al. 2014). Decursin and its derivatives, prepared by chemical synthesis, were investigated for their antileukemic activities in human K562 erythroleukemia and U937 myeloleukemia cells. Decursin, decursinol angelate, decursinol tiglate, decursinol butyrate, and decursinol hexanoate showed potent cytotoxicity in both the cell lines. Decursin, decursinol angelate, and decursinol tiglate were the only compounds which showed cytotoxicity on TUR cells, a protein kinase C (PKC)βII-deficient variant of U937 cells, indicating that their cytotoxicity was PKCβII-independent, while the other derivatives were PKCβII dependent. Furthermore, the cytotoxic mechanism of decursin and decursinol angelate was related to their abilities to induce the downregulation of PKCα and PKCβII in K562 cells and the production of ROS in U937 cells. Coumarin nucleus and the side chain were an important structure determinant of the antileukemic activity, the PKC activation, and the cytotoxic mechanism in leukemia cells (Kim et al. 2005). Jung et al. (2014) confirmed that decursin could potentiate the cytotoxic activity of doxorubicin on three multiple myeloma cell lines, viz., U266, RPMI8226, and MM.1S. The combined therapy enhanced the activation of caspase-9 and caspase-3, the cleavage of PARP, and the sub-G1 population, downregulated the phosphorylation of mTOR and its downstream S6K1, and activated the phosphorylation of ERK. Moreover, the combination reduced mitochondrial membrane potential; suppressed the phosphorylation of JAK2, STAT3, and Src; activated SHP-2; and attenuated the expression of cyclin-D1 and survivin in U266 cells.

15.6 *Angelica dahurica*

Angelica dahurica (holy ghost and wild *Angelica* or *bai zhi* in Chinese) is a wildly grown species in Siberia, Russia Far East, Mongolia, Northeastern China, Japan, Korea, and Taiwan. It is formulated with a plant called *Kanion corydalis* in a commercial product, "Yuan Hu," that is used for the treatment of blood diseases (Thanh et al. 2004). The roots of *A. dahurica* are widely used for its medicinal properties (Fujiwara et al. 1980). Methylene chloride fraction of the methanol extract of *A. dahurica* was proved to have a potent cytotoxic activity in L1210 leukemia cells. Coumarin compounds, pangelin and oxypeucedanin hydrate (Fig. 15.1f, g) acetonide, showed a potent cytotoxic activity with the IC_{50} values of 8.6 and 14.6 pg/mL, respectively (Thanh et al. 2004).

15.7 *Angelica sinensis*

Angelica sinensis commonly known as "*dong quai*" or "female ginseng" is indigenous to China and grows in cool high-altitude mountains in China, Japan, and Korea (Ye et al. 2001). It is commercially incorporated with other herbs in a formulation known in Chinese medicine as "dang gui" for the treatment of gynecological imbalances (Lin et al. 2012). The aqueous extract of *A. sinensis* roots exhibited cytotoxic activities on mouse melanocyte (Raman et al. 1996). Another study examined the antitumor effects of the chloroform extract of the roots on human glioblastoma multiforme (GBM) DBTRG-05MG and rat RG2 GBM brain tumors cells (Tsai et al. 2005). The chloroform extract could potently suppress the growth of malignant brain tumor cells due to the induction of cell cycle arrest and apoptosis. It could upregulate the expression of cyclin-dependent kinase (cdk) inhibitors, including p21, and decrease the phosphorylation of Rb proteins, which resulted in the cell arrest at the G0-G1 phase in DBTRG-05MG and RG2 cells. Moreover, the apoptosis-associated proteins significantly increased and activated in DBTRG-05MG cells and RG2 cells as a result of the treatment with *A. sinensis* extract. The extract triggered both p53-dependent and p53-independent pathways for apoptosis. In in vivo studies, the extract could suppress the growth of malignant brain tumors of rat and human origin and shrink the volumes of in situ GBM and significantly prolong the survival.

15.8 *Angelica pubescens*

Angelica pubescens is a herbal medicine, native to Japan and East Asia. It is widely known in Japanese as "shishiudo" and in Chinese as "usiang tu-huo" (fragrant *Angelica*) (Chen et al. 2007). Osthole (Fig. 15.1g) is an O-methylated pyranocoumarin isolated from the roots of *A. pubescens* (Ko et al. 1992; Hung et al. 2011; Zhang et al. 2012). Several studies have reported the in vitro and in vivo anticancer activities of osthole and its N-hydroxycinnamide derivatives (Lin et al. 2010). Osthole derivatives could induce multinucleation and polyploidy in human colon adenocarcinoma cells. Furthermore, they significantly activated ataxia telangiectasia and rad3-related (ATR) kinase, which in turn triggered activation of the checkpoint kinase 2 (Chk2) signaling pathway and then downregulated Cdc25 phosphatase and Cdc2/cyclin B kinase activities. The other antitumor mechanism included the inhibition of phosphorylation of aurora A kinase, which is important during mitosis, DNA damage, and induction of apoptosis (Liu et al. 2013).

15.9 *Angelica shikokiana*

Angelica shikokiana (Japanese name "inutouki") is an endemic species and perennial herb cultivated in Japan and is used as a substitute drug for ginseng roots (Kimura and Okuda 1989). It is reported to be included among food and drug preparations for protection against cancer (Okuda 1985). The cytotoxic activities of *A. shikokiana* methanol extract (AME) and its isolated compounds were investigated in several types of cancer and normal cell lines. Methanol extract and some of its isolated coumarins and flavonoids showed potent, selective cytotoxicity against cancer cell lines. The cytotoxic mechanism was related to the inhibition of tubulin polymerization, especially by angelicin (Fig. 15.1h) and kaempferol-3-O-rutinoside (Mira and Shimizu 2015a). Docking studies confirmed that the isolated phenolic compounds and furanocoumarins had a strong binding affinity to colchicine-binding site rather than the vinblastine-binding site of tubulin microtubules. Isolated compounds, quercetin, kaempferol, luteolin, chlorogenic acid, and methyl chlorogenate, exhibited a strong activity against histone deacetylase 8 (HDAC8) and a high affinity to trichostatin A-binding site (Mira and Shimizu 2015a). In vivo hepatoprotective activities of the methanol extract of the aerial parts of *A. shikokiana* (AME) and its major coumarin, isoepoxypteryxin were investigated on thioacetamide (TAA)-induced hepatocellular carcinoma (HCC) in male Sprague Dawley rats (Mira and Shimizu 2015b). AME and isoepoxypteryxin significantly reduced the levels of alanine transaminase (ALT), aspartate transaminase (AST), gamma-glutamyl transpeptidase (GGT), and total bilirubin and nitric oxide production when compared with hepatocellular carcinoma group. Besides, the hepatoprotective mechanism was related to the ability of AME and isoepoxypteryxin to induce a significant increase in the level of caspase-3 protein, and a significant decrease in the inducible nitric oxide synthase (iNOS), and vascular endothelial growth factor C (VEGF-C) levels.

15.10 Conclusions

The updated research has revealed the efficiency of *Angelica* species, or their active agents, either alone or in combination with known cytotoxic drugs, as selective and potent therapeutic antitumor agents. Furthermore, the recent research provides the scientific evidence of the traditional use of *Angelica* species among antitumor preparations. Future clinical studies are still needed to complete the activity profile and to begin the applicable activities of *Angelica* species as antitumor agents.

References

Ahn K, Sim W, Lee I, Seu Y, Kim I (1997) Decursinol angelate: a cytotoxic and protein kinase C activating agent from the root of *Angelica gigas*. Planta Med 63:360–361. doi:10.1055/s-2006-957701

Akihisa T, Motoi T, Seki A, Kikuchi T, Fukatsu M, Tokuda H, Suzuki N, Kimura Y (2012) Cytotoxic activities and anti-tumor-promoting effects of microbial transformation products of prenylated chalcones from *Angelica keiskei*. Chem Biodivers 9:318–330. doi:10.1002/cbdv.201100255

Bhat Z, Kumar D, Shah M (2011) *Angelica archangelica* Linn is an angel on earth for the treatment of diseases. Int J Nutr Pharmacol Neurol Dis 1:36–50. doi:10.4103/2231-0738.77531

Chen Y, Tsai H, Wu T (2007) Anti-inflammatory and analgesic activities from roots of *Angelica pubescens*. Planta Med 61:2–8. doi:10.1055/s-2006-957987

Fujiwara H, Yokoi T, Tani S, Saiki Y, Kato A (1980) Studies on constituents of *Angelicae dahuricae* radix. I. On a new furocoumarin derivative. Yakugaku Zasshi 100:1258–1261. PMID:7252783

Haranaka K, Satomi N, Sakurai A, Haranaka R, Okada N, Kobayashi M (1985) Antitumor activities and tumor necrosis factor producibility of traditional Chinese medicines and crude drugs. Cancer Immunol Immunother 20:1–5. doi:10.1007/BF00199765

Hung C, Kuo D, Chou C, Su Y, Ho C, Way T (2011) Osthole suppresses Hepatocyte Growth Factor (HGF)-induced epithelial-mesenchymal transition via repression of the c-Met/Akt/mTOR pathway in human breast cancer cells. J Agric Food Chem 59:9683–9690. doi:10.1021/jf2021489

Jung D, Jeong S, Kwon H, Jung J, Sohn H, Sin E, Kim J, Cho S, Yun M, Lee H, Kim S (2014) Decursin and doxorubicin are in synergy for the induction of apoptosis via STAT3 and/or mTOR pathways in human multiple myeloma cells. Cancer Res 73:836. doi:10.1158/1538-7445.AM2013-836

Kim H, Sik Bang S, Seok Choi J, Han H, Kim I (2005) Involvement of PKC and ROS in the cytotoxic mechanism of anti-leukemic decursin and its derivatives and their structure-activity relationship in human K562 erythroleukemia and U937 myeloleukemia cells. Cancer Lett 223:191–201. doi:10.1016/j.canlet.2004.10.025

Kim J, Noh E, Kim M, Hwang J, Hwang H, Ryu D, Kim H, Yu H, You Y, Kim J, Youn H, Kwon K, Jun S, Lee Y (2014) Decursin prevents TPA-induced invasion through suppression of PKCα/p38/NF-kB-dependent MMP-9 expression in MCF-7 human breast carcinoma cells. Int J Oncol 44:1607–1613. doi:10.3892/ijo.2014.2327

Kimura Y, Okuda H (1989) Effects of active compounds isolated from *Angelica shikokiana* on lipid metabolism in fat cells. J Ethnopharmacol 25:269–280. doi:10.1016/0378-8741(89)90033-0

Ko F, Wu T, Liou M, Huang T, Teng C (1992) Vasorelaxation of rat thoracic aorta caused by osthole isolated from *Angelica pubescens*. Eur J Pharmacol 219:29–34. doi:10.1016/0014-2999(92)90576-P

Lin V, Chou C, Lin Y, Lin J, Yu C, Tang C, Lin H, Way T (2010) Osthole suppresses fatty acid synthase expression in HER2-overexpressing breast cancer cells through modulating Akt/mTOR pathway. J Agric Food Chem 58:4786–4793. doi:10.1021/jf100352c

Lin P, Liu P, Lin S, Harn H (2012) *Angelica sinensis*: a Chinese herb for brain cancer therapy. BioMedicine 2:30–35. doi:10.1016/j.biomed.2011.12.003

Liu L, Huang W, Lin R, Lin S, Liang Y (2013) N-Hydroxycinnamide derivatives of osthole presenting genotoxicity and cytotoxicity against human colon adenocarcinoma cells *in vitro* and *in vivo*. Chem Res Toxicol 26:1683–1691. doi:10.1021/tx400271n

Martin W, Hutchins R (1980) A flora of New Mexico, J. Cramers, Vaduz, vol 1, USA

Mira A, Shimizu K (2015a) *In vitro* cytotoxic activities and molecular mechanisms of *Angelica shikokiana* extract and its isolated compounds. Pharmacogn Mag 11:564–569. doi:10.4103/0973-1296.172962

Mira A, Shimizu K (2015b) *In vivo* and *in vitro* hepatoprotective activities of the methanol extract of *Angelica shikokiana* and its major coumarin: isoepoxypteryxin against hepatocellular carcinoma. Phytochem Soc Asia 64

Moon H, Lee J, Lee Y, Kim K (2011) Antiplasmodial and cytotoxic activity of coumarin derivatives from dried roots of *Angelica gigas* Nakai *in vitro*. Immunopharmacol Immunotoxicol 33:663–666. doi:10.3109/08923973.2011.559248

Nishimura R, Tabata K, Arakawa M, Ito Y, Kimura Y, Akihisa T, Nagai H, Sakuma A, Kohno H, Suzuki T (2007) Isobavachalcone, a chalcone constituent of *Angelica keiskei*, induces apoptosis in neuroblastoma. Biol Pharm Bull 30:1878–1883. doi:10.1248/bpb.30.1878

Ogawa H, Nakashima S, Baba K (2003) Effects of dietary *Angelica keiskei* on lipid metabolism in stroke-prone spontaneously hypertensive rats. Clin Exp Pharmacol Physiol 30:284–288. doi:10.1046/j.1440-1681.2003.03830.x

Okuda H (1985) Pharmaceutical composition comprises extract of *Angelica shikokiana* Makiao belonging to Umbelliferae. In: JP 60260520 A, 23 December. JP 60260520 A

Raman A, Lin Z, Sviderskaya E, Kowalska D (1996) Investigation of the effect of *Angelica sinensis* root extract on the proliferation of melanocytes in culture. J Ethnopharmacol 54:165–170. doi:10.1016/S0378-8741(96)01458-4

Sarker SD, Nahar L (2004) Natural medicine: the genus *Angelica*. Curr Med Chem 11(11):1479–1500. http://dx.doi.org/10.2174/0929867043365189

Sigurdsson S, Ögmundsdottir H, Gudbjarnason S (2004) Antiproliferative effect of *Angelica archangelica* fruits. Z Naturforsch C 59:523–527. doi:10.1515/znc-2004-7-813

Sigurdsson S, Ogmundsdottir H, Gudbjarnason S (2005a) The cytotoxic effect of two chemotypes of essential oils from the fruits of *Angelica archangelica* L. Anticancer Res 25:1877–1880

Sigurdsson S, Ogmundsdottir H, Gudbjarnason S (2005b) Antitumour activity of *Angelica archangelica* leaf extract. In Vivo (Brooklyn) 19:191–194

Thanh P, Jin W, Song G, Bae K, Kang S (2004) Cytotoxic coumarins from the root of *Angelica dahurica*. Arch Pharm Res 27:1211–1215. doi:10.1007/BF02975883

Tsai N, Lin S, Lee C, Chen S, Su H, Chang W, Harn H (2005) The antitumor effects of *Angelica sinensis* on malignant brain tumors in vitro and in vivo. Clin Cancer Res 11:3475–3484. doi:10.1158/1078-0432.CCR-04-1827

Ye Y, Koo M, Li Y, Matsui H, Cho C (2001) *Angelica sinensis* modulates migration and proliferation of gastric epithelial cells. Life Sci 68:961–968. doi:10.1016/S0024-3205(00)00994-2

Zhang L, Jiang G, Yao F, He Y, Liang G, Zhang Y, Hu B, Wu Y, Li Y, Liu H (2012) Growth inhibition and apoptosis induced by Osthole, a natural coumarin, in hepatocellular carcinoma. PLoS One 7, e37865. doi:10.1371/journal.pone.0037865

Chapter 16
Advances in Chinese Herbal Medicine for Rheumatoid Arthritis: Clinical Utilization and Efficacy, Mechanism of Action, and Safety

Xia Mao, Qiuyan Guo, Aiping Lu, Yanqiong Zhang, and Na Lin

Abstract Rheumatoid arthritis (RA) is a debilitating, systemic autoimmune disease that affects people around the world. The disease is characterized by chronic inflammation of the joints, which eventually results in cartilage and bone damage. An increasing number of patients with RA worldwide are seeking help from complementary and alternative medicine (CAM) to alleviate the severity of the disease and to improve physical conditions. Among these treatments, traditional Chinese medicine (TCM) is regarded as a powerful treatment option, and it has been used for RA therapy for thousands of years in China. TCM is characterized by a holistic theory that emphasizes maintaining the balance of the patient's whole body using Chinese herbal medicines (CHMs) with multiple bioactive ingredients. Some studies have revealed that many antiarthritic CHMs may exert anti-inflammatory and immunomodulatory effects by regulating the production of pro-inflammatory cytokines and immuno-related pathways. However, the precise molecular mechanisms underlying the anti-RA activities of CHMs have not been fully elucidated. Moreover, safety issues have also blocked the development of CHMs; therefore, it is of great significance for clinicians, researchers, and pharmaceutical companies to share responsibility by regulating the clinical use of CHMs, strengthening the basic toxicology research, and establishing a strict quality control system to ensure the safe use of CHMs and decrease the number of toxic cases. The present chapter illustrates the clinical utilization of CHMs acting on RA, elucidates their mechanisms of action, analyzes their limitations and problems, and discusses their development and application prospects.

Keywords Chinese herbal medicine • Clinical utilization • Molecular mechanism rheumatoid arthritis • Traditional Chinese medicine

X. Mao • Q. Guo • Y. Zhang (✉) • N. Lin (✉)
Institute of Chinese Materia Medica, China Academy of Chinese Medical Sciences, Beijing 100700, China
e-mail: yqzhang@icmm.ac.cn; linna888@163.com

A. Lu
School of Chinese Medicine, Hong Kong Baptist University, Kowloon, Hong Kong

H.-S. Tsay et al. (eds.), *Medicinal Plants - Recent Advances in Research and Development*, DOI 10.1007/978-981-10-1085-9_16

Abbreviations

BGD	Baihu Guizhi decoction
CAM	Complementary and alternative medicine
CHMs	Chinese herbal medicines
COX-2	Cyclooxygenase-2
CRP	C-reactive protein
DJD	Duhuo Jisheng decoction
DMARDs	Disease-modifying antirheumatic drugs
ESR	Erythrocyte sedimentation rate
FHD	Fangji Huangqi decoction
FLS	Fibroblast-like synoviocytes
GSZD	Guizhi Shaoyao Zhimu decoction
HLXL	Huo-Luo-Xiao-Ling Dan
IL	Interleukin
iNOS	Inducible nitric oxide synthase
JNK	Jun N-terminal kinases
MAPKs	Mitogen-activated protein kinases
MMPs	Matrix metalloproteinases
MTX	Methotrexate
NF-kB	Nuclear transcription-kB
NO	Nitric oxide
NSAIDs	Nonsteroidal anti-inflammatory drugs
OPG	Osteoprotegerin
PGE2	Prostaglandin E2
RA	Rheumatoid arthritis
RANK	Receptor activator of NF-kB
RANKL	Receptor activator of NF-kB ligand
SIN	Sinomenine
TCM	Traditional Chinese medicine
TGP	Total glucosides of paeony
TIMPs	Tissue inhibitor of metalloproteinases
TLR	Toll-like receptor
TNF	Tumor necrosis factor
TRAF	TNF receptor-associated factor
TRP	Transient receptor potential
TRPA	TRP ankyrin type 1
TRPM8	TRP melastatin type 8
TRPV1	TRP vanilloid type 1
TS	Total saponin
TSW3	Triterpenoid saponin W3
TWHF	*Tripterygium wilfordii* Hook F
VEGF	Vascular endothelial growth factor
WLY	Wen Luo Yin

WTD	Wutou decoction
YLB	Yi Shen Juan Bi pill

16.1 Introduction

Rheumatoid arthritis (RA) is a chronic systemic inflammatory disease of unknown etiology. The overall prevalence of RA is approximately 0.5–1 % worldwide (Gibofsky 2014). Substantial variation existed regarding the incidence and prevalence of RA, implying its dynamic characteristic of the epidemiology (Gabriel and Michaud 2009), which is also a reflection of a well-accepted hypothesis that RA is caused by an environmental exposure or "trigger" in individuals who are genetically susceptible (Gibofsky 2012a). RA is manifested by synovial inflammation, insidious pain, morning stiffness, and joint swelling. It could result in erosion of the cartilage and bone and eventually cause joint deformity (Doan and Massarotti 2005). Moreover, patients with RA are more likely to suffer from myocardial infarction, atherosclerosis, stroke, and other complications (Avina-Zubieta et al. 2008), and they also tend to suffer from disability without appropriate and regular treatment. The management of RA should be commensurate with the characteristics and degree of the disease, which demands an integrated approach including both rheumatologists and orthopedic surgeons (Longo et al. 2015). Currently, the therapeutic strategies for the treatment of RA can be mainly divided into two categories. One category contains disease-modifying antirheumatic drugs (DMARDs), which consist of conventional DMARDs, including leflunomide, methotrexate (MTX), and sulfasalazine (Gibofsky 2012a). Extensive clinical experience has indicated that compared to leflunomide, the application of MTX, the "anchor" DMARD, may improve a patients' ability to remain on long-term therapy (Donahue et al. 2012). The second category contains biological agents, including tumor necrosis factor (TNF) inhibitors and non-TNF biologics (abatacept, rituximab, tocilizumab), which target the underlying pathophysiology of the disease and may alter disease progression (Gibofsky 2012b). Although these remedies have benefited many RA patients, their poor efficacies, high prices, and adverse effects are of common concern (Li et al. 2015a). Currently, an increasing number of RA patients have been seeking help from complementary and alternative medicine (CAM) for the alleviation of severe diseases and improvement of physical conditions.

Traditional Chinese medicine (TCM), which is based on empirical applications of experience from thousands of years, has become a crucial component of the modern medical system and has been extensively used as CAM in clinical practice. TCM is characterized by a holistic theory, which emphasizes maintaining the balance of the patients' whole body using Chinese herbal medicines (CHMs) that consist of multiple bioactive ingredients based on syndrome (ZHENG in Chinese) differentiation (Li et al. 2015b; Zhang et al. 2010). According to the theory of TCM, RA is an impediment disease ("Bi" syndrome), which is a group of diseases caused by the invasion of wind, cold, dampness, or heat pathogen into the human body (He et al. 2014). Compared to western medicine, CHMs excel in their lower prices, higher security, and feasibility of long-term administration (Jiang et al. 2012). Regarding the administration, CHMs used in clinics are usually classified into drugs for external use, such as CHM bath, acupoint application therapy, and ion introduction, and for internal use, such as single herbs, herbal formulae, and Chinese patent medicines (Qi et al. 2010). These have been regarded as indispensable strategies to alleviate the conditions and to enhance the life quality of RA patients.

The present chapter illustrates the clinical utilization of CHMs for RA, elucidates their mechanisms of action, analyzes their limitations and problems, and discusses their development and application prospects.

16.2 Clinical Utilization and Efficacy of CHMs in the Treatment of RA

TCM has been applied in the treatment of RA since ancient times in China. Several pathological factors, including weakness, the insufficiency of vital qi, and the invasion of cold and dampness, are always related to RA (Liu and Liu 2011). TCM practitioners determine the appropriate therapeutic schedules based on the categorization of patients through syndrome differentiation, such as a hot pattern, cold pattern, and deficiency pattern (Yuan et al. 2015). Moreover, the anti-inflammatory and antiarthritic activities of multiple CHMs applied in the treatment of RA have been proven through animal experiments and clinical trials (Venkatesha et al. 2011). In this section, we would like to illustrate the clinical utilizations and efficacies of CHMs in the treatment of RA.

16.2.1 Application of CHMs for External Use in the Treatment of RA

16.2.1.1 CHM Bath

A CHM bath is an ancient therapeutic approach in RA therapy and has been regarded as a supplementary method to assist with conventional therapeutic strategies, such as medications, acupuncture, etc., to enhance their therapeutic effectiveness. A CHM bath is a combination of heat therapy and medication, that is, to immerse the whole body or limbs of RA patients in a CHM soup, which exerts antiarthritic effects by dilating the capillaries in the skin and subcutaneous tissues, accelerating blood circulation, and improving drug absorption and infiltration into the lesion site. Moreover, when patients are soaking in the CHM liquid, their active or passive exercises may contribute to the improvement of their physical functions (Li 2001). In TCM theory, CHM baths can expel evil wind, remove dampness, and disperse cold by warming the meridians, which coincides well with the pathogenesis of RA (Zhu et al. 2011). Clinically, there are no fixed formulae for CHM baths. Clinicians make prescriptions according to the physical condition and disease severity of the RA patients. Herbs that have functions of dispelling wind and dampness, stimulating blood circulation, removing blood stasis, and causing muscle and joint relaxation are frequently used (Li 2001). Among them, *Carthami flos* (Honghua), *Angelicae pubescentis* Radix (Duhuo), *Lonicerae japonicae* Caulis (Rendongteng), and *Salviae miltiorrhizae* Radix et Rhizoma (Danshen) are representative herbs (Zhu et al. 2011). Interestingly, growing evidence shows that CHM baths can improve patients' joint function and quality of life and enhance the efficiency of conventional medical treatment in combined applications (Christie et al. 2007). However, although there are no obvious toxicities or side effects, patients suffering from dermatosis, acute inflammation, malignant tumors, severe cardiac insufficiency, and hypertension should avoid CHM baths (Li 2001).

16.2.1.2 Acupoint Application Therapy

Acupoint application therapy prevents and treats diseases under the guidance of a unique theory of TCM, the so-called preventive treatment of diseases. This therapy is performed by applying a mixed herbal cone cake into acupoints on the human body (Yang et al. 2015), and it has been well accepted by RA patients both alone and in combination with other therapies.

Clinically, Qubi analgesic gel paste acupoint application, midnight-noon ebb-flow acupoint application, and Leima adhesive plaster are the three most common therapeutic strategies in the treatment of RA, and they can improve the patients' balance ability and achieve good therapeutic effects (Wang et al. 2013; Gao et al. 2013). Moreover, Chinese herbs such as *Aconite radix* (Chuanwu), *Sinomenii caulis* (Qingfengteng), Moschus (Shexiang), and *Carthami flos* (Honghua) that can dispel wind and dampness can activate blood circulation and remove stasis and are frequently applied to produce herbal cone cakes and act on Dazhui, Tsusanli, Waiguan, Yangliquan, and other related acupoints (Du et al. 2013). It has been shown that acupuncture combined with the acupoint application of CHMs may improve immunologic function, enhance the physical conditions of RA patients, and achieve better therapeutic effects than the sole application of acupuncture (Chen et al. 2014).

16.2.1.3 Ion Introduction of CHM

Ion introduction of CHM electrically transports ionic particles into tissues through the skin by the application of certain devices (Kim et al. 2009). This technique can be used not only to deliver molecules into the body but also to monitor drugs and biomarkers in the clinical environment (Sieg and Wascotte 2009). Ion introduction serves as a complementary therapy in the treatment of many diseases, such as topical anesthesia, endodontics, and temporomandibular joint disorders (Girenes and Ulusu 2014).

Ion introduction of CHM is often applied in combination with other therapies, including conventional medical treatments and acupuncture, in RA therapy. According to TCM theory, the obstruction may cause pain. Chinese medicinal herbs that can promote blood circulation, remove blood stasis, dispel the wind, remove meridian obstructions, dispel cold, and remove dampness are often combined with certain devices to deliver the active ingredients of the herbs deeper into the joints and to achieve better efficacy (Zhang 2002). Growing clinical evidence has shown that ion introduction of CHM may accelerate the functional recovery of RA patients and enhance the therapeutic efficacy of CHM with little side effects (Zhang et al. 2011). However, this therapy is not suitable for patients suffering from severe heart disease, active tuberculosis, hyperpyrexia, and bleeding disorders (Fan and Xia 2007).

16.2.2 Application of CHMs for Internal Use in the Treatment of RA

16.2.2.1 Extracts of CHM Administration

In recent years, numerous studies have been conducted to investigate the prominent active components contained in Chinese medicinal herbs, following that extracts of CHMs have received more popularity due to its better efficacy and fewer adverse effects with the development of pharmaceutical technology. The detailed information about these herbs is provided below.

Tripterygium wilfordii Hook F (TWHF), also known as "Lei Gong Teng" in China, have been widely applied in the treatment of several autoimmune and inflammatory diseases such as ankylosing spondylitis, psoriasis, and especially RA. It has been reported that its ethanol/ethyl acetate extract and chloroform-methanol extract could maximize therapeutic benefit and minimize toxicity (Bao and Dai 2011). Clinical trials assessed by the attainment of ACR20 response criteria had proven that TWHF extract administration resulted in greater therapeutic efficacy than a conventionally used western drug, sulfasalazine, when the treatment for RA lasted for more than 24 weeks (Goldbach-Mansky et al. 2009). Moreover, during a 6-month study, the treatment with ethanol/ethyl alcohol extract (180 mg/day) of TWHF could significantly improve the clinical signs and syndromes of RA patients, such as joint pain, joint swelling, overall well-being as well as in several indicators of inflammation including C-reactive protein (CRP), erythrocyte sedimentation rate (ESR), and pro-inflammatory cytokine interleukin (IL)-6, suggesting great enhancement in active RA achieved by TWHF extracts (Macfarlane et al. 2011). In recent years, several bioactive ingredients extracted from TWHF have been further prepared into capsules, pellets, and powders for the treatment of RA (Lv et al. 2015), including *Tripterygium wilfordii* tablets, tripterygium glycoside tablets, and triptolide tablets, all of which have been extensively applied in clinics for RA therapy.

Sinomenii caulis (Qingfengteng), recorded by the Chinese Pharmacopoeia 2005 (Anonymous 2005), is an ancient Chinese medicinal herb and has been utilized to treat rheumatic diseases for more than a thousand years. It contains many active alkaloids, among which sinomenine (SIN) has been proved to be predominant in anti-inflammatory, antirheumatic, as well as immunosuppressive effects (Zhao et al. 2005). SIN preparation has been recorded in drug standard set by the Chinese Health Ministry and has been widely used in RA therapy in clinics (Chinese Pharmacopoeia 2005). Studies have been performed to compare the efficacy of SIN preparation with those of nonsteroidal anti-inflammatory drugs (NSAIDs). It has been indicated that SIN treatments could be more desirable regarding the total cases of RA patients whose clinical manifestations were significantly enhanced than NSAIDs. Moreover, SIN preparations possess better efficacy in improving the major syndromes of RA such as morning stiffness, joint pain, and swelling and clinical indicators of RA including ESR and CRP with dermato-mucosal adverse effects, which could be completely under the control by antihistamine reagents (Xu et al. 2008). SIN preparations are valuable remedies to treat RA in clinics.

Total glucosides of paeony (TGP) is composed of active components obtained from the roots of a conventional Chinese medicinal herb named *Paeoniae radix* Aiba (Baishao). It has been approved by the State Food and Drug Administration of China as a disease-modifying drug since 1998 and has been utilized in the treatment of RA for centuries (He and Dai 2011). In clinics, it is frequently applied in combination with MTX or leflunomide. It has been indicated that this combination remedy could greatly improve the registered indexes of RA patients such as the time of morning stiffness, the number of swelling joints, grip strength, ESR, CRP, and blood rheumatoid factor when the treatment lasted for 8 weeks (Zhang et al. 2007). Moreover, patients in combination therapy could achieve a better European League against rheumatism response than the sole application of western medicine, and the addition of TGP could significantly alleviate the severe hepatotoxicity resulting from MTX or leflunomide, observed in RA patients who have received therapy for 12 weeks (Chen et al. 2013).

16.2.2.2 Chinese Patent Medicine

Chinese patent medicine refers to CHMs made in the forms of pills, pulvis, emplastrum and unguentum, pellets and capsules, etc., which have the characteristics of being convenient to carry and store and have been regarded as the "essence" of TCM (Gao et al. 2014). In the clinical setting, Chinese patent medicines, such as the Xinfeng capsules, Simiao pills, Yi Shen Juan Bi pill (YLB), and Biqi capsule, are frequently used in the treatment of RA (Zhang et al. 2010).

The Xinfeng capsule is an extensively used commercial antiarthritic Chinese patent medicine and is composed of several herbs, including *Astragali radix* (Huangqi), *Coicis semen* (Yiyiren), TWHF, *Scolopendra* (Wugong), etc., which can invigorate the function of the spleen and resolve dampness. Clinical trials have proved that it could significantly reduce the total score of pain and swelling in involved joints and the level of uric acid and high-sensitivity CRP and also improve RA disease activity index and serum iron reserve (Huang et al. 2013a, b). Also, it has been found to exert amelioration effects in the abarticular pathologic changes in patients with active RA (Liu et al. 2014), indicating that it is a reliable therapeutic remedy.

The Simiao pill, consisting of *Phellodendri chinensis* Cortex (Huangbo), *Atractylodis rhizoma* (Cangzhu), *Achyranthis bidentatae* Radix (Niuqi), and *Coicis semen* (Yiyiren), is a well-known Chinese patent medicine for the treatment of RA with a precise compatibility. In clinics, modified Simiao pills are widely used for accurate individual RA therapies according to the syndromes and signs of RA patients. It has been demonstrated that modified Simiao pill could greatly improve the index of blood uric acid, blood leukocyte count, score of clinical symptoms, etc., and also has been proved to exert a better therapeutic efficacy than western medicine such as MTX with the advantages of low prices and readily availability (Zhao et al. 2013), indicating that it is a promising formula in clinics.

YLB is the pill form of a formula that was first prepared by the National TCM master Zhu Liangchun, and it is composed of 20 herbs including *Zaocys* (Wushaoshe), *Scorpio* (Quanxie), *Scolopendra* (Wugong), and *Corydalis rhizoma* (Yanhusuo), which can expel evil wind and remove dampness. It has been proven to have superior efficacy in the treatment of RA patients with kidney deficiency patterns (Zhou et al. 2007). Clinical trials have demonstrated that YLB could achieve more improvement on life quality and syndromes of RA patients such as arthralgia, joint pain, and joint tenderness than MTX and sulfasalazine after 24-week treatment although with relatively lower ACR20 and ACR50 responses (He et al. 2008).

Biqi capsule is composed of several traditional Chinese medicinal herbs such as *Codonopsis radix* (Dangshen), *Astragali radix* (Huangqi), and *Salviae miltiorrhizae* Radix et Rhizoma (Danshen) and has become an important Chinese patent drug for RA therapy in clinics. Studies have shown that Biqi capsule possesses favorable therapeutic outcomes especially for RA patients with qi deficiency and blood stasis syndrome including alleviating the degree of joint pain, the tender joint number, and the swollen joint number and shortening the morning stiffness time with no obvious adverse reaction (Liu et al. 2006). Moreover, its combination use with western medicine such as MTX showed better clinical efficacy than the application of Biqi capsule and MTX alone, indicating that it could be regarded as an effective treatment program for RA (Jie et al. 2012).

16.2.2.3 TCM Herbal Formulae

The administration of TCM herbal formulae, which are complex mixtures of herbs with multiple bioactive ingredients, is a notable feature of treatment based on holistic principles. The increased efficacy and decreased toxicity of TCM herbal formulae may arise as a result of complex synergistic or antagonistic interactions among different formula components, which meet the requirements of complex disease treatment in a systematic manner (Mao et al. 2015). Regarding the RA treatment, several classic TCM herbal formulae, such as Fangji Huangqi decoction (FHD), Wutou decoction (WTD), Guizhi Shaoyao Zhimu decoction (GSZD), Baihu Guizhi decoction (BGD), and Duhuo Jisheng decoction (DJD), have been extensively used for thousands of years (Liu and Liu 2011). Also, proved recipes prescribed by experienced clinicians have also obtained much recognition due to their satisfactory therapeutic effects. In clinics, the compositions and doses of these formulae are adjusted by clinicians by the syndromes and signs of RA patients.

FHD, composed of *Stephaniae tetrandrae* Radix (Fangji), *Astragali radix* (Huangqi), *Atractylodis macrocephalae* Rhizoma (Baizhu), *Glycyrrhizae radix* et Rhizoma (Gancao), *Zingiberis rhizoma* (Shengjiang), and *Jujubae fructus* (Dazao), is an ancient and effective remedy for the treatment of painful inflammatory disorders, such as RA, as well as the pain and edema caused by abdominal pain. In TCM theory, this remedy can relieve the symptoms of RA patients by replenishing qi to invigorate the spleen and eliminate dampness. Studies have indicated that FHD could significantly shorten the time of morning stiffness, alleviate joint pain, and exert a protective effect on the lung, liver, and kidney, of which mechanisms may be related to peripheral nociceptive pathway such as prostaglandins (Lin et al. 2015). Its components *Stephaniae tetrandrae* Radix (Fangji) and *Astragali radix* (Huangqi) play the most prominent role in this formula.

WTD, originating from the "Synopsis of Golden Chamber" (Chinese name: Jin Gui Yao Lue), is a famous TCM formula for the treatment of RA patients with cold pattern and joint pain with a history of more than a thousand years (Dai et al. 2014). It contains five herbs, including *Aconite radix* (Chuanwu), *Ephedrae herba* (Mahuang), *Astragali radix* (Huangqi), *Paeoniae radix* Aiba (Baishao), and *Glycyrrhizae radix* et Rhizoma (Gancao). In clinics, a prominent effectiveness can be observed when this formula is used for the treatment of RA characterized by an acute onset of severe pain. The formula exerts protective effects against joint destruction, inhibits the swelling of the limbs, and alleviates the severity of the disease. Numerous studies have shown that WTD in a combination of western medicine could enhance the therapeutic efficacy, shorten the treatment course of western medicine, as well as improve the life quality of RA patients in clinics (Hu 2011).

GSZD, originally recorded in the "Synopsis of Golden Chamber," is a classic formula in RA therapy. It comprises the following nine herbs: *Cinnamomi ramulus* (Guizhi), *Paeoniae radix* Aiba (Baishao), *Glycyrrhizae radix* et Rhizoma (Gancao), *Ephedrae herba* (Mahuang), *Zingiberis rhizoma* Recens (Shengjiang), *Atractylodis macrocephalae* Rhizoma (Baizhu), *Anemarrhenae rhizoma* (Zhimu), *Saposhnikoviae radix* (Fangfeng), and *Aconiti lateralis* Radix Preparata (Fuzi). The formula has been regarded as a vital formula in the treatment of chronic RA, which is manifested by joint deformation, body weight loss, dizziness, nausea, vomiting, fatigue, shortness of breath, and pain in multiple joints. Studies have shown that it could significantly alleviate the progression of the disease, improve related indexes such as CRP and ESR, and improve patients' quality of life (Xu et al. 2004).

BGD is another classic formula recorded in the "Synopsis of Golden Chamber" for RA treatment. In contrast to WTD, this remedy is superior to treat RA with the hot pattern, which is characterized by severe pain with hot, red, and inflamed joints, and the joint pain could be alleviated by applying cold to the joints (Lu et al. 2012). It could significantly alleviate syndromes of RA patients with hot pattern and improve their quality of life.

16.3 Pharmacological Mechanisms of CHMs Acting on RA

TCM, as a comprehensive and unique medical system, has excited worldwide interest. As a major component of TCM, CHMs are characterized by their complex nature. Subsequently, the ingredient profiling and molecular mechanisms of CHMs have not been fully elucidated, despite the considerable efforts made by many research groups. These limitations have hindered the application of CHMs in mainstream medicine and the modernization of TCM (Cooper 2007). In this section, we focus on the recent progress in elucidating the underlying mechanisms of CHMs acting on RA.

TWHF has an established history of its application in the treatment of RA. It has been proven to contain more than 70 components. Among them, tripdiolide, triptonide, and triptolide have been reported to exert the immunomodulatory and anti-inflammatory effects validated in both in vitro and in vivo studies (Qiu and Kao 2003). For instance, triptolide functions as an anti-inflammatory agent by inhibiting the production of nitric oxide (NO) and the expression of inducible nitric oxide synthase (iNOS), which lead to the inhibition of nuclear transcription-kB (NF-kB) and Jun N-terminal kinase (JNK) activation, TNF-α-induced cyclooxygenase-2 (COX-2) and prostaglandin E2 (PGE2) production, as well as lymphocyte proliferation (Wang et al. 2004; Zhang et al. 2004; Shao et al. 2004). Then, triptolide may play a protective role on cartilage by interfering with collagen-induced arthritis augmented expression of key enzymes including matrix metalloproteinases (MMPs)-13 and MMPs-3 in the pathological changes of cartilage (Lin et al. 2007). Also, its inhibitory effects on proMMPs-1 and proMMPs-3 and the simultaneous upregulation of tissue inhibitor of metalloproteinases (TIMPs) in IL-1 – treated synovial fibroblasts – may also contribute to the protective role of triptolide on cartilage (Lin et al. 2001). As one of the vital manifestations during RA progression, focal bone destruction within affected joints often severely influences the quality of patients' life. A recent study of Liu et al. has indicated that triptolide may attenuate bone destruction in RA partially by regulating receptor activator of NF-kB ligand/receptor activator of NF-kB/osteoprotegerin (RANKL/RANK/OPG) signal pathway to inhibit osteoclast formation (Liu et al. 2013a, b, c). The authors also indicated that triptolide might exert therapeutic effects on angiogenesis, an essential event in the development of RA, by the downregulation of angiogenic activators and the inhibition of mitogen-activated protein kinase downstream signal pathway activation (Kong et al. 2013). Moreover, Lu et al. found that triptolide could effectively inhibit the bioactivity of IL-18 and its receptor in phorbol 12-myristate 13-acetate-stimulated

RA synovial fibroblasts, highlighting a potential mechanism in RA therapy (Lu et al. 2008).

In addition to triptolide, celastrol is another single compound derived from TWHF with anti-inflammatory and bone protection properties. A recent study has reported that celastrol may directly inhibit the formation and function of osteoclast and be used as a novel medication management of RA in preventing bone destruction (Gan et al. 2015). Li et al. also found that celastrol could suppress the migration and invasion of fibroblast-like synoviocytes (FLS) by inhibiting Toll-like receptor (TLR) 4/NF-kB-mediated MMP-9 expression (Li et al. 2013a, b).

SIN is an alkaloid isolated from the stem of Chinese medicinal plant named *Sinomenii caulis* (Qingfengteng) and has been extensively used to treat RA diseases in China and Japan (Xu et al. 2008). Growing evidence show the anti-inflammatory and arthritis amelioration effects of SIN (Liu et al. 2005). For example, the administration of SIN could suppress inflammation response and joint destruction by targeting myeloid differentiation primary response protein 88 signaling (Mu et al. 2013). SIN could attenuate the formation of osteoclast and *M. tuberculosis* H37Ra-induced bone loss by mediating RANKL signaling pathways (Li et al. 2013). SIN exerts the inhibitory effects on cell invasion and migration abilities in a concentration-dependent manner by suppressing the expression of CD147, MMP-2, and MMP-9 in activated human monocytic THP-1 cells (Ou et al. 2011).

TGP, derived from the root of a Chinese herb *Paeoniae radix* Aiba (Baishao), contains more than 90% paeoniflorin and has received wide popularities in China, Korea, and Japan due to its prominent anti-inflammatory, hepatoprotective, and immunomodulatory effects in the treatment of RA (Chang et al. 2009). It has been proved that TPG exerts anti-inflammatory effects through inhibiting the production of inflammatory mediators and chemokines and leukocyte migration. Also, TGP ameliorates synovitis, a major pathological change occurred in RA patients, by regulating the balance of differentiation and function of Th1/Th2 cells and secretion of pro-inflammatory cytokines originated from lymphocytes, macrophages, and FLS (He and Dai 2011). Notably, TGP also exerts protective effects on joint destruction by reducing the secretion of cartilage degradation enzymes MMPs including MMP-1 and MMP-3, which mainly account for the degradation of cartilage (Zhang and Dai 2012).

Curcumin, derived from the rhizome of *Curcuma longa L.*, Zingiberaceae, has remarkable pharmacological and biological activities such as anti-inflammation and antioxidation against various chronic diseases, including RA (Recio et al. 2012). It has been indicated that curcumin may possess inhibitory effects on the growth and apoptosis of synovial fibroblasts, which was related to the proteolytic activation of caspase-3 and caspase-9 and modulation of poly(ADP-ribose) polymerase protein. Park et al. demonstrated that curcumin could reduce the expression of COX-2, a type of COX prominently involved in the process of inflammatory, partially by inhibiting PGE2 release (Park et al. 2007). Shakibaei et al. also revealed that curcumin might alleviate or reverse the breakdown of degenerative articular chondrocytes stimulated by IL-1β by antagonizing the activation of caspase-3 and matrix production (Shakibaei et al. 2005). Huang et al. also found the anti-inflammatory effects of curcumin, which might be associated with its suppression on the NF-kB and inflammatory loop (Huang et al. 2013). Notably, curcumin is frequently used in combination with other medications in RA therapy including resveratrol and MTX to benefit therapeutic efficiencies in clinical practice. For example, the combination of curcumin with resveratrol could antagonize the destruction of human articular chondrocytes in RA patients by activating the extracellular-regulated protein kinase signaling pathway, which might be associated with the differentiation and survival of chondrocyte (Shakibaei et al. 2011). Curcumin could synergistically support the therapy of MTX through its influence on arachidonic acid, thromboxanes, neutrophils, and lymphocytes. Interestingly, curcumin could also circumvent the hematological toxicities induced by MTX via suppressing the delivery of IL and leukotrienes, sufficiently making up this vital deficiency induced by western medicine (Banji et al. 2011).

The dry root of *Anemone flaccida* Fr. Schmidt (Diwu) is extensively used in clinical prescriptions in RA therapy, mainly excelling in anti-inflammatory, healing fractures and benefiting bone destruction (Han et al. 2013). Triterpenoid saponin W3 (TSW3), the major active ingredient in this plant, exerts anti-inflammatory, immunomodulatory, and analgesia effects (Cheng et al. 2008). Osteoclast plays an important role in the pathogenesis of RA and might result in excessive bone resorption within inflamed joints (Tanaka et al. 2001). The studies have indicated that TSW3 could inhibit the RANKL-induced osteoclast differentiation through downregulating the expression of a signaling adaptor molecule TNF receptor-associated factor (TRAF) 6, leading to the activation of mitogen-activated protein kinases (MAPKs) and NF-kB pathways, as well as the downregulation of two osteoclastogenic transcription factors including c-Fos and nuclear factor of activated T cells (Kong et al. 2015a, b). Also, the total saponin (TS) derived from this plant possesses antiarthritic effects and is undergoing the clinical trial in phase III in the treatment of RA (Huang et al. 2014). It has been clarified that the inhibitory effects of TS on the RANKL-induced osteoclast differentiation and bone destruction were mediated by inhibiting TRAF6 expression and suppressing JNK and p38 MAPKs and NF-kB activation. Subsequently, it downregulated the expression of a c-Fos and nuclear factor of activated T cells, which functions similarly in the way of W3 (Kong et al. 2015). Moreover, Liu et al. indicated that TS could attenuate focal and system bone destruction partially by modulating RANKL/RANK/OPG signal pathway, which has been demonstrated to play a crucial role in the process of bone loss and to inhibit the release of pro-osteoclastogenic cytokines (Liu et al. 2015).

Nobiletin, a citrus polymethoxy flavonoid, possesses numerous pharmacological activities including anti-inflammatory against various arthritic diseases. MMPs, the synthesis and secretion of which are mainly mediated by pro-inflammatory cytokines including IL-1, TNF-α, and IL-6, play an essential role in the destruction of matrix components in the pathological process of RA (Woessner 1991). Recent studies have indicated that nobiletin may suppress the secretion of proMMP-1, proMMP-3, and proMMP-9 partially by inhibiting the production of pro-inflammatory cytokines (Lin et al. 2003). Also, nobiletin also exerts inhibitory effects on PGE2 production in human synovial fibroblasts, which is an important inflammatory mediator due to its upregulation of vascular permeability (Ishiwa et al. 2000). Moreover, nobiletin could prevent cartilage destruction against RA through interfering with the expression of a disintegrin and metalloproteinase with thrombospondin-like motifs (ADAMTS)-4 and ADAMTS-5 (Imada et al. 2008). Murakami et al. proved that nobiletin could inhibit the osteoclastogenesis induced by RANKL possibly by suppressing MAPKs and blocking the differentiation of two key transcription factors including activator protein-1 and NF-kB (Murakami et al. 2007).

WTD is a classic formula for the treatment of RA, especially RA patients with the cold pattern, as previously described (Hu 2011). However, its pharmacological mechanisms have not been fully clarified. Recent studies have indicated that the anti-inflammatory effects of WTD were closely associated with its inhibition of pro-inflammatory cytokines including IL-β and TNF-α and its regulation on the TLR2/TRAF6/Faslg signal pathway (Xu et al. 2010). Accumulating studies have indicated that the pathological process of RA may frequently be accompanied by chronic inflammatory pain (Chiu et al. 2012). Members of the transient receptor potential (TRP) ion channel family especially TRP vanilloid type 1 (TRPV1), TRP ankyrin type 1 (TRPA1), and TRP melastatin type 8 (TRPM8) ion channels are closely involved in the induced inflammatory nociceptive responses (Sousa-Valente et al. 2014). Wang et al. found that WTD could possess antinociceptive property through decreasing mechanical and thermal hypersensitivities, partially resulting from its suppression of the expression of TRPV1, TRPA1, and TRPM8 (Wang et al. 2015). Interestingly, network pharmacology-based approaches have also been used to investigate the underlying mechanisms of WTD. Zhang et al. pointed out that WTD could alleviate RA possibly by reversing the imbalance of the nervous, endocrine, and immune systems, which markedly influence the pathological progression of RA (Zhang et al. 2015).

GSZD excels in the treatment of RA with hot pattern (Xu et al. 2004). An integrative method that combines both network analysis and experimental validation has also been applied to investigate the pharmacological mechanisms of GSZD acting on RA. The authors identified a candidate GSZD-targeted signal axis and found that GSZD plays a role in the treatment of RA partially by regulating inflammation-immune system imbalance (Guo et al. 2016).

Huo Luo Xiao Ling (HLXL) Dan, composed of *Angelicae sinensis* Radix (Danggui), Olibanum (Ruxiang), *Salviae miltiorrhizae* Radix et Rhizoma (Danshen), and *Myrrha* (Moyao), is a well-known herbal formula, and its modified versions have been applied to treat RA (Yu et al. 2013). Pharmacological studies have indicated that HLXL could alleviate the severity of ongoing inflammatory in RA, which might be associated with its alteration in T cells by regulating the antibody response against the RA-associated antigen (the mycobacterial heat-shock protein 65) and the serum level of NO (Yang et al. 2011). HLXL also exerts protective effects against arthritic bone destruction in adjuvant arthritis model, which was mediated by its regulatory effects on the mediators of bone remodeling (Nanjundaiah et al. 2013).

FHD extract possesses antinociceptive, anti-inflammatory, and immune-modulatory activities, some of which was validated in rodents with the acetic acid-induced writhing response, carrageenan-induced edema test, and formalin-induced licking test (Chen 2012). Regarding its antinociceptive effect, a recent study has demonstrated that the pretreatment with FHD extract could decrease acetic acid-induced writhing response and significantly prevent the late phase of formalin-induced licking response (Taber et al. 1969). Also, FHD produces marked antinociceptive activities in a dose-dependent way, which might be associated with the peripheral systems of pain pathway (Lin et al. 2015). Importantly, growing evidence show that almost all herbs in FHD may possess anti-inflammatory effects synergistically by regulating NF-kB, iNOS, and COX-2/prostaglandin pathway and by inhibiting the release of pro-inflammatory cytokines (Lin et al. 2015).

Wen Luo Yin (WLY), originated from the classic formula Guizhi Fuzi decoction and ZhuFu decoction, has been extensively applied in the treatment of RA with a cold pattern in clinics, especially excelling in the RA patients presented by obvious pain. WLY could alleviate pain, inflame joint swelling, and also inhibit the excessive secretion of synoviocytes. These possibly may be due to the decreased number of Golgi apparatus, rough surface endoplasmic reticulum, dense bodies, matrix filaments, and vacuoles, which are involved in the ultrastructures of synoviocytes (Li et al. 2002). Angiogenesis has been considered as an important event in affecting inflammatory and immune responses and further interfering with the pathological process of RA (Thairu et al. 2011). Liu and the group have reported that WLY exerts significant anti-angiogenic effects by suppressing the expression of numerous of angiogenic activators, including TNF-α, IL-1β, IL-17, vascular endothelial growth factor (VEGF), VEGF receptor, angiopoietins, and epidermal growth factor. All these participate in the progression of neovasculature by mutual interaction in the sera of collagen-induced rats in human FLS of RA and human umbilical vein endothelial cells induced by IL-1β, indicating that WLY is a promising therapeutic remedy in RA (Liu et al. 2013).

These previous studies have successfully identified the biological activities and targets of a herb or herbal formula and elucidated their molecular mechanisms for RA treatment. Many research techniques, such as flow cytometry analysis, cell differentiation assays, quantitative real-time PCR, Western blot analysis, cell migration assays, cell proliferation assays, and luciferase assays, have been commonly performed in the experimental studies. With the rapid progress in bioinformatics, system biology, and polypharmacology, network pharmacology has attracted much attention because it can reveal the underlying complex interactions between an herbal formula and cellular proteins, as well as can influence their interactions on the function and behavior of the system. It shifts the "one target, one drug" paradigm to the "network target, multicomponent" strategy (Li et al. 2007). Above all, the combination of the conventional experimental approaches and network pharmacology strategies can provide a powerful means of modern research on TCM in the future.

16.4 Safety and Adverse Effects of CHMs Acting on RA

CHMs perform well in clinical practice and have a bright future in the treatment of RA. However, the safety of CHMs has been a widespread concern due to their complex chemical nature and lack of proper evaluation methods, especially after recent consecutive reports of adverse drug reactions. Because CHMs are often used in humans for an extended period, continuous surveillance of patient safety during CHM treatments may be a convenient and powerful way to detect any potential harm caused by CHM therapy.

HLXL is a TCM formula for the treatment of a variety of immune disorders, including RA. By feeding HLXL to Lewis rats for 6 weeks consecutively, its toxicity was assessed. During the experiments, abnormal behavioral changes and standard manifestations of toxicity were documented (Zhang et al. 2009). Moreover, the blood biochemistry and histopathological variations of the tissues were examined to assess its toxicity. No adverse reactions or toxicity from this formula was observed at normal doses when taking all of the parameters together. Therefore, HLXL is reliable in RA therapy (Yang et al. 2011).

Fuzi (the lateral root of *Aconitum carmichaeli*) is a well-known herb for its bilateral effects, including both effectiveness and toxicity. Aconitine, mesaconitine, and hypaconitine are mainly responsible for its high toxicity. In clinical practice, Fuzi is processed by hydrolyzing toxic components into nontoxic derivatives, resulting in a toxicity reduction (Huang et al. 2007). Studies have been conducted to investigate whether the therapeutic efficacy of Fuzi remains after processing. In adjuvant-induced arthritic rats, it has been proven that processed Fuzi with 120 min decoctions could achieve the same therapeutic efficacy as the products processed for less time (Tong et al. 2013), indicating a non-interdependent relationship between its therapeutic effectiveness and toxicity.

Regarding TWHF, recent studies have reported that the efficacy and toxicity of this CHM are dose dependent (Zhao et al. 2015). The adverse effects of TWHF may be involved in many aspects of human body, such as gastrointestinal tract disturbances, dermatosis, reproductive system malfunction in both males and females, acute hepatotoxicity, and nephrotoxicity (Bao and Dai 2011). It could also lead to the blood system adverse events including the white blood cell decreasing, hemoglobin decreasing, and platelet decreasing (Li et al. 2015c). Triptolide is regarded to be the main contributor to these toxicities. Terpene triptonide and alkaloids in TWHF have been proved to be of no toxicological concern at the dosage of 20-fold of the therapeutic dose. Detoxification of TWHF could be achieved by metabolic eliminations, leading to less reactive metabolites. Moreover, safety issues should be taken into more consideration when it is coadministered with other cyclophosphamide inhibitors or glutathione-depleting agents (Li et al. 2015). TWHF could be a promising drug with the further in-depth investigation of its efficacy and toxicity.

Insect medicine excels in expelling evil wind and removing dampness in RA therapy. *Scorpio*, *centipede*, and *Agkistrodon* are representative insects. However, modern pharmacological studies indicate that homologous proteins such as toxic protein and histamine-like substance in these insects could exert toxic effects or allergic reactions if not properly used (Li and Liao 2011).

Asarum is a key herb in the treatment of arthrodynia. Volatile oils are its active constituent. The safrole containing in volatile oils could lead to several side effects, such as respiratory paralysis or arrhythmias (Li and Liao 2011).

Given all the issues mentioned above, it is clinically important to harness CHM advantages and bypass the disadvantages. The adverse effects of CHMs can be confined within a reasonable range by appropriate methods such as the standardization of CHM production, CHM processing, proper combination, and correct dose administration based on different pathological conditions. Moreover, the more information we obtain about the safety and adverse events of CHM therapies, the greater the likelihood that physicians trained in conventional medicine will be encouraged to use such medicines. Therefore, further studies are required to generate data and information on the adverse effects of existing CHM therapies for the treatment of RA.

16.5 Perspective

CHMs have been recognized extensively to benefit RA patients due to their efficacies, minimal adverse effects, affordable price, therapeutic effects, and possibility of long-term application. However, the following limitations and issues need our consideration.

First, TCM is characterized by individualized treatment. Clinicians make prescriptions according to the syndromes and signs of patients and classify them into different patterns such as hot pattern, cold pattern, deficiency pattern, etc. It is important to clarify the molecular mechanisms of pattern and CHMs. Gaps in knowledge regarding the characteristics and mechanisms of CHMs acting on different patterns of RA still exist due to the complexities of CHMs and limitations of investigative techniques.

Second, cognized consensus standard regarding the safety of CHMs has not been completed. The efficacy of CHMs against RA has long been based on empiricism. Although the appropriate compatibility of herbs could enhance the therapeutic efficacy and reduce toxicity such as the coadministration of *Bupleuri radix* (Chaihu) and *Pinelliae rhizoma* (Banxia) (Liu et al. 2013a, b, c), its safety cannot be fully guaranteed. The application of several modern research tools, such as gene expression microarrays, proteomics, and biological molecular networks, may shed light on a holistic understanding of CHMs. Thus, offer an interface where CHMs and conventional medicine can find common ground to investigate the mechanisms of action of therapeutic products and to enhance their practical use for the ultimate benefit of the patients (Venkatesha et al. 2011).

Finally, efforts are needed to standardize the safety assessment of CHMs, in-depth toxicity-related studies of CHMs, and safer advanced preparation technology, such as nano-drug delivery and other targeted drug delivery. Also, refinement in trial-related issues, such as sample size, explicit inclusion/exclusion criteria, consistent standards for assessing the outcome of therapeutic intervention, and proper statistical analysis are other issues for consideration (Efthimiou and Kukar 2010).

Thus, the future studies that combine the routine application of CHMs and advanced technologies will be essential to explore fully the therapeutic effects, mechanisms, and safety of CHMs acting on RA. There is a promise that CHMs may receive increasing therapeutic approvals and may benefit RA patients in a more effective way.

References

Anonymous (2005) *Caulis sinomenii.* In: Chinese Pharmacopoeia Committee (ed) Chinese pharmacopoeia, vol 1. Chinese Chemical Industry Press, Beijing, p 135

Avina-Zubieta JA, Choi HK, Sadatsafavi M et al (2008) Risk of cardiovascular mortality in patients with rheumatoid arthritis: a meta-analysis of observational studies. Arthritis Rheum 59(12):1690–1697. doi:10.1002/art.24092

Banji D, Pinnapureddy J, Banji OJ et al (2011) Evaluation of the concomitant use of methotrexate and curcumin on Freund's complete adjuvant-induced arthritis and hematological indices in rats. Indian J Pharmacol 43(5):546–550. doi:10.4103/0253-7613.84970

Bao J, Dai SM (2011) A Chinese herb *Tripterygium wilfordii* Hook F in the treatment of rheumatoid arthritis: mechanism, efficacy, and safety. Rheumatol Int 31(9):1123–1129. doi:10.1007/s00296-011-1841-y

Chang Y, Wei W, Zhang L (2009) Effects and mechanisms of total glucosides of paeony on synoviocytes activities in rat collagen-induced arthritis. J Ethnopharmacol 121(1):43–48. doi:10.1016/j.jep.2008.09.028

Chen GX (2012) Clinical trials of the combination of modified Fangji-Huangqi decoction and Duhuo-Jishen decoction in the treatment of RA in 32 cases. Guiding J Trad Chin Med Pharm 18(4):58–59 (In Chinese)

Chen Z, Li XP, Li ZJ et al (2013) Reduced hepatotoxicity by total glucosides of paeony in combination treatment with leflunomide and methotrexate for patients with active rheumatoid arthritis. Int Immunopharmacol 15(3):474–477. doi:10.1016/j.intimp.2013.01.021

Chen ZP, Chen PD, Zhuang LX (2014) Acupoint herbal application combined with electro-acupuncture in the treatment of knee osteoarthritis. J Clin Acupunct Moxibustion 5(5):28–30 (In Chinese)

Cheng S, Du Y, Bing F et al (2008) Synthesis of flaccidoside II, a bidesmosidic triterpene saponin isolated from Chinese folk medicine Di Wu. Carbohydr Res 343(3):462–469

Chiu IM, von Hehn CA, Woolf CJ (2012) Neurogenic inflammation and the peripheral nervous system in host defense and immunopathology. Nat Neurosci 15(8):1063–1067. doi:10.1038/nn.3144

Christie A, Jamtvedt G, Dahm KT et al (2007) Effectiveness of nonpharmacological and nonsurgical interventions for patients with rheumatoid arthritis: an overview of systematic reviews. Phys Ther 87(12):1697–1715

Cooper EL (2007) The immune system and complementary and alternative medicine. Evid Based Complement Alternat Med 4(1):5–8. doi:10.1093/ecam/nem093

Dai PM, Wang Y, Ye L et al (2014) Pharmacokinetic comparisons of benzoylmesaconine in rats using ultra-performance liquid chromatography-tandem mass spectrometry after administration of pure benzoylmesaconine and Wutou decoction. Molecules 19(10):16757–16769. doi:10.3390/molecules191016757

Doan T, Massarotti E (2005) Rheumatoid arthritis: an overview of new and emerging therapies. J Clin Pharmacol 45(7):751–762

Donahue KE, Jonas DE, Hansen RA et al (2012) Drug therapy for rheumatoid arthritis in adults: an update. Agency for Healthcare Research and Quality (US), Rockville. Report No.: 12-EHC025-EF

Du J, Li YJ, Li Q et al (2013) Efficacy of Hanbi powder acupoint application with massage for knee osteoarthritis. J Nurs (12):1–4 (In Chinese)

Efthimiou P, Kukar M (2010) Complementary and alternative medicine use in rheumatoid arthritis: proposed mechanism of action and efficacy of commonly used modalities. Rheumatol Int 30(5):571–586. doi:10.1007/s00296-009-1206-y

Fan DH, Xia B (2007) Observation on therapeutic effect of heat needle combined with herb iontophoresis and western medicine on rheumatoid arthritis. Chin Acupunct Moxibustion 27(10):731–734 (In Chinese)

Gabriel SE, Michaud K (2009) Epidemiological studies of incidence, prevalence, mortality, and comorbidity of the rheumatic diseases. Arthritis Res Ther 11(3):229. doi:10.1186/ar2669

Gan K, Xu L, Feng X et al (2015) Celastrol attenuates bone erosion in collagen-induced arthritis mice and inhibits osteoclast differentiation and function in RANKL-induced RAW264.7. Int Immunopharmacol 24(2):239–246. doi:10.1016/j.intimp.2014.12.012

Gao J, Zhang T, Zhang Q et al (2013) Effect of midnight-noon ebb-flow of acupoint application on balance ability in patients with knee osteoarthritis. Chin J Pract Nurs 29(29):26–28 (In Chinese)

Gao F, Leng J, Fu CM et al (2014) Interpretation of contemporary positioning of traditional Chinese medicine injections and analysis of key problems. China J Chin Mater Med 39(17):3416–3419 (In Chinese)

Gibofsky A (2012a) Comparative effectiveness of current treatments for rheumatoid arthritis. Am J Manag Care 18(13):303–314

Gibofsky A (2012b) Overview of epidemiology, pathophysiology, and diagnosis of rheumatoid arthritis. Am J Manag Care 18(13):295–302

Gibofsky A (2014) Epidemiology, pathophysiology, and diagnosis of rheumatoid arthritis: a synopsis. Am J Manag Care 20(7):128–135

Girenes G, Ulusu T (2014) An *in vitro* evaluation of the efficacy of a novel iontophoresis fluoride tray on remineralization. J Clin Exp Dent 6(4):327–334. doi:10.4317/jced.51376

Goldbach-Mansky R, Wilson M, Fleischmann R et al (2009) Comparison of *Tripterygium wilfordii* Hook F versus sulfasalazine in the treatment of rheumatoid arthritis: a randomized trial. Ann Intern Med 151(4):229–240

Guo QY, Mao X, Zhang YQ, et al (2016) Guizhi-Shaoyao-Zhimu decoction attenuates rheumatoid arthritis partially by reversing inflammation-immune system imbalance. J Trans Med 14(1):165

Han LT, Fang Y, Li MM et al (2013) The antitumor effects of triterpenoid saponins from the *Anemone flaccida* and the underlying mechanism. Evid Based Complement Alternat Med 2013:517931. doi:10.1155/2013/517931

He DY, Dai SM (2011) Anti-inflammatory and immunomodulatory effects of *Paeonia lactiflora* Pall., a traditional Chinese herbal medicine. Front Pharmacol 2:10. doi:10.3389/fphar.2011.00010

He Y, Lu A, Zha Y et al (2008) Differential effect on symptoms treated with traditional Chinese medicine and western combination therapy in RA patients. Complement Ther Med 16(4):206–211. doi:10.1016/j.ctim.2007.08.005

He YT, Ou AH, Yang XB et al (2014) Traditional Chinese medicine versus Western medicine as used in China in the management of rheumatoid arthritis: a randomized, single-blind, 24-week study. Rheumatol Int 34(12):1647–1655

Hu HZ (2011) A clinical analysis of treating rheumatic arthralgia with Wutou decoction. Clin J Chin Med 3(2):20 (In Chinese)

Huang QA, Zhang YM, He Y et al (2007) Studies on hydrolysis of aconitine. China J Chin Mater Med 32(20):2143–2145 (In Chinese)

Huang CB, Liu J, Chen X et al (2013a) Treatment of rheumatoid arthritis by xinfeng capsule: an efficacy observation. Zhongguo Zhong Xi Yi Jie He Za Zhi 33(12):1599–1602 (In Chinese)

Huang G, Xu Z, Huang Y et al (2013b) Curcumin protects against collagen-induced arthritis via suppression of BAFF production. J Clin Immunol 33(3):550–557. doi:10.1007/s10875-012-9839-0

Huang XJ, Tang JQ, Li MM et al (2014) Triterpenoid saponins from the rhizomes of *Anemone flaccida* and their inhibitory activities on LPS-induced NO production in macrophage RAW264.7 cells. J Asian Nat Prod Res 16(9):910–921. doi:10.1080/10286020.2014.954554

Imada K, Lin N, Liu C et al (2008) Nobiletin, a citrus polymethoxy flavonoid, suppresses gene expression and production of aggrecanases-1 and -2 in collagen-induced arthritic mice. Biochem Biophys Res Commun 373(2):181–185. doi:10.1016/j.bbrc.2008.05.171

Ishiwa J, Sato T, Mimaki Y et al (2000) A citrus flavonoid nobiletin suppresses production and gene expression of matrix metalloproteinase 9/gelatinase B in rabbit synovial fibroblasts. J Rheumatol 27(1):20–25

Jiang M, Lu C, Chen G et al (2012) Understanding the molecular mechanism of interventions in treating rheumatoid arthritis patients with corresponding traditional Chinese medicine patterns based on bioinformatics approach. Evid Based Complement Alternat Med 2012:129452. doi:10.1155/2012/129452

Jie HY, Wu QF, Ding ZX (2012) Clinical study of Biqi capsule combined with methotrexate for treatment of rheumatoid arthritis. Zhongguo Zhong Xi Yi Jie He Za Zhi 32(2):195–198 (In Chinese)

Kim HE, Kwon HK, Kim BI (2009) Application of fluoride iontophoresis to improve remineralization. J Oral Rehabil 36(10):770–775. doi:10.1111/j.1365-2842.2009.01992.x

Kong X, Zhang Y, Liu C et al (2013) Anti-angiogenic effect of triptolide in rheumatoid arthritis by targeting angiogenic cascade. PLoS ONE 8(10), e77513. doi:10.1371/journal.pone.0077513

Kong X, Wu W, Yang Y et al (2015a) Total saponin from Anemone flaccidaFr. Schmidt abrogates osteoclast differentiation and bone resorption via the inhibition of RANKL-induced NF-kB, JNK, and p38 MAPKs activation. J Transl Med 13:91. doi:10.1186/s12967-015-0440-1

Kong X, Yang Y, Wu W et al (2015b) Triterpenoid saponin W3 from *Anemone flaccida* suppresses osteoclast differentiation through inhibiting activation of MAPKs and NF-kB pathways. Int J Biol Sci 11(10):1204–1214. doi:10.7150/ijbs.12296

Li ZY (2001) Nursing care of treating rheumatoid arthritis by medicated bath with Chinese traditional herbs. Acta Acad Med Militaris Tertiae 23(6):685–685 (In Chinese)

Li N, Liao YH (2011) Discussion about the application of toxic traditional Chinese medicine in the treatment of rheumatoid arthritis. Guiding J Tradit Chin Med Pharm 17(3):99–100 (In Chinese)

Li S, Lu A, Jia H (2002) Therapeutic actions of the Chinese herbal formulae with cold and heat properties and their effects on ultrastructures of synoviocytes in rats of collagen-induced arthritis. J Tradit Chin Med 22(4):296–302

Li S, Zhang ZQ, Wu LJ (2007) Understanding ZHENG in traditional Chinese medicine in the context of the neuro-endocrine-immune network. IET Syst Biol 1(1):51–60

Li G, Liu D, Zhang Y et al (2013a) Celastrol inhibits lipopolysaccharide-stimulated rheumatoid fibroblast-like synoviocyte invasion through suppression of TLR4/NF-kB-mediated matrix metalloproteinase-9 expression. PLoS ONE 8(7), e68905. doi:10.1371/journal.pone

Li X, He L, Hu Y et al (2013b) Sinomenine suppresses osteoclast formation and Mycobacterium tuberculosis H37Ra-induced bone loss by modulating RANKL signaling pathways. PLoS ONE 8(9), e74274. doi:10.1371/journal.pone.0074274

Li XX, Du FY, Liu HX et al (2015a) Investigation of the active components in *Tripterygium wilfordii* leading to its acute hepatotoxicity and nephrotoxicity. J Ethnopharmacol 162:238–243. doi:10.1016/j.jep.2015.01.004

Li Y, Li R, Ouyang Z et al (2015b) Herb network analysis for a famous TCM doctor's prescriptions on the treatment of rheumatoid arthritis. Evid Based Complement Alternat Med 2015:451319. doi:10.1155/2015/451319

Li ZX, Ma DM, Yang XH et al (2015c) Meta-analysis of blood system adverse events of *Tripterygium wilfordii*. Zhongguo Zhong Yao Za Zhi 40(2):339–345 (In Chinese)

Lin N, Sato T, Ito A (2001) Triptolide, a novel diterpenoid triepoxide from *Tripterygium wilfordii* Hook. f. suppresses the production and gene expression of pro–matrix metalloproteinases 1 and 3 and augments those of tissue inhibitors of metalloproteinases 1 and 2 in human synovial fibroblasts. Arthritis Rheum 44(9):2193–2200

Lin N, Sato T, Takayama Y et al (2003) Novel anti-inflammatory actions of nobiletin, a citrus polymethoxy flavonoid, on human synovial fibroblasts and mouse macrophages. Biochem Pharmacol 65(12):2065–2071

Lin N, Liu C, Xiao C et al (2007) Triptolide, a diterpenoid triepoxide, suppresses inflammation and cartilage destruction in collagen-induced arthritis mice. Biochem Pharmacol 73(1):136–146

Lin YC, Chang CW, Wu CR (2015) Antinociceptive, anti-inflammatory and toxicological evaluation of Fang-Ji-Huang-Qi-Tang in rodents. BMC Complement Altern Med 15:10. doi:10.1186/s12906-015-0527-5

Liu J, Liu RL (2011) The potential role of Chinese medicine in ameliorating extra-articular manifestations of rheumatoid arthritis. Chin J Integr Med 17(10):735–737. doi:10.1007/s11655-011-0872-2

Liu ZQ, Kelvin C, Zhou H et al (2005) The pharmacokinetics and tissue distribution of sinomenine in rats and its protein binding ability in vitro. Life Sci 77(25):3197–3209

Liu W, Zhang L, Xu Z (2006) Clinical observation on treatment of rheumatoid arthritis with the Biqi capsule. Zhongguo Zhong Xi Yi Jie He Za Zhi 26(2):157–159 (In Chinese)

Liu C, Kong X, Li X et al (2013a) Wen Luo Yin inhibits angiogenesis in collagen-induced arthritis rat model and in vitro. J Ethnopharmacol 149(2):478–489. doi:10.1016/j.jep.2013.07.002

Liu C, Zhang Y, Kong X et al (2013b) Triptolide prevents bone destruction in the collagen-induced arthritis model of rheumatoid arthritis by targeting RANKL/RANK/OPG signal pathway. Evid Based Complement Alternat Med 2013:626038. doi:10.1155/2013/626038

Liu CF, Tan SF, Wang DH et al (2013c) Study on efficacy of compatibility between *Aconiti radix* cocta and *Pinelliae rhizoma* by the uniform design method. Zhongguo Zhong Yao Za Zhi 38(13):2169–2175 (In Chinese)

Liu J, Cao Y, Huang C, Wang Y et al (2014) Use of xinfeng capsule to treat abarticular pathologic changes in patients with rheumatoid arthritis. J Tradit Chin Med 34(5):532–538

Liu C, Yang Y, Sun D et al (2015) Total saponin from Anemone flaccida Fr. Schmidt prevents bone destruction in experimental rheumatoid arthritis via inhibiting osteoclastogenesis. Rejuvenation Res 18(6):528–542. doi:10.1089/rej.2015.1688

Longo UG, Petrillo S, Denaro V (2015) Current concepts in the management of rheumatoid hand. Int J Rheumatol 2015:648073. doi:10.1155/2015/648073

Lu Y, Wang WJ, Leng JH et al (2008) Inhibitory effect of triptolide on interleukin-18 and its receptor in rheumatoid arthritis synovial fibroblasts. Inflamm Res 57(6):260–265. doi:10.1007/s00011-007-7128-9

Lu C, Xiao C, Chen G et al (2012) Cold and heat pattern of rheumatoid arthritis in traditional Chinese medicine: distinct molecular signatures identified by microarray expression profiles in CD4-positive T cell. Rheumatol Int 32(1):61–68. doi:10.1007/s00296-010-1546-7

Lv QW, Zhang W, Shi Q et al (2015) Comparison of *Tripterygium wilfordii* Hook F with methotrexate in the treatment of active rheumatoid arthritis (TRIFRA): a randomized, controlled clinical trial. Ann Rheum Dis 74(6):1078–1086. doi:10.1136/annrheumdis-2013-204807

Macfarlane GJ, El-Metwally A, De Silva V et al (2011) Evidence for the efficacy of complementary and alternative medicines in the management of rheumatoid arthritis: a systematic review. Rheumatology (Oxford) 50(9):1672–1683. doi:10.1093/rheumatology/ker119

Mao X, Zhang Y, Lin N (2015) Application perspectives of traditional Chinese medicine in the treatment of liver cancer. Cancer Transl Med 1:101–107

Mu H, Yao RB, Zhao LJ et al (2013) Sinomenine decreases MyD88 expression and improves inflammation-induced joint damage progression and symptoms in rat adjuvant-induced arthritis. Inflammation 36(5):1136–1144. doi:10.1007/s10753-013-9648-5

Murakami A, Song M, Katsumata S et al (2007) *Citrus nobiletin* suppresses bone loss in ovariectomized ddY mice and collagen-induced arthritis in DBA/1J mice: possible involvement of receptor activator of NF-kappaB ligand (RANKL)-induced osteoclastogenesis regulation. Biofactors 30(3):179–192

Nanjundaiah SM, Lee DY, Berman BM et al (2013) Chinese herbal formula Huo-Luo-Xiao-Ling Dan protects against bone damage in adjuvant arthritis by modulating the mediators of bone remodeling. Evid Based Complement Alternat Med 2013:429606. doi:10.1155/2013/429606

Ou Y, Li W, Li X et al (2011) Sinomenine reduces invasion and migration ability in fibroblast-like synoviocytes cells co-cultured with activated human monocytic THP-1 cells by inhibiting the expression of MMP-2, MMP-9, CD147. Rheumatol Int 31(11):1479–1485. doi:10.1007/s00296-010-1506-2

Park C, Moon DO, Choi IW (2007) Curcumin induces apoptosis and inhibits prostaglandin E2 production in synovial fibroblasts of patients with rheumatoid arthritis. Int J Mol Med 20(3):365–372

Qi F, Li A, Inagaki Y et al (2010) Chinese herbal medicines as adjuvant treatment during chemo- or radio-therapy for cancer. Biosci Trends 4(6):297–307

Qiu D, Kao PN (2003) Immunosuppressive and anti-inflammatory mechanisms of triptolide, the principal active diterpenoid from the Chinese medicinal herb *Tripterygium wilfordii* Hook. f. Drugs R D 4(1):1–18

Recio MC, Andujar I, Rios JL (2012) Anti-inflammatory agents from plants: progress and potential. Curr Med Chem 19(14):2088–2103

Shakibaei M, Schulze-Tanzil G, John T (2005) A curcumin protects human chondrocytes from IL-1β-induced inhibition of collagen type II and β1-integrin expression and activation of caspase-3: an immunomorphological study. Ann Anat 187(5–6):487–497

Shakibaei M, Mobasheri A, Buhrmann C (2011) Curcumin synergizes with resveratrol to stimulate the MAPK signaling pathway in human articular chondrocytes in vitro. Genes Nutr 6(2):171–179. doi:10.1007/s12263-010-0179-5

Shao XT, Feng L, Yao HP et al (2004) Effect of triptolide on TNF-α-induced activation of NF-kB and expression of COX-2 and iNOS in human rheumatoid arthritis synovial fibroblasts. Zhejiang Daxue Xuebao Yixue Bao 33(2):160–165 (In Chinese)

Sieg A, Wascotte V (2009) Diagnostic and therapeutic applications of iontophoresis. J Drug Target 17(9):690–700. doi:10.3109/10611860903089750

Sousa-Valente J, Andreou AP, Urban L et al (2014) Transient receptor potential ion channels in primary sensory neurons as targets for novel analgesics. Br J Pharmacol 171(10):2508–2527. doi:10.1111/bph.12532

Taber RI, Greenhouse DD, Rendell JK et al (1969) Agonist and antagonist interactions of opioids on acetic acid-induced abdominal stretching in mice. J Pharmacol Exp Ther 169:29–38

Tanaka S, Nakamura K, Oda H (2001) The osteoclast: a potential therapeutic target of bone and joint destruction in rheumatoid arthritis. Mod Rheumatol Jpn Rheum Assoc 11(3):177–183. doi:10.3109/s101650170001

Thairu N, Kiriakidis S, Dawson P et al (2011) Angiogenesis as a therapeutic target in arthritis in 2011: learning the lessons of the colorectal cancer experience. Angiogenesis 14(3):223–234. doi:10.1007/s10456-011-9208-2

Tong P, Wu C, Wang X et al (2013) Development and assessment of a complete-detoxication strategy for Fuzi (lateral root of *Aconitum carmichaeli*) and its application in rheumatoid arthritis therapy. J Ethnopharmacol 146(2):562–571. doi:10.1016/j.jep.2013.01.025

Venkatesha SH, Rajaiah R, Berman BM et al (2011) Immunomodulation of autoimmune arthritis by herbal CAM. Evid Based Complement Alternat Med 2011:986797. doi:10.1155/2011/986797

Wang B, Ma L, Tao X et al (2004) Triptolide, an active component of the Chinese herbal remedy *Tripterygium wilfordii* Hook F, inhibits production of nitric oxide by decreasing inducible nitric oxide synthase gene transcription. Arthritis Rheum 50(9):2995–303

Wang H, Fang JQ, Fang F et al (2013) Effects of Qubi analgesic gel paste acupoint application on collagen-induced arthritis inflammatory rats. J Zhejiang Chin Med Univ 37(5):613–616 (In Chinese)

Wang C, Liu C, Wan H et al (2015) Wu-Tou decoction inhibits chronic inflammatory pain in mice: participation of TRPV1 and TRPA1 ion channels. Biomed Res Int 2015:328707. doi:10.1155/2015/328707

Woessner JF Jr (1991) Matrix metalloproteinases and their inhibitors in connective tissue remodeling. FASEB J 5(8):2145–2154

Xu Q, Wang J, Liu YH (2004) Clinical observation on treatment of rheumatoid arthritis with United Chinese medicine and western medicine. China J Mod Med 14(8):140–145 (In Chinese)

Xu M, Liu L, Qi C et al (2008) Sinomenine versus NSAIDs for the treatment of rheumatoid arthritis: a systematic review and meta-analysis. Planta Med 74(12):1423–1429

Xu S, Li L, Zhang W, Wang Y (2010) Comparative study of the regulation of TLR/TRAF signaling pathway of classic prescriptions that deals with arthralgia syndrome based on the relevant theory of prescription and syndrome. Zhongguo Zhong Yao Za Zhi 35(8):1025–1029 (In Chinese)

Yang YH, Rajaiah R, Lee DY et al (2011) Suppression of ongoing experimental arthritis by a Chinese herbal formula (Huo-Luo-Xiao-Ling Dan) involves changes in antigen-induced immunological and biochemical mediators of inflammation. Evid Based Complement Alternat Med 2011:642027. doi:10.1155/2011/642027

Yang XC, Yin T, Gao Q et al (2015) The immunomodulatory effect of acupoint application for childhood asthma: a systematic review and meta-analysis. Evid Based Complement Alternat Med 2015:896247. doi:10.1155/2015/896247

Yu H, Lee DY, Nanjundaiah SM et al (2013) Microarray analysis reveals the molecular basis of antiarthritic activity of Huo-Luo-Xiao-Ling Dan. Evid Based Complement Alternat Med 2013:524746. doi:10.1155/2013/524746

Yuan HY, Zhang XL, Zhang XH et al (2015) Analysis of patents on anti-rheumatoid arthritis therapies issued in China. Expert Opin Ther Pat 25(8):909–930. doi:10.1517/13543776.2015.1044972

Zhang LJ (2002) Observation and nursing of treating rheumatoid arthritis by herbal ion infusion. Acta Acad Med Militaris Tertiae 24(7):792–792, 795 (In Chinese)

Zhang W, Dai SM (2012) Mechanisms involved in the therapeutic effects of *Paeonia lactiflora* Pallas in rheumatoid arthritis. Int Immunopharmacol 14(1):27–31. doi:10.1016/j.intimp.2012.06.001

Zhang N, Xu YJ, Zhang ZX (2004) Regulatory function of nuclear factor kappa B on lymphocyte proliferation and apoptosis in bronchial asthmatic rats and effect of triptolide on the regulation. Zhongguo Zhongxiyi Jiehe Zazhi 24(5):435–438 (In Chinese)

Zhang DC, Ye HL, Zhu DB (2007) The prospective efficacy of total glucosides of paeony combined with leflunomide in treating rheumatoid arthritis: a report of 40 cases. J Bengbu Med Coll 32(3):307–308 (In Chinese)

Zhang RX, Fan AY, Zhou AN et al (2009) Extract of the Chinese herbal formula Huo- Luo-Xiao-Ling- Dan inhibited adjuvant arthritis in rats. J Ethnopharmacol 121(3):366–371. doi:10.1016/j.jep.2008.11.018

Zhang P, Li J, Han Y et al (2010) Traditional Chinese medicine in the treatment of rheumatoid arthritis: a general review. Rheumatol Int 30(6):713–718. doi:10.1007/s00296-010-1370-0

Zhang C, Jiang M, Lu A (2011) A traditional Chinese medicine versus western combination therapy in the treatment of rheumatoid arthritis: two-stage study protocol for a randomized controlled trial. Trials 12:137. doi:10.1186/1745-6215-12-137

Zhang Y, Bai M, Zhang B et al (2015) Uncovering pharmacological mechanisms of Wu-tou decoction acting on rheumatoid arthritis through systems approaches: drug-target prediction, network analysis, and experimental validation. Sci Rep 5:9463. doi:10.1038/srep09463

Zhao ZZ, Liang ZT, Zhou H et al (2005) Quantification of sinomenine in *Caulis sinomenii* collected from different growing regions and wholesale herbal markets by a modified HPLC method. Biol Pharm Bull 28(1):105–109

Zhao J, Zha Q, Jiang M et al (2013) Expert consensus on the treatment of rheumatoid arthritis with Chinese patent medicines. J Altern Complement Med 19(2):111–118. doi:10.1089/acm.2011.0370

Zhao QH, Li XY, Feng Q et al (2015) Evaluation of the dosage-based efficacy-toxicity correlation of *Tripterygium wilfordii* against immune inflammation in mice. Zhongguo Zhong Yao Za Zhi 40(6):1139–1143 (In Chinese)

Zhou DH, Zhou ZQ, Zhu WH (2007) Zhu Liangchun's experience prescription in 38 cases treatment of rheumatoid arthritis. J New Chin Med 39(9):71–72 (In Chinese)

Zhu DL, Zeng DZ, Wang BH et al (2011) Observation of curative effects of Chinese medicine bath on rheumatoid arthritis. Mod J Integr Tradit Chin West Med 20(7):816–817 (In Chinese)

Chapter 17
Medicinal Plants Used in the Management of Noncommunicable Diseases in Uganda

Esezah Kakudidi, Claude Kirimuhuzya, Godwin Anywar, Esther Katuura, and Juliet Kiguli

Abstract Noncommunicable diseases (NCDs) are recent and growing health problem in Uganda. The NCDs epidemic is burdening the healthcare systems, which is already under pressure from the high prevalence of communicable diseases. Hypertension is the most common NCD in Uganda; more females than males suffer from NCDs. High blood pressure and heart disease equally affect 5.3 % of the female population, while they affect 2.4 % and 2.6 % males, respectively. Cancers of the prostate and cervix are ranked number one in men and women, respectively. Traditional herbal medicine remains the most utilized form of healthcare. With the emergence of various NCDs, the services of traditional medical practitioners (TMPs) are set to rise. We collated 42 medicinal plants from literature used in the treatment of NCDs, of which 20 (47.6 %) are used in the management of hypertension, an indicator of its prevalence. Seven priority species were also identified for various NCDs by TMPs. The Uganda government realizing the importance of traditional medicine in primary healthcare established the Natural Chemotherapeutic Research Institute to undertake research on medicinal plants used by TMPs with the aim of justifying the therapeutic claims. Research on medicinal plants is still faced with the challenge of funding and collaboration between institutions to harness synergies towards the gradual integration into modern healthcare systems. The Ministry of Health needs to invest in training professional health providers and TMPs and public sensitization using targeted messages on prevention and management of NCDs, as was done for the HIV/AIDS pandemic in Uganda.

E. Kakudidi (✉) • G. Anywar • E. Katuura
Department of Plant Sciences, Microbiology and Biotechnology, College of Natural Sciences, Makerere University, 7062, Kampala, Uganda
e-mail: ekakudidi@cns.mak.ac.ug; esezahk@gmail.com

C. Kirimuhuzya
Western Campus, Department of Pharmacology, Kampala International University, 71, Bushenyi, Uganda

J. Kiguli
Department of Community Health & Behavioral Sciences, College of Health Sciences, Makerere University, 7062, Kampala, Uganda

H.-S. Tsay et al. (eds.), *Medicinal Plants - Recent Advances in Research and Development*, DOI 10.1007/978-981-10-1085-9_17

Keywords Medicinal plants • Noncommunicable diseases • Traditional medical practitioners • Uganda

Abbreviations

ANAMED	Action for natural medicine
BUMETHA	Bushenyi Meditrad Healers' Association
CDL	Chronic diseases of lifestyle
CDs	Chronic diseases
DOB	Department of Botany
HIV/AIDS	Human immunodeficiency virus/acquired immunodeficiency syndrome
IPR	Intellectual property rights
ITFC	Institute of Tropical Forest Conservation
IUPAC	International Union of Pure and Applied Chemistry
MS	Mass spectrometry
NACOTHA	National Council for Traditional Healers and Herbalists Association
NAPRECA	Natural Products Research Network for Eastern and Central Africa
NCDs	Noncommunicable diseases
NCRI	Natural Chemotherapeutic Research Institute
NDA	National Drug Authority
NEMA	National Environmental Management Authority
NMR	Nuclear magnetic resonance
SSA	Sub-Saharan Africa
TCM	Traditional and complementary medicine
THETA	Traditional and Modern Health Practitioners Together Against AIDS
THM	Traditional herbal medicines
TM/CAM	Traditional medical/complementary alternative medicine
TMK	Traditional medicinal knowledge
TMPs	Traditional medical practitioners
UBoS	Uganda Bureau of Standards
UMPG	Uganda Medicinal Plant Growers Ltd.
UNESCO	United Nations Educational Scientific and Cultural Organisation
UNHRO	Uganda National Health Research Organization
WHO	World Health Organization

17.1 Introduction

Uganda is a landlocked country in the eastern part of Africa. It is surrounded by Kenya in the east, Sudan in the north, Tanzania and Rwanda to the south, and the Democratic Republic of Congo to the west. Uganda's climate is shaped by the inter-tropical convergence zone and air currents such as the southeast and northeast monsoons, with two main seasons: the rainy and dry seasons (NEMA 2010).

Uganda covers a total area of 2,415,505.7 Km^2 (UBoS 2014) with an estimated population of 35 million people who are mostly engaged in subsistence agriculture. Uganda also has a high population growth rate of 3.2 % p.a., the second highest in Africa (UBoS 2010). The high population puts pressure on the natural resources, including arable land, which in turn drives up the poverty rate and threatens future gains in agricultural production and food security (NEMA 2010). Uganda has been recovering since 1986 from previous civil, political, and economic turmoil. The Annual Gross National Income is $300 per capita, and mean life expectancy at birth is 50 years on average (UBoS 2015). It is one of the countries in Africa where the HIV epidemic was first reported and was initially most badly affected by HIV (Maher et al. 2011). Furthermore, both medical professionals and lay people assert that noncommunicable diseases (NCDs) particularly hypertension and diabetes are recent, widespread, and growing health problems in Uganda (UBoS 2010; Whyte 2014).

Noncommunicable diseases (NCDs) are sometimes referred to as lifestyle disease or chronic diseases. MedicineNet.com refers to lifestyle disease as being associated with the way a person or group of people lives. Lifestyle diseases have been defined as diseases that are not passed from person to person and are of long duration and slow progression (WHO 2015). The South African Medical Research Council has defined chronic diseases of lifestyle (CDL) as "a group of diseases that share similar risk factors because of exposure, over many decades, to unhealthy diets, smoking, physical inactivity, harmful use of alcohol and possibly stress" (Steyn et al. 2006; Schwartz et al. 2014). The four main types of NCDs are cardiovascular diseases (e.g., heart attacks and stroke), cancers, chronic respiratory diseases (such as chronic obstructed pulmonary disease and asthma), and diabetes (WHO 2013, 2015). Many of these NCDs are associated with urbanization and Western lifestyles, as well as improvements in life expectancy in sub-Saharan Africa (Parkin et al. 2010; Kavishe et al. 2015). Many people are increasingly becoming exposed to high-fat, high-salt, and high-calorie diets, which are low in fiber, coupled with increasingly sedentary lifestyles (Amuna and Zotor 2008; Guthold et al. 2011).

The epidemic of diseases of lifestyle has led to a rapid change in the disease profile of the world, especially in low- and middle-income countries. The World Health Organization (WHO) has reported that NCDs are currently causing more deaths than all other causes combined (WHO 2014). The report indicates that these deaths were projected to increase from 38 million in 2012 to 52 million by 2030 (WHO 2014). The same report further shows that cardiovascular diseases, cancer, chronic respiratory diseases, and diabetes are the major NCDs responsible for 82 %

of the deaths from NCDs globally (WHO 2014) and that low- and middle-income countries take about three quarters (28 million) of the NCD deaths (WHO 2014; Hughes et al. 2015). Cardiovascular diseases account for most NCD deaths with up to 17.5 million people dying annually, followed by cancers (8.2 million), respiratory diseases (4 million), and diabetes (1.5 million) WHO (2015).

The current socioeconomic, cultural, political, and environmental determinants of NCDs that have been reported to have a significant influence are urbanization, globalization, and population aging (WHO 2008). In Uganda, the situation is complicated by lifelong antiretroviral treatment (UBoS 2010).

17.2 Status of Noncommunicable Diseases (NCDs) in Uganda

Data on the burden of NCDs in sub-Saharan Africa (SSA) are limited, and only a few representative community-based studies have been conducted (Dalal et al. 2011). Hypertension is among the world's commonest NCD affecting as much as 20 % of the world's adult population (Osamor and Owumi 2010). Hypertension is the most common NCD in Uganda by various studies including Nuwaha and Musinguzi (2013) and Kavishe et al. (2015). According to UBoS National Household Survey (2009/2010), more females than males suffer from NCDs. High blood pressure and heart disease equally affect 5.3 % of the female population, while they affect 2.4 % and 2.6 %, respectively, in the male population. In Uganda, prostate cancer ranked number one in men (Parkin et al. 2010; Global Burden of Disease Cancer Collaboration 2015). The number of new prostate cancer cases in Uganda increased from 850 in 1990 to 4400 in 2013, whereas cervical cancer increased from 2000 to 3400 in the same period. Parkin et al. (2010) reported an overall increase in cancer in Kampala between 1991 and 2006, with breast and prostate cancers particularly increasing by 4.5 % annually. Prostate and cervical cancers are now the leading causes of cancer morbidity and mortality in men and women. Cervical cancer claimed 2300 lives in women in 2013, up from 1400 in 1990. On the other hand, prostate cancer deaths have nearly tripled, from 390 in 1990 to 1100 in 2013 in Uganda (Global Burden of Disease Cancer Collaboration 2015).

Doctors and government officials in Uganda are becoming increasingly concerned at the country's growing incidence of diabetes, which they attribute to changes in lifestyle and rising obesity. The total number of sufferers is 4 % of the total population or 1,120,000 patients. In 2006, the Uganda Diabetic Association registered 560,000 people with diabetes, but an additional 560,000 patients may have the disease but are unaware of it (Wasswa 2006). There were 693,200 cases of diabetes in Uganda in 2014 with 17,570 adult deaths according to the International Diabetes Federation (http://www.idf.org/membership/afr/uganda accessed January 12th 2016 at 14:06 pm). The threat of cardiovascular disease is serious, yet few reliable data are available. It has resulted in neglecting serious warning signs on the emergence of cardiovascular disease in sub-Saharan Africa (van der Sande 2003). In a cross-sectional population-based survey of cardiovascular disease in southwestern Uganda, the commonest cardiovascular disease risk factor was high blood pressure, with an observed prevalence of 22.5 % in both sexes (Maher et al. 2011). UBoS (2010) report indicated that the rural population is more affected by heart disease than the urban with 4.3 % and 2.3 %, respectively, while 2.6 % of males are affected compared to 5.3 % of females. The report indicated an early start of heart disease under 15-year-olds (0.5 %) to 8.4 % in adults 45+ years.

17.3 The Health Sector Response to NCDs

In 1972, only 254 people in Uganda were given a diagnosis of diabetes. They attended the country's only diabetes clinic, at Mulago, the national referral hospital n Kampala. However, in 2006, ten centers handled diabetic patients countrywide with type 1 diabetes accounting for about 8 % of the total number of patients with diabetes (Wasswa 2006). Due to the realization that diabetes was acute among the cattle-keeping peoples of the southwestern Uganda, 20 hospitals were chosen to handle the growing number of patients (Wasswa 2006). The numbers are on the rise such that three hospitals in major towns in the southwestern region handled between them 2400 patients with diabetes (Wasswa 2006). On the contrary, none of the facilities examined in 54 centers across Uganda met WHO standards for essential tools and medicines needed to implement effective NCD interventions (Schwartz et al. 2014). On the civil society front, the Uganda Diabetes Association has 15 branches throughout the country, with 13,000 registered members (Whyte 2014).

The surveillance and monitoring of NCDs in East Africa have improved, although NCDs are not sufficiently integrated into national health information management systems. Most health personnel are also limited in their capacity for NCD surveillance and data collection (Schwartz et al. 2014). In Uganda, for hypertension and type 2 diabetes, existing interventions are few and modestly funded but linked to the national healthcare system. The Uganda Initiative for Integrated Management of NCDs received support from the World Diabetes Foundation and Yale Global Health Leadership Institute (http://uincd.org). On the other hand, the Uganda NCD Alliance was funded by the Danish Agency for International Development as a collaboration of patient and professional groups. Also, Novo Nordisks' (a pharmaceutical company) Changing Diabetes in Children program provides free insulin to children with type 1 diabetes, while Quality Chemicals in Kampala (a daughter company of Cipla and manufacturer of metformin for diabetes) has supported occasional training of government health workers (Whyte 2014). Additionally since 2014, C_3 Collaborating for Health, another initiative funded by the Burdett Trust for Nursing, has been partnering with the Uganda Diaspora Health Foundation in the UK and the Uganda NCD Alliance to raise awareness among nurses and the public about NCD prevention and to highlight their links with mental health and mental disorders. None of these initiatives is comprehensive in the sense of reaching the majority of people with these conditions, and there is far more attention to diabetes than cardiovascular diseases (Whyte 2014).

17.4 Role of Traditional Medical Practitioners (TMPs) in the Management of NCDs

Traditional herbal medicines, due to their availability, accessibility, acceptability, and adaptability to the community (Anyinam 1987), remain the most utilized form of healthcare in the developing world (WHO/EDM/TRM 2002). The use of alternative and herbal medicine for the management of NCDs is common and on the increase in both developing as well as developed countries (The World Bank Human Development Network 2008; Nuwaha and Musinguzi 2013). Consequently, many of the affected people in these countries depend on alternative therapies such as traditional herbal medicines (THM) (Hughes et al. 2015). There are also reports of some physicians recommending herbal treatment to their patients, especially in cases where allopathic medicine has been found to be ineffective (Nugent 2008; Kiringe 2006). It is not surprising because various uses of many traditional herbal products have been scientifically validated (Hughes et al. 2015). More than 30 % of modern medicines including those that are used for the treatment of NCDs have their origin from plants used by tribal people. Those valued in modern medicine include anticancer medicines such as vincristine and vinblastine; antihypertensive agents such as reserpine, deserpidine, ajmalicine, and rescinnamine; and decongestants such as ephedrine (Farnsworth and Soejarto 1991; Trease and Evans 2009). Many other standardized herbal medicines and galenicals are continually being used.

17.5 Medicinal Plant Use in the Management of NCDs in Uganda

There is evidence of increasing number of people using herbal medicines to treat NCDs across many low-, medium-, and even high-income countries (Osamor and Owumi 2010). With the emergence of various NCDs in Uganda, the use of medicinal plants and services is set to rise correspondingly. Further evidence comes from the ever-increasing number of herbalists in both the print and nonprint media advertising their products for treating various diseases inclusive of hypertension and diabetes (The New Vision http://www.newvision.co.ug/new_vision/news/1083181/nda-warns-herbalists accessed on 17th January 2016 at 18:05 pm) and cancers (Parkin et al. 2010). There are also many herbal products both local and imported used in the management of NCDs. Presented in Table 17.1 are medicinal plants that

Table 17.1 Documented medicinal plants traditionally used in the management of NCDs in Uganda

Medicinal plant	Family	Methods of preparation	Reference
Diabetes			
Steganotaenia araliacea	Apiaceae	Leaf decoction drunk	Ssegawa and Kasenene (2007)
Heart disease			
Brachiaria decumbens	Poaceae	Leaf decoction drunk/chew fresh leaves	Namukobe et al. (2011)
Ficus natalensis	Moraceae	Leaf decoction drunk/chew fresh leaves	Namukobe et al. (2011)
Hydnora abyssinica	Hydnoraceae	Whole plant	Oryem-Origa et al. (2001, 2003)
Euphorbia cotinifolia	Euphorbiaceae	Leaves	Nawaggi et al. (2014)
Hypertension			
Amaranthus spinosus	Amaranthaceae	The whole plant is steamed and eaten	Tabuti et al. (2003)
Citrus limon	Rutaceae	Root bark	Tugume et al. (2016)
Chamaecrista nigricans	Leguminosae	Root eaten with *Sesamum indicum*	Tabuti et al. (2003)
Gomphocarpus physocarpus	Apocynaceae	Eaten with *Arachis hypogea* paste	Tabuti et al. (2003)
Hibiscus sabdariffa	Malvaceae	Leaves	Katende et al. (1999) and Asiimwe et al. (2013)
Hoslundia opposita	Lamiaceae	Roots and leaves	Oryem-Origa et al. (2001, 2003)
Mentha aquatica	Lamiaceae	Dried leaf taken as tea	Namukobe et al. (2011)
Milicia excelsa	Moraceae	Eaten with *Arachis hypogea/Sesamum indicum* paste	Tabuti et al. (2003)
Musa sp.	Musaceae	Fresh leaf/flower decoction drunk	Namukobe et al. (2011)
Persea americana	Lauraceae	Fresh leaf/seed infusion/decoction	Namukobe et al. (2011)
Senna didymobotrya	Leguminosae	Root or whole plant infusion or decoction	Tabuti et al. (2003)
Sesamum calycinum subsp. *angustifolium*	Pedaliaceae	*Arachis hypogaea* L. paste	Tabuti et al. (2003)

(continued)

Table 17.1 (continued)

Medicinal plant	Family	Methods of preparation	Reference
Tinnea aethiopica	Lamiaceae	Eaten with *Arachis hypogaea* L./*Sesamum indicum* L. paste	Tabuti et al. (2003)
Tylosema fassoglensis	Leguminosae	Flower infusion drunk	Tabuti et al. (2003)
Oxalis corniculata	Oxalidaceae	Leaves are chewed	Tugume et al. (2016)
Sesbania sesban	Leguminosae		Tugume et al. (2016)
Vangueria apiculata	Rubiaceae		Tugume et al. (2016)
Zanthoxylum gilletii	Rutaceae	Fresh stem bark infusion	Namukobe et al. (2011)
Platycerium elephantotis	Polypodiaceae	Whole plant	Kakudidi et al. (2000)
Anemia			
Justicia laxa	Acanthaceae	Leaf	Oryem-Origa et al. (2001, 2003)
Hibiscus sabdariffa	Malvaceae	Leaves	Katende et al. (1999) and Asiimwe et al. (2013)
Epilepsy			
Cussonia arborea	Araliaceae	Bark	Oryem-Origa et al. (2001, 2003)
Physalis peruviana	Solanaceae	Leaf powder infusion	Anywar et al. (2014)
Physalis minima	Solanaceae	Leaf powder infusion	Anywar et al. (2014)
Mental illness			
Albizia glaberrima	Leguminosae	Roots and leaves	Oryem-Origa et al. (2001, 2003)
Coccinia sp.	Cucurbitaceae	Roots and leaves	Oryem-Origa et al. (2001, 2003)
Erythrina exelsa	Leguminosae	Roots and bark	Oryem-Origa et al. (2003)
Indigofera arrecta		Leaves and roots	Oryem-Origa et al. (2001, 2003)
Secamone africana	Apocynaceae	Roots	Oryem-Origa et al. (2001, 2003)

(continued)

Table 17.1 (continued)

Medicinal plant	Family	Methods of preparation	Reference
Cancer			
Markhamia lutea	Bignoniaceae	Flowers	Oryem-Origa et al. (2001, 2003)
Cyperus alatus	Cyperaceae	Rhizomes	Oryem-Origa et al. (2001, 2003)

have been documented in the treatment of various NCDs in Uganda. Many of these species used have other ethnomedicinal uses (Lye et al. 2008).

Of the 42 documented Ugandan medicinal plant species used in the management of NCDs, 20 (47.6 %) are used in the management of hypertension, an indicator of how common the disease is. Hypertension is followed by mental illness (five species), epilepsy (three species), heart diseases (four species), cancer (two species), and diabetes and anemia (one species each). In 2010, the National Conservation Assessment and Management Planning Workshop for Priority Medicinal Plants in Uganda (CAMP) identified 50 species of medicinal plants that were selected for conservation action (CAMP 2010). Out of the 50, 7 are used in the management of different NCDs, with documented evidence of pharmacological activity and bioactive phytochemical constituents (Table 17.2).

17.6 Challenges in the Use of TM in the Management of NCDs

Awareness among people with hypertension and other chronic diseases (CDs), particularly in rural areas in Uganda, is low (Kavishe et al. 2015), and as such, people do not get proper treatment in time. Researchers have documented instances where patients combine herbal medicines with Western medicines while treating different lifestyle diseases (Nuwaha and Musinguzi 2013). Some patients do not believe TM to be a complete treatment of various diseases including lifestyle diseases. For instance, Nuwaha and Musinguzi (2013) reported up to 50 % of the 258 patients in one study in Mukono and Buikwe districts in central Uganda using both herbal and Western medicine to treat hypertension. This practice has also been observed at Uganda's national referral hospital and other health units. However, some health workers may not be aware that some patients use alternative medicine in conjunction with the Western medicines prescribed (Eisenberg et al. 1993; Amira and Okubadejo 2007). Like in many parts of the world, Uganda policy makers, health professionals, and the public are still skeptical about the use of TM. They wrestle with questions of safety, efficacy, quality, preservation, and further development. To a large extent, it is associated with the lack of evidence-based efficacies and safety

Table 17.2 Priority medicinal plants in the management of NCDs in Uganda and their phytochemical composition and pharmacological activities

S. no	Plant species (family)	NCD treated	Phytochemical constituents	Pharmacological activities
1	*Prunus africana* (Rosaceae)	Cancer	Phytosterols, e.g., β-sitosterol, tannins, linear aliphatic alcohols, ferulic acid esters, etc. (van Wyk and Wink 2004)	Benign prostatic hyperplasia (BPH) (Edgar et al. 2007), urinary symptoms (Wilt et al. 2002)
2	*Canarium schweinfurthii* (Burseraceae)	Cancer, anemia	Phenolic compounds such as limonene, sabinene, α-pinene, terpenoids such as canarene (Engonga et al. 2012; Atawodi 2010; Edou et al. 2012)	Antioxidant, radical scavenging, anticancer, and antidiabetic activity (Atawodi 2010; Kamtchouing et al. 2006)
3	*Kigelia africana* (Bignoniaceae)	Diabetics	Naphthoquinones, iridoids, fatty acids, sterols, lignans, terpenoid, and flavonoids (Olatunji and Atolani 2009)	Antimicrobial activity (Jeyachandra and Mahesh 2007), antidiarrheal, antimalarial, anti-inflammatory, and anticancer activity (Atawodi and Olowoniyi 2015)
4	*Solanum anguivi* (Solanaceae)	Hypertension	Solamargine, steroid alkaloid glycosides, anguivine and isoanguivine (Ripperger and Himmelreich 1994). Saponins (Olusola et al. 2013)	Antiproliferative activity (Gandhiappan and Rengasamy 2012), hypoglycemic (Grubben and Denton 2004), antioxidant and radical scavenging activities (Olusola et al. 2013)
5	*Pseudospondias microcarpa* (Anacardiaceae)	Diabetes, hypertension	Saponins, phenols, terpenoids, flavonoids, cardiac glycosides, and coumarins (Adongo et al. 2015)	Antimalarial (Lacroix et al. 2011; Chinsembu et al. 2015)
6	*Toddalia asiatica* (Rutaceae)	Asthma, rheumatism	Coumarins such as toddaculin (Vazquez et al. 2012), alkaloids (Hu et al. 2014)	Antimicrobial and antifungal activities (Hu et al. 2014)
7	*Piptadeniastrum africanum* (Leguminosae)	Mental disorders	Triterpenoid saponins, alkaloids, tannins, glycosides, flavonoids, sterols (Note et al. 2013; Brusotti et al. 2013)	Antimicrobial and anthelmintic (Brusotti et al. 2013)

associated with herbal medicine or the complexity of its validity measures (Tilburt and Kaptchuk 2015).

Increased use of TM in the management of NCDs has not been accompanied by adequate clinical or empirical research-based findings in Uganda to prove the therapeutic effects of the herbal drugs. However, there are ongoing efforts through Traditional and Modern Health Practitioners Together Against AIDS (THETA), a nongovernment organization, Natural Chemotherapeutic Research Institute (NCRI), and the academia to bridge the gap between TMPs and modern health providers to collaborate in healthcare provision. Other local organizations involved in mobilizing traditional healers for proper management of medicinal plant resources include "Action for Natural Medicine" (ANAMED), Uganda N'edaggala Lyaayo, and National Council for Traditional Healers and Herbalists Association (NACOTHA), PROMETRA-Uganda, and Bushenyi Meditrad Healers' Association (BUMETHA), plus many others.

17.7 Research on the Management of NCDs with Traditional Medicine

In Uganda, before colonialism, traditional medicine (TM) was the only source of healthcare for the people (Mugumya 1997). The society had many ways of combating disease by the use of different practices, such as herbalists, spiritualists, diviners and magicians, bone-setters, and traditional birth attendants (Anokbonggo 1992), all of whom were highly respected. TM was neglected, outlawed, and relegated as witchcraft (Uganda Legal Information Institute, Witchcraft Act 1957, Chapter 124 (http://www.ulii.org/ug/legislation/consolidated-act/124 accessed on Mon 18 January 2016 14:00 PM) during the colonial times, until the Uganda Constitutional Court decision declared the colonial Witchcraft Act Cap.124 unconstitutional in 1997 (Kyomugisha 2008). This Act, which sought to regulate practices involving the use of supernatural powers, did not properly define witchcraft since it never differentiated between genuine traditional healers and witches.

Biomedicine came into the picture in Uganda at the beginning of the twentieth century, when it immediately developed a strong holding, especially in the urban. Biomedical doctors from the onset made little attempt to liaise with TMPs. Still today there is very little interaction between TM and biomedicine. The two systems operate independently. However, TM plays a pivotal role in the healthcare of more than 80 % of Uganda's population that live in remote areas where they cannot access orthodox medical services (Kakudidi et al. 2000; WHO/EDM/TRM 2002; Kamatenesi and Oryem-Origa 2005). TM is quite popular in Uganda because the majority of the rural and a sizeable urban population rely on it. They believe that it is sometimes more effective than biomedicine, and it had added the advantage of precipitating fewer side effects and cost less. It was against this that the WHO argued nation-states to promote research on TM and make it accessible and affordable (WHO/EDM/TRM 2002).

African participants at the 14th International Union of Pure and Applied Chemistry's (IUPAC) International Symposium on the Chemistry of Natural Products in Poland in 1984 expressed the need for research in natural products in Eastern and Central Africa. After that, the Natural Products Research Network for Eastern and Central Africa (NAPRECA) became a reality with a coordinating office in Addis Ababa, Ethiopia, with branches in seven other countries including Uganda (NAPRECA 1995). NAPRECA received financial support from UNESCO. The main objectives were (i) to initiate, develop, and promote research in natural products in Eastern and Central Africa subregion, (ii) to coordinate and maintain inter- and intra-regional links among different research groups, (iii) to disseminate information pertaining to natural products research, and (iv) to foster and maintain links with scientists in other parts of the world who are actively working in the specific areas of natural products that are pertinent to Africa. To ameliorate problems of isolation and lack of contact with each other as well as with peers elsewhere, NAPRECA has been organizing symposia since 1988 in different member states, with Uganda hosting one in 1995 in Kampala. Like other NAPRECA branches, Uganda branch, coordinated from Makerere University, has continued to produce quality research involving both academia and young researchers in modern nuclear magnetic resonance (NMR) techniques, mass spectrometry (MS), and organic synthesis.

NAPRECA, however, was not addressing the concerns raised on accessibility and affordability, efficacy, and safety of TM in developing countries, and yet TM seekers are increasing (Kakudidi et al. 2000; WHO/EDM/TRM 2002; Hamilton 2008). The Uganda government realized the importance of TM as a key contributor in providing primary healthcare (MoH 2010). Uganda started to promote research and conservation of medicinal plants under the Natural Chemotherapeutic Research Institute (NCRI) since 1964. The institute is mandated to undertake research in natural products used by traditional medicine practitioners in Uganda with the view of justifying therapeutic claims and developing quality natural products and services for improved healthcare. Related research is mostly done at Makerere University and Mbarara University of Science and Technology. The research involves documentation of uses, collecting of herbarium voucher specimens for reference, analysis to establish biological/pharmacological activity of the crude extracts, and molecular characterization.

Once a traditional health practitioner has identified herbal medicine that has proven effective, the TMP may take it to NCRI, providing information on contents and effects (CCFU 2008). The unprocessed product is then tested. The samples are all analyzed for standard therapeutic phyto-components and toxic substances. The formula is also subjected to toxicity tests in rats or mice. In cases where the validation is positive, TMPs are encouraged once their products are notified by National Drug Authority (NDA).

17.8 Challenges in TM Research

The custodians of traditional knowledge are mainly the uneducated that still hold information in secrecy, which in turn has had a negative effect on standardization of the products and services. The protection of intellectual property rights (IPR) is another major constraint. It is because TM has been disregarded and even repressed during colonial and postcolonial periods. TMPs are suspicious of the motive of those who have now rather suddenly developed an interest in their knowledge. A limited understanding of patenting and IPR also intensifies the reluctance to disclose knowledge for fear of exploitation that this may be stolen (CCFU 2008). Characterization of the main active ingredients of medicinal plants is minimal; and yet the active compound modifications are needed towards drug development. Highly trained scientists in areas such as safety, biopharmaceuticals, clinical trials, and pharmacovigilance that would promote the underexploited TM and TMK systems are needed to assist in the integration of TM/CAM into conventional medicine. Working in isolation is a problem within and between institutions; therefore, planned research agenda would help to bring out priority areas and synergies in generating relevant information. Funding is also a major constraint as well as the lack of infrastructure including accredited laboratory facilities and modern equipment.

17.9 Challenges in Marketing Herbal Products for NCDs in Uganda

In the past, TMPs would collect herbal medicines from the wild, prepare, and administer to patients without the need to package. In the recent past where herbals have entered trade, and therefore administered far from the point of collection, it has necessitated them to be packaged. Many TMPs use locally available materials for packaging so that they can keep the prices low. These materials are often of very poor quality and unhygienic making the products spoil after a short time. Consequently, this made many people shun TM. It was worsened by the unspecified dosages. However, some companies have come up with new and innovative ways of processing and packaging herbal medicines, addressing some of these challenges. These include Green Herbs, Kazire Health Products, Sefa Organics, Aloesha, and Quality Chemicals, among others. However, the marketability of many herbal medicines is still faced with many challenges:

(i) The low socioeconomic status of most developing countries including Uganda in which the majority of the population does not understand the causes, symptoms, and management of NCDs (White 2002) makes it problematic to market products that would prevent or treat these diseases.

(ii) Conflicting scientific information on epidemiological, toxicological, and other therapeutic and safety data and the verification of the herbal materials being sold.
(iii) Drug interactions and antagonistic and synergistic harmful works are not clear both to the consumers and scientists for most herbs. It makes marketability of such products cumbersome.
(iv) The packaging of many locally made natural health products in Uganda is still substandard partly because of inadequate technology and lack of locally available and affordable standard packaging materials, thus making the products quality and presentation less attractive than imported products and affecting the general demand. Moreover, even then, artemisinin which is the herbal medicine derived drug is packed using imported materials.
(v) There are many illegal and counterfeit herbal products since not all the products on the market are regulated by NDA as they do not satisfy the standards set by WHO. The registration process is cumbersome and expensive for many TMPs.
(vi) For many Ugandans who believe in the effectiveness of the herbal products (Tabuti 2008), they prefer buying them from trusted sources or persons rather than obtaining them from the open market.
(vii) There are excessive claims and advertisement in mass media. It is not unusual in Uganda to see an advert of a "doctor" offering the most outlandish services (Nassaka 2014). Some TMPs give false information and end up misleading the public or are not able to satisfactorily explain the therapeutic benefit of the products to the clients. As a result, TM is a gray area in which the public remains suspicious. Some TMPs who are seen as charlatans and sorcerers have been lynched by angry mobs.
(viii) Some of the herbal products that would otherwise be efficacious are adulterated with prohibited materials/chemical substances.
(ix) Counterfeit products are also presented in the market. Since most herbal medicine clients are by referrals, failure of a product to treat the claimed condition would result in future clients shunning the product.
(x) There is a lack of organized market structure and limited market information for many medicinal plants in Uganda, as well as limited knowledge of chemical properties and appropriate technology to add value.

17.10 Conservation of Medicinal Plants

Uganda has a rich diversity of plant species estimated at 5000 (IUCN 1990). The interest of medicinal plants in Uganda's local communities is strong. As noted by Hoareau and DaSilva (1999), the future of medicinal plant species is threatened by complacency concerning their conservation. Hoareau and DaSilva (1999) assert that the reserves of many medicinal plants in developing countries are diminishing and in danger of extinction due to growing demands and preference for new plant-based therapeutic products instead of expensive target-specific drugs and biopharmaceuticals. Many medicinal plants are becoming scarce in Uganda given their ever-dwindling natural habitats. The existence of TM depends on the medicinal plants' availability (Kakudidi et al. 2000; WHO/EDM/TRM 2002; Hamilton 2008; Agea et al. 2008).

In spite of the desire of many TMPs to source material from the wild, projects linking conservation and people with particular reference to medicinal plants have had several important ex situ initiatives. Entebbe Botanic Garden maintains an herbal garden of 122 spp.; Makerere University Botanic Garden, 100+spp.; Tooro Botanic Gardens, 200 spp.; and the Institute of Tropical Forest Conservation (ITFC) – located in Bwindi National Park, under Mbarara University of Science and Technology – 100 spp.. Uganda's herbal medicine genetic resources require more attention than currently given, in particular concerning ex situ conservation. Some organizations, such as PROMETRA-Uganda, SEFA Organic, Tooro Botanic Gardens, BUMETHA, Joint Ethnobotanical Research and Advocacy, Nature Uganda, Uganda Medicinal Plant Growers Ltd. (UMPG), etc., that grow medicinal plants for products formulation are also involved in preserving the environment as well as reducing wild resources overexploitation. Ex situ conservation of medicinal plants, which was the concern of the Uganda National Conservation Assessment and Management Planning (CAMP) workshop (DOB 2010), has not been fully embraced by many TM providers who still believe that the medicines from the wild are more effective (Kamatenesi-Mugisha and Oryem-Origa 2005). A natural response to the unavailability of local herb species is to cultivate them for easy access. Cultivation would also be a long-term remedy to conserve the rare and commonly used species and the environment.

The lack of data on the conservation status and the volumes involved in the trade of many medicinal plant species that would inform on the impact of plant species concerned is a serious deficiency in their conservation in Uganda (Mujuni 2014). Already *Citropsis articulata*, the male sex enhancement tree, is overharvested from the wild (Okeowo 2007). The loss of this and many others would not only do irreversible damage to their habitats but would also deprive scientists of the opportunity to study the plants' possible medicinal properties (Kamatenesi-Mugisha and Orem-Origa 2005). The wild resources of *M. whytei*, one of the old traded medicinal plant in most parts of Kampala city with estimated volume of over 1 ton of roots consumed every month, are dwindling (Agea et al. 2008).

In situ initiatives targeting TMPs exist countrywide that seek to ensure sustainable supplies of medicinal plants for the benefit of healthcare and livelihoods (Hamilton 2008). TMPs are sensitized on the management of wild medicinal plants, such as harvesting techniques, collection of seeds, and raising seedlings for nurseries, propagation, and cultivation in home gardens or wild enrichment (Hamilton 2008). There is increased pressure on biodiversity globally including medicinal plants due to overexploitation, poor harvesting techniques, climate change, increased population, urbanization, and modernization, among many other issues (Hawkins 2008). Many collectors continue to harvest medicines from the wild, a situation that is common in Uganda. A number of medicinal plant users have reported scarcity of medicinal plants that used to be common in the past (Katuura et al. 2007). For example, *Prunus africana* for treating prostate cancer is endangered in the wild. It is already on the CITES listings due to the risk of overharvesting (Lange 1997). Increased demand of herbal products on the Ugandan market has aggravated the problem. It is a challenge since there is no specific government conservation intervention targeting medicinal plants except in protected areas when other key species of conservation importance such as the mountain gorilla in Bwindi Impenetrable National Park and the chimpanzees and elephants in Kibale National Park are targeted.

17.11 Policy Issues

Since independence, TM has gradually gained increasing recognition in East Africa, especially with the declaration by the African Union of 2001–2010 as "The Decade of Traditional African Medicine." An East African Network of Traditional Medicine and Medicinal Plants was established in 2007 that is hosted in the Lake Victoria Basin Commission for the East African Community. While a traditional healer policy was established for Tanzania in 2002, Uganda has lagged behind since all her policies on the regulation of TM have not yet been passed by parliament. However, the TMPs are required to register with the Ministry of Health under the Traditional and Complementary Medicine (TCM) Council that promotes, controls, and regulates TCM practice as established by the relevant law.

The National Drug Authority (NDA) has guidelines for the regulation of traditional herbal medicines in Uganda. Between 2005 and 2007, NDA carried out sensitization workshops in the country focusing on the regulation requirements of herbal medicine products, but not all TMPs were sensitized. Consequently, many herbal medicinal products are not notified by NDA. Recorded data about the operations of herbalists was greatly missing throughout the African continent (Sofowora 1993). In Uganda, TMPs have formed associations to represent their collective interests for legitimate practice. However, not all are registered due to the registration fees charged, suspicion, and uncertainty of the benefits they would get by registering and collaborating with other stakeholders (Kakudidi et al. 2015). In liaison with NCRI, NDA regulates use, promotion, and protection of indigenous knowledge and intellectual property rights of TM under the 1993 statute that provides minimum requirements for the registration of herbal medicine products for introduction into the market/commercial use. The statute influenced the development of the 2012 Public Private Partnership for Health Policy that provides for promotion and regulation of authentic, acceptable, safe, and ethical traditional and complementary medicine practice.

17.12 Discussion

The NCD epidemic is of great significance in the healthcare systems in the affected countries which are already under considerable pressure from the high prevalence of communicable diseases such as malaria, tuberculosis, and HIV/AIDS. On top of that, these have to contend with the burden of NCDs (Miranda et al. 2008; Mayosi et al. 2009). NCDs increase with age (UBoS National Household Survey 2009/2010) but go undetected especially in children, since many health workers do not consider this age group to suffer from them. People below 35 years often consider themselves not at risk from NCDs; by the time they are detected, the damage has already been done.

The health sector has made several interventions in surveillance and monitoring of NCDs. THETA's engagement of TMPs in the management of HIV/AIDS is commendable and should be emulated in the management of NCDs. However, TM and traditional medicinal knowledge (TMK) systems have not been fully exploited to improve the quality of life either as value-added TM or biopharmaceuticals in disease management. With an estimated ratio of 1 TMP for 200–400 Ugandans compared to 1 trained doctor per 20,000 Ugandans, TMPs are more available and accessible to patients than Western-trained doctors (CCFU 2008). The services offered by TMPs are holistic in nature within the community setting. Secondly, when patients go to the hospital, they do not have a choice of the type of medicine after diagnosis, other than the orthodox medicine prescribed. If TM and orthodox medicine were integrated within the hospital settings, the patients would have a choice of treatment but also be properly monitored. The effectiveness of medicine whether perceived or actual has been shown to be the main predictor of use of alternative medicine among patients with, for example, hypertension in Uganda (Nuwaha and Musinguzi 2013). The fact that hypertension is the commonest NCD in Uganda is reflected in Table 17.1 from the number of species 20/42 (47.6 %) used to treat it. The main groups of phytochemicals identified in Table 17.2 often work in synergy as opposed to working singly (Trease and Evans 2009). The positive interaction between the phytochemicals results in the overall therapeutic effect on the patients.

Diagnosis of disease by TMPs in Africa is often based on symptoms that are related to the disease (Iwu 1993). Sometimes patients also make their diagnosis alone or with the help of their neighbors. TMPs can recognize several symptoms of the disease such as behavioral changes, fever, jaundice, weakness, pain, bleeding, and diarrhea, among others. Iwu (1993) noted that the recognition of diabetes is fairly recent in Africa, and for many communities, there is no vernacular name for it. In Uganda, for lack of a better name, most indigenous communities refer to it as "sukali" translating into sugar. TMPs in Uganda rely on the knowledge of glucosuria for evidence of diabetes. They may wait to see if ants gather or not at a spot where an individual has urinated.

Many patients are often diagnosed in modern health facilities by Western methods by the time they go to TMPs. One example of such a case has been reported in Kenya by Ochwang'i et al. (2014) for patients seeking cancer treatment. Because the TMPs have no accurate methods of diagnosis, they can only reliably depend on the Western diagnosis. They sometimes ask their patients to go to the hospital for a proper diagnosis before they give them treatment. The fact that TM has not been accorded the same status as Western medicine in Uganda unlike in some Himalayan countries and China (Hamilton 2008) makes its regulation problematic since different standards have to be applied. However, the attempts to regulate TMP practice in Uganda can contribute towards solving this problem. As has been noted by Kakudidi et al. (2015), the registration of herbalists would not eliminate quacks, but it is an important starting point for the process of screening them. A participatory screening exercise by local authorities involving community input can be of paramount importance.

17.13 Conclusion and Recommendations

The rich heritage of TM practice in Uganda developed from her diverse cultural and ethnic background. There is a lot to be gained by tapping into the indigenous knowledge considering that the communicable and NCD burden is on the rise that stretches the modern health sector. There is a need by the Uganda National Health Research Organization (UNHRO) to coordinate research across TMP and orthodox health practitioners, researchers, and academia to develop useful systems of traditional medicine and promote the gradual integration into the modern healthcare systems. Research in TM would introduce systems of drug formulation with standard dosages. The Ministry of Health needs to invest more in training professional health providers and TMPs and in sensitizing the general public using targeted messages on the prevention and management of NCDs, as was done for the HIV/AIDS pandemic in Uganda.

References

Adongo DW, Kukuia KKE, Mante PK et al (2015) Antidepressant-like effect of the leaves of *Pseudospondias microcarpa* in mice: evidence for the involvement of the serotoninergic system, NMDA receptor complex, and nitric oxide pathway. BioMed Res Int 2015:15. doi:10.1155/2015/397943

Agea JG, Katongole B, Waiswa D et al (2008) Market survey of *Mondia whytei* (Mulondo) roots in Kampala City, Uganda. Afr J Trad Complement Alter Med 5(4):399–408

Amira O, Okubadejo N (2007) Frequency of complementary and alternative medicine utilization in hypertensive patients attending an urban tertiary care center in Nigeria. BMC Complement Alternat Med 7:30

Amuna P, Zotor FB (2008) Epidemiological and nutrition transition in developing countries: impact on human health and development. Proc Nutr Soc 67:82–90. doi:10.1017/s0029665108006058

Anokbonggo WW (1992) Preliminary pharmacological experimental approach to Ugandan traditional medicines. Planta Med 21:26–375

Anyinam C (1987) Availability, accessibility, acceptability and adaptability: four attributes of African ethnomedicine. Soc Sci Med 25:803–811

Anywar G, Oryem-Origa H, Kamatenesi-Mugisha M (2014) Wild nutraceutical plants from Nebbi district, Uganda. Eur J Med Plant 4(6):641–660. doi:10.9734/EJMP/2014/7634

Asiimwe S, Kamatenesi-Mugisha M, Namutebi A et al (2013) Ethnobotanical study of nutri-medicinal plants used for the management of HIV/AIDS opportunistic ailments among the local communities of western Uganda. J Ethnopharmacol 150(2):639–648, doi: http://dx.doi.org/10.1016/j.jep.2013.09.017

Atawodi SE (2010) Polyphenol composition and *in vitro* antioxidant potential of Nigerian *Canarium schweinfurthii* Engl Oil. Adv Biol Res 4(6):314–322

Atawodi SE, Olowoniyi DO (2015) Pharmacological and therapeutic activities of *Kigelia africana* (Lam.) Benth. Ann Res Rev Biol 5(1):1–17. doi:10.9734/ARRB/2015/8632

Brusotti G, Tosi S, Tava A et al (2013) Antimicrobial and phytochemical properties of stem bark extracts from *Piptadeniastrum africanum* (Hook f.) Brenan. Ind Crop Prod 43:612–616, doi: http://dx.doi.org/10.1016/j.indcrop.2012.07.068

Chinsembu KC, Hijarunguru A, Mbangu A (2015) Ethnomedicinal plants used by traditional healers in the management of HIV/AIDS opportunistic diseases in Rundu, Kavango East Region, Namibia. South Afr J Bot 100:33–42. doi:10.1186/1746-4269-6-25

Dalal S, Beunza JJ, Volmink J et al (2011) Non-communicable diseases in sub-Saharan Africa: what we know now. Int J Epidemiol 40:885–901. doi:10.1093/ije/dyr050

CAMP(2010) The National Conservation Assessment and Management Planning (CAMP) workshop for prioritization of medicinal plants of Uganda. Workshop Draft Technical Report. Organized by DOB, 24–25 June 2010 at Makerere University

Edgar AD, Levin R, Constantinou CE (2007) A critical review of the pharmacology of the plant extract of *Pygeum africanum* in the treatment of LUTS. Neurol Urodynamic 26(4):458–463. doi:10.1002/nau.20136

Edou EP, Abdoul-Latif FM, Obame LC et al (2012) Volatile constituents of *Canarium schweinfurthii* Engl. essential oil from Gabon. Int J Agric Sci 2(3):200–203

Eisenberg DM, Kessler RC, Foster C et al (1993) Unconventional medicine in the United States – prevalence, costs, and patterns of use. N Engl J Med 328:246–252. doi:10.1056/nejm199301283280406

Engonga EP, Abdoul-Latif FM, Obame LC et al (2012) Volatile constituents of Canarium schweinfurthii Engl. essential oil from Gabon. Int J Agric Sci 2(3):200–203

Farnsworth NR, Soejarto DD (1991) The Global importance of medicinal plants. In: Akerere O, Heywood V, Synge H (eds) Conservation of medicinal plants. Cambridge University Press, Cambridge

Gandhiappan J, Rengasamy R (2012) Antiproliferative activity of *Solanum anguivi* against cancer cell lines. Der Pharm Lett 4(3):875–880. doi:10.4103/0250-474X.108421

Global Burden of Disease Cancer Collaboration (2015) The global burden of cancer 2013. JAMA Oncol 1(4):505–527. doi:10.1001/jamaoncol.2015.0735

Grubben GJH, Denton OA (2004) Plant resources of tropical Africa 2. Vegetables. PROTA Foundation, Wageningen, PROTA Foundation, Wageningen

Guthold R, Louazani SA, Riley LM et al (2011) Physical activity in 22 African countries: results from the World Health Organization STEPwise approach to chronic disease risk factor surveillance. Am J Prev Med 41:52–60. doi:10.1016/j.amepre.2011.03.008

Hamilton AC (ed) (2008) Medicinal plants in conservation and development: case studies and lessons learnt. Plantlife International, Salisbury

Hawkins B (2008) Plants for life: medicinal plant conservation and botanic gardens. Botanic Gardens Conservation International, Richmond

Hoareau L, DaSilva EJ (1999) Medicinal plants a re-emerging health aid. Electron J Biotechnol 2(2):56–70

Hu J, Shi X, Chen J et al (2014) Alkaloids from *Toddalia asiatica* and their cytotoxic, antimicrobial and antifungal activities. Food Chem 148:437–444. doi:10.1016/j.foodchem.2012.12.058

Hughes GD, Aboyade OM, Beauclair R et al (2015) Characterizing herbal medicine use for non-communicable diseases in urban South Africa. Evid Based Complement Alternat Med. doi:10.1155/2015/736074

IUCN (1990) Biodiversity in Sub-Saharan Africa and its Islands: conservation, management, and sustainable use. Occasional Papers of the IUCN Species Survival Commission No. 6. IUCN-The World Conservation Union: Gland, Switzerland

Iwu MM (1993) Handbook of African medicinal plants. CRC Press LLC, Boca Raton

Jeyachandra R, Mahesh A (2007) Antimicrobial evaluation of *Kigelia africana* (Lam). Res J Microbiol 2(8):645–649. doi:10.3923/jm.2007.645.649

Kakudidi EK, Bukenya-Ziraba R, Kasenene JM (2000) The medicinal plants in and around Kibale Park in western Uganda. LIDIA Nor J Bot 5(4):109–124

Kakudidi E, Anywar G, Ayorekire F et al (2015) Antifungal medicinal plants used by communities adjacent to Bwindi Impenetrable National Park, South-Western Uganda. Eur J Med Plant 7(4):184–192. doi:10.9734/EJMP/2015/16237

Kamatenesi-Mugisha M, Oryem-Origa H (2005) Traditional herbal remedies used in the management of sexual impotence and erectile dysfunction in Western Uganda. Afr Health Sci 5(1):40–49

Kamtchouing P, Kahpui SM, Dzeufiet PDD et al (2006) Anti-diabetic activity of methanol/methylene chloride stem bark extracts of *Terminalia superba* and *Canarium schweinfurthii* on streptozotocin-induced diabetic rats. J Ethnopharmacol 104(3):306–309. doi:10.1016/j.jep.2005.08.075

Katende AB, Birnie A, Tangnas BP et al (1999) Wild food plants and mushrooms of Uganda. Regional Land Management Unit (RELMA), Nairobi

Katuura E, Waako P, Ogwal-Okeng J et al (2007) Traditional treatment of malaria in Mbarara District, Western Uganda. Afr J Ecol 45(s1):48–51. doi:10.1111/j.1365-2028.2007.00737.x

Kavishe B, Biraro S, Baisley K et al (2015) High prevalence of hypertension and risk factors for non-communicable diseases (NCDs): a population based cross-sectional survey of NCDS and HIV infection in Northwestern Tanzania and Southern Uganda. BMC Med 13(1):1–21. doi:10.1186/s12916-015-0357-9

Kiringe JW (2006) A survey of traditional health remedies used by the Maasai of Southern Kajiado District, Kenya. Ethnobot Res Appl 4:61–73

Kyomugisha H (2008) Traditional medicine: oversight of practitioners and practices in Uganda (April 2, 2008). Available at SSRN: http://ssrn.com/abstract=1322093 or http://dx.doi.org/10.2139/ssrn.1322093

Lacroix D, Prado S, Kamoga D et al (2011) Antiplasmodial and cytotoxic activities of medicinal plants traditionally used in the village of Kiohima, Uganda. J Ethnopharmacol 133(2):850–855, doi: http://dx.doi.org/10.1016/j.jep.2010.11.013

Lange D (1997) Trade in plant material for medicinal and other purposes: a German case study. Traffic Bull 17(1):20–32

Lye KA, Bukenya-Ziraba R, Tabuti JRS et al (2008) Plant-medicinal for East Africa. Makerere Herbarium handbook no. 2. Makerere University Herbarium, Kampala

Maher D, Waswa L, Baisley K et al (2011) Distribution of hyperglycaemia and related cardiovascular disease risk factors in low-income countries: a cross-sectional population-based survey in rural Uganda. Int J Epidemiol 40:160–171. doi:10.1093/ije/dyq156

Mayosi BM, Flisher AJ, Lalloo UG et al (2009) The burden of non-communicable diseases in South Africa. Lancet 374(9693):934–947. doi:10.1016/S0140-6736(09)61087-4

Ministry of Health (MoH) (2010) Health sector strategic & investment plan: promoting people's health to enhance socio-economic development 2010/2011–2014/2015. http://www.unicef.org/uganda/HSSIP_Final.pdf

Miranda JJ, Kinra S, Casas JP et al (2008) Non-communicable diseases in low- and middle-income countries: context, determinants and health policy. Trop Med Int Health 13(10):1225–1234. doi:10.1111/j.1365-3156.2008.02116.x

Mugumya F (1997) Attitudes and beliefs of people towards hot springs as sources of traditional healing in Hamurwa Sub County, BAA dissertation, Makerere University

Mujuni N (2014). The role of botanic gardens in conservation of medicinal plant species: a case study of Entebbe Bontanic Gardens. Retrieved from http://www.eoearth.org/view/article/51d7084c0cf2d912ea59feb8. Accessed 27 Dec 2015, 1.44 pm

Namukobe J, Kasenene JM, Kiremire BT et al (2011) Traditional plants used for medicinal purposes by local communities around the Northern sector of Kibale National Park, Uganda. J Ethnopharmacol 136(1):236–245, doi: http://dx.doi.org/10.1016/j.jep.2011.04.044

NAPRECA 6th Symposium on Natural Products (1995) Natural products for development chemistry, botany and pharmacology. 10–15 Sept 2015

Nassaka F (2014) Putting an end to enlarged buttocks and tied lovers. http://www.independent.co.ug/features/features/9163-putting-an-end-to-enlarged-buttocks-and-tied-lovers. 29 Dec 2015

National Environment Management Authority (NEMA) (2010) State of the environment report for Uganda 2010. National Environment Management Authority, Kampala

Nawaggi P, Bbosa GS, Kavuma M et al (2014) Diuretic activity of *Euphorbia cotinifolia* extracts in adult Wistar Albino rats: a herb used locally in the management of cardiovascular diseases. Res Rev Biosci 8(6):220–227. doi:10.3923/ajdd.2012.135.142

Noté OP, Tapondjou AL, Mitaine-offer AC et al (2013) Triterpenoid saponins from *Piptadeniastrum africanum* (Hook. f.) Brenan. Phytochem Lett 6(4):505–510, doi: http://dx.doi.org/10.1016/j.phytol.2013.06.006

Nugent R (2008) Chronic diseases in developing countries: health and economic burdens. Ann N Y Acad Sci 1136:70–79

Nuwaha F, Musinguzi G (2013) Use of alternative medicine for hypertension in Buikwe and Mukono districts of Uganda: a cross sectional study. BMC Complement Alternat Med 13:301. doi:10.1186/1472-6882-13-301

Ochwang'i DO, Kimwele CN, Oduma JA et al (2014) Medicinal plants used in treatment and management of cancer in Kakamega County, Kenya. J Ethnopharmacol 151(3):1040–1055, doi: http://dx.doi.org/10.1016/j.jep.2013.11.051

Okeowo A (2007) "Sex Tree," other medicinal plants near extinction in Uganda in Mabira Forest Reserve, Uganda. National Geographic News August 3, 2007

Olatunji AG, Atolani OA (2009) Comprehensive scientific demystification of *Kigelia africana*: a review. Afr J Pure Appl Chem 3(9):158–164

Olusola OE, Kamdem JP, Kade IK et al (2013) Saponins from *Solanum anguivi* Lam. fruit exhibit in vitro and in vivo antioxidant activities in alloxan-induced oxidative stress. Asian J Pharm Clin Res 6:252–257

Oryem-Origa H, Katende AB, Kakudidi EKZ (2001) Ethnobotanical studies of Mabira Forest area, Central Uganda. Discovery and Innovations (special edition). Afr Acad Sci, pp 169–181

Oryem-Origa H, Katende AB, Kakudidi EKZ (2003) Some medicinal plants used in Mukono District. Ug J 49:56–65

Osamor PE, Owumi BE (2010) Complementary and alternative medicine in the management of hypertension in an urban Nigerian community. BMC Complement Altern Med 10(36):1472–6882. doi:10.1186/1472-6882-10-36

Parkin DM, Nambooze S, Wabwire-Mangen F et al (2010) Changing cancer incidence in Kampala, Uganda, 1991–2006. Int J Cancer 126:1187–1195. doi:10.1002/ijc.24838

Ripperger H, Himmelreich U (1994) Anguivine and isoanguivine steroid alkaloid glycosides from *Solanum anguivi*. Phytochemistry 37(6):1725–1727, doi: http://dx.doi.org/10.1016/S0031-9422(00)89600-4

Schwartz J, Guwatudde D, Nugent R et al (2014) Looking at non-communicable diseases in Uganda through a local lens: an analysis using locally derived data. Glob Health 10:77. doi:10.1186/s12992-014-0077-5

Sofowora A (1993) Medicinal plants and traditional medicine in Africa. Spectrum Books, Ibadan

Ssegawa P, Kasenene JM (2007) Medicinal plant diversity and uses in the Sango bay area, Southern Uganda. J Ethnopharmacol 113(3):521–540, doi: http://dx.doi.org/10.1016/j.jep.2007.07.014

Steyn K, Fourie J, Temple N (eds) (2006) Chronic diseases of lifestyle in South Africa: 1995–2005, Technical Report. South African Medical Research Council, Cape Town

Tabuti JRS (2008) Herbal medicines used in the treatment of malaria in Budiope County, Uganda. J Ethnopharmacol 116(1):33–42. doi:http://dx.doi.org/10.1016/j.jep.2007.10.036
Tabuti JRS, Dhillion SS, Lye AK (2003) Traditional medicine in Bulamogi County, Uganda: its practitioners, users and viability. J Ethnopharmacol 85:119–129
The Cross-Cultural Foundation of Uganda (CCFU) (2008) Promoting herbal medicine in Uganda: traditional health practitioners and Government Working Together. www.crossculturalfoundation.or.ug. Accessed 14 Dec 2015
The World Bank Human Development Network (2008) The growing danger of non-communicable diseases. Acting now to reverse course Conference edition
Tilburt JC, Kaptchuk TJ (2015) Herbal medicine research and global health: an ethical analysis. Bull World Health Organ. http://www.who.int/bulletin/volumes/6/8/07-042820/en/. Accessed 28 Dec 2015
Trease ECW, Evans D (2009) Pharmacognosy, 15th edn. Elsevier Ltd, London
Tugume P, Kakudidi EK, Mukadasi B et al (2016) Ethnobotanical survey of medicinal plant species used by communities around Mabira Central Forest Reserve, Uganda. J Ethnobiol Ethnomed 12:5. doi:10.1186/s13002-015-0077-4
Uganda Bureau of Statistics (UBoS) (2010). Uganda National Household Survey Findings 2009/2010. Chapter 5
Uganda Bureau of Statistics (UBoS) (2014) Statistical abstract 2015
van der Sande (2003) Cardiovascular disease in sub-Saharan Africa: a disaster waiting to happen. Neth J Med 61(2):32–36
Van Wyk BE, Wink M (2004) Medicinal plants of the world. Timber Press, Oregon
Vázquez R, Riveiro ME, Vermeulen M et al (2012) Toddaculin, a natural coumarin from *Toddalia asiatica*, induces differentiation and apoptosis in U-937 leukemic cells. Phytomedicine 19(8–9):737–746. doi:10.1016/j.phymed.2012.03.008
Wasswa H (2006) Uganda struggles to cope with rise in diabetes incidence. BMJ 333(7570):672. doi:10.1136/bmj.333.7570.672-c
White K (2002) An introduction to the sociology of health and illness. Sage, London, pp 4–5
WHO (2008) Interventions on diet and physical activity. What works? Geneva, Switzerland
WHO (2013) Global action plan for the prevention and control of non-communicable diseases, 2013–2020. WHO, Geneva
WHO (2014) Global status report on non-communicable diseases 2014. World Health Organization, Geneva
WHO (2015) Noncommunicable diseases progress monitor, 2015. World Health Organization, Geneva
WHO/EDM/TRM (2002) WHO traditional medicine strategy 2002–2005. World Health Organization, Geneva
Whyte SR (2014) Timelines and chronic medication: knowledge about hypertension and diabetes in Uganda. SPP working paper no 7
Wilt T, Ishani A, Mac Donald R et al (2002) *Pygeum africanum* for benign prostatic hyperplasia". Cochrane Database Syst Rev(1): CD001044. DOI:10.1002/14651858.CD001044

Chapter 18
Plant-Derived Agents in Modulation of Rheumatoid Arthritis

Pathirage Kamal Perera

Abstract Rheumatoid arthritis (RA) is a chronic systemic inflammatory disease of unknown cause. The characteristic feature of this condition is persistent polyarthritis that affects hands and feet, although any joint lined by a synovial membrane may be involved. Many of the conventional disease-modifying antirheumatic drugs (DMARD) therapies are immunosuppressive in nature, but lead to serious side effects. In this chapter, we have explored possible immune molecular targets of disease-modifying antirheumatic herbal agents and their place in arthritic management. Considerable past research findings have demonstrated that herbal drug molecules can modulate the major inflammatory cytokine expression in synovial cells. Further studies have clearly demonstrated that certain herbal drugs can significantly inhibit the expression of iNOS and are involved in the regulation of the inflammatory immune process, as a potent PGE2 and COX-2 expression inhibitors. Herbal drug molecules potently induce the apoptosis of synoviocytes through upregulating the caspase cascade and downregulating the NF-kB expression. This chapter concludes that the herbal drugs can be developed as new clinical pharmaceutical agents, which can be used in cost-effective manner to treat arthritic conditions with minimal side effects.

Keywords Apoptosis • Arthritis • Caspases • Cytokines • Herbal drug agents

Abbreviations

Bax	Bcl-2-like protein 4
Bid	BH3 interacting-domain death agonist
Bcl-2	B-cell lymphoma 2
CAM	Complementary and alternative medicine
COX-2	Cyclooxygenase-2
DMARD	Disease-modifying antirheumatic drugs

P.K. Perera (✉)
Institute of Indigenous Medicine, University of Colombo, Rajagiriya, Sri Lanka
e-mail: drkamalperera@yahoo.com

H.-S. Tsay et al. (eds.), *Medicinal Plants - Recent Advances in Research and Development*, DOI 10.1007/978-981-10-1085-9_18

GM-CSF	Granulocyte-macrophage colony-stimulating factor
HLA-DR	Human leukocyte antigen DR4
IFN-g	Interferon gamma
IL	Interleukin
iNOS	Inducible nitric oxide synthase
NF-kB	Nuclear factor kappa B
NOS2	Nitric oxide synthase 2
PGE2	Prostaglandin E2
PBMC	Peripheral blood mononuclear cells
RA	Rheumatoid arthritis
RANKL	Receptor activator of nuclear factor kappaB legend
TCM	Traditional Chinese medicines
TNF	Tumor necrosis factor
TGF	Transforming growth factor
VIP	Vasoactive intestinal peptide

18.1 Introduction

Rheumatoid arthritis (RA) is a chronic disease with multiple pharmacotherapies involved. RA leads to enormous suffering in patients by targeting multiple organs. Conservative symptomatic treatment is often associated with adverse reactions and side effects. Therefore many RA patients seek Complementary and Alternative Medical Systems (CAM) to manage the disease (Park and Ernst 2005). Indian study showed that 40 % of RA patients use either Ayurveda or other traditional medical systems along with allopathic medicine (Sangha 2000). Further in the USA, 60–90 % RA patients seek CAM to manage the disease progression (Soeken et al. 2003).

In RA, mechanism of autoimmunity can be seen either as the induction of cell cycle arrest or uncontrolled cell division or as the induction of apoptosis in stressed cells. Some plant-derived agents have been explored to be effective regulators of the cell cycle by targeting specific cell signaling molecules leading to immune modulations or apoptosis or cellular senescence. Therefore many plant-derived agents potentially employed for treatment of RA as molecular level targets. The plant-derived agents' main action on the suppression of autoimmunity is by upregulating the key signaling molecules like Bax and Bid, a proapoptotic member of the Bcl-2 family, and subsequent downregulation of expression of various other key signaling molecules such as NF-kB and Bcl-2. This leads to activation of caspases in cells which eventually induces apoptosis in target cells. Also most plant-derived agents respond to inflammatory mediators including interleukins, TNF-alpha, TGF-h, IFN-g, VIP, iNOS and cyclooxygenase-2, and prostaglandin E2 (Perera et al. 2011) (Fig. 18.1). This chapter discusses how plant-derived agents are capable of modulating RA and their possible molecular mechanism.

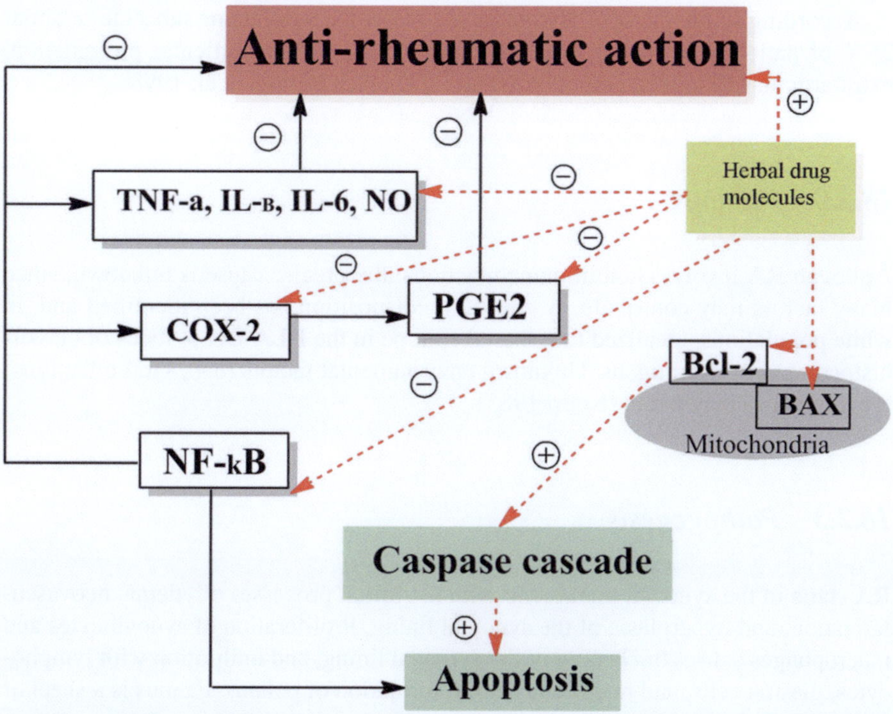

Fig. 18.1 The schematic presentation – molecular level mechanism of plant-derived herbal agents in RA ("+," positive effects; "–," negative effects)

18.2 Features of RA

18.2.1 Clinical Features and Classification

RA is a chronic inflammatory disease characterized by progressive damage to synovial-lined joints and variable extra-articular manifestations. RA affects 0.5–1 % of the population in the industrialized world. It is two to three times more frequent in women than men and can lead to disability and reduced quality of life (Arnett et al. 1988).

Tendon and bursal involvement are often clinically dominant in the early stage of RA. It can affect any joint and prominent in metacarpophalangeal, proximal interphalangeal, and metatarsophalangeal joints, as well as in wrist and knees. Articular and periarticular manifestations including joint swelling and tenderness to palpation, morning stiffness, and motion impairment in joints are common to RA. Insidious onset of pain with symmetric swelling of small joints has been observed as the most frequent symptom. Malaise, fatigue, weight loss, and fever are common symptoms in RA.

According to the onset of RA, it can be classified as acute or subacute in about 25 % of patients, but patterns are palindromic onset, monoarticular presentation, extra-articular synovitis, and polymyalgic-like onset (Grassi et al. 1998).

18.2.2 Etiology

Although RA involves autoimmune reactions, the precise cause is unknown, since many factors may contribute. A genetic predisposition has been identified and, in white populations, localized to a shared epitope in the HLA-DR β_1 locus of class II histocompatibility antigens. Unknown environmental factors (e.g., viral infections) are thought to play a role (Kouri 1985).

18.2.3 Pathogenesis

RA starts in the synovial membrane with the initial processes of edema, neovascularization, and hyperplasia of the synovial lining. Proliferation of synoviocytes and macrophages causes thickening of the synovial lining, and infiltration with lymphocytes, plasma cells, and mast cells caused formation of pannus. Pannus is a sheet of invasive cellular tissue that is continuous with the synovial lining. Pannus causes erosion of the bone and cartilage at the margin of joints due to invaded macrophages and synoviocytes. Synovial villous formation leads to inflammatory joint effusion. It causes capsule distention and stretching of the ligamentous tissues which results in laxity of the capsule. With the progression of the disease, the joint becomes unstable and begins to deform (Otero and Goldring 2007).

At the cellular level, cytokines stimulate synoviocytes to produce cartilage-degrading enzymes. Hyperplastic synovium and pannus produce several enzymes which are capable in degrading bone and cartilage components. Matrix metalloproteinase is a group of enzymes secreted from synoviocytes and chondroblasts in response to cytokines. Genetically RA has been shown to be associated with a positive human leukocyte antigen DR4. Further RA has been shown a strong association with human leukocyte antigen (Stuart et al. 1984).

18.3 The Inflammatory and Immune Response in RA

The immune system is our defense system. It is designed to perform several tasks, for example, (i) to fight off intruders such as bacteria, viruses, and parasites, (ii) to clear transformed cells, (iii) to remove dead cells, and (iv) to heal injured tissues. The immune system is divided into two interacting compartments: the innate immune system and the adaptive immune system. The innate immune system is ancient, being present in all multicellular organisms. In general, innate immunity is a nonspecific, inducible response to pathogens. It is immediate in action, yet short-lived. Cells that contribute to innate responses are dendritic cells, macrophages, neutrophils, mast cells, and natural killer cells. Conversely, the adaptive immune system is much more specific, but takes longer to become activated. T cells and B cells of the adaptive immune system can recognize specific antigens through specific receptors. These undergo clonal proliferation and maturation following activation. Both systems work together to provide protection against a diverse and rapidly evolving array of pathogens. The ability to make immune response is tightly regulated to ensure that when pathogens are eliminated, the immune response is tuned down to avoid unwanted damage. However, the immune system can occasionally attack self-tissues and cause autoimmunity. Autoimmunity is caused by an adaptive immune response against "self"-antigens resulting in tissue damage and organ dysfunction. In RA, the cartilage of the joint is damaged due to massive inflammation in the affected tissues (Stuart et al. 1984).

18.4 The Cytokine Network in RA

Cytokine network imbalance favors the disease progression in RA. In RA excess production of pro-inflammatory cytokines leads to induction of autoimmunity, chronic inflammation, and joint destruction. Still it is a gray area to find the best target cytokines for drug therapy (Perera et al. 2011). Protein and mRNA analyses revealed that many pro-inflammatory cytokines such as TNF-alpha, IL-1, IL-6, GM-CSF, and chemokines are abundant in RA. Also it is compensated to some degree by the increased production of anti-inflammatory cytokines such as IL-10 and TGF-beta and cytokine inhibitors such as IL-1ra and soluble TNF-R (Feldmann et al. 1996).

From the RA joint cell cultures, it was revealed that IL-1 is produced spontaneously and TNF-alpha is the dominant regulator of IL-1. Subsequently, other pro-inflammatory cytokines were also inhibited if TNF-alpha is neutralized. This led to the new concept that the pro-inflammatory cytokines were linked in a network with TNF-alpha at its apex (Feldmann et al. 1996). This hypothesis was tested in animal and clinical trials of anti-TNF-alpha therapy. Re-treatment studies have also explored benefit in repeated relapses indicating RA remains TNF-alpha dependent (Iain et al. 2007).

18.4.1 Effect of Herbal Agents on the Cytokine Expression in RA

Experimental findings have shown that herbal drugs significantly ameliorate symptoms and prevent severe development through modulating cytokines in arthritis (Perera et al. 2010a; Peng et al. 2010). TNF-alpha and IL-1β are key mediators in the joint inflammation and destruction of the cartilage and bone in patients of RA (Firestein 2003). IL-6 is a pro-inflammatory cytokine with a wide range of biological activities in immune regulation, inflammation, and oncogenesis (Kitamura et al. 2004). Increase of serum g-globulin and the emergence of rheumatoid factors are modulated by IL-6 (Naka et al. 2002).

Recent studies have further demonstrated that herbal compounds significantly target the production of IL-1β, IL-6, and TNF-α (Perera et al. 2010b). Further researches have also confirmed that herbal compounds could significantly target on mRNA expression in synovial cells (Perera et al. 2011). Clinical researches have also suggested that traditional herbal medicines not only suppress the expression of pro-inflammatory cytokines but also induce the expression of cytokines with anti-inflammatory and immunomodulatory effects (Liu and Sun 2013). Therefore cytokines are very important targets in antiarthritic herbal drug development.

18.5 Nitric Oxide (NO) and RA

Nitric oxide (NO) is constitutively produced by endothelial or neuronal NO synthases or in higher concentrations by inducible NO synthase (iNOS; NOS2) after stimulation of bacterial products and cytokines. A study by Stefanovic-Racic et al. (1993) has indicated that NO may be important in the pathogenesis of RA, while research on animal models of arthritis has suggested that excessive level of NO can promote tissue injury and contribute to progression of the disease. Likewise, it has been shown that peripheral blood mononuclear cells (PBMC) from RA patients have increased NOS2 expression and enhanced formation of NO that correlates with the disease activity. NO has been shown as a key mediator of apoptosis in the RA joints, as well as an important regulator of the Th1/Th2 balance in autoimmune diseases. The final piece of evidence comes from the finding that administration of NOS2 inhibitors has provided beneficial effects on animal models of RA (van't Hof et al. 2000).

18.5.1 Effect of Herbal Agents on iNOS Expression in RA

Expression of the inducible isoform of NOS (iNOS) and a raised concentration of NO may play key roles in the pathogenesis of RA as a mediator of apoptosis. The transcription of iNOS may be involved in the induction of RA by augmentation of inflammation (McCartney-Francis et al. 1993). Experimental findings have confirmed that herbal drugs significantly inhibit the expression of iNOS in blood serum (Perera et al. 2010a).

18.6 Prostaglandin-E and COX-2 Expression in RA

Recent research explored that prostaglandin subfamily including PGE2 has a pro-inflammatory role in RA. In response to cytokines, PGE2 is produced and it has negative impact on regulating both IL-17 and TNF-alpha expression. This leads to activation of fibroblast-like synoviocytes through EP2/EP4 receptors which results in the modulation of pro-inflammatory cascades (Akaogi et al. 2006). Experimental findings have confirmed that herbal agents can be effectively involved in the regulation of the inflammatory process as a potent PGE2 inhibitor (Perera et al. 2010a).

COX-2 is a key rate-limiting enzyme in the biosynthesis of prostaglandins PGE2. Cyclooxygenase (COX) is the rate-limiting enzyme in the conversion of arachidonic acid (AA) to prostaglandins (PGs). There are two isoforms of COX: constitutively expressed cyclooxygenase 1 (COX-1) and cyclooxygenase 2 (COX-2), an inducible isoform of the enzyme that predominates in pathologies associated with inflammation, such as cancer and RA. PGs produced by COX-1 are thought to mediate housekeeping functions, while COX-2 expression is selectively expressed in some tissues by growth factors, oncogenes, and cytokines, and PGs produced by this isoform contribute to cellular processes such as angiogenesis (Alexanian et al. 2014). Normal cells express limited COX-2, while, in the presence of inflammatory cytokines such as TNF-α, IL-1β is stimulated to affect rapid expression of COX-2, which in turn promotes the production and release of large quantity of PGE2. COX-2 and PGE2 can promote fibroblast cell division and proliferation and induce capillary formation. Therefore, these are important target molecules for inflammation and immunological disorders (Anderson et al. 2009; Noguchi et al. 2008). Western blot determination of COX-2 protein expression revealed that herbal drug molecules could downregulate the COX-2 protein expression and lead to improved experimental arthritic conditions (Peng et al. 2010).

18.7 Induction of Apoptosis by Herbal Agents in RA

Synovial macrophages, fibroblasts, and lymphocytes are critical to the pathogenesis of this disease, in which apoptosis may play divergent roles. In the joints of patients with active RA, few apoptotic cells have been detected, and experimental data suggest that enhanced apoptosis in the joint might be therapeutically beneficial. Research findings suggest that the synovium may undergo apoptosis or proliferate depending upon the actions of common nodal point regulators, one of which may be the apoptotic antagonist Bcl-2 (Perlman et al. 2000). Forced Bcl-2 expression blocks cytochrome *c* release (Kluck et al. 1997) induced by either withdrawal of the growth factor or Bax overexpression. Additionally, overexpression of Bcl-2 may inhibit apoptosis following cytochrome *c* release (Marzo et al. 1998). Previous investigations have demonstrated a less number of apoptotic cells in the synovium or the pannus in RA (Hong et al. 2005), suggesting an antiapoptotic Bcl-2 upregulation in RA. Therefore, in vivo downregulation of Bcl-2 has potential as a novel therapeutic agent, which may enhance apoptosis in the RA synovium and potentially limit disease progression (Perlman et al. 2000). Consistent with these research findings, antirheumatic herbal drugs significantly downregulate the expression of Bcl-2 in synovium in vivo, suggesting that these drugs may inhibit apoptosis in the synovium which leads to amelioration of the symptoms of RA (Perera et al. 2010a).

Hilbers et al. (2003) observed that expression of Bax was higher in patients with RA compared to healthy controls. Western blot results showed that herbal drugs have significant influence on the Bax protein (Perera et al. 2011). Bcl-2 and Bax are related to apoptosis in which Bcl-2 binds to Bax to generate heterodimers, and a decreased Bcl-2/Bax ratio could lead to induction of apoptosis in RA (Szodoray et al. 2003). Caspase, a protease, plays an important role in the cell apoptosis.

Current research indicates that there are two main apoptotic pathways: the extrinsic or death receptor pathway and the intrinsic or mitochondrial pathway. The extrinsic and intrinsic pathways converge on the same terminal or execution pathway. This pathway is initiated by the cleavage of caspase-3 (Kaubuyama et al. 2008; Schulze-Osthoff et al. 1998). This results in fragmentation of DNA, degradation of cytoskeletal and nuclear proteins, cross-link of proteins, formation of apoptotic bodies, and expression of ligands for phagocytic cell receptors and leads to uptake by phagocytic cells (Huang et al. 2008). Perera et al. (2010b) in their experiments observed a significant decrease in procaspase-3 in RA treatment with herbal drugs by increase of apoptosis in cells improved the excessive proliferation.

18.8 Nuclear Factor *k*B (NF-*k*B) and Herbal Agents

Nuclear factor *k*B (NF-*k*B) has been well recognized as a pivotal regulator of inflammation in RA. Studies have revealed a broad involvement of NF-*k*B in progress of T helper 1 response that leads to activation of apoptosis and proliferation of fibroblast-like synovial cells in RA. NF-*k*B activates transcription of a gene encoding for inhibitor of apoptosis protein (IAP). This protein in turn contributes to downregulation of activity of the caspase cascade which forms the core of the apoptotic pathway (Danial and Korsmeyer 2004; De Jong et al. 2004; Eissing et al. 2004). Therefore, gene delivery of NF-kB inhibitors may have distinct advantages in arthritis treatment (Huang et al. 2008). Further research has confirmed that herbal drugs could lower the levels of NF-kBp65 subunit and inhibit the expression of NF-kB (Peng et al. 2010).

18.9 Evidence for Antiarthritic Activity of Plant-Based Herbal Agents (Table 18.1)

18.9.1 Alstonia scholaris *Linn*

A. scholaris commonly known as saptaparni or "devil's tree" is an important medicinal plant in the various folk and traditional systems of medicine in Asia, Australia, and Africa. *A. scholaris* is a medium to large tree, about 40 m high with a somewhat tessellated corky gray to gray-white bark (Meena et al. 2011). Traditionally, bark of

Table 18.1 Mode of antiarthritic activity of plant-based herbal agents

Plant name	Part used	Active compounds	Mode of action	References
Alstonia scholaris Linn.	Bark	Alkaloids especially echitamine. Glycosides of venoterpene, nareline, lagunamine, 19-epischolaricine, butamine, dobutamine, and alschomine	Antiarthritic activity by the reduction in total leukocyte, lymphocytes, and monocytes/macrophages migration	Arulmozhi et al. (2001)
Aristolochia bracteata Lam.	Whole plant	Aristolochic acid	Maintaining the synovial membrane and vascular permeability and inhibiting cytokines and leukotriene infiltration	Chitme and Patel (2009)
Boerhaavia diffusa Linn.	Whole plant	Punernavoside, boeravinones	Anti-inflammatory and antioxidative mechanisms	Dapurkar et al. (2013)
Boswellia serrata Roxb.	Oleo gum resin	β-boswellic acid, acetyl-β-boswellic acid, 11-keto-β-boswellic acid, and acetyl-11-keto-β-boswellic acid	By decreasing the function of membrane marker enzymes and inhibition of leukocyte migration in inflamed tissues	Mishra et al. (2011)
Caesalpinia sappan Linn.	Heartwood	Sappanone A, protosappanin E, and neoprotosappanin	By modulating the levels of cytokines and PGE2	Wang et al. (2011)
Cannabis sativum Linn.	Dried flowering or fruiting tops	Cannabinoid	By induction of apoptosis, inhibition of cell proliferation, suppression of cytokine production, and induction of T-regulatory cells	Malfait et al. (2000)
Cedrus deodara (Roxb)	Heartwood	Deodarone, atlantone, deodarin, deodardione	By modulating the levels of IL-1β, IL-6, TNF-α, and prostaglandin E2 (PGE2)	Uma et al. (2011)
Cinnammomum zeylicanium Blume.	Bark	Cinnamaldehyde, coumarin	Mediating through inhibition of leukocyte emigration and prostaglandin synthesis	Vetal et al. (2013)

(continued)

Table 18.1 (continued)

Plant name	Part used	Active compounds	Mode of action	References
Curcuma longa Linn.	Rhizome	Curcumin	Preventing local activation of NF-kappaB and the subsequent expression of NF-kappaB-regulated genes mediating joint inflammation and destruction, including chemokines, cyclooxygenase 2, and RANKL	Funk et al. (2010)
Ferula asafetida Linn.	Gum resin	Ferulic acid, umbelliferone, asaresinotannols, farnesiferols A, B, and C	By reducing the cytokine level, MMP and increased SOX9 gene expression	Chen et al. (2010)
Phyllanthus amarus Schum and Thomm	Whole plant	Phyllanthin, hypophyllanthin	Reducing the levels of aspartate transaminase and alanine transaminase	Malia et al. (2011)
Piper longum Linn.	Fruit/seeds	Piperine, rutin	Modulating the TNF-α and NF-kB	Yende et al. (2010)
Saussurea lappa Clarke.	Roots	Quercetin, cynaropicrin	Inhibiting TNF-α and NO	Gokhale et al. (2002)
Sida rhombifolia Linn.	Roots	Vasicinol, asicinone, and N-methyl tryptophan	By modulating the levels of cytokines and PGE2	Gupta et al. (2009)
Terminalia chebula Retz.	Fruit	Gallic acid, chebulagic acid, punicalagin, chebulanin, corilagin, neochebulinic acid, ellagic acid, chebulinic acid	Modulating the levels of TNF-α, IL-6, and IL-1β	Nair et al. (2010)
Trigonella foenum-graecum Linn.	Seeds	4-hydroxyisoleucine (4-OH-Ile), fat, diosgenin, iron, phenolic acids, protein, and protodioscin	Modulating the TNF-α and NF-kB and blood parameters	Sindhu et al. (2012)
Vitex negundo Linn.	Leaves, roots	P-cymene, valencene, caryophyllene epoxide, and (E)-nerolidol	Modulating the cytokines, PGE2, and interferon gamma	Pandey et al. (2012)

A. scholaris is used in the treatment of rheumatism (Kalaria et al. 2012). The ethanolic extract of *A. scholaris* leaves showed antiarthritic effect in rat models by reduction in total leukocyte, lymphocytes, and monocytes/macrophages migration (Arulmozhi et al. 2001).

18.9.2 Aristolochia bracteata *Lam*

The extracts of *A. bracteata* have been shown to possess properties like antipyretic (Rajamanickam et al. 2009), anti-allergic (Chitme et al. 2010), anti-inflammatory, and antiarthritic (Chitme and Patel 2009). The different extracts of whole plant showed antiarthritic by maintaining synovial modulation of cytokine and leukotriene infiltration (Chitme and Patel 2009).

18.9.3 Boerhaavia diffusa *Linn*

B. diffusa is valued for its diuretic, laxative, and respiratory disorders in traditional systems of medicines (Rajpoot and Mishra 2011). Investigations have revealed that *B. diffusa* possesses an anti-inflammatory action (Bhalla et al. 1968; Gupta et al. 1969). The root extracts showed antiarthritic effect in rat model (Dapurkar et al. 2013).

18.9.4 Boswellia serrata *Roxb*

B. serrata has been shown to possess an anti-inflammatory (Gupta et al. 1992), analgesic (Menon and Kar 1971), immunomodulatory (Pungle et al. 2003), and anti-osteoarthritic activities (al-Awadi et al. 1991). The gum resin extract of *B. serrata and extract of Glycyrrhiza glabra* exhibited antiarthritic activity in rat models (Mishra et al. 2011). This antiarthritic action was mainly due to decreased function of membrane marker enzymes and inhibition of leukocyte migration in inflamed tissues (Mishra et al. 2011).

18.9.5 Caesalpinia sappan *Linn*

C. sappan is reported to have anti-inflammatory activity (Hikino et al. 1977; Dhawan et al. 1980). *C. sappan* caused an inhibition of phosphodiesterase (Nikaido et al. 1981) and stimulation of glutamate pyruvate transaminase (Lee and Kim 1999) and tyrosinase enzymes (Liu et al. 1992). The ethanol wood extract of *C. sappan* showed antiarthritic action on rat models by modulating serum IL-1β and IL-6, TNF-α, and PGE2 (Wang et al. 2011).

18.9.6 Cannabis sativum *Linn*

Cannabinoid component of *C. sativum* showed antiarthritic action (Silveira and Tufik 1981; Zuardi et al. 1993) and used as anxiolytic antidepressant in schizophrenia (Moreira and Guimarães 2005). Cannabidiol active moiety of *C. sativum* significantly decreased the arthritic scores and inhibited the anti-inflammatory mediator on collagen-induced rat model (Malfait et al. 2000).

18.9.7 Cedrus deodara *(Roxb)*

C. deodara has been excessively used for antiarthritic, anti-inflammatory, and analgesic activities in Ayurveda and traditional medical practice. Also, it was shown that the anti-inflammatory activity of *C. deodara* wood oil could be attributed to its mast cell stabilizing activity and the inhibition of leukotriene synthesis (Shinde et al. 1990).

In another study, the petroleum ether, chloroform, and alcoholic extracts of the heartwood of *C. deodara* were prepared by a Soxhlet extractor and examined for its external antiarthritic activity in rats using Freund's adjuvant method. Application of all the three extracts exhibited a significant inhibition of complete Freund's adjuvant (CFA)-induced rat paw edema when compared with the arthritic control group. These findings seem to justify the use of the plant in traditional Indian medicine in the treatment of inflammation, including arthritic conditions (Uma et al. 2011).

18.9.8 Cinnammomum zeylicanium *Blume*

C. zeylicanium plant is reported to have analgesic, antipyretic (Valero and Salmeron 2003), anti-inflammatory (Valero and Salmeron 2003; Mancini et al. 1999), and antioxidant activities (Rani et al. 2010; Taker et al. 2007). The bark extracts of *C. zeylicanium* showed antiarthritic effect by improving body weight and level of serum C-reactive proteins in rat models (Vetal et al. 2013).

18.9.9 Curcuma longa *Linn*

C. longa is extensively used as anti-inflammatory agent (Mukhopadhyay et al. 1982; Srimal and Dhawan 1973). The antiarthritic activity was reported in essential oils of rhizomes in experimental arthritic animal models (Funk et al. 2010).

18.9.10 Ferula asafetida *Linn*

F. asafetida is reported to exhibit anti-inflammatory, antioxidant, antiarthritic, and immunostimulant activities (Poonam and Shradha 2012). The effect of *F. asafetida* and its effective concentration on porcine chondrocytes were evaluated by real-time PCR, and it was revealed that *F. asafetida* reduced the hydrogen peroxide-induced IL-1 beta and TNF-alpha, MMP-1, and MMP-13 and increased SOX9 gene expression (Chen et al. 2010). The protective effect of *F. asafetida* on joints appears to occur through multiple ways due to the various antiarthritic effects exerted by its constituent bioactive compounds.

18.9.11 Phyllanthus amarus *Schum and Thomm*

P. amarus was used for its antiarthritic activities (Kassuya et al. 2005) and antioxidant activities (Harikumar and Kuttan 2007). The aqueous extract of *P. amarus* showed antiarthritic activity in rat models by reducing levels of aspartate transaminase and alanine transaminase (Malia et al. 2011).

18.9.12 Piper longum *Linn*

P. longum finds its importance due to its anti-inflammatory (Sharma and Singh 1980) and immunomodulatory activities (Mananvalan and Singh 1979). The aqueous extract of *P. longum* seeds showed antiarthritic effect on Freund's complete adjuvant-induced arthritis in rat (Yende et al. 2010). This study further explored that antiarthritic effect was due to modulating the TNF-alpha and NF-kβ in cellular level.

18.9.13 Saussurea lappa *Clarke*

The ethanolic extracts of *S. lappa* exhibited anti-inflammatory and antiarthritic effects by suppressing nitric oxide (NO) production in rat models (Gokhale et al. 2002).

18.9.14 Sida rhombifolia *Linn*

S. rhombifolia is used as nutritive and tonic in arthritic conditions (Nadkarni & Indian 2009). Research revealed that *S. rhombifolia* has cytotoxic (Islam et al. 2003), anti-inflammatory, antipyretic (Alam et al. 1991), and antiarthritic effects.

The aqueous and ethanol extracts of *S. rhombifolia* showed potent antiarthritic effects on rat models (Gupta et al. 2009).

18.9.15 Terminalia chebula *Retz*

Terminalia chebula is reported to possess the immunomodulatory (Sohni and Bhatt 1996) and cytoprotective activities (Hamada et al. 1997). The hydroalcoholic extract of *T. chebula* showed antiarthritic activity by significant modulating effect of TNF-alpha, IL-6, and IL-1β (Nair et al. 2010).

18.9.16 Trigonella foenum-graecum *Linn*

T. foenum-graecum possessed anti-inflammatory activities (Pandian et al. 2002). The T. foenum-graecum showed antiarthritic activities by reducing the levels of SGOT, SGPT, CRP, nitrites, ESR, and white blood cell count (Sindhu et al. 2012).

18.9.17 Vitex negundo *Linn*

In traditional and Ayurveda medicine, *V. negundo* has been used for antiarthritic and anti-inflammatory (Tandon and Gupta 2005; Ravishankar et al. 1985, 1986; Telang et al. 1999), antioxidant (Tandon and Gupta 2005; Munasinghe et al. 2001), and antirheumatic effects (Tandon 2005). The active compounds of *V. negundo* has proven for antiarthritic action by decreasing the levels of ESR, leukotriene B4, PGE2, cytokines IL-17, TNF-alpha, and interferon (Pandey et al. 2012).

18.10 Conclusion

The pro-inflammatory cytokines in RA has been validated in recent preclinical and clinical research. Recent research showed that plant-derived herbal agents significantly reduced the production of IL-1β, IL-6, and TNF-α in RA. Expression of the inducible isoform of NOS (iNOS) and a raised concentration of NO may play key roles in the pathogenesis of RA as a mediator of apoptosis. These data indicate that plant-derived herbal agents may have the modulatory power to ameliorate the arthritic conditions.

Apoptosis is closely involved in the initiation and progression of RA. Recent research has advanced to study the effect of herbal agents on programmed cell death machinery via experiments. The mechanism of apoptosis is remarkably conserved across species and is executed with a cascade of sequential activation of initiation and effector caspases. Caspase-3, which is the ultimate executioner caspase that is essential for the nuclear changes associated with apoptosis, is synthesized as an inactive precursor. Proteolytic processing of this zymogen of 32 kDa (procaspase-3) generates active enzyme (caspase-3), and this caspase-3 activation is an early and irreversible point in the development of apoptosis. Apoptosis mediated through the intrinsic pathway involves the release of cytochrome c from mitochondria, an essential cofactor required for the activation of caspases, and is regulated by the Bcl-2 family of proteins upstream of caspase activation. A major characteristic of these proteins is their frequent ability to form homo- and heterodimers suggesting a neutralizing competition between these proteins. Briefly, Bcl-2 is an apoptosis-suppressing factor that heterodimerizes with Bax and neutralizes the effects of the latter. When Bcl-2 is present in excess, cells are protected against apoptosis. In contrast, when Bax is in excess and the homodimers of Bax dominate, cells are susceptible to programmed death. Therefore, the ratios of Bcl-2/Bax determine the fate of a cell rather than the absolute concentration of either. The treatment of herbal agents in RA shows significant descent in the Bcl-2/Bax ratios. Therefore herbal agents' therapies confirmed that induce apoptosis through caspase-3 pathway.

Further research confirmed that herbal agents are effective in RA, by downregulating the antiapoptotic Bcl-2 and upregulating the apoptotic Bax expression in the synovium. The main products of the cyclooxygenase pathway, namely, prostaglandin E2 (PGE2), play a key role in the erosion of the cartilage and juxta-articular bone in RA. NF-kB activation is a common feature of human RA synovium. Activation of NF-kB has also been detected in different animal models of RA, including adjuvant arthritis in rats. Further these results indicate that mechanism of anti-rheumatoid arthritis action of herbal agents is related to inhibition of NF-kBp65 subunit and COX-2 expression. Research confirmed that herbal agents are effectively involved in the regulation of the inflammatory process as a potent PGE2 inhibitors. Also the mechanism of curing RA by herbal agent is related to inhibiting NF-kB signal transduction pathway with associated protein, and the COX-2 expression factors are closely related. Taken together, these results suggest that herbal agents can be effectively applied at the level of inflammatory cytokine pathways and as mediator regulators in cellular mechanism pathways in RA. These agents may be used as novel targets for clinical antirheumatic drug development by reverse pharmacological methods in the future.

References

Akaogi J, Nozaki T, Satoh M, Yamada H (2006) Role of PGE2 and EP receptors in the pathogenesis of rheumatoid arthritis and as a novel therapeutic strategy. Endocr Metab Immune Disord Drug Targets 6(4):383–394

Alam M, Joy S, Ali SU (1991) Screening of Sida cordifolia Linn, Sida rhombifolia and Triumfetta rotundifolia for anti-inflammatory and anti-pyretic drugs. Indian Drugs 28(2):397–399 al-Awadi F, Fatania H, Shamte U (1991) The effect of a plants mixture extract on liver gluconeogenesis in streptozotocin induced diabetic rats. Diabetes Res 18(4):163–168

Alexanian A, Miller B, Chesnik M, Mirza S, Sorokin A (2014) Post-translational regulation of COX2 activity by FYN in prostate cancer cells. Oncotarget 5(12):4232–4243. doi:10.18632/oncotarget.1983

Anderson GD, Keys KL, De Ciechi PA, Masferrer JL (2009) Combination therapies that inhibit cyclooxygenase-2 and leukotriene synthesis prevent disease in murine collagen induced arthritis. Inflamm Res 58(2):109–117. doi:10.1007/s00011-009-8149-3

Arnett FC, Edworthy SM, Bloch DA, McShane DJ, Fries JF, Cooper NS (1988) The American Rheumatism Association 1987 revised criteria for the classification of rheumatoid arthritis. Arthritis Rheum 31(3):15–24

Arulmozhi S, Mazumder PM, Sathiyanarayanan L, Ashok P (2001) Anti-arthritic and antioxidant activity of leaves of Alstonia scholaris Linn R Br. EuJIM 3(2):83–90

Bhalla TN, Gupta MB, Sheth PK, Bhargava KP (1968) Anti-inflammatory activity of *Boerhaavia diffusa*. Indian J Physiol Pharmacol 12(1):37–38

Chen MP, Yang SH, Chou CH, Yang KC, Wu CC, Cheng YH, Lin FH (2010) The chondroprotective effects of ferulic acid on hydrogen peroxide-stimulated chondrocytes: inhibition of hydrogen peroxide-induced pro-inflammatory cytokines and metalloproteinase gene expression at the mRNA level. J Inflamm Res 59(8):587–595

Chitme RH, Patel PN (2009) Antiarthritis activity of Aristolochia bracteata extract in experimental animals. Open Natl Pro J 2(1):6–15

Chitme HR, Malipatil M, Chandrashekhar VM, Prashant PM (2010) Antiallergic activity of Aristolochia bracteolata Lank in animal model. Indian J Exp Biol 48(1):46–52

Danial NN, Korsmeyer SJ (2004) Cell death: critical control points. Cell 116(2):205–216

Dapurkar KV, Sahu KG, Sharma H, Meshram S, Rai G (2013) Anti-arthritic activity of roots extract of *Boerhaavia Diffusa* in adjuvant induced arthritis rats. Sch Acad J Pharm 2(2):107–109

De Jong H, Gouz´e JL, Hernandez C, Page M, Sari T, Geiselmann J (2004) Qualitative simulation of genetic regulatory networks using piecewise linear models. Bull Math Biol 66(2):301–340

Dhawan BN, Dubey MP, Mehrotra BN, Rastogi RP, Tandon JS (1980) Screening of Indian plants for biological activity: part IX. Indian J Exp Biol 8(6):594–606

Eissing T, Conzelmann H, Gilles ED, Allg¨ower F, Bullinger E, Scheurich P (2004) Bistability analysis of a caspase activation model for receptor-induced apoptosis. J Biol Chem 279(2):36892–36897

Feldmann M, Brennan FM, Maini RN (1996) Role of cytokines in rheumatoid arthritis. Annu Rev Immunol 14(1):397–440

Firestein GS (2003) Evolving concepts of rheumatoid arthritis. Nature 423(6937):356–361

Funk JL, Frye JB, Oyarzo JN, Zhang H, Timmermann BN (2010) Anti-arthritic effects and toxicity of the essential oils of turmeric (*Curcuma longa L.*). J Agric Food Chem 58(2):842–849

Gokhale AB, Damre AS, Kulkami KR, Saraf MN (2002) Preliminary evaluation of anti-inflammatory and anti-arthritic activity of *S. lappa, A. speciosa and A. aspera*. Phytomedicine 9(5):433–437

Grassi W, De Angelis R, Lamanna G, Cervini C (1998) The clinical features of rheumatoid arthritis. Eur J Radiol 27(1 Suppl):S18–S24

Gupta MB, Bhalla TN, Gupta GP, Mitra CR, Bhargava KP (1969) Anti-inflammatory activity of natural products. I. Triterpenoids. Eur J Pharmacol 6(1):67–70

Gupta OP, Sharma N, Chand D (1992) A sensitive and relevant model for evaluating anti-inflammatory activity-papaya latex-induced rat paw inflammation. J Pharmacol Toxicol Methods 28(1):15–19

Gupta SR, Nirmal SA, Patil RY, Asane GS (2009) Anti-arthritic activity of various extracts of Sida rhombifolia aerial parts. Nat Prod Res 23(8):689–695. doi:10.1080/14786410802242778

Hamada S, Kataoka T, Woo JT, Yamada A, Yoshida T, Nishimura T (1997) Immunosuppressive effects of gallic acid and chebulagic acid on CTL-mediated cytotoxicity. Biol Pharm Bull 20(9):1017–1019

Harikumar KB, Kuttan R (2007) Protective effect of *Phyllanthus amarus* against radiation-induced changes in the intestine and mouse chromosomal damage. J Radiat Res 48(1):469–476

Hikino H, Taguchi T, Fujimura H, Hiramatsu Y (1977) Anti-inflammatory principles of *Caesalpinia sappan* and *Haematoxylon campechianum* wood. Planta Med 31(3):214–220

Hilbers I, Hansen T, Petrow PK, Gaumann A, Bräuer R, Salzmann G, Gay RE, Kosmehl H, Kirkpatrick CJ, Gay S, Kriegsmann J (2003) Expression of the apoptosis accelerator Bax in rheumatoid arthritis synovium. Rheumatol Int 23(2):75–81

Hong SJ, Rim GS, Yang HI, Yin CS, Koh HG, Jang MH, Kim CJ, Choe BK, Chung JH (2005) Bee venom induces apoptosis through caspase-3 activation in synovial fibroblasts of patients with rheumatoid arthritis. Toxicon 46(1):39–45

Huang XY, Chen FH, Li J, Xia LJ, Liu YJ, Zhang XM (2008) Mechanism of fibroblast-like synoviocyte apoptosis induced by recombinant human endostatin in rats with adjuvant arthritis. Anat Rec 291(8):1029–1037

Iain B, Schett M, Schett G (2007) Cytokines in the pathogenesis of rheumatoid arthritis. Nat Rev Imm 7(1):429–442. doi:10.1038/nri2094

Islam ME, Haque ME, Mosaddik MA (2003) Cytotoxicity and antibacterial activity of Sida rhombifolia (Malvaceae) grown in Bangladesh. Phytother Res 17(8):973–975

Kalaria P, Gheewala P, Chakraborty M, Kamath J (2012) A phytopharmacological review of Alstonia scholaris: a panoramic herbal medicine. IJRAP 3(3):367–371

Kassuya CA, Leite DF, de Melo LV, Rehder VL, Calixto JB (2005) Anti-inflammatory properties of extracts, fractions and lignans isolated from *Phyllanthus amarus*. Planta Med 71(8):721–726

Kaubuyama Y, Kitamura T, Yamaki J, Homma MK, Kikuchi S, Homma Y (2008) Involvement of thioredoxin reductase 1 in the regulation of redox balance and viability of rheumatoid synovial cells. Biochem Biophys Res 367(1):491–496

Kitamura K, Nakamoto Y, Kaneko S, Mukaida N (2004) Pivotal roles of interleukin-6 in transmural inflammation in murine T cell transfer colitis. J Leukoc Biol 76(6):1111–1117

Kluck RM, Bossy-Wetzel E, Green DR (1997) The release of cytochrome c from mitochondria: a primary site for Bcl-2 regulation of apoptosis. Science 275(5303):1132–1136

Kouri T (1985) Etiology of rheumatoid arthritis. Experientia 41(4):434–441

Lee KT, Kim JH (1999) Brazilin as a new sunless tanning agent. Sci Conf Asian Soc Cosmet Sci 3(1):33–36

Liu J, Sun Y (2013) How does Chinese medicine target cytokine imbalance in rheumatoid arthritis? Chin J Integr Med 19(11):874–880

Liu XR, Han WQ, Sun DR (1992) Treatment of intestinal metaplasia and atypical hyperplasia of gastric mucosa with Xiao wei yan powder. Zhongguo Zhong Xi Yi Jie He Za Zhi 12(10):602–603

Malfait AM, Gallily R, Sumariwalla PF, Malik AS, Andreakos E, Mechoulam R (2000) The non-psychoactive cannabis constituent cannabidiol is an oral anti-arthritic therapeutic in murine collagen-induced arthritis. Proc Natl Acad Sci U S A 97(17):9561–9566

Malia SM, Sinnathambia A, Kapasea CU, Bodhankara SL, Mahadik KR (2011) Anti-arthritic activity of standardised extract of *Phyllanthus amarus* in freund's complete adjuvant induced arthritis'. Biomed Aging Pathol 1(1):85–190

Mananvalan G, Singh J (1979) Chemical and some pharmacological studies on leaves of P. longum Linn. Indian JPharm Sci 41(1):190–193

Mancini DA, Dias AL, Pinto JR, Mancini FJ (1999) Antioxidant aqueous extract from cinnamon (Cinnamomum zeylanicum, Blume) as inhibitors of influenza virus. Rev Bras Cien Farma 35(1):155–160

Marzo I, Brenner C, Zamzami N, Susin SA, Beutner G, Brdiczka D, Rémy R, Xie ZH, Reed JC, Kroemer G (1998) The permeability transition pore complex: a target for apoptosis regulation by caspases and Bcl-2-related proteins. J Exp Med 187(8):12611271

McCartney-Francis NAJ, Mizel DE, Albina JE, Xie QW, Nathan CF, Wahl SM (1993) Suppression of arthritis by an inhibitor of nitric oxide synthase. J Exp Med 178(2):49–54

Meena AK, Garg N, Nain J, Meena RP, Rao MM (2011) Review on ethanobotany, phytochemical and pharmacological profile of Alstonia scholaris. IRJP 2(1):49–54

Menon MK, Kar A (1971) Analgesic and psychopharmacological effects of the gum resin of *Boswellia serrata*. Planta Med 19(4):333–341

Mishra NK, Bstia S, Mishra G, Chowdary AK, Patra S (2011) Anti-arthritic activity of *Glycyrrhiza glabra, Boswellia serrata* and their synergistic activity in combined formulation studied in freund's adjuvant induced arthritic rats. IJPER 2(1):92–98

Moreira FA, Guimarães FS (2005) Cannabidiol inhibits the hyperlocomotion induced by psychotomimetic drugs in mice. Eur J Pharmacol 512(2–3):199–205

Mukhopadhyay A, Basu N, Ghatak N, Gujral PK (1982) Anti-inflammatory and irritant activities of curcumin analogues in rats. Agents Actions 12(4):508–515

Munasinghe TC, Seneviratne CK, Thabrew MI, Abeysekera AM (2001) Anti-radical and antilipoperoxidative effect of some plant extracts used by Sri Lankan traditional medical practitioner for cardioprotection. Phytother Res 15(6):519–523

Nadkarni KM (2009) Indian materia medica. Poupular Prakashan Pvt. Ltd, Bombay

Nair V, Singh S, Gupta YK (2010) Anti-arthritic and disease modifying activity of Terminalia chebula Retz. in experimental models. J Pharm Pharmacol 62(12):1801–1806. doi:10.1111/j.2042-7158.2010.01193.x

Naka T, Nishimoto N, Kishimoto T (2002) The paradigm of IL-6: from basic science to medicine. Arthritis Res 4(Suppl 3):S233–S242

Nikaido T, Ohmoto T, Noguchi H, Kinoshita T, Saitoh H, Sankawa U (1981) Inhibitors of cyclic AMP phosphodiesterase in medicinal plants. Planta Med 43(1):18–23

Noguchi M, Kimoto A, Sasamata M, Miyata K (2008) Micro-CT imaging analysis for the effect of celecoxib, a cyclooxygenase-2 inhibitor, on inflammatory bone destruction in adjuvant arthritis rats. J Bone Miner Metab 26(5):461–468. doi:10.1007/s00774-008-0855-3

Otero M, Goldring MB (2007) Cells of the synovium in rheumatoid arthritis. Chondrocytes. Arthritis Res Ther 9(5):220

Pandey A, Bani S, Satti NK, Gupta BD, Suri KA (2012) Anti-arthritic activity of agnuside mediated through the down-regulation of inflammatory mediators and cytokines. Inflamm Res 61(4):293–304. doi:10.1007/s00011-011-0410-x

Pandian RS, Anuradha CV, Viswanathan P (2002) Gastroprotective effect of fenugreek seeds (Trigonella foenum graecum) on experimental gastric ulcer in rats. J Ethnopharmacol 81(3):393–397

Park J, Ernst E (2005) Ayurvedic medicine for rheumatoid arthritis: a systematic review. Semin Arthritis Rheum 34(5):705–713

Peng C, Perera PK, Li Y, Teng Q, Han C (2010) Experimental study on secondary affection of rat's collagen arthritis treated with Yi Shen Juan Bi pill. Chin J Trad MedSci Tech 17(1):389–390

Perera PK, Peng C, Li Y, Fang W, Han C (2010a) Immunomodulatory activity of a Chinese herbal drug Yi Shen Juan Bi (YJB) in adjuvant arthritis. Indian J Pharmacol 42(2):65–69. doi:10.4103/0253-7613.64489

Perera PK, Peng C, Xue L, Li Y, Fang W, Han C (2010b) Effects of Yi Shen Juan Bi (YJB) pill on experimental rheumatoid arthritis. CJNM 8(1):57–61

Perera PK, Peng C, Xue L, Li Y, Han C (2011) *Ex vivo* and *in vivo* effect of Chinese herbal Pill Yi Shen Juan Bi (YJB) on experimental arthritis. J Ethnopharmacol 134(1):171–175

Perlman H, Georgans C, Pagliari LJ, Koch AE, Haines K, Pope RM (2000) Bcl-2 expression in synovial fibroblasts is essential for maintaining mitochondrial homeostasis and cell viability. J Immunol 164(10):5227–5235

Poonam M, Shradha B (2012) Ferula asafoetida: traditional uses and pharmacological activity. Pharmacogn Rev 6(12):141–146. doi:10.4103/0973-7847.99948
Pungle P, Banavalikar M, Suthar A, Biyani M, Mengi S (2003) Immunomodulatory activity of boswellic acids of *Boswellia serrata Roxb*. Indian J Exp Biol 41(12):1460–1462
Rajamanickam V, Rajasekaram A, Jesupillai M, Darlin Q, Sabitha R (2009) Anti pyretic activity of Aristolochia bracteolate. IJAM 8(1):4
Rajpoot K, Mishra RN (2011) *Boerhaavia diffusa* roots (Punarnava mool) – review as rasayan (rejuvenator/antiaging). IJRPC 2(24):1451–1460
Rani P, Venkatesan M, Binilraj J, Sasidha SS, Amma P (2010) Antioxidant and cytotoxic potential of acetone and methanolic extract of *C. zeylanicum* dry bark. Cell Tissue Res 10(1):2131–2138
Ravishankar B, Bhaskaran NR, Sasikala CK (1985) Pharmacological evaluation of *Vitex negundo* (Nirgundi) leaves. Bull Med Ethno Bot Res 6(1):72–92
Ravishankar B, Bhaskaran NR, Sasikala CK (1986) Pharmacology of *Vitex negundo* Linn (Nirgundi) root. JDRAS 7(1):62–77
Sangha O (2000) Epidemiology of rheumatic diseases. Rheumatology (Oxf) 39(Suppl2):3–12
Schulze-Osthoff K, Ferrari D, Los M, Wesselborg S, Peter ME (1998) Apoptosis signaling by death receptors. Eur J Biochem 254(3):439–459
Sharma A, Singh R (1980) Screening of anti-inflammatory activity of certain indigenous drugs on carrageenan induced hind paw oedema in rat. Bull Med Ethnobot Res 2(3):262
Shinde UA, Phadke AS, Nair AM, Mungantiwar AA, Dikshit VJ, Saraf MN (1990) Studies on the anti-inflammatory and analgesic activity of Cedrus deodara (Roxb.) Loud. Wood oil. J Ethnopharmacol 65(1):21–27
Silveira FN, Tufik S (1981) Comparative effects between cannabidiol and diazepam on neophobia, food intake and conflict behavior. Res Comm Psychol Psychiatr Behav 6(3):25–26
Sindhu G, Ratheesh M, Shyni GL, Nambisan B, Helen A (2012) Anti-inflammatory and antioxidative effects of mucilage of Trigonella foenum graecum (Fenugreek) on adjuvant induced arthritic rats. Int Immunopharmacol 12(1):205–211. doi:10.1016/j.intimp.2011.11.012
Soeken KL, Miller SA, Ernst E (2003) Herbal medicines for the treatment of rheumatoid arthritis: a systematic review. Rheumatology (Oxf) 42(5):652–659
Sohni YR, Bhatt RM (1996) Activity of a crude extract formulation in experimental hepatic amoebiasis and immunomodulation studies. J Ethnopharmacol 54(2–3):119–124
Srimal RC, Dhawan BN (1973) Pharmacology of diferuloylmethane (curcumin), a non-steroidal anti-inflammatory agent. J Pharm Pharmacol 25(6):447–452
Stefanovic-Racic M, Stadler J, Evans CH (1993) Nitric oxide and arthritis. Arthritis Rheum 36(8):1036–1044
Stuart JM, Townes AS, Kang AH (1984) Collagen autoimmune arthritis. Annu Rev Immunol 2(1):199–218
Szodoray P, Jellestad S, Nakken B, Brun JG, Jonsson R (2003) Programmed cell death in rheumatoid arthritis peripheral blood T-cell subpopulations determined by laser scanning cytometry. Lab Invest 83(12):1839–1848
Taker M, Deny S, Mohamad SR, Fadzilah A, Abdul M, Hasnah SM, Ahmad F (2007) Antioxidant activity of cinnamtannin B1 from *Cinnamomum zeylanicum Blume*. Phytomedicine 16(1):601–608
Tandon VR (2005) Medical uses and biological activities of *Vitex negundo*. Nat Prod Radiance 4(3):162–165
Tandon V, Gupta RK (2005) Effect of *Vitex negundo* on oxidative stress. Indian J Pharmacol 37(1):38–40
Telang RS, Chatterjee S, Varshneya C (1999) Studies on analgesic and anti-inflammatory activities of *Vitex negundo* Linn. Indian J Pharmacol 31(5):363–366
Uma C, Shashidhar S, Chandrasekar SB, Narasimha Rao M (2011) Studies of preliminary phytochemical and Anti-arthritic activity of heart wood of Cedrus deodara (Roxb.). Res J Pharm, Biol Chem Sci 2(3):654–660

Valero M, Salmerón MC (2003) Antibacterial activity of 11 essential oils against Bacillus cereus in tyndallized carrot broth. Int J Food Microbiol 85(1):73–81

van't Hof RJ, Hocking L, Wright PK, Ralston SH (2000) Nitric oxide is a mediator of apoptosis in the rheumatoid joint. Rheumatology 39(9):1004–1008

Vetal S, Subhash LB, Vishwaraman M, Prasad AT (2013) Anti-inflammatory and anti-arthritic activity of type-A procyanidin polyphenols from bark of *Cinnamomum zeylanicum* in rats. Food Sci Hum Wellness 2(2):59–67

Wang YZ, Sun SQ, Zhou YB (2011) Extract of the dried heartwood of *Caesalpinia sappan* L. attenuates collagen-induced arthritis. J Ethnopharmacol 136(1):271–278. doi:10.1016/j.jep.2011.04.061

Yende SR, Sannapuri VD, Vyawahare SN, Harle UN (2010) Antirheumatoid activity of aqueous extract of *Piper longum* on freund's adjuvant-induced arthritis in rats. IJPSR 1(9):129–133

Zuardi AW, Cosm RA, Graef FG, Guimarãe FS (1993) Effects of ipsapirone and cannabidiol on human experimental anxiety. J Psychopharmacol 7(1Suppl):82–88

Chapter 19
Ayurvedic Plants with Antidiabetic Potential

Manasi Malgaonkar, Amey Shirolkar, Soma Narasimha Murthy, and Sharad Pawar

Abstract Since ancient period in India, *Madhumeha* (diabetes mellitus) has been treated using indigenous medicines. A rapid increase in the occurrence of diabetes is a serious threat to the humankind. The drugs originated from new bioactive plants have shown more efficacy than oral hypoglycemic agents used in the clinical therapy. Recently attention has been directed toward the selection and confirmation of plants with antidiabetic potential. A number of studies have confirmed the benefits of Ayurvedic plants with hypoglycemic effects in the management of this disorder. These plants usually delay the diabetic complications and rectify the metabolic abnormalities. The present chapter reviews the Ayurvedic perspective of diabetes and the various plants used in diabetes treatment with their mechanism of action and pharmacological test results. Also, the article emphasizes the rational use of traditional as well as indigenous natural medicines.

Keywords Ayurvedic perspective • Diabetes • Madhumeha • Medicinal plants • Prameha • Pathophysiology

Abbreviations

+OH	Hydroxide ion
DM	Diabetes mellitus
DPPIV	Dipeptidyl peptidase 4
GIP	Glucose-dependent insulinotropic polypeptide
GLP-1	Glucagonlike peptide-1
IDDM	Insulin-dependent diabetes mellitus
IDF	International Diabetes Federation
kg	Kilogram
mg/dL	Milligram per deciliter

M. Malgaonkar • A. Shirolkar • S.N. Murthy • S. Pawar (✉)
National Research Institute of Basic Ayurvedic Sciences, Pune, India
e-mail: sd_pawar@yahoo.com

H.-S. Tsay et al. (eds.), *Medicinal Plants - Recent Advances in Research and Development*, DOI 10.1007/978-981-10-1085-9_19

mg/kg	Milligram per kilogram
mg	Milligram
NIDDM	Non-insulin-dependent diabetes mellitus
PPAR	Peroxidase proliferator-activated receptor
PPAR-gamma:	Peroxisome proliferator-activated receptor gamma
SGOT	Serum glutamic-oxaloacetic transaminase
SGPT	Serum glutamic-pyruvic transaminase
STZ	Streptozotocin
TBARS	Thiobarbituric acid-reactive substances
WHO	World Health Organization

Abbreviations Used Ayurveda Part

CSCi6	Caraka Samhita Chikitsasthana 6th Chapter
Caraka Nidana 24	Caraka Nidana 24th Chapter
Caraka Nidana 36	Caraka Nidana 36th Chapter
Caraka Nidana 4/3	Caraka Nidana 4th Chapter Verse 3
Sushruta	*Nidana* 6/4 : Sushruta Nidana 6th Chapter Verse 4
AHNi 6/8	Astang Hrudaya Nidanasthana 6th Chapter Verse 8
Caraka Samhita Indriyasthana 9/8-9	Caraka Samhita Indriyasthana 9th Chapter Verse 8, 9

19.1 Introduction

Diabetes mellitus is one of the common endocrine-metabolic disorders and non-communicable disease. It is a heterogeneous group of disorders characterized by abnormalities in carbohydrate, protein, and lipid metabolism. In practical terms, it is a situation in which cells are starving in the sea of glucose (Uma Devi et al. 2006). It is the fourth leading causes of death in the most of the developed countries and is an epidemic in many developing nations (Arumugam et al. 2013). This dreadful disease is spread in all parts of the world and is becoming a serious threat to human health (Dwivedi and Daspaul 2013).

Diabetes is caused by the abnormality of carbohydrate metabolism which is linked to low blood insulin level or insensitivity of target organs to insulin which in long term affects the quality of human life (Maiti et al. 2004). Insulin is a hormone that is required for the conversion of sugar starches and other food into energy. Glucagon plays an important role in the release of glucose from liver cells into the blood for the production of energy. The human body has to modulate the blood glucose levels at a very narrow range which is done with insulin and glucagon. For a healthy adult, normal fasting blood sugar level should be 70–100 mg/dL, and the value should be less than 140 mg/dL in postprandial (2 h after consuming glucose) (Guyton and Hall 2006; Kumar et al. 2010; Powers 2012). Specifically, the symptom of diabetes mellitus is characterized by constant high levels of blood glucose (sugar), thirst, and polyuria. Also, diabetes mellitus can be considered by hyperglycemia, hyperlipidemia, hyperaminoacidemia, and hypoinsulinemia, which leads to decrease in insulin secretion and insulin action (Altan 2003; Rao et al. 2010).

According to the reports of International Diabetes Federation and WHO, diabetes is growing alarmingly in India and gaining the status of a potential epidemic; more than 65.1 million people are suffering from this disease, compared to 50.8 million in 2010 (Anonymous 2013; Kaveeshwar and Cornwall 2014). The prevalence of diabetes is predicted globally from 171 million in 2000 to 366 million in 2030 with a maximum increase in India. It is predicted that by 2030, diabetes mellitus may afflict up to 79.4 million individuals in India (Wild et al. 2004; Boon et al. 2006; Whiting et al. 2011). The data of the International Diabetes Federation (IDF) indicated the statistics of diabetes mellitus in the age group from 20 to 79 years as following: Bangladesh (9.85 %), India (8.31 %), Sri Lanka (7.77 %), Pakistan (6.72 %), and Nepal (3.03 %).

There are two major types of diabetes mellitus: Type I diabetes (formerly known as insulin-dependent diabetes mellitus or IDDM) and Type II diabetes (formerly known as non-insulin-dependent diabetes mellitus or NIDDM). Type I diabetes occurs in young people usually below 35 years of age, while type II occurs in older people usually above 35 years of age and often overweight. Type II diabetes is characterized by insulin resistance, hyperglycemia, and often hyperlipidemia (Cai et al. 2006; Dong et al. 2012).

Type I diabetes leads to an inability to release insulin resulting in a low rate of glucose uptake into the muscles and adipose tissue (Lehninger et al. 2010). In type I, the pancreas cannot synthesize insulin so the patient must be treated with insulin. In the absence of insulin, the patient cannot survive, and insulin cannot be orally administered. Thus, the patients must receive insulin injections once or twice per day (Cai et al. 2006).

Type II diabetes usually occurs in obese individuals, and it is associated with hypertension and dyslipidemia. In type II diabetes, the pancreas does synthesize insulin, but the body cannot use the insulin properly (Cai et al. 2006; Dong et al. 2012). Untreated or uncontrolled diabetes can lead to several chronic complications such as retinopathy, nephropathy, neuropathy, and cardiovascular diseases as well as morbidity and mortality (Port and Schwartz 1996; Rizvi and Mishra 2013). Many peroxidase proliferator-activated receptor (PPARs) agonists, such as clofibrate, bezafibrate, pioglitazone, and rosiglitazone, have been used in the treatments of type II diabetes (Ding et al. 2005; Cai et al. 2006).

Obesity is one of the major risk factors for diabetes. A recent report confirmed that increasing obesity in South Asia is primarily driven by nutrition, lifestyle, and demographic transitions, increasingly faulty diets and physical inactivity, in the background of genetic predisposition (Misra and Shrivastava 2013). The etiology of diabetes in India is multifactorial and includes genetic factors coupled with environmental influences such as obesity associated with rising standard of living and sedentary lifestyle.

Though oral hypoglycemic agents like sulfonylureas and biguanides are the major means of management of the disease; however, as on date, there is no complete cure in the modern system of medicine. Ayurveda, a system of Indian medicine, mentions the details of diabetes and its management by altering the lifestyle coupled with administration of Ayurvedic drugs. Due to side effects associated with the hypoglycemic agents of modern medicine, there is a growing interest globally in herbal remedies, particularly in Ayurvedic drugs.

Various in vitro models designed to simulate specific therapeutic targets have been used to screen plant extracts for potential antidiabetic activity. These assays include alpha-glucosidase inhibition, DPPIV inhibition, antioxidant activity, glucose uptake, and PPAR-gamma agonist activity. Also, potential hepatotoxicity was assessed using HepG2/C3A cells (Koekemoer et al. 2008).

This article presents a review of some reported antidiabetic medicinal plants (with their botanical name, common name, constituent, and mechanism of action for antidiabetic action) and plant-based marketed polyherbal formulations.

19.2 Causes of Diabetes

1. Insufficient production of insulin (either absolutely or about the body's needs).
2. Production of defective insulin.
3. The inability of the cells to use insulin properly and efficiently leads to hyperglycemia.

The primary problem in type II diabetes is the inability of the cells to use insulin properly affecting the fat tissues and muscles, which finally results in the condition known as "insulin resistance." Beta cells of the pancreas produce a hormone known as insulin (Colbert 1999). The main disorder in type I diabetes is the complete absence of insulin because of the destruction process of beta cells. This decline of beta cells does occur in type II, but the process is very slow (Gupta and De 2012).

The food contains a simple sugar (glucose) which is an essential nutrient for normal function and metabolic activities as it provides energy. After the breakdown of carbohydrates in the small intestine, glucose gets absorbed into the bloodstream and then utilized as per the need. The excess glucose is stored in the liver in the form of glycogen. However, glucose cannot enter the cells alone and needs insulin to aid the transport. In the absence of insulin, cells get starved in spite of the presence of glucose in the blood. The abundant, unutilized glucose is wastefully excreted in the urine.

19.3 Pathophysiology of Diabetes

As mentioned above in type I diabetes, insulin is almost absent, and glucose amount in plasma is elevated. Along with this, beta cells cannot respond to all insulin secretary stimuli; therefore, type I diabetes is known as a catabolic disorder. Further insulin deficiency is caused by lymphocytic infiltration and destruction of beta cells of islets of Langerhans. In such conditions insulin has to be given externally to reverse this catabolic condition, prevent ketosis, decrease elevated glucose levels, and normalize lipid and protein metabolism. Currently, autoimmunity is considered as the major factor in the pathophysiology of type I diabetes. It is now known that amino acid profiling assessment helps in predicting the diabetes risk as amino acid metabolism plays an essential role in the pathogenesis of diabetes (Wang et al. 2011). It was stated by Joergensen et al. (2011) that deficiency of vitamin D in type I diabetes predicts the causes of mortality but not the development of microvascular conditions (Joergensen et al. 2011; Young et al. 2011).

Type II diabetes is characterized by the combination of peripheral insulin resistance and inadequate insulin secretion by pancreatic beta cells. The improper function of the beta cells is the major factor across the spectrum of prediabetes to diabetes. In the progression from normal glucose tolerance to abnormal glucose tolerance, postprandial blood glucose levels increase first; eventually, fasting hyperglycemia develops as suppression of hepatic gluconeogenesis fails (Gupta and De 2012). Insulin resistance also occurs because of a high-calorie diet, steroid administration, or physical inactivity. While increased glycogen levels and increased glucose-dependent insulinotropic polypeptide (GIP) levels accompany glucose intolerance; however, postprandial glucagonlike peptide-1 (GLP-1) response remains unaltered. Therefore, GLP-1 can be used in target therapy of type II diabetes.

The pathophysiology of the disease differs between the types of diabetes, but most of the complications, including microvascular, macrovascular, and neuropathic, are similar. Hyperglycemia determines metabolic and microvascular problems, however, a macrovascular disease in less related to glycemia.

19.4 Diabetes: Perspective of Ayurveda

Ayurvedic medicine, originating from the holy book of *Atharvaveda* in the Hindu tradition in India, has been followed for centuries. Ayurveda has a long history of aiding people suffering from various physical and psychological diseases. Ayurvedic medicine is a highly recommended form of medicine in India, and it is currently being accepted as evidence-based medicine in the Western world. Ayurveda is one of the most ancient traditional medicinal systems which have now spread beyond India to other territories like Europe, Japan, Malaysia, Mauritius, North America, Russia, South Africa, and Sri Lanka (Elder 2004; Hankey 2005).

The description related to *Prameha* is available in many Ayurvedic textbooks like, *Caraka Samhita* (CSCi6) and *Sushruta Samhita* (found in sixth chapter *Nidana Sthana* and *Chikitsa Sthana's* 11th, 12th, and 13th chapter), while *Madhavakara* described it in *Madhavanidana*, 33rd chapter, which is *Prameha Nidana* (Sushruta 1915; Madhavanidana 1999). The Sanskrit word *Prameha* consists of *Pra*, upsarga (prefix) and *Meha*, watering. Therefore, the literal meaning of it is passing of urine profusely, both in quantity and frequency (Agnivesha 1941).

In Ayurveda, 20 types of *Prameha* have been described, *viz*., *Kaphaja Prameha* (10 types), *Pittaja Prameha* (6 types), and *Vataja Prameha* (4 types), mainly by physical examination of urine (*Caraka Nidana* 4/3, 24, 36; *Sushruta Nidana* 6/4). Among these *Kaphaja Prameha* types are mentioned as completely curable; *Pittaja Prameha* is described as difficult to cure. The last four are *Vataja* in origin, not only incurable but also drug dependent or to be maintainable. *Madhumeha* is one among the four types of *Prameha* mentioned above, which shows many clinical similarities with today's diabetes mellitus. *Madhumeha* is composed of "*madhu* means sweetness and *meha* mean urination" indicating the disease having urine quality similar to *Madhu* (honey) in its color, taste, smell, and consistency called along with the pathognomonic features of *Prameha* (i.e., increased frequency and quantity of urine) is *Madhumeha. Vagbhata* has mentioned clearly two types of *Prameha* based on their etiology, *Sahaja* (hereditary – genetic origin) and *Apathyanimittaja* (acquired due to faulty diet and lifestyle errors) (AHNi 6/8) (Arunadutta 1933). Akin to this, "diabetes mellitus" is composed of "diabetes (Greek word) – meaning excessive urine discharge" and "mellitus (Latin word) – meaning sweetness." Therefore, it can be said that *Madhumeha* and "diabetes mellitus" are similar in their literal meaning (Shastri and Chaturvedi 2005). As described by Acharyas (scholars) in *Caraka* (*Caraka Samhita Indriya Sthana* 9/8-9), *Madhumeha* is included in *Ashta Maha Roga* (eight major disorders) indicating the severity of the disease.

Polyuria (frequent urination), polyphagia (increased hunger), polydipsia (increased thirst), and loss of weight are major symptoms of untreated diabetes. Nausea, rash gas, and weight gain are the most common side effects of antidiabetic medicine, and further serious side effects include low blood sugar, liver, and kidney damages as well as heart issues. *Nidanparivarjan* (Ayurvedic principles of prevention) and *Shodhan Chikitsa* (treatment measures) along with the help of *aushadhi*, i.e., single or polyherbal formulation drugs, and diet (pathya-apathya) management are proven to be significant in treating *Madhumeha* patients.

Madhumeha management with the help of Ayurveda not only helps in prevention and health maintenance but also stands with the healthy benefits for any individual. Plants have always been a great source of drugs, which are the part of Ayurvedic treatment. In the treatment of *Madhumeha*, various drugs of herbal and mineral origin are already mentioned in Ayurvedic texts (Sipika and Agarwal 2015).

Sedentary lifestyle and stressful mental conditions are the cause of many diseases, and diabetes is the perfect example of such lifestyle disorder, which can be controlled by Ayurvedic treatment. Ayurvedic medicine is oriented toward prevention, health maintenance, and treatment of diseases. A variety of plant preparations has been mentioned in Ayurveda and another indigenous system of medicine, which is claimed to be useful in the treatment of diabetes mellitus.

19.5 Ayurvedic Plants with Antidiabetic Potential

The herbal drugs or medicinal plants with antidiabetic activity are yet to be commercially formulated as modern medicines. The plants are often used as adjunctive therapy with conventional medications, enhancing the possibility for synergistic interaction, and even they are acclaimed for their therapeutic properties in the traditional systems of medicine (Koekemoer et al. 2008; Wadkar et al. 2008). Plants provide a potential source of hypoglycemic drugs. Several medicinal plants have potential and have been used in the Indian System of Medicines, including Ayurveda and Siddha. In modern medicine, there are varieties of glucose-lowering agents available for the treatment of type II diabetes with many side effects, including weight gain and the risk of hypoglycemia (Hassan et al. 2010). Among the best existing alternative therapies are herbal remedies, which have been used since ancient times for the treatment of diabetes mellitus (Rang and Dale 1999). About 90 % of the population in rural areas of developing countries rely solely on traditional medicines for their primary health care, and 80 % of the world population use the herbal medicines or natural products as per WHO report (Sheth 2005).

India is known as the botanical garden of the world due to the occurrence of various medicinal plants (Patil et al. 2011). Plant derivatives with antidiabetic potentials have been traditionally used around the world (Yeh et al. 2003). Plants provide a potential source of hypoglycemic drugs and are widely used in several traditional systems of medicine (Shokeen et al. 2008). The flora is a renewable non-exhaustive source of bioactive compounds. Preparation of medicinal plant inventories is necessary for the preservation of medicinal plant information and knowledge. Such inventories will help the society in the identification of medicinal plants and generate economic awareness.

Herbal medicines have been the highly esteemed source of medicine throughout human history. In recent years these are widely used, as these are affordable to many, indicating that herbs are a growing part of modern and advanced medicine (Rao et al. 2010). Over the past decade, herbal medicines have been accepted universally, and they have an impact on world health and international trade. The current scenario of the market reveals that Ayurvedic and herbal medicines are estimated to be more than 100 billion USD (Sheth 2005). Medicinal plants have natural therapeutic values against various diseases. Traditional medicine is used for the treatment of diabetes in developing countries where the cost of conventional medicines is a burden to the population (Saravanan and Pari 2008). Many indigenous Indian medicinal plants have been found to be successful in the management of diabetes. The ethnobotanical information reports about 800 plants that may possess antidiabetic potentials (Alarcon et al. 1998). Of these medicinal herbs like *Fructus corni* (*Cornus officinalis* Sieb. et Zucc.), *Fructus schisandrae* (Turcz.) Baill., *Rhizoma alismatis* (*Alisma plantago-aquatica* L.), and *Rhizoma dioscoreae* (*Dioscorea opposita* Thunb.) are commonly used in traditional Chinese system of medicine to treat diabetes (Lau et al. 2008). These types of herbs usually lower the blood glucose levels. Antidiabetic drugs usually have antioxidant properties, and such drugs (herbal formulations) have a potent effect on blood sugar levels with lesser side effects. Antidiabetic activity of medicinal plants is mainly due to the presence of phenolic compounds, flavonoids, terpenoids, coumarins, and other constituents which show a reduction in blood glucose levels (Saravanan and Pari 2008).

Therefore, many researchers have studied the medicinal properties of different plant parts individually. Some of the plant species and their parts used as medicines have been listed in Table 19.1 (Kawalekar 2011; Dwivedi and Daspaul 2013; Borokini et al. 2013; Krup et al. 2013). New oral hypoglycemic compounds can be provided by antidiabetic plants for many rural populations in developing countries.

Approximately 21,000 plant species are used as the basis for the preparation of medicinal drugs, out of which 2500 species are available in the Indian continent, from these 800 plants are well known for their antidiabetic activity (Modak et al. 2007; Patil et al. 2011; Dwivedi and Daspaul 2013). Ayurvedic drugs are prescribed widely due to their effectiveness, fewer side effects, and relatively low cost. Various plant-derived compounds have been used in the treatment of diabetes to control the blood sugar of patients.

19.5.1 Working Principle of Antidiabetic Plants

The specific ingredient which is responsible for antidiabetic effect and therapeutic effect is not known for most of the plants. In Ayurvedic preparations, many plants or plant parts are mixed, which is likely to work together to produce the desired therapeutic effect. There are many factors which may affect the quality and desire effect. Those can be summarized as the type of environment (climate, soil), healthiness of the herb at the time of collection, the age of the plant, and harvesting/processing techniques (Patel et al. 2012; Dwivedi and Daspaul 2013).

19.5.2 General Mechanism of Action of Antidiabetic Ayurvedic Plants

The mechanism of action of antidiabetic plants can broadly be grouped as following (Modak et al. 2007; Dwivedi and Daspaul 2013):

1. Adrenomimeticism, pancreatic beta cell potassium channel blocking, cAMP stimulation
2. Cortisol lowering activities
3. Increasing the size and number of cells in the islet of Langerhans

Table 19.1 List of plants having antidiabetic properties

S. No.	Plant part used	Name of plants
1.	Aerial parts	*Artemisia pallens* Wall. ex. DC.; *Bidens pilosa* L.; *Bixa orellana* L.; *Teramnus labialis* (L.f.) Sprang.
2.	Bark	*Albizia odoratissima* (L.f.) Benth.; *Anogeissus leiocarpus* (DC.) Guill and Perr.; *Aristolochia ringens* Vahl.; *Azadirachta indica* A.Juss.; *Cinnamomum zelanicum* Blume.; *Croton cajucara* Benth.
3.	Bulb	*Allium cepa* L.; *Allium sativum* L.; *Allium ascalonicum* L.
4.	Flower	*Ageratum conyzoides* L.; *Senna auriculata* (L.) Roxb.; *Gentiana olivieri* Griseb.; *Musa sapientum* L.; *Senna alata* L. Roxb.;
5.	Fruit	*Aegle marmelos* (L.) Corr.; *Aframomum melegueta* (Rose.) K. Schum.; *Ananas cosmos* (L.) Merrill.; *Carica papaya* L.; *Carum carvi* L., *Citrullus lanatus* (Thunb.) Matsum and Nakai; *Citrus aurantiifolia* (Christm.) Swingle; *Citrus sinensis* (L.) Osbeck.; *Coriandrum sativum* L.; *Phyllanthus emblica* L.; *Eugenia aromatica* (L.) Baill.; *Juniperus communis* L.; *Momordica charantia* L.; *Xanthium strumarium* L.
6.	Leaves	*Abrus precatorius* L.; *Aegle marmelos* (L.) Corr.; *Ageratum conyzoides* L.; *Alangium lamarckii* Miq ex Clarke; *Aloe barbadensis* (L.) Burm.f.; *Aloe vera* L.; *Axonopus compressus* (Sw.) P. Beauv.; *Azadirachta indica* A. Juss.; *Brassica oleracea* L.; *Carica papaya* L.; *Cinnamomum tamala* (Hamm.) Nees and Eberm.; *Citrus aurantifolia* (Christm.) Swingle; *Costus speciosus* (Koem.) Sm.; *Cucurbita maxima* Duch.; *Gossypium hirsutum* L.; *Gymnema sylvestre* R.Br.; *Hibiscus acetosella* Welw ex. Hiern.; *Holarrhena antidysenterica* (L.) R.Br.; *Jatropha curcas* L.; *Jatropha gossypifolia* L.; *Mangifera indica* L.; *Momordica charantia* L.; *Moringa oleifera* Lam.; *Murraya koienigii* (L.) Spreng.; *Pterocarpus marsupium* Roxb.; *Ricinus communis* L.; *Senna alata* L. Roxb.; *Sida acuta* Burm.f.; *Solanum aethiopicum* L.; *Syzygium cumini* L.; *Wattakaka volubilis* (L.f.) Stapf.
7.	Rhizome	*Curcuma longa* L.; *Rheum emodi* L.
8.	Roots	*Aegle marmelos* (L.) Corr.; *Catharanthus roseus* L.; *Cucurbita maxima* Duch.; *Dialium guineense* Willd.; *Sida acuta* Burm.f.
9.	Seed	*Acacia Arabica* (L.) Willd.; *Acacia nilotica* (L.) Willd. ex Del.; *Azadirachta indica* A. Juss.; *Brassica juncea* L.; *Carica papaya* L.; *Holarrhena antidysenterica* (L.) R.Br.; *Pterocarpus marsupium* Roxb.; *Syzygium cumini* L.; *Terminalia chebula* Retz.; *Trigonella foenum-graecum* L.
10.	Stem	*Aegle marmelos* (L.) Corr.; *Pterocarpus marsupium* Roxb.; *Tinospora cordifolia* (Willd.) Hook.f.and Thorn.
11.	Whole plant	*Ageratum conyzoides* L.; *Hybanthus enneaspermus*; *Phyla nodiflora* (L.) Greene.; *Ocimum sanctum* L., *Viscum album* L.

4. Inhibition in renal glucose reabsorption
5. Inhibition of β-galactosidase and α-glucosidase
6. Inhibition of α-amylase
7. Improvement in digestion along with a reduction in blood sugar and urea
8. Providing certain necessary elements like calcium, zinc, magnesium, manganese, and copper for beta cells

9. The protective effect on the destruction of the beta cells
10. Prevention of pathological conversion of starch to glucose
11. Reduction in insulin resistance
12. Regenerating and/or repairing pancreatic beta cells
13. Stimulation of insulin secretion by beta cells of islets and/or inhibition of insulin denaturation processes
14. Stimulation of insulin secretion
15. Stimulation of glycogenesis and hepatic glycolysis

19.5.3 Conventional Medicinal Plants Used in Ayurvedic Preparations

The use of herbs in the diabetic treatment is known since ages and has been prevalent in Indian society. Several medicinal plants are reported to possess potential hypoglycemic activity. Plants native to India having antidiabetic potential are *Acacia arabica* (babul), *Aegle marmelos* (bael), *Agrimonia eupatoria* (church steeples), *Allium cepa* (onion), *Allium sativum* (garlic), *aloe vera* (Ghrit kumari), *Azadirachta indica* (neem), *Beta vulgarus* (beetroot), *Benincasa hispida* (ash gourd), *Caesalpinia bonducella* (fever nut), *Citrullus colocynthis* (bitter apple), *Coccinia indica* (ivy gourd), *Eucalyptus globulus* (eucalyptus), *Ficus benghalensis* (banyan tree), *Gymnema sylvestre* (gudmar), *Hibiscus rosa chinensis* (Gurhal/Jaswandi), *Ipomoea batatas* (sweet potato), *Jatropha curcas* (purging nut), *Mangifera indica* (mango), *Momordica charantia* (karela), *Morus alba* (mulberry), *Mucuna pruriens* (kiwach), *Ocimum sanctum* (tulsi), *Pterocarpus marsupium* (Beejsar/Vijaysar), *Punica granatum* (anar), *Syzygium cumini* (jamun), *Tinospora cordifolia* (giloy), and *Trigonella foenum-graecum* (methi) (Rizvi and Mishra 2013).

19.5.4 Antidiabetic Herbal Products Marketed in India

Some of the manufacturers have designed the products using herbs, and these are marketed in India under different brand names such as Ayurveda alternative herbal formula to diabetes (Chakrapani Ayurveda), bitter gourd powder (Garry and Sun natural remedies), Dabur Madhu Rakshak (Dabur), Diabecon (Himalaya), Diabecure (Nature beaute sante), Diabeta (Ayurvedic cure Ayurvedic Herbal Health Products), Diabetes Daily Care (Nature's Health Supply), Dia-care (Admark Herbals limited), Diasulin, Epinsulin (Swastik Formulations), Gurmar powder (Garry and Sun natural remedies), Hyponidd (Charak Pharma), Madhumeha Kusumkara Rasa (Shree Dhootapapeshwar Limited), Madhumehari Granules (Baidyanath), Ojamin (Tates remedies), Pancreatic tonic 180 cp (Ayurvedic herbal supplement), Synedrex (Plethico Laboratories), and Zpter (Om Pharmaceuticals Limited) (Jonnalagadd and

Selkar 2013; Gupta and De 2012; Jegede et al. 2011). Many researchers have concluded in their studies that there is a need for authentic clinical trials along with the molecular mechanism of the herbs possessing antidiabetic potential.

19.6 Plants with Antidiabetic Properties

19.6.1 Aegle marmelos *(L.) Correa. (Bael) (Family: Rutaceae)*

The chief constituent of the drug is marmelosin which is chemically furocoumarin. It also has alkaloids like O-methylhalfordinol and isopentylhalfordinol which are isolated from fruits and are active components. The aqueous extract of *A. marmelos* leaves significantly reduces the blood glucose level and is known to have antioxidative activity in alloxan diabetic rats (Upadhya et al. 2004; Sabu and Kuttan 2004). When leaves of bael are mixed with fenugreek seeds, a similar effect was seen in non-insulin-dependent diabetes mellitus patients (Mohammad 2009). Leaf and callus extract have also shown antidiabetic effect.

19.6.2 Alangium lamarckii *Thwaites. (Family: Alangiaceae)*

Alcoholic extract (250 and 500 mg/kg) of *A. lamarckii* has a significant antidiabetic activity which was recorded in STZ-nicotinamide-induced diabetic rat (Kumar et al. 2011b).

19.6.3 Albizia odoratissima *(L.f.) Benth. (Shirisha) (Family: Fabaceae)*

The bark of this plant is known to possess antidiabetic properties. In specific, methanolic extract of *A. odoratissima* bark showed an antidiabetic effect in alloxan-induced diabetic mice through a significant reduction in the serum cholesterol, triglycerides, serum glutamic-oxaloacetic transaminase (SGOT), serum glutamic-pyruvic transaminase (SGPT), alkaline phosphatase, and total protein (Kumar et al. 2011a).

19.6.4 Allium cepa *L. (Palandu) (Family: Liliaceae)*

A. cepa (also called onion) is known to have antioxidant and hypolipidemic activity. Raw and boiled onion extracts are used to treat diabetes. Aqueous extract of *A. cepa* showed hypoglycemic and hypolipidemic activity in alloxan-induced diabetic animals (Ozougwu 2011; Ogunmodedee et al. 2012). It contains essential amino acids like arginine, histidine, lysine, tryptophan, phenylalanine, methionine, threonine, leucine, and isoleucine. The bulb contains several phenolic acids, such as p-hydroxybenzoic acid, vanillic acid, caffeic acid, etc.; along with these citric, abietic, oxalic, and malic acids are also present. It also contains several oligosaccharides (Kokate et al. 2009). The active ingredient of *A. cepa* is called allyl propyl disulfide, which blocks the breakdown of insulin by the liver and possibly to stimulate insulin production by the pancreas, thus increasing the amount of insulin and reducing the sugar levels in the blood.

19.6.5 **Allium sativum** *L. (Garlic, Rasona) (Family: Liliaceae)*

Garlic bulbs contain 29 % carbohydrates, 56 % proteins (albumin), 0.1 % fat, and 0.1 % volatile oil. It also contains phosphorous, iron, and copper. Volatile oil of the drug is a chief active constituent and contains allyl propyl disulfide, diallyl disulfide, allin, and allicin. Garlic cloves lower blood sugar levels significantly (Kokate et al. 2009). Disulfides present in the garlic are responsible for the antidiabetic effect (Thomson et al. 2007). The antidiabetic activity of ethanolic extract of *A. sativum* was confirmed in normal and streptozotocin-induced rats (Eidi et al. 2006; Thomson et al. 2007).

19.6.6 **Aloe vera** *(L.) Burm.f. (Ghritkumari) (Family: Xanthorrhoeaceae)*

Various extracts of *Aloe vera* showed hypoglycemic activity in hyperglycemic rats (Tanaka et al. 2006; Afaf et al. 2008). *Aloe vera* gel and its isolated compounds are also known to have the same activity in non-insulin-dependent diabetes mellitus mice (Rehman et al. 2011).

19.6.7 **Andrographis paniculata** *(Burm.f.) Wall ex Nees. (Kalmegh) (Family: Acanthaceae)*

This plant is known to have hypoglycemic activity and beta cell protective effects in alloxan-induced diabetic mouse model because of the presence of "andrographolide" compound (Zhang et al. 2009). Aqueous leaf extract is reported to have antioxidant properties and helps to reduce the blood glucose level in STZ-induced diabetic rats (Dandu and Inamdar 2009).

19.6.8 **Annona squamosa** *L. (Sugar Apple, Sitaphal) (Family: Annonaceae)*

Aqueous extract of *A. squamosa* seeds was found useful when supplemented during an experiment with diabetic rats. It was noticed that it controls blood glucose level and improves the plasma insulin and lipid metabolism (Kaleem et al. 2006a; Bhuyan 2015).

19.6.9 **Artemisia sphaerocephala** *Krasch. (Family: Asteraceae)*

In the studies conducted to check the antioxidant effects on STZ-induced diabetic rat, administration of extracts of *A. sphaerocephala* lowers the levels of thiobarbituric acid-reactive substances (TBARS) and +OH in serum and liver tissue. The activity levels of the liver and serum superoxide dismutase were found to decrease significantly. It was concluded from their studies that *A. sphaerocephala* possesses very good antioxidant activity.

19.6.10 **Azadirachta indica** *(Neem) (Family: Meliaceae)*

Aqueous extract of neem leaves showed noticeable antihyperglycemic activity in male albino rats of Wistar strain (Bajaj and Srinivasan 1999).

19.6.11 **Berberis vulgaris** *L. (Barberry) (Family: Berberidaceae)*

Hypoglycemic effect of *B. vulgaris* in STZ-induced diabetic rats was studied. Results indicated that water extract and saponins show a significant hypoglycemic effect (Meliani et al. 2011).

19.6.12 **Boerhaavia diffusa** *L.* **nom. cons.** *(Punarnava) (Family: Nyctaginaceae)*

Extract of *Boerhaavia diffusa* leaves was reported by Rao et al. (2004) and Bhatia et al. (2011) to have a significant antidiabetic activity in STZ-induced hyperglycemic rats.

19.6.13 Brassica juncea *(L.) Vas. (Mustard) (Family: Brassicaceae)*

This is a traditional plant commonly used in Indian spice, especially in Tamil Nadu, a southern state of India. The aqueous seed extract of *B. juncea* (250, 350, 450 mg/kg) showed hypoglycemic activity which was investigated in STZ-induced diabetic male albino rat (Thirumalai et al. 2011). Also, the methanol extract of leaves showed hypoglycemic activity in glucose-induced mice (Rahmatullah et al. 2010).

19.6.14 Calotropis gigantea *(L.) R.Br. (Mandar) (Family: Apocynaceae)*

The chloroform extract of *C. gigantea* flowers showed antihyperglycemic potential in alloxan- induced hyperglycemic rats (Choudhary et al. 2011). In a separate study, the chloroform extract of flowers and leaves was reported to have a significant antidiabetic activity in STZ-induced diabetic rats (Rathod et al. 2011).

19.6.15 Calotropis procera *(Aiton) Aiton. (Arka) (Family: Apocynaceae)*

Investigation of the antihyperglycemic effect on STZ-induced diabetic rats carried out using *C. procera* extracts confirmed the antihyperglycemic potential (Bhaskar and Ajay 2009). Latex of *C. procera* was found to have an effect on the antidiabetic activity and decreased the blood glucose level in alloxan-induced diabetic rats (Preethi et al. 2012).

19.6.16 Cassia fistula *L. (Golden Shower Tree) (Family: Fabaceae)*

The bark of *C. fistula* is known to have antidiabetic potential. Ethyl acetate extract exhibited a significant antihyperglycemic effect in alloxan-induced diabetic rats as well it is known to lower the lipid profile (Malpani and Manjunath 2012). Similar results were noticed in hexane extract of stem bark on STZ diabetic rats (Nirmala et al. 2008).

19.6.17 **Cassia grandis** *L. (Pink Shower, Carao) (Family; Fabaceae)*

The aqueous and ethanolic extracts of *C. grandis* showed significant antidiabetic activity in alloxan-induced diabetic rats (Lodha et al. 2010).

19.6.18 **Cassia occidentalis/Senna occidentalis** *(L.) Link. (Kasamarda, Kasari) (Family: Fabaceae)*

Methanolic extract of *C. occidentalis* leaves exhibited significant antidiabetic activity on alloxan-induced diabetic rats and STZ-induced diabetic rats (Emmanuel et al. 2010).

19.6.19 **Catharanthus roseus** *(L.) G. Don. (Nithyakalyani, Sadabahar) (Family: Apocynaceae)*

Methanolic leaf extract has shown to have a hypoglycemic effect on alloxan-induced diabetic rats. The blood glucose level significantly decreased when compared with control rats. This result was more potent than glibenclamide and metformin (Kaleem et al. 2005; Ohadoma and Michael 2011).

19.6.20 **Chaenomeles sinensis** *(Thouin) Koehne (Family: Rosaceae)*

Ethyl acetate fraction of fruits of *C. sinensis* is a good antidiabetic agent. The extract (50 and 100 mg/kg) has shown a significant antidiabetic activity (Sancheti et al. 2009; Sancheti et al. 2011).

19.6.21 **Cinnamomum tamala** *(Buch. Ham.) Nees and Eberm. (Indian Bay Leaf) (Family: Lauraceae)*

The aqueous extracts exhibited antihyperglycemic as well as antioxidant activities in STZ-induced diabetic rats (Chakraborty and Das 2010).

19.6.22 Cinnamomum zeylanicum *Blume. (Cinnamon) (Family: Lauraceae)*

Cinnamon bark contains volatile oil, tannins, mucilage, calcium oxalate, starch, and mannitol. The cinnamon oil contains cinnamaldehyde and terpenes like phellandrene, pinene, cymene, etc. (Kokate et al. 2009). Alcoholic extract of cinnamon leaves has shown antidiabetic activity. Also, cinnamon improves glucose and lipid of the people (Khan et al. 2003).

19.6.23 Coccinia grandis *(L.) Voigt. (Ivy Gourd) (Family: Cucurbitaceae)*

Coccinia grandis has synonyms *Coccinia indica* and *Cephalandra indica*. Alcoholic extract of leaves is shown to have hypoglycemic activity (Jose and Usha 2010). Oral administration of an alcoholic extract of leaves (600 mg/kg) showed the significant hypoglycemic effect on blood glucose level in normal fasted rats (Ajay 2009). It was reported that the active compounds in plant inhibit glucose-6-phosphatase (Shibi et al. 2015), a key enzyme in the liver involved in regulating the sugar metabolism.

19.6.24 Curcuma longa *L. (Turmeric, Haldi) (Family: Zingiberaceae)*

Rhizome powder of turmeric along with amla juice and honey is commonly used in *Madhumeha* (Acharya 1994). *Curcuma longa* increases the insulin level but does not usually affect the plasma glucose levels in healthy subjects (Wickenberg et al. 2010). The active principles in rhizome are curcuminoids which maintain antioxidant enzymes like superoxide dismutase, catalase, and glutathione peroxidase (Faizal et al. 2009).

19.6.25 Cynodon dactylon *(L.) Pers. (Family: Poaceae)*

An aqueous extract has been demonstrated to have antidiabetic activity in alloxan-induced diabetic rats (Jarald et al. 2008).

19.6.26 Dalbergia sissoo *Roxb. (Shisham) (Family: Fabaceae)*

Ethanolic extract of *D. sissoo* bark is known to possess significant antidiabetic activity in alloxan-induced diabetic rats (Pund et al. 2012).

19.6.27 **Ferula asafoetida** *L. (Asafoetida) (Family: Apiaceae)*

Various extracts of this plant have been reported to have antidiabetic activity in alloxan-induced diabetic rats (Saber and Zaiton 2010).

19.6.28 **Ficus benghalensis** *L. (Indian Banyan) (Family: Moraceae)*

The aqueous extract of *F. benghalensis* exhibits significant antidiabetic activity in streptozotocin-induced diabetic rats (Mahalingam and Krishnan 2008a), and hot water extract is reported to be beneficial in diabetes mellitus induced by alloxan in rabbits (Shukla et al. 1994).

19.6.29 **Ficus glomerata** *Roxb. (Indian Fig) (Family: Moraceae)*

The aqueous extract of stem bark is known to reduce blood glucose level in streptozotocin-induced diabetic rats (Ahmed and Urooj 2008).

19.6.30 **Ficus hispida** *L. (Hairy Fig) (Family: Moraceae)*

The ethanol extract of bark has shown significant antidiabetic activity in diabetic rats (Ghosh et al. 2004).

19.6.31 **Gymnema sylvestre** *R. Br. (Gurmari, Meshasringa) (Family: Apocynaceae)*

Gymnema sylvestre has been called the "sugar killer" due to its ability to reduce the blood sugar level. It helps the pancreas with insulin production in type II diabetes and also improves the ability of insulin to lower blood sugar level. Traditionally leaf extracts are used to treat the diabetic patients. Gymnemic acids have been isolated for the treatment. The plant has been able to normalize blood sugar in animals treated with agents that destroy beta-cell function. It is frequently included in Ayurvedic formulations and often used as a folk treatment for diabetes (Khajuria and Thomas 1992). The extracts of *G. sylvestre* leaves had been investigated for antidiabetic activity and noticed to have a significant effect in alloxan-induced diabetic rats (Sathya et al. 2008; Mall et al. 2009). Water extract is seen to increase the effect of exogenous insulin in normal and hypoglycemic rats (Chattopadhyay et al. 1993).

19.6.32 Hemidesmus indicus *(L.) R.Br. (Indian Sarsaparilla, Sariwa) (Family: Apocynaceae)*

Aqueous extracts of roots demonstrated antidiabetic activity on STZ-induced diabetic rats (Mahalingam and Krishnan 2008b).

19.6.33 Madhuca longifolia *(J. Konig) Macbr. (Family: Sapotaceae)*

Aqueous extract of the roots exhibited potential as antidiabetic agent on STZ-induced diabetic rats (Dahake et al. 2010).

19.6.34 Mimosa pudica *L. (Touch Me Not Plant) (Family: Fabaceae)*

The ethanolic extract of the leaves showed significant antidiabetic activity (Sutar et al. 2009).

19.6.35 Momordica charantia *L. (Bitter Gourd) (Family: Cucurbitaceae)*

This plant is commonly known to have antidiabetic as well as antihyperglycemic activity and used as an antidiabetic agent in India and other Asian countries. Extracts of fruit pulp, seeds, leaves, and whole plant are commonly used in the *Ayurveda*. It is used as a fresh juice extract, tincture, and powdered leaf. An aqueous extract of seeds showed antidiabetic activity in STZ-induced diabetic rats (Sekar et al. 2005). In a separate study, crude juice extract of *M. charantia* showed antidiabetic activity in rats (Matheka et al. 2011). This plant contains lectin which has insulin-like activity due to its nonprotein-specific linking with insulin receptors. This lectin lowers blood glucose level by acting on peripheral tissues (Kokate et al. 2009). Triterpenoids, such as momordicin, momordicilin, and momordol, were isolated and showed antidiabetic activity (Tan et al. 2008).

19.6.36 Nigella sativa *L. (Kalonji) (Family: Ranunculaceae)*

The crude aqueous extract of seeds exhibited the antidiabetic activity and showed effects on intestinal glucose absorption in the in vitro studies (Ogunmodedee et al. 2012). The ethanol extract exhibited antidiabetic activity on STZ-induced diabetic rats (Kaleem et al. 2006b; Meddah et al. 2009).

19.6.37 Ocimum sanctum *L. (Tulsi, Holy Basil) (Family: Lamiaceae)*

This whole plant is known to have medicinal properties. Leaf extracts showed antidiabetic activity in experimental models (Hannan et al. 2006; Khan et al. 2010).

19.6.38 Opuntia ficus-indica *(L.) Mill. (Indian Fig Opuntia) (Family: Cactaceae)*

The main chemical constituent of this plant includes methoxytyramine, candicine, hordenine, tyramine, etc. Boiled stems are used in the treatment of type II diabetes (Lopez et al. 2014).

19.6.39 Oroxylum indicum *(L.) Benth ex Kurz. (Shyonaka, Kampong) (Family: Bignoniaceae)*

The aqueous and ethanolic extract of roots was examined in alloxan- and dexamethasone-induced diabetic rats and showed a significant decrease in plasma glucose levels (Tamboli et al. 2011).

19.6.40 Panax ginseng *Baill and Oken (Ginseng) (Family: Araliaceae)*

This plant contains a mixture of several saponins, such as ginsenosides, panaxosides, and chikusetsusaponin. These saponins are known to reduce the blood glucose levels and are used as the hypoglycemic agent.

19.6.41 Piper longum *L. (Pepper) (Family: Piperaceae)*

The ethanolic extract of dried fruits showed antihyperglycemic activity on alloxan-induced diabetic rats (Manoharan et al. 2007).

19.6.42 Piper nigrum *L. (Black Pepper) (Family: Piperaceae)*

The aqueous extract of *P. nigrum* seeds and *Vinca rosea* flowers evaluated in alloxan-induced diabetic rats showed antidiabetic activity (Kaleem et al. 2005).

19.6.43 Phyllanthus emblica *L. (Amla) (Family: Euphorbiaceae)*

Dried and fresh fruits are used in the treatment of diabetes. Amla is a rich natural source of vitamin C. It contains 0.5 % fat, phyllemblin, 5 % tannin, pectin, phosphorous, iron, and calcium. *P. emblica* has shown antidiabetic activity in animal models (Kokate et al. 2009; Mehta et al. 2009).

19.6.44 Pterocarpus marsupium *Roxb. (Indian Kino, Vijaysar) (Family: Fabaceae)*

The gum resin of this plant looks like a dried blood is occasionally used in Ayurveda. The clinical studies also proved the blood sugar balancing property. Epicatechin, a flavonoid extracted from the bark of this plant, is responsible for beta cell protection. It is the only herb which has beta cell rejuvenating property. Crude alcohol extract have been shown to regenerate functional pancreatic beta cells (Dash 1987; Devaraj and Nirmala 2012).

19.6.45 Pongamia pinnata/Millettia pinnata *(L.) Pani. (Family: Fabaceae)*

The ethanolic extract showed hypoglycemic activity on alloxan-induced animals (Lanjhiyana et al. 2011).

19.6.46 Punica granatum *L. (Pomegranate) (Family: Lythraceae)*

Aqueous extract of fruits showed its antidiabetic potentials in alloxan-induced diabetic rats (Khalil 2004; Radhika et al. 2011; Jain et al. 2012).

19.6.47 Rheum emodi *L. (Rhubarb) (Family: Polygonaceae)*

It is also known as Himalayan rhubarb consists of active constituents such as emodin, aloe-emodin, chrysophanol, physcion, and their glycosides. The ethanolic extract of rhizome exhibits antidiabetic activity (Radhika et al. 2010, 2012).

19.6.48 Rubia cordifolia *L. (Indian Madder, Manjistha) (Family: Rubiaceae)*

Roots of this plant contain important active principles. Alcoholic extract of roots and leaves were found to have promising antidiabetic activity against animal models. The extract of roots reduced the blood sugar level in alloxan-treated diabetic rats (Patil et al. 2006). The aqueous extract is usually known to normalize hyperglycemia, hypertriglyceridemia, enhanced transaminases of the liver and kidney, hypochromic microcytic anemia, and loss of body weight in STZ-induced diabetic rat models (Baskar et al. 2006). Serum cholesterol and triglyceride level decreased, whereas serum high-density lipoprotein and protein levels increased in diabetic rats (Viswanathaswamy et al. 2011). A mixed aqueous extract of *R. cordifolia* and Ashwagandha is found to be effective for treating diabetes and for the dressing of diabetic foot (Murthy 1996).

19.6.49 Rubus fruticosus *(Blackberry) (Family: Rosaceae)*

It is commonly known as blackberry. Fruit powder is used to treat diabetes. The principal compounds isolated from blackberry leaves are hydrolysable tannins (Nonaka 1982). Common flavonoids are isolated from leaves, such as rutin, kaempferol, quercetin, and decanal (Khabibullaeva 1972; Maga 1992; Gudej and Rychlinska 1996).

19.6.50 Syzygium cumini *(L.) Skeels. (Jambul, Jamun) (Family: Myrtaceae)*

This plant is commonly known as Jamun or Indian blackberry and has been indicated in Ayurveda for use in the treatment of diabetes mellitus. This plant has been reported to have hypoglycemic effects in experimental as well as in clinical studies (Ravi et al. 2004). Different extracts of *S. cumini* possessed antidiabetic potentials against STZ-induced diabetic rats (Saravanan and Leelavinothan 2006; Kumar et al. 2008).

19.6.51 Tinospora cordifolia *(Thunb.) Miers. (Gulvel) (Family: Menispermaceae)*

Stem and roots of gulvel are commonly used in the treatment. The active adaptogenic constituents are diterpene compounds. Various extracts of the stem were found to be possessing potent antidiabetic activity in STZ-induced diabetic rats (Rajalakshmi et al. 2009).

19.6.52 Trigonella foenum-graecum *L. (Fenugreek) (Family: Fabaceae)*

Fenugreek has components like nicotinic acid, alkaloid trigonelline, and coumarin which are active ingredients for its antidiabetic properties. The leaves and seeds are used in therapeutic purpose. Soluble dietary fractions were evaluated which exhibited antidiabetic effects (Hannan et al. 2007). The aqueous extract of leaves possessed a hypoglycemic effect in normoglycemic and alloxan-induced hyperglycemic rats (Renuka et al. 2009).

19.6.53 Zingiber officinale *L. (Ginger) (Family: Zingiberaceae)*

Ginger, an underground rhizome produced a significant hypoglycemic effect in alloxan-induced rats (Abd-Elraheem et al. 2009; Jafri et al. 2011). Similarly, ethanolic extract showed antidiabetic property in alloxan-induced diabetic rats (Venkata et al. 2011). An aqueous extract of raw ginger was reported to possess antidiabetic in STZ-induced diabetic rats (Zainab et al. 2006).

19.7 Conclusion

Diabetes (*Madhumeha*) is a universal and stern metabolic disorder of the malfunctioning of carbohydrates, fats, and proteins metabolism. This disorder specifically results from defects in insulin secretion, action, or both. Though synthetic oral hypoglycemic drugs/insulin are primarily used in the treatment, these have several noticeable side effects. Therefore, there is an increasing demand for the alternative treatment. Traditional herbal products have been used all over the world for treating *Madhumeha*. Several plants have shown the antidiabetic potential in controlling and curing diabetes with minimal or negligible side effects. Therefore, it is essential to include plants with antidiabetic potentials in the regular diet for healthy living. The combined efforts of ethnobotanists, phytochemists, pharmacognosists, pharmacologists, and Ayurvedic practitioner are desired to document and appraise the efficacy and safety of the natural drugs.

References

Abd-Elraheem AE, Muhammad MAS, Mahrous MAM (2009) Effect of Ginger extract consumption on levels of blood glucose, lipid profile and kidney functions in alloxan induced diabetic rats. Egypt Acad J Biol Sci 2(1):153–162. doi:http://www.academia.edu/8955140/

Acharya YT (1994) Charaka Samhitha of Agnivesh with the Ayurveda Dipika commentary, 4th edn. Chaukambha Sanskrit Samstha, Varanasi, p 447

Afaf IA, Osman MKM, Elmahdi B (2008) Effects of *Aloe vera* (Elsabar) Ethanolic extract on blood glucose level in wistar albino rats. J Appl Sci Res 4(12):1841–1845

Agnivesha (1941) Charaka Samhita. Revised by Charaka Chakrapani, ed by Yadavaji, Niranaya Sagar Press, Bombay

Ahmed F, Urooj A (2008) Antihyperglycaemic activity of *Ficus glomerata* stem bark in streptozotocin-induced diabetic rats. Glob J Pharm 2(3):41–45. doi:http://www.academia.edu/1355176

Ajay SS (2009) Hypoglycemic activity of *Coccinia indica* (Cucurbitaceae) leaves. Int J Pharm Technol Res 1(3):892–893. doi:http://sphinxsai.com/PTVOL3/PT=78,AJAY%20SINGH%20(892–893)

Alarcon AFJ, Roman RR, Perez GS, Aguilar CA, Contreras WCC, Flores SJL (1998) Study of the anti-hyperglycemic effect of plants used as antidiabetics. J Ethnopharmacol 61:101–110. doi:10.1016/S0378-8741(98)00020-8

Altan VM (2003) The pharmacology of diabetic complications. Curr Med Chem 10:1317–1327. doi:10.2174/0929867033457287

Anonymous (2013) International Diabetes Federation Diabetes Atlas, 6th edn. International Diabetes Federation, Brussels

Arumugam G, Manjula P, Paari N (2013) A review: antidiabetic medicinal plants used for diabetes mellitus. J Acute Dis 2:196–200. doi:10.1016/S2221-6189(13)60126-2

Arunadutta (1933) Sarvanga Sundara Comm. on Astanga Hridya, Nirnaya Sagar Press, Bombay

Bajaj S, Srinivasan BP (1999) Investigations into the anti-diabetic activity of *Azadirachta indica*. Indian J Pharm 31(2):138–141. doi:http://www.ijp-online.com/temp/IndianJPharmacol312142-2297269_062252.pdf

Baskar R, Bhakshu LM, Bharathi GV, Reddy SS, Karuna R, Reddy GK, Saralakumari D (2006) Antihyperglycemic activity of aqueous root extract of Rubia cordifolia in Streptozotocin-induced diabetic rats. Pharm Biol 44:478–479. doi:10.1080/13880200600798593

Bhaskar VH, Ajay SS (2009) Evaluation of antihyperglycemic activity of extracts of *Calotropis procera* (Ait.) R.Br on streptozotocin induced diabetic rats. Glob J Pharmacol 3(2):95–98. doi:10.1.1.491.3214

Bhatia V, Kinja K, Bishnoi H, Savita S, Gnaneshwari D (2011) Antidiabetic activity of the alcoholic extract of the aerial part of *Boerhaavia diffusa* in rats. Recent Res Sci Technol 3(7):4–7. doi:http://scienceflora.org/journals/index.php/rrst/article/viewFile/727/712

Bhuyan SI (2015) Medicinal plants used for diabetes treatment by the different tribes in Bongaigaon district, North East India. Int J Adv Res Sci Eng Technol 2(1):376–379. doi:http://www.ijarset.com/upload/2015/january/8_IJARSET_bhuyan2.pdf

Boon NA, Colledge NR, Walker BR (2006) Diabetes mellitus in Davidson's principles and practice of medicine (Hunter JAA, ed), Elsevier, Edinburg, pp 805–847

Borokini TI, Ighere DA, Clement M, Ajiboye TO, Alowonale AA (2013) Ethnobotanical survey of traditional medicine practice for the treatment of piles and diabetes mellitus in Oyo State. J Med Plant Stud 1(5):30–40. doi:http://www.plantsjournal.com/vol1Issue1/Issue_sep_2013/2.1.pdf

Cai ZF, Liu Q, Li PP, Guo ZR, Shen ZF (2006) Synthesis and anti-diabetic activity of (RS)-2-ethoxy-3-{4-[2-(4-trifluoro-8methanesulfonyloxy-phenyl)-ethoxy]-phenyl}-propionic acid. Acta Pharmacol Sin 27(5):597–602. doi:10.1111/j.1745-7254.2006.00295.x

Chakraborty U, Das H (2010) Antidiabetic and antioxidant activities of Cinnamomum tamala leaf extracts in STZ-treated diabetic rats. Glob J Biotechnol Biochem 5(1):12–18

Chattopadhyay RR, Medda C, Das S (1993) Hypoglycemic and antihyperglycemic effect of *Gymnema sylvestre* leaf extract in rats. Fitoterapia 64(5):450–454

Choudhary NK, Jha AK, Sharma S, Goyal S, Dwivedi J (2011) Anti-diabetic potential of chloroform extract of flowers of *Calotropis gigantea*: an *in vitro* and *in vivo* study. Int J Green Pharm 5:296–301. doi:10.4103/0973-8258.94351

Colbert D (1999) Bible cure for diabetes. Published by Strang Communications Company, p 96

Dahake AP, Chakma CS, Chakma RC, Bagherwal P (2010) Antihyperglycemic activity of methanolic extract of *Madhuca longifolia* bark. Diabetol Croat 39–41. doi:http://www.idb.hr/diabetologia/10no1-1.pdf

Dandu AM, Inamdar NM (2009) Evaluation of beneficial effects of antioxidant properties of aqueous leaf extract of *Andrographis paniculata* in STZ-induced diabetes. Pak J Pharm Sci 22(1):49–52. doi:http://www.ncbi.nlm.nih.gov/pubmed/19168420

Dash VB (1987) Materia medica of Indo-Tibetan medicine. MDIA Classics India Publication, New Delhi

Devaraj AR, Nirmala ARR (2012) Herbal solution for the treatment of diabetes mellitus. J Pharm Sci Inno 1(1):27–31. doi:http://jpsionline.com/admin/php/uploads/20_pdf.pdf

Ding SY, Shen ZF, Chen YT, Sun SJ, Liu Q, Xie MZ (2005) Pioglitazone can ameliorate insulin resistance in low-dose streptozotocin and high sucrose-fat diet-induced obese rats. Acta Pharmacol Sin 26:575–580. doi:10.1111/j.1745-7254.2005.00090.x

Dong H, Wang N, Zhao L, Lu F (2012) Berberine in the treatment of type 2 diabetes mellitus: a systemic review and meta-analysis. Evid Based Complement Alternat Med 2012:1–12. doi:10.1155/2012/591654

Dwivedi C, Daspaul S (2013) Antidiabetic herbal drugs and polyherbal formulations used for diabetes: a review. J Phytopharmacol 2(3):44–51. doi:http://www.phytopharmajournal.com/V2issue308.pdf

Eidi A, Eidi M, Esmaeili E (2006) Antidiabetic effect of garlic (*Allium sativum* L.) in normal and streptozotocin-induced diabetic rats. Phytomedicine 13:624–629. doi:10.1016/j.phymed.2005.09.010

Elder C (2004) Ayurveda for diabetes mellitus: a review of the biomedical literature. Altern Ther Health Med 10:44–50. doi:http://www.ncbi.nlm.nih.gov/pubmed/14727499

Emmanuel S, Rani MS, Sreekanth MR (2010) Antidiabetic activity of *Cassia occidentalis* Linn. In streptozotocin-induced diabetic rats: a dose dependent study. Int J Pharm Bio Sci 1(4):B14–B25. doi:http://www.ijpbs.net/issue-4/Bio-2.pdf

Faizal IP, Suresh S, Satheesh Kumar R, Augusti KT (2009) A study on the hypoglycemic and hypolipidemic effects of an Ayurvedic drug Rajanyamalakadi in diabetic patients. Indian J Clin Biochem 24:82–87. doi:10.1007/s12291-009-0014-1

Ghosh R, Sharatchandra K, Rita S, Thokchom IS (2004) Hypoglycemic activity of *Ficus hispida* (bark) in normal and diabetic albino rats. Indian J Pharmacol 36(4):222–225

Gudej J, Rychlinska I (1996) Flavonoid compounds from the leaves of *Rubus idaeus* L. Herba Pol 42(4):257–261

Gupta AD, De A (2012) Diabetes mellitus and its herbal treatment. Int J Res Pharmaceut Biomed Sci 3(2):706–721. doi:http://ijrpbsonline.com/files/42-3262.pdf

Guyton AC, Hall JE (2006) Textbook of medical physiology, 11th edn. Elsevier Saunders, Philadelphia

Hankey A (2005) The scientific value of Ayurveda. J Altern Complement Med 11:221–224. doi:https://www.online.liebertpub.com/doi/abs/10.1089/acm.2005.11.221

Hannan JMA, Marenah L, Ali L, Rokeya B, Flatt PR, Abdel Wahab YHA (2006) *Ocimum sanctum* leaf extracts stimulate insulin secretion from perfused pancreas, isolated islets, and clonal pancreatic β-cells. J Endocrinol 189:127–136. doi:PMID: 16614387

Hannan JMA, Ali L, Rokeya B, Khaleque L, Akhter M, Flatt PR, Abdel-Wahab YHA (2007) Soluble dietary fibre fraction of *Trigonella foenumgraecum* (fenugreek) seed improves glucose homeostasis in animal models of type 1 and type 2 diabetes by delaying carbohydrate digestion and absorption, and enhancing insulin action. Br J Nutr 97:514–521. doi:PMID: 17313713

Hassan Z, Yam MF, Ahmad M, Yusof PM (2010) Antidiabetic properties and mechanism of action of *Gynura procumbens* water extract in streptozotocin-induced diabetic rats. Molecules 15:9008–9023. doi:10.3390/molecules15129008

Jafri SA, Abbas S, Qasim M (2011) Hypoglycemic effect of ginger (*Zingiber officinale*) in alloxan induced diabetic rats (*Rattus norvagicus*). Pak Vet J 31(2):160–162. doi:http://www.pvj.com.pk/pdf-files/31_2/160-162.pdf

Jain V, Viswanatha GL, Manohar D, Shivaprasad HN (2012) Isolation of antidiabetic principle from fruit rinds of *Punica granatum*. Evid Based Complement Altern Med 2012:1–11. doi:http://dx.doi.org/10.1155/2012/147202

Jarald EE, Joshi SB, Jain DC (2008) Antidiabetic activity of aqueous extract and non polysaccharide fraction of *Cynodon dactylon* Pers. Int J Experiment Biol 46:660–667. doi:http://nopr.niscair.res.in/bitstream/123456789/4578/1/IJEB%2046(9)%20660-667.pdf

Jegede A, Oladosu P, Ameh S, Kolo I, Izebe K, Builders P (2011) Status of management of diabetes mellitus by traditional medicine practitioners in Nigeria. J Med Plan Res 5(27):6309–6315

Joergensen C, Hovind P, Schmedes A, Parving HH, Rossing P (2011) Vitamin D levels, microvascular complications, and mortality in type 1 diabetes. Diabetes Care 34(5):1081–1085. doi:10.2337/dc10-2459

Jonnalagadd VG, Selkar N (2013) Antidiabetic herbal products marketed in India: an update. J Med Plant Stud 1(6):24–26. doi:http://www.plantsjournal.com/vol1Issue1/Issue_nov_2013/6.1.pdf

Jose E, Usha PTA (2010) Evaluation of the antidiabetic efficacy of *Coccinia indica* in rats. Indian J Anim Res 44(3):168–172. doi:http://www.indianjournals.com/ijor.aspx?target=ijor:ijar1&volume=44&issue=3&article=002

Kaleem M, Sheema SH, Bano B (2005) Protective effects of *Piper nigrum* and *Vinca rosea* in alloxan induced diabetic rats. Indian J Physiol Pharmacol 49(1):65–71. doi:PMID: 15881860

Kaleem M, Asif M, Ahmed QU, Bano B (2006a) Antidiabetic and antioxidant activity of *Annona squamosa* extract in streptozotocin-induced diabetic rats. Singap Med J 47(8):670–75. doi:PMID: 16865205

Kaleem M, Kirmani D, Asif M, Ahmed Q, Bano B (2006b) Biochemical effect of *Nigella sativa* L seed in diabetic rats. Indian J Exp Biol 44:745–748. doi:http://nopr.niscair.res.in/bitstream/123456789/6595/1/IJEB%2044(9)%20745-748.pdf

Kaveeshwar SA, Cornwall J (2014) The current state of diabetes mellitus in India. Australas Med J 7(1):45–48. doi:10.4066/AMJ.2013.1979

Kawalekar JS (2011) Survey and documentation of some of the antidiabetic plants by local vaidyas in Gavanal, village, at Hukkeri taluk, Belgaum district. Plant Sci Feed 1(3):34–38

Khabibullaeva L (1972) Phytochemical study of raspberry leaves. Mater Yubileinoi Resp Nauchn Konf Farm Chem Abs 83:4960

Khajuria S, Thomas J (1992) Traditional Indian beliefs about the dietary management of diabetes – an exploratory study of the implications for the management of Gujarati diabetes in Britain. J Hum Nutr Diet 5(5):311–321

Khalil EMA (2004) Antidiabetic effect of an aqueous extract of Pomegranate (*Punica granatum* L.) peels in normal and alloxan diabetic rats. Egyptian J Hosp Med 16:92–99. doi:https://doaj.org/article/bd9d4c4b9aaa440087e4554a79433374

Khan A, Safdar M, Ali Khan MM, Khattak KN, Anderson RA (2003) *Cinnamon* improves glucose and lipids of people with type 2 diabetes. Diabetes Care 26:3215–3218. pmid:14633804. doi:10.2337/diacare.26.12.3215

Khan MRI, Islam MA, Hossain MS, Asadujjaman M, Wahed MII, Rahman BM, Anisuzzaman ASM, Shaheen SM, Ahmed M (2010) Antidiabetic effects of the different fractions of ethanolic extracts of *Ocimum sanctum* in normal and alloxan induced diabetic rats. J Sci Res 2(1):158–168

Koekemoer T, Van de Venter M, Roux S (2008) A comparative study on the *in vitro* anti-diabetic properties of four plants frequently used to treat diabetes in South Africa. Afri J Tradi, Comple Alt Med (AJTCAM), Abstracts of the World Congress on Medicinal and Aromatic Plants, Cape Town

Kokate CK, Purohit AP, Gokhale SB (2009) Textbook of pharmacognosy, 44th edn. Nirali Prakashan, New Delhi

Krup V, Prakash LH, Harini A (2013) Pharmacological activities of Turmeric (*Curcuma longa linn*): a review. J Homeop Ayurv Med 2:133. doi:10.4172/2167-1206.1000133

Kumar A, Ilavarasan R, Jayachandran T, Deecaraman M, Aravindan P, Padmanabhanand M, Krishan V (2008) Anti-diabetic activity of *Syzygium cumini* and its isolated compound against streptozotocin-induced diabetic rats. J Med Plant Res 2(9):246–249. doi:http://www.academicjournals.org/article/article1380381651_Kumar%20et%20al.pdf

Kumar V, Abbas AK, Fausto N, Aster JC (2010) Robbins and Cotran pathologic basis of disease, 8th edn. Sauders Elsevier Publication, Philadelphia, pp 1131–1146

Kumar D, Kumar S, Kohli S, Arya R, Gupta J (2011a) Antidiabetic activity of methanolic bark extract of *Albizia odoratissima* Benth. in alloxan induced diabetic albino mice. Asian Pac J Trop Med 4(11):900–903. doi:10.1016/S1995-7645(11)60215-0

Kumar R, Pate DK, Prasad SK, Sairam K, Hemalatha S (2011b) Antidiabetic activity of alcoholic leaves extract of *Alangium lamarckii* Thwaites on streptozotocin-nicotinamide induced type 2 diabetic rats. Asian Pac J Tropical Med 4(11):904–909. doi:10.1016/S1995-7645(11)60216-2

Lanjhiyana S, Garabadu D, Ahirwar D, Bigoniya P, Rana AC, Patra KC, Lanjhiyana SK, Karuppaih M (2011) Hypoglycemic activity studies on aerial leaves of *Pongamia pinnata* (L.) in alloxan-induced diabetic rats. Der Pharmacia Lettre 3(1):55–70. doi:http://scholarsresearchlibrary.com/DPL-vol3-iss1/DPL-2011-3-1-55-70.pdf

Lau C, Cham CM, Chan YW, Lau KM, Lau TW, Lam FC, Che CT, Leung PC, Fung KP, Ho YY, Lau CBS (2008) *In vitro* antidiabetic activities of five medicinal herbs used in Chinese medicinal formulae. Phytother Res 22(10):1384–1388. doi:10.1002/ptr.2513

Lehninger AL, Nelson DL, Cox MM (2010) Principle of biochemistry. Worth Publishers, New York

Lodha SR, Joshi SV, Vyas BA, Upadhye MC, Kirve MS, Salunke SS, Kadu SK, Rogye MV (2010) Assessment of the antidiabetic potential of *Cassia grandis* using as *in vivo* model. J Adv Pharm Technol Res 1(3):330–333. doi:10.4103/0110-5558.72429

Lopez RP, Pichardo OE, Avila NA, Vazquez MN, Tovar AR, Pedraza CJ, Torres N (2014) The effect of nopal (*Opuntia ficus-indica*) on postprandial blood glucose, incretins and antioxidant activity in Mexican patients with type 2 diabetes after consumption of two different composition breakfasts. J Acad Nutr Diet 114(11):1811–1818. doi:10.1016/j.jand.2014.06.352

Madhava-Nidanam of Sri Madhavakara with the 'Madhukosa' Sanskrit commentary by Srivijayaraksita and Srikanthadatta with The 'Vidyotini' Hindi commentary by Shri Sudarsana Sastri, and revised and edited by Prof. Yadunandana Upadhyaya, Part-I (1999) and II (2009 reprint), Chaukhambha Prakashan, 221001, Varanasi (India)

Maga J (1992) Bramble leaf volatiles. Dev Food Sci 29:145

Mahalingam G, Krishnan K (2008a) Antidiabetic and ameliorative potential of *Ficus bengalensis* bark extract in streptozotocin induced diabetic rats. Indian J Clin Biochem 23(4):394–400. doi:10.1007/s12291-008-0087-2

Mahalingam G, Krishnan K (2008b) Hypoglycemic activity of *Hemidesmus indicus* R. Br. on streptozotocin-induced diabetic rats. Int J Diabetes Dev Ctries 28(1):6–10. doi:10.4103/0973-3930.41979

Maiti R, Jana D, Das UK, Ghosh D (2004) Antidiabetic effect of aqueous extract of seed of *Tamarindus indica* in streptozotocin induced diabetic rats. J Ethnopharmacol 92:85–91. doi:PMID: 15099853

Mall GK, Mishra PK, Veeru P (2009) Antidiabetic and hypolipidemic activity of *Gymnema sylvestre* in alloxan induced diabetic rats. Glob J Biotechnol Biochem 4(1):37–42. doi:http://idosi.org/gjbb/gjbb4(1)09/6.pdf

Malpani SN, Manjunath SK (2012) Antidiabetic activity and phytochemical investigations of *Cassia fistula* L. bark. Int J Pharmac Sci Res 3(6):1822–1825. doi:http://ijpsr.com/bft-article/antidiabetic-activity-and-phytochemical-investigations-of-cassia-fistula-linn-bark

Manoharan S, Silvan S, Vasudevan K, Balakrishnan S (2007) Antihyperglycemic and antilipidperoxidative effects of *Piper longum* (Linn.) dried fruits in alloxan induced diabetic rat. J Biol Sci 7:161–168. doi:http://docsdrive.com/pdfs/ansinet/jbs/2007/161-168.pdf

Matheka DM, Kiama TN, Alkizim FO, Bukachi F (2011) Glucose-lowering effects of *Momordica charantia* in healthy rats. Afr J Diabetes Med 19(2):15–19

Meddah B, Ducroc R, Faouzi MEA, Eto B, Mahraoui L, Benhaddou-Andaloussi A, Martineau LC, Cherrah Y, Haddad PS (2009) *Nigella sativa* inhibits intestinal glucose absorption and improves glucose tolerance in rats. J Ethnopharmacol 121:419–424. doi:10.1016/j.jep.2008.10.040

Mehta S, Singh RK, Jaiswal D, Rai PK, Watal G (2009) Anti-diabetic activity of *Emblica officinalis* in animal models. Pharm Biol 47(11):1050–1055. doi:10.3109/13880200902991532

Meliani N, Amine Dib ME, Allali H, Tabti B (2011) Hypoglycaemic effect of *Berberis vulgaris* L. in normal and streptozotocin induced diabetic rats. Asian Pac J Trop Biomed 6:468–471. doi:10.1016/S2221-1691(11)60102-0

Misra A, Shrivastava U (2013) Obesity and dyslipidemia in South Asians. Nutrients 5(7):2708–2733. doi:10.3390/nu5072708

Modak M, Dixit P, Londhe J, Ghaskadbi S, Devasagayam TPA (2007) Indian herbs and herbal drugs used for the treatment of diabetes. J Clin Biochem Nutr 40:163–173. doi:10.3164/jcbn.40.163

Mohammad YMI (2009) Clinical evaluation of antidiabetic activity of *Trigonella* seeds and *Aegle marmelos* Leaves. World Appl Sci J 7(10):1231–1234

Murthy SN (1996) Study of Manjishtha and Ashwagandha with special reference to its clinical and immuno-pathological effect of diabetic foot ulcer. MD thesis submitted to Department of Dravyaguna, Institute of Medical Sciences, Banaras Hindu University, Varanasi, India

Nirmala A, Eliza J, Rajalakshmi M, Priya E, Daisy P (2008) Effect of hexane extract of *Cassia fistula* barks on blood glucose and lipid profile in streptozotocin diabetic rats. Int J of Pharm 4(4):292–296

Nonaka G (1982) A dimeric hydrolyzable tannin, Sanguin H-6 from *Sanguisorba officinalis* L. Chem Pharm Bull 30(6):2255

Ogunmodedee OS, Saalu LC, Ogunlade B, Akunna GG, Oyewopo AO (2012) An evaluation of hypoglycemic, antioxidant and hepatoprotective potential of onion (*Allium cepa* L.) on alloxan induced diabetic rabbit. Int J of Pharm 8(1):21–29. doi:http://docsdrive.com/pdfs/ansinet/ijp/2012/21-29.pdf

Ohadoma SC, Michael HU (2011) Effects of co-administration of methanol leaf extract of *Catharanthus roseus* on the hypoglycemic activity of metformin and glibenclamide in rats. Asian Pac J Trop Med 4(6):475–477. doi:10.1016/S1995-7645(11)60129-6

Ozougwu JC (2011) Anti-diabetic effects of *Allium cepa* (onions) aqueous extracts on alloxan-induced diabetic *Rattus novergicus*. Pharmacologyonline 1:270–281. doi:http://pharmacologyonline.silae.it/files/archives/2011/vol1/027.ozougwu.pdf

Patel P, Harde P, Pillai J, Darji N, Patel B (2012) Antidiabetic herbal drugs: a review. Int Res J 3:18–29

Patil RA, Jagdale SC, Kasture SB (2006) Antihyperglycemic, antistress and nootropic activity of roots of *Rubia cordifolia* Linn. Ind J Exp Biol 44:987–992. doi:PMID: 17176672

Patil R, Patil R, Ahirwar B, Ahirwar D (2011) Current status of Indian medicinal plants with antidiabetic potential: a review. Asian Pacific J Trop Biomed S291–S298. doi:http://www.apjtb.com/zz/2011s2/33.pdf

Port DJ, Schwartz MW (1996) Diabetes complications: why is glucose potentially toxic? Science 27:699–700. doi:10.1126/science.272.5262.699

Powers AC (2012) Diabetes mellitus. In: Longo DL et al (eds) Harrison's principles of internal medicine, vol II, 18th edn. McGraw-Hill Medical Publishing Division, New York

Preethi PJ, Bindusree K, Pavan Kumar K, Rajavelu R, Sivakumar T (2012) Antidiabetic and antioxidant activity of latex of *Calotropis procera*. Int J Adv Pharm Res 3(6):975–980

Pund KV, Vyawahare NS, Gadakh RT, Murkute VK (2012) Antidiabetic evaluation of *Dalbergia sissoo* against alloxan induced diabetes mellitus in Wistar albino rats. J Nat Prod Plant Resour 2(1):81–88. doi:http://scholarsresearchlibrary.com/archive.html

Radhika R, Kumara K, Sudarsanam D (2010) Antidiabetic activity of *Rheum emodi* in Alloxan induced diabetic rats. Int J of Pharma Sci and Res 1(8):296–300. doi:http://www.ijpsr.info/docs/IJPSR10-01-08-08.pdf

Radhika S, Smila KH, Muthezhilan R (2011) Antidiabetic and hypolipidemic activity of *Punica granatum* Linn on alloxan induced rats.World. J Med Sci 6(4):178–182. doi:http://www.idosi.org/wjms/6(4)11/3.pdf

Radhika R, Ragavan B, Pawar SD, Sudarsanam D (2012) Action of marker enzymes of *Rheum emodi* in Alloxan induced diabetic rats. Asian J Exp Biol Sci 3(2):420–423

Rahmatullah M, Shefa TF, Hasan L, Hossain MT, Ahmed S, Al Mamun A, Islam MR, Rahman S, Chowdhury MH (2010) A Study on antinociceptive and anti-hyperglycemic activity of methanol extract of *Brassica juncea* (L.) Czern. leaves in mice. Adv Nat Appl Sci 4(3):221–225

Rajalakshmi M, Eliza J, Priya CE, Nirmala A, Daisy P (2009) Antidiabetic properties of *Tinospora cordifolia* stem extracts on streptozotocin- induced diabetic rats. Afr J Pharm Pharmacol 3(5):171–180. doi:http://www.academicjournals.org/journal/AJPP/article-full-text-pdf/AA123FE34034

Rang HP, Dale MM (1999) The endocrine system pharmacology, 2nd edn. Longman Group Ltd., London

Rao KN, Boini KM, Nammi S (2004) Effect of chronic administration of *Boerhaavia diffusa* Linn. leaf extract on experimental diabetes in rats. Trop J of Pharm Res 3(1):305–309. doi:http://dx.doi.org/10.4314/tjpr.v3i1.14614

Rao MU, Sreenivasulu M, Chengaiah B, Reddy KJ, Chetty CM (2010) Herbal medicines for diabetes mellitus: a review. Int J of Pharm Technol Res 2(3):1883–1892. doi:http://sphinxsai.com/july-sept_2010_vol2.3/pharmtech/pharmtechvol2.3july-sept210/PT=36%20(1883–1892).pdf

Rathod NR, Chime HR, Irchhaiya R, Chandra R (2011) Hypoglycemic effect of *Calotropis gigantea* Linn. leaves and flowers in streptozotocin-induced diabetic rats. Oman Med J 26(2):104–108. doi:10.5001/omj.2011.26

Ravi K, Ramachandran B, Subramanian S (2004) Effect of *Eugenia jambolana* seed kernel on antioxidant defense system in streptozotocin induced diabetes in rats. Life Sci 75(22):2717–2731. doi:PMID: 15369706

Rehman SU, Jafri SA, Hassan S, Ahmed I, Naim M (2011) Study on antidiabetic effect of *Aloe vera* extract on alloxan induced diabetic rats. Libyan Agric Res Center J Int 2(1):29–32. doi:http://www.aloe9.com/image/users/215470/ftp/my_files/Aloe%20studies/Aloe-Diabetic-2.pdf?id=10509668

Renuka C, Ramesh N, Saravanan K (2009) Evaluation of the antidiabetic effect of *Trigonella foenumgraecum* seed powder on alloxan induced diabetic albino rats. Int J PharmTechnol Res 1(4):1580–1584. doi:http://sphinxsai.com/PTVOL4/pdf_vol4/PT=99%20(1580–1584).pdf

Rizvi SI, Mishra N (2013) Traditional Indian medicines used for the management of diabetes mellitus. J Diabetes Res 1–11. doi:10.1155/2013/712092

Saber A, Zaiton A (2010) Antibiotic activity of *Ferula asafoetida* extract in normal and alloxan induced diabetic rats. Pak J Biol Sci 13(2):97–100. doi:PMID: 20415145

Sabu MC, Kuttan R (2004) Antidiabetic activity of *Aegle marmelos* and its relationship with its antioxidant properties. Indian J Physiol Pharmacol 48(1):81–88. doi:http://www.ijpp.com/IJPP%20archives/2004_48_1/vol48_no1_orgnl_07.pdf

Sancheti S, Sancheti S, Seo S (2009) *Chaenomeles sinensis*: a potent α- and β-Glucosidase Inhibitor. Am J Pharmacol Toxicol 4:8–11. doi:10.3844/ajptsp.2009.8.11

Sancheti S, Sancheti S, Seo SY (2011) Antidiabetic and antiacetylcholinesterase effects of ethyl acetate fraction of *Chaenomeles sinensis* (Thouin) Koehne fruits in streptozotocin-induced diabetic rats. Exp Toxicol Pathol 65(1–2):55–60. doi:10.1016/j.etp.2011.05.010

Saravanan G, Leelavinothan P (2006) Effects of *Syzygium cumini* bark on blood glucose, plasma insulin and C-peptide in streptozotocin induced diabetic rats. Int J Endocrinol Metab 4:96–105

Saravanan G, Pari L (2008) Hypoglycemic and antihyperglycemic effect of *Syzygium cumini* bark in streptozotocin-induced diabetic rats. J Pharmacol Toxicol 3:1–10. doi:http://www.docsdrive.com/pdfs/academicjournals/jpt/2008/1-10.pdf

Sathya S, Kokilavani R, Gurusamy K (2008) Hypoglycemic effect of *Gymnema sylvestre* (Retz.) R.Br leaf in normal and alloxan induced diabetic rats. Anc Sci Life 28(2):12–14. doi:PMC3336356

Sekar DS, Sivagnanam K, Subramanian S (2005) Antidiabetic activity of *Momordica charantia* seeds on streptozotocin induced diabetic rats. Pharmazie 60:383–387. doi:http://www.ncbi.nlm.nih.gov/pubmed/15918591

Shastri PK, Chaturvedi GN (2005) Charaka Samhita of Agnivesa elaborated 'Vidyotini' Hindi commentary, Part-I (Reprint year: 2005) and Part-II (Reprint year: 2006), Varanasi, Chaukhambha Bharati Academy, 221001, India

Sheth PP (2005) Global opportunities and challenges for medicinal uses of Ayurveda, herbal products, nutraceuticals, and alternatives. Chapter- 17 of Health administrator XIX(1):74–75

Shibi BA, Khan LA, Rahman R (2015) Hypoglycemic activity of *Coccinia indica* and Momordica charantia in diabetic rats. Biochem J 293:267–270. doi: PMCID: PMC1134299

Shokeen P, Anand P, Murali YK, Tandon V (2008) Antidiabetic activity of 50 % ethanolic extract of *Ricinus communis* and its purified fractions. Food Chem Toxicol 46:3458–3466. doi:10.1016/j.fct.2008.08.020

Shukla R, Anand K, Prabhu KM, Murthy PS (1994) Hypoglycaemic effect of the water extract of *Ficus bengalensis* in alloxan recovered, mildly diabetic and severely diabetic rabbits. Int J Diabetes Dev Countries 14:78–81. http://diabetes.org.in/journal/1994_july-sept/original_article2.pdf

Sipika S, Agarwal P (2015) Diabetes mellitus: an Ayurvedic view. J Sci Innov Res 4(4):193–196. http://www.jsirjournal.com/Vol4_Issue4_08.pdf

Sushruta (1915) Sushruta Samhita. Comm by Dalhana, ed. by Yadavaji, Nirnaya Sagar Press Bombay

Sutar NG, Sutar UN, Behera BC (2009) Antidiabetic activity of the leaves of *Mimosa pudica* Linn. in albino rats. J of Herb Med and Tox 3(1):123–126

Tamboli AM, Karpe ST, Shaikh SA, Manikrao AM, Kature DV (2011) Hypoglycemic activity of extracts of *Oroxylum indicum* (L.) vent roots in animal models. Pharmacol Online 2:890–899. doi:http://pharmacologyonline.silae.it/files/archives/2011/vol2/091.tambolii.pdf

Tan MJ, Ye JM, Turner N, Hohnen-Behrens C, Ke CQ, Tang CP, Chen T, Weiss HC, Gesing ER, Rowland A, James DE, Ye Y (2008) Antidiabetic activities of Triterpenoids isolated from bitter melon associated with activation of the AMPK pathway. Chem Biol 15(3):263–273. doi:10.1016/j.chembiol.2008.01.01

Tanaka M, Misawa E, Ito Y, Habara N, Nomaguchi K, Yamada M, Toida T, Hayasawa H, Takase M, Inagaki M, Higuchi R (2006) Identification of five phytosterols from Aloe vera gel as antidiabetic compounds. Biol Pharm Bull 29(7):1418–1422. doi:http://www.ncbi.nlm.nih.gov/pubmed/16819181

Thirumalai T, Theresa SV, Elumalai EK, David E (2011) Hypoglycemic effect of *Brassica juncea* (seeds) on streptozotocin induced diabetic male albino rat. Asian Pac J Trop Biomed 1(4):323–325. doi:10.1016/S2221-1691(11)60052-X

Thomson M, Zainab MAA, Khaled KAQ, Lemia HS, Ali M (2007) Anti-diabetic and hypolipidaemic properties of garlic (*Allium sativum*) in streptozotocin induced diabetic rats. Int J Diabetes Metab 15(3):108–115. doi:http://ijod.uaeu.ac.ae/iss_1503/f.pdf

Uma Devi P, Selvi S, Suja S, Selvam K, Chinnaswamy P (2006) Antidiabetic and hypolipidemic effect of *Cassia auriculata* in alloxan induced diabetic rats. Int J Pharmacol 2:601–607. doi:http://docsdrive.com/pdfs/ansinet/ijp/2006/601-607.pdf

Upadhya S, Shanbhag KK, Suneetha G, Naidu BM, Upadhya S (2004) A study of hypoglycemic and antioxidant activity of aegle marmelos in alloxan induced diabetic rats. Indian J Physiol Pharmacol 48(4):476–480. doi:http://www.ijpp.com/IJPP%20archives/2004_48_4/a_study_of_hypo_476.pdf

Venkata KSN, Santhosh D, Narasimha Rao D, Sanjeeva Kumar A, Charles Martin A (2011) Preliminary phytochemical screening and antidiabetic activity of *Zingiber officinale* rhizomes. Int J Pharm Life Sci 2(12):1287–1292. doi:http://www.ijplsjournal.com/issues%20PDF%20files/dec%202011/4.pdf

Viswanathaswamy AHM, Koti BC, Singh AK, Thippeswamy AHM (2011) Antihyperglycemic and antihyperlipidemic effect of *Rubia cordifolia* leaf extract on Alloxan-induced Diabetes. RGUHS J Pharm Sci 1:49–52. doi:10.5530/rjps.2011.1.10

Wadkar KA, Magdum CS, Patil SS, Naikwade NS (2008) Antidiabetic potential and Indian medicinal plants. J Herb Med Toxicol 2:45–50. doi:10.1.1.453.4973

Wang TJ, Larson MG, Vasan RS, Cheng S, Rhee EP, McCabe E (2011) Metabolite profiles and the risk of developing diabetes. Nat Med 17(4):448–453. doi:10.1038/nm.2307

Whiting DR, Guariguata L, Weil C, Shaw J (2011) IDF Diabetes atlas: global estimates of the prevalence of diabetes for 2011 and 2030. Diabetes Res Clin Pract 94:311–321. doi:10.1016/j.diabres.2011.10.029

Wickenberg J, Ingemansson S, Hlebowicz J (2010) Effects of *Curcuma longa* (turmeric) on postprandial plasma glucose and insulin in healthy subjects. Nutr J 9:43. doi:10.1186/1475-2891-9-43

Wild S, Roglic G, Green A, Sicree R, King H (2004) Global prevalence of diabetes estimates for the year 2000 and projections for 2030. Diabetes Care 27(3):1047–1053. doi:10.2337/diacare.27.5.1047

Yeh GY, Eisenberg DM, Kaptchuk TJ, Phillips RS (2003) Systematic review of herbs and dietary supplements for glycemic control in Diabetes. Diabetes Care 26(4):1277–1294. doi:10.2337/diacare.26.4.1277

Young KA, Snell-Bergeon JK, Naik RG, Hokanson JE, Tarullo D, Gottlieb PA, Garg SK, Rewers M (2011) Vitamin D deficiency and coronary artery calcification in subjects with type 1 diabetes. Diabetes Care 34(2):454–458. doi:10.2337/dc10-0757

Zainab MA, Thomson M, Khaled KAQ, Shalaby RP, Ali M (2006) Antidiabetic and hypolipidemic properties of ginger (*Zingiber officinale*) in streptozotocin-induced diabetic rats. Br J Nutr 96:660–666. doi:http://www.ncbi.nlm.nih.gov/pubmed/17010224

Zhang Z, Jiang J, Yu P, Zeng X, Larrick JW, Wang Y (2009) Hypoglycemic and beta-cell protective effects of andrographolide analogue for diabetes treatment. J Transl Med 7:62. doi:10.1186/1479-5876-7-62

Chapter 20
Herbal Medicines: Boon or Bane for the Human Liver?

Dipita Bhakta-Guha and Thomas Efferth

Abstract Since ages, medicine is the most consistent companion of man. While in primeval time, disease was cured through natural preparations, onset of technology has made today's formulations synthetic or nature derived. Across the ages, the plausibility of the "savior-turned-slayer" functionality of these drugs remained constant. With the increment in documentation, it becomes evident that many of the commonly used drugs are associated with toxicities. Thus, any drug, irrespective of its origin, needs to be thoroughly assessed. Rampant use of herbal drugs has often been a threat to human health owing to the scarcity in quality assessment. What needs to be understood is that everything natural is not safe. They may pose a threat and thus need to be verified before being touted as a healer. Herein, we discuss the toxicity quotient of herbal medicine in reference to the liver – the metabolic controller of our body. Liver holds a prime position owing to its multitudinal functionality. We attempt to present a brief overview of the mechanisms of hepatotoxicity and highlight the nexus between herbal medicines and liver injury. We discuss the potential ways that can assure quality check of herbal medicines through imposition of regulatory laws, databases, and resorting to toxicogenomics. While constant endeavors need to be made in the quest for novel and more effective drugs, at the same time, assuring its safety is of paramount significance. Thus, plants, which are a huge reservoir of drug leads, necessitate the need to examine products derived from them and subsequently propose their efficacies in human health.

Keywords Herbal medicine • Hepatotoxicity • Liver • Toxicogenomics

---

D. Bhakta-Guha
Cellular Dyshomeostasis Laboratory (CDHL), SASTRA University, Thanjavur 613401, Tamil Nadu, India

T. Efferth (✉)
Institute of Pharmacy and Biochemistry, Johannes Gutenberg University, Mainz 55128, Germany
e-mail: efferth@uni-mainz.de

H.-S. Tsay et al. (eds.), *Medicinal Plants - Recent Advances in Research and Development*, DOI 10.1007/978-981-10-1085-9_20

Abbreviations

ABC	ATP-binding cassette
ADP	Adenosine diphosphate
Akr7a3	Aldo-keto reductase family 7, member A3
AMPK	5′ adenosine monophosphate-activated protein kinase
APAP	Acetaminophen
ATP	Adenosine triphosphate
CCl_4	Carbon tetrachloride
CD	Cluster of differentiation
CYP	Cytochrome P450
CYP1A2	Cytochrome P450, family 1, subfamily A, polypeptide 2
CYP2D6	Cytochrome P450, family 2, subfamily D, polypeptide 6
CYP2E1	Cytochrome P450 2E1
CYP3A4	Cytochrome P450, family 3, subfamily A, polypeptide 4
CYP7A1	Cholesterol 7 alpha-hydroxylase
DILI	Drug-induced liver injury
FDA	Food and Drug Administration
Fe	Iron
FoxO1	Forkhead box protein O1
Gclc	Glutamate-cysteine ligase catalytic subunit
Gpx2	Glutathione peroxidase 2
GSH	Glutathione
GSH/GSSG	Glutathione/glutathione disulfide
Gsr	Glutathione reductase
Gstp1	Glutathione S-transferase pi gene
H3	Histone 3
HCV	Hepatitis C virus
HepG2	Liver hepatocellular carcinoma
HH	Herbal hepatotoxicity
Hmox1	Heme oxygenase (decycling) 1
ICAM-1	Intercellular adhesion molecule 1
IFN-γ	Interferon gamma
IGF-1	Insulin-like growth factor 1
IL	Interleukin
LI	Liver injury
Mac-1	Macrophage-1 antigen
MRP2	Multidrug-resistant protein 2
NAFLD	Nonalcoholic fatty liver disease
NAPQI	N-acetyl-p-benzoquinone imine
NFATc4	Nuclear factor of activated T cells, cytoplasmic, calcineurin-dependent 4
NF-kB	Nuclear factor kappa-light-chain-enhancer of activated B cells
NK	Natural killer

NKT	Natural killer T
NO	Nitric oxide
Nqo1	NAD(P)H dehydrogenase [quinone] 1
PEP	Phosphoenolpyruvic acid
PGC-1α	Peroxisome proliferator-activated receptor gamma coactivator 1-alpha
PPARα	Peroxisome proliferator-activated receptor-alpha
RNS	Reactive nitrogen species
ROS	Reactive oxygen species
SIRT1	Silent mating-type information regulation 2 homologue 1
SLE	Systemic lupus erythematosus
SNP	Single nucleotide polymorphism
SREBP-1	Sterol regulatory element-binding transcription factor 1
TCM	Traditional Chinese medicine
TNF-α	Tumor necrosis factor-α
VCAM-1	Vascular cell adhesion molecule 1
WHO	World Health Organization

20.1 Herbal Medicines: The Edge Over Synthetic Drugs

Wherever the art of medicine is loved, there is also a love of humanity ~ Hippocrates

Medicines for long have been man's best friend in disease. Despite differences in the process of manufacture and source from which they are derived, all medicines serve the common goal – relieving the consumer from a pertinent or plausible ailment. Over the decades, improvements in research and technology have helped pharmaceutical industries to translate research effectively from "bench to bedside." It has been a constant endeavor of researchers, to come up with novel or alternative forms of existing medicines that are more efficient, target specific and less toxic. For years, different sources (natural and synthetic) have been utilized to achieve the said goal.

Out of the plethora of sources that serve as the reservoir, plants pose as a significant source of drug leads. Out of the 252 drugs that have been earmarked by the World Health Organization (WHO) as essential medicine, 11 % of them are derived exclusively from plants. Undoubtedly, development of diverse drugs from plants has benefitted human health. Isolation of opioids, alkaloids, polyphenols, and other secondary metabolites as leads and their subsequent development into drugs have proved beneficial in mitigating myriad diseases in human. Surprisingly though, and quite contrary to the brighter side, without concrete scientific validation, the general consensus supports the belief that a complementary or alternative medicine derived from any natural source especially from plant is nontoxic. This belief has, in fact, snowballed into the rampant use of natural products (herbal drugs/medicines or dietary supplements) without asserting their scientific credibility.

Owing to the purported safety assurance that initially got tagged along with phytocompounds, herbal products etched an imperative niche in the pharmaceutical sector. The early publicity of herbal drugs was laden with safety promises due to their origin from natural sources and soon proved to be a great competition to the contemporary synthetic versions. Unfortunately, apart from the approved herbal products, there are many more of their counterparts available in the market that are consumed and prescribed without caution. It often leads to severe health complications, and thus judicious consumption of the same needs to be monitored.

20.2 The Why(s) and How(s) of the Preferred Cult!

Through the ages, herbal medicines have amassed attention and prominence as a consequence of traditional knowledge transferred across generations. A large chunk of the population in developing countries still relies more on traditional or alternative medicines, particularly those derived from plants. According to Wachtel-Galor and Benzie (2011), in such regions, herbal medicine constitutes the major component of primary health care. Africa boasts of a mammoth use of herbal drugs by around 80 % of its total population. Despite lack of evidence, equating all herbal products with nontoxicity has led these supplements to be grossly used as devices for holistic improvement of generic well-being all across the globe. Medicines and dietary supplements such as *Aloe vera*, *Ginseng*, black cohosh, and several others have forayed deep into daily consumption. The scenario in developed countries equally favors such medicines. Thirty-eight percent of US population use herbal drugs. The country alone contributed to an overwhelming revenue of around $6,032 million through the sale of herbal medicines. This preferential consumption is further exploited by several pharmaceutical setups and standalone drug providers who advertise and subsequently indoctrinate the use of herbal medicine. A report estimates that going by the current trend, by 2017, the global expenditure on herbal drugs is estimated to reach a staggering $107 billion (Navarro and Lucena 2014).

20.3 Do Benefits Come at a Price?

However, despite enormous advances in chemically synthesized drugs and stringent policies of regulatory bodies, unaccounted sales of over-the-counter herbal drugs are rampant. The majority of the people exposed to these medicines use them without approval or prescription from doctors. The primary problem that allures concern is the fact that not all herbal products are safe for human use. Multiple studies have indicated to the verity that many herbal drugs are endowed with severe toxic effects which might be acute or chronic. Many of the abundantly prevalent drugs have been reported to be carcinogenic, nephrotoxic, hepatotoxic and also posed a risk of inducing unwanted drug interactions.

For instance, cat's claw (*Uncaria tomentosa*) used as an anti-inflammatory agent, *Tripterygium wilfordii* exploited as a remedy against rheumatoid arthritis, and *Artemisia sp.*-derived essential oil commonly used to stimulate digestion have been reported to induce acute renal failure and severe nephrotoxicity (Hilepo et al. 1997; Asif 2012; Li et al. 2015b). Products from well-known plants such as *Ginkgo biloba*, *Euphorbia tirucalli*, and *Pteridium sp.* have been reported to induce or aggravate tumors (Gomes et al. 2012; CSPI 2013; Machado et al. 2016). To add to the woe, the latest studies suggest that innumerable herbal drugs are laden with potential to afflict serious injuries to the liver. Their predisposition as causative agents of severe hepatotoxicity has come to the forefront in the recent decade and thus demands more attention and caution.

20.4 Unfurling Hepatotoxicity

The liver plays a crucial role in human health since it acts as the metabolic hub of our body. It is the primary organ which is entrusted with the task of carbohydrate, fat, and protein metabolisms through varied mechanisms. In addition to these, the liver also governs the biotransformation of xenobiotics. In case, any foreign substance enters the body, the liver takes up the duty of converting the foreign material into metabolites that can be readily excreted out of the system. Subsequently for a healthy system, it becomes imperative that proper functioning of the liver is ensured.

Corroborating to this fact, if the liver is inflicted with injuries (which might be physical, chemical, or genetic), it leads to severe malformation and flawed metabolism. There are innumerable ways and mechanisms through which plethora of substances can induce injury or toxicity to the liver. The few prominent ones are enlisted below. Having said that it is important to mention that these processes are interrelated to an extent and one can give rise to the other.

20.4.1 Oxidative Stress-Induced Injury

Homeostasis helps maintain a balance between the rate of reactive oxygen species (ROS) production and innate responses to counter their toxicity via antioxidant defenses. A skewed balance between the two parameters often resorts to oxidative stress, a phenomenon that is known to induce toxicity and is labeled responsible for generating several anomalies and subsequent pathologies. There are several factors that contribute to oxidative stress. While each of those factors has their characteristic mechanisms, all of them lead to a common consequence – liver injury (LI). For example, hepatitis C virus (HCV) augments ROS and reactive nitrogen species (RNS) generation, increased lipid peroxidation products, and elevated oxidized glutathione (GSH) (Choi and Ou 2006; Muriel 2009). In addition, HCV contributes to chronic inflammation, an overload of iron and subsequently LI manifesting into cirrhosis, hepatocellular carcinoma, and other HCV-mediated hepatotoxicities (Muriel 2009). In an another scenario, in nonalcoholic fatty liver disease (NAFLD), extensive calorie-rich diet promotes accumulation of excess fatty acid that gets converted into triacylglycerol and gets stored in liver cells as oil droplets (Zivkovic et al. 2007). It leads to an enhanced oxidation of fatty acid generating surplus amount of ROS, a major cause of steatohepatitis.

20.4.2 Mitochondrial Dysfunction-Mediated Hepatotoxicity

Different types of steatosis arise owing to mitochondrial dysfunctionality. One of the primary implications is defective fatty acid oxidation which promotes deposition of lipid vesicles in hepatocyte cytosol. Further, energy production is disrupted depleting delivery of hepatic ketone bodies to peripheral tissue, decline in energy substrates, which eventually leads to multiple organ failure (Fabbrini et al. 2010; Begriche et al. 2011). Several drugs inhibit β-oxidation of fatty acid and subsequently impair mitochondria-mediated homeostasis.

20.4.3 Role of CYP in Inducing Hepatotoxicity

The fundamental purpose of drug metabolism is to ensure that the toxic nonpolar metabolites are converted into their polar forms which can be easily excreted out of the system. This process is facilitated by a number of players, one of them is CYP450 family of enzymes. One of the members CYP2E1 is closely associated with metabolism of xenobiotics, for example, carbon tetrachloride (CCl_4), ethanol, or drugs such as acetaminophen (APAP). A number of CYP isoforms (CYP2D6, CYP1A2, CYP3A4, and CYP2E1) regulate APAP metabolism. In particular, metabolism of APAP by CYP2E1 generates a toxic metabolite, N-acetyl-p-benzoquinone imine (NAPQI) that can selectively bind to organelles inducing apoptosis or necrosis. Alternatively, NAPQI can bind to glutathione causing severe depletion of the latter. Also, the former can covalently get attached to proteins to promote formation of acetaminophen-cysteine adduct. It can co-localize along with nitrated tyrosine (produced by peroxynitrite that in turn gets manufactured by NO and superoxide) thus all the more augmenting the process of necrosis. In corroboration to this, it has been observed that mice treated with macrophage inactivators could mitigate APAP-mediated LI. This property was attributed to the fact that the inactivators could deplete tyrosine nitration and subsequently reduce the degree of co-localization (Jaeschke et al. 2014). However, it is interesting to note that nitric oxide may pose harmful or beneficial depending on the GSH level. In case of already depleted GSH by APAP, NO would contribute and finally magnify the extent of toxicity.

Similarly, in CYP2E1-overexpressing HepG2 cells, it was observed that the presence of APAP/CCl_4/ethanol enhanced ROS production. In the presence of Fe, stronger ROS were produced that inflicted cellular damages through protein oxidation and lipid peroxidation. In fact, some of the CYP2E1-derived diffusible ROS leave the cell where they are produced and spread to neighboring stellate cells, which results in overproduction of collagen subsequently unleashing a fibrotic response (Jaeschke et al. 2014).

20.4.4 Bile Acid-Induced Apoptosis

Bile constitutes bile salts, phospholipids, water, electrolyte, pigments, and trace amounts of metals and acts as emulsifier which is essential for digestion of fatty acids especially in cholesterol metabolism. One of the end products of cholesterol metabolism is bile acid which governs the homeostasis of the gastrointestinal tract and also monitors proliferation and inflammatory responses of the liver. *CYP7A1* gene regulates the synthesis of bile acids, and the produced salts are known to operate a feedback mechanism wherein they repress the same gene thereby regulating their synthesis (Chiang 2009). Bile is produced in the liver and transported off to the gallbladder for further concentration. However, flawed localization of bile from the liver to the duodenum causes cholestasis – a disease characterized by pruritus and jaundice. This disease on the hindsight is an indicator of a probable mechanical obstruction, defective pump functioning, or genetic anomaly. This aberration conversely promotes retention of bile acids in the liver, which subsequently drive the liver cells (hepatocytes) to apoptosis. They are also associated with inducing oxidative stress and necrosis. Consequently, prolonged retention of bile acids induces cirrhosis, LI, and ultimately liver failure, which in the absence of a replacement is fatal.

20.4.5 Role of Polymorphism in Hepatotoxicity

Polymorphism refers to the presence of varied phenotypes within the same population. Cases where variation in a single nucleotide contributes to different phenotypes within members of the same species are known as single-nucleotide polymorphism (SNP). The presence of SNP often renders the bearer predisposed to certain conditions. In context to hepatotoxicity, a number of reports pose testimony to the fact that several cases of LI are primarily due to the presence of SNPs. A very recent study by Zazuli et al. (2015) reported that in Indonesian patients of tuberculosis, the presence of SNP {25385C>T (rs3814055)} in pregnane X receptor made them more susceptible to antituberculosis drug-induced LI. Similarly CYP1A2 1C polymorphism in a small Chinese population was reported to make them predisposed to greater risk of *P. multiflorum*-induced LI (Ma et al. 2014). Multidrug-resistant protein 2 (MRP2) is a type of ABC transporter that is expressed in hepatocyte and is responsible for biliary transport. Undoubtedly, mutation and polymorphism in MRP2 gene are associated with defective transport of bile, which might augment the occurrence of LI. Corroborating to this mechanism, a cohort study on a group of Korean people exhibited that the variation g.-1774delG conglomeration of variant g.-1549G>A and g.-24C>T rendered MRP2 anomalous and thus made those patients predisposed to HH and also DILI (Choi et al. 2007).

20.4.6 Neutrophil-Mediated Liver Injury

As a consequence of injury in the liver via microbial attack, alcohol, ischemia-reperfusion, or any other mode, the inflammatory response is initiated. It recruits the macrophages and neutrophils to the site of injury who primarily remove dead or injured cells produced as a result of the injury and subsequently prepare the tissue for regeneration. However, quite to the dismay, the savior often turns slayer here since mediators produced by phagocytes are toxic and exasperate the already present LI. As a result of the primary injury, CXC chemokine, TNF-α, and IL are generated which upregulate β_2 integrin situated on neutrophil and move them in at the location of injury. The latter aggregate in the liver sinusoid and get adhered to venular endothelial cells with the help of adhesion molecules such as ICAM-1, VCAM-1, etc. Complement receptor Mac-1 (CD11b/CD18) present on the surface of neutrophil helps in the latter's adhesion to the target cell. Mac-1 subsequently releases protease and perpetuates the ROS production. It further facilitates transmigration of neutrophils to hepatocytes where the presence of chemokines and lipid peroxidation-derived oxidative stress prompts the neutrophils to degranulate and hence induce necrotic liver cell injury which can occur as fast as in one hour in human.

20.4.7 Alcohol-Induced Hepatotoxicity

Incessant exposure to ethanol (alcohol) progresses into alcoholic fatty liver disease (AFLD), one of the most common forms of LI. It can lead to cirrhosis and eventually HCV. The mechanism through which alcohol can inflict damage to the liver is regulated by a network of multiple signaling molecules, namely, SIRT1-AMPK axis. SIRT1 is a class III histone deacetylase which plays an imperative role in the liver. It administers lipid metabolism and simultaneously regulates inflammation in the liver. Usually, SIRT1 regulates the acetylation status of its downstream targets and thus presides over their regulation. Ethanol is known to inhibit SIRT1 as a result of which the downstream targets of the SIRT1-AMPK axis such as SREBP-1, H3, lipin-1, PGC-1α, PPARα, β-catenin, FoxO1, NF-kB, or NFATc4 are severely affected (You et al. 2015). Depletion in lipin-1α, β-catenin further depletes PGC-1α/PPARα which decreases the rate of fatty acid oxidation, thereby facilitating the accumulation of fatty acid in hepatocytes. Alternatively, inhibition of SIRT1 leads to acetylation of FoxO1 because of which FoxO1 nuclear protein is downregulated that renders processes of lipid metabolism and oxidative stress-response dysfunctional (Daitoku et al. 2011; Boutant and Cantó 2014; You et al. 2015). Under normal condition, SIRT1 deacetylates the transcription factor NF-kB thereby promoting an anti-inflammatory effect. On the contrary, alcohol-induced SIRT1 inhibition leads to the acetylation of NF-kB which results in the overproduction of TNF-α (pro-inflammatory cytokine) which is proposed to be a prime contributor of alcoholic liver disease. Similarly, nuclear localization, accumulation, and activation of NFATc4 are induced by alcohol-mediated SIRT1 inhibition. This in turn generates copious amounts of TNF-α and IL-6 that favor inflammation.

20.4.8 Drug-Induced Liver Injury (DILI)

The liver owes its pragmatic position of significance because, on one hand, it is at the intersection of absorption and systemic circulation, while on the other hand, it is the prime site for metabolism and elimination of foreign material. DILI is one of the greatest culprits causing hepatotoxicity, which further translates into a number of liver diseases and congruently is responsible for approximately 50 % of all cases of acute liver failure. Nonsteroidal anti-inflammatory drugs, amoxicillin, nefazodone, trovafloxacin, and many other drugs are known to inflict LI. However, the deadliest of them all is acetaminophen (APAP), which single handedly poses as the reason for 39 % of acute liver failures. The primary modes of action pursued by DILI are either through direct hepatotoxicity or an indirect way through immune reactions. In one mode, bioactivation of drug metabolites interacts with cellular macromolecules such as lipid, protein, etc. leading to their peroxidation or dysfunction, finally amassing oxidative stress. Additionally, these reactive metabolites might disrupt ion gradients disturbing intracellular Ca^{2+} levels, which in turn render erratic mitochondrial function, subsequently depleting energy production and impairing liver function. This hepatic stress produced can trigger another mode of action, the initiation of immunological response which might stimulate an innate as well as an adaptive response. As a result, NK, NKT, and Kupffer cells produce pro-inflammatory factors and secrete cytokines (TNF-α, IFN-γ, etc.) that promote liver tissue damage (Gao et al. 2009). Another school of scientists proposes DILI to proceed through three steps – (a) direct stress-mediated, mitochondrial inhibition-mediated, or immune response-mediated primary injury; (b) initial injury proceeds to permeability transition of mitochondria which might trigger the apoptotic cascade through activation of Bax and sensitization of TNF-α and FasL binding to death receptors; and (c) finally, based on the availability of ATP, cells proceed to either apoptosis or necrosis (Holt and Ju 2006; Yuan and Kaplowitz 2013). The situation worsens owing to idiosyncratic DILI, a complicated occurrence wherein doses of an otherwise safe drug can elicit severe LI in certain patients. It might be due to a polymorphism in the genetic makeup of an individual, which renders him predisposed to DILI.

Talking about APAP, the most touted reason of DILI, high doses of the drug are closely associated with the fatal disease centrilobular hepatic necrosis. On consumption, the reactive metabolite of APAP produced by cytochrome P450 metabolism binds to proteins, reduces glutathione, and increases ROS and RNS. This elevated oxidative stress alters Ca^{2+} levels and promotes loss of mitochondrial membrane potential. Consequently, ATP production is hampered which drives the hepatocytes toward necrosis (Choi and Ou 2006). Similarly, herbal medicines and supplements also can induce liver injury. In such a case, the injury caused to the liver is known as herb-induced liver injury (HILI) or herbal hepatotoxicity (HH).

20.5 Close Nexus Between Herbal Medicine and Hepatotoxicity

Herb-induced liver injury is one of the pragmatic reasons that shouts out for more stringent regulations in context of prescription and consumption of herbal medicines. About 15 % of all DILI cases are herb mediated. Secondary metabolites are usually the defensive components of a plant and thus, many of them are toxic too. Therefore, it becomes quite difficult to discriminate them from pharmacologically active secondary metabolites. Subsequently, these herbal derivatives act as one of the prominent reasons for toxicity. In discrete, independent cohort studies across the globe, occurrence of HILI substantiates the potential threat of rampant usage of herbal derivatives without appropriate measures to ensure their safety.

In an attempt to consolidate the hepatotoxicity quotient of various Traditional Chinese medicine (TCM) based on available literature, Teschke (2010) published an article that spoke in length about the present scenario of the safety of such preparations. The paper also asserted the fact that the increase in number of hepatotoxicity cases linked to these preparations makes it imperative to set strict rules for evaluating the safety of herbal medicine. In another discrete study by the same group, 60 herbal drugs were reported to inflict HH as revealed through the screening of 185 publications. In 2015, they also reported that 28 out of 57 TCM caused liver diseases. Multiple other articles by Abdualmjid and Sergi (2013), Bunchorntavakul and Reddy (2013), Navarro and Lucena (2014), and many other groups have reviewed the current status of HH highlighting the current challenges that make the field murky without appropriate regulation. The total number of cases that report HH is massive, and putting them all in a single chapter is beyond the scope of this book. Therefore, we mention a few case reports as a representation of the mammoth proportion of safety issues that still lie unaddressed in context to herbal medicines.

20.5.1 Case I, Reported by Li and Coworkers: (Li et al. 2015a)

Qibao Meiran Wan (Chinese herbal product) is often used as a mutipotent remedy to a bunch of issues ranging from premature graying of hair to constipation. This inflicted acute liver injury to a 26-year-old man who contracted symptomatic hepatitis within 1 month of starting this drug.

20.5.2 Case II, Reported by Yellapu et al. (2011)

"Fat burners" comprising green tea and *Commiphora wightii* extracts and usnic acid (from *Usnea* sp.) are commonly employed to reduce weight by proposed enhancement of fat loss. However, this initiated LI in a healthy, 28-year-old female who developed jaundice and within a week exhibited extremely elevated levels of liver damage markers. She could recover only after a liver transplant.

20.5.3 Case III, as per Langrand et al. (2014)

Tinospora crispa, often used as antispasmodic, antidiabetic-induced chronic lower back pain in a 49-year-old male. On further evaluation of his liver profile, serious increment in bilirubin and aspartate aminotransferase levels was observed. The patient developed acute hepatitis which normalized after stopping the herbal medicine.

The above cases are merely a micro-tip of the huge iceberg that poses a threat to health in the form of herbal hepatotoxicity. It goes without saying that more are the number of incidences of indiscriminate use of herbal products, the greater are the chances of them inducing HILI. However, there are a few touted groups that are notorious for contributing to the maximum number of cases of HILI. They are the following:

Black cohosh (*Actaea racemosa*): Rhizomes of this plant are commonly used by the Native Americans as a remedy to cold, menstrual aberrations, fatigue, and constipation. It is rich in actein (terpene glycoside). It has been reported to elicit autoimmune hepatitis augmented with cholestasis and diffuse cell necrosis in a number of people (Cohen et al. 2004; Oh et al. 2013).

Blue thistle (*Atractylis gummifera*): Abundant in the Mediterranean regions, root extract of this plant is consumed as a diuretic, emetic, and antipyretic. Blue thistle can inhibit mitochondrial oxidative phosphorylation. This is attributed to the presence of diterpenoid glucosides – carboxy-atractyloside and atractyloside, which interact with translocator responsible for ATP/ADP antiport and associated with membrane permeability of mitochondria (Georgiou et al. 1988; Daniele et al. 2005). Consequently, this manifests into LI, cholestasis, cyanosis, and internal hemorrhages and has been reported to be fatal, particularly in children.

Chaparral (*Larrea tridentata*): Widespread in Southwest America and Mexico, chaparral is consumed as a tea to relieve people from rheumatic pain and bronchitis. It is also ingested in other forms as a curative of cold, diarrhea, chronic skin anomalies, and also as antibiotic. The active compound nordihydroguaiaretic acid inhibits cyclooxygenase and lipoxygenase-mediated pathways consequently asserting its role as a potent antibiotic (Abdualmjid and Sergi 2013). This, however, induced hepatitis, cirrhosis, and liver failure. In a 22-year-old, breast-feeding female, ingestion of chaparral led to the development of hepatic fibrosis.

Impila (*Callilepis laureola*): It is a perennial plant that is prevalent in Natal region of South Africa. It is consumed as a cure for gastric ailments, infertility, worm infestations, and common cold. The presence of atractyloside renders it extremely toxic that subsequently gets translated into hepatitis and nephrotoxicity often proving to be fatal (Abdualmjid and Sergi 2013). The toxic effect of impila is closely associated with the occurrence of hepatic dysfunction, renal tubular necrosis, and hypoglycemia.

Kava kava or kava (*Piper methysticum*): Kava has been used as a beverage for centuries in Pacific Islands. It is employed as an antipsychotic and is widely consumed for its touted potential to relieve one from anxiety and depression. The roots are rich in kavalactones and kavapyrones, which render the plant pharmacologically active. However, this ancient herbal medicine has been the cause of many severe LI cases necessitating liver transplant. People across all age groups have exhibited symptoms of kava toxicity. In a previous review article (Teschke 2010), comprehensive documentation of cases of kava hepatotoxicity suggests the potential havoc; this otherwise common herbal medicine can create, provided stern steps to ensure the quality of raw material required for production are not adopted.

Pyrrolizidine alkaloid (PA)-containing plants: These compounds are known to be predominant in members of the families Compositae, Boraginaceae, and Leguminosae. Comfrey (*Symphytum officinale*), one of the members of Boraginaceae family, is rich in PA and is also the constituent of common herbal tea popular all across the globe. Apparently, comfrey relieves inflammation, enhances wound healing, and cures thrombophlebitis. However, owing to the presence of PA, comfrey has been reported to inflict hepatic veno-occlusive disease, annihilation of smaller hepatic venules, and liver failure that turned fatal in many cases (Stickel and Seitz 2000).

Saw palmetto (*Serenoa repens*): Available as the dried fruit of American dwarf palm, this is perhaps one of the most commonly used herbal medicines to counter urinary tract infections as well as benign prostatic hyperplasia in certain cases. It is widely available in West Indies and Southeastern America. The dried fruit is laden with phytosterols and is known to harbor estrogenic as well as androgenic properties. This, in turn, is one of the major reasons of HILI. Several case reports reveal that this herbal supplement often leads to cholestatic hepatitis, acute liver failure, acute pancreatitis, and also a combinatorial occurrence of all the aforesaid (Jibrin et al. 2006; Wargo et al. 2010; Navarro and Lucena 2014). In another separate report, a patient with Gilbert's syndrome (elevated unconjugated bilirubin in the blood) was found to develop acute LI (Lapi et al. 2010).

20.6 "Taming" the "Dragon"!

The current scenario of indiscriminate use of herbal medicines is one that of "feeding" the "dragon." What remains to be the need of the hour is to "tame" that "dragon." The list in Table 20.1 is an alarm for the tremendous safety concern that the herbal medicines generate. The references enlisted in the Table 20.1 talk about the hepatotoxicity that the corresponding products have induced. It goes without saying that it becomes all the more essential to ensure safety of these products since their usage is so rampant across the globe and specifically across all age groups. The main intent of any scientific discussion is to ensure that the vices can be converted into virtues. Consolidated efforts to ensure safety need to be devised. Also, serious efforts need to be made so that the existing regulations are revisited, updated, and adhered to, under all circumstances. The pressing situation that needs to be addressed right now is that, how can it be ensured that the herbal medicines available as over-the-counter drugs are in the best of interest for generic consumption? It is to be ensured that the raw materials, the techniques to be used for production, must totally comply with the safety guidelines. The downstream processing of the drugs/supplements need to be dealt with utmost care. Special emphasis should be laid on the need of quality checks to ensure that no lacunae creep up while commercializing or manufacturing an herbal formulation intended for human consumption.

At this juncture, it becomes mandatory to lay down few strict, rational, and achievable legal instructions that must be abided by one and all in pertinence to the safety assurance of herbal medicines. Failure to do so could lead to severe implications amounting to levying hefty fines, retraction of manufacture/sale license, and potential imprisonment. Stricter the policies, lesser will be the loopholes which otherwise would have helped the noncompliers to escape.

20.7 Need for Legal Intervention to Assess and Ensure Safety

Undoubtedly, one of the major ways of ensuring quality check of a commodity is to implement a certain set of regulatory principles especially those with a legal functionality. This very moment, several lacunae in defining the regulations of herbal medicine production and sale plague the problem of its safety. A few of such loopholes are highlighted in this section which talks about the general lackadaisical

Table 20.1 Common herbal products that are known to inflict hepatotoxicity

Product	Source	Commonly used as/in	Reference
Aloe	*Aloe vera*	Wound healing, antitumor, antiseptic	Yang et al. (2010)
Bajiaolian	*Dysosma pleianthum*	Dysmenorrheal, lumbago, snake bite	Kao et al. (1992)
Camphor	*Cinnamomum camphora*	Mucolytic	Uc et al. (2000)
Cang Er Zi	*Xanthium strumarium*	Headache, sinusitis, arthritis, urticaria	Wang et al. (2011)
Cascara sagrada	*Rhamnus purshiana*	Constipation	Nadir et al. (2000)
Chinese herbal mixture	*Dictamnus dasycarpus*	Rheumatism, jaundice, skin disease, cough	Jang et al. (2008)
Copalchi	*Coutarea latiflora*	Antidiabetic	Bruguera et al. (2007)
Gan cao	*Glycyrrhiza uralensis*	Viral hepatitis	Yuen et al. (2006)
Germander	*Teucrium chamaedrys* L.	Antiseptic, antidepressant, cholerectic	Gori et al. (2011)
Greater celandine	*Chelidonium majus* L.	GIT disorders, spasmolytic, cholelithiasis	Stickel et al. (2003)
Green tea	*Camellia sinensis*	Weight reduction	Bonkovsky (2006)
He Shou Wu	*Polygonum multiflorum*	Antioxidant, laxative	Navarro and Lucena (2014)
Jin Bu Huan	*Lycopodium serratum*	Analgesic, sedative, sleep inducer	Woolf et al. (1994)
Ma Huang	*Ephedra sinica*	Bronchitis, slimming agent	Bajaj et al. (2003) and Teschke et al. (2012)
Margosa oil	*Azadirachta indica*	Itching, inflammation, worm infestation	Abdualmjid and Sergi (2013)
Mistletoe	*Viscum album*	Insomnia, epilepsy, urinary ailments	Teschke et al. (2012)
Mountain germander	*Teucrium polium*	Gastrointestinal tract ailments, enhance breast milk	Dag et al. (2014)
Noni juice	*Morinda citrifolia*	Depression, hypertension	Stadlbauer et al. (2005)
Pennyroyal oil	*Hedeoma pulegioides*	Abortifacient, insect repellent	Bakerink et al. (1996)
Purple coneflower	*Echinacea purpurea*	Wound healing, psoriasis, eczema	Navarro and Lucena (2014)
Sassafras	*Sassafras albidum*	Anti-inflammatory, renal diseases	Dietz and Bolton (2011)
Senna	*Cassia angustifolia*	Laxative	Sonmez et al. 2005
Shou Wu Pian	*Polygonum multiflorum*	Spermatorrhea, lumbago, chronic prostatitis	Mazzanti et al. (2004)

(continued)

Table 20.1 (continued)

Product	Source	Commonly used as/in	Reference
Skullcap	*Scutellaria baicalensis*	Anti-inflammatory, sedative, hysteria	Burke et al. (2014)
St. John's wort	*Hypericum perforatum*	Insomnia, bronchitis, hypothyroidism, scabies	Domínguez Jiménez et al. (2007)
TWHF	*Tripterygium wilfordii*	Immunosuppressant, anti-inflammatory	Xue et al. (2012)
Usnic acid	*Usnea dasypoga*	Weight loss	Durazo et al. (2004)
Valerian	*Valeriana officinalis*	Sedative, aches, palpitation, hypertension	Cohen and Toro (2008)
Wild germander	*Teucrium chamaedrys*	Bronchitis, depression, asthma, fever	Pauwels et al. (1992) and Nencini et al. (2014)
Would sacaca	*Croton cajucara*	Obesity, hypercholesterolemia	de Freitas et al. (2013)

attitude toward the need for safety parameter of herbal drugs. As per the results of a global survey conducted by WHO, a mere 32 % of the 141 member countries were found to have a national policy on regulation of traditional/alternative/herbal medicines. The United Kingdom legislative provisions corroborated by the Medicines Act 1968 exempt herbal medicines from licensing "provided that they are supplied after a one-to-one consultation." According to the laws in the USA, herbal products are defined as "dietary supplement" and thus do not necessitate meeting safety and efficacy checks contrary to pharmaceutical drugs. However, the FDA regulates the production of herbal derivatives through its "The Dietary Supplement Health and Education Act of 1994" (DSHEA). It is essential to follow strict compliance with good manufacturing practices (GMP) and good agricultural practices (GAP).

An increment in the number of HILI cases must make us ponder on the current laws and also motivate the scientific community to suggest ways in which these laws can be updated so as to uphold the integrity of herbal medicines and ensure their "safe for consumption" agenda. It is necessary to ensure that noncompliance of the law can lead violators toward adverse consequences.

20.8 Need for Appropriate Documentation or Assessment

The presence of compounding factors makes the diagnosis of HILI a lot more complex. These anomalies are compound specific and often patient specific also. To worsen the situation, there is a serious lack of a unanimous procedure/algorithm that can systematically analyze or assess the cause of HILI. As of now, it is the jurisdiction and judgment of a physician that determines the diagnosis and subsequent treatment. Thus, it is imperative to lay down a common, unambiguous protocol/algorithm/evaluation sheet that will be unanimously used by all clinicians and scientists. It would enable a correct, universal assessment of a beneficial/harmful effect of herbal medicine, thereby eradicating all judgmental/ad hoc biases. As examples, The Council for International Organizations of Medical Sciences (CIOMS) scale, WHO introspection method, etc. must be applied for assessing HILI. Also, a universal database can be created which might store all updates on case reports that showed signs of HILI. It would give a sneak peek into the potential threats and simultaneously indicate toward a plausible target that can help alleviate the malaise.

20.9 The Emerging Role of Toxicogenomics

The comprehensive study of toxic effects of environmental or pharmaceutical chemicals on human health through the application of holistic genomic tools is known as toxicogenomics. This is perhaps the future of toxicology research. It is a high-throughput approach which enables quick analysis of multiple subjects by combining information pertinent to toxicology and genomic technologies such as genetics, sequence analysis, proteomics, gene expression profiling, metabolomics, and others. Further, this provides an insight into risk assessment of a particular compound and also predicts the potential hazard associated with it. The advantage of this technology over the traditional toxicity studies lies in the fact that first and foremost, a massive amount of information can be generated out of a single experiment. Secondly, computational techniques permit multidirectional information synthesis as well as analyze them in more novel ways. The results obtained through this upcoming mode of assessment are primarily determined by the experimental design. For example, applying means of transcriptomics, microarray-assisted techniques help to evaluate the change in expression at the mRNA level on exposure to different agents. It also helps to assess the short-/low-level exposure, which so far had been a distant reality with conventional toxicity profiles. The best part of this technique is that it discloses pattern changes involving multiple molecules. This resolution is far superior to the assessment generated out of the pattern studied for a single molecule (current methods). The overall success of toxicogenomics depends on the reliability and reproducibility of results obtained from a multiple study/studies. It is ensured by the process of validation – it asserts the efficacy of the test in consistently measuring the decided end points taking into account both the technical and biological quantification. Platform validation, software/data analysis validation, biological validation, and regulatory setting validation are the different tiers of validation method that are employed to assess the reliability of the data/information generated out of toxicogenomic analyses (National Research Council (US) Committee on Applications of Toxicogenomic Technologies to Predictive Toxicology 2007).

In the recent years, several studies via the tools of toxicogenomics have helped unravel many novel mechanisms of hepatotoxicity inflicted by different physical/

chemical/biological agents. Herein, we cite a few examples that stand testimony to the above statement.

20.9.1 Trovofloxacin-Inflicted Hepatotoxicity

Trovofloxacin (TFX) is a broad-spectrum antibiotic that inhibits DNA uncoiling in bacteria. In studies carried out by (Liguori et al. 2005, 2008), they reported the mechanism of TFX-mediated LI by employing microarray analysis as a part of the toxicogenomic study. Hepatocytes [HepG2 and primary human hepatocytes (PHH)] isolated from donors were employed in the studies. Despite wide variation that existed among the donors regarding weight, lifestyle, and history of medications, changes in gene expression profile were prominently massive and uniform across all the four donor cells exposed to TFX. Several genes are particularly regulated by TFX, which until this study was attributed to idiosyncratic hepatotoxicity mechanisms. Out of the many genes regulated, mitofusin-1 (maintains morphological features of mitochondria) was observed to be downregulated by the drug which in turn resulted in dysfunction of mitochondria (Liguori et al. 2005). Likewise, Bax, another mitochondrial morphology-maintenance gene, was downregulated by TFX. The latter also upregulated several genes associated with oxidative stress and thus holistically contributed to mitochondria toxicity which subsequently induced LI in the isolated hepatocytes (HepG2). TFX also interfered in transcription by downregulating genes associated with the same and processing of RNA exhibiting its cross-reactivity with eukaryotic polymerase II system (Liguori et al. 2005; Reymann and Borlak 2008). Additionally, genes associated with organogenesis and phosphorylation were also seen to be regulated by TFX administration, which was attributed to being involved in the mechanism of hepatotoxicity. However, differential expression was observed in TFX-exposed PHH and HepG2 cells in context to inflammation-associated genes (e.g., IL-8). While the former was downregulated in PHH, the same gene was overexpressed by almost ninefold in HepG2 cells (Liguori et al. 2005).

20.9.2 Pentamethylchromanol (PMCol)-Induced Hepatotoxicity

PMCol is a potent anticancer agent that is quite effective in the therapy of androgen-dependent cancers. Parman and coworkers in 2011, by the help of toxicogenomic studies aided with metabolomic analyses, assessed the effect of PMCol in inducing LI in rats. Gene set enrichment analysis revealed that genes associated with glutathione metabolism were affected by PMCol treatment. The chemical was able to upregulate genes associated with overexpression of Gpx2, Gsr, and Gclc in the liver of rats under study. It also upregulated Akr7a3 and Gstp1 – biomarkers of conjugation-type reduction of glutathione. Also, indicators of oxidative stress (Nqo1, Hmox1) were upregulated by PMCol too. The latter comprehensively inhibited biosynthesis of glutathione and effectively depleted liver glutathione through both nonconjugative and conjugative modi operandi. The metabolomic study revealed that PMCol competes with campesterol and vitamin E to be taken up by a cell. It also reduced the levels of several hydrophilic and lipophilic dietary antioxidants in hepatocytes thereby facilitating LI (Parman et al. 2011). PMCol tampered lipid biosynthesis in the liver through depletion of malonyl-CoA and acetyl-CoA. To add to this, hepatic XO, which is associated with the production of H_2O_2 by oxidizing hypoxanthine to allantoin, is also affected by PMCol. Comprehensively, PMCol reduces glutathione either through direct inhibition of its biosynthesis or by promoting flawed xenobiotic transcriptional response that subsequently generates ROS and thereby augments the consumption of glutathione (Parman et al. 2011). Now, since the latter is already in less abundance due to PMCol, the antioxidants are fast consumed ultimately leading to their scarcity depicted as erroneous physiology and functionality of the liver and subsequently progressing into LI.

20.9.3 Triptolide-Induced Hepatotoxicity

Triptolide is a diterpenoid isolated from *Tripterygium wilfordii*. It is an important constituent of TWHF, a traditional herbal supplement commonly used to alleviate a number of autoimmune disorders, namely, SLE, rheumatoid arthritis, and other diseases such as leprosy and nephritis. In 2013, Wang and the group carried out a study to determine the hepatotoxic effect of triptolide in Wistar rats by analyzing their gene expression profiles, 14 days after administration of the drug. Microarray studies exhibited that female rats were more sensitive to triptolide than their male counterparts (Wang et al. 2013). In the former population, 3,329 genes were seen to be differentially expressed. The majority of these genes played crucial roles in significant pathways of normal liver function. In other words, pathways mediating glucose metabolism, insulin signaling, ROS-mediated cell stress, cell-cycle progression, and apoptosis were seen to be affected owing to the altered expression of these genes. Triptolide significantly enhanced expression of caspase-3 and reduced GSH/GSSG ratio, serum glucose, liver glycogen, activities of gluco-6-phospatase, and PEP carboxykinase. On the hindsight, these augmentations and depletions could get translated to altered redox balance in the liver, serum glucose levels, and increase in hepatocyte apoptosis. This study for the first time indicated to the fact that IGF-1 and glucocorticoids might be associated with triptolide-induced LI (Wang et al. 2013).

The above examples are a few among the plethora of applications that have been generated out of the field of toxicogenomics. This conglomeration of techniques is the future of toxicology research where in the wedding of conventional techniques to updated computational approaches can work wonders in understanding and predicting the toxicity quotient of a particular drug/agent under question.

20.10 How to Strike the Right Balance: The "Take-Home Message"!

Like the "head" and "tail" are the two sides of the same coin, herbal medicines are endowed with their indispensable two sides too – the pros and the cons. On one hand herbal medicines have dire effects on the liver, and on the other hand, they do have certain beneficial aspects too. Therefore, it is necessary to augment the positives and deplete the negatives as much as possible. The essence is to strike a perfect balance between use and disuse of herbal formulations. The liver is one of the most vital organs of the body, which administers metabolism as well as detoxification. Thus, it becomes all the more imperative to assess the effect of each and every herbal supplement on the liver, before allowing them for consumption. Employment of toxicogenomic studies can help take into consideration different parameters at a given point in time. This subsequently caters to us a map of probable multifactorial effects because of a particular drug. This in turn would enable to devise a more effective prophylactic or preventive measure to curb the plausible toxicity of the drug. Strict measures to ensure that before these drugs are released into the market, they are subjected to thorough quality control steps. It will ensure not only reduced toxicity problems attached to herbal medicine but also enable the end-point users to reap maximum benefits, which otherwise could not have been catered by medicines from other sources. Universal hepatotoxicity assessment criteria shall boost the endeavor to frame an assessment of HILI extent of a particular herbal drug. Thus, a conglomeration of proper assessment, strict regulations, and adherence to law and overall judicious use of herbal supplement would enable us to enjoy the true benefits of herbal medicines without making the liver unhappy!

References

Abdualmjid RJ, Sergi C (2013) Hepatotoxic botanicals – an evidence-based systematic review. J Pharm Pharm Sci 16:376–404

Asif M (2012) A brief study of toxic effects of some medicinal herbs on kidney. Adv Biomed Res 1:44. doi:10.4103/2277-9175.100144

Bajaj J, Knox JF, Komorowski R, Saeian K (2003) The irony of herbal hepatitis: Ma-Huang-induced hepatotoxicity associated with compound heterozygosity for hereditary hemochromatosis. Dig Dis Sci 48:1925–1928

Bakerink JA, Gospe SM, Dimand RJ, Eldridge MW (1996) Multiple organ failure after ingestion of pennyroyal oil from herbal tea in two infants. Pediatrics 98:944–947

Begriche K, Massart J, Robin MA, Borgne-Sanchez A, Fromenty B (2011) Drug-induced toxicity on mitochondria and lipid metabolism: mechanistic diversity and deleterious consequences for the liver. J Hepatol 54:773–794. doi:10.1016/j.jhep.2010.11.006

Bonkovsky HL (2006) Hepatotoxicity associated with supplements containing Chinese green tea (*Camellia sinensis*). Ann Intern Med 144:68–71

Boutant M, Cantó C (2014) SIRT1 metabolic actions: integrating recent advances from mouse models. Mol Metab 3:5–18. doi:10.1016/j.molmet.2013.10.006

Bruguera M, Herrera S, Lazaro E, Madurga M, Navarro M, De Abajo FJ (2007) Hepatitis aguda asociada al consumo de Copalchi. A propósito de 5 casos. Gastroenterol Hepatol 30:66–68

Bunchorntavakul C, Reddy KR (2013) Review article: herbal and dietary supplement hepatotoxicity. Aliment Pharmacol Ther 37:3–17. doi:10.1111/apt.12109

Burke DW, Zakhary B, Pinelis E (2014) Hepatotoxicity associated with baikal skullcap (*Scutellaria baicalensis*). In: D47. Case vignettes in critical care I. American Thoracic Society, pp A6118–A6118

Chiang JYL (2009) Bile acids: regulation of synthesis. J Lipid Res 50:1955–1966. doi:10.1194/jlr.R900010-JLR200

Choi J, Ou JHJ (2006) Mechanisms of liver injury. III. Oxidative stress in the pathogenesis of hepatitis C virus. Am J Physiol Gastrointest Liver Physiol 290:G847–G851. doi:10.1152/ajpgi.00522.2005

Choi JH, Ahn BM, Yi J, Lee MG (2007) MRP2 haplotypes confer differential susceptibility to toxic liver injury. Pharmacogenet Genomics 17:403–415. doi:10.1097/01.fpc.0000236337.41799.b3

Cohen DL, Toro YD (2008) A case of valerian-associated hepatotoxicity. J Clin Gastroenterol 42:961–962. doi:10.1097/MCG.0b013e3180500348

Cohen SM, O'Connor AM, Hart J, Merel NH, Te HS (2004) Autoimmune hepatitis associated with the use of black cohosh: a case study. Menopause 11:575–577

CSPI (2013) FDA urged to prohibit sale of Ginkgo in wake of cancer study. http://www.cspinet.org/new/201306031.html. Accessed 28 Mar 2016

Dag M, Ozturk Z, Aydnl M, Koruk I, Kadayfc A (2014) Postpartum hepatotoxicity due to herbal medicine *Teucrium polium*. Ann Saudi Med 34:541–543. doi:10.5144/0256-4947.2014.541

Daitoku H, Sakamaki J, Fukamizu A (2011) Regulation of FoxO transcription factors by acetylation and protein–protein interactions. Biochim Biophys Acta BBA – Mol Cell Res 1813:1954–1960. doi:10.1016/j.bbamcr.2011.03.001

Daniele C, Dahamna S, Firuzi O, Sekfali N, Saso L, Mazzanti G (2005) *Atractylis gummifera* L. poisoning: an ethnopharmacological review. J Ethnopharmacol 97:175–181. doi:10.1016/j.jep.2004.11.025

de Freitas RB, Rovani BT, Boligon AA, de Brum TF, Piana M, da Silva Jesus R, Veloso CF, Kober H, Moresco RN, da Costa AIC, de Freitas BL, Athayde ML (2013) Hepatotoxicity evaluation of aqueous extract from *Scutia buxifolia*. Molecules (Basel Switz) 18:7570–7583. doi:10.3390/molecules18077570

Dietz BM, Bolton JL (2011) Biological reactive intermediates (BRIs) formed from botanical dietary supplements. Chem Biol Interact 192:72–80. doi:10.1016/j.cbi.2010.10.007

Domínguez Jiménez JL, Pleguezuelo Navarro M, Guiote Malpartida S, Fraga Rivas E, Montero Alvarez JL, Poyato González A (2007) Hepatotoxicity associated with *Hypericum* (St. John's wort). Gastroenterol Hepatol 30:54–55

Durazo FA, Lassman C, Han SHB, Saab S, Lee NP, Kawano M, Saggi B, Gordon S, Farmer DG, Yersiz H, Leonard R, Goldstein I, Ghobria M, Busuttil RW (2004) Fulminant liver failure due to usnic acid for weight loss. Am J Gastroenterol 99:950–952. doi:10.1111/j.1572-0241.2004.04165.x

Fabbrini E, Sullivan S, Klein S (2010) Obesity and nonalcoholic fatty liver disease: biochemical, metabolic and clinical implications. Hepatology (Baltimore, MD) 51:679–689. doi:10.1002/hep.23280

Gao B, Radaeva S, Park O (2009) Liver natural killer and natural killer T cells: immunobiology and emerging roles in liver diseases. J Leukoc Biol 86:513–528. doi:10.1189/jlb.0309135

Georgiou M, Sianidou L, Hatzis T, Koutselinis A (1988) Hepatotoxicity due to *Atractylis gummifera*-L. J Toxicol Clin Toxicol 26:487–493. doi:10.3109/15563658809038564

Gomes J, Magalhães A, Michel V, Amado IF, Aranha P, Ovesen RG, Hansen HC, Gärtner F, Reis CA, Touati E (2012) *Pteridium aquilinum* and its paquiloside toxin induce DNA damage response in gastric epithelial cells, a link with gastric carcinogenesis. Toxicol Sci 126:60–71. doi:10.1093/toxsci/kfr329

Gori L, Galluzzi P, Mascherini V, Gallo E, Lapi F, Menniti-Ippolito F, Raschetti R, Mugelli A, Vannacci A, Firenzuoli F (2011) Two contemporary cases of hepatitis associated with *Teucrium chamaedrys* L. decoction use: case reports and review of literature. Basic Clin Pharmacol Toxicol 109:521–526. doi:10.1111/j.1742-7843.2011.00781.x

Hilepo JN, Bellucci AG, Mossey RT (1997) Acute renal failure caused by "Cat's Claw"; herbal remedy in a patient with systemic *Lupus erythematosus*. Nephron 77:361–361. doi:10.1159/000190304

Holt MP, Ju C (2006) Mechanisms of drug-induced liver injury. AAPS J 8:E48–E54. doi:10.1208/aapsj080106

Jaeschke H, Xie Y, McGill MR (2014) Acetaminophen-induced liver injury: from animal models to humans. J Clin Transl Hepatol 2:153–161, 10.14218/JCTH.2014.00014

Jang JS, Seo EG, Han C, Chae HB, Kim SJ, Lee JD, Wang JH (2008) Four cases of toxic liver injury associated with *Dictamnus dasycarpus*. Korean J Hepatol 14:206–212. doi:10.3350/kjhep.2008.14.2.206

Jibrin I, Erinle A, Saidi A, Aliyu ZY (2006) Saw palmetto-induced pancreatitis. South Med J 99:611–612. doi:10.1097/01.smj.0000215642.76198.44

Kao WF, Hung DZ, Tsai WJ, Deng JF (1992) Podophyllotoxin intoxication: toxic effect of Bajiaolian in herbal therapeutics. Hum Exp Toxicol 11:480–487. doi:10.1177/096032719201100607

Langrand J, Regnault H, Cachet X, Bouzidi C, Villa AF, Serfaty L, Garnier R, Michel S (2014) Toxic hepatitis induced by a herbal medicine: *Tinospora crispa*. Phytomedicine 21:1120–1123. doi:10.1016/j.phymed.2014.04.031

Lapi F, Gallo E, Giocaliere E, Vietri M, Baronti R, Pieraccini G, Tafi A, Menniti-Ippolito F, Mugelli A, Firenzuoli F, Vannacci A (2010) Acute liver damage due to *Serenoa repens*: a case report. Br J Clin Pharmacol 69:558–560. doi:10.1111/j.1365-2125.2010.03618.x

Li X, Qu C, He Q, Chen W, Zhang X, Liu X, Liu Y, Tang Y (2015a) Acute hepatitis induced by a Chinese herbal product Qibao Meiran Wan: a case study. Int J Clin Exp Med 8:11624–11627

Li XX, Du FY, Liu HX, Ji JB, Xing J (2015b) Investigation of the active components in Tripterygium wilfordii leading to its acute hepatotoxicity and nephrotoxicity. J Ethnopharmacol 162:238–243. doi:10.1016/j.jep.2015.01.004

Liguori MJ, Anderson MG, Bukofzer S, McKim J, Pregenzer JF, Retief J, Spear BB, Waring JF (2005) Microarray analysis in human hepatocytes suggests a mechanism for hepatotoxicity induced by trovafloxacin. Hepatology 41:177–186. doi:10.1002/hep.20514

Liguori MJ, Blomme EAG, Waring JF (2008) Trovafloxacin-induced gene expression changes in liver-derived in vitro systems: comparison of primary human hepatocytes to HepG2 cells. Drug Metab Dispos 36:223–233. doi:10.1124/dmd.107.017608

Ma KF, Zhang XG, Jia HY (2014) CYP1A2 polymorphism in Chinese patients with acute liver injury induced by Polygonum multiflorum. Genet Mol Res 13:5637–5643. doi:10.4238/2014.July.25.19

Machado MM, de Oliveira LFS, Zuravski L, Souza RO, Fischer P, Duarte JA, Rocha MO, Güez CM, Boligon AA, Athayde ML (2016) Evaluation of genotoxic and cytotoxic effects of hydroalcoholic extract of Euphorbia tirucalli (Euphorbiaceae) in cell cultures of human leukocytes. An Acad Bras Cienc 88:17–28. doi:10.1590/0001-3765201520140076

Mazzanti G, Battinelli L, Daniele C, Mastroianni CM, Lichtner M, Coletta S, Costantini S (2004) New case of acute hepatitis following the consumption of Shou Wu Pian, a Chinese herbal product derived from *Polygonum multiflorum*. Ann Intern Med 140:W30

Muriel P (2009) Role of free radicals in liver diseases. Hepatol Int 3:526–536. doi:10.1007/s12072-009-9158-6

Nadir A, Reddy D, Van Thiel DH (2000) Cascara sagrada-induced intrahepatic cholestasis causing portal hypertension: case report and review of herbal hepatotoxicity. Am J Gastroenterol 95:3634–3637. doi:10.1111/j.1572-0241.2000.03386.x

National Research Council (US) Committee on Applications of Toxicogenomic Technologies to Predictive Toxicology (2007) Applications of toxicogenomic technologies to predictive toxicology and risk assessment. National Academies Press (US), Washington, DC

Navarro VJ, Lucena MI (2014) Hepatotoxicity induced by herbal and dietary supplements. Semin Liver Dis 34:172–193. doi:10.1055/s-0034-1375958

Nencini C, Galluzzi P, Pippi F, Menchiari A, Micheli L (2014) Hepatotoxicity of *Teucrium chamaedrys* L. decoction: role of difference in the harvesting area and preparation method. Indian J Pharmacol 46:181–184. doi:10.4103/0253-7613.129313

Oh HJ, Mok YM, Baek MS, Lee JK, Seo BS, Kim TH, Choi KH, Hwang IK, Ra JE, Oh YR, Kim YS, Cho EY, Kim HC, Sohn YW (2013) Co-development of autoimmune hepatitis and Sjögren's syndrome triggered by the administration of herbal medicines. Clin Mol Hepatol 19:305–308. doi:10.3350/cmh.2013.19.3.305

Parman T, Bunin DI, Ng HH, McDunn JE, Wulff JE, Wang A, Swezey R, Rasay L, Fairchild DG, Kapetanovic IM, Green CE (2011) Toxicogenomics and metabolomics of pentamethylchromanol (PMCol)-induced hepatotoxicity. Toxicol Sci 124:487–501. doi:10.1093/toxsci/kfr238

Pauwels A, Thierman-Duffaud D, Azanowsky JM, Loiseau D, Biour M, Levy VG (1992) Acute hepatitis caused by wild germander. Hepatotoxicity of herbal remedies: two cases. Gastroenterol Clin Biol 16:92–95

Reymann S, Borlak J (2008) Topoisomerase II inhibition involves characteristic chromosomal expression patterns. BMC Genomics 9:324. doi:10.1186/1471-2164-9-324

Sonmez A, Yilmaz MI, Mas R, Ozcan A, Celasun B, Dogru T, Taslipinar A, Kocar IH (2005) Subacute cholestatic hepatitis likely related to the use of senna for chronic constipation. Acta Gastroenterol Belg 68:385–387

Stadlbauer V, Fickert P, Lackner C, Schmerlaib J, Krisper P, Trauner M, Stauber RE (2005) Hepatotoxicity of NONI juice: report of two cases. World J Gastroenterol 11:4758–4760

Stickel F, Seitz HK (2000) The efficacy and safety of comfrey. Public Health Nutr 3:501–508

Stickel F, Pöschl G, Seitz HK, Waldherr R, Hahn EG, Schuppan D (2003) Acute hepatitis induced by Greater Celandine (*Chelidonium majus*). Scand J Gastroenterol 38:565–568. doi:10.1080/00365520310000942

Teschke R (2010) Kava hepatotoxicity – a clinical review. Ann Hepatol 9:251–265

Teschke R, Wolff A, Frenzel C, Schulze J, Eickhoff A (2012) Herbal hepatotoxicity: a tabular compilation of reported cases. Liver Int Off J Int Assoc Study Liver 32:1543–1556. doi:10.1111/j.1478-3231.2012.02864.x

Uc A, Bishop WP, Sanders KD (2000) Camphor hepatotoxicity. South Med J 93:596–598

Wachtel-Galor S, Benzie IFF (2011) Herbal medicine: an introduction to its history, usage, regulation, current trends, and research needs. In: Benzie IFF, Wachtel-Galor S (eds) Herbal medicine: biomolecular and clinical aspects, 2nd edn. CRC Press, Boca Raton

Wang Y, Han T, Xue LM, Han P, Zhang QY, Huang BK, Zhang H, Ming QL, Peng W, Qin LP (2011) Hepatotoxicity of kaurene glycosides from *Xanthium strumarium* L. fruits in mice. Pharmazie 66:445–449

Wang J, Jiang Z, Ji J, Wang X, Wang T, Zhang Y, Tai T, Chen M, Sun L, Li X, Zhang L (2013) Gene expression profiling and pathway analysis of hepatotoxicity induced by triptolide in Wistar rats. Food Chem Toxicol Int J Publ Br Ind Biol Res Assoc 58:495–505. doi:10.1016/j.fct.2013.04.039

Wargo KA, Allman E, Ibrahim F (2010) A possible case of saw palmetto-induced pancreatitis. South Med J 103:683–685. doi:10.1097/SMJ.0b013e3181e1e3ee

Woolf GM, Petrovic LM, Rojter SE, Wainwright S, Villamil FG, Katkov WN, Michieletti P, Wanless IR, Stermitz FR, Beck JJ, Vierling JM (1994) Acute hepatitis associated with the Chinese herbal product jin bu huan. Ann Intern Med 121:729–735

Xue M, Jiang Z, Wu T, Li J, Zhang L, Zhao Y, Li XJ, Zhang LY, Yang SY (2012) Anti-inflammatory effects and hepatotoxicity of Tripterygium-loaded solid lipid nanoparticles on adjuvant-induced arthritis in rats. Phytomedicine Int J Phytother Phytopharmacol 19:998–1006. doi:10.1016/j.phymed.2012.06.006

Yang HN, Kim DJ, Kim YM, Kim BH, Sohn KM, Choi MJ, Choi YH (2010) Aloe-induced toxic hepatitis. J Korean Med Sci 25:492–495. doi:10.3346/jkms.2010.25.3.492

Yellapu RK, Mittal V, Grewal P, Fiel M, Schiano T (2011) Acute liver failure caused by "fat burners" and dietary supplements: a case report and literature review. Can J Gastroenterol 25:157–160

You M, Jogasuria A, Taylor C, Wu J (2015) Sirtuin 1 signaling and alcoholic fatty liver disease. Hepatobiliary Surg Nutr 4:88–100. doi:10.3978/j.issn.2304-3881.2014.12.06

Yuan L, Kaplowitz N (2013) Mechanisms of drug induced liver injury. Clin Liver Dis 17:507–518. doi:10.1016/j.cld.2013.07.002

Yuen MF, Tam S, Fung J, Wong DKH, Wong BCY, Lai CL (2006) Traditional Chinese medicine causing hepatotoxicity in patients with chronic hepatitis B infection: a 1-year prospective study. Aliment Pharmacol Ther 24:1179–1186. doi:10.1111/j.1365-2036.2006.03111.x

Zazuli Z, Barliana MI, Mulyani UA, Perwitasari DA, Ng H, Abdulah R (2015) Polymorphism of PXR gene associated with the increased risk of drug-induced liver injury in Indonesian pulmonary tuberculosis patients. J Clin Pharm Ther 40:680–684. doi:10.1111/jcpt.12325

Zivkovic AM, German JB, Sanyal AJ (2007) Comparative review of diets for the metabolic syndrome: implications for nonalcoholic fatty liver disease. Am J Clin Nutr 86:285–300

MIX
Papier aus verantwortungsvollen Quellen
Paper from responsible sources
FSC® C105338

If you have any concerns about our products,
you can contact us on
ProductSafety@springernature.com

In case Publisher is established outside the EU,
the EU authorized representative is:
Springer Nature Customer Service Center GmbH
Europaplatz 3, 69115 Heidelberg, Germany

Printed by Libri Plureos GmbH
in Hamburg, Germany